DEVELOPMENTS IN **QUATERNARY SCIENCE** **2**
SERIES EDITOR: JIM ROSE

QUATERNARY GLACIATIONS EXTENT AND CHRONOLOGY

PART I: EUROPE

Developments in Quaternary Science
(Series editor: Jim Rose)

Volumes in this series

1. The Quaternary Period in the United States
Edited by A.R. Gillespie, S.C. Porter, B.F. Atwater,
0-444-51470-8 (hardbound); 0-444-51471-6 (paperback) – 2004

2. Quaternary Glaciations – Extent and Chronology
Part I Europe; Part II North America; Part III South America, Asia, Africa, Australia, Antarctica
Edited by J. Ehlers, P.L. Gibbard
0-444-51462-7 (hardbound + CD-ROM) – 2004

3. Ice Age Southern Andes – A Chronicle of Paleoecological Events
Authored by C.J. Heusser
0-444-51478-3 (hardbound) – 2003

4. Spitsbergen Push Moraines – Including a translation of K. Gripp: Glaciologische und geologische Ergebnisse der Hamburgischen Spitzbergen-Expedition 1927
Authored by J. van der Meer
Forthcoming – 2004
0-444-51544-5

5 Tropical West Africa – Marine and Continental Changes during the Late Quaternary
Authored by P. Giresse
Forthcoming – 2004

For further information as well as other related products, please visit the Elsevier homepage (http://www.elsevier.com)

Developments in Quaternary Science, 2
Series editor: Jim Rose

QUATERNARY GLACIATIONS EXTENT AND CHRONOLOGY

PART I: EUROPE

Editors:

J. Ehlers
Geologisches Landesamt,
Hamburg, Germany

P.L. Gibbard
University of Cambridge
Cambridge, United Kingdom

ELSEVIER

2004
Amsterdam – Boston – Heidelberg – London – New York – Oxford – Paris
San Diego – San Francisco – Singapore – Sydney – Tokyo

ELSEVIER B.V.
Sara Burgerhartstraat 25
P.O. Box 211, 1000 AE Amsterdam
The Netherlands

ELSEVIER Inc.
525 B Street, Suite 1900
San Diego, CA 92101-4495
USA

ELSEVIER Ltd
The Boulevard, Langford Lane
Kidlington, Oxford OX5 1GB
UK

ELSEVIER Ltd
84 Theobalds Road
London WC1X 8RR
UK

First edition 2004

Library of Congress Cataloging in Publication Data
A catalog record is available from the Library of Congress.

British Library Cataloguing in Publication Data
A catalogue record is available from the British Library.

ISBN: 0 444 51462 7
ISSN: 1571 0866 (series)

♾ The paper used in this publication meets the requirements of ANSI/NISO Z39.48-1992 (Permanence of Paper).
Printed in The Netherlands.

Foreword

"Quaternary Glaciations – Extent and Chronology, Part I: Europe", edited by Jürgen Ehlers and Philip Gibbard is the second volume in the *Elsevier Book Series* on "*Developments in Quaternary Science*". This book represents the first volume of three covering Quaternary glaciations across the Earth. Part II will cover North America and Part III: South America, Asia, Africa, Australia, and Antarctica.

These books represent the activities of Working Group 5 of the INQUA Commission on Glaciation, which had the mission of bringing together up-to-date information about the extent, volume and timing of Quaternary glaciations. This programme was established at the 1995 INQUA Congress in Berlin and early results were presented to members of the Quaternary community at the INQUA Congresses at Durban in 1999 and Reno in 2003. A brief preview of the work and the background to its development is given in a Viewpoint Article in *Quaternary Science Reviews* (Ehlers and Gibbard, 2003, Volume 22, Issue 15-17, pp 1561-1568).

The books contain contributions from over 200 scientists working in more than 80 countries and territories around the world, and will provide the most complete survey of the evidence for Quaternary glaciation ever attempted.

Initially, publication of this book was discussed in terms of a special issue of *Quaternary Science Reviews* or *Quaternary International*, following the path of other output from INQUA commissions. However, as discussion progressed, it became apparent that the size and scope of the book was such that it would swamp the journal in its year of publication, and little space would be available for normal research-and-review-papers or other special issues. It was partly for this reason that the book series: *Developments in Quaternary Science* was instigated. This series is designed to provide an outlet for topics that require substantial space, are linked to major scientific events, require special production facilities (i.e. use of interactive electronic methods), or are archival in character. The subjects covered by the series will consider Quaternary science across different parts of the Earth and with respect to the diverse range of Quaternary processes. The texts will cover the response to processes within the fields of geology, biology, geography, climatology archaeology and geochronology. Particular consideration will be given to issues such as the Quaternary development of specific regions, comprehensive treatments of specific topics such as global scale consideration of patterns of glaciation, and compendia on timely topics such as dating methodologies, environmental hazards and rapid climate changes. This series will provide an outlet for scientists who wish to achieve a substantial treatment of major scientific concerns and a venue for those seeking the authority provided by such an approach.

"Quaternary Glaciations – Extent and Chronology, Part I: Europe", edited by Jürgen Ehlers and Philip Gibbard not only reflects many of the aspirations of the Book Series in terms of size and substantial treatment of a major Quaternary topic, it also reflects the requirement for special production facilities. With the need to represent such large quantities of cartographic information about different-age ice limits and the effects of glaciation, the editors have chosen to use the ArcView GIS package based on the 1:1,000,000 Digital Chart of the World, and to provide this information on a CD included in each volume. The use of this electronic format means that users can interrogate the maps with GIS software (ArcView, ArcInfo), and the maps can then be used as part of future research.

I know from personal experience that the use of digital representation has caused the Editors considerable work. Just before publication, Jürgen Ehlers found that the original version of the Digital Chart of the World could not be accessed by the latest version of ArcInfo, and so he had to reformat the entire document in order to make it accessible to present and near future users. The Editors have also had to deal with the diverse views of the many contributors, with some areas being represented by quite contrasting interpretations. This gives the work outstanding value as it allows long-held traditional, or even individual idiosyncratic views to be included and compared with the results of recent research based on new methodologies and incorporating contemporary Quaternary concepts.

The publication of this book gives me great pleasure, and reflects outstanding efforts by the editors over the last eight years. I look forward to the appearance of Volumes II and III, which are already with the publishers and should be in press later in 2004. The Editors are to be admired and congratulated on their achievement. These volumes will be a major archive for the future, and because of the use of the interactive digital imagery will be documents that are part of future work on Quaternary glaciations.

Jim Rose,
Series Editor

Preface

Jürgen Ehlers [1] and Philip L. Gibbard [2]

[1] *Geologisches Landesamt, Billstr. 84, D-20539 Hamburg, Germany*
E-mail: jehlersqua@aol.com

[2] *University of Cambridge, Godwin Institute of Quaternary Research, Department of Geography,*
Downing Place, Cambridge CB2 3EN, UK
E-mail: plg1@cus.cam.ac.uk

At the XIV Congress of the International Quaternary Association (INQUA) in Berlin in 1995 the INQUA Commission on Glaciation decided to form a new working group entitled Extent and Chronology of Glaciations. The aim of this project was to provide a comprehensive overview of the extent and chronology of Quaternary glaciation as currently known throughout the world.

The initial results of the working group were presented at the XV INQUA Congress in Durban, South Africa in 1999. Here these results were further discussed and open questions resolved. Now, before the XVI INQUA Congress in Reno, Nevada (2003), the first part of our results are published. This first volume is dedicated to the glaciation of Europe. The second volume will present the evidence from North America, whilst the third will present the results from the rest of the World. Those volumes will be published shortly.

The greatest challenge in a project of this size and scope is how best to display the information. For this purpose we chose ArcView for the Geographical Information System (GIS). As a base map we have used the 'Digital Chart of the World' at a scale of 1:1,000,000. The compilers, ESRI kindly allowed us to publish the relevant layers of information including rivers, lakes, glaciers, settlements, contours (in feet), roads and railway lines together with the glacial map.

For the purposes of this project, we have attempted to compile three sets of glacial limits from all areas (if at all possible):

1. the maximum limit of Pleistocene glaciation
2. the Late Weichselian / Wisconsinan / Valdaian / Würmian Glacial Maximum (LGM)
3. the Early (or Middle) Weichselian glacial maximum

These three limits are the foundation of the maps and the text. In areas where more information is available, additional glacial limits, are included, such as

- different deglaciation timelines
- various pre-Weichselian glacial limits

The digital maps were compiled from topographic maps showing the glacial limits and other features mentioned above, then digitized by us. Several contributors, who had access to ArcInfo or ArcView, provided their data in digitized form so that we could include them directly into our system.

Because of the size of the area covered by the project, it is simply impractical to publish printed maps. Instead the maps are provided as ArcView projects, which can either be used by a GIS (ArcView, ArcInfo) or viewed through a browser (ArcExplorer, available free-of-charge through the Internet). Printed versions of the maps can therefore be produced as required by the user.

The following original information is included in the digital maps:

1. Glacial limits.
2. Morphologically-expressed end moraines.
3. Ice-dammed lakes connected to certain ice-marginal positions.
4. Glacially-induced drainage diversions.
5. Location of key sections through which the glacial limits are defined and dated.

If the original ArcView projects are used to read the digital maps, the glacial limits are shown as thick lines in different colours, continuous lines for certain boundaries and stippled lines for uncertain boundaries. The end moraines are shown as red line features or green polygons. The ice-dammed lakes are shown as blue hatched areas. Glacially-induced drainage diversions are shown as thick blue lines, and key sections are shown as different symbols.

The text contributions include a brief overview of the glacial history of the area, discuss glacial limits plotted on the digital maps (quality of data, alternative interpretations) and discuss the dating of the glacial limits and any open questions.

The eastern and southeastern boundaries of Europe (the Urals, the Caucasus) run directly across formerly glaciated areas. In both cases, we have decided to include the complete ice sheets in this volume. Contributions are presented by country in alphabetical order, starting with Austria and ending with the Ukraine. Some countries (e.g. Britain, France, Germany, Ireland, Italy, Russia, Spain) are represented by more than one contribution. Where no recent information was available (e.g. in some parts of the Balkans), published sources were used in combination with evaluation of satellite imagery to identify formerly-glaciated areas.

So far, the most complete coverage of the glacial limits in Europe had been the 'International Quaternary Map of Europe', 1: 2,500,000, published in the 1970s and 80s. We have been working at a scale of 1: 1,000,000, allowing more detail to be included.

As can be seen from the maps, information on many of the features is very detailed. However, other things are much more poorly known, including the pre-LGM Weichselian glacial limits in the circum-Baltic region, the extent of all pre-Weichselian glaciations in western Siberia, the extent of the Don Glaciation beyond the Don lobe, and the glacial limits in the offshore areas around Britain.

We hope that this compilation will prove to be useful to the broad Quaternary community, will stimulate further research to close the gaps and will allow a better reconstruction of the earth's glacial history.

This work would not have been possible without the substantial and enthusiastic support and co-operation of all our colleagues who have contributed to this volume. It would not even have been begun without the encouragement of Professors Jan Mangerud, Jim Rose and Dave Mickelson. To all these people, and anyone we have forgotten, we express our sincere thanks.

Jürgen Ehlers & Philip Gibbard

Cambridge, 4.10.02

Contents

Quaternary Glaciations – Extent and Chronology
Editors J. Ehlers and P.L. Gibbard

Quaternary glaciations in Austria

Dirk van Husen

Institut für Ingenieurgeologie, Technische Universität, Karlsplatz 13, A- 1040 Wien, Austria

E-mail: dirk.van-husen@telering.at

Abstract

From the onset of rhythmic loess accumulation, at the turn from the Gauss to the Matuyama Epoch, a frequent variation from humid warm to dry cool climate with loess and occasional gravel accumulation occurred until the end of the Matuyama Epoch. Remnants of four glaciations (Günz, Mindel, Riss, Würm) within the Eastern Alps and their foreland have long since been known. More recently evidence for an additional cold stage between the two older ones was found. As a result of a major cooling and build up of piedmont glaciers in the foreland, the four glaciations show a complete succession of terminal moraines with glaciofluvial terraces connected to them. The last interglacial/glacial cycle can easily be reconstructed climatologically and by sediment development. It may serve as a model for understanding the climatic conditions of the older ones, which had very similar successions.

The course of the Quaternary

Around the Neogene/Quaternary boundary (Aguirre & Pasini, 1985), at about 1.8 Ma BP (Fig. 1), the drainage pattern of the Eastern Alps was fully developed. Loess accumulation began at the beginning of the Upper Pliocene (Frank *et al.*, 1997), at the Gauss and Matuyama Epoch boundary, some 2.6 Ma ago. According to palaeontological investigations, the Neogene to the Quaternary boundary period was characterised by a moderately warm, humid climate alternating with cool, dry periods. There were frequent repeated changes of climate, but lacking any reference to glacial events. Within the Eastern Alps no sediments from this early phase of the Quaternary have been preserved, due to the high relief energy and the later glaciations. Sediments from this period, mostly loess, are found only in the Alpine foreland along the Danube and its tributaries.

The loess section at Krems shooting range (Fink *et al.*, 1976) reveals a repeated change of dry-cold (*Pupilla* fauna), warm dry (*Striata* fauna) and warm humid (*Chilostoma* fauna) conditions at the beginning of the Quaternary (Frank & van Husen, 1995; Fink *et al* , 1976). The sequence contains evidence of 17 interglacials. They post date the Olduvai Event and thus are younger than Marine Isotope Stage (MIS) 63. The loess was certainly deposited under dry, cold conditions. However, no evidence was found for any climatic deterioration strong enough to generate glaciation.

The four Alpine glaciations

Following younger interpretations of the marine oxygen isotope record, MIS 24 at about 0.9 Ma represents a clear transition to a different style of climatic regime characterized by more extreme glaciations than the previous period (Shackleton, 1995). This may explain why within the Eastern Alps and their foreland in Austria no glacial deposits older than the Günz Glaciation, *sensu* Penck & Brückner (1909/11), have been found.

In the Salzach, Traun and Krems river areas, terminal moraines of this glaciation are connected to terrace bodies (Weinberger, 1955; Kohl, 1974). These *'Ältere Deckenschotter' sensu* Penck & Brückner (1909/11) form part of a widespread gravel cover that occurs between the rivers Traun and Enns. In terms of gravel composition and age, this unit is a genetically polymict body (van Husen, 1980, 1981). It probably accumulated over a long time spanning several cold stages. Reworking and lateral erosion by the rivers resulted in the incorporation of older deposits into what appears now as a single terrace accumulation.

In the Alpine foreland east of the Salzach River (Figs. 2 and 3), remnants of three glaciations younger than the oldest represented above can be easily identified (Weinberger, 1955; Del Negro, 1969; Kohl, 1976; van Husen, 1977, 1996; Sperl, 1984). Knowledge regarding the extent, development of tills, terminal moraines, glaciofluvial sediments and weathering is good enough to allow the reconstruction of the individual ice streams and tongues of the piedmont glaciers (Penck & Brückner, 1909/11; Sperl, 1984).

Palaeomagnetic investigations (Fink, 1979) suggest that the tills of these glaciations and their connected outwash terraces all fall into the Brunhes Epoch. The marine oxygen isotope record suggests that four major glaciations affected the northern hemisphere during the Brunhes Epoch (Shackleton, 1987; Raymo, 1997), generally correlated with MIS 2, 6, 12 and 16 (Raymo, 1997). According to radiometric dating, stratigraphical position, as well as

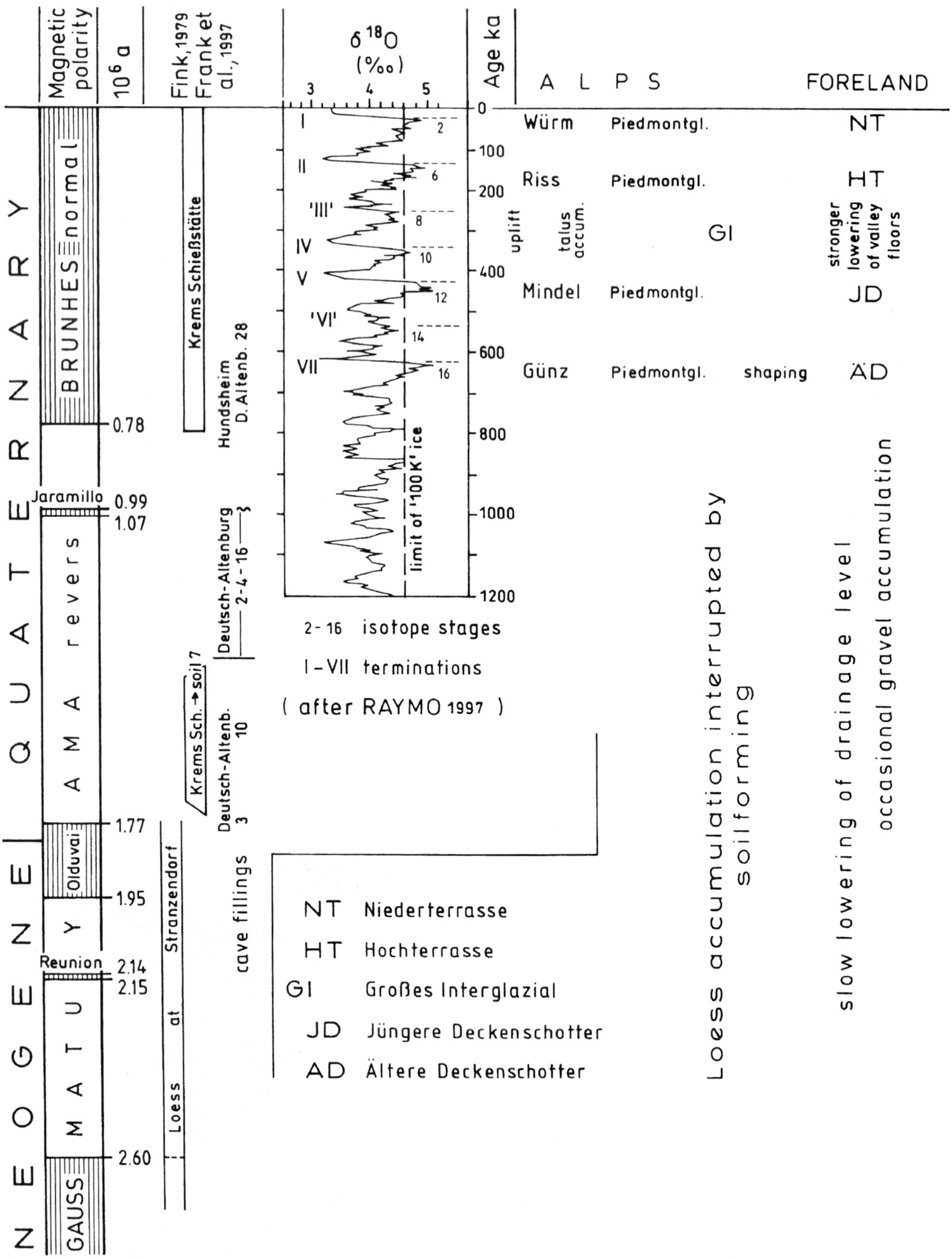

Fig. 1. Geological time scale of the Quaternary events in Austria.

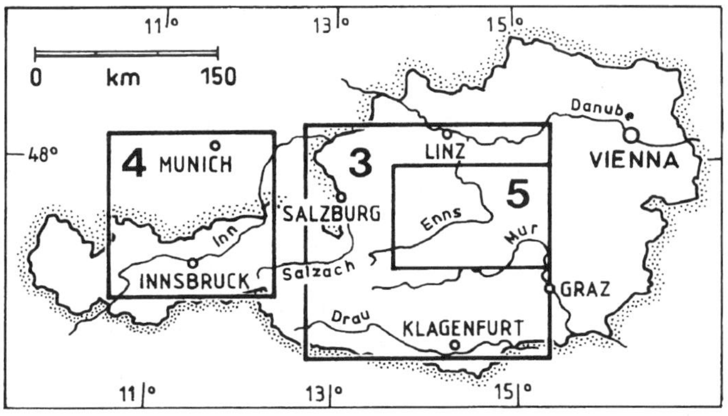

Fig. 2. Location map showing the main drainage systems of Austria. Squares mark figures 3, 4 and 5.

weathering, the two youngest Alpine glaciations are correlated with MIS 2 and 6. However, the much stronger weathering, cementation and periglacial modification of all the Mindelian deposits, in comparison to those of the Rissian Stage, indicates that a longer time span must have separated these glaciations. This *'Großes Interglazial'* (great interglacial) was first postulated by Penck & Brückner (1909/11). Thus, the Mindelian probably correlates with MIS 12 and the Günzian with MIS 16 (Fig. 1).

Remnants of another major cold period with gravel deposition in the Alpine foreland that fall between depositions of the Günzian and Mindelian stages have been found more recently (Kohl, 1976). They are separated from both by weathering. This cold stage probably correlates with MIS 14.

Tectonic activity

The incision of the circum-alpine drainage system as far as the level of the *'Ältere Deckenschotter'* and beyond seems to have been of an intensity similar to that observed in the Danube and its tributaries (Graul, 1937; Fischer, 1977; Fuchs, 1972). This suggests that no differential uplift has occurred and that there are no traces of other tectonic activity, such as faulting within the northern foreland of the Alps. However, age control of the remnants of terrace systems older than the *'Ältere Deckenschotter'* is not yet possible

On the basis of the loess sequence near Krems (Fink *et al.*, 1976), the amount of erosion along the Danube system was about 50 m during the Quaternary period (Fig. 1). This supposedly undisturbed development along the northern edge of the Alps ends where the Danube enters the Vienna Basin. Clear evidence of tectonic activity that continues well into the Holocene is recognisable here (Plachy, 1981). The glaciofluvial terraces dating from the last two glaciations (*Hochterrasse, Niederterrasse*, as well as Holocene deposits), show clear evidence of tectonic displacement; indeed parts of the *'Hochterrasse'* are displaced by as much as *c.* 10 m along major faults (Kröll *et al.*, 1993).

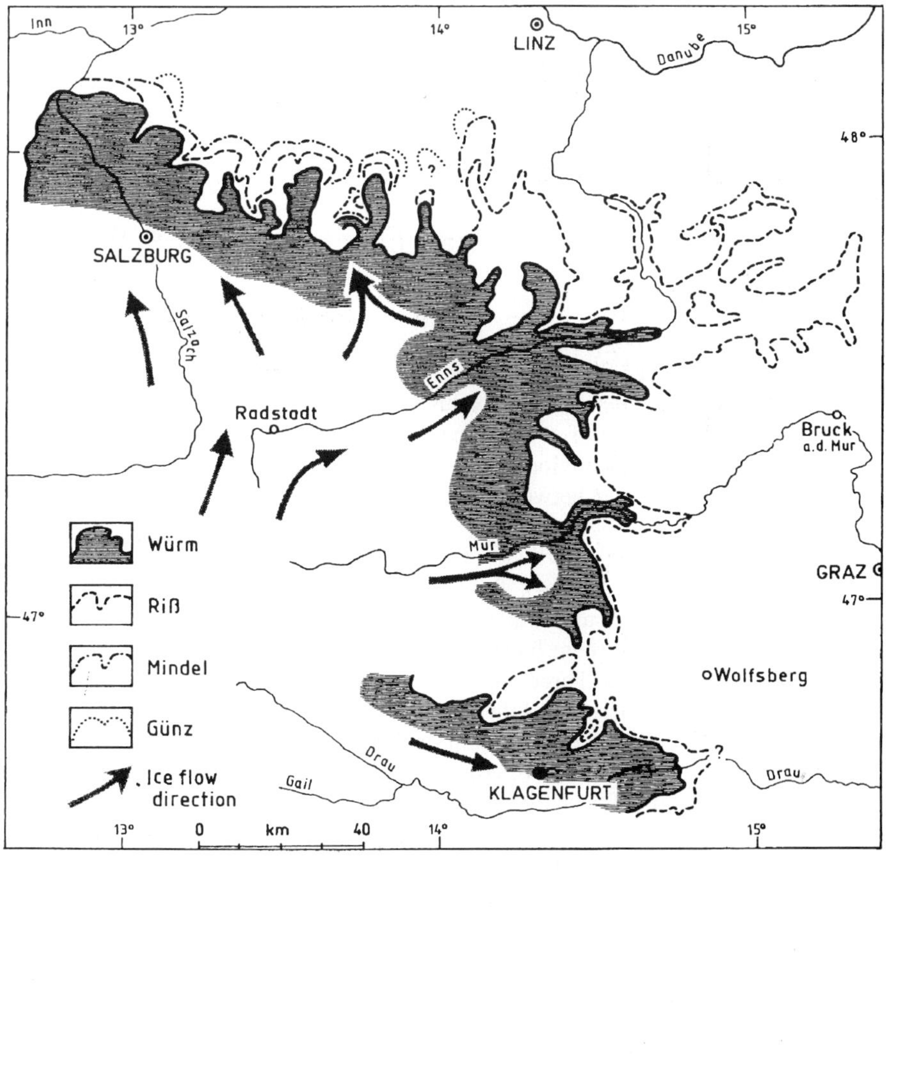

Fig. 3. The extent of the glaciers of the four glaciations known in Austria. Only glaciers of the main valleys are indicated. Ice flow direction is rather similar during each glaciation

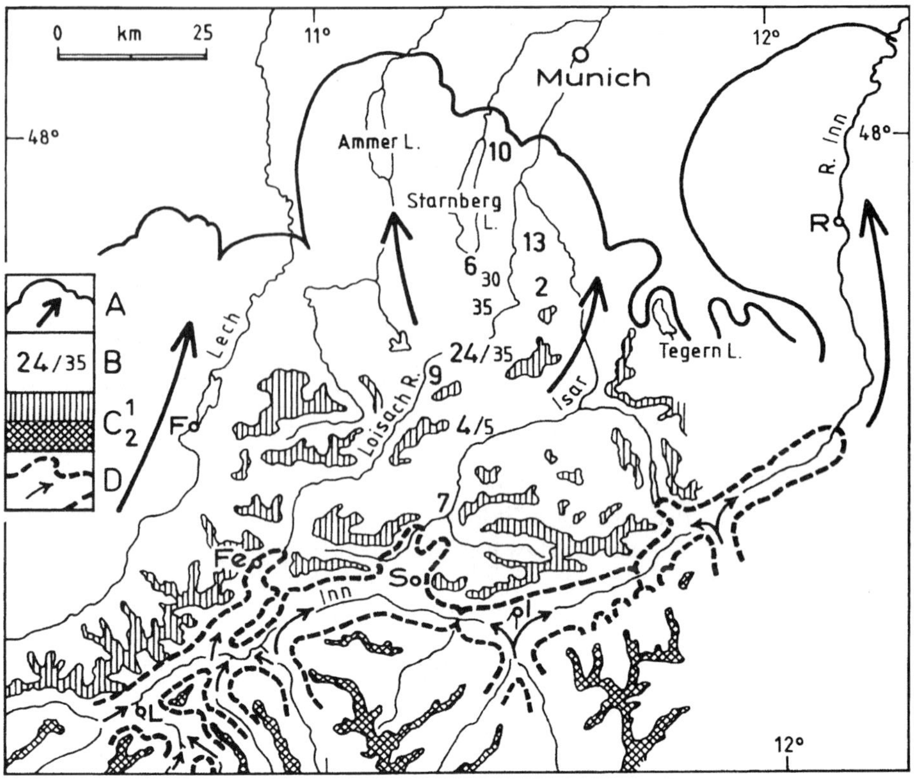

Fig. 4. Ice streams of the Inn, Lech, Isar and Loisach valleys. (A) Glacier extent during the Würm MIS 2. (B) Percentage of crystalline boulders. The larger numbers in basal till. (C) Nunataks formed of sedimentary rocks (C1) and crystalline rocks (C2). (D) Probable ice extent in the Inn valley around the beginning of the final ice build up during MIS 2. F: Füssen, Fe: Fernpass, I: Innsbruck, L: Landeck, R: Rosenheim, S: Seefeld.

Development of glaciers

The Alpine drainage system was filled by dendritic glaciers during all four major glaciations. The topography of the longitudinal valleys (e.g. the Inn, Salzach, Enns, Mur and Drau; Fig. 2) controlled glacier build-up in their tributaries and surrounding areas. On the one hand this explains the outstanding differences in glacier extent between some of the catchment areas. On the other it explains the strongly accelerated ice build-up towards the glacial maximum. Two examples are given here:

Inn valley

The Inn valley is the most extensive drainage system in the Eastern Alps. Its source areas are located south of the valley and in the Engadine, areas almost exclusively underlain by crystalline rocks. North of the Inn valley, the Northern Calcareous Alps have no major valley tributary of the Inn, which, follows the border between these two tectonic units east of Landeck (Fig. 4).

The valleys in the Northern Calcareous Alps are orientated to the north (Fig. 4), forming the heads of the rivers Isar and Loisach and other small rivers. The Inn valley is separated from them by mountain chains rising up to 2,500 m a.s.l. or more. Three gaps in the mountain chain are found at Fernpass, Seefeld and Achensee, with watersheds 400-600 m above the Inn valley. During the major glaciations, these cols were crossed by ice (Fig. 4).

Till composition reflects the petrographic composition of the catchment areas. In the basal tills of the Inn valley, crystalline components are dominant; only on the northern slope is there a considerable admixture of limestones, derived from the local bedrock. However, up to 35% of crystalline erratics (Fig. 4) are found in the glacial and glacigenic material of the Würmian Isar and Loisach glaciers (Dreesbach, 1983). Most of the crystalline material is fresh and unweathered except for a small amount affected by multiple reworking. Such a large amount of unweathered crystalline rocks has to be attributed to extensive ice transport from the Inn valley to the Isar and Loisach systems, across the Fernpass and the Seefeld passes (Penck & Brückner, 1909/11). This transfluence reflects the gradient of the ice surface. The ice filling the Inn valley reached a much higher elevation than in the valleys in the north. The extraordinarily high crystalline boulder content in the sediments in the advance phase of the ice streams suggests a strong and rapid ice build-up in the Inn valley between Landeck and Innsbruck. This would allow strong transfluence from the Inn valley even at an early phase of the glaciation, suppressing the influence of the local glaciers and affecting both till and outwash gravel composition.

Such an early overflow may result from the internal ice flow conditions within the Inn valley system. The large tributaries from the south, with their extensive source areas at high altitudes, carried huge glaciers during the final approach of the last glaciation, gradually filling the whole Inn valley. Hence, five glaciers finally merged around Landeck. At the same time, the ice flowing out of the Sill and Ziller valleys reached the main valley and thus the glaciers may have blocked each other creating a

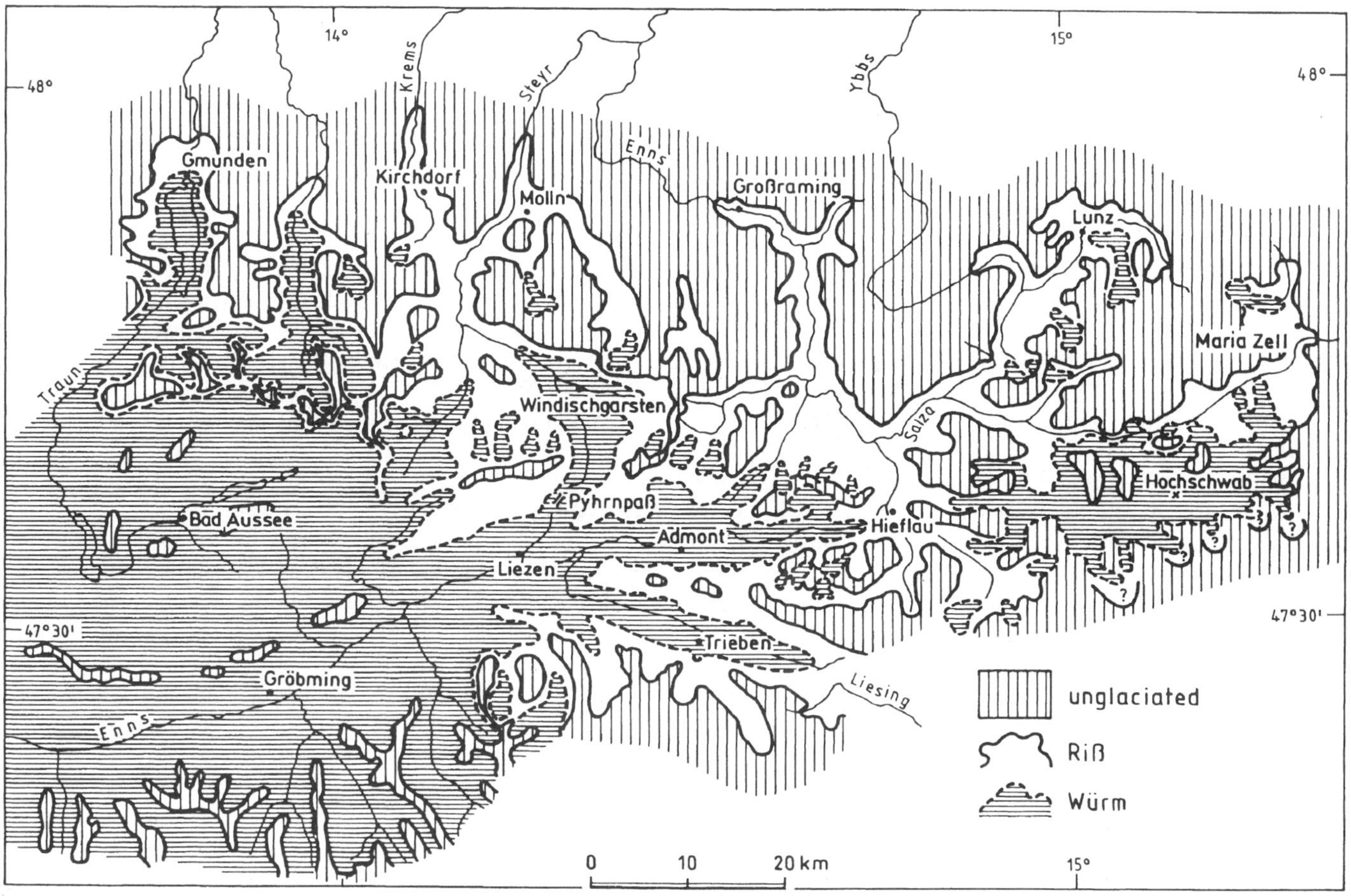

Fig. 5. Glacier extent during the Würmian (MIS 2) and Rissian (MIS 6) in the Enns valley and environs and in the Steyr and Ybbs valleys.

conspicuous rise in the ice surface. Therefore, crossing the watersheds the ice of the Inn valley system reached the drainage systems of the Isar and Loisach (with smaller catchment areas at lower altitudes) early enough before these valleys were filled with local ice strong enough to hinder this process.

On the other hand, ice congestion in the narrow Inn valley also resulted in a high ice-table leading to a rapid expansion of the feeding area, which again favoured rapid ice-buildup. Thus the high gradient to the north persisted throughout the entire glaciation. The volume of ice discharge north across the watersheds (crystalline material included) can be judged from the size of the Isar and Loisach piedmont glaciers as compared with their neighbours, which were fed only by their own source areas within the Northern Calcareous Alps (Fig. 4).

Enns valley

The Enns valley glacier (Fig. 2) was connected with the Salzach glacier in the west and the Traun glacier in the north (Fig. 5). The Enns valley follows the border between the crystalline zone in the south and the Northern Calcareous Alps in the north. The topography differs slightly from that of the Inn valley. In the western part, the tributaries from the south originate in extensive areas of high elevation, and the wall-like Northern Calcareous Alps drain mainly towards the north. Further to the east, however, the elevation of the crystalline zone declines abruptly over a short distance, and the Northern Calcareous Alps change from continuous chains and plateaux to more isolated mountains surrounded by much lower terrain.

During the Würmian, the Enns valley was occupied by a glacier extending down to the area of more isolated high mountains. The latter supported local glaciers (Fig. 5). Both the local glaciers and the valley glacier were in contact, but had little influence on one another with regard to the ice discharge. To the south, a 20 km long tongue of the valley glacier entered the otherwise unglaciated valley around Trieben. To the north, ice crossed the Pyhrn Pass, entered the drainage system of the Steyr river and filled the small basin at Windischgarsten. The content of crystalline boulders in the tills is around 2-3% in this area.

During the penultimate glaciation (Rissian), the equilibrium line of the glacier system was about 100 m lower than during the Würmian (Penck & Brückner, 1909/11). This led to more extended piedmont glaciers on the northern rim of the Alps and longer valley glaciers within the mountains. The difference in length was in most cases about 5-6 km (Fig. 3). The situation was very different in the Enns and Salza valleys and the Steyr-Krems drainage system (Fig. 5). In the former, the end moraines at Grossraming (van Husen, 1999) show a glacier extending *c.* 40 km further than during the Würmian. The tills throughout the valley and in the immediate vicinity north of Hieflau contain abundant crystalline boulders, indicating that the Enns valley was filled by a much larger Enns glacier with little influence from the local glaciers.

With respect to the lower position of the equilibrium line during the Riss glaciation, both the local glaciers and the valley glacier grew much larger. Thus, the local glaciers were powerful enough to block the valley glacier in the very narrow portion of the valley west of Hieflau. This impediment to ice flow caused a higher ice surface in the Enns valley west of Admont. This area, that formed part of the ablation area during the Würmian, consequently became an accumulation area in the Rissian Stage. This considerable addition to the accumulation area, together with the ice supply from the local glaciers, affected the extent of the glacier, as explained above.

The glaciers on the north slope of Hochschwab grew in the same way (Kolmer, 1993: Fritsch, 1993), filling the Salza valley, which became part of the accumulation area. As a result an extended ice stream formed here. This process may have been accelerated by the additional cooling effect of the glaciated Enns valley (Fig. 5).

The same feedback mechanism occurred in the Windischgarsten basin. The stronger transfluence of ice from the Enns valley in the south caused a larger glacier tongue to develop and to extend further and further into the basin. In the same way the larger local glaciers contributed to the overall ice volume. Thus, the whole basin became part of the accumulation area. This greatly enlarged feeding area enabled the glacier to advance down valley over 30 km beyond its Würmian limits. It was even powerful enough to cross the watershed into the Krems river system (at Kirchdorf on the Krems), generating a much greater ice stream in that valley (Kohl, 1976).

Overdeepened valleys

Since the beginning of research on ancient glaciers and their palaeogeographical distribution, overdeepened tongue basins have been known (Fig. 6). They seem to form predominantly in the ablation area, where the higher ice velocity increases the basal debris load and where basal meltwater drainage under hydrostatic pressure occurs (van Husen, 1979). They were shaped successively during each

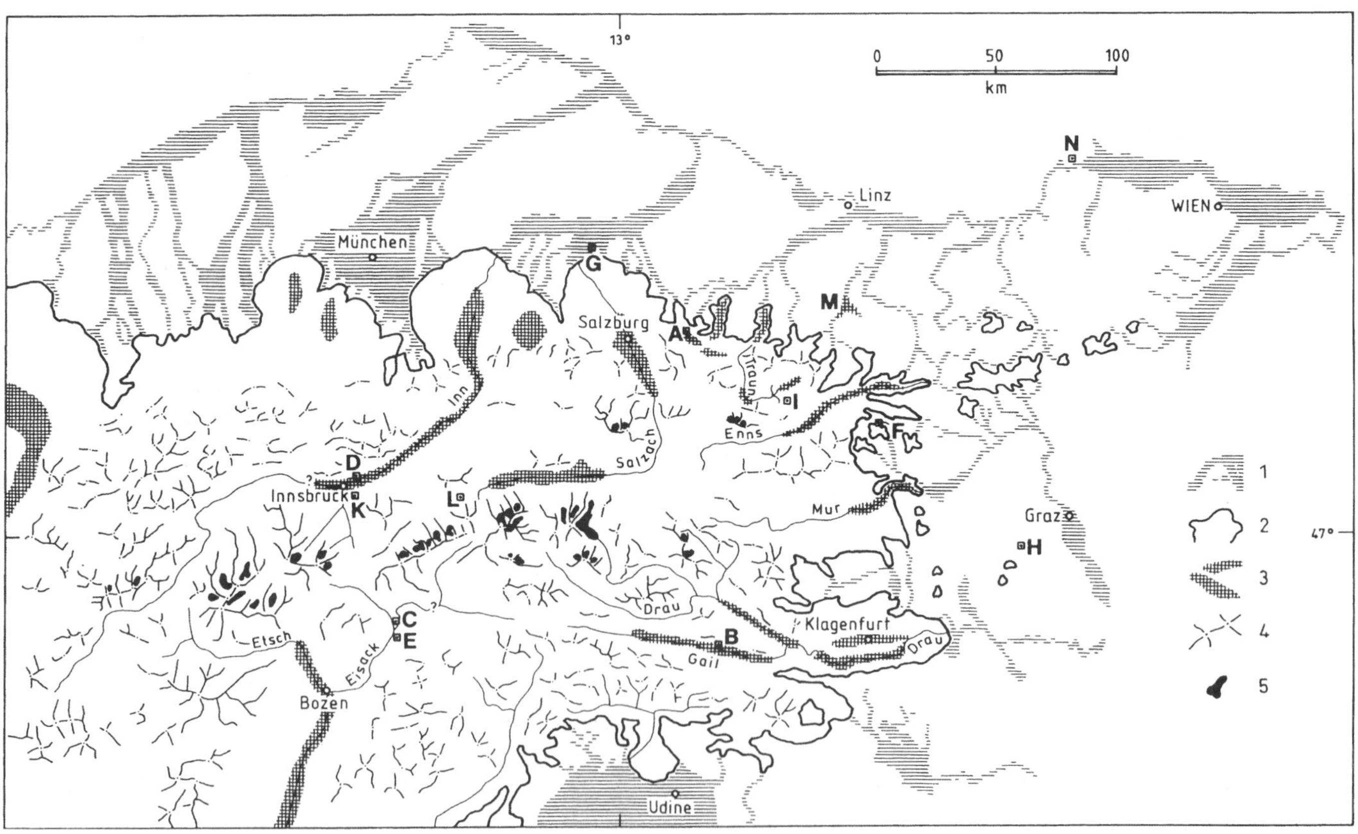

Fig. 6. Sketch map of the Eastern Alps during Würmian (MIS 2); 1 Terrace 'Niederterrasse'; 2 Maximum extent of glaciers; 3 Overdeepened parts of the valleys; 4 Nunataks. 5; Glacier extent of the Holocene. Localities mentioned in the text: A: Mondsee; B: Nieselach; C: Schabs; D Baumkirchen; E: Albeins; F: Hohentauern; G: Duttendorf; H: Neurath; I: Mitterndorf; K: Lans; L: Gerlos; M: overdeepened area at Molln.

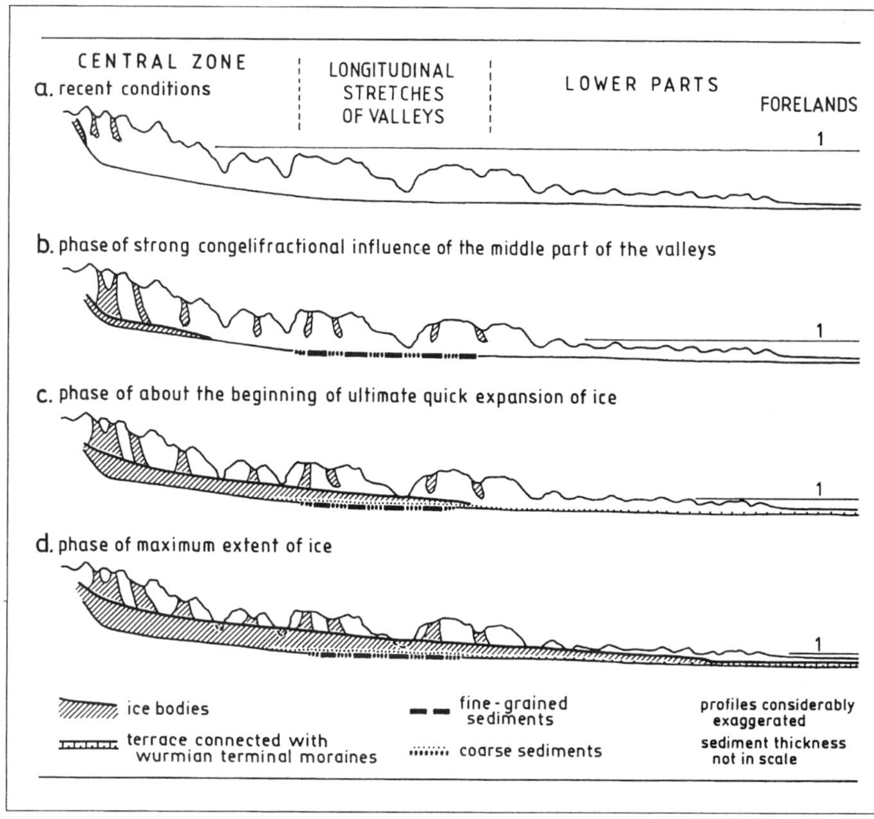

Fig. 7. Sketch profiles showing the consequences of gradual climatic deterioration indicated by the estimated lower limit (l) of strong congelifraction. It shows that only a little lowering of this limit and the equilibrium line (c to d) is necessary to cause a very rapid and substantial expansion of the valley glaciers, such as during the Würmian maximum (see Fig. 8).

glaciation because the tongue areas always developed more or less in the same positions (*cf.* Fig. 3). Hydrogeological investigations in the longitudinal valleys provide new data on the shape and depth of these basins. Geophysical investigations, together with boreholes also provide good evidence on the sediment filling of the basins and the position of the underlying bedrock.

Thus, on the one hand, the underlying bedrock in the tongue basins of the Salzach glacier was cored repeatedly between 160-340 m, on the other the base in the Gail and Drau valleys was found at 200-240 m (Kahler, 1958). These investigations did show that the overdeepening may be limited to about 400 m in this part of the Austrian Alps. Similar depths are also found in some lakes (e.g. Traunsee) which have a depth of nearly 200 m and a thick sediment fill at the bottom. A depth of *c.* 200 m was also determined for the Steyrtal which was affected by the most extensive glaciations (Fig. 6, M) (Enichlmaier, pers. comm.).

Stronger overdeepening is reported from the Inn valley. East of Innsbruck, for example, the pre-Quaternary basement lies at *c.* 180 m above sea level, about 400 m below the valley floor (Aric & Steinhauser, 1977). Further to the east, seismic investigations revealed that the bedrock lies at about 500 m below sea level (1000 m below the valley floor). The latter was proved by a drill hole 900 m deep that passed through unconsolidated gravel, silt and clay without reaching the bedrock (Weber *et al.*, 1990).

This excessive amount of erosion and overdeepening may result from the stronger linear ice and meltwater discharge along valleys with extensive catchment areas. It is in close accordance with the erosion of about 600-1000 m, recently reported from the longitudinal valleys of Rhine and Rhône in Switzerland (Pfiffner *et al.*, 1997).

The filling of the overdeepened valleys depends strongly on the relation of the major rivers to their tributaries, in terms of water and debris discharge, as well as on the size of the basin. Large basins, with a strong main river and small tributaries, were often filled with a thick sequence of fine-grained bottom-set beds that interfinger with coarse delta deposits (e.g. Salzachgletscher: Brandecker, 1974). Strong input of coarse gravel and sand from major tributaries creates a more inhomogeneous valley fill with alternating layers of gravel, sand and silt all over the area.

Last Interglacial-Glacial Cycle

According to oxygen isotope records, all climatic cycles within the Brunhes Epoch and in the Late Matuyama Epoch show a similar pattern. This is particularly true of the four major cycles before the Terminations I, II, V and VII, following the major glaciations at MIS 2, 6, 12 and 16 (Raymo, 1997) each of which is characterised by a step by step cooling, interrupted by short phases of climatic amelioration, leading eventually to the very cold short true glaciation period (Fig. 1). The last, Eemian/Würmian, climatic cycle is relatively well investigated. Therefore, it may serve as a model for the reconstruction of the other cycles in terms of climatically-induced sedimentation and

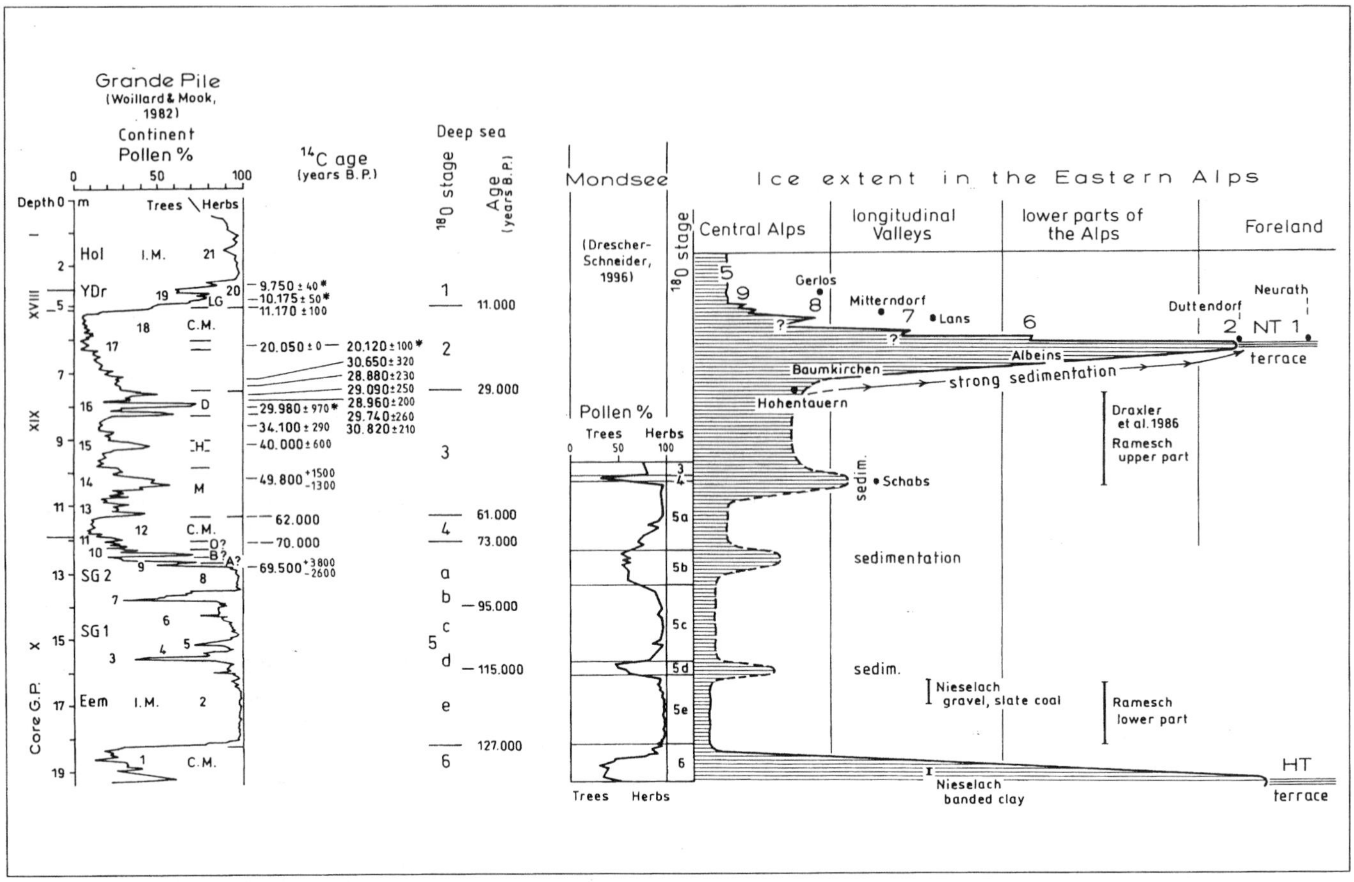

Fig. 8. Temporal development of the ice extent in the Eastern Alps over the last 140,000 years (after van Husen, 2000).

facies diversification in the Eastern Alps (van Husen, 1989).

On the northern edge of the Eastern Alps (Fig. 6, A), the Mondsee sedimentary sequence has given good, continuous evidence on the climatic development from Termination II, through the Eemian, and well into the first half of the Würmian Stage. The fine-grained sediments, north of the shoreline of the present lake Mondsee, were first investigated by Klaus (1987), and believed to represent a complete sequence covering the time between the Rissian and Würmian glaciations. Recent investigations, based on three long cores have revealed a delta structure in an ancient lake with a water surface around 50 m above the present lake level. Bottom set, foreset and thin topset beds of a classical Gilbert delta structure were covered by the till of the last glaciation (Krenmayr, 1996).

The results of palynological investigations (Drescher-Schneider, 1996) suggest that the Eemian in this area was a warm period with temperatures averaging 2-3 °C above the current Holocene values. The valleys and foreland of the Alps were densely forested at this time with well-developed mixed oak forest with a high content of *Abies* (fir). This phase ended with an abrupt climatic deterioration that affected the forest on the northern edge of the Alps and brought coarser sediments to the delta.

During the Early Würmian, forests recovered twice, showing some elements of mixed oak forest, with a cold stadial intervening between these periods. During the cold stadial the treeline descended close to the lake level. The cold period at the beginning of the Middle Würmian saw the treeline more or less at the level of the Alpine foreland. This was followed by a slight warming, allowing a forest dominated by *Larix* (larch) and *Picea* (spruce) to grow around the lake. This series of events corresponds closely to the Samerberg sequence from east of the Inn valley (Grüger, 1979), as well as with that at La Grande Pile in the Vosges (Woillard & Mook, 1982).

Chronology

The final climatic deterioration and glacier advance phase is represented at the Würmian Albeins and Baumkirchen (Fig. 6 E, D) sites where till overlies gravel and lake deposits respectively. Radiometric dating indicates that the glaciers of the tributary valleys reached the main longitudinal valleys at about 25-24,000 years BP. The rate at which further ice build-up occurred is unknown, because of the total lack of chronological evidence. During this final climatic decay (Fig. 8) the valleys were filled with coarse gravel, "Vorstoßschotter" at many places to high elevations, as a result of progressive overloading of the main river with debris (van Husen, 1989). The "Vorstoßschotter" extend laterally into the terraces in the foreland, which accumulated at the same time as the thick gravel bodies along the rivers (Fig. 7). These were developed also in non glaciated areas by periglacial activity (congelifraction). In only two places, Duttendorf and Neurath (Fig. 6 G, H), in the Eastern Alps can radiocarbon dates constrain the climax of the climatic deterioration and the maximum extent of glaciers and periglacial activity, as well as strong congelifraction and periglacial downwash around 20,000 years BP (van Husen, 1989).

Around the Eastern Alps, detailed mapping of terminal moraines and outwash terraces (e.g. the Traun, Enns, Mur, Drau) has shown that most glaciers behaved in a similar way. First of all, the greatest extent of the glacier tongues is marked by small morainic ridges connected to outwash fans. After this, the glacier fronts retreated some hundreds of metres and formed distinctive, high and wide end moraines, also connected to outwash, which grades morphologically into the downstream fluvial terraces. These outwash fans, and their transition into downstream terraces, have a lower gradient than the earlier ones. Within 1-2 km, they are on the same level, merging to form the *'Niederterrasse'* (Lower Terrace) which continues downstream, and can also be traced along the Danube to Vienna (van Husen, 1987). The terraces also correspond to those of the unglaciated tributary valleys.

No evidence of weathering has been found between the sediments of these different Würmian terminal moraines, nor within the outwash gravel sequence. This suggests that very little time elapsed between these two events, the *'Maximalstand'* and the *'Hochstand'* (van Husen, 1977). It is not known how long the glaciers were in place to form the 20-40 m high moraines of the *'Hochstand'*, since no datable material has yet been found.

The first retreat from these terminal moraines was in the order of some hundreds of metres to several kilometres, depending on the size of the glacier. This generally led to a drainage concentration to only one or two outlets, and an initial minor incision into the outwash fan and terrace. This stage is characterized by small morainic ridges and kettle holes, indicating continuing permafrost conditions. Large blocks of ice were preserved below the sediments during the whole time span from the glacial maximum to its early retreat phase (van Husen, 1977). The kettle holes formed after the downmelting of the glacier tongue, when a single deeply-incised outlet drained the overdeepened glacier basin.

The duration of the maximum extent of glaciation and climatic deterioration during the LGM (Last Glacial Maximum) can only be tentatively estimated. After the ice advance around 21 ka BP, the glacier front may have remained in these three maximal positions for about 4000 years, on the basis of data from the subsequent deglaciation phases.

Phase of ice decay

Following the LGM, large-scale retreat and downwasting of the glacier tongues began in the Alpine foreland, as well as in the valleys. In all the great valleys of the Eastern Alps (Inn, Salzach, Drau, Mur, Enns and Traun) no sequences of end moraines or equivalent evidence of former ice margins have been found from this phase. Only kames and ice-marginal terraces, that formed in temporary lakes have been identified. These ice-contact sediments, developed especially around the overdeepened parts of the valleys (Fig. 8), indicate continuous downmelting, without glacier stillstands or readvances. With respect to the distribution and internal structure of the sediments which formed in temporary lakes, downmelting was rapid. Some hundreds to one thousand years may have been all that was necessary for the loss of about 50% of the glacier lengths in the Eastern Alps. Ice lakes first formed at the glacier margins, and then extended over the entire area of the overdeepened basins. This probably resulted in glacier calving, which would have enhanced the rate of ice recession.

Knowledge on the further recessional phases is mainly based on investigations in two valley systems. One is the Traun valley, where the complete sequence of retreat and readvance phases from the LGM to the beginning of the Holocene was mapped in this relatively short valley and investigated by sediment analysis, palynology and radiocarbon dating (Draxler, 1977; van Husen, 1977). A second is the Inn valley where all the type localities of these events are situated (Mayr & Heuberger, 1968), and where intensive investigations have recently been undertaken by S. Bortenschlager and G. Patzelt. The close correspondence of the sequences in both areas, in terms of sediment and vegetation development and radicarbon dating, allows the use of classical terms for easier understanding. The following paragraphs describe key sites that provide some evidence about the deglaciation of the Eastern Alps.

The Bühl Phase

The first sign of a halt in the downmelting of the glaciers is marked around the intramontane basin of Bad Ischl. Here extensive kame deposits, partly covered by a thin layer of till, are connected to small morainic ridges, and this assemblage suggests a stillstand of the glacier margin with minor oscillations (van Husen, 1977). This phase is comparable to the 'Bühl Stage' of Penck & Brückner (1909/11), as shown by the more detailed investigation of the type locality by Mayr & Heuberger (1968). The lithology of pebbles and boulders in the till suggests that a dendritic ice stream in the main valley was still connected to glaciers in all of the tributary valleys at this time. Large kame terraces also exist in other valleys, such as the Drau valley, but they have not been mapped and studied in detail. Nevertheless, their general distribution suggests that they

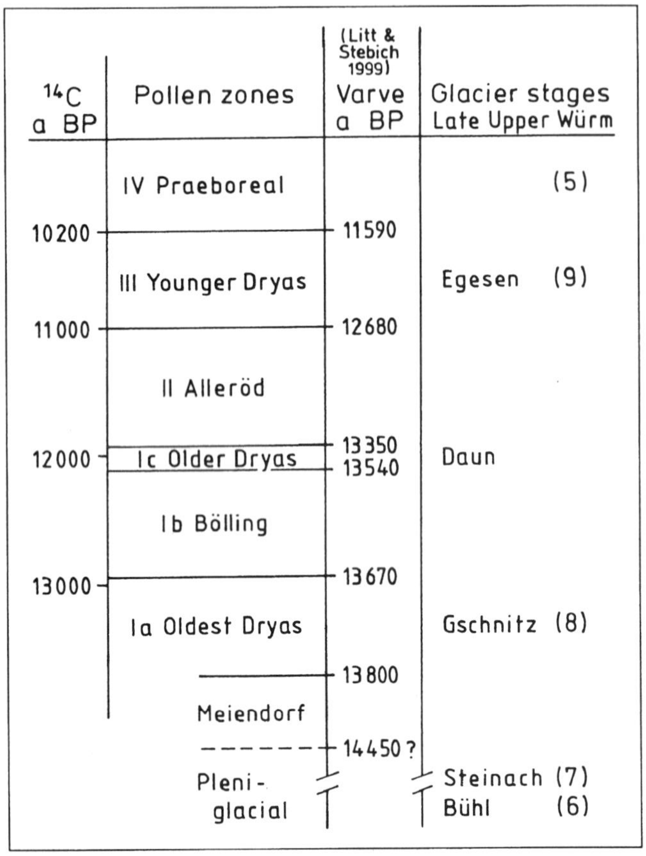

Fig. 9. Temporal position of the late Upper Würmian phases (Termination 1). For explanation see Fig. 8).

were associated with ice streams comparable to those typical of the Bühl Phase (Fig. 8).

The Steinach Phase

The phase of deglaciation that followed the Bühl was characterized by a minor readvance of the by then much smaller glaciers, again linked with kame terraces and inactive ice masses. Thus, the glacier tongue in the Traun valley near Bad Goisern had advanced over lacustrine and fluvial sediments deposited high above the valley bottom, apparently formed when drainage in the valley to the north was still blocked by stagnant ice masses (van Husen, 1977). A similar situation was described for the Steinach Phase (a term introduced by Senarclens-Grancy, 1958) from the Sill valley, south of Innsbruck (Mayr & Heuberger, 1968). Here, a thick sequence of gravels was also deposited in contact with stagnant ice masses and was covered by the till of an ice readvance. Thus, after the Bühl Phase, the main valleys had become free of active ice and the glaciers had retreated into the tributary valleys. The time interval between the two phases could not have been very long, since large inactive ice masses lingered in the valleys, despite climatic amelioration and large meltwater lakes (Figs. 8 and 9).

The Gschnitz Phase

The next phase in the deglaciation sequence is marked by well-developed blocky end moraines. In the Traun valley, these are present around Bad Goisern and in all of the other source areas of the Würmian ice stream. The morainic ridges are connected to outwash gravels almost everywhere. In the Traun valley north of Bad Goisern and around Bad Aussee, these deposits form terraces extending about 10 km downstream. This indicates that the valleys were free of dead-ice, permitting free drainage along the valley bottoms (van Husen, 1977). This suggests a comparatively long period of time, probably climatic amelioration between the Gschnitz and Steinach Phases, and considerable ice recession (Fig. 8). The Gschnitz moraines are relatively unmodified by slope processes suggesting that little or no solifluidal shaping has occurred since their deposition. The moraines of the Steinach Phase, in contrast, were clearly smoothed by solifluction during the following (Gschnitz) glacial event. This Gschnitz Phase is also well developed at Trins, south of Innsbruck (Mayr & Heuberger, 1968). Moreover, similar distinctive moraines can be recognized in many of the large tributary valleys draining from higher parts of the central Eastern Alps, as well as in the high cirques to the north and south. This implies that the glacial event was regionally extensive, reflecting a uniform lowering of the equilibrium line (ELA) to a position about 600 m lower than that of the Little Ice Age (Gross *et al.*, 1978).

Chronology

These glacial phases have been indirectly dated by palynological studies of bogs in the Traun valley (Draxler, 1977, 1987). During the early phase, after the main LGM deglaciation, some depressions were slowly filled with varved clay. The pollen record of this time is dominated by *Artemisia, Helianthemum, Ephedra, Hippophaë* and *Juniperus*, in addition to *Pinus*; the *Juniperus* becomes important towards the end of the sequence. This vegetational assemblage, especially the high content of *Artemisia*, is typical of the pioneering phase under dry, cold conditions (Draxler, 1987). The same feature of this early phase has also been described from the western part of the Eastern Alps (Bortenschlager, 1984). The phase terminates with the rapid increase of *Pinus* pollen to values of 70-80%. This interval is well dated in the Traun area to around 12,300 ^{14}C years BP. The following dates are reported in van Husen (1977):

Moos Alm: 730 m a.s.l., 12,520 ±180 years BP. VRI-431;

Ödensee: 770 m a.s.l., 12,220 ±180 years BP. VRI-433;
Plakner: 550 m a.s.l., 12,410 ±190 years BP. VRI 430;
Ramsau: 515 m a.s.l., 11,970 ±200 years BP. VRI-432;
Rödschitz: 790 m a.s.l., 12,440 ±420 years BP. VRI-485.
Dates for the equivalent interval from the Tyrol, reported by Bortenschlager (1984), include:
Lanser See: 840 m a.s.l., 13,230 ±190 years BP. HV 5269, and
Gerlos: 1590 m a.s.l., 12,155 ±210 years BP. HV 5284.

The difference of some 100 years between sites may be due to the contrasting rates of soil-forming processes on limestone and crystalline bedrock, as well as to differences in plant immigration. However, the dates suggest that this event occurred during, or at the end of the Bølling Ib Chronozone (Fig. 9). Bogs documenting this event in the Traun valley lie both outside and inside end moraines of the Gschnitz Phase. Thus, this glacial advance occurred no later than the Oldest Dryas.

This event can be dated more precisely in the pollen profile from Rödschitz in front of the Gschnitz moraines, in which the cooling event is marked by a strong increase in *Artemisia* at 6.40 m depth. Radiocarbon dates from gyttja at 5.40 m depth (12,420 ±440 years BP. VRI485) and organic detritus (pieces of shrub and herbs) at 7.20-7.00 m depth, yield an age of 15,400 ±470 years BP. VRI-484, suggest that the Gschnitz cold phase had occurred around 14 ka BP, assuming an approximately constant rate of sedimentation in the lake. A similar estimate was made by Patzelt (1975).

Recently it was attempted to date this event by surface exposure dating (^{10}Be und ^{26}Al) at the type locality (Ivy Ochs *et al.* 1997, 2000) indicating a final forming of the terminal moraine before 16,000 years. This probably would push it into a temporal position before the Meiendorf Stadial at the end of the Pleniglacial (Fig. 9).

The periglacial modification of end moraines of the earlier Steinach advance, and the lack of such reshaping on the Gschnitz moraines, agrees well with an Oldest Dryas age for the Gschnitz event, immediately preceding the climatic improvement at Termination I. After the Bølling Interstadial, there is no evidence for permafrost conditions on the floors of the main valleys in the Eastern Alps.

Based on evidence from the Rödschitz site (basal date of *c.* 15,400 ^{14}C years BP), the Steinach event may have occurred at around 16 ka BP (Fig. 9). Thus, the earlier Bühl Phase possibly culminated shortly before this date. During the warmer conditions of the Bølling Chron, the valley bottoms became ice free. Only high parts of the limestone plateaux of the Northern Calcareous Alps and valleys at higher elevations in the Central Alps remained covered with ice.

The Daun Phase

During the subsequent short, cold Older Dryas Chron, enhanced ice accumulation stimulated small glaciers on the higher limestone plateax, like the Dachstein Plateau. These small ice masses formed blocky end moraines, including boulders the size of houses. However, in general, this and the following (Egesen) event are not well-marked on the comparatively low terrain of the limestone plateaux. By contrast, in the higher parts of the Eastern Alps, moraines of Older and Younger Dryas age are well developed. The older event shows the ELA at a position more than 300 m lower than the A.D. 1850 (Little Ice Age) snowline (Gross *et al.*, 1978).

The Egesen phase

This event is marked by well-developed end moraines, which, according to palynological records and radiocarbon (Patzelt & Bortenschlager, 1978) and exposure dating (Ivy Ochs *et al.*, 1996), are believed to have formed during the Younger Dryas Chron of NW Europe. The ELA was lowered by about 300 m (Gross *et al.*, 1978) at this time, but arising from precipitation differences across the mountains, this value varied between 280 m in the drier continental part, to 400 m in the more oceanic northern ranges (Kerschner, 1980). Generally the Younger Dryas was characterized by drier, more continental conditions (around 70% of modern precipitation), with a lowering of the mean annual temperature by some 2.5-4°C). The climatic deterioration was felt most strongly in the drier central parts of the Alps. Many rock glaciers were reactivated under these cold and dry conditions (Kerschner, 1980).

With the onset of the Holocene, the glaciers receded behind their recent limits beginning with a sequence of readvances and retreats (Patzelt, 1995). The last significant ice advance was that during the Little Ice Age.

References

Aguirre, E. & Pasini, G. (1985). The Pliocene-Pleistocene boundary. *Episodes*, **8**, 116-120.

Aric, K & Steinhauser, P. (1977). Geophysikalische Untersuchungen des Inntaluntergrundes bei Thaur östlich von Innsbruck. *Zeitschrift für Gletscherkunde und Glazialgeologie*, **12**, 37-54.

Bortenschlager, S. (1984). Beiträge zur Vegetationsgeschichte Tirols I. Inneres Ötztal und unteres Inntal. *Berichte des naturwissenschaftlich-medizinischen Vereins Innsbruck*, **71**, 19-56.

Brandecker, H. (1974). Hydrogeologie des Salzburger Beckens. *Steirische Beiträge zur Hydrogeologie*, **26**, 26-39.

Del Negro, W. (1969). Bemerkungen zu den Kartierungen L. Weinbergers im Traungletschergebiet (Attersee- und Traunseebereich). *Verhandlungen der Geologischen Bundesanstalt*, 112-115.

Draxler, I. (1977). Pollenanalytische Untersuchungen von Mooren zur Spät- und postglazialen Vegetationsgeschichte im Einzugsgebiet der Traun. *Jahrbuch der Geologischen Bundesanstalt*, **120**, 131-163.

Draxler, I. (1987). Zur Vegetationsgeschichte und Stratigraphie des Würmspätglazials des Traungletschergebietes. *In:* van Husen, D. (ed.), Das Gebiet des Traungletschers, O.Ö. Eine Typregion des Würm-Glazials. *Mitteilungen der Kommission für Quartärforschung der österreichischen Akademie der Wissenschaften*, **7**, 19-35.

Dreesbach, R. (1983). Sedimentpetrographische Untersuchungen zur Stratigraphie des Würmglazials im Bereich des Isar-Loisachgletschers. Dissertation, München: 176 pp.

Drescher-Schneider, R. (1996). Die Klimaentwicklung des Eems in den nördlichen Voralpen an neuen Daten der Interglazialablagerung von Mondsee. DEUQUA '96' Alpine Gebirge im Quartär p 8, Hannover.

Fink, J. (1979). Stand und Aufgaben der österreichischen Quartärforschung. *Innsbrucker Geographische Studien*, **5**, 79-104.

Fink, J , Fischer, H., Klaus, W, Koci, A., Kohl, H., Kukla, J., Lozek, V., Piffl, L & Rabeder, G. (1976). Exkursion durch den österreichischen Teil des nördlichen Alpenvorlandes und den Donauraum zwischen Krems und Wiener Pforte. *Mitteilungen der Kommission für Quartärforschung der Österreichischen Akademie der Wissenschaften*, **1**,113 pp.

Fischer, H. (1977). Tal- und Hangentwicklung in pleistozänen Schotterterrassen (Enns Ybbs-Platte). *Zeitschrift für Geomorphologie, N F, Supplement-Band*, **28**, 161-180.

Frank, C. & van Husen, D. (1995). Eastern Alps Traverse. *In:* Schirmer, W (ed.), *Quaternary field trips in Central Europe International Union for Quaternary Research, XIV International Congress Berlin*, **1**, 417-418, München, Friedrich Pfeil.

Frank, Ch., Nagel, D. & Rabeder, G. (1997). Chronologie des österreichischen Plio-Pleistozäns. *In:* Döppes, D & Rabeder, G. (eds), *Pliozäne und pleistozäne Faunen Österreichs*. 359-374, Wien, Österreichische Akademie der Wissenschaften.

Fritsch, A. (1993). Das Quartär der westlichen Hochschwab-Nordabdachung unter Berücksichtigung des Bergsturzes von Wildalpen. Unpublished Dipl.-Arbeit, 122 pp., Wien.

Fuchs, W. (1972). Tertiär und Quartär am Südostrand des Dunkelsteiner Waldes. *Jahrbuch der Geologischen Bundesanstalt*, **115**, 205-245.

Graul, H. (1937). Untersuchungen über Abtragung und Aufschüttung im Gebiet des unteren Inn und des Hausruck. *Mitteilungen der Geographischen Gesellschaft München*, **30**, 181-256.

Gross, G., Kerschner, H. & Patzelt, G. (1978). Methodische Untersuchungen über die Schneegrenze in alpinen Gletschergebieten. *Zeitschrift für Gletscherkunde und Glazialgeologie*, **18**, 249-286.

Grüger, E. (1979). Spätriß, Riß/Würm und Frühwürm am Samerberg in Oberbayern - ein vegetationsgeschichtlicher Beitrag zur Gliederung des Jungpleistozäns. *Geologica Bavarica*, **80**, 5-64.

Ivy Ochs, S., Schlüchter, Ch., Kubik, P., Synal, H.A., Beer, J. & Kerschner, H. (1996). The ^{10}Be, ^{26}Al und ^{36}Cl exposure age of an Egesen moraine at Julier Pass, Switzerland, measured with the cosmogenic radionuclides. *Eclogae Geologicae Helvetiae*, **89**, 1049-1063.

Ivy Ochs, S., Kerschner, H., Kubik, P.W., Synal, H.-A., Patzelt, G. & Schlüchter, Ch. (1997). Moraine formation in the European Alps mirrors North Atlantic Heinrich events. *Lunar and Planetary Institute Contribution* 921 (Seventh annual V.M. Goldschmidt Conference, Houston Ivy Ochs, S., H. Kerschner, P.W. Kubik, TX), 103.

Ivy Ochs, S., Kerschner, H., Kubik, P.W. & Schlüchter, Ch. (2000). Expositionsalter und paläoklimatische Interpretation der Gschnitz-Moräne in Trins, Stubaier Alpen. DEUQUA 2000 (Abstract Band) Bern.

Kahler, F. (1958). Die Tiefe der Felsoberfläche in den Senken des Klagenfurter Beckens. *Carinthia*, **11, 68**, 5-8.

Kerschner, H. (1980). Outlines of the Climate during the Egesen Advance (Younger Dryas, 11,000-10,000 years BP) in the Central Alps of the Western Tyrol, Austria. *Zeitschrift für Gletscherkunde und Glazialgeologie*, **16**, 229-240

Klaus, W. (1987). Das Mondsee-Profil; R/W Interglazial und vier Würm-Interstadiale in einer geschlossenen Schichtfolge. *Mitteilungen der Kommission für Quartärforschung der österreichischen Akademie der Wissenschaften*, **7**, 3-17.

Kohl, H. (1974). Die Entwicklung des quartären Flußnetzes im Bereich der Traun Enns-Platte/Oberösterreich. *Heidelberger Geographische Arbeiten, H.-Graul-Festschrift*, 31-44.

Kohl, H. (1976). In: J. Fink (ed.), Exkursion durch den österreichischen Teil des nördlichen Alpenvorlandes und den Donauraum zwischen Krems und Wiener Pforte. *Mitteilungen der Kommission für Quartärforschung der österreichischen Akademie der Wissenschaften*, **1**, 37-48.

Kolmer, Ch. (1993). Die quartäre Landschaftsentwicklung der östlichen Hochschwab Nordabdachung. Diplom-Arbeit, Naturwissenschaftliche Fakultät der Universität Wien, 111 pp.

Krenmayr, H G. (1996). Sedimentologie der letztinterglazialen lim nischen Deltasedimente von Mondsee als Basis für die paläoklimatologische Interpretation. DEUQUA '96' Alpine Gebirge im Quartär p. 26, Hannover

Kröll, A., Gnojek, I., Heinz, H., Jirecek, R., Meurers, B., Seiberl, W, Steinhauser, P., Wessely, G. & Zych, D. (1993). Karten über den Untergrund des Wiener Beckens und der angrenzenden Gebiete (+ Erläuterungen). Geologische Bundesanstalt, Wien.

Mayr, F. & Heuberger, H. (1968). Type Areas of Late Glacial and Post-Glacial Deposits in Tyrol, Eastern Alps. *In: University of Colorado Studies: Glaciation of the Alps*, 143-165.

Patzelt G. (1975). Unterinntal-Zillertal Pinzgau, Kitzbühel. Spät- und postglaziale Landschaftsentwicklung *Innsbrucker geographische Studien*, **2**, 309-329.

Patzelt, G. (1995), Eastern Alps Traverse. *In:* Schirmer, W. (ed.), *Quaternary Field Trips in Central Europe. International Union for Quaternary Research, XIV International Congress,* **1**, 385. Berlin.

Patzelt G. & Bortenschlager, S. (1978). Spät- und nacheiszeitliche Gletscher- und Vegetationsentwicklung im inneren Otztal. Führer zur Tirol-Exkursion 'Innsbrucker Raum und Ötztal', DEUQUA, 13-25, Innsbruck

Penck, A. & Brückner, E. (1909/11). *Die Alpen im Eiszeitalter* 1199 pp., Leipzig, Tauchnitz.

Pfiffner, O. A., Heitzmann, P., Lehner, P., Frei, W., Pugin, A. & Felber, M. (1997). Dynamic Alps. Incision and backfilling of Alpine valleys: Pliocene, Pleistocene and Holocene processes *In:* Pfiffner, O.A., Lehner, P., Heitzmann, P., Mueller, St. & Steck, A. (eds), *Deep structure of the Swiss Alps*. 265-288, (Birkhäuser), Boston, Berlin.

Plachy, H. (1981). Neue Erkenntnisse zur Tektonik im Wiener Raum. *Mitteilungen der Österreichischen Geologischen Gesellschaft*, **74/75**, 231-243.

Raymo, M.E. (1997). The Timing of Major Climatic Terminations. *Paleoceanography*, **12**, 577-585.

Senarclens-Grancy, W. (1958). Zur Glazialgeologie des Ötztales und seiner Umgebung. *Mitteilungen der Geologischen Gesellschaft Wien*, **49**, 275-314.

Shackleton, N. J. (1987). Oxygen Isotopes, Ice Volume and Sea Level. *Quaternary Science Reviews*, **6** (183-190.

Shackleton, N. J. (1995). The Deep Sea Sediment Record and the Pliocene-Pleistocene Boundary. *Quaternary International*, **40**, 33-35.

Sperl, H. (1984). Geologie und Sedimentologie des Quartärs im Attergau/Oö. Dissertation, naturwissenschaftliche Fakultät der Universität Wien, 251 pp.

van Husen, D. (1977). Zur Fazies und Stratigraphie der jungpleistozänen Ablagerungen im Trauntal. *Jahrbuch der Geologischen Bundesanstalt*, **120**, 1-130.

van Husen, D. (1979). Verbreitung, Ursachen und Füllung glazial übertiefter Talabschnitte an Beispielen in den Ostalpen. *Eiszeitalter und Gegenwart*, **29**, 9-22.

van Husen, D. (1980). Bericht 1979 über Aufnahmen im Quartär auf Blatt 66 Gmunden *Verhandlungen der Geologischen Bundesanstalt* **1980**, A42-A43.

van Husen, D. (1981). Geologisch-sedimentologische Aspekte im Quartär von Österreich. *Mitteilungen der Österreichischen Geologischen Gesellschaft*, **74/75** (197-230.

van Husen, D. (1987). Die Entwicklung des Traungletschers während des Würm-Glazials. *In:* van Husen, D., (ed.). Das Gebiet des Traungletschers, O.Ö. Eine Typregion des Würm-Glazials. *Mitteilungen der Kommission für Quartärforschung der Österreichischen Akademie der Wissenschaften*, **7**: 19-35.

van Husen, D. (1989). The last interglacial glacial cycle in the Eastern Alps. *Quaternary International*, **3/4**, 115-121.

van Husen, D. (1996). LGM and Late Glacial Fluctuations in the Eastern Alps. *Quaternary International*, **38**, 109-118.

van Husen, D. (1999). *In:* Egger, H.; *Geologische Karte der Republik Österreich 1:50,000, Blatt 69 Großraming*. Geologische Bundesanstalt, Wien.

van Husen, D. (2000). Geological Processes during the Quaternary. *Mitteilungen der Österreichischen Geologischen Gesellschaft*, **92**: 135-156.

Weber, F., Schmid, Ch. & Figala, G. (1990). Vorläufige Ergebnisse reflexionsseismischer Messsungen im Quartär des Inntals/Tirol. *Zeitschrift für Gletscherkunde und Glazialgeologie*, **26**, 121-144.

Weinberger, L. (1955). Exkursion durch das österreichische Salzachgletschergebiet und die Moränengürtel der Irrsee- und Attersee-Zweige des Traungletschers. *Verhandlungen der Geologischen Bundesanstalt* **1955**, 7-34.

Woillard, G.M. & Mook, W.G. (1982). Carbon-14 dates at Grande Pile: Correlation of land and sea chronologies. *Science*, **215**, 159-161.

Quaternary Glaciations – Extent and Chronology
Editors J. Ehlers and P.L. Gibbard

The main glacial limits in Belarus

A.K.Karabanov, A.V.Matveyev and I.E.Pavlovskaya

Institute of Geological Sciences of the Academy of Sciences of Belarus, Zhodinskaya Str., 7 Minsk, 220141, Belarus

E-mail: karab@ns.igs.ac.by

Introduction

The present interpretation of glacial limits in Belarus is based on data from the Geomorphological Maps of the Belarussian SSR (Gurski, 1980, 1986) drafted in accordance with the stratigraphical scheme of the Pleistocene which was accepted in the past by the majority of geologists (Gurski, 1974; Makhnach, 1971; Matveyev, 1995; Matveyev *et al.*, 1988). The aim of this paper is to present some of the latest modifications to the glacial pattern in Belarus in comparison to the 1980 Geomorphological Map. These modifications are based on the results of geological and geomorphological investigations undertaken during the last decade. The present authors follow the new stratigraphical scheme of the Pleistocene in Belarus as elaborated by Velichkevich and co-authors (1996) and the digital map 1: 1,000,000 presented is an attempt to integrate the revised ice-marginal positions into the previously-mapped glacial limits.

The present-day landscape of Belarus was shaped by repeated Pleistocene glaciations. In order to understand the glacial pattern in Belarus it is necessary first of all to summarize the glacial history. Five major ice sheets affected the country during the Pleistocene: the Narev, Berezina, Dnieper, Sozh and Poozerian glaciations. The sequence of glacial/interglacial deposits reaches a maximum thickness of 325 m, with an average thickness of 75-80 m (here and below: data on the thickness of glacial deposits cited after Matveyev, 1995).

Narev Glaciation

The oldest-known glaciation in Belarus is the Narev Glaciation. Deposits of this glaciation have been mostly eroded and disturbed during the subsequent ice advances. Today, Narev glacial sediments are nowhere exposed; they underlie the Berezina deposits.

Berezina Glaciation

The Berezina ice sheet covered almost all of Belarus (except the southernmost part of the country). Generally, the Berezina glacial sediments are 10-15 m thick in the northern part of the country, 50-70 m in the centre and 15-25 m in the southern part. One of the peculiarities of the Berezina deposits is the abundance of glaciolacustrine sediments which are widespread, especially in western and eastern Belarus. The other is the presence of an intricate network of glacial channels which are deeply-incised into the pre-Quaternary bedrock (to 125 m b.s.l.). After deglaciation, the landscape was dominated by relatively flat surfaces with very few highlands and with large morainic plateaux dissected by deep glacial valleys.

Dniepr Glaciation

The main morainic highlands of Belarus were originally formed during the Dniepr Glaciation. The Dniepr Stage saw the most extensive Pleistocene glaciation which advanced well beyond the limits of the country. The thickness of glacial deposits reaches over 100 m and is generally 40-50 m. The most remarkable result of this glaciation was the formation of the cores of the Grodno, Minsk, Oshmiany, Mozyr and other morainic highlands. In many respects these uplands controlled the dynamics of the last (Poozerian) glaciation, as well as peculiarities of the last ice advance during the Sozh Stage.

Sozh Glaciation

The Sozh glacial complex is 10-20 m thick on average with a maximum thickness of about 100 m (in the Minsk highland area). The distribution of the Sozh Till is irregular. In the northern part of Belarus it has been almost completely eroded. Some patches of the original till cover were preserved within the largest glacial depression in Belarus - the Polotsk glacial depression. Recent investigations have shown that the Sozh deposits are absent from eastern Belarus (Pavlovskaya *et al.*, 1997). This discovery has significantly changed opinions concerning the age of the upper till in that area. These results confirm the view of Velichkevich *et al.* (1996) that there was only the Dniepr and the Sozh deposits represent glacial stage sediments in the strict stratigraphic sense of this term.

Poozerian Glaciation

Deposits of the subsequent Poozerian Glaciation cover the northern and north-western part of Belarus. The thickness of the glacial sequence varies from a few metres to 70 m; on average they reach about 20-30 m. During the Poozerian

Analysis of the pebble fraction (2-5 cm) in both tills indicates their different composition. Clasts of sedimentary rocks (mainly dolomite, limestone, sandstone, marl, flint) prevail in the lower till, whilst crystalline rocks dominate in the upper till (Gurski *et al.,* 1990).

Since the same characteristics have been found in other sites within the Poozerian limits, it has been possible to identify the upper reddish-brown unit as the Poozerian Till. Therefore, these data suggest that the Upper Berezina lowland, above Borisov, was overridden by the Poozerian Glaciation. The maximum limit of the glaciation can be identified 40-50 km to the south along the Berezina river valley up to its bend at Bolshoye Stakhovo (some 5-7 km north of Borisov). This area was previously attributed to the Sozh Glaciation.

Further downstream, in the section of the Berezina river valley between Bobruisk and Parichi and also in the Sozh river valley downstream of Slavgorod, investigations of the till clast petrography (fractions of 1-2, 2-3, 3-5, 5-7, 7-10 cm) have been carried out (Gurski *et al.,* 1990). The results of these investigations suggest that separate glacial tongues might have extended significantly farther to the south (30-40 km in this area) then previously assumed.

Conclusions

On the basis of the results of recent investigations, the glacial pattern in Belarus was more complicated than previously thought. The most significant recent modifications of this pattern concern the limits of the Poozerian Glaciation in the northwestern and northern parts of the country. The maximum limit of this glaciation in NW Belarus was located south of the Vilia valley and the glacier also spread further south within the Upper Berezina lowland.

The modified limits of the older glaciations (the Dniepr Glaciation and the Sozh Substage) are mainly based on the results of till lithological studies and palaeobotanic investigations (Gurski, 1974; Makhnach, 1971; Pavlovskaya *et al.,* 1997; Savchenko & Pavlovskaya, 1999; Velichkevich, 1982; Velichkevich *et al.,* 1993). The main modifications concern the position of the glacial limit of the Sozh Substage in eastern Belarus and the maximum extent of the Dniepr Glaciation in southern Belarus.

References

Baltrunas, V., Krupickas, R. & Jurgaitis, A. (1985). Possibility of stadial and phasial till correlation by means of lithological and geochemical methods. *Nauchnye trudy VUZ Litovskoi SSR. Geologia,* **6**, Vilnius, 102-111 [*in Russian*].

Gaigalas, A. (1995). Glacial history of Lithuania. *In*: Ehlers, J., Gibbard, P.L. & Kozarski, S. (eds), *Glacial deposits in north-east Europe.* Rotterdam / Brookfield, Balkema, 127-155.

Guobyte, R. (1999). *Quaternary geological map of Lithuania. Scale 1:200 000.* Geological Survey of Lithuania, Vilnius.

Guobyte, R. & Pavlovskaya I. (1998). The stadial limits and maximum extent of the Weichselian glaciation in north-eastern Lithuania and north-western Belarus. *In:* Field Symposium on glacial processes and Quaternary environment in Latvia, 25-31 May, 1998. Abstracts. Riga. p. 23.

Gurski, B.N. (ed.) (1980). *Geomorphological Map of Belarussian SSR, Scale 1:500 000.*

Gurski, B.N. (ed.) (1986). *Geomorphological Map of Belarussian SSR, Scale 1:500 000.*

Gurski, B.N., Levkov, E.A., Karabanov, A.K. & Bessarab, D.A. (1990). New data about glaciations limits within the Belarus territory. *Doklady Akademii Nauk Belarusi.* **34, 4**, 345-348 [*in Russian]*.

Gursky, B.N. (1974). *Lower and Middle Antropogene of Belarus.* Minsk, Nauka i tekhnika [*in Russian*]. 144 pp.

Karabanov, A.K. (1987). *Grodno highland.* Minsk, Nauka i tekhnika [*in Russian*]. 108 pp.

Makhnach, N.A. (1971). *Stages of the Quaternary evolution of flora in Belarus.* Minsk, Nauka i tekhnika [*in Russian*]. 212 pp.

Matveyev, A.V. (1995). Glacial history of Belarus. *In*: Ehlers, J., Gibbard, P.L. & Kozarski, S. (eds), *Glacial deposits in north-east Europe.* Rotterdam, Brookfield: Balkema, 267-276.

Matveyev, A.V., Gurski B.N. & Levitskaya R.I. (1988). *Relief of Belarus.* Universiteskoye, Minsk. 320 pp.

Pavlovskaya, I. (1994). *The Polotsk glaciolacustrine basin: structure, relief and history.* Minsk, Nauka i tekhnika [*in Russian*]. 128 pp.

Pavlovskaya I., Simakova G. & Yakubovskaya T. (1997). New sections with Alexandrian interglacial deposits in eastern Belarus. *Doklady Akademii Nauk Belarusi,* **41, 5**, 95-99 [*in Russian with English summary*];

Sanko, A.F. (1987). Neopleistocene of north-eastern Belarus and adjacent areas. Minsk, Nauka i tekhnika [*in Russian*]. 178 pp.

Savchenko, I. & Pavlovskaya, I. (1999). Muravian (Eemian) and Early Poozerian (Weichselian) deposits at Azarichi (eastern Belarus). - In: *Proceedings of the Fifth European Palaeobotanical and Palynological Conference*, 523-527. Cracow.

Velichkevich, F.Yu. (1982). *Pleistocene flora of glaciated areas of the East-European Plain.* Minsk, Nauka i tekhnika [*in Russian*]. 239 pp.

Velichkevich, F.Yu., Rylova, T.B., Sanko, A.F. & Fedenia B.M. (1993). *The Pleistocene Bereza stratoregion in Belarus.* Minsk, Nauka i technika [*in Russian*]. 147 pp.

Velichkevich, F.Yu., Sanko, A.F., Rylova, T.B., Nazarov, V.I., Khursevich, G.K. & Litviniuk G.I. (1996). Stratigraphic scheme of the Pleistocene deposits of Belarus. *Stratigraphiya. Geologicheskaya korrelatsiya.* **4, 6**, 75-87. Moscow [*in Russian*].

Quaternary Glaciations – Extent and Chronology
Editors J. Ehlers and P.L. Gibbard

Glacial history of the Croatian Adriatic and Coastal Dinarides

Ljerka Marjanac[1*] and Tihomir Marjanac[2]

[1] *Croatian Academy of Science and Arts, Institute of Quaternary paleontology and geology, 10000 Zagreb, Ante Kovačića 5, Croatia*

[2] *Faculty of Science, Department of Geology, 10000 Zagreb, Zvonimirova 8, Croatia*

1. Introduction

The first concept that even Mediterranean (and consequently the Adriatic Sea) was affected by Pleistocene glaciation dates back to 1840 when Louis Agassiz briefly discussed the idea, but provided no supporting evidence. Recent comprehensive study of Quaternary sediments along the northern Croatian Adriatic coasts, and on Krk and Pag islands, included deposits which were attributed to the Middle and Late Pleistocene. Former biostratigraphical studies of Quaternary deposits based on ostracods from lacustrine deposits (Malez & Sokač, 1969; Šimunić, 1970) in Žegarsko polje (KS 13, Table 1), Erveničko polje (KS 14, Table 1) and Kninsko polje (KS 15, Table 1) yielded Mindel and Riss age of glaciations.

The view that at least the highest peaks of the Croatian Dinarides were glaciated dates from 1899 and Albrecht Penck's journey to the Dinarides. Hranilović (1901), Gavazzi (1903a,b) and Schubert (1909) also promoted the idea of glaciation on the Velebit Mountain, but it was not accepted by other researchers (*i.e.* Gorjanović, 1902). Later, the geomorphologist Milojević (1922) described what he interpreted as glacial deposits in the Paklenica canyon on the Velebit Mountain, as also did Bauer (1934/35) and Degen (1936). Their conclusions were rejected by Poljak (1947) who regarded all of the evidence quoted by previous authors as misinterpreted. In contrast, he favoured a fluvial origin of extremely coarse-grained sediments, first described by Milojević (*ibid.*), and here illustrated in Figures 3 and 4. Malez (1968) made a general statement that the mountain tops of the coastal Dinarides, including the Velebit Mountain, were glaciated during the Pleistocene, but provided no documentation. Nikler (1973) first described a well-preserved 'terminal' moraine ridge on the Velebit Mountain, at 920 m a.s.l. Later a more complex geomorphological study of the same part of the Velebit Mountain was undertaken by Belij (1985) who reconstructed the path and extent of the corresponding valley glacier. During the past ten years new evidence on the glaciation of the northern Velebit Mountain were presented by Faivre (1991), and Bognar *et al.* (1999).

2. Key sections

The data presented on the digital map are based on published results and recent sedimentological study of the Quaternary deposits in the coastal region and on the islands of the Croatian Adriatic (Hrvatsko Primorje and Northern Dalmatia) and partly its hinterland (Ravni Kotari and Bukovica regions) (Fig. 1). Fifteen locations with key sections (Table 1) have been selected and marked on the map. The study of Quaternary deposits at these sites has enabled interpretation of the glacial maxima during Pleistocene. Unfortunately, the inland part of Croatia has not been recently studied in detail and thus any data on glacial limits for this area are unavailable.

3. Glacial limits

The following information about glacial maxima has been included on the map:

A) Glacial limits: Early/Middle Würmian glacial maximum.
B) Sites which might be related to the Pleistocene glacial maximum.
C) Morphologically-expressed end moraines.
D) Ice-dammed lakes of Early/Middle Pleistocene age.

The boundaries of glacial maxima are still uncertain due to lack of sufficient stratigraphical data, and the fact that the exploration is still in progress.

Würmian Glaciation

The boundary of the Würmian glacial maximum probably corresponds to the Early or Middle Würmian glaciation which cannot yet be distinguished. There are indications of a possible Late Würmian glacial advance (LGM) on the Risnjak Mountain (KS 2, Table 1) (Bognar & Prugovečki, 1997). Because of the local character of the evidence, no 'glacial limit' could be drawn. The glaciated areas were certainly above the present altitude of 800 m a.s.l.. The known end moraines on the Risnjak Mountain are located at

* Corresponding author. Fax: +385+1+4856211; E-mail address: ljerka@hazu.hr

Table 1. Key sections presented on the map.

NO	KEY SECTION	LITHOLOGY	STRATIGRAPHY	ALTITUDE
1	GOMANCE	- glaciofluvial sands and gravels - glaciolacustrine siltstone	ca 17000 y BP (C14) Late Weichselian and older	900 m
2	RISNJAK	- terminal moraine	Weichselian	940 m and above
3	JURANDVOR Krk island	kame-terrace - coarse-grained stratified breccia with large boulders	pre-Weichselian (Middle Pleistocene)	190 m
4	GAJEVI Krk island	kame-terrace / terminal moraine - coarse-grained stratified breccia - glaciofluvial sands and gravels	pre-Weichselian (Middle Pleistocene)	80 m
5	BOŠANE Pag island (Fig. 6)	kame-terrace - glaciofluvial sands and gravels	pre-Weichselian (Middle Pleistocene)	75 m
6	GRADAC Pag island (Fig. 5)	kame-terrace - coarse-grained stratified breccia with large boulders	pre-Weichselian (Middle Pleistocene)	80 m
7	VELIKO RUJNO Mt. Velebit (Fig. 2)	medial moraine	Weichselian	840 - 950 m
8	VELIKA PAKLENICA (Figs. 3, 4)	terminal moraine with extra large boulders	Early or Middle Weichselian, possibly older	270 m
9	SELINE Starigrad-Paklenica	glaciolacustrine deposits - varved-like siltstones with dropstones	Middle Pleistocene	at sea level and below
10	NOVIGRADSKO MORE (Figs. 8, 9)	terminal moraine 1 and 2 glaciolacustrine deposits (varved-like siltstone with dropstones and rippled sands)	Middle Pleistocene	at sea level and below
11	KARINSKO MORE	glaciolacustrine siltstones + gravels with possible glacial push effect	Middle Pleistocene	at sea level and below
12	SMILČIĆ	proglacial outwash - braided stream sands and gravels	Middle/Upper Pleistocene	190 m
13	ŽEGARSKO POLJE	glaciolacustrine deposits	Mindel + Riss/Würm	63 m
14	ERVENIČKO POLJE	glaciaolacustrine deposits	Mindel + Riss/Würm	127 m
15	KNIN-GOLUBIĆ	moraine (diamictite) glacial lake deposits - varved siltstones glaciofluvial sands and gravels	Mindel + Riss Riss	300 m

940 - 950 m a.s.l. and are attributed to the Late Würmian (Bognar & Prugovečki, 1997). The moraine at Veliko Rujno (NW from Velika Paklenica canyon - KS 8), which lies between 830 and 950 m a.s.l., was interpreted as a terminal feature by Nikler (1973), Belij (1985) and Perica (1993). However, it shows all characteristics of a medial

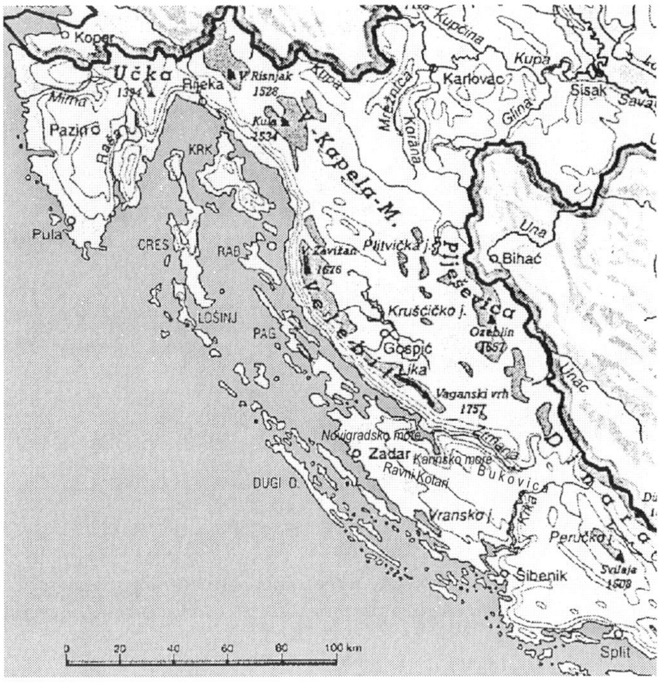

Fig. 1. Location map of the areas studied in Croatia.

moraine (Fig. 2) and is considered as such in this paper. Consequently, the corresponding glacial tongue proceeded further downslope towards lower altitudes leaving behind an extensive lag of coarse debris. An erosional remnant of end moraine also occurs in the Velika Paklenica canyon (KS 8, Table 1) (Figs. 3, 4) at about 250 m a.s.l. (Marjanac & Marjanac, 1999) and probably represents an Early or Middle Würmian glaciation. Thereafter, the boundary of Early or Middle Würmian glacial maximum was set approximately at an altitude of 800 m although some glacial tongues extended down to 250 m, as in the Paklenica canyon.

The Pleistocene glacial maximum

Whilst there is a general agreement on the extent of the Late Würmian glaciers, there has been no consensus on the extents of older glaciations. Traces of earlier glaciations are scarce and their interpretation remains controversial in this area considering the present overall knowledge about the extent of Alpine glaciations. The investigations during the past ten years have provided possible evidence of a very extensive glaciation in the Croatian Adriatic (Marjanac *et al.*, 1990). There are a number of features that could only be interpreted as of glacial origin after critical evaluation of all other possible sedimentary origins. These include, for example, kame-terraces on the Krk and Pag islands (Table 2) and glacial and periglacial deposits at Novigradsko More (KS 10, Table 1) and Karinsko More (KS 11, Table 1).

Features interpreted as kame-terraces occur on both slopes of the Baška valley (Krk island) and Pag valley (Pag island), and two types are recognised: a) slopeward-dipping, very coarse-grained and thick bedded sedimentary bodies with large boulders (Figs. 5) and b) sedimentary bodies composed of glaciofluvial sands and gravels with rare boulders. These clearly indicate longitudinal transport along the valley (Figure 6). A reconstruction of a possible glacial environment on the Pag island is given in Figure 7.

Along the southern coast of Novigradsko More there are exposures of various facies which are interpreted as representing glaciofluvial, glaciolacustrine and glacial (end moraines) environments. These deposits are extensively eroded. There are two levels of morainic deposits exposed along the coast: the older moraine is eroded and mainly represented by piles of very large boulders (up to 8 m in diameter). The younger unit is composed of pebbles, cobbles and rare large boulders supported in sandy matrix. The moraines contain predominantly rounded, but also faceted and striated boulders (Fig. 9). Other features exposed at Novigradsko More section include cryoturbations and sediment wedges which are either filled with fine-grained sediment or coarse gravel and are

Fig. 2. Medial moraine of Würmian age at Veliko Rujno (KS 7, table 1). A) seaward view of the composite moraine ridge and B) detail of the medial moraine ridge. The direction of ice movement was seawards, i.e. away from the camera.

Fig. 3. An end moraine of Early/Middle Würmian age in the Velika Paklenica canyon (KS 8, Table 1) in the southeastern part of the Velebit Mountains (270 m a.s.l.). The moraine consists mainly of very large limestone pebbles and boulders (more than 20 m in diameter), locally supported in a fine-grained matrix. The sediment is well-cemented here.

interpreted as possible ice-wedge casts. Together they clearly indicate deposition in a periglacial environment

The Karinsko More section (KS 11, Table 1) also shows glaciolacustrine deposits and one gravelly interval with cryoturbations, which are laterally equivalent to those in the Novigrad section.

The lack of stratigraphical evidence makes it impossible to distinguish individual boundaries of either the Mindelian nor Rissian-age glacial maxima. The only stratigraphical marker is known from the Žegarsko polje, the Erveničko polje (Malez & Sokač, 1969), and the Kninsko polje (Šimunić, 1970) where ostracod faunal assemblages found in glacial lake deposits indicate a Mindelian glaciation and Riss/Würm (Eemian) interglacial age. Equivalent lacustrine sediments over 90 m thick have also been recognised in the

Fig. 4. Detail of the end moraine in Velika Paklenica canyon (KS 8, Table 1). Boulders have rounded edges and range in size from less than 1 m up to tens of cubic metres. Interspices are filled with fine and coarse gravel, partly cemented.

Fig. 5. A possible kame-terrace on Pag island (KS 6, Tables 1 and 2). Stratified and cemented coarse-grained breccia dip towards the former ice-contact slope and contain a) rounded limestone boulders and b) rare extra-large boulders more than 10 m^3 in size (b). The scale on (a) is 25 cm.

Fig. 6. A possible kame-terrace on Pag island (KS 5, Tables 1 and 2) consists of glaciofluvial sands and gravels (predominantly carbonate in composition). In the centre of the picture rather steeply-dipping (ca 45°) Mesozoic limestone slope is exposed by sand exploitation and erosion.

borehole near Obrovac (Fritz, 1977). Varved-like silts with dropstones in the Novigrad section (Fig. 8) also contain rare ostracods which only indicate a period of cold climate, because the taxa found are of limited stratigraphical significance (Marjanac *et al.*, 1990). The lacustrine siltstones in the Seline section (KS 9, Table 1) near Starigrad-Paklenica, and Ražanac, on the opposite side of the Velebit channel, are the probable time-equivalent of those in the Novigradsko More section. At both sites they are associated with end moraines which form lateral barriers.

If the sites with lacustrine sediments represent ice-dammed lakes, they may have marked the retreat path of a large piedmont glacier, over 50 km long. On the basis of the debris provenance this glacier originated from the mountains north of the city of Knin. Since the Würmian glaciation was limited to areas over 800 m a.s.l., this large glacier, together with the associated periglacial features, are attributed to an Early or Middle Pleistocene glaciation.

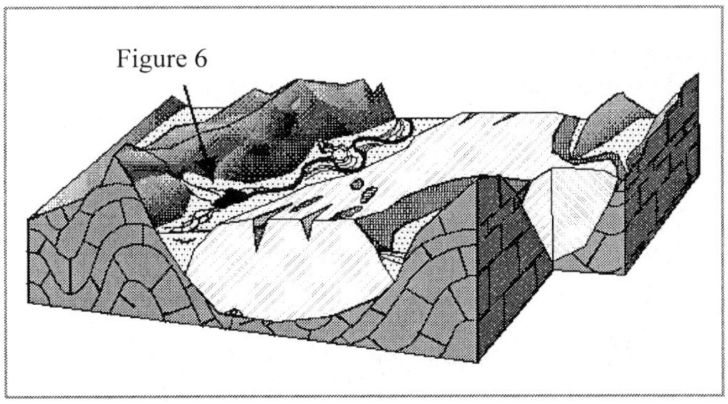

Fig. 7. Reconstruction of a possible Middle Pleistocene glacial environment on Pag island, with a glacier occupying the Pag valley and coarse and fine-grained sediment accummulating between the glacier walls and valley slopes (kame-terraces).

The possibility of such an extensive former glaciation may seem contradictory at present. However, it should be remembered that Cvijić (1917) had interpreted the Boka Kotorska Bay on the southern Adriatic coast (Montenegro) as a fjord, which might support the idea of a major Adriatic glaciation.

4. Dating of glacial limits

The age of the Pleistocene maximum glaciation in Croatia is not known. The ostracods identified in lacustrine sediments in the Novigrad section have a limited stratigraphical significance, but those found in lacustrine sediments in Žegarsko polje (KS 13, Table 1) and Erveničko polje (KS 14, Table 1) (situated *c.* 25 km eastward from the Novigrad section) indicate a Middle Pleistocene age (Malez & Sokač 1969).

Fig. 8. Glaciolacustrine siltstone with rhythmic bedding (varves?) and a dropstone at Novigrad (KS 10, Table 1).

The Rujno moraine (KS 7, Table 1) on the Velebit Mountain was attributed to the Würmian glaciation by Nikler (1973), and this correlation was also adopted by subsequent researchers (e.g. Belij 1985).

The boundaries presented on the map are still uncertain due to lack of stratigraphical evidence. The stratigraphy of the Quaternary deposits at various sites is also tentative. A scapula of *Bos primigenius* excavated in the Gomance valley gravel pit (KS 1, Table 1) was dated as 17 ky BP by radiocarbon (Marjanac *et al.*, 2001). These deposits are therefore Late Würmian in age and were accumulated on the outwash plain of the Snežnik glacier in Slovenia.

Radiometric dating (^{14}C dating) has also been performed on samples from the Krk and Pag islands. One bed with charcoal particles found in the Gajevi sand pit (KS 4, Table 1) on the Krk island yielded a date of *c.* 28 ky BP. The

Fig. 9. Various types of boulders are exposed along the Novigradsko More coast (KS 10, Table 1) washed-out from the possible moraine deposits: a) and b): large, rounded boulders more than 1 m in diameter as the most common, c) facetted large boulder and d) striated small boulder.

dating of other two strata with charcoal debris (Gajevi and Bošane; KS 5, Table 1) on Pag island, yielded ages of over 40 ky BP which may be infinite.

5. Open questions

The stratigraphy of the Quaternary deposits is a constant problem since most of dating methods other than ^{14}C have not been applied in Croatia. Radiocarbon dating is not appropriate for most of the sediment samples, whilst the age of the sediments exceeds that of the ^{14}C range. Lacustrine deposits should be submitted to palynological analysis but no specialists are available in Croatia. There are very good paleosol profiles that could be dated, but again detailed investigations have yet to be undertaken. Future international co-operation may help to solve the dating problems and clarify some of the open questions.

6. Acknowledgements

The study of Quaternary sediments along eastern Adriatic coasts was originally funded by INA-Naftaplin, the Croatian petroleum company. The project was conducted under the auspices of INA-Geological Consulting from 1988 to 1992. We thank INA for supporting the scientific research and granting us permission to publish the data. Since 1992 our research has been supported by the Croatian Ministry of Science and Technology through research projects 1-09-084 and 119303.

The ostracods have been studied by Ana Sokač of the Faculty of Mining, Geology and Petroleum Engineering of Zagreb, and ^{14}C analyses were undertaken in the Rugjer Bošković Institute of Zagreb by Nada Horvatinčić. We thank them all for their efforts and expertise.

7. References

Agassiz, L. (1840). Evidence of a Glacial Epoch. *In*: Mather, K.F. & Mason, S.L.(eds), *A Source Book in Geology 1400-1900.* 329-335, Harvard University Press., Cambridge (1967).

Bauer, B. (1934/35). Über die Landformen des Nördlichen Velebit. Sonderabdruck Jahresberichte des Bundes-Realgymnasiums Knittelfeld, 49 pp. Knittelfeld.

Table 2. Kame-terraces studied on the Krk and Pag islands (3 = location number on the map).

KAME-TERRACE	ALTITUDE	ESTIMATED HIGHT/WIDTH	VALLEY SLOPE DIP	TERRACE BEDDING DIP	LITHOLOGY/FEATURES
Krk island - Jurandvor (3)	140 - 190 m	45 m / 200 m	230/40	76/20	very coarse-grained, poorly sorted, stratified breccia
Krk island - Batomalj	80 - 120 m	40 m / 200-300 m	locally subvertical	200-220/10-20	very coarse-grained, poorly sorted, stratified breccia
Krk island - Gajevi (4)	50 - 80 m	55 m / 250 m	40/35	250/30, 200/13	coarse-grained, poorly sorted, stratified breccia with large boulders + well sorted, trough-cross stratified breccia + coarse-grained calcarenite
Pag island - Metajna	50 m	20 m / 100 m	220/25	30/25	coarse-grained stratified breccia
Pag island - Bošane (5)	0 - 70 m	80 m / up to 500 m	50/40-80	random	glaciofluvial sand/gravel body
Pag island - Gradac (6)	30 - 100 m	60 m /50 -150 m	30/40	226/36, 302/32	coarse-grained to fine-grained, poorly to well sorted breccia with extra-large boulders (> 10 m^3)
Pag island - Gorica	70 - 80 m	10 m	220/30	30/25	coarse-grained, medium sorted, stratified breccia + calcarenite

Belij, S. (1985). Glacijalni i periglacijalni reljef južnog Velebita. *Posebno izdanje Srpskog geografskog društva,* **61**, 5-68, Beograd.

Bognar, A., Faivre, S. & Pavelić, J. (1991). Tragovi oledbe na Sjevernom Velebitu (Glaciation traces on the Northern Velebit). *Geografski glasnik,* **53**, 27-39, Zagreb.

Bognar, A. & Prugovečki, I. (1997). Glaciation Traces in the Area of the Risnjak Mountain Massif. *Geologia Croatica,* **50/2**, 269-278, Zagreb.

Cvijić, J. (1917): L'Epoque glaciaire dans le Péninsule balkanique. *Annales Géographique,* **26/141**, 189-218.

Degen, A. (1936). *Flora Velebitica*, vol. 1. 7-202, Ungarische Akademie der Wissenschaften, Budapest.

Frakes, L.A. (1979). *Climate Through Geologic Time*. New York, Elsevier Scientific Publications, 310 pp.

Fritz, F. (1977). Litostratigrafska i tektonska osnova hidrogeoloških odnosa Ravnih Kotara i Bukovice (Lithostratigraphical and tectonical framework to hydrogeology of Ravni Kotari and Bukovica). PhD Thesis, University of Zagreb, 148 pp.

Gavazzi, A. (1903a). Tragovi oledbe u našem kršu. *Glasnik hrvatskog naravoslovnog društva*, **14**, 174-175, Zagreb.

Gavazzi, A. (1903b). Trag oledbe na Velebitu? *Glasnik hrvatskog naravoslovnog društva*, **14**, 459, Zagreb.

Gorjanović, D. (1902). Geomorfološki problemi iz hrvatskog krasa. *Glasnik hrvatskog naravoslovnog društva*, **13/4-6**, 193-196, Zagreb.

Hranilović, H. (1901). Geomorfološki problemi iz hrvatskog krasa. *Glasnik hrvatskog naravoslovnog društva*, **13/1-3**, 93-133, Zagreb.

Malez, M. (1968). O razvoju kvartara na području vanjskih Dinarida. I kolokvij o geologiji Dinarida, Ljubljana 1966, I, 203-210.

Malez, M. & Sokač, A. (1969). O starosti slatkovodnih naslaga Erveničkog i Žegarskog polja. III Simpozij dinarske asocijacije 1, 81-93, Zagreb.

Marjanac, Lj. & Marjanac, T. (1991). Pleistocene alluvial deposits on the Krk island (Northern Adriatic). Abstracts IAS 12th Regional Meeting Bergen, 31.

Marjanac, Lj. & Marjanac, T. (1994). Slopeward-dipping terraces on NE Adriatic islands, Pleistocene low altitude kame-terraces? Abstracts IAS 15th Regional Meeting Ischia, 271-272.

Marjanac, Lj., Poje, M. & Marjanac, T. (1992/93). Pleistocene marine and terrestrial sediments with striata fauna on the Island of Krk. *Rad Hrvatske akademije znanosti i umjetnosti, Razred prirodnih znanosti,* **26 (463)**, 49-62, Zagreb.

Marjanac, Lj. & Marjanac, T. (1999). Polyphase kame terraces on the Island of Pag (Eastern Adriatic, Croatia). Field Symposium on Pleistocene Stratigraphy and Glacial Chronology, Southern Estonia 1999, Abstract volume, 34-35, Tartu.

Kneblová-Vodičková, V. (1961). Entwicklung der Vegetation im Elster/Saale -Interglazial im Suchá-Stonava-Gebiet (Ostrava Gebiet). *Anthropozoikum*, **9**, 129-174.

Kočí, A., Růžičková, E. & Kadlec, J. (1991). On dating of Labe river terraces in the environs of Ústí n.Labem. *Věstník Ústředního Ústavu Geologického*, **66**, **1**, 43-49.

Králík, F. (1989). Nové poznatky o kontinentálních zaledněních severních Čech. *Sborník geologických věd, Antropozoikum*, **19**, 9-74.

Macoun, J. (1962). Stratigrafie sprašových pokryvů na Opavsku (Stratigraphie der Lössdecken im Opava-Troppauer Gebiet). *Přírodovědný časopis slezský*, **23**, 15-24.

Macoun, J. (1980). Paleogeografický a stratigrafický vývoj Opavské pahorkatiny v pleistocénu. *Časopis Slezského Muzea Opava*, **A 29**, 113-132, 193-222.

Macoun, J. (1985). Stratigraphie des Mittelpleistozäns in Mähren in Bezug auf das Quartär Europas. *Časopis Slezského Muzea Opava*, **A 34**, 125-143, 219-237.

Macoun, L. (1989). Die kontinentale Vereisung in der Mährischen Pforte. *Sborník geologických věd, Antropozoikum*, **19**, 75-104.

Macoun, J. & Králík, F. (1995). Glacial history of the Czech Republic. *In:* Ehlers, J., Kozarski, S. & Gibbard, P.L. (eds), *Glacial Deposits in North-East Europe*, 384-405. Rotterdam, Brookfield, Balkema.

Macoun, J. & Růžička, M. (1967). The Quaternary of the Upper Moravian Basin in the relation to the sediments of Continental glaciation. *Sborník geologických věd*, **A 4**, 125-168.

Macoun, J. & Šibrava, V. (1958). K otázce zalednění Hlučínska (Zur Vereisung des Troppauer und Hultschiner Gebiets). *Anthropozoikum*, **7**, 241-260.

Macoun, J., Šibrava, V., Tyráček, J. & Vodičková V. (eds) (1965). *Kvartér Ostravska a Moravské brány* (Quartär im Gebiet von Ostrava und in der Mährischen Pforte). *Ústřední ústav geologický, Academia Praha*, 419 pp.

Mojski, J.E. (1995). Pleistocene glacial events in Poland. *In:* Ehlers, J., Kozarski, S. & Gibbard, P.L. (eds), *Glacial Deposits in North-East Europe*, 287-292. Rotterdam / Brookfield, Balkema.

Opletal M. *et al.*, (2000). Vysvětlivky k základním geologickým mapám České republiky O2-222 Jiříkov a 02-224, 03-113 Varnsdorf. MS Český Geologický Ústav Praha.

Prosová, M. (1981). Oscilační zóna kontinentálního ledovce. Jesenická oblast. *Acta Universitatis Carolinae, Geologica*, **3**, 265-294.

Růžička, M. (1980). Sedimenty sálského zalednění na Opavsku a Hlučínsku (Saale Glaciation deposits in the Opava and Hlučín areas). *Sborník geologických věd.*, **A 13**, 127-144.

Růžička, M. (1986). Till fabric studies in the Ostrava region. *Sborník geologických věd.*, **A 17**, 43-67.

Růžička, M. (1995). Genesis and petrography of glacial deposits in the Czech Republic. *In:* Ehlers, J., Kozarski, S. & Gibbard, P.L. (eds), *Glacial Deposits in North-East Europe*, 406-420. Rotterdam / Brookfield, Balkema.

Růžičková, E., Růžička, M., Zeman, A. & Kadlec, J. (2001). Quaternary clastic sediments of the Czech Republic. Praha: Český Geologický Ústav. 160 pp.

Sekyra, J. (1968). Geomorfologická mapa Krkonoš. in Chaloupský, J. *et al.*, Geologická mapa krkonošského národního parku. Praha: Ústřední Ústav Geologický.

Šibrava, V. (1965). The Pleistocene lacustrine sediments of the České středohoří Mountains: *Sborník geologických věd*, **A 3**, 101-122.

Šibrava, V. (1967). Study on the Pleistocene glaciated and non-glaciated areas of the Bohemian Massif. *Sborník geologických věd*, **A 4**, 7-38.

Tyráček, J. (1961). Nové názory na rozšíření maximálního zalednění v Moravské bráně (Neue Ansichten über die maximale Vereisung in der Mährischen Pforte). *Přírodovědní časopis slezský*, **22**, 247-254.

Tyráček, J. (1963). On the problem of the parallelization of the continental and alpine glaciation on the territory of Czechoslovakia. *Report of the VIth INQUA Congress, Warsaw 1961*, **3**, 375-384. Lodz.

Woldstedt, P., (1955). *Norddeutschlands und angrenzende Gebiete im Eiszeitalter*, 2nd ed., 500 pp. Stuttgart, Koehler

Žebera, K. (1961). Altpleistozäne Bändertone (Warven) in Ctiněves am Říp. *Věstník Ústředního Ústavu Geologického*, **36**, 457-460.

Žebera, K. (1974). Kvartér Podřipska (III.část). *Sborník geologických věd*, **A 10**, 23-40.

Quaternary Glaciations – Extent and Chronology
Editors J. Ehlers and P.L. Gibbard

The Pleistocene of Denmark: a review of stratigraphy and glaciation history

Michael Houmark-Nielsen

Geological Institute, University of Copenhagen, Øster Voldgade 10, DK-1350 København K, Denmark

E-mail: michaelh@geo.geol.ku.dk

Abstract

Climatically dependent environmental changes and the advance and retreat of glaciers have controlled the development of sedimentary successions in Denmark during the Middle and Late Pleistocene. Over the last four glacial-interglacial cycles radical alternations of marine and terrestrial environments, combined with eustatic and isostatic response to glaciation, has lead to the following geological history. The first Quaternary glaciation reached the country during the Menapien or the early Cromerian. The Cromerian II interglacial is represented by single finds of lacustrine deposits. During the Elsterian glaciation three major glacial advances from different parts of Scandinavia are recognized. By the end of the Elsterian and during the Holsteinian interglacial infilling of incised valleys took place. In local lake basins, the Holsteinian is characterised by calcareous gyttja and diatomite with a flora characteristic of leached soils. The Drenthe phase of the Saalian glaciation began with an ice advance from southern Norway, it culminated with glaciation from Middle Sweden and terminated during the Warthe stadial with ice flow from the Baltic depression. During deglaciation, arctic to boreo-arctic marine conditions and tundra vegetation was restored. The Eemian interglacial includes high sea levels with boreo-lusitanian marine faunas and dense temperate forests. In the Early Weichselian shallow and boreal to boreo-arctic seas prevailed along with open forest and tundra vegetation. The Middle Weichselian began with arctic seas accompanied by ice rafting and followed by major glacier cover from the Baltic and possibly preceded by glaciation from southern Norway. Most of the remaining Middle Weichselian witnessed boreo-arctic conditions with ice rafting and the establishment of a shrub tundra vegetation. Glaciers calved in high arctic Kattegat and Skagerrak accompanied by surging glaciers in the Baltic. The Middle Weichselian ended with ameliorated interstadial environments. Shrub tundra and boreo-arctic seas prevailed. The Late Weichselian, Jylland stadial comprises complex conditions with several glacier advances and retreats. An ice stream from southern Norway was followed by glaciation from central Sweden during the maximum extension of Weichselian glaciers and the glaciations were terminated by ice streaming from the Baltic while arctic marine conditions were restored in Skagerrak and Kattegat.

Introduction

The stratigraphy, lithology and distribution of Pleistocene deposits in Denmark is highly controlled by the geological setting of the Pre-Quaternary bedrock. Permian rifting and regional subsidence connected with the braking up of Pangaea laid down the foundation to the present day bedrock geology. The Fennoscandian border zone separates the Baltic shield from the Mesozoic Danish and German basins, the sediments of which constitutes the soft and readily eroded and deformed substratum for Quaternary glaciers. During the Neogene, tilting, uplift and erosion of the Danish area took place (Håkansson & Surlyk, 1997; Japsen & Chalmers, 2000).

A dense network of incised valleys now filled with Quaternary sediments is reported from the Danish North Sea area as well as the mainland (Huuse & Lykke-Andersen, 2000). In Mid-Pleistocene ice-free periods, much of these valleys were occupied by fjords and lakes (Knudsen, 1987). Valleys were smoothed out and filled up during the Holsteinian interglacial. In the Late Pleistocene buried valleys were re-excavated and the regional tectonic activity in the Danish basin shifted eastward controlled by movements along the Fennoscandian border zone. Thus, the Kattegat region were subject to intensified subsidence, gradually bringing more of northern and eastern Denmark below sea level (Lykke-Andersen, 1987), and eventually, the Alnarp-Esrum valley linking the Baltic Sea and the Norwegian channel through the Kattegat basin was established. The so-called tunnel valleys in east Jylland were carved out along the path of older buried valleys during deglaciation after the Late Weichselian glacial maximum.

The growth of glaciers and ice caps influenced eustatic sea level and subsequent spreading of ice from central domes in Scandinavia into the north European lowland caused the formation of water-filled proglacial depressions. This favoured the development of outlet surges and the formation of ice streams that could be channelled through the Baltic depression and the Norwegian trench (Boulton *et al.*, 2001; Lagerlund, 1987; Erlingsson, 1994; Larsen *et al.*, 2000).

Key and reference sites

The following account,that covers the glacial history of the Pleistocene of Denmark, is based on a multitude of

Fig.1. Sketch map of Denmark with localities mentioned in the text.

classical literature summarised by Sjørring (1983) and more recent papers, summarised by Houmark-Nielsen (1987, 1989, 1999), Kronborg *et al.*, (1990) and Larsen & Kronborg (1994). Regions of neighbouring countries are also considered in order to present a more comprehensive outline. Much of the cited references deals with the stratigraphy of key sites containing a complex stratigraphy with a long depositional history, where as reference sites holds evidence of more specific events (Fig. 1). The following key sites may be highlighted:

Trelde Næs comprises Elsterian and Saalian glacial successions separated by lacustrine deposits of Holsteinian age (Houmark-Nielsen 1987).

Røgle Klint includes Elsterian, Saalian and Weichselian glacigenic successions and glaciomarine deposits probably of Late Elsterian age (Madsen & Nordmann, 1940; Sjørring, 1983; Houmark-Nielsen, 1987).

Ristinge Klint exhibits a complex succession of lacustrine and marine Eemian strata and Weichselian interstadial and glacigenic deposits strongly glaciotectonised as detached thrust slices (Sjørring, 1983; Kristensen *et al.*, 2000).

The Skærumhede boring and several other drill sites in the northern part of Jylland contains a Late Saalian, Eemian and Early-Middle Weichselian marine sequence (Lykke-Andersen & Knudsen 1991, Seidenkrantz & Knudsen, 1993; Kristensen *et al.*, 1998.

Lønstrup Klint contains Middle- and Late Weichselian marine deposits, freshwater strata, glacigenic deposits and the cliff exhibits large scale glaciotectonic thrust and load structures (Jessen 1931, Houmark-Nielsen *et al.*, 1996; Richardt, 1996; Sadolin *et al.*, 1997; Houmark-Nielsen, 2002).

Klintholm, Møn comprises Weichselian glacigenic deposits of three separate ice advances and interstadial, lacustrine deposits of the Middle and Late Weichselian (Houmark-Nielsen, 1994).

Bovbjerg Klint is a cliff site showing a cross section through the Main Stationary Line (M in Fig.7) which marks the classical outer limit of Weichselian glaciers in Denmark (Pedersen *et al.*, 1988, Larsen & Kronborg, 1994; Houmark-Nielsen, 2002).

The following reference sites comprise:

Harreskov, a Middle Pleistocene interglacial (Cromerian) lacustrine deposit overlying glacial strata and Ølgod where the same interglacial is found, and overlain by an

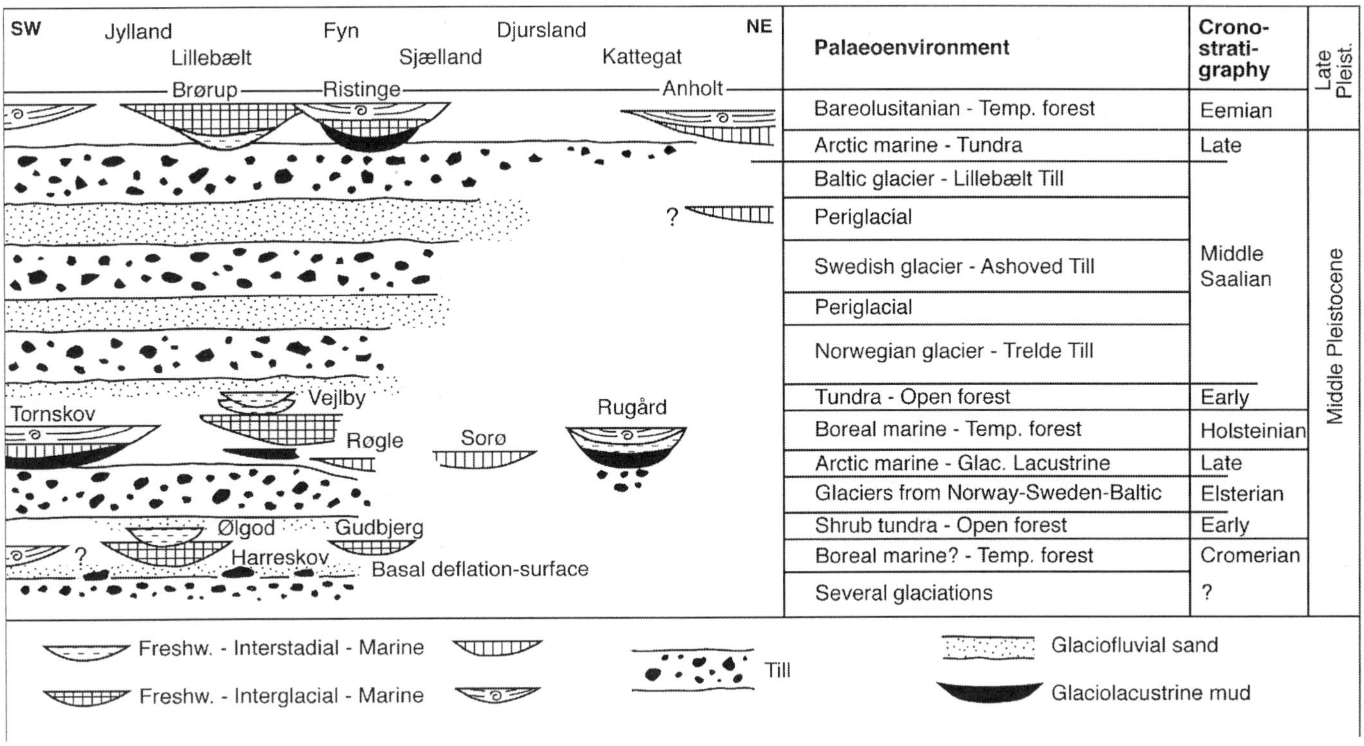

Fig. 2. Event-stratigraphic compilation of the Middle Pleistocene in a SW-NE cross- section through Denmark.

Early Elsterian sequence, including two interstadials (Andersen, 1965; Kronborg *et al.* 1990).

In the Rands-Vejlby area, close to Trelde Næs limnic deposits of the Holsteinian are found (Andersen 1965).

A boring at Tornskov contains marine Holsteinian deposits characterised by an arctic to boreal foraminifer succession. The palynology is equivalent to the Holsteinian pollen record (Andersen, 1963; Knudsen, 1987).

At Stensigmose in south Denmark, Eemian freshwater and marine deposits are underlain by Late Saalian glaciofluvial and lacustrine sand and mud and covered by Weichselian glacial deposits (Jessen, 1945; Konradi, 1976).

The 'Interglaical freshwater deposits' in central Jylland (e.g. Brørup, Herning and Hollerup) contain limnic deposits from the Late Saalian, the Eemian and the Early Weichselian (Jessen & Milthers, 1928; Andersen, 1957, 1961, 1965; Björck *et al.* 2000).

Pleistocene stratigraphy and glacial history of Denmark

The present outline of the Pleistocene stratigraphy in Denmark includes compromises on the issues of glaciation chronology and provenance. This has been done in order to combine opposing views, especially those differences expressed in the works of Houmark-Nielsen (1987, 1988, 1989, 1999) and Kronborg *et al.*, (1990) and Larsen & Kronborg (1994). However, consensus of the broad outline of this stratigraphic model is readily reached, even though lack of published results and reliable datings has partly been responsible for some discrepancies, other factors include the quality of exposures. An effort to solve some of the age problems has been initiated (Houmark-Nielsen 2002). The presence of well-defined interglacial strata with floras and faunas provide important markers in the model. Problems in correlation of glacigenic deposits, especially tills, arise when such beds are absent, and more indirect evidence has to be taken into account. These include provenance-dependent clast compositional characteristics, directional properties of tills and the nature and stratigraphic position of glaciotectonic unconformaties.

Traces of pre-Elsterian glaciation

Deposits of pre-Elsterian glaciations have only been recognised sporadically in Denmark (Kronborg *et al.* 1990). Andersen (1965, 1967) argued that the interglacial deposits at Harreskov are older than the Holsteinian; they can possibly be correlated with the Cromerian Interglacial II (Zagwijn, 1996). Between the Harreskovian Interglacial deposits and beds of Tertiary clay, glaciolacustrine and glaciofluvial deposits overlie a clayey glacial diamicton (Andersen 1967). Thus, possibly during the Menapian glaciation or, more likely, the older part of the Cromerian Complex, an ice sheet from Scandinavia penetrated southwards leaving the oldest record of glaciation in Denmark (Fig. 2). It is not unlikely, that the ventifacted, mainly quartzite- and quartzitic sandstone, rock and boulder horizons found between the Tertiary deposits and the older tills in Jylland comprise residual till material dating from this pre-Elsterian glaciation (Houmark-Nielsen, 1987). According to Andersen (1967), pre-Harreskovian till deposition was followed by sedimentation of glaciofluvial material around Harreskov and Ølgod in central Jylland. During the close of this glaciation local depressions were filled with meltwater clay and solifluction material.

The Harreskovian Interglacial

During the Harreskovian Interglacial some of these local depressions were filled with lake marl and diatomaceous gyttja in the area of west and central Jylland. Pollen analysis (Andersen 1965, 1967) indicates that open Vegetation was succeeded by temperate, deciduous forest. As the calcareous soil was leached, coniferous forests took over and by the end of the Harreskovian open Vegetation again replaced the forests. Some uncertainty exists regarding the presence of marine deposits of Cromerian age; thus Knudsen (1994) indicates that marine deposits referred to as Holsteinian in age holds amino acid values and luminescence ages suggesting a greater antiquity.

The Elsterian Glaciation

Elsterian glacigenic deposits comprise very quartz-rich tills of Norwegian, Middle Swedish and Baltic provenance, possibly deposited in this order. At Ølgod, deposits of the Harreskovian Interglacial are overlain by solifluction material interbedded with two lacustrine gyttja horizons representing the Ølgod I- and Ølgod II Interstadials (Andersen 1965, 1967). During the Ølgod I Interstadial, temperate forest reclaimed the area, whereas during the time of the Ølgod II Interstadial, a less pronounced climatic amelioration allowed only open vegetation to re-occur.

Meltwater streams emerging from the first Elsterian ice sheet (Norwegian Elster Advance) probably deposited outwash sand and- gravel, overlying the solifluction material and interstadial deposits. The ice invaded the country from the north and deposited a sandy till, which is characterised by large amounts of Norwegian indicators (Houmark-Nielsen, 1987). This ice advance apparently terminated east of the Netherlands (ter Wee 1983) and left prominent terminal moraines on the north German lowlands, where it deposited till and outwash material rich in Norwegian indicators (Ehlers *et al.* 1984).

Whether or not the Norwegian Elster Advance, and the advance which deposited the Elsterian sandy to clayey till of Middle Swedish composition, were separated by interstadial conditions is uncertain, neither is the distribution of this Middle Elsterian Advance known in any detail (Houmark-Nielsen, 1987; Kronborg *et al.* 1990). The latter advance was succeeded by a Baltic ice stream, which deposited a clayey till in central Denmark (Houmark-Nielsen, 1987; Kronborg *et al.* 1990). The older Baltic tills of southwest Jylland (Sjørring, 1983), and western Limfjord area (Jensen & Knudsen 1984) equate with this till. This Baltic Elster Advance probably terminated in northern Germany and the North Sea. Meltwater clay was washed down into local depressions in Jylland, as well as in the large buried valleys (Andersen 1965, 1967; Jensen 1985; Bruun-Petersen, 1987). These basins and lakes were later to be filled with lacustrine and marine Holsteinian deposits. According to Jensen (1985), the meltwater clay could very well be equivalent to the so-called Lauenburg Clay from northern Germany (cf. Ehlers *et al.*, 1984).

At the end of the Elsterian, a marine transgression reached the North Sea coast and the Limfjord region (Knudsen 1994) and marine clay containing an arctic foraminifera fauna was deposited. Pollen successions (Fig. 3), comparable to the inland lacustrine deposits, can be found throughout the marine sequence at Tornskov and

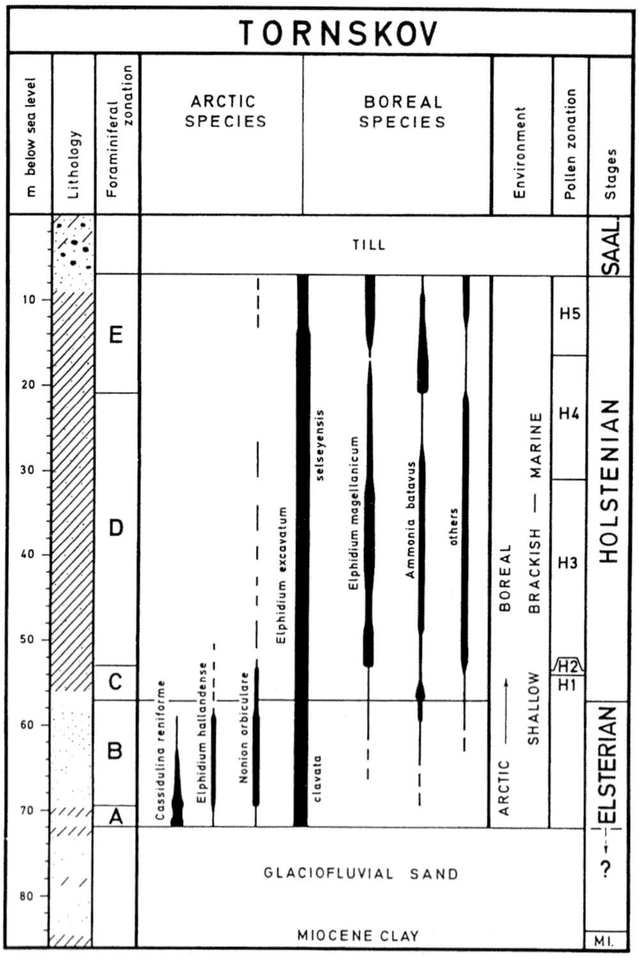

Fig. 3. Foraminiferal analyses, marine palaeoenvironments and pollen zonation of the Late Elsterian and the Holsteinian at the Tornskov boring, SW Jylland. From Knudsen 1987.

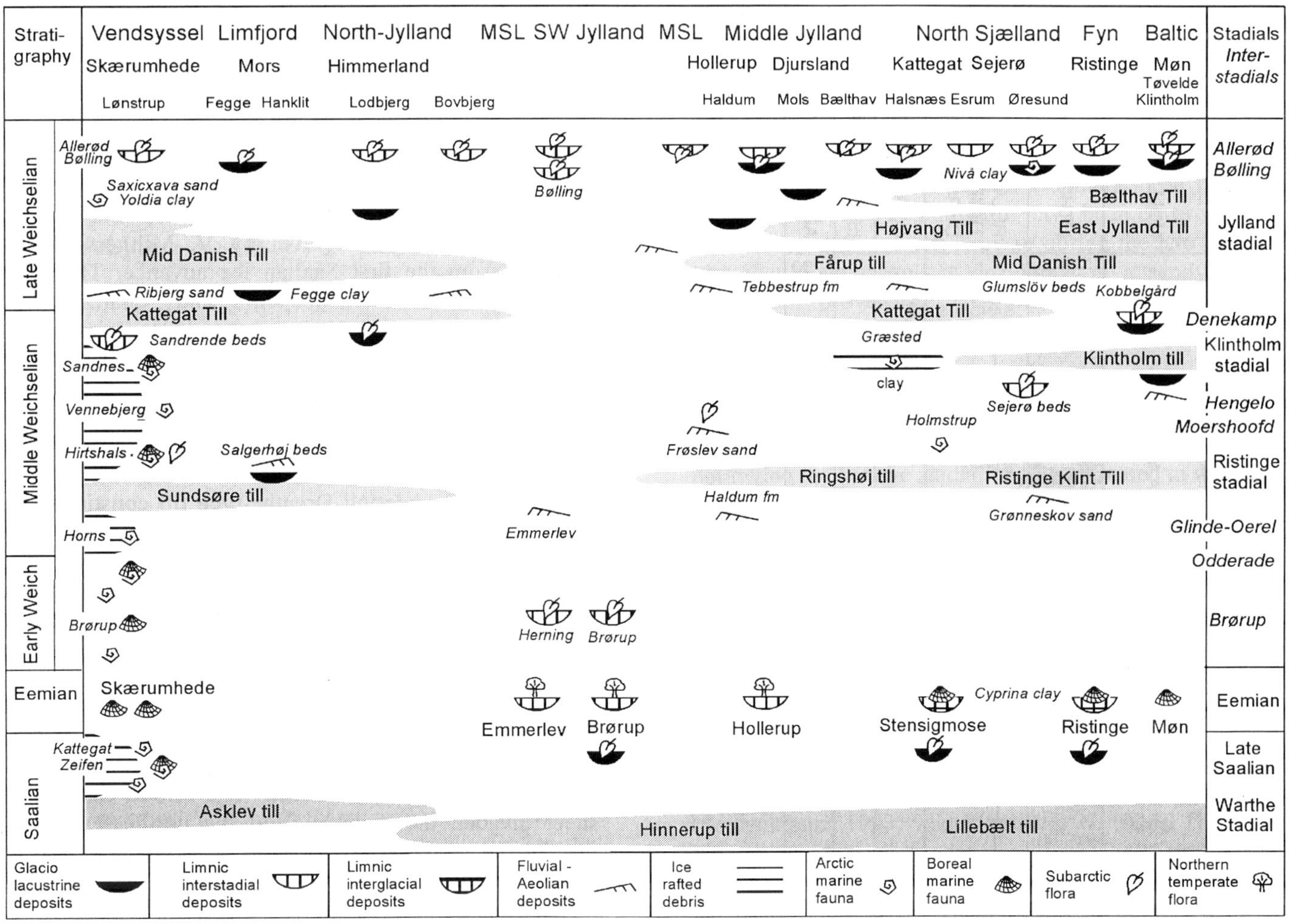

Fig. 4. Event-stratigraphic model for the Late Pleistocene of Denmark. The left side of the cross-section goes southward through northern Jylland crosses the Main Statiorary Line (MSL) and goes eastwards towards the Baltic in the right side of the diagram.

span the late Elsterian and most of the Holsteinian (Andersen 1963; Knudsen 1994). Madsen & Nordmann (1940) suggested that the *Tellina* Clay of Røgle Klint was deposited contemporaneously with the lower part of the (Holsteinian) *Yoldia* Clay at Esbjerg. The marine clay at Røgle has an arctic and poor mollusc fauna deposited in a glacio-marine environment. The assignment of the *Tellina* Clay to the transition between the late Elsterian and the early Holsteinian was favoured by Knudsen (1986); however, there are still insufficient micropaleontological information concerning the *Tellina* Clay, just as luminescence dates and amino-acid correlation makes an Elsterian age somewhat uncertain (Knudsen, 1994).

The Holsteinian Interglacial

The transgression reached its peak during the Holsteinian interglacial and marine clay with a boreal foraminifer fauna was deposited (Knudsen 1994). In central and eastern Jylland local depressions were filled with lacustrine eposits consisting mainly of lake marl and diatomite (Andersen, 1965; Houmark-Nielsen, 1987). Palynological investigations have revealed a characteristic plant succession, apparently developed on leached soils (Andersen 1965, 1967). Herbaceous and open vegetation developed into mostly coniferous, temperate forests which, by the end of the Holsteinian, were again succeeded by herbaceous Vegetation.

The Saalian Glaciation

Three major ice advances, that covered the whole of Denmark and separated in time by deposition of glaciofluvial material, dominate the Saalian. This glaciation was initiated with the final infilling of the previously mentioned lake basins. Solifluction material alternating with two lacustrine beds of the Vejlby I- and Vejlby II Interstadials are recognised in the Rands Fjord area (Andersen, 1965, 1967). Open, almost temperate forest vegetation characterises the interstadials.

Glaciofluvial sand and gravel locally filled the lake basins and formed floodplains deposited by westward-

Nordmann, V. (1928). La position stratigraphiquc des depots d'Eem. *Damnarks Geologiske Undersøgelse* **II, 47**, 81 pp.

Petersen, K.S. (1984). Stratigraphical position of Weichselian tills in Denmark. *In:* Königsson, L.K. (ed.), Ten Years of Nordic Till Research. *Striae,* **20**, 75-78.

Pedersen, S.A.S., Petersen, K.S. & Rasmussen, L.Aa. (1988) Observations on glaciodynamic structures at the Main Stationary Line in western Jutland, Denmark. *In:* D. Croot (ed.), *Glaciotectonics: Forms and processes,* 177-183. Rotterdam, Balkema.

Petersen, K.S. & Kronborg, C. (1991). Late Pleistocene history of the inland glaciation in Denmark. *In:* Frenzel, B. (ed.), *Klimageschichtliche Probleme der letzen 130,000 Jahre. Paläoklimaforschung,* **1**, 331-342.

Påsse, T., Robertsson, A.-M., Miller, U. & Klingberg, F. (1988). A Late Pleistocene sequence in Margreteberg, southwestern Sweden. *Boreas,* **17**, 141-163.

Richardt, N. (1996). Sedimentological examination of the Late Weichselian sea-level history following deglaciation of northern Denmark. *Geological Society of London, Special Publication,* **111**, 261-273.

Sadolin, M., Pedersen, G.K. & Pedersen, S.A.S. (1997). Lacustrine sedimentation and tectonics: an example from the Weichselian at Lønstrup Klint, Denmark. *Boreas,* ***26***, 113-126.

Seidenkrantz, M.-S. (1993). Foreminifera from the Quaternary sequence in the Anholt boring, Denmark. *Boreas,* ***22***, 283-290.

Seidenkrantz, M.-S. & Knudsen, K.L. (1993). Middle Weichselian to Holocene palaeoecology in the eastern Kattegat, Scandinavia: Foraminifera, ostracods and ^{14}C measurements. *Boreas,* **22**, 299-310.

Seidenkrantz, M.-S., Kristensen, P & Knudsen, K.L. (1995). Marine evidence for climatic instability during the last interglacial in fhelf records from NW Europe. *Journal of Quaternary Science,* **10**, 77-82

Seidenkrantz, M.-S., Bornmalm, L., Johnsen, S., Knudsen, K.L., Kuijpers, A, Lauritzen S-E, Leroy S.A.G., Mergeai, I., Schweger, C. & van Vliet-Lanoë, B. (1996). Two-step deglaciation at the oxygen isotope stage 6/5e transition: The Zeifen-Kattegat climate oscillation. *Quaternary Science Reviews,* **15**, 63-75.

Sejrup, H.P., Larsen, E., Landvik, J., King, E.L., Haflidason, H. & Nesje, A. (2000). Quaternary glaciations in southern Fennoscandia: evidence from southwestern Norway and the northern North Sea region. *Quaternary Science Reviews,* **19**, 667-685.

Sjørring, S. (1983). The glacial history of Denmark. *In*: Ehlers, J. (ed.), *Glacial deposits in north-west Europe,* 163-179. Rotterdam, A.A. Balkema.

ter Wee, M.W. (1983). The Saalean glaciation in the northern Netherlands. *In:* J. Ehlers (ed.), *Glacial deposits in north-west Europe,* 405-412. Rotterdam, A.A. Balkema.

Wysota, W. (1999). Ice sheet maximum limit of the Vistulian Glaciation in the mid-eastern Chelmno-Dobrzyń Lakeland, northern Poland. *Geological Quarterly,* **43**, 189-202.

Zagwijn, W.H. (1996). The Cromerian Complex Stage of the Netherlands and correlation with other areas in Europe. *In:* Turner, Ch. (ed.), *The early Middle Pleistocene of Europe.* Rotterdam, AA. Balkema, 145-172.

Quaternary Glaciations – Extent and Chronology
Editors J. Ehlers and P.L. Gibbard

Pleistocene glacial limits in England, Scotland and Wales

Chris D. Clark [1], Philip L. Gibbard [2] and James Rose [3]

[1] *Department of Geography, University of Sheffield, Sheffield, S10 1TN, UK.*
[2] *Godwin Institute of Quaternary Research, Department of Geography, University of Cambridge, Cambridge, CB2 3EN, UK*
[3] *Department of Geography, Royal Holloway, University of London, Egham, Surrey, TW20 0EX, UK.*

1 Introduction

This is a review of evidence for glacial limits in England, Scotland and Wales as understood at the beginning of 2002. It draws on a wide variety of work ranging from early studies such as those by James Geikie (1894) and H. Carvill Lewis (1894), to those based on work very recently published or in progress, such as systematic mapping and lithological studies in eastern England (Hamblin *et al.*, 2000), cosmogenic dating of glacial deposits in western Britain (Bowen *et al.*, 2002) or derivation and interpretation of remote-sensed imagery across the whole glacierized region (Clark, in preparation).

It is perhaps inevitable that the different authors have different interpretations of the evidence, and some are totally at variance one with another. In order that the topic of Pleistocene glacial limits within England, Scotland and Wales is represented fully and fairly the different views are given, and each section is attributed to the relevant author by initials CDC, PLG, JR. The submitted document has been compiled and edited by JR. Locations of sites referred to are shown in Fig. 1 and limits proposed by CDC, PLG and JR are shown on Fig. 2.

2 Context for glaciaton of the British Isles

2.1 Geographical position

As an island off the coast of continental Europe, Great Britain is unique in its geographical setting by comparison to the neighbouring regions. Together with Ireland, Great Britain occupies a position where during interglacial periods it is bathed by the North Atlantic Drift that ensures cool temperate, maritime climates predominate. These conditions provide a stark contrast to those during the cold periods (glacials) when the migration of the polar front, as far south as the Iberian peninsula results in establishment of severe, arctic-type conditions in the British Isles. The consequences of this are the appearance of cold winters, the annual development of sea-ice, periglacial conditions on land and when a sufficient supply of precipitation is available, the build-up of glacial ice in mountain regions. The lowering of global sea-level, leading to the linking of the British Isles to the Continent further augments this development of cold-stage continentality. Whilst during warm periods eustatic sea-level rise ensures the repeated separation of the islands.

The contrasts between the glacial versus interglacial conditions are so strongly marked that the geological record bears a sharply-defined imprint of the combined effects of these changes. This provides a fundamental basis for stratigraphical subdivision that is essential for the differentiation of multiple glaciation events. In addition, the proximity of Britain to the North Atlantic Ocean implies that the region is highly sensitive to even minor variations in marine currents and their accompanying wind patterns. Such changes are seen as environmental responses at a range of scales, particularly glacial advance and retreat **(PLG).**

2.2 Evidence for glaciation.

Evidence for glaciation of England, Scotland and Wales is primarily lithological with glacial episodes being identified by till and glaciofluvial sediments and glacial limits being determined by the extent of these deposits (Rose, 1989). Additionally geomorphological evidence has played an important role in reconstructing the extent of ice masses in younger glaciations. Moraine ridges and ice-contact landforms, including patterns of glacio-isostatically deformed displaced shorelines, have played an important role in the determination of ice limits of these younger glacial events (Geikie, 1894; Charlesworth, 1926, 1956; Sissons, 1967). Biological evidence has played a role in separating glacial events (Rose, 1989b) and in indicating a tendency towards climatic deterioration, or the existence of cold conditions which may be associated with glaciation (West 1980a). In the majority of cases this biological evidence has taken the form of pollen, but mollusca and plant macros have also been used to differentiate different stages of the Quaternary and insect faunas have been used to provide direct evidence for the presence of glacial meltwater (Penny *et al.*, 1967). Soil evidence, usually in the form of permafrost structures, has been used to indicate cold climate conditions, which have been linked with the formation of glacier ice elsewhere in England, Wales and Scotland (Rose *et al.*, 1985a, b; Ballantyne & Harris, 1994).

Hitherto, chemical proxies of temperature have not been used to infer glaciation within the region of this report **(JR)**.

2.3 Methods of dating and correlating glacial events.

Although glaciations in England, Wales and Scotland are primarily identified by rock and landform evidence, and lithology and morphology of glacial deposits has been used to determine glacial coverage and correlate ice marginal positions, they have for historical reasons, been dated primarily by relation to a biologically derived chrono-stratigraphy based on climatostratigraphic events (Mitchell *et al.*, 1973, Bowen, 1999a). This biostratigraphy was based on apparently diagnostic pollen assemblage biozones which provided a method of correlating different sites representing different interglacial stages across England, and to a very minor degree in Scotland and Wales (Mitchell, *et al.*, 1973; West, 1980a, 1989; Ehlers *et al.*, 1991b). Such a method had an important role in the 1950's and 1960's when other geochronometric methods were not available and methodologies for the amplification of the lithostratigraphy and landforms were either little used or not available. Furthermore it was believed that pollen assemblages could be correlated with similar evidence from continental Europe. Thus the glacial stages of England, Scotland and Wales were identified, correlated and ordered in time by relation to sites with apparently diagnostic pollen assemblage biozones (Mitchell *et al.*, 1973; West 1989), and the widely used sequence of Anglian, Wolstonian and Devensian Glacial Stages was established (Mitchell *et al.*, 1973).

However, it is now apparent that this approach is scientifically unsatisfactory and the needs of 50 years ago no longer apply. The reasons are as follows:

i) with increasing understanding of the processes responsible for the formation of glacial and glaciofluvial deposits (Goldthwait, 1971; Goldthwait & Matsch, 1989), it became possible to make sound lithostratigraphic correlations and use the materials produced by the glaciers for lithostratigraphic correlation;

ii) with increasing awareness of the form of glaciated terrain, the application of remote sensed imagery and a new understanding of the mechanisms of formation of glacier bedforms and ice marginal landforms, morphostratigraphic evidence actually produced by glaciers could be used directly to determine glacial flow paths and glacial marginal positions and ice limits (Boulton *et al.*, 1977, 1985, 1991; Boulton,1992);

iii) with the use of the piston corer to obtain ocean cores for the analysis of biological and isotopic sequences, Quaternary scientists have became aware of the complexity and scale of climate change during Quaternary time (Shackleton & Opdyke, 1973), and in consequence equally aware of the inadequacy of the pollen assemblage record in terms of the continuity, resolution and stratigraphic utility (Bowen, 1978);

iv) developments in the study of palaeoecology from pollen evidence, as opposed to using pollen evidence to derive a stratigraphic signal (Birks & Birks, 1980; Delcourt & Delcourt, 1991) highlighted the flaws in the use of pollen to provide a meaningful biostratigraphic signal (Tzedakis & Bennett, 1995);

v) new dating methods, such as radiocarbon, U-series, amino-acid racemisation ratios, luminescence, electron-spin resonance and cosmogenic nuclide exposure dating have been developed over the period. Although all these methods have problems, their application has reinforced the pattern of global climate change represented by ocean records (and now ice-core records), and have provided independent methods of dating the glacial sequences.

Thus, we are now in a position to use primary stratigraphic evidence such as glacial lithostratigraphy and glacial morphostratigraphy to determine the occurrence and extent of glaciation in England, Wales and Scotland. This does introduce new problems as many of the glacial events determined by the lithostratigraphic evidence do not have established chronostratigraphic stage names and must therefore be newly defined. However, in view of the sensitivity of the British Isles to climate forcing from the North Atlantic Ocean as outlined in Section 2.1 above, and the fact that the North Atlantic Ocean is a major factor in the development of global climate, the pattern of glaciation can be referred to the marine Marine Isotope Stages (MIS) and Substages (MISs) derived from ocean cores and to the ice-core stages derived from the Greenland ice-core record. Clearly there will be local peculiarities of glacier expansion and wastage on the British land area relative to the global pattern. This will be caused by factors such as the configuration of the glacier body, changes in relief over time, variations in glacier bed material both spatially and over time, and interaction with deep-water bodies. However it is realistic to see these as minor factors relative to climate forcing which would control both temperature and precipitation. For this reason, and for ease of communication, the glacial events are identified as far as is possible with Marine Isotope Stages (MIS) and Substages (MISs) (cf. Bowen 1978; Ehlers *et al.*, 1991b; Hamblin *et al.*, 2000) **(JR)**

3 The British Pleistocene succession

3.1 Pattern of glaciation.

Evidence for glaciation is found over most of Britain, indeed it was one of the regions from which glaciation was originally identified. However, the island shows marked physiographical contrasts in geology, Pleistocene history, tectonic setting, relief and glacial regimes. Upland northern and western Britain is dominated by erosional evidence, particularly in the mountainous districts of NW Scotland, the English Lake District and N Wales, which formed the main ice accumulation centres and supported their own ice caps. Where depositional sequences occur they tend to

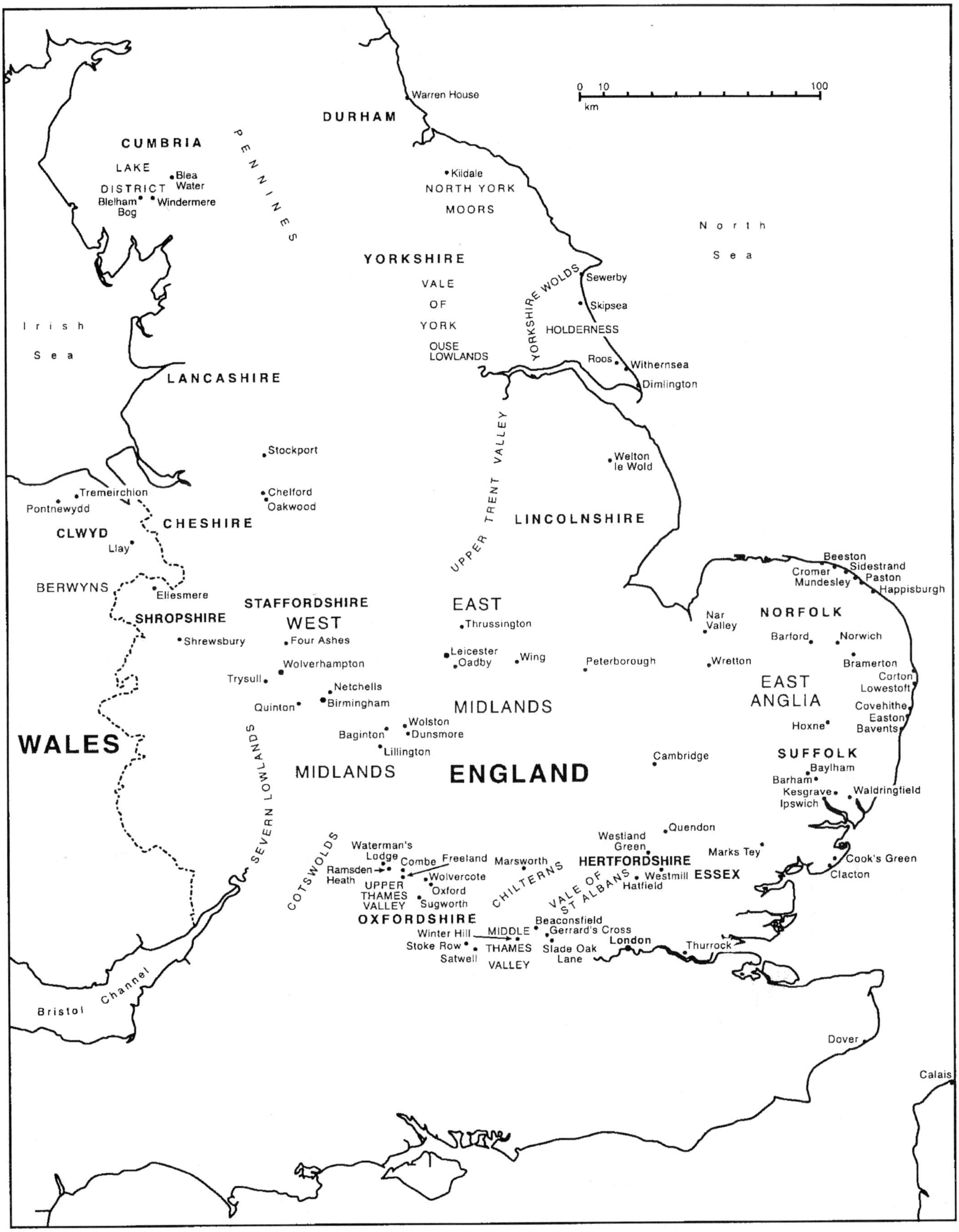

Fig. 1. Location map of sites mentioned in the text, and major towns: a) England, b) Scotland, c) Wales.

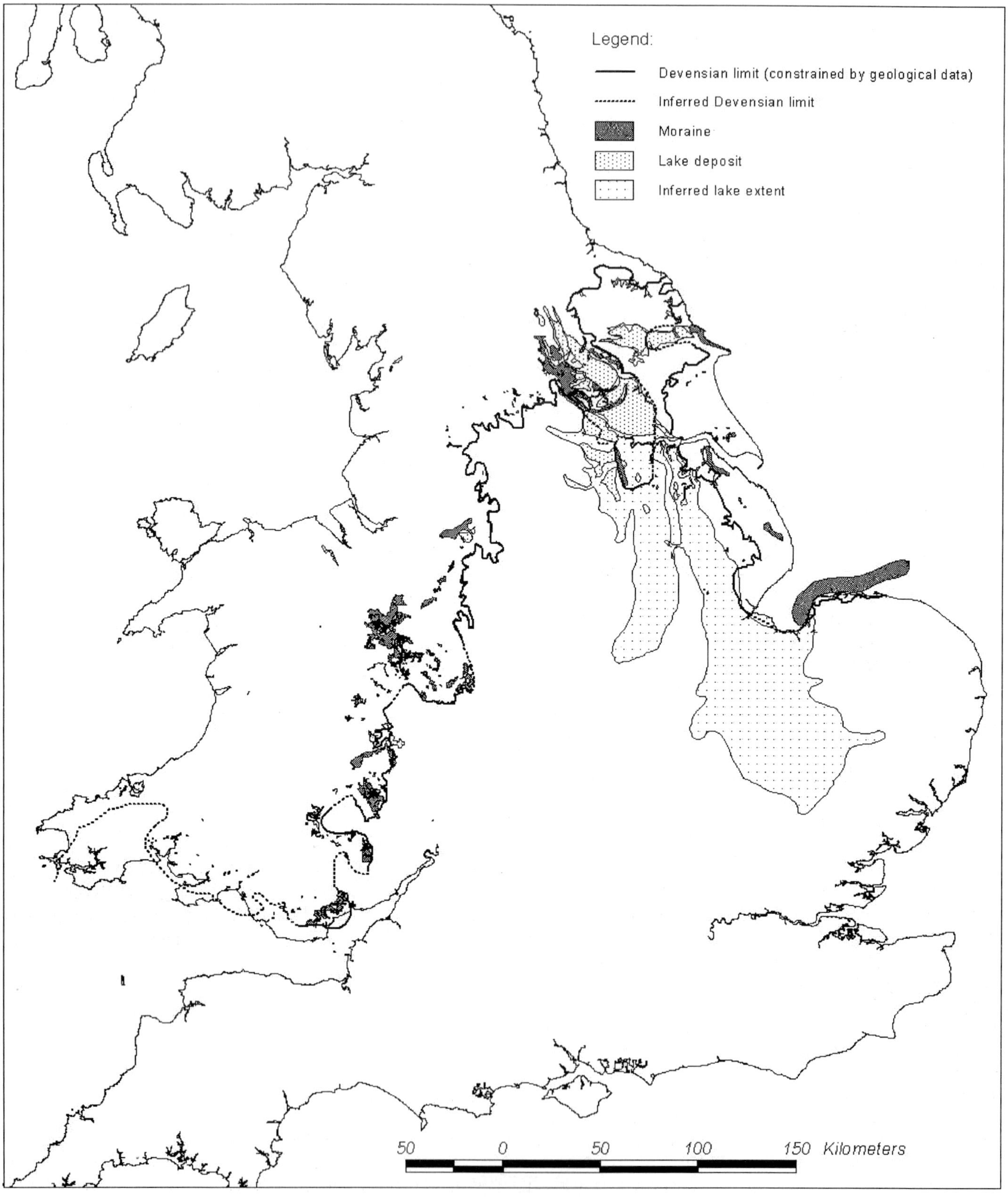

Fig. 2. Glacial limits in England, Scotland and Wales, based on the descriptions given in the text. The Devensian Dimlington Stadial limit, marked as solid where evidence exists and dotted where mostly inferred. Moraines are marked in black; light grey represents inferred extent of ice dammed lakes; dark grey record lake deposits. In places the ice limit is recorded by position of moraines or an ice limit that it required in order to dam a lake, but for most of the area the limit is recorded by drift limits of presumed Dimlington Stadial age. Note that just behind the maximum extent a suite of moraines exist, that likely reflects a major stillstand during retreat.

represent only the most recent glaciation (Devensian = Weichselian, Valdaian, Wisconsinan) and the evidence is often dominated by retreat phase sequences. In lowland south and eastern Britain, by contrast, depositional sequences predominate with relatively subdued landforms but sediments and associated features indicating glacial expansion and retreat of at least three and potentially more glaciation episodes (Anglian = Elsterian, 'Wolstonian' = Saalian, and Devensian). The extraordinary variety of British substrate geology and the consequent topography developed upon it, has given rise to a vast range on locally-variable deposits and landforms across what is, after all, a relative small area. These factors have together resulted in the variety of glacigenic sedimentation, glaciotectonics, erosion and glacial evolution represented in Britain. Finally, southernmost Britain, broadly south of a line from the Thames to the southern shore of the Bristol Channel, remained unglaciated throughout the Pleistocene **(PLG).**

3.2 Background to glaciation of Britain

Taking the island of Great Britain as a whole, the Pleistocene sequence can be broadly divided into two major parts: the pre-glacial Pleistocene and the glacial Pleistocene. With some minor exceptions, the bulk of the evidence for the pre-glacial Pleistocene is restricted to southern and eastern England and particularly to East Anglia, remaining areas having been, for the most part, subjected to erosion through the late Cenozoic. This is because eastern East Anglia forms the western margin of the tectonically-downwarping North Sea basin. Sedimentation here, although in a marginal setting, records mostly sea-level highstand sequences of shallow marine, littoral and sublittoral environments. Rapid infilling of this basin led to retreat of the sea through the later Early Pleistocene (taking the Plio-Pleistocene boundary at 2.4 My) and its replacement by fluvial and terrestrial sedimentation until the later Middle Pleistocene (c. 0.45 My).

No direct evidence for glaciation is known from these sequences although Solomon (1935) invoked glaciation in the North Sea region to explain the occurrence of anolomous frequencies of Scandinavian minerals in heavy mineral analyses from cold stage (Baventian = Tiglian C4c Substage) coastal deposits. This is now interpreted as resulting from tidal transport from the northern Netherlands (cf. Burger in Gibbard *et al.*, 1991). Similar indirect evidence for glaciation is seen as erratic clasts in fluvial sediments, e.g. rhyolitic rocks from the N. Welsh Berwyn Mountains in Thames gravels (Hey & Brenchley, 1977), are seen as evidence for contemporaneous glaciation that entered the uppermost part of the Thames catchment.

The earliest undisputed glacial diamicton known on land in Britain is the Happisburgh Diamicton, which is generally regarded as being of Anglian age, although this has been recently questioned (see below). This marks the arrival of the second glacial part of the Pleistocene sequence. Once again evidence for this glaciation is best developed in eastern and southern central England, although sediments are known to extend as far as the Welsh borderlands region near Hereford and south Wales. Anglian-age glaciogenic sediments also occur offshore in the southern North Sea (see below) and Irish Sea basins (cf. Scourse, this volume).

The second glaciation episode is of late Middle Pleistocene age, intermediate between the Hoxnian (=Holsteinian) and Ipswichian (=Eemian) interglacial Stages. This 'Wolstonian' glaciation was originally recognised in the English Midlands where glacial sediments underlie a large area in the west and central district. However, they are now referred to the Anglian Stage. Despite of this conclusion, a younger Middle Pleistocene glacigenic sequence is known from Midlands, Yorkshire and northern East Anglia, Because it seems to have been of similar extent to that in the subsequent Devensian Stage, this glacial episode is much more poorly represented in the Pleistocene record.

The third, or Devensian glaciation is well-represented throughout most of Britain, although it was possibly the least extensive of the three major events. Although considerable detail is known about its retreat phases, comparatively little evidence is available about the ice-sheet's advance. In general, ice extended outwards from the mountain glacial centres, reaching a maximum at about 20 ky and then retreating in an oscillatory manner until 13 radiocarbon ky. Following the Lateglacial Interstadial 13 - 11 radiocarbon ky the ice readvanced in the Loch Lomond Stadial (=Younger Dryas Stadial) **(PLG).**

3.3 Early and early Middle Pleistocene glaciations

There is no incontrovertable direct evidence of glacial deposits in England, Scotland and Wales older than the Happisburgh Till of East Anglia (as outlined in Section 3.2). Evidence for glaciations prior to the deposition of this deposit takes the form of lithological traces such as heavy minerals, erratic clast lithologies and SEM textures on sand grains, supported by soil and biological evidence of climatic deterioration. There are also diamictons composed of the insoluble residue, and these may be the weathered remains of till.

3.3.1 Early Pleistocene Baventian glaciation

3.3.1.1 Evidence

Evidence for the Baventian glaciation is found in marine deposits at Easton Bavents and Covehithe in East Anglia (Fig. 1), and takes the form of far-travelled, fresh, heavy minerals in shallow water marine clays (Solomon, 1935; Solomon, in Funnell & West, 1962; Burger, in Gibbard *et al.*, 1991b). These minerals are not local to the region, but typical of a high grade metamorphic province such as Scandinavia or Scotland. Thus the case was made that glaciation contributed fine grained material to the seas of the time and this material was incorporated in these sediments. This is the case for the earliest glaciation of Britain. An alternative interpretation has been put forward that these minerals may have been transported to the site by tidal transport (see Section 3.2 above), but this is not likely in view of the weatherable nature of many of the mineral species. Tidal and coastal processes are highly aggressive, leading to the destruction of such minerals and the selective preservation of robust, stable grains. Transportation in suspension in highly turbid sea-water or by ice-rafting is realistic, and the glaciogenic origin proposed by Solomon (1935) is favoured.

Further support for glaciation is provided by the clasts in the sand and gravel units below the marine clays. These are dominated by locally derived flint, but there is about 7 - 10% far travelled materials, especially quartzite, vein quartz, Carboniferous chert and *Rhaxella* chert (Rose, Moorlock & Hamblin, work in progress). This frequency of far travelled rocks is typical of the Dobb's Plantation Member of the Wroxham Crag Formation and reflects the input of river bedload to the western coast of the southern North Sea, a factor that was brought about by either glacial or periglacial erosion in the Welsh mountains (Rose *et al.*, 2001).

Biological evidence, in the form of a grass-heath pollen spectra, a low arctic / high boreal mollusc fauna and a 'semi-glacial' foraminiferal assemblage (meaning that there are likely to have been glaciers in Scandinavia, Scotland and northern England (West *et al.*, 1980, p 7)) also support the interpretation of cold climate conditions at the time that the shallow marine clays were deposited, and are additional evidence for a Baventian Glaciation.

3.3.1.2 Likely age

The evidence for the Baventian glaciation comes from the stratotype of the Baventian Stage of the British Pleistocene at Easton Bavents in East Anglia (Fig. 1). This is correlated with the Tiglian C4c of the Netherlands on the basis of the microtine rodent, freshwater and marine mollusc bio-stratigraphy (Gibbard *et al.*, 1991b; Preece & Parfitt, 2000). On the basis of the cold climate identified to this event Funnell (1995) correlated this event with MIS 68 which is dated to about 1.86 Ma.

3.3.1.3 Extent and limits of glaciation

The deposits described above give no direct evidence of the extent of glaciation other than that there was probably ice in Scandinavia, Scotland and the Welsh mountains all contributing sediment to the rivers and North Sea. Additionally, on the basis of erratic content, Catt (1982) suggested that the remnants of Baventian glacial deposits are to be found on the Yorkshire Wolds **(JR).**

3.3.2 Early and early Middle Pleistocene mountain glaciations

3.3.2.1 Evidence

These Early and early Middle Pleistocene glaciations are grouped together as the evidence for their existence is erratic materials from Wales and west Midland England and glacially fractured sand grains. This evidence is found in Thames river deposits along the Thames valley and over much of East Anglia (Rose *et al.*, 1999c) and shallow marine deposits in northeastern East Anglia (Rose *et al.*, 2001). The sediments that contain this evidence are the Kesgrave Sands and Gravels and the Wroxham Crag, and the differentiation of the glacial events is based on the separation of the river aggradations/terraces and the relationship of these to the shallow marine sediments. Because ten different aggradations (Table 1) have been recognised in the Kesgrave Sands and Gravels it is possible that a maximum of ten different glacial events can be recognised, although as explained below this number should be seen as indicative of the frequency of glaciation rather than a specific number.

The suggestion that glaciation contributed to the aggradations and terraces of the river Thames prior to the Anglian glaciation (MIS 12, see below) was first made by Hey (1965) on the basis of clast content, and was subsequently elaborated by Rose *et al.* (1976) and Rose & Allen (1977) who also described intraformational ice-wedge casts from the deposits demonstrating that permafrost existed in lowland eastern England at the time. These deposits form two formations based on a lithological differentiation, that reflect major changes in the palaeo-geography of the region. The older Sudbury Formation includes six separate aggradations which can be differentiated into six terraces at different elevations (listed in Table 1). It is defined by >23% of quartzite and vein quartz which is predominantly colourless, and a flint: quartzite + vein quartz ratio of <3.3. The younger Colchester Formation can be differentiated into four aggradation (listed in Table 1) and is defined by up to 19% quartzite and vein quartz which is predominantly reddish brown, and a flint: quartzite + vein Quartz ratio of >4.0. Additionally both units contain other far travelled clasts including acid volcanic rocks and greywackes from north and central Wales respectively (Hey & Brenchley, 1977, Green *et al.*, 1980, Green *et al.*, 1982, Whiteman, 1990). Additionally, glacially fractured sand grains (Hey, 1980) have been recognised in sediments of the Sudbury Formation.

The aggradations or terraces have a significant altitudinal and geographical distribution in that those of the Sudbury Formation are linked by a continuous long profile to the west Midlands and Wales (Fig. 3) whereas those of the Colchester Formation are separated from the west Midlands and Wales by the Cotswold Hills and are at elevations that are lower than the lowest col across these hills. This means that the aggradations of the Sudbury

Table 1. Lithostratigraphy and chronostratigraphy of the Kesgrave Sands and Gravels (from Rose et al., 1999, with additional information from Westaway et al., 2002).

Lithostratigraphy			Morphostratigraphy	Chronostratigraphy		Marine Isotope Stage*	Age and Environment
Group	Formations	Members	Terrace	Britain	Netherlands		
							Lowland
	Maidenhead Fm			Anglian	Elsterian	12	glaciation
						12	c. 0.45 Ma
	Colchester Fm	Lower St Osyth	Lower St Osyth	hitherto defined		12*	'Truncated
		Wivenhoe	Wivenhoe	as Cromerian		16*	Thames'
		Ardleigh	Ardleigh	and		18*	river
		Waldringfield	Waldringfield	Beestonian		20*	catchment
Kesgrave Group						21	c. 0.90 Ma
(Kesgrave Sands	Sudbury Fm	Gerrards Cross	Gerrards Cross		'Bavel	22*	
and Gravels)		Beaconsfield	Beaconsfield	hitherto defined	Complex'	34/32*	'Great
		Satwell	Satwell	as	Menapian	46*	Thames'
		Westland Green	Westland Green	pre-Pastonian	Waalian	62*	river
		Waterman's Lodge	Waterman's Lodge		Eburonian		catchment
		Stoke Row	Stoke Row			64*	
							c. 1.70 Ma
	Nettlebed Fm			Pastonian	Tiglian C5-6		Low
							energy
				pre-Pastonian	Tiglian C4c	68	rivers

* from Westaway et al., 2002

Formation could be linked directly to glaciers in the uplands of western Britain by meltwater drainage along the ancestral Thames valley, but that any glacial sediment added to the aggradations of the Colchester Formation require ice to cross the Cotswold Hills and directly enter the present Thames catchment. In East Anglia the Kesgrave Sands and Gravels have been linked with shallow marine offshore sands, muds and sands and gravels known as Wroxham Crag, and the members of this unit show similar lithological and permafrost evidence for cold climate conditions. The Kesgrave Sands and Gravels and Wroxham Crag are reviewed in Rose *et al.* (1999c, 2001).

The erratic clasts and the glacially fractured sand grains, along with the substantial quantities of materials from outwith the present catchment are considered to be evidence for glaciation in the upper parts of the river catchment. On the basis of this interpretation, and adoption of the classic model (Bridgland, 1994) that major aggradations are due to an input of coarse grained sediment, it is suggested that each aggradation may represent a glacial

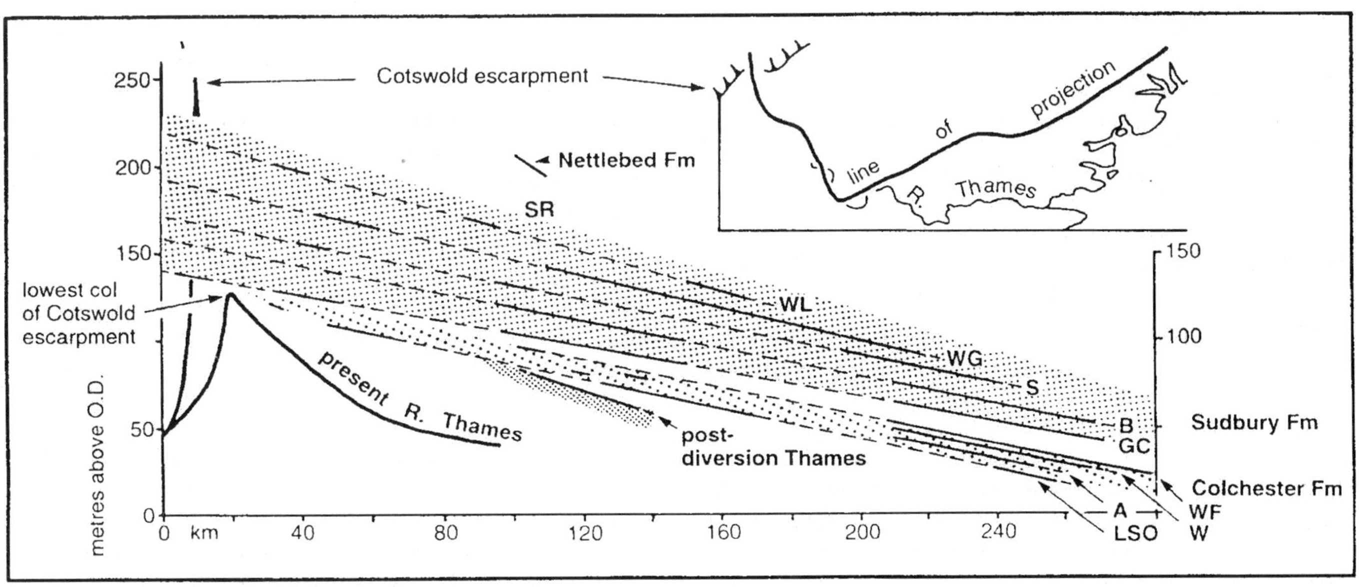

Fig. 3. Terraces of the pre-glacial Thames shown in relation to the elevation of the lowest col through the Cotswold escarpment (from Rose et al., 1999c, published with permission of the Geological Society)

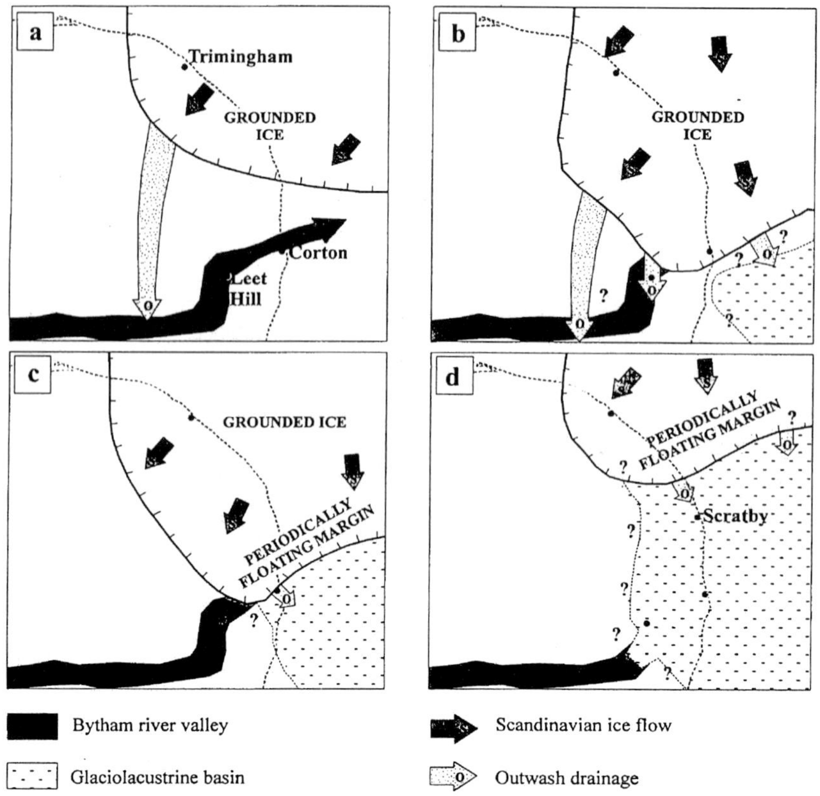

Fig. 4. Patterns of glaciation associated with the deposition of the Happisburgh and Corton Tills in Norfolk, England (from Lee, 2001).

a) Arrival of the ice in Norfolk: subglacial deposition of the Happisburgh Till with outwash reaching the Bytham River.

b) The grounded ice reaches the area of Corton and blocks the Bytham River valley causing southwards diversion and the development of a proglacial lake. The ice extended further west across east Anglia between stages b and c (see Fig. 2).

c) A periodically floating ice margin in the proglacial lake with the deposition of grounding line till at Corton as part of a subaqueous grounding line fan.

d) Retreat of the ice margin and deposition of the Corton Sands.

It is essential to recall, when looking at these figures and Fig. 2, that the Wash and Fen basin did not exist at this time (from Lee, 2001, published with permission of the Geological Society).

episode in which glacial and/or periglacial processes transferred sediment from the slopes to the channels in the headwater regions of the catchment. Following from this assumption it is evident that the number of aggradations need to be defined, in order that the number of glacial episodes can be determined. This requires correlation of the Thames river deposits and terraces throughout the region. However, correlation of the aggradations and terrace landforms across the whole region is difficult and a number of different schemes have been proposed (Whiteman & Rose, 1992; Bridgland, 1994; Rose *et al.*, 1999c). This problem will only be satisfactorily resolved when independent geochronometry is available. However accord-ing to the publications that contributed to Rose *et al.* (1999c) there are six aggradations in the Sudbury Formation (Table 1). Erratic material from Wales has been identified in each of these (Rose *et al.*, 1999c) and it is proposed that six separate cold glacial events occurred in the mountains of Wales and perhaps adjacent lowlands. Erratic material is also recognised in the four aggradations of the Colchester Formation. In three of the units the frequency is lower than the frequencies in the Sudbury Formation and this may reflect reworking of the existing deposits. However it is significant that the percentage erratic material increases in the Ardleigh Sands and Gravels (McGregor & Green, 1978) and this suggests that ice from Wales and the West Midlands entered the Thames catchment, downstream of the Cotswold escarpment, at this time.

A search for glaciogenic deposits to complement this glaciofluvial / cold climate fluvial evidence reveals a number of deposits throughout the region. In the Upper Thames region the Northern Drift is a diamicton that contains far travelled material. This has been interpreted as till (Dines, 1928; Arkell, 1947; Shotton *et al.*, 1980), but careful studies by Hey (1986) and Briggs *et al.* (1975) have shown that the deposit originated as a limestone-rich sorted sediment that has subsequently been decalcified. This process has produced a diamicton in which the fine fraction is the insoluble residue of the limestone, the coarse fraction is the resistant pebbles and the heterogeneous appearance has been produced by collapse and active layer mixing. A similar diamicton, known as the Chiltern Drift, has also been recognised on the north side of the Middle Thames valley and has also been used as evidence for glaciation. As with the Northern Drift this has been shown to be formed by decalcification and mixing of originally sorted sediment (McGregor & Green, 1983). However, in the region around Bruern Abbey in the Evenlode valley north of Oxford (Fig. 1) a clay-rich bouldery diamicton extends across the irregular land surface (Hey, 1986). As with the other deposits there has been much controversy as to whether this deposit is a decalcified sorted sediment or a weathered till, but Hey (1986, p. 294) makes the point that relative to the other deposits within this part of the Thames region this is "the one occurrence that has hitherto been almost universally accepted as a till". In this case the deposit at Bruern Abbey is considered a till deposited by ice that entered the Upper Thames catchment during the formation of the aggradation that deposited the Ardleigh Sand and Gravel (Fig. 2).

Similar river and coastal sediments rich in far-travelled lithologies are known in northern East Anglia and Midland

England. These are associated with the Ancaster river (Clayton, 2000; Rose *et al.*, 2001) which drained the southern Pennine hills and contributed to the formation of the northern part of the Wroxham Crag, and with the Bytham river which also drained the southern part of the Pennine hills and the west Midlands (Rose, 1987, 1989a, 1994, Rose *et al.*, 2001; Lewis, 1993, Lewis *et al.*, 1999) and formed the Bytham Sands and Gravels and contributed to the formation of the Wroxham Crag. Although both of these deposits are dominated by far-travelled materials there is no evidence for rocks that can be traced to the mountains of Wales, and there is no specific evidence to suggest that the far travelled rocks were introduced by glaciation rather than periglacial hillslope processes. Nevertheless it is not unreasonable to consider that glaciation may have occurred in the southern Pennines and the west Midlands (the relief of the west Midlands was much higher at the time). Both of these suites of river sediments are either eroded by glaciation or buried beneath glacial deposits. The result is that the full succession of aggradations has yet to be defined. Thus, these deposits cannot yet be used to identify a number of glacial events in the upper parts of their catchments, as has been done with the Thames.

3.3.2.2 Likely age

The age of the sediment bodies that are taken as proxies of glaciation in the uplands of western Britain is established by the relation of the Kesgrave Sands and Gravels to sediments with meaningful biostratigraphy, as other geochronometrical methods are, as yet, not available. Previously this relationship was made with the Pre-Pastonian, Pastonian Beestonian and Cromerian pollen-defined stages (Rose & Allen, 1977; Rose *et al.*, 1976; Rose in Bowen *et al.*, 1986; Whiteman & Rose, 1992), but after numerous revisions and reversals of this chronotratigraphic scheme (Funnell *et al.*, 1979; Funnell & West, 1977; West & Norton, 1974; West, 1961, 1980b; West *et al.*, 1980; Gibbard *et al.*, 1991) it is now realized that pollen assemblage biozones cannot provide a reliable stratigraphic framework for the Early and early Middle Pleistocene and the method is of no value for chronostratigraphy (Preece & Parfitt, 2000). Thus, at the present time, the only biostratigraphy of value is the association of the beginning of deposition of Kesgrave Sands and Gravels / Wroxham Crag with the appearance of *Macoma balthica* in the North Sea and this is related to the Late Tiglian Substage (TC4c, Meijer, 1993; Lister, 2000) of the Netherlands which is dated to about 1.86 Ma and is correlated with the distinctive global cooling of MIS 68. Formation of these aggradations continued through to the onset of the Anglian glaciation at MIS 12 (Rose, 1994; Rose *et al.*, 1999c). Hitherto the Sudbury Fm has been related to the period from Tiglian C4c to the Dorst Glacial of the Netherlands, and the Colchester Formation has been related to the Cromerian Complex of the Netherlands (Whiteman & Rose, 1992; Rose, 1994; Rose *et al.*, 1999c). Recently, Westaway *et al.* (2002) has attempted to allocate ages to the individual aggradations of the Kesgrave Sands and Gravels and hence to the glacial episodes, using a range of information set out in that paper (see Westaway *et al.*, 2002, p. 567). In this case the glaciations that deposited the aggradations / terraces of the Sudbury Formation are attributed to MIS 64 (Stoke Row), 62 (Westland Green), 46 (Satwell), 34/32 (Bea-consfield), 22 (Gerrards Cross), and those of the Colchester Formation to MIS 20 (Waldringfield), 18 (Ardleigh), 16 (Wivenhoe) and 12 (Lower St. Osyth) (Table 1).

3.3.2.3 Extent and limits of glaciation

Only the 'Bruern Till' can be used to infer a glacial limit and this location has been placed on Fig. 2. All other evidence simply implies cold climate conditions, which may be glaciation, but may only be periglaciation, in the mountains of Scotland, Wales and possibly in the surrounding lowlands including the west Midlands, and in the southern Pennines. It is not possible to draw any precise lines on Fig. 2 **(JR).**

3.4 Later Middle Pleistocene lowland glaciation in Scotland, Wales and England.

3.4.1 Overview of effects of later Middle Pleistocene glaciation on the landscape of Britain

These glacial events are represented by tills and other glaciogenic sediments and the evidence reflects the direct action of glacier ice on the land, rather than evidence that is just a proxy of glaciation. These glaciations are associated with an entirely different landscape from that of the earlier glacial events. At the earlier times the landscape was a well organized fluvial landscape with large river catchments, such as the basins of the Solent, Thames, Bytham and Ancaster rivers, mainly low angle slopes and thick, well developed lowland soils.

With the onset of lowland glaciation, this changed and glacial erosion caused the destruction of the large river valleys in Midland England (Rose, 1994) and the diversion of the river Thames from East Anglia to its present course through London (Gibbard, 1977). In the place of the large river valleys new shorter rivers were established, such as the Trent, Witham, Welland, Ouse and Nene in midland England, and the Lark, Bure, Waveney, Stort, Gipping and Chelmer in East Anglia (Rose *et al.*, 1985a). Many of these rivers occupied the depressions of tunnel valleys caused by subglacial meltwater erosion (Woodland, 1970) with a relief and flow direction caused by the hydrostatic configuration and gradient of the glacier. Thus, the lobe of the Anglian glacier, that fanned-out from the highly deformable bed centred around the impermeable and plastic midland England (Rose *et al.*, 1985a). Glacial erosion also

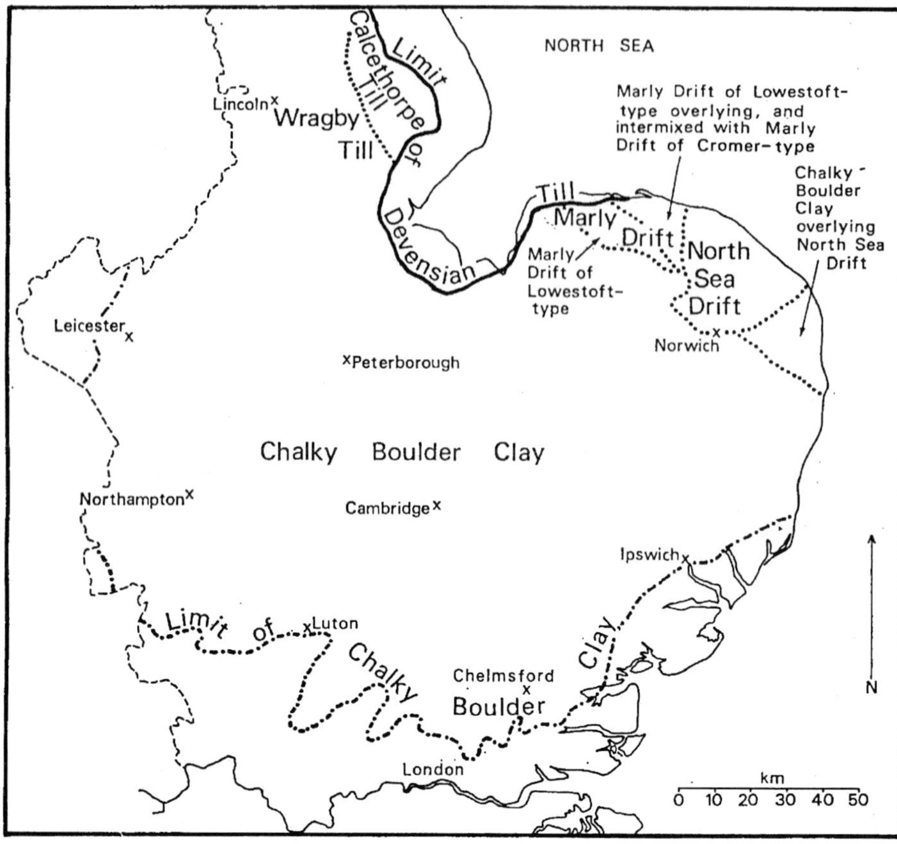

Fig. 5. Till types associated with pre-Devensian Glaciation in eastern England. Chalky Boulder Clay is Lowestoft Till (from Perrin et al., 1979, published with permission of the Royal Society of London).

created large new lowland terrain in regions such as the Mesozoic mudstones of the Wash and Fen basin created a radial pattern of rivers across East Anglia and south Wash and Fen basin (Rose *et al.*, 1985a), the vales between the Pennine Hills and the North York Moors and Yorkshire and Lincolnshire Wolds, and the low area that is now sea, east of the Yorkshire and Lincolnshire coasts. Additionally the glacial landscapes included enclosed basins caused by the melt of isolated blocks of stagnant ice (kettle holes), local glacial scour and localized subglacial meltwater scour. All these enclosed basins provided locations in which sediments such as lake clays, lake muds and calcareous marls could accumulate following glaciation and provide evidence suitable for the reconstruction of the palaeo-environment following ice wastage. It is these locations that provided the sites for some parts of the assemblage-based pollen stratigraphy (Mitchell *et al.*, 1973; West, 1980a, 1989) **(JR).**

3.4.2 Brief history of research and ideas on lowland glaciation in England

Traditionally, British glacial history was divided into four glaciations separated by interglacials (Geikie, 1894) and this was perpetuated by correlation with the scheme proposed for the European Alps (Penck & Brückner, 1909) (see Bowen, 1978). This scheme changed in the 1960's and 1970's to a correlation from Britain with the Alps to Britain with continental Europe (Shotton, 1967, 1968, 1976, 1977; Mitchell *et al.*, 1973) and in so doing the number of glaciations was reduced from four to three. This scheme envisaged, from oldest to youngest, the following:

i) Anglian Glaciation, based on a type site at Corton in East Anglia (Mitchell *et al.*, 1973; Pointon, 1978; Rose, 1989b) was considered the equivalent to the Elsterian of eastern Germany (Eissmann, 2002);

ii) Wolstonian Glaciation, based on a type site at Wolston near Coventry in Midland England (Mitchell *et al.*, 1973, Shotton, 1968, 1976, 1983; Straw, 1979a, 1983a, b) was considered to be equivalent to the Saalian of eastern Germany (Eissmann, 2002);

iii) Devensian Glaciation, based on a type site at Four Ashes near Wolverhampton in west Midland England (Mitchell *et al.*, 1973; Shotton, 1967, 1977) was correlated with the Weichselian of north Germany (Ehlers, 1983).

Subsequently it was realised, on the basis of field mapping, lithological studies and careful stratigraphic work, that the deposits at Wolston were of pre-Anglian (part of the Bytham river system) and Anglian glacial age (Perrin *et al.*, 1979; Sumbler, 1983a, b, 1995, 2001; Rose, 1987, 1989a, b, 1991, 1994). This view has been challenged on the basis of traditional thinking (Gibbard & Turner, 1988; Gibbard, 1991), but there is now general consensus, even by those making the criticism cited above, that only the Anglian glaciation extended across Midland and Eastern England (see Section 3.4.4), and that the Devensian glaciation reached a southern limit that extended from the

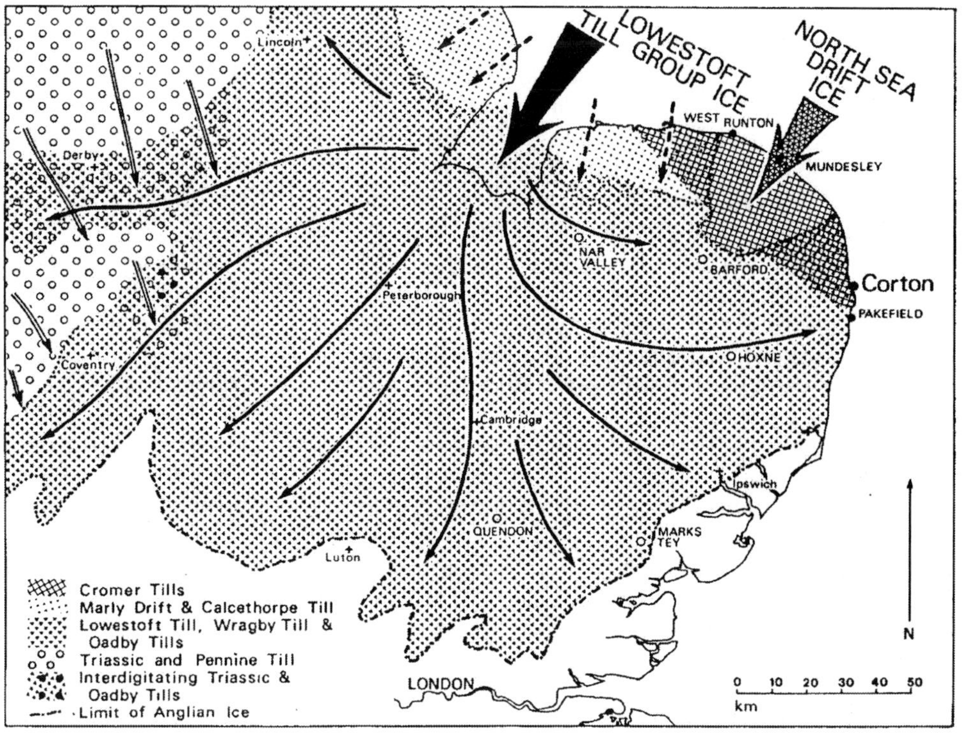

Fig. 6. Distribution of till types and ice flow paths associated with the traditional interpretation of the Anglian Glaciation in Eastern and Midland England. The sites located with a black circle have organic deposits beneath the glaciogenic sediments, and the sites indicated with an open circle have organic deposits defined as Hoxnian Interglacial based on the pollen assemblage biozones (from Rose, 1989).

north Norfolk coast in the east, to the area of Wolverhampton in the west Midlands, around the Welsh Borderland and to the southern coast of Wales and possible the Scilly Isles (Bowen *et al.*, 1986, and see Section 3.5.2).

In Scotland it is assumed that Anglian and Devensian ice covered the whole of the country, although an alternative view suggests that Devensian ice did not cover the distal parts of the northeastern peninsulas (Section 3.5) (Figs 2, 5 and 7).

Thus, the two glaciations were recognised in lowland Midland and Eastern England (Perrin *et al.*, 1979, Sumbler, 1983a, b; Rose in Bowen *et al.*, 1986; Rose, 1989a, b, Ehlers *et al.*, 1991b), although patchy evidence for additional lowland glaciations was described from the area around Birmingham (Maddy *et al.*, 1995) and below Devensian deposits in eastern England (Alabaster & Straw, 1976; Rose in Bowen *et al.*, 1986, for summary). Rose, in Bowen *et al.* (1986, pp 308-309, Tables 1 and 2) suggested that this additional glaciation should be called the Welton Glaciaton based on a stratotype at Welton-le-Wold in Lincolnshire, but this term has not been adopted and the patches of glacial deposits were insufficient to allow limits to be drawn on maps across more than the very smallest areas.

Recently however, as a result of accumulating work on Quaternary sediments in Midland and Eastern England, including regional geological mapping and detailed analyses of critical sections (Sumbler, 1995, 2001; Keen, 1999; Hamblin *et al.*, 2000; Rose *et al.*, 2000), a fundamental revision of lowland glaciation of Britain has been proposed. It is now suggested that the number of Late Middle and Late Pleistocene glaciations of lowland Midland and Eastern England is five! This view has been criticized on the basis of traditional thinking (Banham *et al.*, 2001), but the new scheme takes into consideration new evidence (Hamblin *et al.*, 2001) and resolves some of the long outstanding anomalies of the British glacial history. The original paper (Hamblin *et al.*, 2000) was presented as a preview of current thinking following discussion at a Quaternary Research Association Annual Field Meeting in April 2000, and the five-glaciation scheme is outlined here following under the initials JR, along with the traditional two-glaciation view given under the initials PLG.

The contribution below by JR reflects the outcome of joint work with the Brian Moorlock and Richard Hamblin of the British Geological Survey and Jon Lee of the Department of Geography at Royal Holloway, University of London. This work has not yet been published, but this will be done in the near future. The five-glaciation model does not resolve all the problems that confront the topic of glaciation in England, but it does explain problems that are not explained by the two-glaciation model for lowland England. As one of the people most involved in establishing the two-glaciation model (Perrin *et al.*, 1979; Rose in Bowen, 1986; Rose 1989) I (JR) have clearly given the new five-glaciation model considerable thought before changing my interpretation. I am persuaded to make the case after careful consideration of the new evidence. The critical factors that stimulated this revision are: i) the recognition, as a result of field mapping, that the Lowestoft Till and the 2nd Cromer Till were one and the same deposit; and ii) that the 1st Cromer Till (Happisburgh Till and Corton Till) is to be found as clasts in the 2nd Terrace of the Bytham river which predates the Anglian Stage (MIS 12) by at one, and probably two, temperate episodes. Other reasoning that favours this five-glaciation model are given in the text below. It is important to understand that i) intervening biological deposits or soils have yet to be found between the tills of these five glaciations, as indeed there are no biological deposits between the tills of the two glaciation model, and ii) that chronostratigraphy, based on pollen assemblage biozones is not accepted as a reliable stratigraphic method, although, ofcourse, these assemblages may indicate elements of palaeoecology **(JR).**

3.4.3 Glaciation associated with the deposition of the Happisburgh and Corton Tills (MIS 16)

3.4.3.1 Evidence

The Happisburgh and Corton Tills are found in northern East Anglia (Lee, 2001). They are a sandy diamicton with a clast fraction dominated by flint from Cretaceous Chalk, quartz and quartzite from the underlying Wroxham Crag Formation and shells from the bed of the North Sea. Far travelled erratics include Magnesian Limestone from northern England and igneous and metamorphic lithologies from Scotland (Lee *et al.*, in press), a heavy-mineralogy characteristic of a metamorphic and igneous provenance (Perrin *et al.*, 1979; Lee *et al.*, in press) and Carboniferous palynomorphs from northern England (Lee *et al.*, in press). These tills have also been known as the First Cromer Till (Banham, 1968; West and Banham, 1968; Perrin *et al.*, 1979) of the North Sea Drift group of sediments. Patches of these deposits are found inland in Norfolk and in the area of the Bytham river valley (which is now buried beneath Anglian till). Clasts of the Happisburgh and Corton Tills or erratic lithologies characteristic of these tills are found in the deposits of the second aggradation/terrace of the Bytham Sands and Gravels.

Traditionally, from the classical work of Reid & Woodward (1882) and Reid (1890) onwards, this deposit was considered to have been deposited by ice that moved across the North Sea from Scandinavia (and see Section 3.4.4), but current work on the lithological characteristics of these deposits has shown that they are more likely to have been deposited by ice that moved southwards from eastern Scotland, and not from Scandinavia. The two tills are differentiated on the basis of their structure and particle size distribution and the Happisburgh Till is interpreted as a subglacial lodgement deposit, whereas the Corton Till is a waterlain or grounding-line sub-aquatic deposit (Fig. 5) (Lee, 2001).

3.4.3.2 Likely age

Traditionally this deposit was considered to be deposited during the Anglian glaciation (Mitchell *et al.*, 1973; Perrin *et al.*, 1979; Bowen *et al.*, 1986; Rose, 1989b, Hart & Boulton, 1991; Hart & Peglar, 1990; Lunkka, 1994) because the till overlies the Cromer Forest-bed Formation (West, 1980 b), and is overlain by the Lowestoft Till, which is, in turn overlain by Hoxnian Interglacial deposits (Rose, 1989b, Hart & Pegler, 1990). This is a classic biostrati-graphic constraint for a lithostratigraphic unit. The Anglian glacial deposits being defined by their position between pollen-stratigraphically defined organic deposits.

However, new work on the Cromer Forest-bed shows that most of these deposits are separated from the glacial deposits by a significant hiatus (Preece & Parfitt, 2000) and new work in southern Norfolk shows that the Happisburgh and Corton Tills are contemporaneous with second terrace deposits of the Bytham river (Lewis *et al.*, 1999; Rose *et al.*, 2000; Hamblin *et al.*, 2000, 2001) means that the Happisburgh Till must have been deposited long before to the Anglian glaciation. On present understanding of the history of deposition and erosion of Bytham river sediments (Ashton *et al.*, 1992), it is likely that the aggradation of the second terrace is separated from the Anglian glaciation that overrode and destroyed the Bytham river by two temperate episodes. If the Anglian is accepted to be MIS 12 (see below for reasoning) then the episode of deposition of the Happisburgh Till is considered to be MIS 16. As recognised in Hamblin *et al.* (2000) and re-stated in by Banham *et al.* (2001) this interpretation is not supported by the rodent biostratigraphy. Specifically, the microtine rodent fauna from the Cromer Forest-bed Formation beneath the Happisburgh Till at Sidestrand contains *Arvicola t. cantiana* (Preece & Parfitt, 2000), which is considered to have first appeared in MIS 15. However the correlation between the Marine Isotope Stages and the rodent biostratigraphy is ambiguous (Vandenberghe, 2000) and this issue needs to be resolved before this evidence can be used with confidence.

3.4.3.3 Extent and limits of glaciation

On the basis of the distribution of the Happisburgh and Corton Tills this glaciation is considered to have entered northern East Anglia, reaching the catchment of the Bytham river, but only rarely crossing the Bytham river valley. Ice flow appears to have been from the north and the source of the ice appears to have been central Scotland, and not Scandinavia as is traditionally proposed (Lee *et al.*, in press). This distribution is shown on Fig. 2. Also included is Fig. 4 which is taken from Lee (2001) which shows the development of this glaciation and the deposition of these tills. Of particular importance in the determination of this limit is the presence of Starston Till, which is lithologically equivalent to the Happisburgh Till, at Knettishall (Lewis *et al.*, 1999), the Harleston area (Lawton, 1982) and the Redgrave area (Auton *et al.*, 1985). Deposits known as the Banham Beds (Mathers *et al.*, 1987) in the region around Diss have similar lithological properties, and similar lithologies have been recorded at Barnham, (Lewis, 1998; and JR personal observation and analysis), and Feltwell (JR personal observation). It is essential to recall, when looking at Figs 2 and 4, which show ice limits at this time, that the Wash and Fen basin and the area west of the Chalk was high enough to drain into the Bytham river valley **(JR).**

3.4.4 The Anglian Glaciation (MIS 12)

3.4.4.1 Preliminary comments

The Lowestoft Till is defined in Perrin *et al.* (1979) as a clay rich diamicton with chalk and flint clasts. This is a very distinctive and extensive lithology and its presence has had a profound influence on the interpretation of the Quaternary lithostratigraphy of England as a lithostrati-graphic marker and because of its contribution to the lithostratigraphy of river deposits formed during and following the glaciation in which it was deposited (Whiteman & Rose, 1992).

The lithology that currently comes under the title of Lowestoft Till was initially described in terms of two units (Harmer, 1928; Baden-Powell, 1948; West & Donner, 1956). Of the two units, one is dominated by a dark grey matrix and chalk and flint clasts and was defined by Harmer (1928) as the 'Chalky Boulder Clay with Jurassic matrix' having been derived from the Mesozoic mudstones that outcrop in the region of the Wash and Fen basin; the other is dominated by a light brown (often called 'buff') chalky matrix with chalk clasts. This was defined as the 'Chalky Boulder Clay with Chalk matrix' having been derived from erosion of the Chalk (Harmer, 1928). Reference to these two lithologies persisted until the 1970's and indeed the terms Lowestoft Till and Gipping Till were used to define the Jurassic and Chalky boulder clays, respectively (West & Donner, 1956), each associated with a different ice-flow direction, and with a different age. The Jurassic / Lowestoft Till was ascribed to the Anglian Glaciation and the Chalky / Gipping Till to the Wolstonian Glaciation (Mitchell *et al.*, 1973). Straw (1979a, 1983, 1991) elaborated on this interpretation in Lincolnshire and east Midland England.

Because of the absence of any organic deposits between these two tills the proposal was made that they represented only one glacial event (Cox & Nickless, 1972; Bristow & Cox, 1973 and discussion therein). Subsequently, as a result of a regionally extensive, quantitative study of the particle size properties (sand, silt , clay) and lithological properties (light and heavy minerals, calcium carbonate content), Perrin *et al.* (1979) concluded that the Jurassic and Chalky tills were produced by single glacial event and simply represented the products of different ice flow paths across different lithologies. Although this interpretation was subsequently revised by one of the authors of Perrin *et al.* who suggested two, rather than one ice flow direction (Rose, 1992), the view of a single glacial episode during the Anglian as been maintained to the present, and is described below at the 'traditional model'.

The Happisburgh and Corton Tills, described in Section 3.4.3. are one of the North Sea Drift group of glaciogenic deposits (Boswell, 1914, 1916; Harmer, 1928; Perrin *et al.*, 1979). These North Sea Drift deposits have been given a number of names (Hart & Boulton, 1991; Lunkka, 1994), but for the purpose of this publication the names: 1st, 2nd and 3rd Cromer Tills, proposed by Banham (1968) will be used.

These tills were all studied quantitatively by Perrin *et al.* (1979). They are sandy diamictons, with a low calcium carbonate content, a low percentage of opaque heavy minerals relative to the Lowestoft Till, and an assemblage of non-opaque heavy minerals that are typical of derivation from a metamorphic bedrock region. The quantitative study and statistical analysis showed a distinctive difference from the Lowestoft Till in all the properties studied, and the 1st and 3rd Cromer Tills were recognised as being similar, but the 2nd Cromer Till has a higher calcium carbonate and clay content. Distinctive indicator erratics such as rhomb porphyry and larvikite, both from the Oslo area of Norway, along with the sandy matrix derived from the incorporation of North Sea sediments suggested that the three Cromer Tills / North Sea Drift group of glacial deposits, were deposited by ice that flowed across the North Sea from Scandinavia (Perrin *et al.*, 1979).

The absence of Lowestoft Till from the core region of the North Sea Drift (Figs 5 and 6) suggested that the Scandinavian ice covered northeast Norfolk while the ice that deposited the Lowestoft Till covered the rest of the region. The absence of any organic deposits from between these tills, along with the apparent interdigitation of Lowestoft and North Sea Drift diamictons and sand and gravel deposits (Cox & Nickless, 1972; Perrin *et al.*, 1979) suggested that both were formed in the same glaciation. Following the reasoning used to determine the age of the Lowestoft Till both were therefore attributed to the Anglian Glaciation, between Cromerian and Hoxnian interglacial deposits (Rose, 1989b).

The following description (Section 3.4.5) gives the 'traditional' interpretation of the Anglian Glaciation. The revised model, based on recent discoveries will follow in Section 3.4.6 **(JR).**

3.4.5 Anglian glaciation - the traditional model

The glaciation in the Anglian Stage was the most extensive of the British Pleistocene. However, since for the greater part of its extent its deposits were overridden by younger glaciations, much of the sedimentary record has been lost. There is therefore only fragmentary evidence for reconstructing the glacial history of the Anglian Stage.

For this reason the Anglian glacial sediments are best preserved beyond the margins of later ice advances, and eastern England and the adjacent off-shore region include extensive sequences of deposits and features formed during this event. This review will therefore concentrate on discussion of these areas, but will also consider the evidence for the extent of glacial deposits in the English Midlands, the Welsh borderland and south Wales.

The dating of the Anglian glacial deposits in the type area is based on their relation to temperate stage sediments. In north Norfolk glacial deposits rest on Cromerian Complex Stage deposits and throughout their distribution area are overlain directly by Hoxnian Stage interglacial sediments (Preece & Parfitt, 2000; Banham *et al.*, 2001).

No evidence of interstadial or higher rank climatic oscillations has been identified intervening within the Anglian sequence and therefore it is generally interpreted as representing a single complex glacial event. However, this was recently questioned by Hamblin *et al.* (2000) and Moorlock *et al.* (2000). On the basis of strong lithological and detailed stratigraphical relationships the Anglian is equated with the continental Elsterian Stage.

The considerable antiquity of the Anglian glaciation means that ice-marginal landforms are difficult to find because of their subsequent modification, and in many cases, total removal during later events. Limits must therefore be determined using a variety of means, including primarily the distribution of diamictons and the occurrence of characteristic erosional and deposition features, such as tunnel valleys, marginal ridges, outwash and glaciolacustrine sequences. The following summary is based on this type of combined evidence, but will emphasise those areas for which particularly characteristic features are known.

3.4.5.1 Pre-Devensian (Pre-Weichselian) Pleistocene glaciation limits

3.4.5.1.1 East Anglia

The Anglian Glaciation began as the ice advanced from the NE and overrode the pre-existing sedimentary sequences throughout the region. This advance deposited a complex suite of interbedded tills and associated meltwater sediments, that are termed the North Sea Drift Formation. These sediments, best exposed on the northern and northeastern coasts of East Anglia, from Weybourne to Lowestoft (Fig. 1), can also be traced inland as far as Norwich and south towards Diss (Mathers *et al.*, 1987). Internally this sequence is highly complex, including evidence for a three-fold oscillation of the ice sheet; the intervening phases represented by deltaic outwash and glaciolacustrine sedimentation (Hart 1987; Lunkka 1994). For all but the final still-stand phase that deposited the so-called Cromer Ridge in north-east Norfolk, no precise margins are known for these individual advances.

The ice that deposited the North Sea Drift sediments characteristically contains a suite of igneous and metamorphic erratics of southern Norwegian origin. Some of these erratics, rhomb porphyry and larvikite, were derived from the Oslofjord area and have therefore been generally accepted as evidence that the ice sheet originated in southern Norway and crossed the North Sea basin. Although these erratics are found at all exposures of North Sea Drift diamictons, an indication that Scandinavian ice may have reached further south is the present distribution of Norwegian erratics. Rhomb porphyries and larvikites can be found as far south as Bedford, Hitchin, Ipswich and Cambridge (Ehlers & Gibbard, 1991) where they were probably carried by later ice movement.

During the next phase, ice of British origin advanced through the Vale of York and Lincolnshire into central and western East Anglia. The ice crossed the Fenland Basin, possibly exploiting a pre-existing gap in the Lincolnshire-Norfolk Chalk ridge. On the basis of lithology (Perrin *et al.*, 1979), erratic content (Baden-Powell, 1948), detailed fabric measurements (West & Donner, 1956; Ehlers *et al.*, 1987) and fossil assemblages from erratic chalk pebbles (Fish & Whiteman, 2001), it can be shown that the ice radiated outwards from the Fenland Basin in a fan-like pattern (Fig. 6). It is thought that interaction of this British ice with the Scandinavian ice was responsible for this unusual pattern of ice movement. Moreover, in the eastern part of the area the British ice seems to have expanded close to its maximal position and replaced the Scandinavian ice. The British ice expanded progressively towards the east and northeast, reaching the present east coast at Lowestoft.

Over a large part of East Anglia the British ice deposited by the so-called Lowestoft Formation till. It has a grey to blue-grey clay matrix (up to 45%), is rich in flint and sub-rounded chalk clasts and includes erratics that originated on the British landmass. Grain-size distribution is remarkably uniform across the region, minor changes being attributed to incorporation of local materials, such as North Sea Drift Formation sediments etc. (Perrin *et al.*, 1979).

With the continued withdrawal of the Scandinavian ice, the Lincolnshire coast and adjacent offshore area became open for the expanding British ice advancing south-eastwards into Norfolk. Again the Wash and the Breckland Gap directed the ice stream. Ice that flowed over the relatively high ground of the Chalk escarpment and down into the Fenland Basin led to deposition of the characteristic chalk-rich 'Marly Drift' till facies onto more typical Lowestoft Till near Kings Lynn (Straw, 1991). Ehlers *et al.* (1987) showed that this facies does not represent a separate glaciation but simply locally-derived diamicton deposited during a later phase of the Anglian.

During the 'Marly Drift' phase, chalk-rich till was transported into much of East Anglia, including the Gipping Valley in Suffolk, where it overlies the Jurassic clay-rich tills (Lowestoft facies) of the preceding Lowestoft advance. By this time, however, the ice no longer reached its maximum limits. In East Anglia, the area beyond the Cromer Ridge was no longer covered by ice. As the ice retreated, vast meltwater formations were laid down as sands and gravels, for example those between East Dereham, Swaffham and Fakenham. Here the gravels rest on the chalk-rich till of the last Anglian ice advance. Similarly in the area around the Glaven Valley in North Norfolk kames, dead-ice topography and an esker are associated with the retreat of the chalk-rich till ice (Sparks & West, 1964; Ehlers *et al.* 1987).

Throughout East Anglia a series of deep, steep-sided valleys have been found cutting through the Chalk and associated bedrock. These valleys form a radiating pattern, broadly parallel to the dip of the Chalk; i.e. they trend east-west in the north and north-south in the south of the region. Detailed study of these tunnel valleys, by Woodland (1970) and Cox (1985) among others, indicates that they are normally filled with glacial sediments, predominantly meltwater sands, gravels or fines. Tills also occur but are less frequent. Closely comparable in size to the rinnen of Denmark, Northern Germany, Poland etc., they are undoubtedly of glacial origin and probably result from subglacial drainage under high hydrostatic pressure (Ehlers *et al.*, 1984).

During deglaciation meltwater seems to have become concentrated into large channel-like valleys that formed in a generally radiating pattern again parallel to the dip of the Chalk and Lower Tertiary bedrock. In many places this alignment is determined by the course of the subglacial tunnel valleys. Valleys such as the Stort, the Gipping, the Waveney and the Wensum are all of this type. However, in northern Norfolk, substantial proglacial outwash sandur plains developed particularly at Kelling and Salthouse Heaths and mark retreat-phase ice-front stillstand positions. The radial pattern of meltwater discharge continued until ice withdrew beyond the Chalk escarpment, north-northwest of which drainage became directed by the regional slope into the Fenland Basin.

In the marginal areas, particularly at Corton, the ice front retreated in standing water (Bridge & Hopson, 1985; Hopson & Bridge, 1987; Bridge, 1988). A similar phenomenon is found in the Nar Valley (Ventris, 1986; 1996) where the tunnel valley deepens towards the west i.e. in the direction of the ice retreat. A proglacial ice-dammed lake seems to have formed in the valley and caused the thinning ice initially to float, depositing subaquatic till, and later to retreat completely leaving laminated clays to accumulate. Subsequent deglaciation gave rise to drainage of the lake and establishment of fluvial sedimentation.

3.4.5.1.2 Southeast England and the Thames Valley.

Advance of the British Lowestoft Formation ice into Essex and Hertfordshire brought it into the region influenced by the River Thames and its tributaries. A substantial series of Thames deposits termed the Kesgrave Formation are aligned WSW-ENE to W-E across southern East Anglia and represent the preglacial course of the river. The ice overrode these terrace-like accumulations in all but the southernmost part of the region. Oscillations of the ice-sheet margin have been repeatedly identified in the southern marginal zone, particularly in the Ipswich, Chelmsford and Hertfordshire areas (Allen *et al.*, 1991). Similar frontal movements also occurred in the north and east London areas (Gibbard, 1979; 1994, Cheshire, 1986). These oscillations had profound effects on the River Thames since the Lowestoft ice advanced into and overrode the river's valley in Hertfordshire; the till units interdigitating with the fluvial sediments

Subsequent erosion has largely removed any marginal features left by the ice, the maximum limits being determined principally on distribution of diamicton. The most convincing feature is the marginal moraine-like form in the Finchley-Hendon area of north London. Another landform marking the maximum ice extent is the substantial marginal meltwater-fan complex that occurs south of Chelmsford, Essex; the Danbury-Tiptree ridge. Here over 30 m of meltwater deposits occur on the leeside of the range of Tertiary hills that formed the southern side of the Anglian Thames valley. Although an early oscillation was able to by-pass these hills to the south, they apparently later formed an insurmountable barrier to the Anglian ice which rode up against the hills and discharged vast quantities of meltwater eastwards towards the North Sea (Gibbard and Boreham, in preparation).

Elsewhere, there is evidence for ice extension several kilometres beyond the maximum mapped extent of glacial diamicton. For example, small scale 'tunnel valleys', infilled by glacial sediment have been identified in Suffolk by Mathers *et al.* (1991) and Cornwell & Carruthers (1986). The glacial maximum on the map (Fig. 2) therefore includes these records.

3.4.5.1.3 English Midlands

Following the classic work of Shotton (1953, 1968, 1983a, b), the glacial sequence of the English Midlands in the area around Coventry and Birmingham was considered to be of Wolstonian (= Saalian) age. The sequence comprises a series of glacial sediments identified over a large area in the west and central Midlands (Rice & Douglas, 1991). Of these, diamictons and associated meltwater sediments provide the evidence for an extensive glaciation of the region as far south as Morton-in-Marsh in Gloucestershire. Perrin *et al.* (1979) and Sumbler (1983a, b) questioned the basis for the relative age of the Wolston Formation sequence, pointing to the apparent continuity with the Anglian deposits further to the east, and therefore concluding that the Midlands' sequence should also be assigned to the Anglian Stage.

Undoubted Anglian and post-Anglian (pre-Devensian) glacial sediments do indeed occur in the west Birmingham area. Here two lake basin infillings, the form and situation of which closely resemble kettle holes, have yielded characteristic Hoxnian pollen sequences. Both the Nechells (Kelly, 1964) and Quinton (Horton, 1974) deposits occur interstratified in glacial and associated sediments and therefore potentially hold the key to clarifying some of the questions concerning the Midlands' sequence. However, their occurrence overlying glacial sediments, including the Nurseries Till derived from the NW, filling channel-like depressions in the underlying bedrock (Horton, 1974), emphasises that Anglian events in the Midlands may have been somewhat similar to those in eastern England.

3.4.5.1.4 South Midlands

In the South Midlands area, east of Leominster and Hereford, glacial deposits of the Middle Pleistocene Risby Formation record the extent of an ice lobe that extended westwards from Wales to the foot of the Malvern Hills. These deposits include a till rich in Palaeozoic clastic material of southern central Welsh origin. Apart from this maximal extent, evidence for four additional retreat phases is also represented in the highly-fragmentary sequence. The ice-lobe seems to have adopted virtually the same path as that followed by the Late Devensian ice (Richards, 1998).

The Anglian age of these glacial deposits is indicated by their stratigraphical relationships to interglacial sequences, paralleling those found in East Anglia. They are underlain by late Cromerian Complex-age fluvial sediments of the Mathon Formation and overlain by Hoxnian-age pond sediments in the Cradley Brook Valley (Barclay *et al.*, 1992).

3.4.5.1.5 South Wales

In general, the pre-Devensian glaciation of southern Wales is referred to as the Irish Sea Glaciation (Pringle & George, 1961). This glaciation potentially includes evidence for one or more glacial episode during which complete glaciation of the province occurred by local ice-sheets. The margin of Anglian-age Welsh ice is clearly marked on south and west Gower by the Paviland Moraine, a marginal complex of meltwater gravels and sands resting on a red clay till (Bowen, 1999b), rich in Millstone Grit clasts derived from southern Welsh Carboniferous rocks. At the same time, the glaciation extended across the Bristol Channel as far as the northern coast of the English South-West Peninsula (Bowen *et al.*, 1986), athough there is some dispute over precisely how far it reached. There seems to be no doubt that it did not reach Cornwall or the Scilly Isles, where deposits previously thought to be Anglian are now assigned to the Devensian (Campbell *et al.*, 1999).

3.4.5.1.6 Offshore areas

Many of the marginal areas of the Anglian glaciation ice sheet lie not on land but on the sea floor. Therefore it might be expected that sedimentary basins like the North and Irish Seas would provide excellent sediment traps in which Anglian glaciation sediments would be well preserved. However, because of repeated later erosion this is not the case.

Beneath the St. George's Channel, adjacent to the Welsh landmass, probable Anglian-age diamictons and associated sediments of the Caermarfon Bay Formation have been identified by Tappin *et al.* (1994), Wingfield (1994, 1995) and Cameron & Holmes (1999) infilling glacial valleys. The southern limit of these valleys occurs at about 51°N and, according to Wingfield (1989), marks the stillstand position of a grounded ice-front in the basin. Floating ice may well have extended further south of this line at times.

The occurrence of Anglian tills is also limited to relatively small areas in the southern part of the North Sea (south of Dogger Bank). As in the St. George's Channel, many buried channels occur beneath the southern and central North Sea floor, the oldest of which have been shown to be of Anglian age (Balson & Cameron, 1985; Cameron *et al.*, 1987). They contain tills and meltwater sediments assigned to the Swarte Bank Formation. These channels are remarkably similar in form and scale to the tunnel valleys of East Anglia or the rinnen of the adjacent Continent, already discussed.

This margin is very similar to that suggested by Laban (1995) which is marked by subglacial tongue basins, subglacial tunnel valleys and substantial glaciotectonic deformation structures.

Beyond these areas the evidence for the extent of the Anglian glaciation remains vague **(PLG).**

3.4.6 Anglian glaciation - the revised model

In many respects the revised model is similar to the traditional model described above in Section 3.4.5. For instance, it is associated with the most extensive ice cover in eastern and southwest England, is associated with the formation of the tunnel valleys in eastern England and it is associated with the diversion of the River Thames through London. However, there are significant differences also. First and most important, the revised model equates the Lowestoft Till with the 2nd Cromer Till of the North Sea Drift group of deposits and both are associated with a northern British ice source. Secondly this model does not include much of the deposit that was defined by Perrin *et al.* (1979) as Lowestoft Till - this material is now attributed to both the Anglian and a later stage glaciation (see below). Thirdly, deposits in most of Midland England included in the section above (3.4.5) are now attributed to a later glaciation.

3.4.6.1 Glacial deposits

The glacial deposits attributed to this stage are the Thrussington Till of the west Midlands (Rice, 1968, 1981; Rice & Douglas, 1991), the chalk-free dark grey till with Carboniferous erratics which is found throughout the east Midlands and in parts of the south Midlands (this has not been formally defined, and has previously been incorporated in the Lowestoft group of tills), and the Jurassic and Chalky facies of the Lowestoft till throughout East Anglia.

The Thrussington Till is a reddish brown till with a dominance of quartzite and vein quartz clasts. The lower Triassic facies of the Moreton Drift in the area around Moreton-in-March is the equivalent of this unit in the south Midlands (Sumbler, 2001). These tills are derived from the Triassic rocks of west Midland England and reflects ice flow into midland England from the northwest. The chalk-free dark grey till with Pennine erratics is found in the Trent Valley (Pennine Tills of Deeley, 1886), south of the Vale of Belvoir, in the region around Melton Mowbray, as the

Lower Till of Northamptonshire (and as patches of basal till in Buckinghamshire (JR, personal observation). It is probable that the Heath Till of west Lincolnshire and the Wragby Till of east Lincolnshire (Straw, 1991) are associated with this group (Fig. 5). The Jurassic and Chalky facies of the Lowestoft Till in East Anglia cannot be differentiated on lithological grounds from the same lithologies attributed to a subsequent glaciation (MIS 10, explained below in Section 3.4.7), but they are defined by geochronometry and correlation with dated sequences of river deposits / terraces. On these criteria the Lowestoft Till deposited during the Anglian (MIS 12) exist in East Anglia at Hoxne where the interglacial deposits are dated by U-series / ESR to Stage 11 (Grün & Schwarcz, 2000) and at Marks Tey where the interglacial deposits are dated by U-series to MIS 11 (Rowe *et al.*, 1999). In the Thames valley the deposits of this glacial event are identified by their relation to the Thames river aggradations / terraces that are attributed on the basis of AAR geochronology and terrace stratigraphy to MIS 12 (Bridgland, 1994). It is probable, in view of this distribution, that this part of the Lowestoft Till covers East Anglia east of the Chalk escarpment (Fig. 2), and is the only glacial deposit in this region, except in the north and northeast of the region where earlier and later deposits occur (see elsewhere in this report).

3.4.6.2 Evidence for age

This has been given above, and takes the form of U-series and U-series / ESR determinations on organic materials that overlie the Lowestoft Till at Marks Tey and Hoxne respectively (Rowe *et al.*, 1999; Grün & Schwarcz, 2000). Both of these sites are considered to show continuous sedimentation through from glacial to temperate conditions (West, 1956; Turner, 1970) and both of these age determinations suggest glaciation during MIS 12.

Attribution to the Hoxnian Interglacial is not taken as of stratigraphic significance as pollen assemblage biozones may be repeated on numerous occasions and depend upon circular reasoning for their stratigraphic application. For instance, the pollen assemblages attributed to the early part of the Late Anglian glaciation / Early Hoxnian Interglacial at Trimingham (Hart & Peglar, 1990) are similar to sites from the Late Devensian / Early Holocene, with respect to the 'diagnostic' characteristics (*Hippophaë* peak) (Godwin, 1964), simply reflecting an equivalent ecological niche.

The other key line of evidence used for dating is the relationship of the tills described above to the aggradations / terraces of the Thames river system. This relationship was established by careful mapping by Gibbard (1977) and Cheshire (1986), and shows that glaciation occurred during the formation of river deposits that can be dated by amino acid ratios, terrace stratigraphy, and a model for terrace aggradation / incision (Bridgland 1994) to MIS 12.

All this dating evidence is in line with the initial attribution of the Anglian glaciation to MIS 12 of the marine oxygen isotope curve, based on the correlation of the most extensive British glaciation with the most prominent of the isotopic peaks representing magnitude of ice volume (Shackleton & Turner, 1967).

3.4.6.3 Provenance and extent

All evidence suggests that this glaciation is from ice centres in northern Britain, that moved from the north down the lowlands of eastern and western England into the English Midlands and across the whole of East Anglia. The 2nd Cromer Till facies of the deposit represents a flow path down the North Sea on the eastern side of the British land area. This may have been constrained by Scandinavian ice in the North Sea, but there is no evidence for this, and there is no evidence that Scandinavian ice reached England at this time. It is this glaciation that initiated erosion in the clay vales of the lowland England, including the Wash and Fen Basin, and it is this glaciation that terminated the existence of the Bytham river, destroying its valley in areas of the Fen Basin and burying it beneath glacial deposits in Warwickshire, Leicestershire, Lincolnshire, Norfolk and Suffolk.

The extent of this glaciation is that described in Section 3.4.5 and shown on Fig. 2, with respect to East Anglia, Southeast England and the Thames Valley, South Wales and Offshore areas, but it is not the same for the South Midlands including the upper Thames valley, where it was over-ridden by the ice of MIS 10. In the South Midlands including the upper Thames valley it is not possible to determine the limit of the Anglian Glaciation (**JR**).

3.4.7 Glaciation associated with the deposition of the Bacton Green Till and chalky tills of Midland England, including the Oadby Till (MIS 10)

3.4.7.1 Glacial deposits

The key deposit of this glaciation is the Bacton Green Till or 3rd Cromer Till of the north Norfolk coast. This overlies the 2nd Cromer Till that is a variant of the Lowestoft Till that was deposited in MIS 12 (see Section 3.4.6). The Bacton Green Till is correlated with the Jurassic and Chalky Lowestoft Till of midland England including, especially, the Oadby Till (Rice, 1968, 1981).

The Bacton Green Till is a sandy diamicton that overlies the chalky, clayey diamicton of the 2nd Cromer Till at Bacton Green (Jon Lee, pers. comm.). In many respects it is like the Corton Diamicton (1st Cromer Till), but is separated from this by the glacial deposits of MIS 12. This deposit can be traced further south in Norfolk as the Norwich Brickearth (Perrin *et al.*, 1979), but this is an ambiguous correlation as the Norwich Brickearth is a soil developed on a sandy diamicton (Rose *et al.*, 1999b) and could equally be developed on the 1st Cromer Till if this is exposed at the surface. The Jurassic and Chalky Lowestoft Till extends across midland England from the Wash and

Fen Basin to the Derby area in the west Midlands and to the Moreton in the Marsh region of the south Midlands. To the east the limit is not apparent and can only be determined by U-Series dates on peat overlying glacial deposits at Tottenhill in west Norfolk. The deposit of this glaciation is represented most extensively by the Oadby Till of the Leicester area (Rice, 1968, 1981; Douglas & Rice, 1991) and the Great Chalky Boulder-clay of the Trent valley (Deeley, 1886). Clast content indicates a preponderance of Jurassic mudstones from the Wash and Fen Basin and Chalk from the North Sea east of the Lincolnshire coast (Perrin *et al.*, 1979). Heavy minerals indicate a dominant opaque content, derived from the Jurassic rocks of the North Sea region off the Yorkshire coast and a non-opaque fraction from a metamorphic province. Both heavy mineral suites indicate transport to the region through the Wash and Fen Basin (Perrin *et al.*, 1979) (Fig. 6). Clast fabric studies indicate radial flow from the region of the Fen Basin southwestward across the English Midlands (Fig. 2). At present no other deposits can be attributed to this glaciation.

3.4.7.2 Evidence for age

As with the dating evidence for the revised interpretation of the Anglian, the age for this glaciation is based on geochronometry of overlying non-glacial deposits and correlation with river aggradations / terraces of the River Thames, this time in the Upper Thames region. The critical evidence for dating comes from a number of U-series determinations on peat from sediment that lie directly above, and apparently in continuous succession with the Lowestoft Till at Tottenhill, near Kings Lynn in northwest Norfolk (Rowe *et al.*, 1997). This is the most rigorous U-Series determination carried out for this period of the Quaternary of Britain and results indicate MIS 9 age meaning that the till is likely to have been deposited in MIS 10.

In south Midland England Sumbler (1995, 2001) has mapped the glacial and river deposits and shown that the chalky till facies of the region (Oadby Till) can be correlated with the terrace of the Thames which are dated to MIS 10 (Bridgland, 1994). Throughout midland England the river terrace sequence differs from that of the Thames in that the succession only extends back as far as MIS 10 (Keen, 1999). The logical explanation for this is that ice covered the area during MIS 10 and river development in the region only started as the ice wasted.

3.4.7.3 Provenance and extent

The lithological evidence described above suggests one glacial event that entered the region from northern Britain through the Wash and Fen basin then spread out across the region as a piedmont lobe, on low relief, highly deformable bed material as described in Perrin *et al.*, (1979) (Fig. 2). The main source of this ice was in Scotland and the main flow path was down the western side of the North Sea basin where extensive erosion of Jurassic and Cretaceous rocks took place and in the already existing Wash and Fen Basin which was eroded in the Anglian (MIS 12). It is possible that the ice flow was constrained by Scandinavian ice in the North sea region, but there is no evidence to support this other than the flow-path adopted by this glacier.

The extent of this ice appears to be the region just west of Derby in the Trent valley and around the northern edge of the Cotswold Hills in the south Midlands (Fig. 2). Elsewhere the limit is not clear. This is especially the case at the eastern margin of the ice sheet where the till lithology appears to be similar to that of the preceding Anglian glaciation. However, on the basis of the location of the sites which identify an Anglian Glaciation it is suggested that this MIS 10 ice sheet terminated along the west side of the Chalk escarpment that extends from the region of Luton to north Norfolk (Fig. 2). It is probable that in north Norfolk the chalky till extends eastwards towards the region of Bacton Green where a clear stratigraphic position is observable. The limit of this glaciation south of Bacton Green is the subject of current research, but appears to be near Norwich which is close to the southern limit of the Norwich Brickearth (Fig. 2).

This undefined eastern margin is far from satisfactory, but it is interesting to note that an identical situation has occurred in the past when West (1977, p 280, Fig. 12.3) drew a limit for the 'Wolstonian' glaciation in eastern and midland England which was unsupported by any sedimentary evidence **(JR)**.

3.4.8 Glaciation associated with the deposition of the Warren House Till of County Durham, the Basement Till of east Yorkshire, the Welton Till of Lincolnshire, the Beeston Regis Formation and associated glacial landforms of North Norfolk (MIS 6)

3.4.8.1 Glacial deposits and landforms

The glacial deposits which are considered to be the product of this glacial event are the Warren House Till of County Durham, the Basement Till of east Yorkshire and the Welton Till of east Lincolnshire all of which are located beneath the Devensian deposits of the respective regions. Further evidence is to be found at Tottenhill, in northwest Norfolk (the same site as that which the MIS 9 U-series determination was derived) where outwash has been identified above MIS 9 deposits and beyond the limit of the Devensian Glaciation, and at Shooters Hill near Weybourne, and Beeston Regis near Cromer, both in north Norfolk, where outwash deposits have been mapped with a significant content of Scandinavian erratics.

In addition to the sedimentary evidence, the sites along the north Norfolk coastal region are associated with distinctive topography that has the appearance of ice contact and ice-marginal terrain (this is the same area as described as being of traditional Anglian age in section 3.4.5.1.1). Thus, in the western area a set of kames, an esker and an east-west trending ice-contact slope with distal outwash plain are taken as additional evidence of an MIS 6 ice marginal position and further east this landform continues as the Cromer Ridge which is associated with outstanding glaciotectonic structures.

The Warren House Till (Trechman, 1915; Francis, 1970), the Basement Till (Catt & Penny, 1996, Catt, 1991**b**) and the Welton Till (Alabaster & Straw, 1976) are all distinctive diamictons with a predominance of Scottish, North Sea and Scandinavian erratics, quite different from any pre-existing tills in eastern England. These tills are also quite different from the Devensian tills of the region in terms of colour, particle size distribution and clast lithology. The sands and gravels at Tottenhill, that are interpreted as outwash, a dominated by flint and do not show the Scandinavian clast component described above. However, they form a surface landform, and include plant remains indicative of cold climate conditions in existence during their deposition (Gibbard *et al.*, 1991a, 1992). They also have a palaeosol on their surface which shows only one phase of temperate weathering in addition to the at of the Holocene, and is therefore considered to have started to form in the Last Interglacial (MISs 5e) after the deposition of the outwash gravels (Lewis & Rose, 1991). The Britons Lane Sands and Gravels which are the glaciofluvial deposits with a Scandinavian clast component in north Norfolk are the subject of on-going work. However, they are consistently the youngest glacial deposit of the succession, they are not glaciotectonically deformed, and they can be identified as part of a glaciogenic landform.

The position of the glaciogenic landforms associated with this glaciation is important in a number of ways. Firstly, these are the only glaciogenic landforms that can be recognized with any confidence outside the limit of the Last Glacial Stage. This is an issue that has caused much debate (Sparks & West, 1964) and section 3.4.5.1.1 of this paper these landforms are attributed to the traditional Anglian Glaciation. As pointed out by Sparks & West (1964) it is interesting to note that in northern continental Europe glacial landforms of Saalian age in Northern Europe have similar relief. In particular, the push moraines of the Netherlands and North Germany, which were formed in MIS 6 are recognisible in the landscape as of glaciogenic form, but like the features in north Norfolk, without the subtleties of crenulate ice-contact relief and without the presence of kettle holes. Secondly it has been observed during the process of geological mapping of north Norfolk that these landforms occupy positions in a relief that has been changed significantly since the previous glaciation (Richard Hamblin and Brian Moorlock, pers. comm.). For instance the kames of the Glaven valley fill a large valley that is eroded into pre-existing chalky till. Finally, the scale of the relief is such that it is possible to recognize significant glacial landscapes along the coastal zone between Weybourne and Overstrand in north Norfolk, including an ice-tongue basin with upstanding kames behind outwash fans in the west and below the level of a push moraine ridge and the distal outwash fan in the east.

Evidence for MIS 6 glaciation in the west Midlands is also demonstrated by relation of the Ridgacre Formation to the Kidderminster Station Member of Severn Terraces which is dated by biostratigraphy and AAR (Maddy *et al.*, 1995).

3.4.8.2 Evidence for age

The evidence for age comes from two sources: one within the region, and the other by comparison with northern continental Europe. The critical dating evidence from Britain is to be found at Sewerby in east Yorkshire where the Basement Till has been observed beneath the biostratigraphically and luminescence dated Last Interglacial beach (Catt & Penny, 1966, Bateman & Catt, 1996; Briant, 2002). This means that this deposit is older then MISs5e.

Comparison with the glacial sequence in northern continental Europe is based primarily upon the similarity of the landforms in north Norfolk and those in the Netherlands and northern Germany beyond the Weichselian ice limit (Ehlers *et al.*, 1984). Lithological comparison is not possible as the ice streams that deposited the British and north European glacial deposits are different (Laban, 1995). A link is also proposed based on sea-bed evidence (Oele & Schüttenhelm, 1979, p 194) who also correlated the Dutch glaciogenic deposits with the Welton Till. The glacial deposits in the Netherlands are dated by the incorporation of MIS 7 deposits in the glaciotectonic structures of the push moraines at Franche Kampe, and the burial of the glacial landforms by Eemian (MIS 5e) marine deposits (Ruegg, 1991, Vandenberghe *et al.*, 1993). Thus the glaciation that produced similar landforms in the Netherlands, to those in north Norfolk, is securely dated to MIS 6.

The correlation between MIS 6 deposits in eastern England with those in the Netherlands and Germany is also realistic in terms of the nature of the glacial event. In both cases they are related to a Scandinavian ice source, the only time this is known to have occurred in Britain.

3.4.8.3 Provenance and extent

All the evidence from the tills in County Durham, east Yorkshire and east Lincolnshire and the outwash in north Norfolk suggests that the ice that deposited these sediments came across the North Sea from Scandinavia. This event is similar to that which placed ice over northern continental Europe from the region of Amsterdam eastward into northern Germany. There is sea-bed evidence in the form of subdued ridges and the distribution of tunnel valleys to support this link between northeast Norfolk and the Netherlands (Laban, 1995). In these circumstances it is likely that an ice dammed lake was formed between the ice margin and the Chalk ridge between Dover and northern France and it is possible that the overspilling of this lake contributed to the formation of the Straits of Dover, thus solving some of the problems created by an earlier separation of Britain and Europe (Smith, 1985) suggested by the biological evidence (Meijer & Preece, 1995).

The position of the ice margin in eastern England north of the Devensian ice limit is difficult to determine because of subsequent burial by Devensian glacial deposits. However, in north Norfolk, the margin can be recognized as at the northern end of the outwash deposit at Tottenhill, around the southern margin of the glaciogenic landforms in the Glaven valley, along the ice-contact slope at the western and northern sides of the Kelling and Salthouse outwash fans, then eastwards along the ice-contact slope at the northern side of the Cromer moraine ridge. This ridge extends out to sea in the area of Trimingham, due to the extensive coastal erosion in the area. The ice margin then extended eastwards and southwards across what is now the North Seas to reach the Netherlands in the area of Amsterdam (Fig. 2) **(JR)**.

3.5 Late Pleistocene glaciation limits

3.5.1 The terrestrial record of the maximum extent of the Devensian ice sheet in Britain (MIS 2)

3.5.1.1 Introduction

Here we focus on the maximum terrestrial limit of the Devensian (between 22 – 18 ka BP) Ice Sheet over England, Wales and Scotland. The maximum extent has been a topic of intense, but sporadic, interest over the last hundred years, with a multitude of varying maximal ice limits having been drawn. It is perplexing that academic papers, geological maps and text books contain examples of confidently drawn lines and yet finding the detail of the evidence for each segment of line is hard to achieve. It is within this context that we have tried to ascertain the basis of actual evidence and the best estimate for the maximum ice limit. We do not claim to have exhaustively searched the literature, and so other pieces of the jigsaw may still appear (the findings of a recently published paper that includes significant re-interpretation are given at the end of this section by JR (Section 3.5.2)). The most widely reproduced ice limit is that marked on the Quaternary Map of the United Kingdom (Institute of Geological Sciences, 1977), but unfortunately there is no accompanying text that explains upon what evidence it is based. Taking this line as a starting we point we examine and report evidence along it and modify the line where appropriate. More recent work on Scottish ice margins, and importantly, the vertical extent of the ice sheet, are also reported.

We make no attempt to review the history of changing views regarding the ice limit, but note the following generalizations. The southern limit of Devensian glaciation has been known to a precision of around 40 km for a long time. In fact, the Glacial Map of H. Carvill Lewis (1894, p 27) turns out to be mostly correct and would probably serve as an accurate enough test for any ice sheet modelling experiments. Since this time, the greatest debate and breadth of disagreement occurred for ice extent in Wales, but around three quarters of the southern limit has remained fairly uncontentious. Continued uncertainty has oscillated with regard to ice limits in Scotland, notably over the area of Buchan, the far NE of Scotland and the Hebrides. Views on ice sheet thickness have varied markedly, between complete inundation of the British landscape to reconstructions with much thinner ice. Only until recently has evidence for the vertical extent of the ice sheet been sought, and this is now providing valuable information in the third dimension.

From a glaciological viewpoint, actual palaeo-margin positions are of most interest as these could be used as tests or inputs to numerical models. Due to limitations in dating and scarcity of evidence, however, it has not been possible produce a reliable ice sheet-wide map of an actual palaeo-margin position. Efforts have been more focused on an assessment of the maximum extent of ice and this does not necessarily equate to a palaeo-margin position *at a point in time*. This is because the maximum extent likely occurred at different times in different places, and so the limits drawn here should be treated with this in mind.

In the absence of an extensive end moraine system clearly demarcating the ice limit, other forms of evidence have been enrolled. These include, meltwater channels that due to their topographic position must have been carved at an ice margin; inferred positions of an ice dam that is required to block a proglacial lake; some moraines and ice-marginal fans; weathering trimlines defining nunataks that projected above the ice sheet surface; and for much of Britain the maximum limit is defined by spreads of superficial drift or till. There is no clear diagnostic feature associated with any of these types of evidence that allow us to define the ice limit as *the maximum*, other than an absence of evidence beyond their geographic position. This creates particular problems, especially as the prior glaciation is known to have been more extensive, and herein lies scope for possible confusion. For drift limits this has been problematic with great efforts in the past being made to distinguish between 'newer' and 'older' drifts mainly on the basis of geological composition of drift with regard to the source areas of ice. For the areas without field evidence, judicious interpolation is required in order to define a continuous line for the ice limit. It seems that, quite sensibly, this has been done with due respect to topographic variations, such that lobes may splay out between uplands.

We note that some of the moraines and meltwater channels reported in the literature and entered on the accompanying map have been modified according to our own re-interpretations based on analysis of Ordnance Survey 1:50,000 digital elevation models.

The following notes are to elaborate on the map presented in Fig. 2, and to document the nature and type of evidence used to demarcate the presumed maximal ice limit. For ease of reference the ice limit is split into 'southern margin', 'Scottish limits' and 'vertical ice sheet extent'

3.5.1.2 Southern margin

With reference to Fig. 2 we outline the evidence for the maximum limit working from east to west.

In north Norfolk there is a strong contrast in lithologies of tills that has permitted a reliable splitting of 'newer' from 'older' drifts, that is, Devensian from Anglian tills (Hart & Boulton, 1991). Suggate & West (1959) noted and mapped the Hunstanton Till, which extends for only a short distance (<5 km) onshore, and is bounded by, and banked up against a former sea cliff (Straw, 1979a). The southern limit of Hunstanton drift was defined by Straw (1960) and it is this limit that is marked on Fig. 2, and we take this as the maximum ice limit for this area. Associated with this till are a series of ice-contact landforms such as meltwater channels and kames, and an esker (Straw, 1960, 1979a). Recent work has identified a moraine ridge complex in the Stiffkey area of Norfolk, just west of where the ice margin extends offshore, associated with the formation of an ice dammed lake and subsequent river diversion (Brand *et al.*, in press)[1]. The equivalent of the Hunstanton Till is found in the offshore geology, the Bolders Bank Formation, which is interpreted as a Devensian lodgement till (Balson & Jeffery, 1991), and part of it as a 'probable ice pushed ridge' by Cameron *et al.* (1992).

The Fen Basin, at the west side of Norfolk, is an extremely low-lying area very close to present day sea level. There have been many arguments that if ice impinged on the north Norfolk coast and southern Lincolnshire then it should easily have been able to penetrate the low land that is now the Wash, and it is likely that a lobe existed in this area. Such an argument on topographic grounds and the distribution of Devensian till seams reasonable. Evidence in support of this is as follows. Suggate & West (1959) argued for an ice lobe pushing southwestwards into the Fen Basin on the basis of a small number of till fabrics that possibly demonstrate a splaying pattern of flow. The actual ice limit is probably impossible to find as Holocene sedimentation has masked any landforms that may exist. It has long been regarded that an ice margin in the vicinity of the Wash and Fen Basin would have impounded fluvial drainage seawards and produced a large ice dammed lake. This lake has been called Lake Fenland or more recently Lake Sparks, and whilst it is poorly known and few shorelines have been found, lacustrine sediments do exist. Straw (1979a, 1991) reports sand deposits interpreted as strand-lines of the former lake shoreline, lacustrine clays and sands from its infilling and a fan of glaciofluvial gravels deposited at the lake edge. Investigation of varved sediments at Somersham, Cambridgeshire (TL 375 800) has provided further evidence for the existence of Lake Sparks, and radiocarbon dates confirm the Lake Devensian age (West, 1993; West *et al.*, 1999). We conclude that a lake existed which required an ice margin in the vicinity of the Wash and Fen Basin, although its precise margin location is unknown. It has been argued that this ice margin position was not produced by the ice sheet in steady state, but rather by a short lived surge-event down the east coast of England (*cf.* Evans, *et al.*, 2001)

Along the east coast of Lincolnshire and Yorkshire there are a series of large moraine systems either at the coast or a short distance inland (Straw, 1961, 1969, 1979a; Penny & Rawson, 1969). However, these do not mark the maximum limit of ice moving onshore, but more likely a standstill or minor readvance during overall retreat. The maximum limit of Devensian ice is demarcated by the extent of till that has been dated to have been deposited during the Dimlington Stadial (26 - 13 radiocarbon ka BP) of the Devensian (Rose, 1985). Various limits of this till spread have been reported (e.g. Suggate & West, 1959; Catt & Penny, 1966; Madgett & Catt, 1978), and we use the extent defined by Catt (1991a) and which was derived from Soil Survey mapping. For the 1:250,000 Soil Survey maps of this part of England (Soil Survey, 1983), substrate type was recorded (i.e. chalky till, sands and gravels etc.) and from this data Catt produced the extent of Devensian till, which we take as the maximum ice limit.

Additions or modifications to the drift limit described above are made on the basis of moraines or glaciofluvial ice marginal deposits that were found beyond the edge of the till sheet. In the Vale of Pickering, ice is inferred to have penetrated further inland than previously thought and impounded a glacial lake (Lake Pickering) that is much smaller than previously postulated (Catt, 1987, 1991b; Edwards, 1978; Foster, 1985). Prior views regarded Lake Pickering to have been dammed and the maximum limit to be at the Wykeham Moraine (Penny & Rawson, 1969) or the Flamborough Moraine (cf. Straw, 1979a). In the Vale of York, the impressive York and Escrick Mo-raines were originally taken to mark the maximum ice limit, as this is where the main till sheet terminates. However, some 50 km to the south, Gaunt (1976) found a suite of gravels containing glacial erratics which he interpreted to have been deposited in an ice marginal environment. This has been taken (e.g. IGS, 1977; Catt, 1991b) to make the maximum extent of ice in the area. Gaunt (1976) describes how the area was occupied by the large glacial Lake Humber, which was impounded elsewhere, and that a tongue of ice moved southwards into the lake, when it was at its 'high level'. It reached as far south as Doncaster. When the lake was at its 'lower level' the ice tongue had retreated back to its position at the York and Escrick Moraines. The position of the southern limit as far as Doncaster has been disputed by Straw (1991) who regards the stratigraphic position of Gaunt's (1976) gravels as uncertain and argues that they may be pre-Devensian. The York and Escrick Moraines (cf. Straw, 1979a) are the largest moraines in Britain and presumably represent a major standstill of the ice margin.

From York, the ice limit runs down the backbone of the Pennines and then down their western flack, past the Peak

[1] Inserted by JR.

District and nearly as far south as Birmingham. Much of this limit has in the past been defined by Jowett and Charlesworth's (1929) limit of 'extraneous drift' or drift limits mapped on the 1 inch Geological Survey sheets. The limit we have reproduced is from Catt (1991a) which is based on the Soil Survey mapping.

From Birmingham, across the Welsh borders and into South Wales, there have been a variety of maximal ice limits drawn, mostly based on drift limits. They have not varied by much and indeed the mapping by the IGS (1977) shows a similar pattern to that mapped by Lewis in 1894. We presume that limits of till portrayed on the Quaternary Map of the United Kingdom (IGS, 1977) were a distillation of data from individual geological map sheets (1 inch to the mile and 1:50,000), and use the former as a record of the ice limit across the region. This is modified slightly for the section around Bridgnorth to Wolverhampton, from Worsley (1991) who based his limit on drift extents compiled by Wills (1924) and the Geological Survey. Similar to elsewhere, a sequence of moraines, the Whitchurch moraine system, exists a short distance back from the maximal limit, which are interpreted to mark a standstill or minor advance during overall retreat. These have been mapped and described in Poole & Whiteman (1966), Peake, (1981) and Thomas (1989).

A series of moraines we call the Hereford moraine system, are marked on the map and as with elsewhere this is likely a record of a standstill of the margin during retreat. These landforms are part of the South Wales End Moraine reported by Charlesworth (1929), although we adopt the revised mapping of Luckman (1970).

The precise ice limit around Ludlow and Hereford is hard to define, with at least 8 authors drawing the limit in different places based upon where they have mapped the 'newer drift' to terminate (*cf.* Brandon, 1989, Fig. 19). We adopt the drift limit mapped by the British Geological Survey (Brandon, 1989) and note that it is virtually a return to the limit proposed by Charlesworth (1929).

Around the Abergaveny area the maximum extent, marked by the limit of till, appears to have formed an ice lobe spreading down the Usk Valley. We use the limit as recorded by IGS (1977).

The ice limit across South Wales has been the most debated part of the Devensian southern limit, with limits varying by over 50 km and with some inferring almost complete glacial cover of Wales (Charlesworth, 1929) and others with large swathes of west Wales unglaciated (Mitchell, 1960). Differences have mainly arisen from the contrasting approaches adopted: those based on geomorphology and identification of moraines, and those based on lithostratigraphy and distribution of till sheets (cf. Bowen, 1981). We mark ice limits across South Wales based on the limit of till and hummocky glaciofluvial deposits as defined by Bowen (1973) and marked on the Quaternary Map of the British Isles (IGS, 1977). Elsewhere we have reproduced mapped drift limits, but fail to do so here and simply adopt the ice limit as defined by Bowen. The problem of age of these till sheets has been resolved by elucidating their stratigraphic relationship with dated Last Interglacial raised beaches and periglacial 'head' deposits (Bowen, 1973, 1981; Bowen, *et. al.,* 1986). It is by this means that the tills on the Gower Peninsula for example have been split into Devensian and older. Ice limits are least well known over Pembrokeshire (southern Dyfed), and have in part been demarcated by ice-marginal glacifluvial deposits.

3.5.1.3 Scottish limits

There has been considerable debate about ice limits in parts of Scotland with two basic hypotheses: one of complete glacial cover, against a more restricted ice sheet with the far northeast of Scotland and Buchan remaining ice free and the Outer Hebrides nourishing their own ice cap (Sutherland, 1984). Figs 2 and 8 illustrate the restricted view of glaciation. For the lowland area of Buchan the uncertainty arose due to thin tills that are widely disturbed by periglacial action. This led to the view that Buchan remained as an ice free enclave during the Late Devensian thus allowing intense periglacial modification of a previously deposited till sheet. Critically, the finding of organic deposits at Crossbrae Farm (NJ 753512), dated as Late Devensian, that only had periglacial deposits lying on top of them (Hall, 1984) was taken to confirm the view of an ice free area. This was in spite of a system of meltwater channels in the area as reported in Clapperton & Sugden (1977) and who argued for complete glacial cover. Since Sutherland's (1984) review and hypothesis of the ice limit in this area (marked on Fig. 2), the restricted ice limit has been widely accepted (e.g. Rose, 1985; Bowen *et al.*, 1986; Sejrup *et al.*, 1987). On the basis of the meltwater imprint in the area (Clapperton & Sugden, 1977), and recent revision of the stratigraphic context and dates of the organic deposits at Crossbrae Farm (Whittington *et al.*, 1998) we overturn this view and no longer regard Buchan to have been ice free. We assume that the ice limit lay offshore.

For the far northeast tip of Scotland, around Thurso - Wick, similar arguments to those used above (i.e., strong periglacial imprint on upper till surfaces) have been used to hypothesize for an ice-free enclave during the Late Devensian (*cf.* Sutherland, 1984). Without further work to define the age of the tills in this area it is safest to reject the restricted view of glaciation, and we presume that the ice limit lay offshore.

The western limits of the mainland Devensian ice sheet are also in dispute. One view is that a large ice sheet, with its main ice divide running north-south across Rannoch Moor, should easily have been capable of invading both the inner and outer Hebrides, only some 150 km distance. This is especially so considering that Scottish ice is known to have invaded the north of Ireland, over 200 km away (Colhoun, 1971; Stephens *et al.*, 1975; Clark & Meehan, 2001). The alternative view is that the mainland ice sheet did not reach the Outer Hebrides and that these islands nourished their own independent ice caps (Flinn, 1987;

Sutherland, 1984). Either way, it is likely that the whole of Scotland was covered by ice. What is unresolved, by these methods, is whether ice in the Outer Hebrides emanated from the mainland ice sheet or just from local ice caps (see below).

3.5.1.4 Vertical ice sheet extent

Unlike the outer geographic limit of the ice sheet, the upper limit has until recently, received scant attention. Frost-shattered debris and tors on some mountain summits led some (e.g. Linton, 1955) to conclude that these peaks were not glaciated and formed nunataks above the ice sheet surface. Whilst, others (e.g. Sugden, 1968; Boulton, *et al.*, 1977) used the high altitude carriage of glacial erratics, and that the wide extent of the ice sheet required thick ice, to conclude that all mountain tops were inundated. Preservation of the mountain top tors was ascribed to the protective properties of cold-based ice (Sugden, 1968).

Recent work has shed considerable light on the vertical extent of the ice sheet. By visiting high mountains around the periphery of the ice sheet, it has been possible to distinguish and map upper zones of shattered bedrock, tors and blockfields, from lower zones of ice-moulded bedrock. The dividing line is taken as a weathering limit that records the upper extent of ice cover. Such trimlines have now been identified in the Highlands of NW Scotland, Inner and Outer Hebrides, and the SW part of the English Lake District, and North Wales (McCarroll *et al.*, 1995; Ballantyne, 1997; 1999a, b, c; Ballantyne *et al.*, 1997; 1998a; Dahl *et al.*, 1996: Ballantyne & McCarroll, 1997; Lamb & Ballantyne, 1998; McCarroll & Ballantyne, 2000). From these papers it is possible to define palaeo-nunataks of the ice sheet which form an important part of the limit of glaciation.

Along with the maximum geographical spread of the ice sheet reported in this paper, the vertical extent of the ice sheet may be used as important tests of numerical modelling experiments. Also, using assumptions regarding the likely profile of the ice sheet surface (dependent upon basal shear stress), it possible to use palaeo-nunatak data to empirically reconstruct contours of the ice sheet surface. This approach has been used to good effect to resolve a conflict outlined in the preceding section. Ballantyne *et. al* (1998b) used their trimline data of north-west Scotland to reconstruct ice surface elevation and by doing so have demonstrated that Devensian ice extent completely covered (nunataks excepted) the inner and outer Hebrides and probably for some distance across the continental shelf. They established that independent ice divides existed over the Isle of Skye and the Outer Hebrides.

Palaeo-nunataks are not marked on the ice limit map for reasons of scale. It is also premature to do so as analysis of assumed ice sheet profiles is required to convert these data into meaningful expressions of geographic ice limits and ice sheet volume (CDC).

3.5.2 Timing of glaciation during the Last Glacial Stage (MIS 4 - 2)

3.5.2.1 Introduction

Over the last half century there have been two views about the extent and timing of ice cover in England Scotland and Wales during the Last Glacial Stage (MIS 4 - 2). Traditional views are that the only evidence for glaciation was during what is now known as the Last Glacial Maximum (LGM) during which the ice reached its limit about 20k calendar years ago, in common with much of the rest of the world (Sibrava *et al.*, 1986). The description given above in Section 3.5.1 refers to such a glacial event. However, alternative views exist whereby there was extensive ice cover during an earlier part of the Last Glaciation. One of these views, based on evidence from Scotland, Cheshire and Lincolnshire places the earlier glaciation in the Early Devensian - the equivalent of MIS 4, the other finds evidence for glaciation prior to the LGM either in late MIS 3 or early in MIS 2. This latter view arose initially from offshore evidence in the North Sea region, but has recently been supported by cosmogenic nuclide exposure dating. The following section outlines these different hypotheses.

3.5.2.2 Glaciation during the LGM - the traditional view

The evidence for dating Devensian glaciation of England, Scotland and Wales to the LGM is lithostratigraphy of glaciogenic sediments constrained by radiocarbon and luminescence dates from a designated type area in east Yorkshire. This glaciation is known as the Dimlington Stadial of the Late Devensian Substage and the evidence is outlined in Rose (1985). This evidence is shown diagrammatically in Fig. 7 and this area is linked with the remainder of England, Scotland and Wales using the distribution of glacial deposits and morphostratigraphy, as outlined in Section 3.5.1. These ice limits are the basis of ice sheet modelling by Boulton *et al.* (1977, 1991) deriving both a minimum extent model, which is essentially that based on the evidence described in Section 3.5.1 (Fig. 8) and a maximum extent model (Fig. 9) reflecting some of the areas of debate outlined above. The pattern of ice wastage from these glacial limits is generalised by isochons tied to geochronometrically fixed points and ice flow behaviour derived from glacier bedform lineations and ice marginal landforms (Fig. 10) (Boulton *et al.*, 1991, Boulton, 1992).

3.5.2.3 Early Devensian glaciation (MIS 4)

Contrasted glacial terrain and associated glacial lake sediments in east Lincolnshire and east Yorkshire have caused Straw (1979b, 1991) to suggest that there is evidence for Early Devensian glaciation in this part of eastern England.

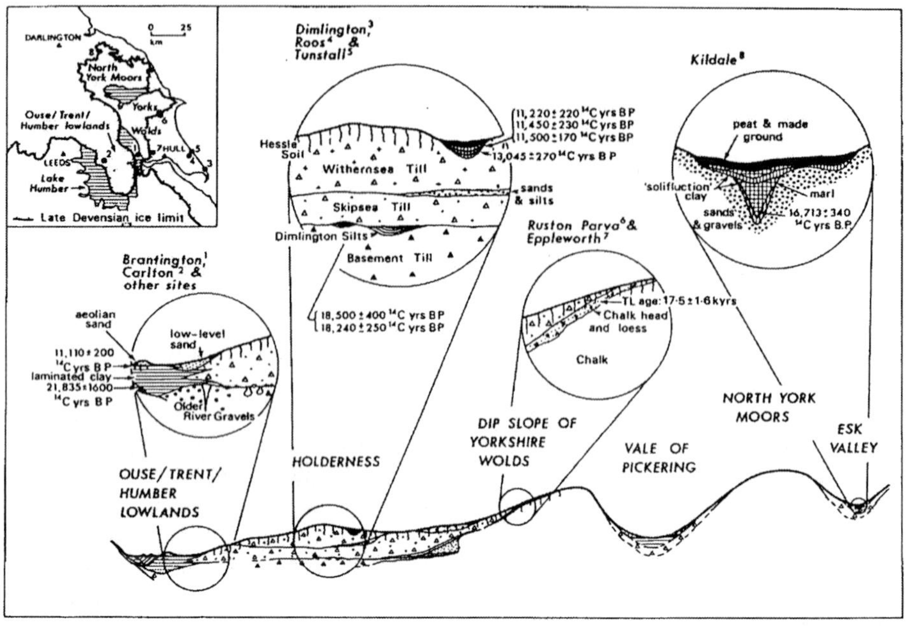

Fig. 7. Diagrammatic representation of the stratigraphic evidence for the Dimlington Stadial of the Late Devensian in eastern England. The site locations are shown on the inset map (from Rose, 1989).

Essentially, the evidence is based on the morphological expression of glacial deposits and glacial meltwater landforms associated with tills that are younger than the Last Interglacial. The reason for attributing the more subdued forms to the Early Devensian rather than the an oscillation of the LGM is the apparent association of the earlier deposits with lacustrine and fluviatile sediments that have been dated to earlier than Middle Devensian.

Quite different evidence, in the form of amino acid determinations on marine shells from glaciogenic sediments in northeast Scotland suggest that these deposits were formed by glaciation during the Early Devensian (Bowen, 1991). 'Amino acid ratios from the shelly deposits both in Caithness and Orkney are characterised by a younger molluscan faunal element ascribed to Oxygen Isotope Substage 5e, unlike other shelly deposits in Scotland which also contain faunal elements of Middle and Late Devensian Substage age (Bowen & Sykes, 1988). It would appear, therefore that the shelly deposits are of Early Devensian Substage age' (Bowen, 1991, p. 8).

3.5.2.4 Two stage glaciation during the later part of the Last Glacial Stage

The view that ice cover during the Last Glacial Stage was of longer duration than the classical LGM model developed from evidence in the North Sea region and has recently been supported by the interpretation of cosmogenic nuclide dates from western Britain and Ireland (Bowen *et al.*, 2002).

Cores taken from the bed of the northern North Sea have yielded *in situ* marine sediments and tills (Sejrup *et al.*, 1994). Radiocarbon dates and amino acid ratios have been obtained from these shells (Fig. 11) and are interpreted as showing that ice covered the northern North sea from Scandinavia to Scotland between about 29 and 24 radiocarbon ka BP. Radiocarbon dates from the Norwegian side indicate renewed glaciation as far as the Norwegian Trench between 18.6 and 15.1 radiocarbon ka BP. This latter readvance is correlated with the Dimlington Stadial of England and Scotland. Thus it is suggested that the LGM had two phases, the earlier of which occurred at the end of MIS 3 and the beginning of MIS 2 when the British and Scandinavian ice sheets were in contact (? the maximum extent model of Boulton *et al.*, 1977, 1991) (Fig. 9), and the latter which is classical LGM and was restricted to the British side of the North Sea basin (? the minimum extent model of Boulton *et al.*, 1977, 1991) (Fig. 8).

Cosmogenic nuclide (^{36}Cl) surface-exposure dates from surfaces in Ireland and D-Aile/L-Ile amino acid ratios from marine shells from marine and glacial deposits in Britain and Ireland are reported by Bowen *et al.* (2002) and provide a basis for a new glacial history in the Middle and Late Devensian of Britain and Ireland, and have a bearing on the validity of the Early Devensian glaciation described above (Section 3.5.2.3).

Evaluation of the age determinations identified a number of clusters which are taken as evidence for the timing of glacier cover and retreat. ^{36}Cl Group 1 indicate large parts of Ireland became ice-free 37 and 32 ka BP, and ^{36}Cl Group 2 indicate ice wastage about 22 ka BP. ^{36}Cl Groups 3 and 4 provide dates for ice wastage across the region and ^{36}Cl Group 5 is evidence for wastage following the Younger Dryas glacier expansion (Bowen *et al.*, 2002, Table 1, Fig. 2). Constraints for the timing of the onset of these glacial episodes is provide by amino acid radios from marine shells in glaciogenic deposits, radiocarbon determinations cited in Section 3.5.2.2 above (Rose, 1985) and correlation with the pattern of sediment flux (as a proxy of glaciation) from Core MD-2006 on the Barra Fan (Bowen *et al.*, 2002). Amino acid ratios from marine shells in glacial deposits are from aminozone 5e suggesting that glaciation is younger than the Last Interglacial, and

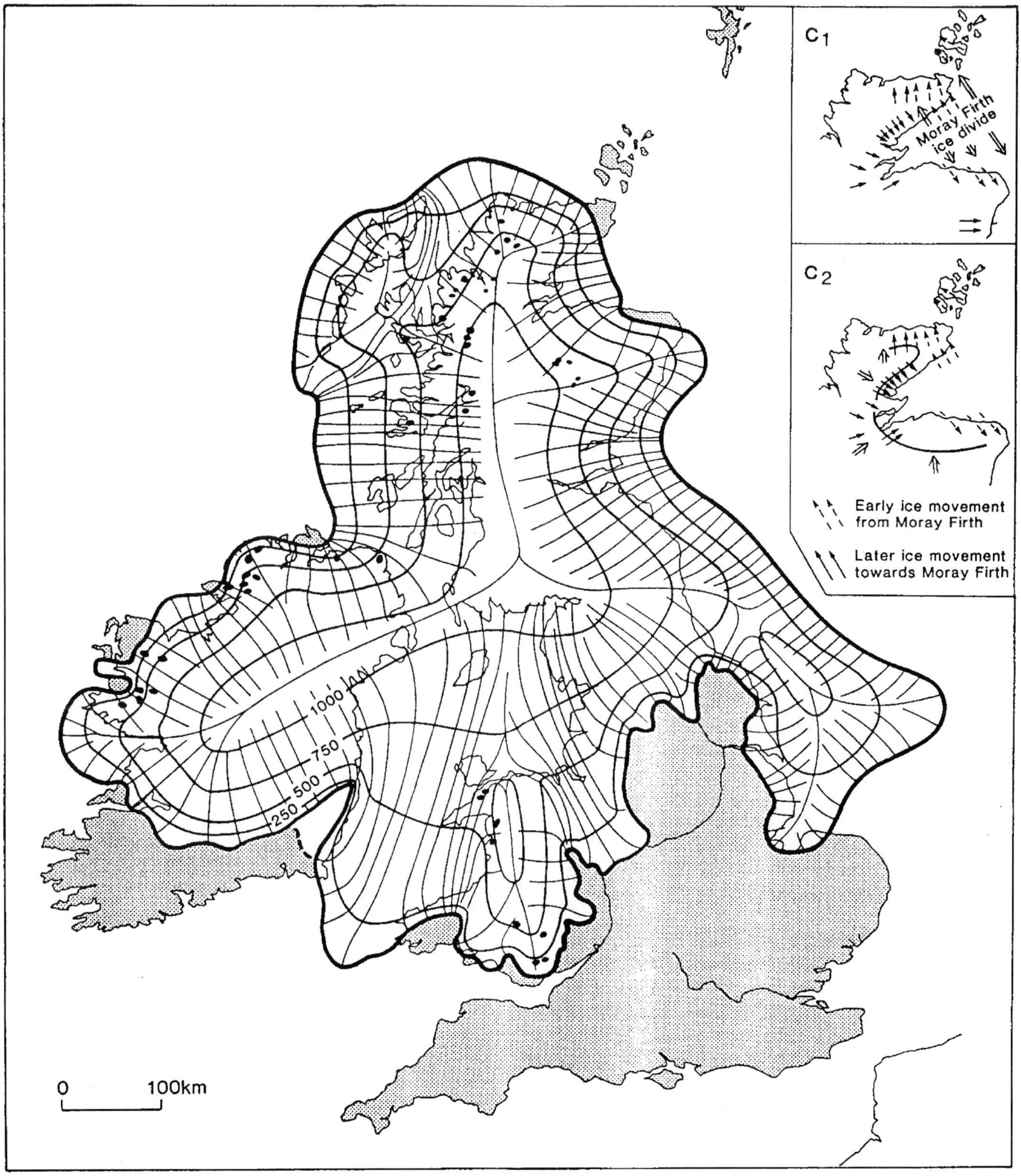

Fig. 8. Reconstruction of the limit and configuration of the ice sheet at the Last Glacial Maximum in Britain and Ireland. This is based on the limited extent model and assumes a basal shear street of 70 kPa on the land area and 30 kPa on the sea area. Principal nunatak areas are shown as dots (from Boulton et al., 1991, published with permission of the Geological Society).

radiocarbon ages fix the glaciation to younger than c. 18 radiocarbon ka BP (21 cal ka BP) at Dimlington and Tremeirchion in the Vale of Clwyd, north Wales.

On this basis Bowen *et al.* (2002) support the Early Devensian glaciation described in Section 3.5.2.3, and suggest a MIS 3 glaciation, possibly associated with Hein-

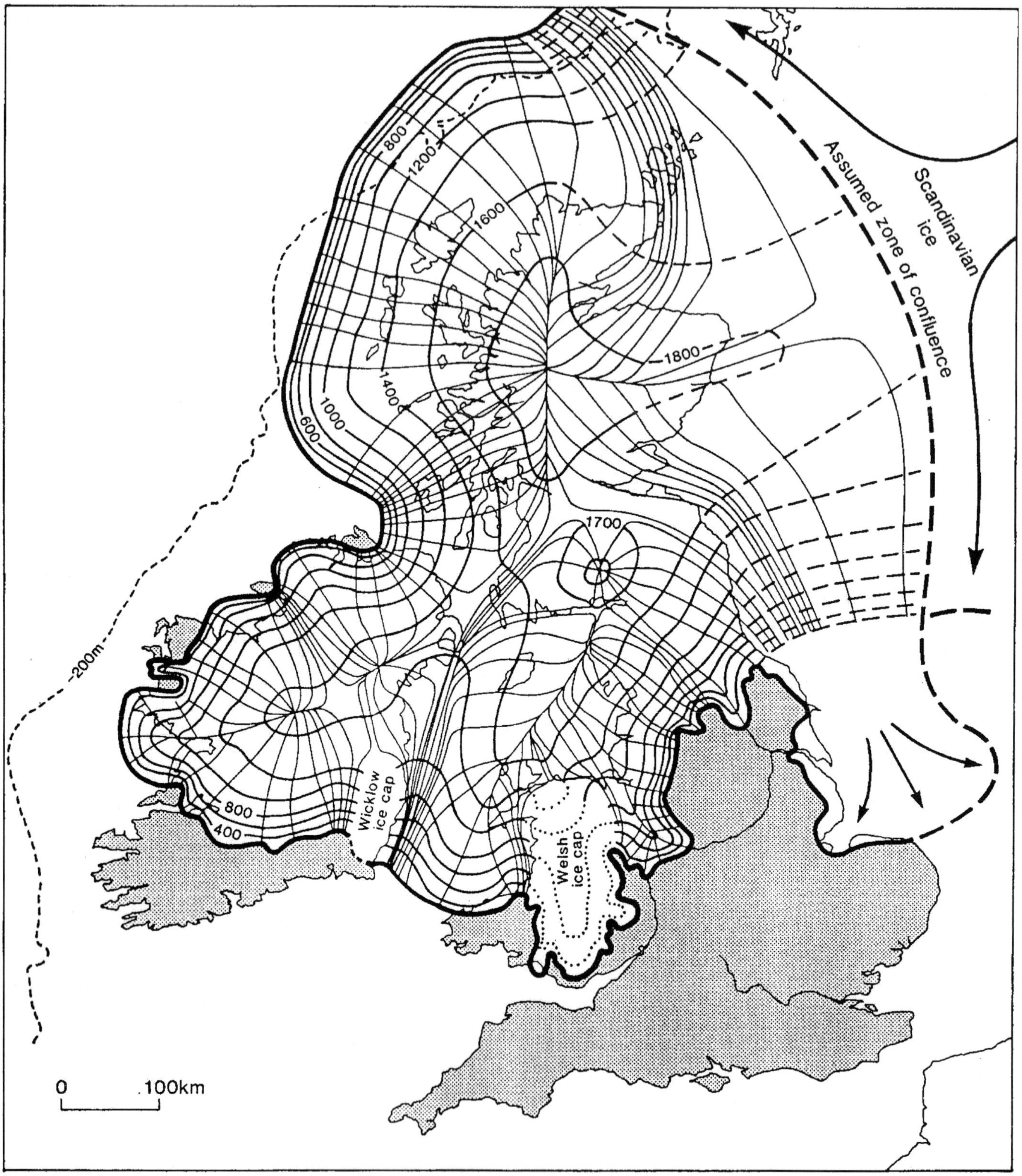

Fig. 9. Reconstruction of the limit and configuration of the ice sheet at the Last Glacial Maximum in Britain and Ireland. This is based on the maximal extent model with Scottish and Scandinavian ice sheets confluent in the North Sea using a 100 kPa basal sheet stress. No significant nunataks occur (from Boulton et al., 1991, published with permission of the Geological Society).

rich Event 4 at about 40 ka, during which the British and Scandinavian ice sheets were in contact. They suggest fluctuations of the ice margin between c. 40 and 25 ka and confirm the timing of the LGM at 22 cal ka BP when the British ice sheet was separate from that in Scandinavia (Fig.12).

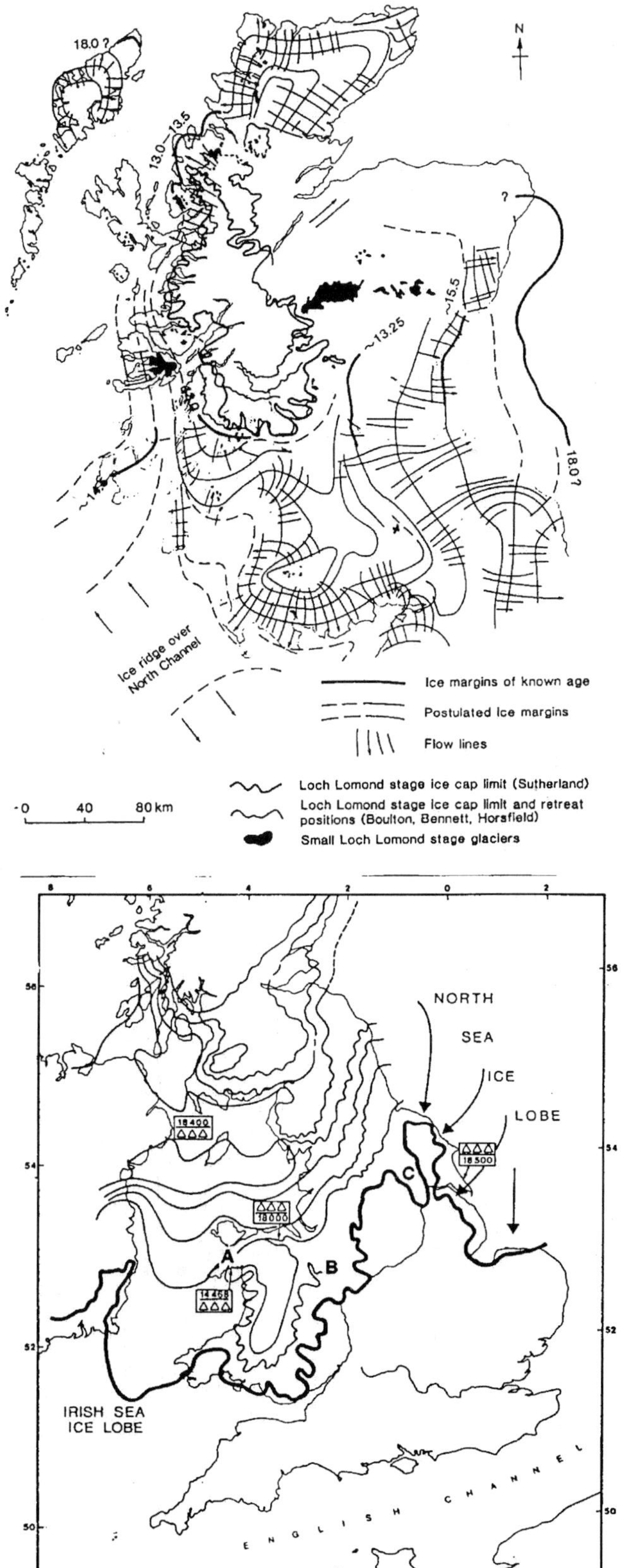

Fig. 10. Inferred pattern of retreat of the Late Devensian ice sheet inferred from geomorphological features. The lines represent isochrons though they cannot be dated, and important dated sites and their relationships to glacial deposits are shown (from Boulton et al., 1991 (upper Fig.) and Boulton, 1992 (lower Fig.), published with permission of the Geological Society).

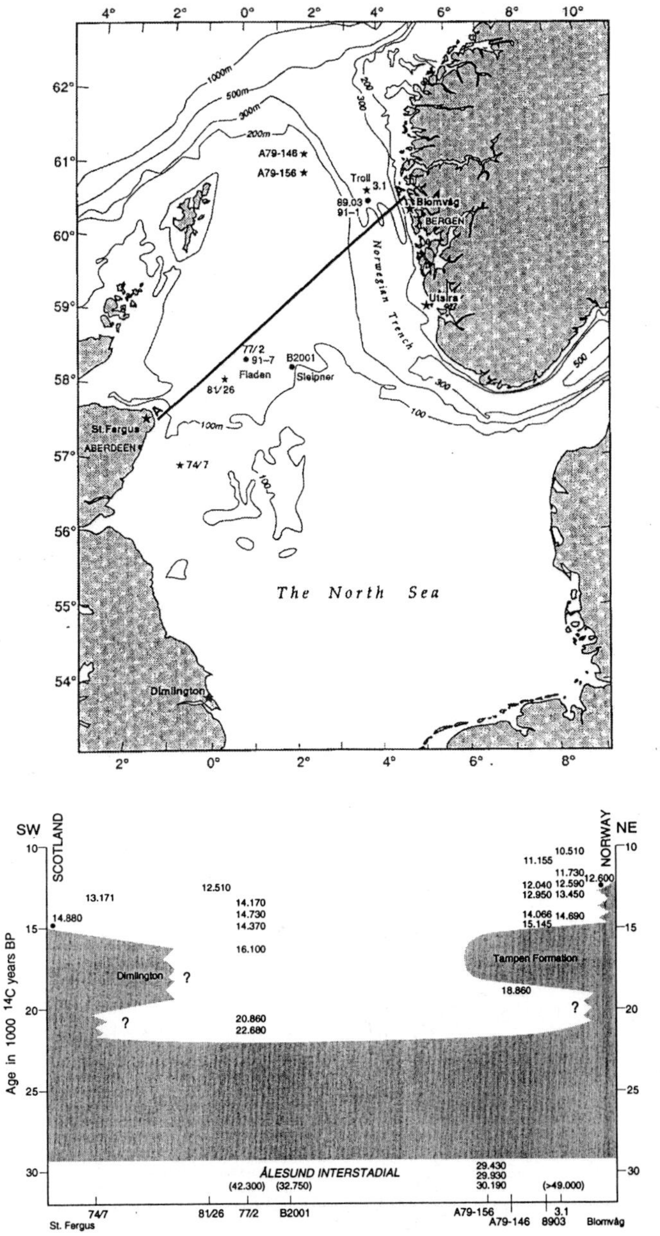

Fig. 11. Ice cover during the later part of the Last Glacial Stage in the North Sea between Britain and Norway. The upper Fig. shows the location of the transect and the lower Fig. shows the core and sites used to obtain dates to constrain the pattern and timing of glaciation. The shaded area represents ice cover and the dates in parentheses were performed on foraminiferal samples from till units (from Sejrup et al., 1994, reproduced by permission of the authors).

Although this paper provides much welcome new information, and clearly has a bearing on problems associated with the understanding of the Devensian glacial history, described above, certain problems remain unresolved. Firstly, the evidence from the Barra Fan provided in Bowen *et al.* (2002, Fig. 5) and in Knutz *et al.* (2001, Figs 3 and 4) indicates only fluctuating sediment flux without any well

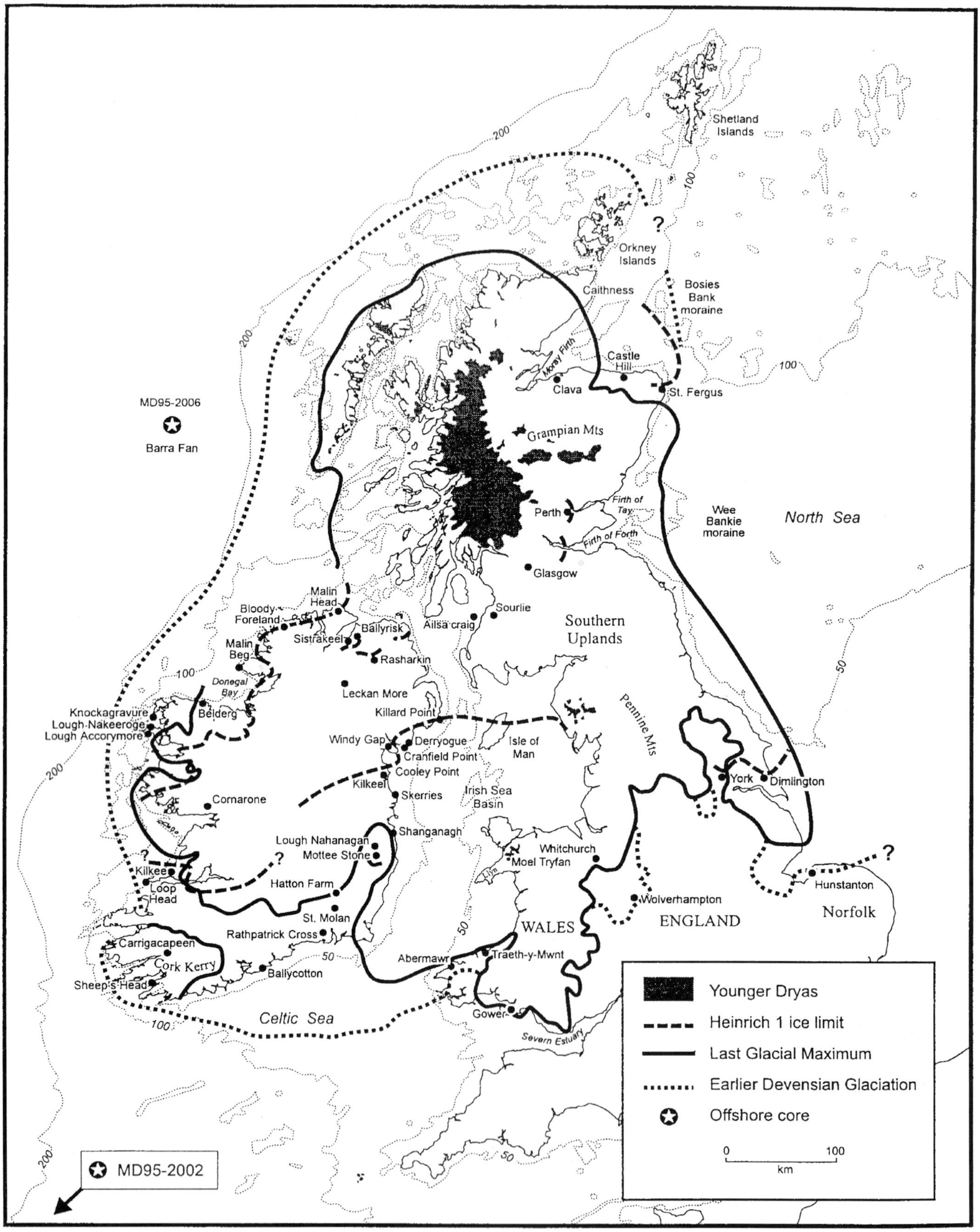

Fig. 12. Ice margins and critical localities for the later part of the Last Glacial Stage in Great Britain and Ireland (from Bowen et al., 2002, reproduced by permission of David Bowen and Elsevier Science Ltd).

defined glacial event other than that associated with Heinrich 1 (c. 17 ka). Secondly, the interpretation of the more extensive glaciation, prior to the LGM, depends upon the interpretation of the Basement Till at Dimlington to LGM age based on amino acid ratios from derived *Macoma* shells (Eyles *et al.*, 1994, Bowen *et al.*, 2002, p 94). This deposit has been observed beneath Last Interglacial sediments at Sewerby, further north in east Yorkshire (Catt & Penny, 1966; Briant, 2002, p 38) which invalidates the amino acid determinations. Finally, it is difficult to accept that ice from Scotland covered the North Sea over the period between 30 and 22 ka when radiocarbon dates on bone, peat, and plant detritus, and U-series ages from speleothem give ages within this range from sites close to the centres of ice accumulation on the mainland of Scotland (Boulton *et al.*, 1991, Fig. 15.8, Table 15.3). These dates and supporting palaeoecology indicate un-glaciated terrain at this time close to these centres of glaciation **(JR)**.

Acknowledgements

JR wishes to thank Brian Moorlock, Richard Hamblin and Jon Lee for permission to cite unpublished information. This will be published jointly in the near future.

References

Alabaster, C. & Straw, A. (1976). The Pleistocene context of faunal remains and artefacts discovered at Welton-le-Wold, Lincolnshire. *Proceedings of the Yorkshire Geological Society*, **41**, 75-94.

Allen, P., Cheshire, D.A. & Whiteman, C.A. (1991). The tills of southern East Anglia. *In:* Ehlers, J., Gibbard, P.L. & Rose, J. (eds), *Glacial Deposits in Great Britain and Ireland*, Rotterdam, Balkema, 255-278.

Arkell, W.J. (1947). *The Geology of Oxford*. Clarendon Press, Oxford, 267 pp.

Ashton, N., Cook, J., Lewis, S.G. & Rose, J. (eds). (1992). *High Lodge: Excavations by G. de G. Sieveking & J. Cook, 1988*. British Museum, London, 192 pp.

Auton, C.A., Morigi, A.N. & Price, D. (1985). The sand and gravel resources of the country around Harleston and Bungay, Norfolk and Suffolk. Description of 1:25,000 resource sheets comprising parts of TM 27, 28, 38 and 39. *Mineral Assessment Report*, **145**, British Geological Survey, Keyworth, 28 pp. + appendices

Baden-Powell, D.F.W. (1948). The chalky bounder clays of Norfolk and Suffolk, *Geological Magazine*, **85**, 279-296.

Ballantyne, C.K. (1997). Periglacial trimlines in the Scottish Highlands. *Quaternary International*, **38**, 119-136.

Ballantyne, C.K. (1999a). A Late Devensian nunatak on the Knoydart Peninsula, NW Scotland: implications for ice sheet reconstruction. *Scottish Geographical Journal*, **115**, 319-328.

Ballantyne, C.K. (1999b). Maximum altitude of Late Devensian glaciation on the Isle of Mull and Isle of Jura. *Scottish Journal of Geology*, **35**, 97-106.

Ballantyne, C.K. (1999c). An Teallach: a Late Devensian nunatak in Wester Ross. *Scottish Geographical Journal*, **115**, 249-259.

Ballantyne, C.K. & McCarroll, D. (1997). Maximum altitude of the Late Devensian ice sheet on the Isle of Rum. *Scottish Journal of Geology*, **33**, 183-186.

Ballantyne, C.K. & Harris, C. (1994). *The Periglaciation of Great Britain*. Cambridge, Cambridge University Press, 330 pp.

Ballantyne, C.K., McCarroll, D., Nesje, A. & Dahl, S.O. (1997). Periglacial trimlines, former nunataks and the altitude of the last ice sheet in Wester Ross, northwest Scotland. *Journal of Quaternary Science*, **12**, 225-238.

Ballantyne, C.K., McCarroll, D., Nesje, A., Dahl, S.O. & Stone, J.O. (1998a). The last ice sheet in north-west Scotland: reconstruction and implications. *Quaternary Science Reviews*, **17**, 1149-1184.

Ballantyne, C.K., McCarroll, D., Nesje, A., Dahl, S.O., Stone, J.O. & Fifield, L.K. (1998b). High-resolution reconstruction of the last ice sheet in NW Scotland. *Terra Nova*, **10**, 63-67.

Balson, P.S. & Cameron, T.D.J. (1985). Quaternary mapping offshore east Anglia. *Modern Geology*, **9**, 221-239.

Balson, P. S. & Jeffery, D.H. (1991). The glacial sequence of the southern North Sea. *In:* Ehlers, J., Gibbard, P.L. & Rose, J. (eds), *Glacial Deposits in Great Britain and Ireland*, 245-253, Rotterdam, Balkema.

Banham, P.H. (1968). A preliminary note of the Pleistocene stratigraphy of northeast Norfolk. *Proceedings of the Geologists' Association*, **79**, 507-512.

Banham, P.H., Gibbard, P.L., Lunkka, J.P., Parfitt, S.A., Preece, R.C. & Turner, C. (2001). A critical assessment of 'A New Glacial Stratigraphy for Eastern England'. *Quaternary Newsletter*, **93**, 5-14.

Barclay, W.J., Brandon, A., Ellison, R.A. & Moorlock, B.S.P. (1992). A Middle Pleistocene palaeovalley-fill west of the Malvern Hills. *Journal of the Geological Society*, **149**, 75-92.

Bateman, M.D. & Catt, J.A. (1996). An absolute chronology for the raised beach and associated deposits at Sewerby, East Yorkshire, England. *Journal of Quaternary Science*, **11**, 389-395.

Birks, H.J.B. & Birks, H.H. (1980). *Quaternary Palaeoecology*, Arnold, London, 289 pp.

Boswell, P.G.H: (1914). On the occurrence of the North Sea Drift (Lower Glacial) and certain other brick-earths in Suffolk. *Proceedings of the Geologists' Association*, **25**, 121-153.

Boswell, P.G.H: (1916). The petrology of the North Sea Drift and Upper Glacial Brick-earths in East Anglia. *Proceedings of the Geologists' Association*, **27**, 79-98.

Boulton, G.S. (1992). Quaternary. *In:* Duff, P.McL.D. & Smith, A.J. (eds), *Geology of England and Wales*, 413-444, London, The Geological Society.

Boulton, G.S., Jones, A.S., Clayton, K.M. & Kenning, M.J. (1977). A British ice sheet model and patterns of glacial erosion and deposition in Britain. *In:* Shotton, F.W. (ed.), *British Quaternary Studies: recent advances*, 231-246, Oxford, Clarendon Press.

Boulton, G.S., Smith, G.D., Jones, A.S. & Newsome, J. (1985). Glacial geology and glaciology of the last mid-latitude ice sheets. *Journal of the Geological Society*, **142**, 447-474.

Boulton, G.S., Peacock, J.D. & Sutherland, D.G. (1991). Quaternary. *In:* Craig, G.Y., (ed.), *Geology of Scotland*, 3rd Edition, 503-543, London, The Geological Society.

Bowen, D.Q. (1978). *Quaternary Geology*, Oxford, Pergamon, 221 pp.

Bowen, D. Q. (1973). The Pleistocene history of Wales and the borderland. *Geological Journal*, **8**, 207-224.

Bowen, D. Q. (1981). The 'South Wales End-Moraine': fifty years after. *In:* Neale, J. & Flenley, J. (eds), *The Quaternary in Britain*, 60-67, Oxford, Pergamon Press.

Bowen, D.Q. (1991). Time and space in the glacial sediment systems of the British Isles. *In:* Ehlers, J., Gibbard, P.L. & Rose, J. (eds), *Glacial Deposits in Great Britain and Ireland.* Rotterdam, Balkema, 3-11.

Bowen, D.Q. (ed.). (1999a). A revised correlation of Quaternary deposits in the British Isles. *The Geological Society Special Report*, **23**, 174 pp.

Bowen, D.Q. (1999b). Wales. *In:* Bowen, D.Q. (ed.), A revised correlation of the Quaternary deposits in the British Isles, *Geological Society Special Report*, **23**, 79-90.

Bowen, D.Q. & Sykes, G.A. (1988). Correlation of marine events and glaciations on the northeast Atlantic margin. *Philosophical Transactions of the Royal Society of London*, **B318**, 619-635.

Bowen, D.Q., Rose, J., McCabe, A.M. & Sutherland, D.G. (1986). Correlation of Quaternary glaciations in England, Ireland, Scotland and Wales. *Quaternary Science Reviews*, **5**, 299-340.

Bowen, D.Q., Phillips, F.M., McCabe, A.M., Knutz, P.C. & Sykes, G.A. (2002). New data for the Last Glacial Maximum in Great Britain and Ireland. *Quaternary Science Reviews*, **21**, 89-101.

Brand, D., Booth, S. & Rose, J. (in press). Late Devensian Glaciation, Ice-Dammed Lake and River Diversion, Stiffkey, North Norfolk, England. *Proceedings of the Yorkshire Geological Society*,

Brandon, A. (1989). *Geology of the country between Hereford and Leominster.* Memoir of the British Geological Survey, sheet 198. Keyworth, HMSO.

Briant, B. (2002). The Quaternary of east Yorkshire and north Lincolnshire. Report on QRA Short Field Meeting, 13-16 September 2001. *Quaternary Newsletter*, **96**, 38-42.

Briant, R.M., Rose, J., Branch, N.P. & Lee, J.A. (1999). 'Pre-glacial' Quaternary sediments from Trimingham, north Norfolk, England. *Bulletin of the Geological Society of Norfolk*, **49**, 15-47.

Bridge, D.M. (1988). Corton Cliffs. *In:* Gibbard, P.L. & Zalasiewicz (eds), *Pliocene-Middle Pleistocene of East Anglia*, Field Guide, Quaternary Research Association, Cambridge, 119-125.

Bridge, D.M. & Hopson, P.M., 1985. Fine gravel, heavy mineral and grain size analysis of mid-Pleistocene deposits in the lower Waveney Valley, East Anglia. *Modern Geology*, **9**, 129-144.

Bridgland, D.R. (1994). *Quaternary of the Thames.* London, Chapman and Hall, 441 pp.

Briggs, D.J., Coope, G.R. & Gilbertson, D.D. (1975). The chronology and environmental framework of early Man in the Upper Thames valet: a new model. *British Archaeological report, British Series*, **137**, 176 pp

Bristow, C.R. & Cox, F.C. (1973). The Gipping Till: a reappraisal of East Anglian glacial stratigraphy. *Journal of the Geological Society*, **129**, 1-37.

Cameron, T.D.J. & Holmes, R. (1999). The continental shelf. *In:* Bowen, D.Q. (ed.). (1999). A revised correlation of Quaternary deposits in the British Isles. *The Geological Society Special Report*, **23**, 125-139.

Cameron, T.D.J., Stoker, M.S. & Long, D. (1987). The history of Quaternary sedimentation in the UK sector of the North Sea basin. *Journal of the Geological Society*, **144**, 43-58.

Campbell, S., Hunt, C.O., Scourse, J.D., Keen, D.H. & Croot, D.G. (1999) Southwest England. *In:* Bowen, D.Q. (ed.) A revised correlation of the Quaternary deposits in the British Isles, *Geological Society Special Report*, **23**, 66-78.

Catt, J.A. (1982). The Quaternary deposits of the Yorkshire Wolds. *Proceedings of the North of England Soils Discussion Group*, **18**, 61-17.

Catt, J.A. (1987). The Quaternary of East Yorkshire and adjacent areas. *In:* Ellis, S. (ed.), *East Yorkshire*, Field Guide, Quaternary Research Association, Cambridge, 1-14.

Catt, J.A. (1991a). Late Devensian glacial deposits and glaciations in eastern England and the adjoining offshore region. *In:* Ehlers, J., Gibbard, P.L. & Rose, J. (eds), *Glacial Deposits in Great Britain and Ireland.* Rotterdam, Balkema, 61-68.

Catt, J.A. (1991b). The Quaternary history and glacial deposits of East Yorkshire. *In:* Ehlers, J., Gibbard, P.L. & Rose, J. (eds), *Glacial Deposits in Great Britain and Ireland.* Rotterdam, Balkema, 185-191.

Catt, J.A. & Penny, L.F. (1966). The Pleistocene deposits of Holderness, East Yorkshire. *Proceedings of the Yorkshire Geological Society*, **35**, 375-420.

Charlesworth, J.K. (1926). The glacial geology of the southern uplands of Scotland, west of Annandale and Upper Clydesdale. *Transactions of the Royal Society of Edinburgh*, **55**, 1-23.

Charlesworth, J.K. (1929). South Wales end moraine. *Quarterly Journal of the Geological Society of London*, **85**, 335-358.

Charlesworth, J.K. (1956). The Late-glacial history of the Highlands and Islands of Scotland. *Transactions of the Royal Society of Edinburgh*, **62**, 769-928.

Cheshire, D.A. (1986). The Lithology and Stratigraphy of the Anglian Deposits in the Lea Basin. Unpublished PhD Thesis, CNAA, 515 pp.

Clapperton, C.M. & Sugden, D.E. (1977). The Late Devensian glaciation of north-east Scotland. *In:* Gray, J.M. & Lowe, J.J. (eds) *Studies in the Scottish Lateglacial Environment*, 1-13, Oxford, Pergamon.

Clark, C.D. & Meehan, R.T. (2001). Subglacial bedform geomorphology of the Irish Ice Sheet reveals major configuration changes during its growth and decay. *Journal of Quaternary Science*, **16**, 483 - 496.

Clayton, K.M. (2000). Glacial erosion of the Wash and Fen basin and the deposition of the chalky till of eastern England. *Quaternary Science Reviews*, **19**, 811-822.

Colhoun, E.A. (1971). The glacial stratigraphy of the Sperrin Mountains and its relation to the glacial stratigraphy of north-west Ireland. *Proceedings of the Royal Irish Academy*, **71B**, 37-52.

Cornwell, J.D. & Carruthers, R.M. (1986). Geophysical studies of a buried valley system near Ixworth, Suffolk. *Proceedings of the Geologists' Association*, **97**, 357-364.

Cox, F.C. (1985). The tunnel valleys of Norfolk, East Anglia. *Proceedings of the Geologists' Association*, **96**, 357-369.

Cox, F.C. & Nickless, E.F.P. (1972). Some aspects of the glacial history of central Norfolk. *Bulletin of the Geological Survey of Great Britain*, **42**, 79-98.

Dahl, S.O., Ballantyne, C.K., McCarroll, D. & Nesje, A. (1996). Maximum altitude of Devensian glaciation on the Isle of Skye. *Scottish Journal of Geology*, **32**, 107-115.

Deeley, R.M. (1886). The Pleistocene Succession in the Trent basin. *Quarterly Journal of the Geological Society of London*, **42**, 437-480.

Delcourt, H.R. & Delcourt, P.A. (1991). *Quaternary Ecology*, London, Chapman and Hall, 242 pp.

Dines, H.G. (1928). On the glaciation of the north Cotteswold area. *Summary of Progress of the Geological Survey of Great Britain*, **2**, 66-71.

Edwards, C.A. (1978). The Quaternary history and stratigraphy of north-east Yorkshire. Unpublished PhD thesis, University of Hull.

Ehlers, J. (1983). The glacial history of north-west Germany. *In:* Ehlers, J. (ed.), *Glacial Deposits in north-west Europe*, 229-238, Rotterdam, Balkema.

Ehlers, J. & Gibbard, P.L. (1991). Anglian glacial deposits in Britain and the adjoining offshore regions. *In:* Ehlers, J., Gibbard, P.L. & Rose, J. (eds), *Glacial Deposits in Great Britain and Ireland*, 17-24, Rotterdam, Balkema.

Ehlers. J., Meyer, K.-D. & Stephan, H.-J. (1984). The pre-Weichselian glaciations of North-West Europe, *Quaternary Science Reviews*, **3**, 1-40.

Ehlers, J., Gibbard, P.L. & Whiteman, C.A. (1987). Recent investigations of the Marly Drift of northwest Norfolk, England. *In:* Van der Meer, J.J.M. (ed.), *Tills and Glaciotectonics*, 39-54, Rotterdam, Balkema.

Ehlers, J., Gibbard, P.L. & Whiteman, C.A. (1991a). The glacial deposits of northwestern Norfolk. *In:* Ehlers, J., Gibbard, P.L. & Rose, J. (eds), *Glacial Deposits in Great Britain and Ireland*, 223-232, Rotterdam, Balkema.

Ehlers, J., Gibbard, P.L. & Rose, J. (1991b). Glacial deposits of Britain and Europe: General overview. *In:* Ehlers, J., Gibbard, P.L. & Rose, J. (eds), *Glacial Deposits in Great Britain and Ireland*, 493-501, Rotterdam, Balkema.

Eissmann, L. (2002). Tertiary and Quaternary geology of the Saale-Elbe region of eastern Germany. *Quaternary Science Reviews*, **21**, 1243-1359

Evans, D.J.A., Thomson, S.A. & Clark, C.D. (2001). Introduction to the Late Quaternary of East Yorkshire and North Lincolnshire. *In:* Bateman, M.D., Buckland, P.C., Frederick, C.D. & Whitehouse, N.J. (eds), *The Quaternary of East Yorkshire and North Lincolnshire*, Field Guide, Quaternary Research Association, London, 1-12.

Eyles, N., McCabe, A.M. & Bowen, D.Q. (1994). The stratigraphic and sedimentological significance of Late Devensian ice sheet surging, Holderness, Yorkshire. *Quaternary Science Reviews*, **13**, 727-759.

Fish, P.R. & Whiteman, C.A. (2001). Chalk micropalaeontology and the provenancing of Middle Pleistocene Lowestoft Formation till in eastern England. *Earth Surface Processes and Landforms*, **26**, 953-970.

Flinn, D. (1987). The glaciation of the Outer Hebrides. *Geological Journal*, **13**, 195-199.

Foster, S. W. (1985). The late Glacial and early Post Glacial history of the Vale of Pickering and northern Yorkshire Wolds. Unpublished PhD thesis, University of Hull.

Francis, E.A. (1970). Quaternary. *In:* Geology of County Durham. *Transactions of the Natural History Society of Northumberland*, **41**, 134-152.

Funnell, B.M. (1995). Global sea-level and the (pen-) insularity of late Cenozoic Britain. *In:* Preece, R.C. (ed.), Island Britain: a Quaternary Perspective. *Geological Society Special Publication*, **96**, 3-13.

Funnell, B.M. & West, R.G. (1962). The Early Pleistocene of Easton Bavents, Suffolk. *Quarterly Journal of the Geological Society of London*, **118**, 125-141.

Funnell, B.M. & West, R.G. (1977). Preglacial Pleistocene deposits of East Anglia. *In:* Shotton, F.W. (ed.), *British Quaternary Studies: recent advances*, 247-265. Oxford, Clarendon Press.

Funnell, B.M., Norton, P.E.P. & West R.G. (1979). The Crag at Bramerton, near Norwich, Norfolk. *Philosophical Transactions of the Royal Society of London*, **B287**, 489-534.

Gaunt, G. D. (1976). The Devensian maximum ice limit in the Vale of York. *Proceedings of the Yorkshire Geological Society*, **40**, 631-637.

Geikie, J. (1894). *The Great Ice Age*, Third Edition, London, Stanford, 850 pp.

Gibbard, P.L. (1977). Pleistocene history of the Vale of St. Albans. *Philosophical Transactions of the Royal Society of London*, **B280**, 445-483.

Gibbard, P.L. (1979). Middle Pleistocene drainage in the Thames valley. *Geological Magazine*, **116**, 35-44.

Gibbard, P.L. (1991). The Wolstonian Stage in East Anglia. *In:* Lewis, S.G., Whiteman, C.A. & Bridgland, D.R.

(eds), *Central East Anglia and the Fen Basin*, Field Guide, Quaternary Research Association, London, 7-14.

Gibbard, P.L. (1994). *Pleistocene History of the lower Thames Valley*. Cambridge, Cambridge University Press, 229 pp.

Gibbard, P.L. & Turner, C. (1988). In defence of the Wolstonian Stage. *Quaternary Newsletter*, **54**, 9-14.

Gibbard, P.L., West, R.G., Andrew, R. & Pettit, M. (1991a). Tottenhill, Norfolk (TF 636115). *In:* Lewis, S.G., Whiteman, C.A. & Bridgland, D.R. (eds), *Central East Anglia and the Fen Basin*, Field Guide, Quaternary Research Association, London, 131-143.

Gibbard, P.L., West, R.G., Zagwijn, W.H., Balson, P.S., Burger, A.W., Funnell, B.M., Jeffery, D.H., De Jong, J., Van Kolfschoten, T., Lister, A.M., Meijer, T., Norton, P.E.P., Preece, R.C., Rose, J., Stuart, A.J., Whiteman, C.A. & Zalasiewicz, J.A. (1991b). Early and early Middle Pleistocene correlations in the southern North Sea Basin. *Quaternary Science Reviews*, **10**, 23-52.

Gibbard, P.L., West, R.G., Andrew, R. & Pettit, M. (1992). The margin of a Middle Pleistocene ice advance at Tottenhill, Norfolk, England. *Geological Magazine*, **129**, 59-76.

Godwin, H. (1964). Late Weichselian conditions in south-eastern Britain: organic deposits at Colney Heath, Hertfordshire. *Proceedings of the Royal Society of London*, **B160**, 258-275.

Goldthwait, R.P. (1971). *Till: A symposium*. Columbus, Ohio State University Press, 402 pp.

Goldthwait, R.P. & Matsch, C.L. (1989). *Genetic Classification of Glacigenic Deposits*, Rotterdam, Balkema, 294 pp.

Green, C.P., Hey, R.W. & McGregor, D.F.M. (1980). Volcanic pebbles in Pleistocene gravels of the Thames in Buckinghamshire and Hertfordshire. *Geological Magazine*, **117**, 59-64.

Green, C.P., McGregor, D.F.M. & Evans, A.H. (1982). Development of the Thames drainage system in Early and Middle Pleistocene times. *Geological Magazine*, **119**, 281-290.

Grün, R. & Schwarcz, H.P. (2000). Revised open system U-series / ESR age calculations for teeth from Stratum C at the Hoxnian Interglacial type locality, England. *Quaternary Science Reviews*, **19**, 1151-1154.

Hall, A.M. (1984). Introduction. *In:* Hall, A.M. (ed.) *Buchan Field Guide*, Quaternary Research Association, Cambridge, 1-26.

Hamblin, R.J.O., Moorlock, B.S.P. & Rose, J. (2000). A new glacial stratigraphy for Eastern England. *Quaternary Newsletter*, **92**, 35-43.

Hamblin, R.J.O., Moorlock, B.S.P. & Rose, J. (2001). A critical assessment of 'A new glacial stratigraphy for Eastern England' - a reply. *Quaternary Newsletter*, **94**, 16-18.

Harmer, F.W. (1928). The distribution of erratics and drift. *Proceedings of the Yorkshire Geological Society*, **21**, 102-150.

Hart, J.K. (1987). The genesis of north eat Norfolk Drift. Unpublished PhD thesis, University of East Anglia.

Hart, J.K. & Boulton, G.S. (1991). The glacial drifts of Norfolk. *In:* Ehlers, J., Gibbard, P.L. & Rose, J. (eds), *Glacial Deposits in Great Britain and Ireland*, 233-243, Rotterdam, Balkema.

Hart, J.K. & Peglar, S.M. (1990). Further evidence for the timing of the Middle Pleistocene Glaciaton in Britain. *Proceedings of the Geologists' Association*, **101**, 187-196.

Hey, R.W. (1965). Highly quartzose pebble gravels in the London Basin. *Proceedings of the Geologists' Association*, **76**, 403-20.

Hey, R.W. (1980). Equivalents of the Westland Green Gravels in Essex and East Anglia. *Proceedings of the Geologists' Association*, **91**, 279-290.

Hey, R.W. (1986). A re-examination of the Northern Drift of Oxfordshire. *Proceedings of the Geologists' Association*, **97**, 291-301.

Hey, R.W. & Brenchley, P.J. (1977). Volcanic pebbles from the Pleistocene gravels in Norfolk and Essex. *Geological Magazine*, **114**, 219-225.

Hopson, P.M. & Bridge, D.M.C., 1987. Middle Pleistocene stratigraphy in the lower Waveney valley, East Anglia. *Proceedings of the Geologists' Association*, **98**, 171-185.

Horton, A. (1974). The sequence of Pleistocene deposits proved during the construction of Birmingham motorways. *Institute of Geological Sciences Report*, **74/22**, 21 pp.

IGS. (1977). *Quaternary Map of the United Kingdom*. Institute of Geological Sciences, Map at 1:625,000 scale.

Jowett, A. & Charlesworth, J.K. (1929). The glacial geology of the Derbyshire Dome and the S.W. Pennines. *Quarterly Journal of the Geological Society of London*, **85**, 307-334.

Keen, D.H. (1999). The chronology of Middle Pleistocene ('Wolstonian') event in the English Midlands. *In:* Andrews, P. & Banham, P. (eds), *Late Cenozoic Environments and Hominid Evolution: a tribute to Bill Bishop*, 159-168, London, Geological Society.

Kelly, M.R. (1964). The Middle Pleistocene of north Birmingham. *Philosophical Transactions of the Royal Society of London*, **B247**, 533-592

Knutz, P.C., Austin, W.E.N. & Jones, E.J.W. (2001). Millenial-scale depositional cycles related to the British Ice Sheet variability and North Atlantic palaeocirculation since 45 kyr B.P., Barra Fan, U.K. margin. *Palaeoceanography*, **16**, 53-64.

Laban, C. (1995). The Pleistocene glaciations in the Dutch sector of the North Sea. PhD Thesis, University of Amsterdam, 194 pp.

Lamb, A.L. & Ballantyne, C.K. (1998). Palaeonunataks and the altitude of the last ice sheet in the SW Lake District, England. *Proceedings of the Geologists' Association*, **109**, 305-316.

Lawton, T.E. (1982). *Geological notes and local details for 1:10,000 sheets TM 28 NW, NE, SW, SE (Harleston, Norfolk)*. Keyworth, Institute of Geological Sciences.

Lee, J.R. (2001). Genesis and palaeogeographical significance of the Corton Diamicton (basal member of the

North Sea Drift Formation), East Anglia, UK. *Proceedings of the Geologists' Association*, **112**, 29-43.

Lee, J.R., Rose, J., Riding, J.B., Moorlock, B.S.P. & Hamblin, R.J.O. (in press). Testing the case for a Middle Pleistocene Scandinavian glaciation in Eastern England: evidence for a Scottish ice source for tills within the Corton Formation of East Anglia. *Boreas*

Lewis, H.C. (1894). Papers and notes on the glacial geology of Great Britain and Ireland. London, Longman, Green and Co., 469 pp.

Lewis, S.G. (1993). The status of the Wolstonian Glaciation in the English Midlands and East Anglia. Unpublished PhD thesis, University of London.

Lewis, S.G. (1998). Quaternary geology of East Farm brick pit, Barnham and the surrounding area. *In:* Ashton, N., Lewis, S.G. & Parfitt, S. (eds), Excavations at the Lower Palaeolithic Site at East Farm, Barnham, Suffolk, 1989-94. *British Museum Occasional Paper*, **125**, 23-78.

Lewis, S.G. & Rose, J. (1991). Tottenhill, Norfolk (TF 639 120). *In:* Lewis, S.G., Whiteman, C.A. & Bridgland, D.R. (eds), *Central East Anglia and the Fen Basin*, Field Guide, Quaternary Research Association, London, 145-148.

Lewis, S.G., Rose, J. & Davies, H. (1999). Pre-Anglian fluvial and Anglian glaciogenic sediments, Knettishall, Suffolk, England. *Proceedings of the Geologists' Association*, **109**, 17-32.

Linton, D.L. (1955). The problem of tors. *Geographical Journal*, **121**, 470-487.

Lister, A.M. (2000). Pliocene to Early Pleistocene mammal faunas from the Norwich Crag and Cromer Forest-bed Formations. *In:* Lewis, S.G., Whiteman, C.A. & Preece, R.C. (eds). *The Quaternary of Norfolk and Suffolk*, Field Guide, Quaternary Research Association, London, 47-48.

Luckman, B.B. (1970). The Hereford Basin. *In:* C.A. Lewis (ed.), *The glaciations of Wales and adjoining regions*, 175-196, London, Longman.

Lunkka, J.P. (1994). Sedimentology and lithostratigraphy of the North Sea Drift and Lowestoft Till Formations in the coastal cliffs of NE Norfolk. *Journal of Quaternary Science*, **9**, 209-233.

Maddy, D., Green, C.P., Lewis, S.G. & Bowen, D.Q. (1995). Pleistocene Geology of the lower Severn valley, U.K. *Quaternary Science Reviews*, **14**, 209-222.

Madgett, P. A. & Catt, J.A. (1978). Petrography, stratigraphy and weathering of Late Pleistocene tills in East Yorkshire, Lincolnshire and north Norfolk. *Proceedings of the Yorkshire Geological Society*, **42**, 55-108.

Mathers, S.J., Zalasiewicz, J.A., Bloodworth, A.J. & Morton, A.C. (1987). The Banham Beds: a petrologically distinct suite of Anglian glacigenic deposits from central East Anglia. *Proceedings of the Geologists' Association*, **98**, 229-240.

Mathers, S.J., Zalasiewicz, J.A. & Wealthall, G.P. (1991). Styles of ice-marginal channel sedimentation: as revealed by a conductivity meter and extendable augers. *In:* Ehlers, J., Gibbard, P.L. & Rose, J. (eds), *Glacial Deposits in Great Britain and Ireland*, 405-414, Rotterdam, Balkema.

McCarroll, D., Ballantyne, C.K., Nesje, A. & Dahl, S.O. (1995). Nunataks of the last ice sheet in northwest Scotland. *Boreas*, **24**, 305-323.

McCarroll, D. & Ballantyne, C.K. (2000). The last ice sheet in Snowdonia. *Journal of Quaternary Science*, **15**, 765-778.

McGregor, D.F.M. & Green, C.P. (1978). Gravels of the River Thames as a guide to Pleistocene catchment changes. *Boreas*, **7**, 197-203.

McGregor, D.F.M. & Green, C.P. (1983). Post-depositional modification of Pleistocene terraces of the River Thames. *Boreas*, **12**, 23-33.

Meijer, T. (1993). Stratigraphical notes on Macoma (Bivalvia) in the southern part of the North Sea Basin and some remarks on the arrival of Pacific species. *Scripta Geol., Special Issue*, **2**, 297-312.

Meijer, T. & Preece, R.C. (1995). Malacological evidence relating to the insularity of the British isles during the Quaternary. *In:* Preece, R.C. (ed.), Island Brita*In:* a Quaternary Perspective. *Geological Society Special Publication*, **96**, 89-110.

Mitchell, G. F. (1960). The Pleistocene history of the Irish Sea. *The advancement of Science*, **17**, 313-325.

Mitchell, G.F., Penny, L.F., Shotton, F.W. & West R.G. (1973). A Correlation of Quaternary deposits in the British Isles. *Geological Society of London Special Report*, **4**, 99 pp.

Moorlock, B.S.P., Booth, S., Fish, P., Hamblin, R.J.O., Kessler, H., Riding, J., Rose, J. & Whiteman, C.A. (2000). In Lewis, S.G., Whiteman, C.A. & Preece, R.C. (eds). *The Quaternary of Norfolk and Suffolk*, Field Guide, Quaternary Research Association, London, 53-54.

Norton, P.E.P. Marine molluscan assemblages in the Early Pleistocene of Sidestrand, Bramerton and the Royal Society Borehole at Ludham, Norfolk. *Philosophical Transactions of the Royal Society of London*, **B253**, 161-200.

Oele, E. & Schüttenhelm, R.T.E. (1979). Development of the North Sea after the Saalian glaciation. *In:* Oele, E., Schüttenhelm, R.T.E. & Wiggers, A.J. (eds), *The Quaternary history of the North Sea, Acta Universitatis Upsaliensis, Symposia Universitatis Upsaliensis Annum Quingentesimum Celebrantis*, **2**, 191-215.

Peake, D. S. (1981). The Devensian glaciation on the north Welsh border. *In:* Neale, J & Flenley, J. (eds), *The Quaternary in Britain*, 49-59. Oxford, Pergamon Press.

Penck, A. & Brückner, E. (1909). *Die Alpen im Eiszeitalter*. Leipzig, 1199 pp.

Penny, L.F., Coope, G.R. & Catt, J.A. (1967). Age and insect fauna of the Dimlington Silts, east Yorkshire. *Nature*, **224**, 65-67.

Penny, L. F. & Rawson, P.F. (1969). Field Meeting in East Yorkshire and North Lincolnshire. *Proceedings of the Geologists' Association*, **80**, 193-218.

Perrin, R.M.S., Rose, J. & Davies, H. (1979). The distribution, variations and origins of pre-Devensian tills in eastern England. *Philosophical Transactions of the Royal Society of London*, **B287**, 536-570.

Pointon, K.W. (1978). The Pleistocene succession at Corton, Suffolk. *Bulletin of the Geological Society of Norfolk*, **30**, 55-76.

Poole, E. G. & Whiteman, A.J. (1966). *The geology of the country around Nantwich and Whitchurch*. Memoir of the Geological Survey of Great Britain. London, Her Majesty's Stationery Office, 154 pp.

Preece, R.C. & Parfitt, S.A. (2000). The Cromer Forest-bed Formation: new thoughts on an old problem. *In:* Lewis, S.G., Whiteman, C.A. & Preece, R.C. (eds), *The Quaternary of Norfolk and Suffolk*, Field Guide, Quaternary Research Association, London, 1-27.

Pringle, J & George, T.N. (1961). British Regional Geology, South Wales, London, Her Majesty's Stationary Office, 152 pp.

Reid. C. (1890). *The Pliocene deposits of Britain*. Memoirs of the Geological Survey, England and Wales, 362 pp.

Reid, C. & Woodward, H.B. (1882). *The geology of the country around Cromer*. Memoirs of the Geological Survey, England and Wales, 143 pp.

Rice, R.J. (1968). The Quaternary deposits of the central Leicestershire. *Philosophical Transactions of the Royal Society of London*, **A262**, 459-509.

Rice, R.J. (1981). The Pleistocene deposits of the area around Croft in south Leicestershire. *Philosophical Transactions of the Royal Society of London*, **B293**, 385-418.

Rice, R.J. & Douglas, T. (1991). Wolstonian glacial deposits and glaciation in Britain. *In:* Ehlers, J., Gibbard, P.L. & Rose, J. (eds), *Glacial Deposits in Great Britain and Ireland*, Rotterdam, Balkema, 25-36.

Richards, A.E. (1998). Re-evaluation of the Middle Pleistocene stratigraphy of Herefordshire. *Journal of Quaternary Science*, **13**, 115-136.

Rose, J. (1985). The Dimlington Stadial / Dimlington Chronozone: a proposal for naming the main glacial episode of the Late Devensian in Britain, *Boreas*, **14**, 225-230.

Rose, J. (1987). Status of the Wolstonian glaciation in the British Quaternary, *Quaternary Newsletter*, **53**, 1-9.

Rose, J. (1989a). Tracing the Baginton-Lillington Sands and Gravels from the West Midlands to East Anglia. *In:* Keen, D.H. (ed.), *West Midlands Field Guide*, Quaternary Research Association, London, 102-110.

Rose, J. (1989b). Stadial type sections in the British Quaternary. *In:* Rose, J & Schlüchter, Ch. (eds), *Quaternary Type Sections: Imagination or Reality*, 45-67, Rotterdam, Balkema.

Rose, J. (1991). Stratigraphic basis of the 'Wolstonian Glaciation', and retention of the term 'Wolstonian' as a chronostratigraphic Stage name -a discussion. *In:* Lewis, S.G., Whiteman, C.A. & Bridgland, D.R. (eds), *Central East Anglia and the Fen Basin*, Field Guide. Quaternary Research Association, London, 15-20.

Rose, J. (1992). High Lodge - regional context and geological background. *In:* Ashton, N., Cook, J., Lewis, S.G. & Rose, J. (eds), *High Lodge: Excavations by G. de G. Sieveking and J. Cook, 1988*, 13-24, London, British Museum.

Rose, J. (1994). Major river systems of central and southern Britain during the Early and Middle Pleistocene, *Terra Nova*, **6**, 435-443.

Rose, J. & Allen, P. (1977). Middle Pleistocene stratigraphy in southeast Suffolk. *Journal of the Geological Society*, **133**, 83-102.

Rose, J., Allen, P., Kemp, R.A., Whiteman, C.A. & Owen, N. (1985a). The Early Anglian Barham Soil of eastern England. *In:* Boardman, J. (ed.). *Soils and Quaternary Landscape Evolution*, 197-230, Chichester,Wiley.

Rose, J., Boardman, J., Kemp., R.A. & Whiteman, C.A. (1985b). Palaeosols and the interpretation of British Quaternary stratigraphy. *In:* Richards, K.S., Arnett, R.R. & Ellis, S. (eds), *Geomorphology and Soils*, 348-375, London, Allen and Unwin.

Rose, J., Allen, P. & Hey, R.W. (1986). Middle Pleistocene stratigraphy in southern East Anglia. *Nature*, **263**, 492-494.

Rose, J., Lee, J.A., Candy, I. & Lewis, S.G. (1999a). Early and early Middle Pleistocene river systems in eastern England. Evidence from Leet Hill. *Journal of Quaternary Science*, **14**, 347-360.

Rose, J., Lee. J.A., Moorlock, B.S.P. & Hamblin, R.J.O. (1999b). The origin of the Norwich Brickearth: micromorphological evidence for palaeopedological alteration of sand Anglian Till in northeast Norfolk. *Proceedings of the Geologists' Association*, **110**, 1-8.

Rose, J., Whiteman, C.A., Allen, P. & Kemp, R.A. (1999c). The Kesgrave Sands and Gravels: 'pre-glacial' Quaternary deposits of the River Thames in East Anglia and the Thames valley. *Proceedings of the Geologists' Association*, **110**, 93-116.

Rose, J., Candy, I. & Lee, J.R. (2000). Leet Hill (TM 3849260): pre-glacial and glaciofluvial river deposits - with possible evidence for a major glaciation prior to the deposition of the Lowestoft Till. *In:* Lewis, S.G., Whiteman, C.A. & Preece, R.C. (eds). *The Quaternary of Norfolk and Suffolk*, Field Guide. Quaternary Research Association, London, 207-217.

Rose, J., Moorlock, B.S.P. & Hamblin, R.J.O. (2001). Pre-Anglia fluvial and coastal deposits in Eastern England: lithostratigraphy and palaeoenvironments. *Quaternary International*, **79**, 5-22.

Rowe, P.J., Atkinson, T.C. & Turner, C. (1999). U-series dating of Hoxnian interglacial deposits at Marks Tey, Essex, England. *Journal of Quaternary Science*, **14**, 693-702.

Rowe, P.J., Richards, D.A., Atkinson, T.C., Bottrell, S.H. & Cliff, R.A. (1997). Geochemistry and radiometric dating of a Middle Pleistocene peat. *Geochimica et Cosmochimica Acta*, **61**, 4201-4211.

Ruegg, G.H.J. (ed.) (1991). Geology and archaeology of ice-pushed Pleistocene deposits near Wageningen (the Netherlands). *Mededelingen Rijks Geologische Dienst*, **46**, 99 pp.

Sabine, P.A. The source of some erratics from north-eastern Northamptonshire and adjacent parts of Huntingdonshire, *Geological Magazine*, **113**, 255-260.

Sejrup, H.P., Aarseth, I., Ellingsen, K.L., Reither, E., Jansen, E., Lovlie, R., Bent, A., Brigham-Grette, J., Larsen, E. & Stoker, M. (1987). Quaternary Stratigraphy of the Fladen area, central North Sea: a multidisciplinary study. *Journal of Quaternary Science*, **2**, 35-58.

Sejrup, H.P., Haflidason, H., Aarseth, I., King, E., Forsberg, C.F., Long, D. & Rokoengen, K. (1994). Late Weichselian glaciation history of the northern North Sea. *Boreas*, **23**, 1-13.

Shackleton, N.J. & Opdyke, N.D. (1973). Oxygen isotope and palaeomagnetic stratigraphy of Equatorial Pacific core V28-238: oxygen isotope temperatures and ice volumes on a 10^5 year and 10^6 year scale. *Quaternary Research*, **3**, 39-55.

Shackleton , N.J. & Turner, C. (1967). Correlation between marine and terrestrial Pleistocene successions, *Nature*, **216**, 1079-1082.

Shotton, F.W. (1953). The Pleistocene deposits of the area between Coventry, Rugby and Leamington, and their bearing on the topographic development of the Midlands. *Philosophical Transactions of the Royal Society of London*, **B237**, 209-260.

Shotton, F.W. (1967). Age of the Irish Sea Glaciation. *Nature*, **215**, 1366.

Shotton, F.W. (1968). The Pleistocene succession around Brandon, Warwickshire. *Philosophical Transactions of the Royal Society of London*, **B254**, 387-400.

Shotton, F.W. (1976). Amplification of the Wolstonian Stage of the British Pleistocene. *Geological Magazine*, **113**, 241-250.

Shotton, F.W. (1977). The Devensian stage: its development, limits and substages. *Philosophical Transactions of the Royal Society of London*, **B280**, 107-118.

Shotton, F.W. (1983a). The Wolstonian Stage of the British Pleistocene in and around its type area of the English Midlands. *Quaternary Science Reviews*, **2**, 261-280.

Shotton, F.W. (1983b). Observations on the type Wolstonian glacial sequence. *Quaternary Newsletter*, **40**, 28-36.

Shotton, F.W., Goudie, A.S., Briggs, D.J. & Osmaston, H.A. (1980). Cromerian interglacial deposits at Sugworth near oxford, England, and their relation to the Plateau Drifts of the Cotswolds and the terrace sequence of the Upper and Middle Thames. *Philosophical Transactions of the Royal Society of London*, **B289**, 55-86.

Sibrava, V., Bowen, D.Q. & Richmond, G.M. (eds). (1986). Quaternary glaciations of the Northern Hemisphere. *Quaternary Science Reviews*, **6**, 1-510.

Sisson, J.B. (1967). *The Evolution of Scotland's Scenery*. Edinburgh, Oliver and Boyd, 289 pp.

Smith, A.J. (1985). A catastrophic origin for the palaeovalley system of the eastern English Channel. *Marine Geology*, **64**, 65-75.

Soil Survey (1983). *Soil map of England and Wales 1:250,000. Sheet 1, Northern England.* Harpenden, Soil Survey of England and Wales.

Solomon, J.D. (1935). The Westleton Series of East Anglia: its ages, distribution and relations. *Quarterly Journal of the Geological Society of London*, **91**, 216-238.

Sparks, B.W. & West, R.G. (1964). The drift landforms around Holt, Norfolk. *Transactions of the Institute of British Geographers*, **35**, 27-35.

Stephens, N., Creighton, J.R. & Hannon, A. (1975). The late Pleistocene Period in north-eastern Ireland: an assessment, 1975. *Irish Geography*, **8**, 1-23.

Straw, A. (1960). The limit of the 'Last' glaciation in north Norfolk. *Proceedings of the Geologists' Association*, **71**, 378-390.

Straw, A. (1961). Drifts, meltwater channels and ice-margins in the Lincolnshire Wolds. *Transactions of the Institute of British Geographers*, **29**, 115-128.

Straw, A. (1969). Pleistocene events in Lincolnshire: a survey and revised nomenclature. *Transactions of the Lincolnshire Naturalists' Union*, **18**, 85-98.

Straw, A. (1979a). Eastern England. *In:* A. Straw, A. & Clayton, K.M. (eds*), The geomorphology of the British Isles: eastern and central England*, 3-139, London, Methuen.

Straw, A. (1979b). An Early Devensian glaciation in eastern England. *Quaternary Newsletter*, **28**, 18-24.

Straw, A. (1983). Pre-Devensian glaciation of Lincolnshire (Eastern England) and adjacent areas. *Quaternary Science Reviews*, **2**, 239-260.

Straw, A. (1991). Glacial deposits of Lincolnshire and adjoining areas. *In:* Ehlers, J., Gibbard, P.L. & Rose, J. (eds), *Glacial Deposits in Great Britain and Ireland*, 213-221, Rotterdam, Balkema,.

Sugden, D.E. (1968). The selectivity of glacial erosion in the Cairngorm Mountains, Scotland. *Transactions of the Institute of British Geographers*, **45**, 79-92.

Suggate, R.P. & West, R.G. (1959). On the extent of the last glaciation in eastern England. *Proceedings of the Royal Society of London*, **B150**, 263-283.

Sumbler, M.G. (1983a). A new look at the type Wolstonian Glacial deposits. *Proceedings of the Geologists' Association*, **94**, 23-31.

Sumbler, M.G. (1983b). The type Wolstonian sequence - some further observations. *Quaternary Newsletter*, **40**, 36-39.

Sumbler, M.G. (1995). The terraces of the rivers Thame and Thames and their bearing on the chronology of glaciation in central and eastern England. *Proceedings of the Geologists' Association*, **106**, 93-106.

Sumbler, M.G. (2001). The Moreton Drift: a further clue to glacial chronology in central England. *Proceedings of the Geologists' Association*, **112**, 13-27.

Sutherland, D.G. (1984). The Quaternary deposits and landforms of Scotland and neighbouring shelves: a review. *Quaternary Science Reviews*, **3**, 157-254.

Tappin, D.R., Chadwick, R.A., Jackson, A.A., Wingfield, R.T.R. & Smith, N.J.P. (1994). *The geology of Cardigan Bay and the Bristol Channel*. British Geological Survey, UK Offshore Regional report, London, HMSO, 107 pp.

Thomas, G.S. P. (1989). The Late Devensian glaciation along the western margin of the Cheshire - Shropshire lowland. *Journal of Quaternary Science*, **4**, 167-181.

Trechmann,. C.T. (1915). The Scandinavian Drift of the Durham coast and the general glaciology of south-eat Durham. Quarterly *Journal of the Geological Society* of London, **71**, 53-82.

Turner, C. (1970). The Middle Pleistocene deposits at Marks Tey, Essex. *Philosophical Transactions of the Royal Society of London*, **B257**, 373-440.

Tzedakis, P.C. & Bennett, K.D. (1995). Interglacial vegetation succession: a view from southern Europe, *Quaternary Science Reviews*, **14**, 967-982.

Vandenberghe, J. (2000). A global perspective of European chronostratigraphy for the past 650 ka. *Quaternary Science Reviews*, **19**, 1701-1707.

Vandenberghe, J., Roebroeks, W. & van Kolfschoten, T. (eds) (1993). Maastricht-Belvédère: stratigraphy, palaeo-environment and archaeology of the Middle and Late Pleistocene deposits; Part ii, *Mededelingen Rijks Geologische Dienst*, **47**, 91 pp.

Ventris, P.A. (1986). The Nar Valley. *In:* West R.G. & Whiteman, C.A. (eds), *The Nar Valley and North Norfolk*, Field Guide, Quaternary Research Association, Coventry, 6-55.

Ventris, P.A. (1996). Hoxnian Interglacial freshwater and marine deposits in northwest Norfolk, England and their implications for sea-level reconstruction, *Quaternary Science Reviews*, **15**, 437-450.

West, R.G. (1956). The Quaternary deposits at Hoxne, Suffolk. , *Philosophical Transactions of the Royal Society of London*, **B239**, 265-356.

West, R.G. (1961). Vegetational history of the Early Pleistocene of the Royal Society Borehole at Ludham, Norfolk, *Proceedings of the Royal Society of London*, **B155**, 437-453.

West, R.G. (1977). *Pleistocene Geology and Biology*, 2nd edition. London,Longman, 440 pp.

West, R.G. (1980a). Pleistocene forest history in East Anglia. *New Phytologist*, **85**, 571-622.

West, R.G. (1980b). *The pre-glacial Pleistocene of the Norfolk and Suffolk coasts*, Cambridge, Cambridge University Press, 203 pp.

West, R.G. (1989). The use of type localities and type sections in the Quaternary, with especial reference to East Anglia. *In:* Rose, J & Schlüchter, Ch. (eds), *Quaternary Type Sections: Imagination or Reality*, 3-10, Rotterdam, Balkema.

West, R. G. (1993). On the history of the Late Devensian Lake Sparks in southern Fenland, Cambridgeshire, England. *Journal of Quaternary Science*, **8**, 217-234.

West, R.G. & Banham, P.H. (1968). Short field meeting on the north Norfolk Coast. *Proceedings of the Geologists' Association*, **79**, 493-512.

West, R.G. & Donner, J.J. (1956). The glaciations of East Anglia and the East Midlands: a differentiation based on stone-orientation measurements of the tills. *Quarterly Journal of the Geological Society of London*, **112**, 69-91.

West, R.G. & Norton, P.E.P. (1974). The Icenian Crag of southeast Suffolk, *Philosophical Transactions of the Royal Society of London*, **B269**, 1-18.

West R.G., Funnell, B.M. & Norton, P.E.P. (1980). An Early Pleistocene cold marine episode in the North Sea: pollen and faunal assemblages at Covehithe, Suffolk, England . *Boreas*, **9**, 1-10.

West, R.G., Andrew, R., Catt, J.A., Hart, C.P., Hollin, J.T., Knudsen, K.-L., Miller, G.F., Penney, D.N., Pettit, M.E., Preece, R.C., Switsur, V.R., Whiteman, C.A. & Zhou, L.P. (1999). Late and Middle Pleistocene deposits at Somersham, Cambridgeshire, U.K.,: a model for reconstructing fluvial / estuarine depositional environments, *Quaternary Science Reviews*, **18**, 1247-1314.

Westaway, R., Maddy, D. & Bridgland, D. (2002). Flow in the lower continental crust as a mechanism for the Quaternary uplift of south-east England: constraints from the Thames terrace record. *Quaternary Science Reviews*, **21**, 559-603.

Whiteman, C.A. (1990). Early and Middle Pleistocene stratigraphy and soils in central Essex, England. Unpublished PhD Thesis, University of London.

Whiteman, C.A. (1992). The palaeogeography and correlation of pre-Anglian-Glaciation terraces of the River Thames in Essex and the London Basin. *Proceedings of the Geologists' Association*, **103**, 37-56.

Whiteman, C.A. & Rose, J., 1992. Thames River sediments of the British Early and Middle Pleistocene. *Quaternary Science Reviews*, **11**, 363-375.

Whittington, G., Connell, R.E., Coope, R.G., Edwards, K.J., Hall, A.M., Hulme, P.D. & Jarvis, J. (1998). Devensian organic interstadial deposits and ice sheet extent in Buchan, Scotland. *Journal of Quaternary Science*, **13**, 309-324.

Wills, L.J. (1924). The development of the Severn Valley in the neighbourhood of Iron-Bridge and Bridgnorth. *Quarterly Journal of the Geological Society of London*, **80**, 274-314.

Wingfield. R.T.R. (1989). Glacial incisions indicating Middle and Upper Pleistocene ice limits in Britain. *Terra Nova*, **1**, 538-548.

Wingfield, R.T.R. (1994). Pleistocene and Holocene. *In:* Tappin, D.R., Chadwick, R.A., Jackson, A.A., Wingfield, R.T.R. & Smith, N.J.P. *The geology of Cardigan Bay and the Bristol Channel*, 76-93, British Geological Survey, UK Offshore Regional report, London, HMSO.

Wingfield, R.T.R. (1995). Pleistocene and Holocene. *In:* Jackson, D.I., Jackson, Evans, D., Wingfield, R.T.R., Barnes, R.P. & Arthur, A.A., M.J. (eds), *The Geology of the Irish Sea*, 85-102. British Geological Survey, UK Offshore Regional report, London, HMSO.

Woodland, A.W. (1970). The buried tunnel-valleys of East Anglia. *Proceedings of the Yorkshire Geological Society*, **37**, 521-578.

Worsley, P. (1991). Glacial deposits of the lowlands between the Mersey and Severn rivers. *In:* Ehlers, J. Gibbard, P.L. & Rose, J. (eds), *Glacial Deposits in Great Britain and Ireland*. Rotterdam, Balkema, 203-211.

Quaternary Glaciations – Extent and Chronology
Editors J. Ehlers and P.L. Gibbard

Pleistocene glaciations in Estonia

Anto Raukas[1], Volli Kalm[2], Reet Karukäpp[1] and Maris Rattas[2]

[1] *Institute of Geology, Tallinn Technical University, 7 Estonia Ave, 10143 Tallinn, Estoniaėėėėė*

[2] *Institute of Geology, University of Tartu, Vanemuise 46, 51014 Tartu, Estonia*

Introduction

The possibility that more than one glaciation had covered Estonia was first proposed by Grewingk (1879). He described two separate till beds, the superficial reddish-brown and the lowermost grey till in the buried valleys in Tartu, SE Estonia (Fig.). The Rõngu (Eemian) section was the first occurrence of fossiliferous deposits, the interglacial nature of which was confirmed palynologically (Orviku, 1939; Thomson, 1939, 1941). On the basis of this discovery, the first stratigraphical scheme of Pleistocene deposits in Estonia was presented by Orviku (1939). In 1938, Tammekann compiled a map of glacial landforms (Tammekann, 1938). Together with glacial landforms of different origin, he presented five zones of ice-marginal formations. A first attempt to date the ice-marginal formations and to correlate them with the neighbouring countries was undertaken by Serebrjannyi and Raukas (1966). Several stratigraphical schemes have been compiled for Estonia (in 1956, 1957, 1961, 1963, 1970, 1976). There were mainly correlative parts of the schemes of the European part of the former Soviet Union or the Baltic States and Belarus (Orviku 1956, 1960; Raukas 1978). In the scheme compiled by Kajak *et al.* (1976) local geographical names were for the first time used to denote stratigraphic units. In 1993 a new regional stratigraphical scheme for the Quaternary deposits in the Baltic States as a whole, and local stratigraphical schemes for each of the three republics (Estonia, Latvia, and Lithuania) were agreed. All stratigraphical units are based on stratotype sections (Kondratienė, 1993; Raukas *et al.*, 1993; Raukas & Kajak, 1995). Medium-scale (1:200,000) geological mapping of Estonia was completed in 1978 and large-scale mapping is in progress at present.

From the dense network of boreholes available, it is known that glacial, glaciofluvial and glaciolacustrine deposits comprise about 95% of the Quaternary sediments in Estonia. The areas of glacial accumulation and erosion remained relatively stable through time. Glacial erosion predominated in the lobate depressions and plateau-like uplands where bedrock is exposed (Aboltinš *et al.*, 1989; Tavast & Raukas, 1982). Glacial accumulation was concentrated on the insular accumulative heights and interlobate and ice-marginal formations (Aboltinš *et al.*, 1989; Raukas & Karukäpp, 1979). The thickness of the Quaternary deposits is less in northern and western Estonia, where it does not usually exceed 5 metres. The Quaternary deposits are at their thickest (50–200 m) in the Fore-Klint Lowland, in the Saadjärve Drumlin Field, in the heights of southeastern Estonia and in the buried valleys. In the Baltic offshore area, the thickness of the Quaternary deposits usually varies between 10 and 20 m. Five till beds, often of great thickness, are more or less continually traceable. They are separated in several places from each other by Prangli (Eemian, Mikulino) and Karuküla (Holsteinian, Likhvin) interglacial deposits or interstadial beds (Kajak, 1995; Liivrand, 1991; Raukas, 1976, 1978, 1995; Raukas & Kajak, 1995, 1997b).

Early and Middle Pleistocene

Early Pleistocene glaciations, to which the Lower Pleistocene tills in the southern Peribaltic region belong, have not been identified in Estonia. Strong erosion by subsequent glaciations has destroyed all evidence. On the basis of present knowledge, deposits of Lower Pleistocene age are absent from Estonia and even the Middle Pleistocene sequence is rather incomplete.

Sangaste (Elsterian, Oka) Glaciation

On the basis of its position underlying the Karuküla (Holsteinian) interglacial deposits, the lowermost compact till (Sangaste Till), located in some buried valleys of central and southern Estonia (Keskküla, Saadjärv, Sudiste, Mägiste), is correlated with the Elsterian Glaciation (Raukas, 1978). This brownish or sometimes greenish till, rich in clasts from the Fennoscandian Shield (25–95%) rests directly on bedrock. Within a sequence of Sangaste Till at a depth of 143,3–169,2 m, clayey sands were discovered in the Otepää buried valley in southern Estonia (Kajak, 1995). These sands were correlated by Kajak with the Middle Elsterian Turgeljai and Belovežje beds in Lithuania and Belarus, based on some similarities of the pollen spectra (Kajak & Liivrand, 1967). Glacigenic deposits below these beds were consequently correlated (Kajak, 1995) with the Early Elsterian glaciation. However, direct evidence in favour of this conclusion has

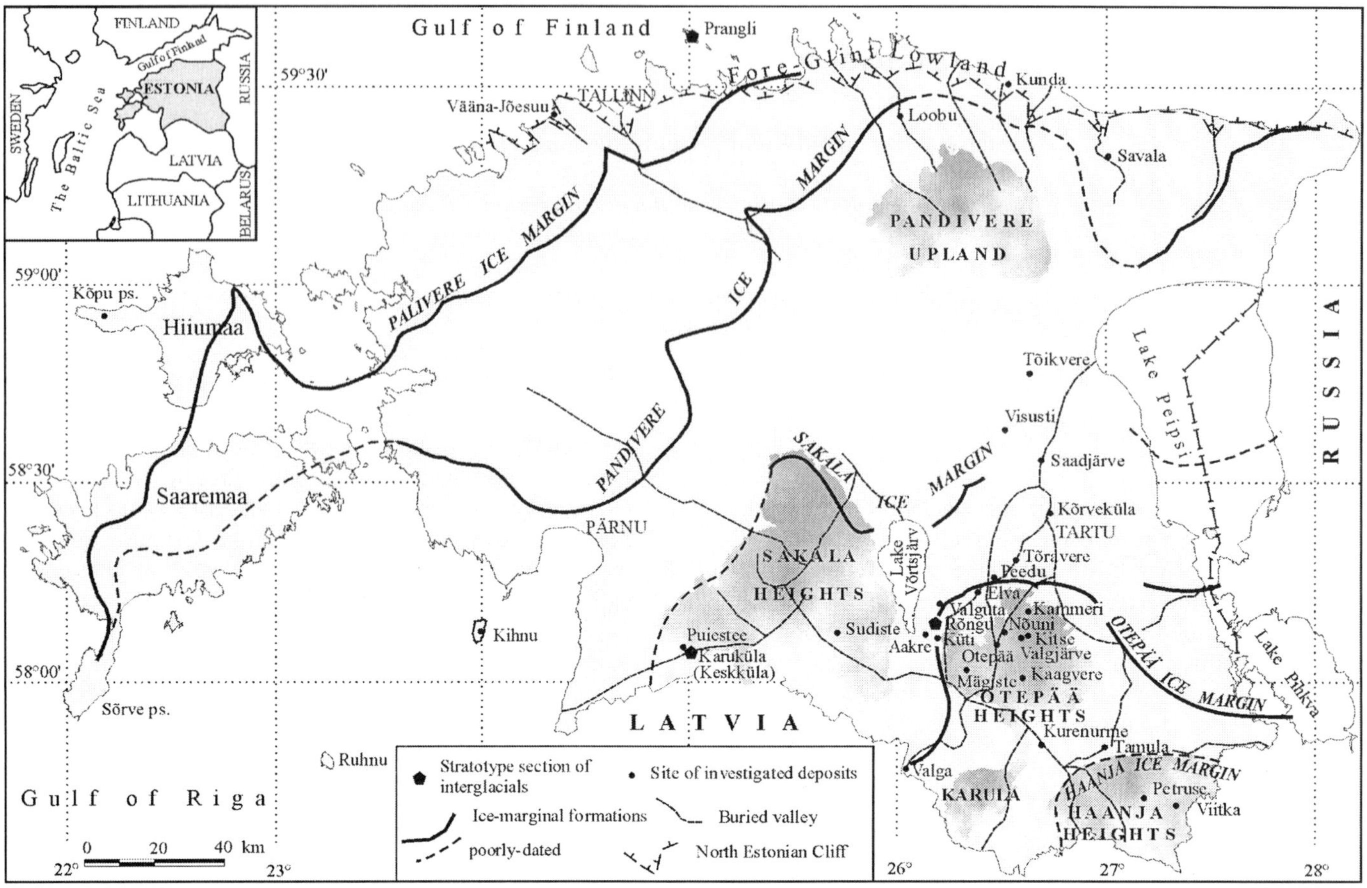

Fig. 1: Stratigraphically important sections of Pleistocene and ice-marginal formations in Estonia.

not been found. Liivrand (1991) assigned all these sediments to the Valdai (Weichselian) Glaciation, what is also disputable.

Karuküla (Holsteinian, Likhvin) Interglacial

The two known occurrences of Karuküla (Holsteinian) Interglacial deposits in Estonia (at Karuküla and Kõrveküla) are found in allochthonous position (Levkov & Liivrand, 1988; Liivrand, 1991; Liivrand & Saarse, 1983), however this conclusion was argued by Kajak (1995). The Karuküla section in SW Estonia was first described by Orviku (1944) and originally correlated with the upper climatic optimum of the Riss–Würm Interglacial (Orviku, 1944, 1960) in Denmark (Jessen & Milthers, 1928). Later it was referred to the Brørup Interstadial (Orviku & Pirrus, 1965). The Middle Pleistocene (Holsteinian, Likhvin) age of these deposits was first suggested by Danilans (1966). Radiocarbon dates from the 1960s and early 1970s (33,450±800, Ta-99; 49,100±1799, Ta-100; 48,100±1650, Ta-101; >45,000, Ta-106; 40,800±700, Ta-275; 47,800±1100, Ta-276; 48,800±1,200, Ta-277; >52,780, LU-44; >53,240, LU-123; >48,750, Birm-249; >46,000, GSC-1975; 48,600±1,600, GSC-1976; 47,760±11,000, Tln-384), however, seemed to suggest a Middle Weichselian (Middle Valdai) age. Based on the detailed pollen and karpological studies Liivrand (1972, 1984, 1991) and Velichkevich & Liivrand (1976, 1984) confirmed the Holsteinian age of the Karuküla and Kõrveküla sediments and therefore the Karuküla Interglacial was included in the Estonian (Raukas, 1978) and regional East-Baltic (Kondratienė, 1993) stratigraphical schemes. The radiocarbon ages of the deposits were discarded as unreliable or being beyond the limits of the method. Both Karuküla Interglacial sections in Estonia are continental deposits of forest peat, gyttja, and lake or river sands (Liivrand, 1984, 1991) which suggest that the sea in the Holsteinian most likely did not reach Estonia (Raukas & Kajak, 1997a). Pollen analysis of the Karuküla interglacial deposits demonstrates widespread conifer forests and a very limited occurrence of *Corylus* (Liivrand, 1984, 1991).

Ugandi (Saalian) Glaciation
Early Ugandi (Saalian I, Drenthe, Dniepr) Substage

Lower Ugandi till occurs in buried valleys in both southern and northern Estonia or in sheltered positions in the cores of the southern Estonian heights. However, at present there are no sites where the Lower Ugandi till directly overlies the Karuküla interglacial deposits. The Early Saalian (Early Ugandi) age of the till is deduced from its position. The lithologically easily traceable reddish-brown, compact till bed is up to 50 m thick. It is the 4th from the surface and the second complex of glacial deposits below the Prangli (Eemian) interglacial deposits. Its clast composition (the till is rich in Vyborg rapakivi) indicates that the Lower Ugandi till was deposited by a southward moving ice sheet (Raukas, 1978; Raukas & Gaigalas, 1993; Raukas & Kajak, 1997b).

Middle Ugandi (Saalian) Interstadial

Possible interstadial deposits are found at a few places (Puiestee, Valguta, Elva, Aakre and Nõuni) in southern Estonia between the two Ugandi (Dniepr and Moscow) till strata (Kajak, 1995; Raukas, 1995; Raukas & Kajak, 1997b). However, Liivrand (1991) has shown that the Middle Ugandi (Odintsovo) organic sands and sandy loams contain redeposited pollen and their stratigraphical position is not firmly established.

Late Ugandi (Saalian II, Warthe, Moscow) Substage

The till lithology (lacking Vyborg rapakivi and Suursaari quartz-porphyries) and clast fabrics suggest that the ice flow during this glaciation was mostly from the northwest to southeast (Raukas, 1978; Raukas & Gaigalas, 1993; Raukas & Kajak, 1997a). In the foreclint (coastal) area of the Gulf of Finland, the clasts in the till originate from the Fennoscandian Shield and from the Baltic Sea floor. Elsewhere in Estonia carbonate clasts predominate (65–80%) in this till (Raukas, 1978). The Upper Ugandi till is up to 70 m thick and has been identified in buried valleys in southern and northern Estonia or in the 'cores' of the heights. In the Valga borehole, in southern Estonia, the third till unit from the top (the Upper Saalian or Moscow Till) was dated using in this case arguable TL method (Punning & Raukas, 1983), which gave an age of 216,000±10,000 TL years (Tln-TL-115) (Kajak *et al.*, 1981).

Late Pleistocene

Prangli (Eemian, Mikulino) Interglacial

Both marine and continental Prangli Interglacial deposits have been found in Estonia. The terrestrial deposits at Rõngu were investigated in detail over half a century ago (Orviku, 1939; Thomson, 1939, 1941). In the Gulf of Finland on the Prangli Island (Kajak, 1961), and in the Gulf of Riga on Kihnu Island (Liivrand, 1976; Raukas, 1978; Liivrand, 1991) the two upper (Valdai, Weichselian) and two middle (Ugandi) tills are separated from each other by marine clayey deposits of the last (Eemian) interglacial. Diatom analyses of the authochtonous interglacial deposits in the Prangli section (Cheremisinova, 1961; Znamenskaya & Cheremisinova, 1962) have revealed two diatom assemblages that represent two stages in the development of the Eemian Baltic Sea (Raukas, 1991). In some places in SE Estonia (Rõngu, Küti, Kitse, Peedu), peat-sapropelic bog-lacustrine deposits occur (Liivrand, 1977) at the same stratigraphical level. Although most of the Prangli (Eemian) interglacial deposits in Estonia are found in ice-dislocated positions (Liivrand, 1991), both marine and continental sediments may easily be correlated using pollen assemblage zones (Kondratiene, 1993; Liivrand, 1991).

Järva (Weichselian, Valdai) Glaciation

The beginning of the Weichselian Glaciation is characterised by sediments containing pollen representing periglacial vegetation (Liivrand, 1991; Raukas & Kajak; 1995, 1997b). These sediments immediately overlie Prangli interglacial sediments (e.g. on Prangli Island) or rest between Upper Ugandi and Lower Valdai till beds (e.g. at Otepää, Tõravere, Valga and Valguta) belong to the Kelnase substage (Raukas & Kajak, 1995, 1997b). The Tõravere site in southern Estonia demonstrates the existence of a till bed between the Lower and Middle Valdai (Weichselian) periglacial strata (Liivrand, 1995). Comprehensive study of these sites indicates that in Estonia there is sufficient evidence to recognize two Valdai glaciation cycles, each lasting some 50-55 thousand years (Raukas, 1992a).

Valgjärv (Early Weichselian, Early Valdai) Substage

The second complex of glacial deposits from the surface, the so-called purplish-grey till in southern Estonia, is referred to the Early Valdai Stadial, as is the grey till overlying the Prangli interglacial deposits in northern Estonia (Raukas, 1978). In the Kitse borehole in SE Estonia at a depth of 4.2–31.1 m the Lower Weichselian till overlies Prangli (Eemian) interglacial deposits (Kajak, 1995). The till is widely-distributed in central and southern Estonia but is absent from the areas of bedrock highs and other regions of intense glacial erosion in northern and north-western parts of the country. The purplish-grey, i.e. the second till unit from the surface, in two boreholes from southern Estonia has yielded thermoluminescence (TL) ages of 75,000 (Tln-TL-40, Peedu borehole) and 153,000±10,000 (Tln-TL-114, Valga borehole) (Kajak *et al.*, 1981). Even if the first date is in agreement with the widely-accepted age-model of the Valdai Glaciation, the TL method, however, is arguably suitable for the dating of tills.

Savala (Middle Weichselian, Middle Valdai) Interstadial

Palynological investigations have revealed Middle Valdai terrestrial interstadial deposits that yield pollen assemblages representing cold and dry periglacial conditions at several sites (e.g. Savala, Tõravere, Valguta, Peedu, Vääna-Jõesuu and Tõikvere) (Liivrand 1974, 1990, 1991, 1995; Raukas, 1978, Raukas & Liivrand, 1971). In southern Estonia the lower part of the Middle Valdai sandy sediments at the Valguta site have given TL ages of 66,500 (Tln-TL-67) and 62,400 (Tln-TL-63) (Kajak *et al.*, 1981; Liivrand, 1991). At Peedu radiocarbon dating of the interstadial layers has given ages of 39,180±1,960 (Ta-136), 39,700±850 (Ta-254) and 31,200±800 (Ta-254A) (Punning, 1970; Liivrand & Saarse, 1976; Kajak *et al.*, 1981). These dates are in a good agreement with those obtained from the other Middle Weichselian sites in the region, such as Lejasčiems (32.2–36.1 ka) and Talsi (56.1 ka) in NW Latvia (Arslanov *et al.*, 1975; Dreimanis & Zelčs, 1995; Meirons, 1986) or Graždanskii Prospekt (40.4 ka) in St.Petersburg (Arslanov *et al.*, 1981).

Võrtsjärv (Late Weichselian, Late Valdai) Substage

The Late Valdai ice sheet deposited tills of different colour, and meltwater sediments both above and beneath the till. On the Gulf of Finland coast, the till is bluish-grey, rich in clasts from the Fennoscandian Shield and almost carbonate-free where it overlies the Cambrian 'blue' clays. However, on the Ordovician and Silurian carbonate bedrock, the till is very much enriched in local limestone and dolomite clasts. On the Devonian sandstones of southern Estonia the till is reddish-brown in colour and has a high sand content (Raukas, 1978). Moreover, at several sites intermorainic interstadial or other sediments occur within this till unit. In a number of places (see above) the Upper Valdai till also rests directly on the Savala (Middle Valdai) interstadial deposits.

Deglaciation history

During the course of thinning of the ice sheet, the movement of its individual lobes was controlled by the underlying bedrock topography (Tavast & Raukas, 1982). Former ice-marginal positions are represented in the modern topography by discontinuous chains of end moraines and glaciofluvial formations (Raukas & Kajak, 1997a). Five ice-marginal zones (in order of decreasing age: Haanja, Otepää, Sakala, Pandivere and Palivere) are usually distinguished in Estonia (Raukas et al., 1971, Raukas & Karukäpp, 1979; Raukas, 1986) (Fig.). They were formed either as the result of standstills of the ice margin or in some cases as a result of readvances. Sites where till-covered interstadial-type sediments (e.g. Viitka, Petruse, Kurenurme and Kaagvere) occur and lithostratigraphical observations provide evidence for events of ice front oscillations (Raukas & Rähni, 1966; Karukäpp & Miidel, 1972; Kajak *et al.*, 1976; Raukas, 1986; Pirrus & Raukas, 1996). The Haanja, Otepää, Sakala and Palivere phases are characterized by a dominant southeasterly ice movement direction, the Pandivere Stadial by its southerly or even southwesterly ice movement direction (Raukas & Karukäpp, 1979; Raukas, 1978, 1992a).

The deglaciation history in Estonia has been dated using the conventional ^{14}C method, varve chronology, thermoluminescence (TL) and optical stimulated luminescence (OSL) methods. The main conclusion is that Estonia and the Gulf of Finland south of the Salpausselkä ice-marginal formations became ice-free at about 13,500-11,000 ^{14}C years BP (Raukas, 1986; Pirrus & Raukas, 1996; Raukas & Kajak, 1997a). In the light of pollen analytical interpretations, the retreat of the ice margin from the Haanja zone (the oldest in Estonia) began in the Bølling, whereas Estonia was finally ice-free in the second half of the Allerød Chron (Pirrus & Raukas, 1969, 1996).

Haanja ice-marginal zone

The age of the Haanja ice-marginal zone (13,430–13,630 ^{14}C years BP, Pirrus & Raukas, 1996; Raukas & Kajak, 1997a) was derived from interstadial deposits in the Raunis section, south of Haanja, in north-central Latvia. These Raunis interstadial deposits are dated to 13,390±500 (Mo-296), 13,250±160 (Ta-177), 13,320±250 (Ri-39) ^{14}C years BP (Dreimanis & Zelčs, 1995; Punning *et al.*, 1968). The peat bed sampled for the ^{14}C dates in the Raunis section is, however, thoroughly penetrated by recent rootlets and therefore the ^{14}C determinations are questionable (Dreimanis & Zelčs, 1995). It will be possible, that organic deposits in Raunis section have accumulated in Early-Holocene and covered with pseudotill (slope deposits). Haanja Heights, and later also Otepää Heights, formed ice divides between the individual peripheral ice streams. In the central part of the heights, both active and passive ice played a significant role in landscape formation. Glacio-karst and slope processes later caused redeposition of Late Glacial organic deposits. Therefore the ^{14}C dates from intermorainic or submorainic sequences from the distal side of the Haanja zone are often younger than one would expect (Petruse: 12,670±200 and 12,080±120 BP; Viitka: 10,950±80 BP). On the basis of magnetostratigraphical investigations and correlations, Hang and Sandgren (1996) demonstrated that sedimentation in Tamula Lake immediately to the north of the Haanja ice-marginal zone began some 12,700–12,800 varve years ago. This coincides well with the assumed time of ice recession from southern Estonia. The correlative of the Haanja ice-marginal zone in Latvia is the North Latvian and in the St. Petersburg region the Luga ice-marginal zone (Donner, 1995; Raukas *et al.*, 1971).

Otepää ice-marginal zone

The age of the Otepää ice-marginal zone (*c.* 12,600 ^{14}C years BP, Pirrus & Raukas, 1996, Raukas & Kajak, 1997b) has been determined at the Kurenurme section, where a sandy loam with organic detritus and wood overlies the Haanja Till. Radiocarbon dating of wood (12,650±520, Ta-57) and organic detritus (12,420±100, Tln-35) indicates that these deposits accumulated at the beginning of the Bølling (Ilves *et al.*, 1974, Raukas & Kajak, 1997b). In the Kaagvere section southeast of Otepää at the distal side of the Otepää ice-marginal zone, the ^{14}C dates obtained from the organic layers on the surface till (15,150±575, Ta-50, and >30,000, Ta-36) suggest redeposition of older interglacial material. Similarly to the Haanja Heights, the Otepää Heights also served as an ice divide between the Võrtsjärve ice lobe to the west and the Peipsi ice lobe to the east.

Sakala ice-marginal zone

The Sakala zone is the most difficult to trace and most poorly-dated amongst the ice-marginal formations in Estonia. End moraines, eskers and kame fields referred to the Sakala zone are described from the Saadjärve Drumlin Field area between the depressions of Lake Peipsi and Lake Võrtsjärv and on Sakala Heights, SW Estonia (Lõokene, 1961; Raukas *et al.*, 1971). The extension of the Sakala ice in the Lake Peipsi and Lake Võrtsjärv depressions is rather unclear because of the lack of ice-marginal formations. Pollen analysis from the Visusti section, which is located between the Sakala and Pandivere ice-marginal zones in eastern Estonia, has revealed that accumulation of lacustrine silts and sands on the surface till began during the Older Dryas (12,200–11,800 BP). Thus the Sakala phase is estimated to date between the Otepää and Pandivere phases (Pirrus & Raukas, 1996).

Pandivere ice-marginal zone

The Pandivere ice-marginal zone is dated according to varve chronology to about 12,500 BP (Raukas 1992a; Raukas & Kajak, 1997b). The correlative of the Pandivere ice-marginal zone in the St. Petersburg region is considered to be the Neva zone (Donner, 1995; Raukas *et al.*, 1971). Judging by the possible mean rates of ice recession and the varve correlations with the St. Petersburg District, Karukäpp *et al.* (1992) assumed, that the Pandivere zone was formed during the Older Dryas. Later, on the basis of clay-varve chronologies of the eastern Baltic and NW Russia, Hang (1997) concluded that the ice sheet would have retreated from the Pandivere (Neva) ice-marginal zone about 12,330 varve years BP. On the west Estonian Islands, the Pandivere Till, and the till of the following Palivere phase, are separated from each other by the lacustrine Kõpu Sands found in a few boreholes on the Kõpu and Sõrve peninsulas (Kadastik & Kalm, 1998).

The Palivere ice-marginal zone

Apart from some end moraines on the Island of Saaremaa, the Palivere ice-marginal zone is represented by a curved belt of so-called marginal eskers about 60 km in length. The marginal eskers, which represent a specific type of glaciofluvial delta, are the most clearly outlined landforms of this zone. In contrast to the older oscillatory phases in Estonia (Haanja, Otepää, Sakala) the Palivere is clearly of readvance character, as reflected by the distribution of push moraines and by the fact that the older glaciolacustrine sediments are overlain by glaciofluvial material (Raukas & Rähni, 1966). The Palivere ice-marginal zone has been dated to 11,630 BP (Raukas, 1992a, 1992b; Pirrus & Raukas, 1996), correlated to the ice recession line of 11,965 BP in Sweden (Eriksson, 1992) or dated to about 11,800 varve years BP (Hang & Sandgren, 1996). The composition of the pollen assemblages in the deposits overlying till at Kunda in NE Estonia suggests that the Palivere zone became ice-free no later than the beginning of the Allerød (Pirrus & Raukas 1969; Raukas, 1992b). In southern Finland, south of Salpausselkä I, the correlative of the Palivere Till is possibly the Espoo Till (Salonen & Glückert, 1992). From lake sediments overlying the Pandivere Till south of the Palivere ice-marginal zone in northeastern Estonia some radiocarbon dates have been obtained: Kunda 11,690±150 BP (Ta-194), Loobu 14,725±260 (Ta-138) and 13,970±115 BP (Ta-137) (Ilves *et al.*, 1974; Punning *et al.*, 1968). The dates from Loobu were interpreted as undoubtedly too old as a result of the hard water effect on the organic deposits (Pirrus & Raukas, 1996; Raukas, 1986).

Conclusions

Taking into account all the lithostratigraphical, biostratigraphical and geochronological data available, the existence of two Pleistocene interglacials – Karuküla (Holsteinian) and Prangli (Eemian) and three glacials – Sangaste (Elsterian), Ugandi (Saalian) and Järva (Weichselian) have been recognized. Probably most of the Prangli and all the known Karuküla interglacial deposits in Estonia are found in allochthonous situations. During the various glaciations, the ice-movement directions differed, enabling till units to be correlated on the basis of their lithological composition. According to its position, the lowermost till in the SE Estonian buried valleys (Sangaste Till) is correlated with the Elsterian Glaciation and is probably the oldest till in Estonia. Marginal positions of the Late Weichselian ice sheet are marked in the present topography by interrupted chains of end moraines and glaciofluvial formations. Five ice-marginal zones that formed during the deglaciation have been distinguished in Estonia and four of them indirectly dated (in the scale of ^{14}C years): Haanja (ca 13,500 BP), Otepää (12,800 – 12,600 BP), Pandivere (12,480 – 12,230 BP) and Palivere (11,800 – 11,630 BP).

References

Aboltinš, O., Aseyev, A., Vonsavičius, V., Isachenkov, V., Možhayev, B. & Raukas, A. (1989). Development and use of glacial accumulative insular heights. *Proceedings of the Estonian Academy of Sciences, Geology*, **38 (1)**, 23-33. (in Russian with English summary).

Arslanov, H.A., Velichkevich, F.Y., Kondratienė, O.P. & Krukle, M.J. (1975). Novye dannye o khronologii i paleogeografii sredne-valdaiskogo mezhstadialnogo kompleksa razreza Leyastšiems na reke Gauya [New data on chronology and palaeogeography of Middle-Valdaian interstadial complex of Lejasziems section on the Gauja River]. *Doklady Akademii Nauk SSSR*, **223 (6)**, 1421-1424. (in Russian).

Arslanov, H.A., Breslav, S.L., Zarrina, E.P., Znamenskaja O.M., Krasnov, I.I., Malakhovski, D.B. & Spiridonova, E.A. (1981). Klimat, stratigrafiya i khronologiya srednego valdaya na severo-vostoke i tsentre Russkoi ravniny. [Climate, stratigraphy and chronology of the Middle-Valdai in the North-West and Centre of Russian Plain]. *In*: Velichko, A.A. & Faustova, M.A. (eds) *Pleistotsenovye oledeneniya na Vostochno-Evropeiskoi ravnine.* Moskva, Nauka: 12-27 (in Russian).

Cheremisinova, E.A. (1961). Diatomovye morskikh mezhlednikovykh otlozhenii Estonskoi SSR [Diatoms in the marine interglacial deposits of the Estonian SSR]. *Doklady Akademii Nauk SSSR*, **141 (3)**, 698-700 (in Russian).

Danilans, I. (1966). Pyltsevye zony mindel-risskikh otlozhenii basseina r. Letiža i ikh sopostavlenie s analogichnymi zonami v drugikh raionakh [Pollen zones of the Mindel-Riss aged deposits in the catchment of river Letiža, and their correlation to analogues in other regions]. *In:* Springis, K. (ed.), *Palynologiya v geologicheskikh issledovaniyakh Pribaltiki* [Palynology in geological research in the Baltic Soviet Republics], Riga, Zinatne, 36-44 (in Russian).

Donner, J. (1995). *The Quaternary history of Scandinavia.* Cambridge University Press, 200 pp.

Dreimanis, A. & Zelčs, V. (1995). Pleistocene stratigraphy of Latvia. *In:* Ehlers, J., Kozarski, S. & Gibbard, P. (eds.), *Glacial Deposits in North-East Europe.* Rotterdam, Balkema, 105-113.

Eriksson, J.A. (1992). Natural history of xerotherm vegetation and landscapes on Stora Karlsö, an island in the Western Baltic Basin, Sweden. *Striae*, **35**, 78 pp.

Grewingk, C. (1879). Erläuterungen zur zweiten Ausgabe der geognostischen Karte Liv-, Est- und Kurlands. *Archiv für Naturkunde Liv-, Est- und Kurlands*, **1 (8)**,: 343-466.

Hang, T. (1997). Clay varve chronology in the Eastern Baltic area. *Geografiska Föreningens i Stockholm Förhandlingar*, **119**: 295-300.

Hang, T. & Sandgren, P. (1996). Magnetostratigraphy of varved clays. *In*: Meidla, T., Puura, I., Nemliher, J., Raukas, A. & Saarse, L. (eds), *The Third Baltic Stratigraphicalal Conference. Abstracts. Field Guide*, Tartu, 152-154.

Ilves, E., Liiva. A. & Punning, J.-M. (1974). *Radiouglerodnyi metod i yego primenenie v chetvertichnoi geologii i arheologii Estonii.* [Radiocarbon method of dating and its usage in Quaternary geology and archaeology of Estonia]. Academy of Sciences of Estonia, Tallinn. 231 pp. (in Russian).

Jessen, K. & Milthers, V. (1928). Stratigraphical and palaeontological studies of interglacial freshwater deposits in Jutland and North-West Germany. *Danmarks Geologiske Undersøgelse*, **2, 48**, 1-379.

Kajak, K. (1961). Kvaternaarsete setete Prangli saare tugiprofiil. [Reference profile of the Quaternary deposits from the Island of Prangli]. *VI Eesti Looduseuurijate päeva ettekannete teesid*, Tartu: 20-21. (in Estonian).

Kajak, K. (1995). *Eesti kvaternaarisetete kaart. Mõõtkava 1:2 500 000.* [Map of the Quaternary deposits of Estonia. In scale 1:2 500 000] Eesti Geoloogiakeskus. Tallinn, 26 pp. (in Estonian).

Kajak, K., Kessel, H., Liivrand, E., Pirrus, R., Raukas, A. & Sarv, A. (1976). Mestnaja rabotchaja stratigrafitcheskaja skhema chetvertichnykh otlozhenii Estonii [Local stratigraphical cheme of Quaternary deposits in Estonia]. *In*: Vaitiekunas, P.P. & Gaigalas, A.J. (eds.), *Stratigrafiya chetvertichnykh otlozhenii Pribaltiki* [Stratigraphy of Quaternary deposits in the Baltic countries]. Vilnius, Pergale, 4-52 (in Russian).

Kajak, K. & Liivrand, E. (1967). O nižne- i verhnepleistotsenovykh otlozheniyakh Estonii. [About Lower and Upper Pleistocene deposits of Estonia]. *In:* Goretskij, G.I. & Kriger, N.I. (eds), *Nizhnii Pleistotsen lednikovyh raionov Russkoi ravniny.* Moscow, 149-157 (in Russian).

Kajak, K., Raukas, A. & Hütt, G. (1981). Opyt issledovaniya raznovozrastnykh estonskikh moren pri pomoshchi termoluminestsentnogo metoda. [Experience of investigation of Estonia's different-aged tills with thermoluminescence method]. *In*: Jevzerov, V.J. (ed.), *Geologiya pleistotsena Severo-Zapada SSSR*, 3-11, Apatity (in Russian).

Kadastik, E. & Kalm, V. (1998). Lithostratigraphy of Late-Weichselian tills on West-Estonian Islands. *Bulletin of the Geological Society of Finland*, **70, 1-2**, 5-17.

Karukäpp, R., Raukas, A. & Hyvärinen, H. (1992). Deglaciation of the area. *In*: Raukas, A. & Hyvärinen, H. (eds), *Geology of the Gulf of Finland. Academy of Sciences of Estonia*, Tallinn: 112-136 (in Russian with English summary).

Karukäpp, R. & Miidel, A. (1972). O strojenii i ocobennostjah formirovaniya nekotorykh tipov predfrontalnykh krajevykh obrazovanii. [About geological structure and formation of some genetic types of glacial marginal formations]. *In*: O.Aboltinš (ed.), *Materialy IV Vsesojuznogo mezvedomstvennogo soveštšaniya po izutsheniju krajevykh obrazovanii materikovogo oledeneniya* [Materials of the 4th All-Union conference on investigation of ice-marginal formations of continental glaciations]. Riga, Latvian State University, 86-103 (in Russian).

Kondratienė, O. (ed.) (1993). *Catalogue of Quaternary Stratotypes of the Baltic Region.* Baltic Stratigraphical Association, Vilnius, 56 pp. (in Russian with English summary).

Levkov, E. & Liivrand, E. (1988). On glaciotectonic dislocations of interglacial deposits in the Karuküla and Kõrveküla sections. *Proceedings of the Estonian Academy of Sciences, Geology*, **37 (4)**, 161-167 (in Russian with English summary).

Liivrand, E. (1972). Die palynologische Charakteristik und Korrelation der Interglazialablagerungen des Profils von Karuküla. *Proceedings of the Estonian Academy of Sciences, Geology & Chemistry*, **21 (4)**, 358-367 (in Russian with German summary).

Liivrand, E. (1974). Zur Altersbestimmung der Ablagerungen des Bohrprofils von Valguta nach Pollenanalyse. *Proceedings of the Estonian Academy of Sciences, Geology & Chemistry*, **23 (1)**, 54-58 (in Russian with German summary).

Liivrand, E. (1976). Rebedded pollen and spore in Pleistocene deposits and their role in stratigraphy. *In:* Bartosh, T., Kabailiene, M. & Raukas, A. (eds), *Palynology in continental and marine geologic investigations.* Riga, 166-178 (in Russian with English summary).

Liivrand, E. (1977). The bedding of Mikulian Interglacial deposits in South-East Estonia. *Proceedings of the Estonian Academy of Sciences, Geology & Chemistry*, **26 (4)**, 289-303 (in Russian with English summary).

Liivrand, E. (1984). Biostratigraficheskaya osnova raschleneniya pleistotsenovykh otlozhenii Estonii [Biostratigraphical base of the Pleistocene deposits subdivision in Estonia]. *In*: Kondratienė, O.P. & Mikalauskas, A.P. (eds), *Paleogeografiya i stratigrafiya chetvertichnogo perioda Pribaltiki i sopredelnykh raionov* [Palaeogeography and Stratigraphy of the Quaternary in the Peribaltic and Adjacent Regions]. Academy of Sciences of Lithuania, Vilnius, 136-139 (in Russian).

Liivrand, E. (1990). Type section of Lower- and Middle-Weichselian interstadial deposits at Tõravere (South-East Estonia). *Proceedings of the Estonian Academy of Sciences, Geology*, **39 (1)**, 12-17. (in Russian with English summary).

Liivrand, E. (1991). *Biostratigraphy of the Pleistocene deposits in Estonia and correlations in the Baltic Region.* Stockholm University. Department of Quaternary Research. Report 19, Doctoral Thesis, 114 pp.

Liivrand, E. (1995). South East Estonia as a key area in solution of disagreements in the Pleistocene stratigraphical scheme. *In*: Meidla, T., Jõeleht, A., Kalm, V. & Kirs, J. (eds), *Liivimaa Geoloogia*: 107-112. Tartu. (in Estonian with English summary).

Liivrand, E. & Saarse, L. (1976). Geological and palynological characterization of Upper-Pleistocene deposits in Peedu section. *Proceedings of the Estonian Academy of Sciences, Geology & Chemistry*, **25 (4)**, 334-342 (in Russian with English summary).

Liivrand, E. & Saarse, L. (1983). Mezhlednikovye otlozheniya razreza Kyrvekyla (Yuzhnaya Estonia) i yikh stratigraficheskoje znachenie [Interglacial deposits in Kõrveküla section (South Estonia) and their stratigraphical importance]. *In*: Bartosh, T. (ed.), *Palynologic researches in geologic studies of the Baltic Region and the Baltic Sea.* Riga, 41-50 (in Russian).

Lõokene, E. (1961). Mandrijää servamoodustistest, fluvioglatsiaalsetest setetest ja mandrijää taandumisest Sakala kõrgustiku põhja- ja keskosas.[About glacial marginal formations, glaciofluvial deposits and retreat of the continental glacier from the northern and central part of Sakala Upland]. *In*: Orviku, K. (ed.) *Geoloogiline Kogumik.* Tartu, Eesti NSV Teaduste Akadeemia Loodusuurijate Selts, 84-105 (in Estonian).

Meirons, Z.V. (1986). Stratigrafiya pleistotsenovykh otlozhenii Latvii [Stratigraphy of Pleistocene deposits in Latvia]. *In*: Kondratienė O.P. & Mikalauskas, A.P. (eds), *Issledovaniya lednikovykh obrazovanii Pribaltiki.* Vilnius, 69-81 (in Russian).

Orviku, K. (1939). Rõngu interglatsiaal – esimene interglatsiaalse vanusega organogeensete setete leid Eestist. [Rõngu Interglacial – the first find of the interglacial organogenic deposits in Estonia]. *Eesti Loodus*, **1**, 1-21 (in Estonian).

Orviku, K. (1944). Jäävaheaegade geoloogiast Eestis. [About geology of interglacials in Estonia]. *Eesti Sõna*, **138**, 4 (in Estonian).

Orviku, K. (1956). Stratigraficheskaya skhema antropogenovykh (chetvertichnykh) otlozhenii territorii Estonskoi SSR. [The stratigraphical scheme of the Antropogene (Quaternary) deposits of the Estonian SSR]. *Trudy Instituta geologii Akademii nauk Estonskoi SSR*, **1**, 105-112 (in Russian).

Orviku, K. (1960). Eesti geoloogilisest arengust antropogeenis. [Geological development of the Estonian territory in the Anthropogene]. *Eesti Loodus*, **1**, 6-16 (in Estonian).

Orviku, K. & Pirrus, R. (1965). Mezhmorennye organogennye otlozhenia v Karukyla (Estonskaya SSR) [Intermorainic organogenic deposits of Karuküla (Estonian S.S.R.)]. *In*: Orviku, K.K. (ed.), *Litologiya i stratigrafiya chetvertichnykh otlozhenii Estonii.* Tallinn: 3-21 (in Russian with English summary).

Pirrus, R. & Raukas, A. (1969). O kharaktere i vremeni osvobozhdenia territorii Estonii ot lednikov poslednego oledenenia [About the character and time of deglaciation of Estonia from the glaciers of last glaciation]. *Voprosy chetvertichnoi geologii* [Problems of Quaternary Geology], **4**, 47-57. Riga, Zinatne (in Russian).

Pirrus, R. & Raukas, A. (1996). Late-Glacial stratigraphy in Estonia. *Proceedings of the Estonian Academy of Sciences, Geology*, **45 (1)**, 34-45.

Punning, J.-M. (1970). On possible errors in the determination of sample age by means of radioactive carbon (^{14}C) and on the checking of results. *Proceedings of the Estonian Academy of Sciences, Geology & Chemistry*, **19 (3)**, 238-243 (in Russian with English summary).

Punning, J.-M. & Raukas, A. (1983). The age of tills: Problems and methods. *In*: Evenson, E.B., Schlüchter, Ch., Rabassa, J. (eds), *Tills and Related Deposits*: 357-364. Rotterdam, Balkema.

Punning, J.-M., Liiva, A. & Ilves, E. (1968). Spisok radiouglerodnyh datirovok Instituta Zooloogii i Botaniki Akademii Nauk Estonskoi SSR. [List of radiocarbon dates obtained in the Institute of Zoology and Botany, Academy of Sciences of Estonian SSR]. *Proceedings of the Estonian Academy of Sciences, Biology,* **17 (4)**, 417-426 (in Russian).

Raukas, A. (1978). *Pleistocene deposits of the Estonian SSR*. Tallinn, Valgus Publishers, 310 pp. (in Russian with English summary).

Raukas, A. (1986). Deglaciation of the Gulf of Finland and adjoining areas. *Bulletin of the Geological Society of Finland*, **58 (2)**, 21-33.

Raukas, A. (1991). Eemian interglacial record in the northwestern European part of the Soviet Union. *Quaternary International*, **10-12**, 183-189.

Raukas, A. (1992a). Late- and Postglacial Geological Development and Human Impact in Estonia. *In:* Hackens, T., Lang, V. & Miller, U. (eds.): *Estonia: Nature, Man and Cultural Heritage*, (PACT 37), 23-34, Rixensart, Belgium.

Raukas, A. (1992b). Ice marginal formations of the Palivere zone in the eastern Baltic. *Sveriges Geologiska Undersökning*, **Ca 81**, 277-284.

Raukas, A. (1995). Properties, origin and stratigraphy of Estonian tills. *In*: Ehlers, J., Kozarski, S. & Gibbard, P.L. (eds), *Glacial Deposits in North-East Europe*. Rotterdam, Balkema, 93-101.

Raukas, A. & Gaigalas, A. (1993). Pleistocene glacial deposits along the eastern periphery of the Scandinavian ice sheets – an overview. *Boreas*, **22**, 214-222.

Raukas, A. & Kajak, K. (1995). Quaternary Stratigraphy in Estonia. *Proceedings of the Estonian Academy of Sciences, Geology*, **44 (3)**,149-162.

Raukas, A. & Kajak, K. (1997a). Ice ages. *In*: A.Raukas, A. Teedumäe (eds.), *Geology and Mineral Resources of Estonia*. Tallinn, Estonian Academy Publishers, 256-262.

Raukas, A. & Kajak, K. (1997b). Quaternary cover. *In*: Raukas, A.& Teedumäe, A. (eds.), *Geology and Mineral Resources of Estonia*. Tallinn, Estonian Academy Publishers, 125-136.

Raukas, A. & Karukäpp, R. (1979). Eesti liustikutekkeliste akumulatiivsete saarkõrgustike ehitus ja kujunemine. [Geology and formation of the glacial accumulative insular heights of Estonia]. *In*: Raukas, A. (ed.) *Eesti NSV saarkõrgustike ja järvenõgude kujunemine*. Tallinn, Valgus, 9-28 (in Estonian).

Raukas, A. & Liivrand, E. (1971). Pleistocene deposits in the sequence of Vääna-Jõesuu borehole (North Estonia). *Proceedings of the Estonian Academy of Sciences, Geology & Chemistry*, **20 (1)**, 60-72 (in Russian with English summary).

Raukas, A., Liivrand, E. & Kajak, K. (1993). Quaternary stratotypes of Estonia (in Russian with English summary). *In*: Kondratienė, O. (ed.), *Catalogue of Quaternary stratotypes of Baltic region*, Vilnius: 42-53.

Raukas, A. & Rähni, E. (1966). K voprosu ob otstupaniy materikovogo l'da poslednego oledeneniya s territorii Estonskoi SSR [On the question about deglaciation history of the last continental glaciation on the territory of Estonian SSR]. *In*: Kivila, H. (ed.), *Eesti Geograafia Seltsi aastaraamat 1964/1965*. Tallinn, Valgus, 5-17. (in Russian).

Raukas, A., Rähni, E. & Miidel, A. (1971). *Marginal glacial formations in North Estonia*. Tallinn, Valgus, 226 pp. (in Russian with English summary).

Salonen, V.-P. & Glückert, G. (1992). Late-Weichselian glacial activity and sediments in southwestern Finland. *Sveriges Geologiska Undersökning*, **Ca 81**, 313-318.

Serebrjannyi, L. & Raukas, A. (1966). Transbaltiiskie korreljatsii kraevykh lednikovykh obrazovanii pozdnego pleistotsena [Trans-Baltic correlations of ice-marginal formations of Late Pleistocene]. *In*: Gritchuk, V., Ivanova, I., Kind, N. & Ravskij, E. (eds), *Upper Pleisto-cene. Stratigraphy and Chronology*. Moscow, Nauka,12-28 (in Russian).

Znamenskaya, O.M. & Cheremisinova, E.A. (1962). Rasprostraneniye mginskogo mezhlednikovogo morya i osnovnye cherty paleogeografii [The distribution of the Mga interglacial sea and its main palaeogeographical characteristics]. *In:* Tokareva, T.N. (ed.) *Voprosy stratigrafii chetvertichnykh otlozhenii severo-zapada evropeiskoi chasti SSSR*. Leningrad, 140-160 (in Russian).

Tammekann, A. (1938). Die Endmoränen in Estland. *Comptes Rendus du Congrés International de Géographie Amsterdam. Union Géographique Internationale,* 251-256. Leyden.

Tavast, E. & Raukas, A. (1982). *The bedrock relief of Estonia*. Tallinn, Valgus, 192 pp. (in Russian with English summary).

Thomson, P. (1939). Eem-interglatsiaali metsade ajalugu Eestis ning Rõngu interglatsiaali stratigraafiline asend. [History of the forest development in Estonia during the Eemian interglacial and the stratigraphical position of the Rõngu interglacial]. *Eesti Loodus*, **1**, 21-24 (in Estonian).

Thomson, P. (1941). Die Klima- und Waldentwicklung des von K.Orviku entdeckten Interglazials von Ringen bei Dorpat (Estland). *Zeitschrift der Deutschen Geologischen Gesellschaft*, **93 (6)**, 274-281.

Velichkevich, F. & Liivrand, E. (1976). New investigations of the interglacial flora from the Karuküla section in Estonia. *Proceedings of the Estonian Academy of Sciences, Geology & Chemistry,* **25**, 215-221 (in Russian with English summary).

Velichkevich, F. & Liivrand, E. (1984). O vozraste Karukyulaskikh otlozhenii [On the age of Karuküla deposits]. *In:* Kondratienė, O.P. & Mikalauskas, A.P. (eds), *Paleogeografiya i stratigrafiya chetvertichnogo perioda Pribaltiki i sosednikh raionov.* Vilnius: 140-148 (in Russian).

Quaternary Glaciations – Extent and Chronology
Editors J. Ehlers and P.L. Gibbard

Glaciation of Finland

Juha Pekka Lunkka [1], Peter Johansson [2], Matti Saarnisto [1]
and Olli Sallasmaa [1]

[1]*Geological Survey of Finland, P. O. Box 96, FIN-02151 Espoo, Finland*
[2]*Geological Survey of Finland, P. O. Box 77, FIN-96101 Rovaniemi, Finland*

1. A brief glaciation history of Finland

1.1 Pre-Weichselian cold stages

The Scandinavian Ice Sheet (SIS), the centre of which was situated in the Scandinavian mountain range, covered Finland and the NW Russian Plain several times during the Quaternary cold stages. It is not exactly known how many times Finland and adjacent areas were covered by ice during the Quaternary. This is because the area is situated close to the glaciation centre and each ice advance eroded most of the previously deposited interglacial and glacial sediments. In most cases therefore only the sediments deposited during the last cold stage (Weichselian) rest on the Pre-Cambrian bedrock. Except for some scattered remnants of Saalian-age esker ridges in the ice-divide zone of northern Finland (Finnish Lapland), there are no distinct geomorphological landforms related to pre-Weichselian glaciations. However, there are a number of sites where pre-Weichselian organic and glacial sediments have been preserved, particularly in central Lapland and western Finland. These sites provide the basis for the Quaternary stratigraphy of Finland.

According to the Finnish till stratigraphy, there are six, stratigraphically-significant till beds in Finnish Lapland. The uppermost three are thought to represent Weichselian tills (Till Beds I-III). The so-called Till Bed IV was laid down during the Saalian glaciation, and the two lowermost till beds, that underlie a Holsteinian peat stratum (Hirvas & Eriksson, 1988) may represent Elsterian or pre-Elsterian tills (*cf.* Hirvas & Nenonen, 1987, Hirvas, 1991). Although Elsterian or pre-Elsterian tills are preserved at scattered localities in northern Finland, there is no conclusive evidence for pre-Saalian tills in southern Finland. On the other hand, Saalian glacigenic sediments are well preserved in the Pohjanmaa (Ostrobothnia) area in southwestern Finland. In that area, Eemian interglacial deposits are underlain by Saalian till and glaciofluvial sediments at several sites (*cf.* Eriksson *et al.*, 1980, Forsström *et al.*, 1988, Hirvas & Niemelä, 1986, Aalto *et al.*, 1989, Gibbard *et al.*, 1989, Saarnisto & Salonen, 1995, Nenonen, 1995). There is only one site south of the Salpausselkä end moraine zone in Helsinki where three till beds have been encountered (Hirvas *et al.*, 1995). The lowermost of these might have been laid down prior to the Weichselian Stage.

1.2 Weichselian cold stage

Southern and northern Finland experienced a rather different glacial history during the Weichselian and therefore these areas are dealt with separately in this brief overview. Basically, the main difference between these two areas is that northern Finland was already covered by the Scandinavian Ice Sheet during the Early Weichselian while there is no evidence of Early Weichselian ice cover in the southern part of Finland at this time. This is probably a result of the development of two separate ice-dome areas in the Scandinavian Mountains that behaved semi-independently in space and time during the Weichselian. There is also evidence that the behaviour of the ice-streams over northern and southern Finland differed during the late Middle Weichselian (*cf.* Lunkka *et al.* 2001, Svendsen *et al.* 2001).

1.2.1 Early and Middle Weichselian substages in the southern part of Finland

In southern Finland there are no end-moraines related to the Early or Middle Weichselian ice advances. Information on the extent of the ice streams is limited and based on the distribution of interstadial organic remnants and till stratigraphy.

It is generally assumed that after the Eemian Interglacial ice began building up in the Scandinavian Mountains and spread into the adjacent areas twice during the Early Weichselian stadials. However, during the Early Weichselian interstadials (Brørup, MIS 5c and Odderade, MIS 5a), the ice melted completely even in the Scandinavian Mountains (*cf.* Anderson & Mangerud, 1990, Saarnisto & Salonen, 1995, Donner, 1995). Lithostratigraphical and biostratigraphical evidence, combined with a number of optically-stimulated luminescence (OSL) dates, suggest that the Scandinavian Ice Sheet (SIS) did not cover the western and southern parts of Finland in the Early Weichselian (Hütt *et al.*, 1993, Saarnisto & Salonen, 1995, Nenonen, 1995). It is suggested that the maximum ice extent in the Early Weichselian was reached during Marine Isotope Substage (MIS) 5d (*cf.* Lundqvist, 1992, Donner, 1995) and at that time ice covered areas north of Oulu (for location see Fig. 1), including the present coast of the northern Gulf of Bothnia(Saarnisto & Salonen, 1995). It has

Sollid, J.L., Andersen, S., Hamre, M., Kjeldsen, O., Salvigsen, O., Sturød, S., Tveitå, T. & Wilhelmsen, A. (1973). Deglaciation of Finnmark, North Norway. *Norsk Geografisk Tidskrift,* **27**, 233-325.

Strömberg, B. (1990). A connection between the clay varve chronologies in Sweden and Finland. *Annales Academiae Scientiarum Fennicae,* **A III 154**, 32 pp.

Sutinen, R. (1984). On the glacial stratigraphy in Pudasjärvi area, Peräpohjola. *Striae,* **20**, 91-94.

Sutinen, R. (1992). Glacial deposits, their electrical properties and surveying by image interpretation and ground penetrating radar. *Geological Survey of Finland, Bulletin,* **359**, 123 pp.

Svendsen, J.I., Astakhov, V.I., Demidov, I., Bolshiyanov, D.Yu., Dowdeswell, J.A., Gataullin, V., Grosfjeld, K., Henriksen, M. Hermichen, W.-D., Hjort, C., Hubberten, H.W., Larsen, E., Lunkka, J. P., Lyså, A., Mangerud, J., Maslenikova, O., Matioushkov, A., Mellers, M., Möller, P., Saarnisto, M., Siegert, M.J., & Wisher, F. (2001). The Weichselian Glaciations in the Eurasian Arctic. *In:* Svendsen (ed.), *Eurasian Ice Sheets - Final Report for the EU. Environment and Climate Research Programme (Contract No ENV4-CT97-0563). Part B - a Scientific Report-a Synthesis*, 63-93. European Commission.

Tanner, V. (1915). Studier öfver kvartärsystemet i Fennoskandias nordliga delar III. (Résumé en français: Études sur le système quaternaire dans les parties septentrionales de la Fennoscandie). *Bulletin de la Commission géologique de Finlande,* **38**, 815 pp.

Ukkonen, P., Lunkka, J.P., Jungner, H. & Donner, J. (1999). New radiocarbon dates from Finnish mammoths indicating large ice-free areas in Fennoscandia during the Middle Weichselian. *Journal of Quaternary Science*, **14**, 711-714.

Quaternary Glaciations – Extent and Chronology
Editors J. Ehlers and P.L. Gibbard

The palaeogeography of the last two glacial episodes in France: the Alps and Jura

Jean-Francois Buoncristiani and Michel Campy

Centre des Sciences de la Terre, 6, Bd gabriel
F-21000 Dijon, France
E-mail: Jean-Francois.Buoncristiani@u-bourgogne.fr

Today of the mountainous massifs in France only the Alps are glaciated. They support glaciers of 600 km^2 (Vivian, 1975). However, both the Jura and the Alps were covered by major glaciers during the last two cold periods of the Quaternary. The maps presented here show the tentative limits of the last two glaciations derived both from evaluating the literature and field investigations. As a result of the pre-existing morphology two different types of glacier developed in the two different mountainous regions (Alps and Jura). The Jura has a massif morphology only slightly dissected by valleys, only supported an ice-caps which are represented by abundant morainic deposits. This contrasts with the situation in the Alps which are transversed by numerous deep valleys (e.g. Rhône and the Arve in the North, the Durance and the Var in the South) which allowed the development of large valley glaciers.

Nowhere in the Alps have Quaternary deposits been completely preserved. Therefore the basic stratigraphic subdivision is combined from various regional stratigraphies, though in many cases the determination is difficult because age control is lacking. Indications of Early Quaternary cold stages have long since been reported from the Italian and French Alps. In both areas an abrupt change from fine-grained sedimentation to the deposition of coarse gravel and conglomerates is manifested in what was regarded as the 'Upper Pliocene'. However, the evidence is ambiguous. The change in sedimentary environment might either be explained by tectonic causes or by climatic changes (Billard & Orombelli, 1986).

On the Chambaran Plateau west of Grenoble, in a sequence of strata thought to be older than 1.6 million years, a clayey sediment with striated boulders has been found. Bourdier (1961) could not determine with certainty whether this diamicton represented a slope deposit or a till. On the western Chambaran Plateau the coarse gravels are overlain by strongly cemented loess layers that contain, apart from a fossil soil, a rich mammal fauna (Viret 1954). The faunal composition suggests that the age of the loess is ca. 2.2 million years (Guérin, 1980).

1. The Alpine chain and the Jura

Since Penck & Brückner (1909) two morainic complexes have been recognised in the marginal zone around the Alpine chain and the Jura mountains. These sedimentary complexes form a zone about 400 km long extending from south of Grenoble to the North of the Jura (Fig. 1) (Bourdier, 1961; Monjuvent, 1978; Campy, 1982; Mandier, 1984; Monjuvent, 1984):

The sedimentary record of the glacial advances is discontinuous. The complexity of the readvance phases caused only the most extensive to be preserved, and there are problems to date these phases (cf. 3.1 Chronology). Thus it was decided to present the palaeogeography that corresponds to the maximum glaciation at each stage.

- the 'External Moraine Complex' (CME) indicates the maximum glacial extension (MEG) of Pleistocene glaciation to the west and northwest where it reaches the western margin of the Jura, the Lyon region and where it covers the region from the Dombes between Bourg-en-Bresse and Lyon. It has been attributed to the last-but-one glaciation, (i.e. the Riss *s.l.* after Penck & Brückner, 1909)
- the 'Internal Moraine Complex' (CMI = last glacial maximum, LGM) can be traced 10 to 40 km inward of the preceding limits. It has been correlated with the last maximum glaciation (i.e. Würmian after Penck & Brückner, 1909).

The original interpretation was that the glacial advances that laid down the deposits of these two complexes were both derived from inside the Alps. This interpretation is evident for the southern half of the region (as far as approximately the latitude of Geneva) because in this zone there is no mountainous massif separating the Alps from the piedmont. However, further north, the situation is more complex because the Jura formed an obstacle to the advance of the Alpine glaciers.

Until the 1980s, most authors, e.g. Tricart (1961, 1965) and Jäckli (1962, 1970), considered that also the two morainic complexes present west of the Jura were emplaced under the dominant influence of Alpine glaciers. This hypothesis was based upon the occurrence of erratic boulders of Alpine origin (granite, gneiss etc.) in certain morainic deposits in the Jura. It was also thought that the Jura mountains were too small to have supported their own ice-cap. Thus it was thought that the Alpine glaciers overrode the entire Swiss plain, occupied the Lake Geneva basin and penetrated the Jura through depressions on their eastern side.

This question has been partly re-examined in the course of precise mapping and detailed investigation of the glacial deposits on the margin of the Jura massif: The morainic complexes of the western slopes (Campy, 1982; 1992), the glacial deposits of the Geneva Basin and the eastern slope (Arn, 1984; Campy & Arn, 1991) and the moraines south of the Jura (Sbaï, 1986; Monjuvent, 1988). The problem of the Internal Moraine Complex shall be discussed first, followed by the External Moraine Complex.

1.1 The Internal Moraine Complex (CMI)

1.1.1 Mapping and petrography

From south of the Alps to the extreme northeast of the Jura chain, the CMI has been traced with great precision for over 400 km. It comprises a series of frontal moraines, particularly north of Grenoble. The limit of glacier deposits marking the maximum position of the associated glacial advance is clearly defined. In the southern Alps the morainic deposits in the middle Durance valleybelonging to the CMI, have yielded two samples of fossil wood that have deen dated using ^{14}C: 18,600 ±200 B.P. (LY 6338) and 17,680 ±180 B.P. (LY 6387). These indicate that the CMI dates to the Weichselien Upper Pleniglacial (MIS2) in this region (Jorda *et al.*, 2000).

The petrography of these deposits differs between the northern and southern zones. From the extreme north to the Cluse des Hôpitaux, the deposits only comprise material derived from the Jura massif (Jurassic and Cretaceous limestones). By contrast, from the Cluse des Hôpitaux further south the petrography is much more varied (Mandier, 1984) and includes material of alpine origin (granite, gneiss, greenstones etc.). It seems that two ice advances from different sources were supplying the CMI: one solely from the Jura on the western side of the Jura massif, and the other of Alpine origin at the southern side of the Jura.

1.1.2 Glacial deposits on the eastern flank of the Jura

Two till types are present in this region. The younger till contains only Jurassic limestone material transported by the glacial advances from the west. The older till contains mostly Alpine material (Helvetic and Briançon limestones, granite, gneiss and micaschist), brought in by ice advances from the east. The two formations were mapped by Arn (1984), whose petrographical investigations helped to clarify the regional Quaternary stratigraphy.

An example section

The synthetic section from the Nozon valley (Arn & Aubert, 1984) shows the relations of the two complexes (Fig. 2). Resting on tectonised Upper Jurassic and Cretaceous bedrock, the glacial formations reach from an altitude of 650 to 1000 m. In stratigraphical order from the base upwards four main formations can be seen:

- First horizontally-bedded clayey silts with rare stones which are interpreted as glaciolacustrine sediments.
- Then a basal till of essentially Alpine material, very thick in the central part of the section (over 50 m) and thinning up-valley.
- In the upper part of the section from 950 - 1000 m this is overlain by a basal till of exclusively Jura material, 20 m thick, ending in a small morainic ridge at Plan de la Sagne.
- A number of terraces composed of mixed Jura and Alpine material spread between 950 - 700 m a.s.l. They are interpreted as ice-contact landforms (kame terraces) emplaced during the progressive retreat of the Alpine glacier (Arn, 1984).

This sequence is repeatedly found on the eastern side of the Jura. It demonstrates that at the time of LGM the two ice flows were in contact: Alpine ice occupied the Swiss plain up to a certain height (see below). A second ice flow from the Jura brought materials that were deposited onto the Alpine sediments. However, the Jura ice did not pass beyond the foot of the Jura chain. The stratigraphical relationship of the two till types shows that the Jura mountains did indeed support their own ice-cap during the LGM, confirming the views of Agassiz (1843), Nussbaum & Gygax (1935) and Aubert (1965). The Jura glaciation was independent of the Alpine glaciation.

1.1.3 The Alpine glacier at the LGM maximum in SW Switzerland

The configuration of the Alpine glacier in the Geneva Basin at the LGM was investigated by Swiss geologists since the end of the 19th century. The size and surface level of the ice were clearly determined by Jäckli (1962) and by his map *'la Suisse pendant la derniére période glaciaire'* (Switzerland during the last glacial period) (Jäckli, 1970). His interpretation is still valid today. The maximum height reached by the Alpine ice at its contact with the Jura during the LGM was about 1200 m a.s.l.. It descended gradually from this maximum to *c.* 400 m a.s.l. at the terminal moraines in the Rhône valley in the southern Jura. In the north and northeast the ice thinned towards the terminal moraines in the Soleure region to an altitude of about 600 m a.s.l.

The level of the Alpine ice surface only very rarely exceeded that of the Jura massif. Only the valleys were lower than the ice surface. However, their entrance was blocked by morainic deposits which prevented the Alpine ice from penetrating the Jura. It was never able to gain sufficient strength to supply sediment to the CMI on the western side of the Jura.

1.2 The External Moraine Complex (CME)

The CME is recognised all around the peripheral Alpine piedmont (Penck & Brückner, 1909). In the NW of the Alps

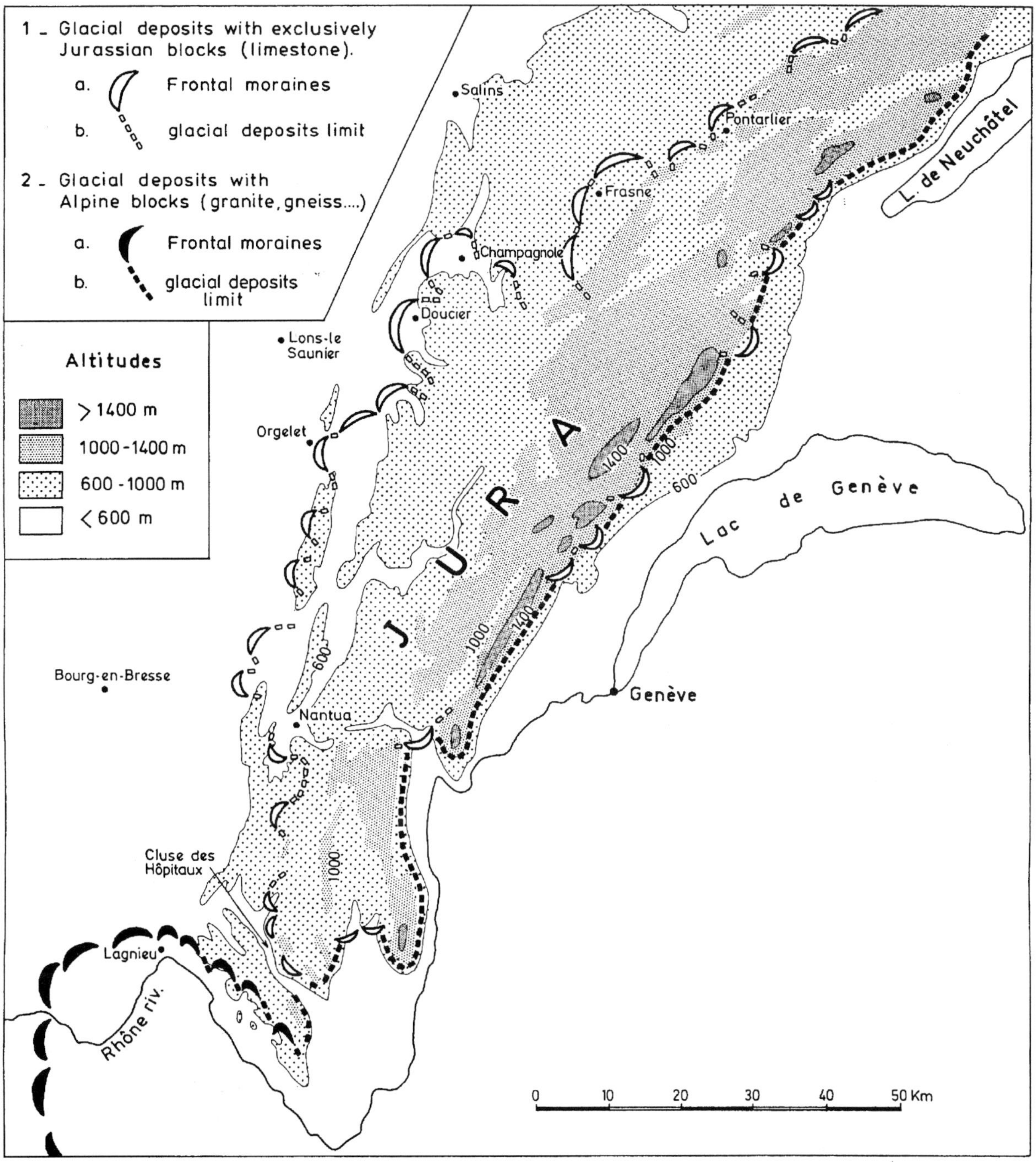

Fig. 1. Glacial deposits from the Alps and from the Jura: Internal Morainic Complexes (CMI).

(Fig. 1) it is represented by the lines of end moraines some 10 - 40 km west of the CMI. Because of their older age they had been more strongly eroded. Consequently, the features are more poorly preserved, than the CMI. However, the deposits are sufficiently well preserved to indicate a maximum glacial limit for the contemporaneous ice sheet.

1.2.1 Mapping

The CME can be traced over a distance of 200 km from the northern Jura to the Lyon region (Fig. 3).

Between Ornans in the North and Bourg-en-Bresse in the South, it follows the western slopes of the Jura at altitudes of around 500 m a.s.l.. The main deposits of these moraines are basal tills, ablation tills and glaciolacustrine deltas identified from outcrops and borehole information (Campy, 1982). The limit of these glacial deposits occurs as lobes developed towards the west opposite the main outer Jura valleys, in the Ornans, Salins, Poligny, Voiteur and Lons-le-Saunier regions.

In the extreme north of the Jura, outcrops are much less common so the deposits' limit is much less precise. In keeping with the work here by Hantke (1978) it can be

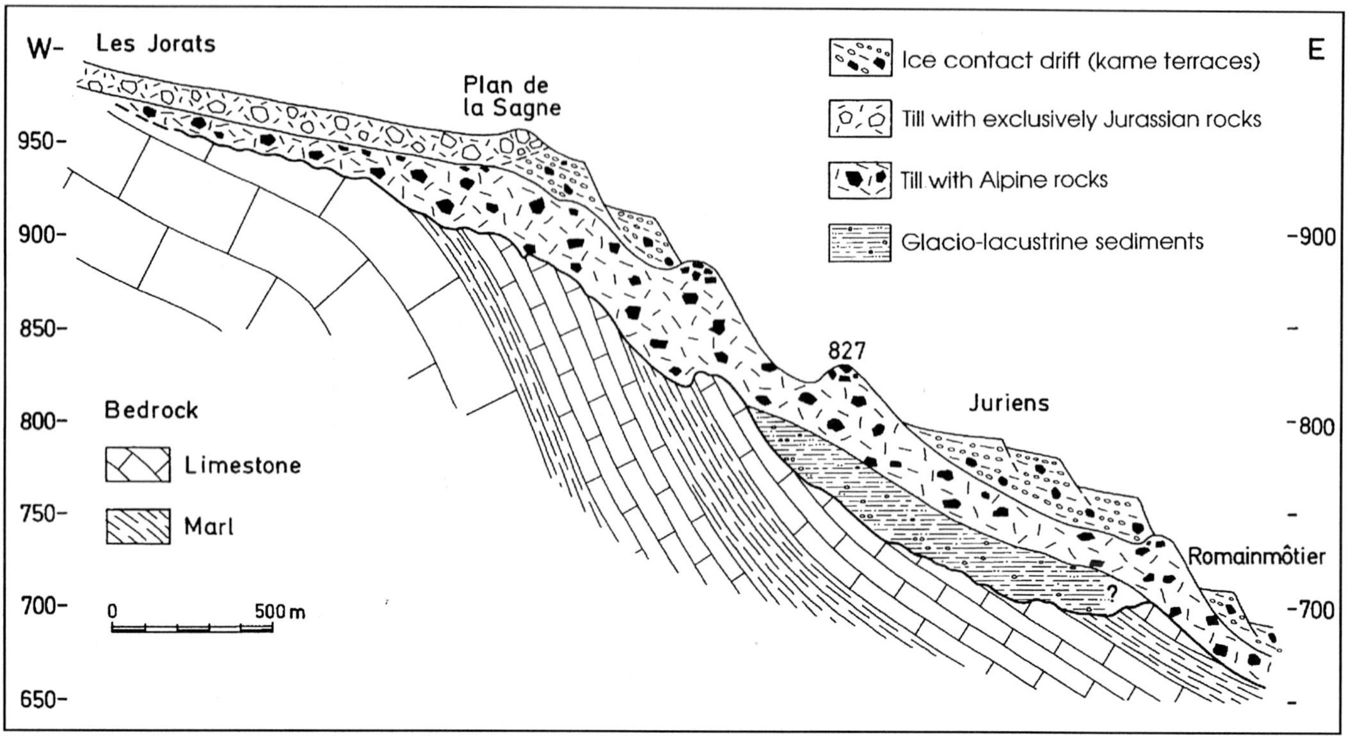

Fig. 2. Relationship between Jura and Alpine glacial deposits along the vallée du Nozon.

shown that the ice limit occurred at heights of about 800 m, bending towards the east.

In the central area, between Bourg-en-Bresse and Lyon, the CME glacial deposits are very common. There, they have been mapped and recognised by many previous authors, e.g. Falsan & Chantre (1879), Delafond & Deperet (1893), Penck & Brückner (1909) and classically termed *Glaciaire de la Dombes*. They consist of sediment several metres thick overlying Pliocene alluvium of the River Bresse (Fleury & Monjuvent, 1984). Recent studies (Mandier, 1984; Monjuvent, 1984, Fleury & Monjuvent, 1984) have shown that these deposits were laid down by a vast ice lobe from the Alps. It is possible that the CME was formed during several glacial episodes (Billard & Derbyshire, 1985), but for the most part the deposits date from the period immediately before Eemian Interglacial (de Beaulieu & Reille, 1984, 1989). It therefore seems that the classic '*Complexe des moraines externes*' is of Middle Pleistocene (= Rissian *s.l.*) age.

In the south, the Durance glacier advanced as far as Sisteron (Tiercelin, 1974). In the *Alpes provençales* the glaciers failed to leave very well developed frontal morainic systems. The main valleys were glaciated, however. For example, traces of glaciation are found in the valleys of the Var, the Bléone and the Verdon.

1.2.2 Petrography

The petrography of the CME deposits varies from north to south.

- In the North, in the Ornans and Pontarlier regions, a dozen oucrops have been studied (Campy, 1982). They expose basal till, ablation till and glaciolacustrine delta deposits. They all contain limestone material from the Jura (40 to 90%), associated with boulders of Alpine origin (10 to 60%).
- In the central zone between Salins and Bourg-en-Bresse, over a distance of about 100 km the CME contains only material derived from the Jura (Campy, 1982).
- South of Bourg-en-Bresse, the Dombes glacial deposits contain Jura limestone material with or without an admixture of Alpine material (20 to 60%) (Mandier, 1984; Fleury & Monjuvent, 1984).

The results of field mapping and the petrography of the CME from the northern Jura to the Lyon region is shown schematically in Fig. 3. It shows several features: 1. In the northern and southern zones, where Alpine ice deposited the CME, it must have crossed the Jura. 2. The central Jura, between Salins in the North and Bourg-en-Bresse in the South was not overridden by Alpine ice. 3. However, the occurrence of clear frontal moraines in this zone demonstrates that glaciers were present here. These glaciers must have been local glaciers from the Jura.

The topography of the Jura does explain this separation of glacier flows (Fig. 3): In the central zone of the Jura the eastern slope rises to a height of 1250 m which must have formed a barrier to the Alpine glaciers. However, in the southern and northern zones the altitudes are lower. Some peaks are over 1250 m high, but there are many cols

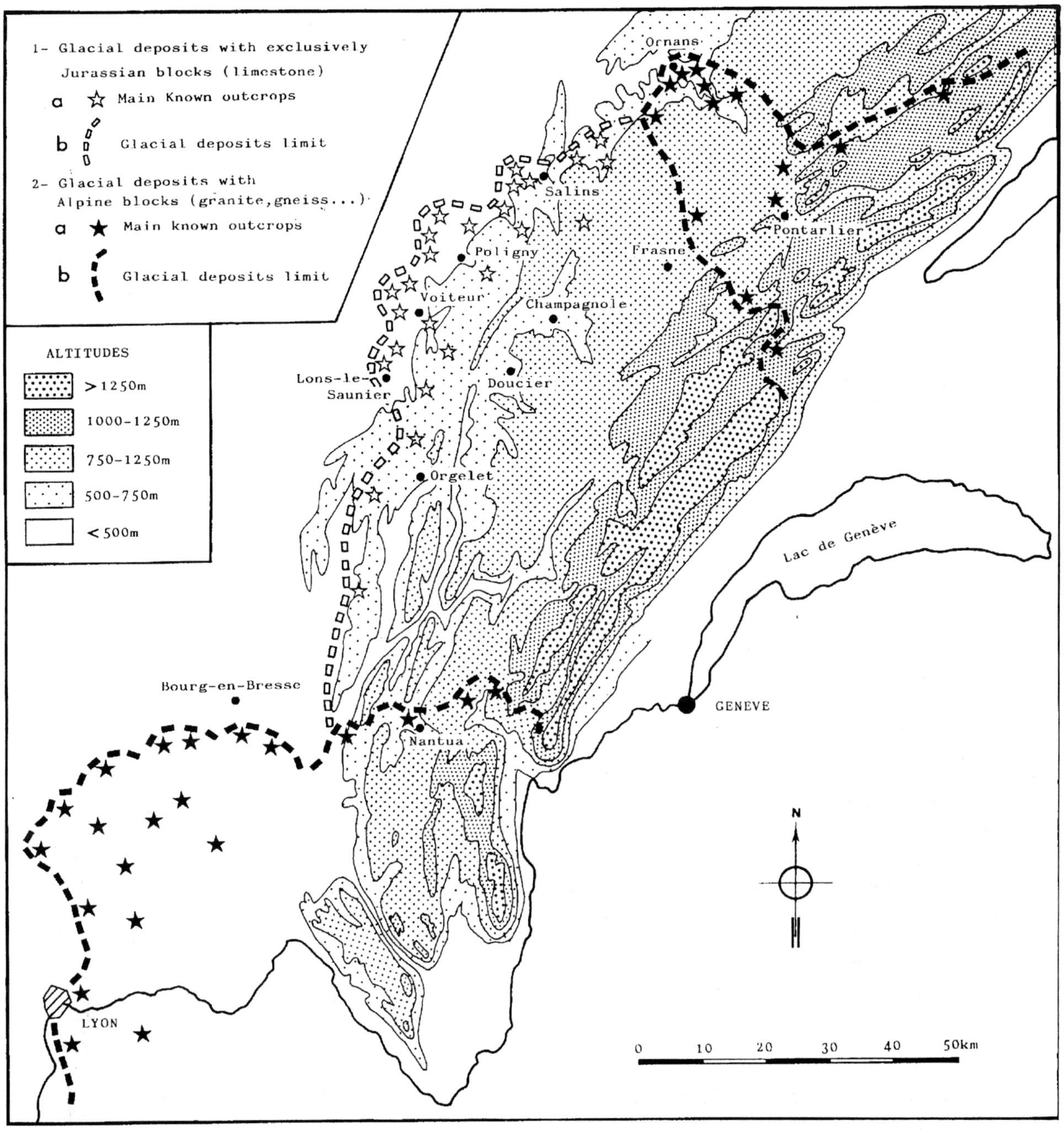

Fig. 3. Glacial deposits from the Alps and from the Jura: External Morainic Complexes (CME).

between them at altitudes of less than 1000 m. In these zones, the Jura would not represent a significant barrier to the advance of the Alpine ice.

1.3 Palaeogeographical reconstruction of the last two glaciations in the NW Alps

1.3.1 Emplacement of the Internal Moraine Complex (CMI)

It is consisted a series of terminal moraines well individualized especially in the north of Grenoble. In the Alps of North all the valleys and transfluences were filled by glaciers which received flow from local glaciers, forming a continuous network. At the glacial maximum, the ice flows from the central Alpine zone took the form of a vast piedmont glacier when it entered the Swiss plain. This lobe collided with the Jura at the latitude of Lake Geneva and was forced to flow towards both to the north and south. The southern glacial front stabilised as a piedmont lobe about 20 km from Lyon.

At the maximum advance of the LGM, the Alpine ice did not enter the Jura mountains, which were covered by a local ice-cap (Fig. 4a). The assumption of a discrete Jura ice-cap is based on the following evidence:

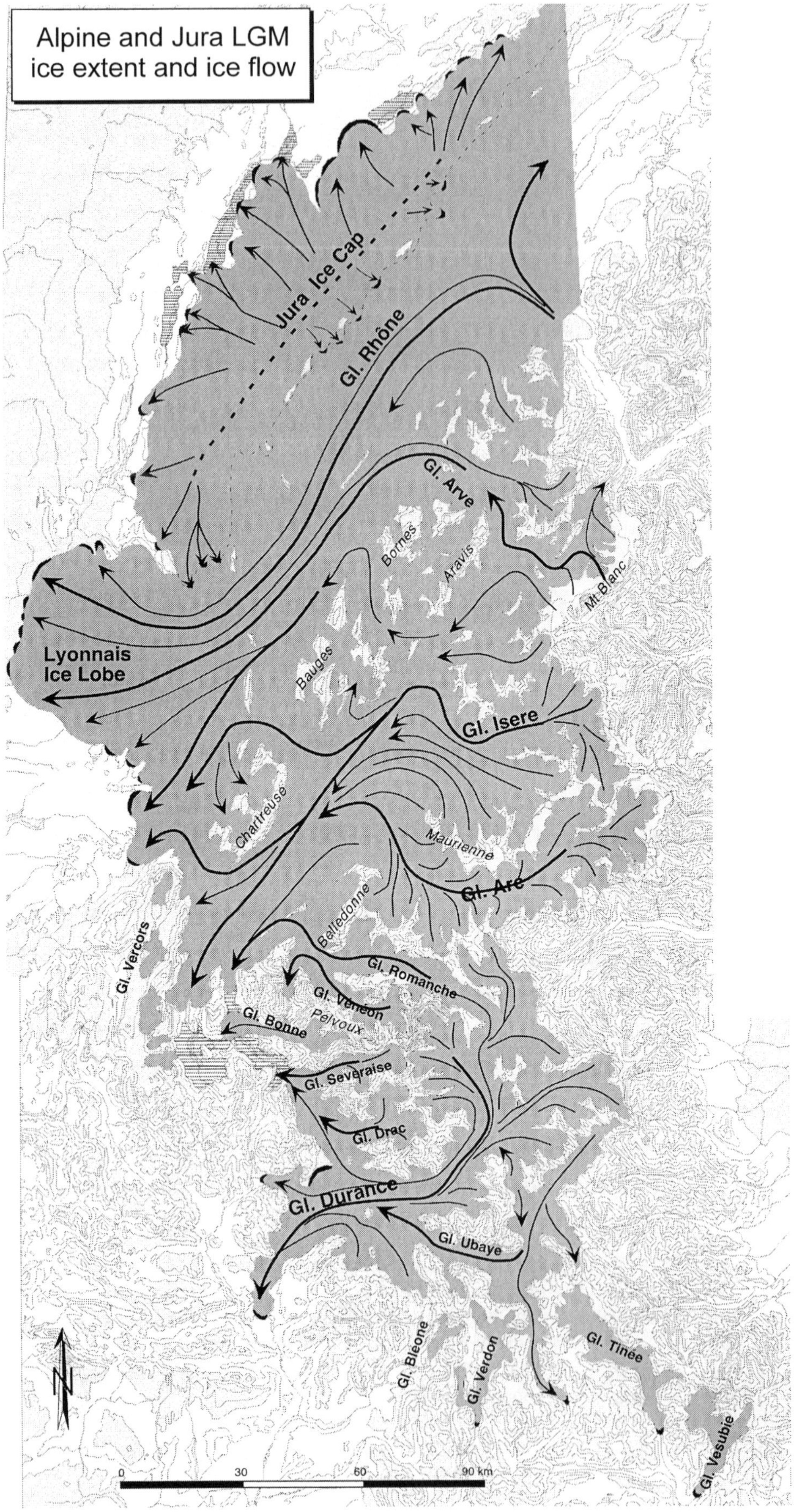

Fig. 4. Alpine and Jura ice sheet

a - Alpine and Jura ice sheet extent and ice flow directions during the Last Glacial Maximum (LGM).

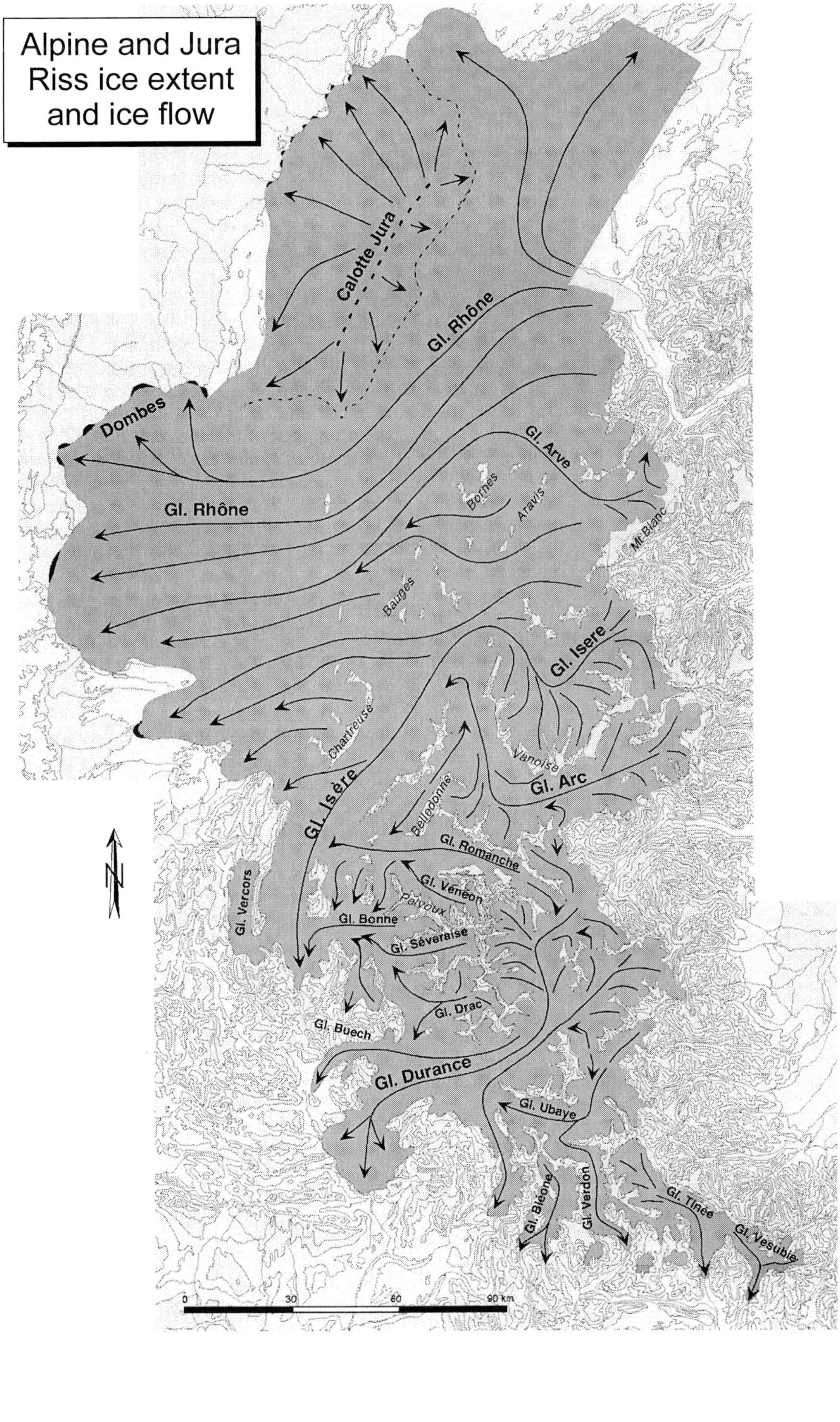

b - Alpine and Jura ice sheet extent and ice flow directions during the MGE Stage

1. The presence of exclusively local material in the morainic complex on the western slope of the Jura (Fig. 3) demonstrates that the associated ice flow originated only in the Jura. This morainic complex is particularly developed in the central zone of the western slopes right behind the highest parts of the Jura massif. This is the region where the Alpine ice would have met the greatest obstacle to crossing the mountains.
2. Next, the stratigraphical relationships of the Alpine and Jura tills at the eastern border of the mountains (Figs 1 and 3) clearly shows that two opposing ice flows occurred in this zone.
3. The surface level of the ice that overrode the Swiss Basin during the glacial maximum did not allow it to advance far into the Jura massif even though the Jura was not buried by ice at the same time.

It must be considered, however, that the general form of the Jura ice-cap shown in Fig. 4a is more certain in the central zone than further to the north and south. In the latter regions, the moraines are not sufficiently clear to allow a reliable reconstruction but there is no evidence conflicting with the general outline shown in the maps (Campy, 1982; Sbaï, 1986).

In the Isère valley the CMI correspond to the Bank moraines which result from the confluent glacier of l'Isère glacier and Arc glacier. While progressing towards the South (Fig. 4a), the valleys become gradually ice field and we find only little glacier glacier (Bléone, the Verdon, Var).

1.3.2. Emplacement of the External Moraine Complex (CME)

Reconstruction of the CME is more difficult than for the CMI complex because the External Moraine Complex is more eroded and degraded. However, the available information allows a coherent reconstruction of the MGE that formed the CME deposits (Fig. 4b).

The external morainic front passes further to the west than the Internal Morainic Complex. This suggests that the MGE ice thickness was much greater. In the Geneva basin the upper limit of the MGE ice was indeed *c.* 200 m higher than during the LGM (Monjuvent, 1984). Therefore the equilibrium line must have been much lower than during the LGM.

As a result of this greater ice thickness the glacier was able to partially override the Jura in the relatively lower areas, i.e. to the north towards the Ornans and in the south as far as Bourg-en-Bresse as far as Lyon. However, in the central part of the Jura the petrography of the CME deposits indicates that this sector had only been occupied by Jura ice (Fig. 3). By comparison to the LGM Jura ice-cap, the authors think that the level reached by the MEG Jura ice-cap was c. 200 m higher and that it terminated at 2000 m altitude in the central part.

2. Discussion and Conclusions

2.1 Chronology

The palaeogeographic reconstruction shown in the maps is largely based on the External and Internal Moraine Complexes (CME and CMI) (cf. Penck & Brückner, 1909). However, each morainic complex corresponds to several glacial advances out of the mountain massifs, and has been rearranged each time in the course of numerous climatic fluctuations. Consequently, the maps represent a synthesis of various climatic oscillations during the cold periods. Also, it must be taken into account that the maxima of the glacial advances may not neccessarily have been synchronous in all the valleys. The CME can be attributed to the last-but-one glaciation (MEG *s.l.*) whilst the CMI is equivalent to the last glaciation maximum (LGM).

Because of the 'freshness' of the deposits and the large number of exposures available, identification of the LGM ice advance is easily achieved. On the basis of the ocean core sediments and the ice-core sequences it is now known that the last cold stage includes two major glacial advances; the first during the Early Pleniglacial (60,000 B.P. = MIS 4) and the second in the Late Pleniglacial (20,000 B.P. = MIS 2). The problem of the apparent diachroneity of the maximum ice advance is still unsolved. In the Jura it occurred at 20,000 B.P. (Campy & Richard, 1988), in the Rhône Glacier region of Lyon prior to 26,000 B.P., with a readvance in the northern Alps at 20,000 B.P. (Monjuvent & Nicoud, 1988), and in the southern Alps it is also dated to 20,000 B.P. (Jorda *et al.*, 2000).

Initially, some authors correlated the Rhône glacier maximum with MIS 4, based only on comparison with the ocean isotope curve (Monjuvent & Nicoud, 1988; Blavoux, 1988). However, resolution of the oceanic signal is not sufficiently fine enough to distinguish the weak influence of the Alpine Glaciation in the geochemistry of the oceans. It is therefore preferable to correlate with the temperature curve from the Greenland ice-cap (Grootes *et al*, 1993). Regarding the dates available for the last glacial maximum of the principle Alpine glaciers (Rhône, Linth and Rhine) two series of results available give different ages. Therefore there are two solutions possible: a long record with the LGM before 27,000 cal B.P. and a short solution with the LGM near 22,000 cal B.P. (Schoeneich, 1998). The shorter variant is in accord with the chronology proposed for the Jura by Campy & Richard (1988) and with that proposed for the southern Alps by Jorda *et al.* (2000). And whichever solution may be correct, there can be no doubt that the LGM correlates with MIS 2.

However, in all correlation attempts the erosional effects of the glaciations must be taken into account. Complete sequences of all ice advances are seldom preserved, often only traces of the most extensive phases are found. This is especially true for the older, pre-LGM glaciations. Once the glaciers completely retreat, the rivers of the large Alpine valleys continue to erode and remove some or all of the evidence. This implies that the stratigraphical sequences in the large Alpine valleys are very incomplete. It is therefore reasonable to conclude that the preserved traces of glacial advances will not always be synchronous for each reconstructed advance phase in both the Alps and the Jura.

2.2 Jura ice sheet Palaeogeography

In the North-West of the Alps (fig. 3), the two Morainic Complexes (CMI and CME) were not everywhere set up by ice of alpine origin. The Jura had a determining influence on glacial flows and the contents of the Morainic Complexes. It constituted, a topographic stopping to the glacial flows resulting from the Alps, and it was itself an ice cap during various glaciations of Pleistocene.

It may seem surprising that low mountains like the Jura would be capped during the LGM by a thick ice cover. In reality, only a few peaks of the mountain chain are over 1500 m high and the elevations over 1000 m are concentrated in a narrow zone 15 - 30 km wide and 120 km long (Fig. 1). There are two reasons for this important glacial build-up: climate and morphology.

Climatic perspective

The Jura is a particularly cold region because of its position adjacent to the mountains. Even at present the lowest temperatures in France are always found there. This is because the orientation of the major landforms leaves the Jura open to winds from the north-east derived from the Central European anticyclone. It is also a region of very heavy precipitation. Today, there is 1800 mm of precipitation annually in the zone above 1000 m which falls as snow between September and June. It seems reasonable to assume that these characteristics also applied during the recent glacial periods.

Morphological perspective

The Jura is of a form that retains snow by slowing its melting (Aubert, 1965). Therefore, the central part of this massif is not incised by deep valleys. The highest zone is characterised by long, synclinal valleys at heights of 900 - 1000 m, flanked by smooth higher ground at altitudes of 1200 - 1600 m. The poorly drained valley form efficient traps in which great thicknesses of snow can accumulate. The spring melt is considerably slowed in these closed passings even at the present day. One can therefore imagine that during glacial times the balance between accumulation and ablation would have been positive here so that an ice cap could grow. When the level of the ice surface passed that of the surrounding rock, outlet glaciers would form, particularly on the western and eastern flanks of the Jura chain.

2.3 Alpine glacier Palaeogeography

On the Alps, a topographic gradient between the North and the South with high mountains in North where are still present glaciers (Mont Blanc, Vanoise..) and lower topographic mountain in the South cut by steepsided valleys (the Verdon, Var, Bléone). This North-South topographic gradient can partly explain of the extention aog glacier in the Northern zone of the Alps during great Pleistocene ice ages.

However, the different extention of the glacier in the North of the Alps between the LGM and the MGE are a also function of the variations of the glacial whatershed variations between these two periodes.During the last cold period of the LGM, the alpine glaciers were developed than for the cold period of MGE as illustrated by the theoretical calculation of the Isere glacial topography in Grenoble which was 1100mduring LGM and 1500 m asl during MGE (Monjuvent, 1978). The topographic barrier formed by the principal solid masses of the Western Alps: Chartreuse, the Bauges, the Bornes and the Aravis probably had a significant impact on the limits of glacial diffluences during two last glaciations.

During the LGM, the Isère glacier was a large confluent glacier with the Arc glacier, and the topograpic mountain barrier stop the Isère ice flow in direction of the Rhône glacier, this mountains just allowing only weak diffluences between this two glaciers. On the other hand, with a higher development of the glaciers during the MEG, the influence of the mountain barrier decrease and allowed the Isère glacier and the Arc glacier to flow away more easily towards the Rhône glacier. This increase in the whatershed of the Rhône glacier would thus imply an additional alimentation, which could explain the development of the lobe of Piedmont resulting from the Rhône glacier of the Rhone during the MGE.

Acknowledgement

This article was translated by P.L. Gibbard.

References

Agassiz, L. (1843). Le Jura a eu ses glaciers propres. *Actes Sociètè Helvétique des Sciences Naturelles. 28° session, Lausanne*, 284-285

Arn, R. (1984). Contribution à l'étude stratigraphique du Pléistocène de la région lémanique. Université de Lausanne, thesis, 307 pp.

Arn, R. & Aubert, D. (1984). Les formations quaternaires de l'Orbe et du Nozon, au pied du Jura. *Bulletin de la Société Vaudoise de Sciences Naturelles*, **76 (2)**, 203-214.

Aubert, D. (1965). Calotte glaciaire et morphologie jurassienne. *Eclogae Geologicae Helvetiae*, **58 (1)**, 555-578.

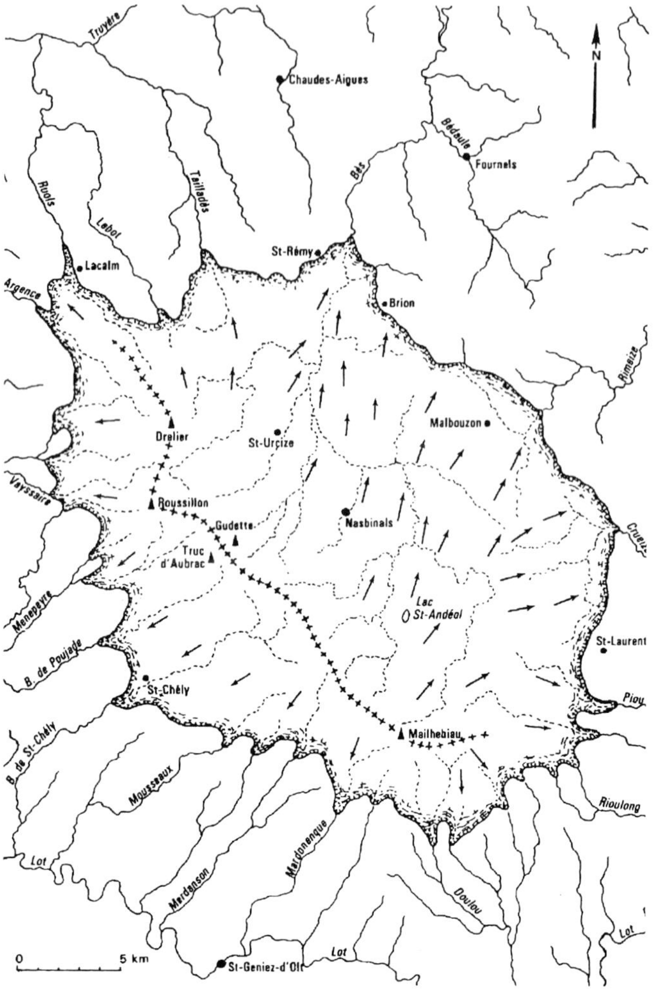

Fig. 2. Würmian-age ice-cap at Aubrac (after Poizat & Rousset, 1975).

cap of 520 km^2 and 200 m thick (Rousset, 1970). This ice cap is asymmetrical, but its central ice-divide, displaced towards the SW, corresponds to the crest line of the massif.

Le Forez

The 'internal and external morainic complexes' and their morainic deposits are well represented on this massif. This is attributed to the occurrence of an ice-cap centred on the Haute Chaume, that probably covered the whole massif, from which several glaciers flowed outwards. Calculations suggest an LGM equilibrium line altitude for this ice-cap at 1280 m (Etlicher, 1985).

The Eastern massifs

The eastern part of the Massif Central comprises small massifs and associated valleys, including Mont Lozère, Mezenc, Tanargue and the valleys of Congues, of Crénades or of Entrayques. The remains of moraines in this area indicate only limited glaciation (Valadas, 1984).

Conclusions

The glaciation of the Massif Central is for most of the area represented by two morainic complexes. The 'external moraines' which are the oldest and the poorest preserved, are of Rissian age. The 'internal moraines' which have a well-preserved morphology, date from the Last Glacial Maximum (Würmian). Therefore the climatic variations of the two glacial periods are represented in the Massif Central by glaciations of different massifs where several types of glacier occurred. They include ice-caps, valley and piedmont glaciers. Because of the strongly marked topographic gradient, the largest ice mass occurred in the west of the massif, whilst only small examples existed in the east.

References

de Goër, H. (1972). La Planèze de Saint Flour: formes et dépots glaciaires. *Annales de la faculté des sciences de clermont-ferrand*, **48**, 204 pp.

de Goër, H. & Veyret, Y. (1976). Les glaciers quaternaires dans le massif central. *Préhistorie française*, **1**, 50-51. Paris, Centre National de la Recherche Scientifique.

Etlicher, B. (1985). Corrélations glaciares-periglaciare dans l'Est du Massif Central Français. *Bulletin de l'Association Française pour l'Etude du Quaternaire,* **2/3**, 117-124.

Etlicher, B. & de Goër, H. (1988). La déglaciation wurmeinne dans le Massif Central francais: le point sur les travaux recent. *Bulletin de l'Association Française pour l'Etude du Quaternaire*, **2/3**, 103-111.

Poizat, M. & Rousset, C. (1975). Les calottes de glaces quaternaires des monts d'Aubrac (Massif Central, France): caractéristiques, contexte paléoclimatique. *Revue de Geographie physique et de geologie Dynamique*, **2/17**, 171-190.

Rousset, C. (1970). Détermination géomorphologique et climatique de l'évolution des calottes glaciaires quaternaires dans les monts d'Aubrac (Massif Central, France). *Compte rendue de Académie des Sciences Paris*, **D 270**,: 464-467.

Veyret, Y. (1981). Les modelés et formations d'origines glaciaire dans le Massif Central français : probmlème de distribution et de limites dans le milieu de la moyenne montagne. Thèse d'état de Paris I, Presse universitaire de Lille, 2 Vol.: 783 pp.

Valadas, B. (1984). Les hautes terres du Massif Central francais. Contribution à l'étude des morphodynamiques récentes sur versants cristallins et volcaniques. Thèse de l'Université de Paris: 927 pp.

Quaternary Glaciations – Extent and Chronology
Editors J. Ehlers and P.L. Gibbard

The glacial history of the Vosges Mountains

Jean-Luc Mercier and Natacha Jeser

Université Louis Pasteur, Faculté de Géographie, 3 Argonne
67083 Strasbourg cedex, France

E-mail: jlm@equinoxe.u-strasbg.fr

Abstract

The map of the end moraines of the Vosges massif presents a summary of 162 years of contributions to the Quaternary geology of central Europe. The investigations in this area have produced two main results: 1) the occurrence of dated Holocene end moraines at low altitude on the eastern side; 2) ^{10}Be-dated evidence of a moraine predating la Grande Pile bog on the western side.

1. Introduction

The Vosges are a mid-latitude (48° N, 7° E) Hercynian massif that has been uplifted since the Pliocene. This ridge between the sedimentary Paris Basin, to the west, and the Rhine Graben to the east, the Vosges shows a very pronounced asymmetry: a long, gentle western slope and a smooth convex upland; opposed to a short steep eastern side towards the graben. The main ridge of this massif is orientated NNE-SSW, with elevations ranging from 1424 m at the *Grand Ballon* in the south, to a 700 m high plateau in the north.

During the Quaternary, the Vosges were completely covered by a local ice cap that occurred between the Scandinavian and the Alpine ice sheets, with valley glaciers descending on both sides of the mountains. The lengths of these glaciers were different: 40 km on the western side, as compared to a mere 15 km on the eastern side. At the end of the last cold stage, the latitudinal position together with the amplification effect of the landscape, made these glaciers very sensitive to climatic variations; their response times being very short. Some valleys include as many as 7 successive Holocene-aged end moraines.

2. History

Modern Quaternary research in the Vosges mountains began in 1838 with Leblanc. In 1837 and 1838, Leblanc published twice papers on the same section with two different interpretations, the second being that which is still valid today. The same year he went to Neufchâtel to visit Agassiz. Historically, the studies conducted in the Vosges contributed to the principles which define Quaternary Geology. In his paper he gave a general description of some of the western Vosges moraines. Two years later, in 1840, Hogard produced a precise description of the southern moraines. Collomb (1846) and Grad (1858) soon afterwards described the major glacial cirques of the eastern side and Meyer (1913) presented the first general map of the glacial features in this area. Théobald & Gardet (1935) identified 3 moraines in the south-east. Tricart (1963) was the first to recognize Holocene moraines at low altitude. More recently, Woillard (1978) described the Grande Pile peat bog sequence. The number of publications on the glaciation of the Vosges mountains including these by pioneers, is 300 - 400. A partial historical overview has been presented by Seret (1967) and Flageollet (1982).

3. Principles

3.1. *Approaches to glacial stratigraphy and correlation*

Morphostratigraphy, lithostratigraphy, rock and soil weathering and sometimes absolute age determination (^{14}C in peat bogs or ^{10}Be on morainic boulders or roches moutonnées) are the classical tools used by the present authors.

3.1.1. *Morphostratigraphy and spatial continuity*

The first step in the cartographic process is to determine the geometry and the spatial continuity of either the morainic ridges (lateral or end moraines) or of other glacial features. However, to be efficient, this approach requires application of a relative (or sometimes absolute) dating technique. In the western Vosges, both continuity of the features and sufficient dates are available (Woillard *et al.* 1982) for the oldest glacial limit (Seret & Mook, 1990). Because of the steeper slopes in the eastern part, the continuity does not occur and thus interpretation not supported by dates.

3.1.2. *Sedimentological study*

Morphological and sedimentological arguments (mineralogical composition, size, orientation, shape and roundness of pebbles, sorting of sands, maturity index (quartz or feldspar concentration) soil weathering) and variations in rock weathering characteristics have been used by the authors to identify and establish a relative

Table 1. Exposure ages (k ^{10}Be yr BP) on morainic boulders (B), roches moutonnées (RM) and erratics (E) from Mercier et al. (1999).

Samples	Cold events (morainic ridges)		Warm (erratics, roches moutonnées)	
	Exposure age	Relative error	Exposure age	Relative error
A 6			(E) 15.3	± 1.2
M 5a	(B) 11.5	± 0.9		
M 5c	(B) 11.5	± 1.0		
M 5b	(B) 10.6	± 1.1		
M 5d	(B) 10.6	± 1.0		
W 10			(RM) 10.1	± 1.3
Ln 2	(B) 9.7	± 1.1		
Ta 4			(E) 9.3	± 1.6
W 15			(RM) 9.3	± 1.1
W 11			(RM) 9.2	± 1.0
W 7	(B) 9.0	± 1.1		
W 14	(B) 9.0	± 1.0		
W 13			(RM) 8.8	± 1.0
W 8	(B) 8.3	± 1.1		
W 9			(RM) 7.6	± 1.9
W 12			(RM) 6.3	± 0.8
Ln 1			(RM) 5.1	± 1.2

Table 2. Last Glacial Maximum and Younger Dryas, cirques and nivation hollows orientations.

	Orientation	Number	%
Donon Last Glacial Maximum (LGM) and Younger Dryas Cirques Nivation hollows	N	27	24,3
	NE	14	12,6
	E	36	32,4
	SE	5	4,5
	S	0	0
	SW	4	3,6
	W	9	8,1
	NW	16	14,4
Champ du Feu Younger Dryas Nivation hollows	N	5	11
	NE	12	26
	E	5	11
	SE	3	6,6
	S	1	2
	SW	6	13
	W	6	13
	NW	7	15
Main Vosges Last Glacial Maximum (LGM) and Younger Dryas Cirques Nivation hollows	N	56	11,57
	NE	111	22,93
	E	113	23,34
	SE	56	11,57
	S	16	3,30
	SW	22	4,54
	W	34	7,02
	NW	76	15,70

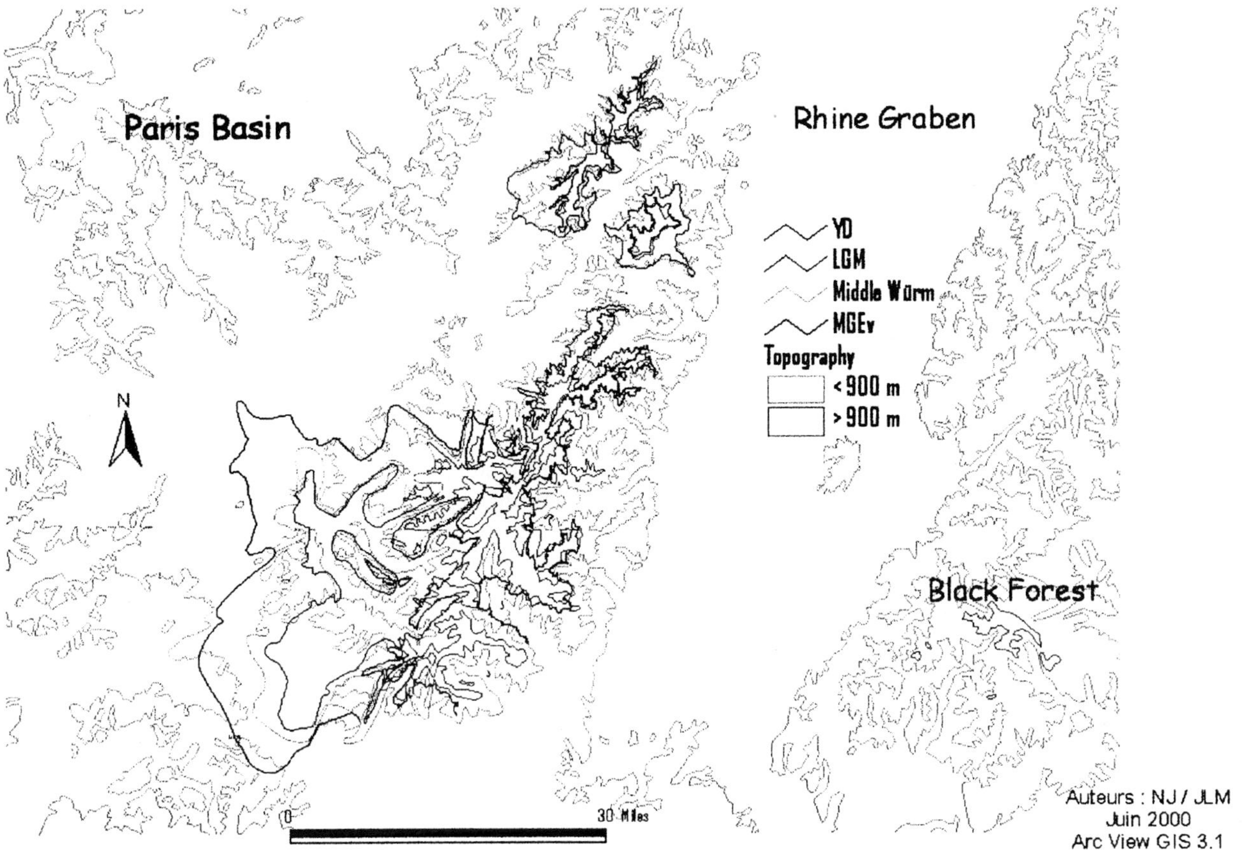

Fig. 1. Glacial limits in the Vosges.

chronology between the deposits of different areas. For pebbles, they used the Cailleux's roundness: 1000 (2r/L) and shape (L+l)/2e indices.

Shape and roundness

Because of the combination of features involved, there is no unequivocal interpretation of the shape and roundness values. In the southern Vosges, the roundness is: 142 to 192, shape 0.62 for MGE_V, as opposed to a roundness of 130 to 200, and a shape 0.8 to 1.82 for the middle Würmian, and an roundness 55 for Last Glacial Maximum (LGM).

In the central Vosges, the weathering features on pebbles are the same, but for geomorphological reasons, the roundness is lower and the shape value higher than in the southern Vosges: roundness 85 to 130, shape 1.9 for the Last Glacial Maximum (LGM); and for the Younger Dryas, roundness 70, and shape 1.5-1.6

Weathering characteristics of morainic boulders

Differential development of weathering over time in the Vosges area.

- Maximum Glacial Extent for the Vosges (MGE_V) = weathering rinds on locally-indurated pebbles (sandstones, quartzites, granites). The granites are decomposed and can be disaggregated by hand. When broken, the feldspar grains produce a fine white powder.
- Middle Würmian = surface weathering on granitic pebbles. The feldspars in the sand fraction show the same degree of bleaching as the pebbles.
- Last Glacial Maximum (LGM) = few visible weathering features on pebbles. The potassium feldspars are fresh; only the plagioclase shows some etching. Sometimes, biotite has undergone partial alteration to chlorite.
- Younger Dryas = no visible weathering on pebbles or the sand.

Soil profiles

Soils show a deeper profile and thicker soil horizons with increasing age, but they are generally weakly-developed and show very few diagnostic properties. The clay illuviation (cutans) and concentration in the B horizons can be used to identify the Maximum Glacial Extent for the Vosges (MGE_V) and the colour of these horizons is orange to reddish brown (5YR 4/8 to 6/8). In these B horizons, the clay mineral composition is dominated by illite, with few illite-vermiculite intergrades (Vogt 1992). On the Middle Würmian moraines, the B horizon is light brown in colour (7.5 YR 5/6 to 6/8). On younger deposits, the soils can be classified as cambisols (according to the FAO classification), and the soil horizons colour is directly related to the parent material. On the eastern side, some of the Younger Dryas moraines show the development of podzols (FAO) on dated granitic parent material (9.7 ± 1.1 k ^{10}Be yr BP.; Mercier *et al.*, 1999)

3.2. *Relative and absolute dates*

The Vosges mountains are covered by 250 peat bogs that occur on the main ridge, in some cirques and nivation hollows and in the valleys. Numerous pollen profiles have been obtained and dated (Reille, 1990) from these sequences. Palynological investigations began very early with the works of Hatt (1937), Firbas *et al.* (1948) and Lemée (1963). The results of those early studies continue to be important and corroborate the absolute dates obtained on morainic boulders and roches moutonnées by the exposure dating method (Mercier *et al.*, 1999).

3.3. *Age determination and stratotypes*

Because of the asymmetry of the massif, each side must be discussed separately. On the western part, the stratotype of MGE_V (V for Vosges) is the till that underlies the Grande Pile peat sequence (Seret *et al.,* 1990) and the morainic ridge beyond the bog (Linexert Moraine). This moraine predates the last interglacial (Woillard, 1979). Between the Vosges highlands and Grande Pile depression all the morainic ridges and till accumulations are related to the last glaciation (= Würmian Stage). The glacier length (over 20 km) and differences in soil weathering have been used to separate the chronology into two parts: Middle Würmian to Older Dryas (OD) and Younger Dryas (YD) to Holocene moraines.

On the eastern side, some authors at the beginning of the century had assumed that valley glaciers had advanced out of the mountain valleys into the Rhine Graben (e.g. Baulig, 1922). However, since the middle of the last century, there is no evidence for the Maximum Glacial Extent in this area. All morainic ridges older than a certain erratic (dated at 15.3 ± 1.2 k ^{10}Be yr BP) are referred to as Last Glacial Maximum (LGM). Indeed, no secure remnants of a Middle Würmian substage glaciation have been found.

The stratotype of the deglaciation of the Vosges is the Wormsa valley. Exposure dating (Table 1) has allowed the identification of five oscillatons (boulders on morainic ridges, dated at 11.5; 10.6; 9.7; 9.0; 8.3 k ^{10}Be yr) in this area, separated by warmer periods (erratics and roches moutonnées) during Younger Dryas and Holocene. A sixth cold event is estimated to have occurred at 7.8 k ^{10}Be yr.

The deglaciation of the Vosges highlands is dated from pollen profiles associated with ^{14}C and roches moutonnées. Exposure dates of 5.1 ± 1.2 k ^{10}Be yr BP and 6.3 ± 0.8 k^{10}Be yr BP have been obtained within cirques (Table 1). It appears that a permanent ice cover disappeared from the main crest of the Vosges around 6.0 k ^{10}Be yr BP.

4. Results and mapping limits

From north to south three areas have been mapped: Donon, Champ du Feu and the Main Vosges. During the Maximum Glacial Extent an ice sheet covered the entire Vosges massif. The moraines of this glaciation are still present on the SW slope; this ice sheet degraded during the Middle Würmian and only three small ice caps remained to expand later during the Last Glacial Maximum.

4.1. *The Donon*

During the Last Glacial Maximum (LGM), both cirques and valley glaciers existed on the western side. Their maximum extent as indicated by that associated with La Maix cirque is 3 km. On the eastern side only three cirques were glaciated (Table 2).

The ice during the Younger Dryas was limited to cirques and nivation hollows (Table 2). The mean height of cirque crests and floors are 825 and 702 m respectively. A residual ice cap occurred near Le Rocher de Mutzig at 1010 m and near Schneeberg at 990 m. Pollen sequences (Hatt, 1937; Becker & Sittler, 1952) indicate that deglaciation began during the early Holocene Preboreal period.

4.2. *The Champ du Feu*

During the Last Glacial Maximum an ice cap covered the whole massif and was surrounded by five outlet glaciers; the two largest, Schirgoutte and Andlau were each 4 km long. The petrography and morphology of the Champ du Feu area did not favour the development of cirques.

A small independent ice cap covered the top (960-1100 m) of the Champ du Feu in Younger Dryas time, and two short ice streams extended from this summit towards the NW: the Schirgoutte and Serva (Table 2) which are represented by two and three recessional moraines respectively.

4.3. *The main Vosges*

The limits of the Middle Würmian and the Maximum Glacial Extent show the same characteristics. The longest valley glacier during these periods, that in the Moselle valley, is marked by an end moraine at Epinal (some 40 km from the main ridge).

During the Würmian maximum an ice cap was present on the main upland (1250-1300 m) from which valley glaciers extended down valley (5-6 km in the eastern side; 10 km in the west). After the maximum, (Do you mean after the maximum or since the maximum?) three sectors became more separated; the extents differed between the east and the west sides, because of climatic and topographic differences. In the South, an ice cap was probably located on the Plateau de la Haute Saône, and some ice streams were derived from the Moselle glacier tongue. The diffluences mapped are: col des Luthiers (890 m), col des Croix (690 m), col du mont de Fourche (644 m), col du Grand Faing (780 m), col des Sarrazins (730 m), col de la Source de la Combeauté (672 m), col de Raon (537 m) and col du Rond Caillou (546 m).

During the Younger Dryas cirques and nivation hollows facing E (23%), NE (23%), NW (11.5%) (Table 2) are found in the upper parts of the valleys. These typical cirques have their steep sides and backwalls from 1240 m to 985 m. At this time an ice cap existed on the main upland similar to that in the LGM, but the valley glaciers were smaller (2-3 km), reaching down to only 510 m (end moraines in the Rougigoutte valley).

5. Remaining problems

This mid-latitude, relatively small mountain range shows a number of very impressive glacial features. A regional glacier covered the area during the last glaciation. This ice sheet was divided into three individual minor ice caps before the Last Glacial Maximum (LGM). On the western side, the Maximum Glacial Extent (MGE_V) and the Last Glacial Maximum (LGM) can be clearly delimited and dated. The glaciers built end moraines up to 40 km from the main mountain ridge in the deeply incised valleys; and 15 km on the eastern flank. Deglaciation occurred slowly during the Holocene, each cold oscillation giving rise to a readvance and the formation of a new end-moraine ridge in the valleys (cf. Wormsa) in which the two last cold events are dated at 8.3 and 7.8 kyr. Permanent ice cover seem to have disappeared around 6 kyr.

The main unanswered questions deal with the glaciology or glacial geomorphology; it remains to understand the relationship between the unknown pre-glacial landscape and oscillating climate.

Geomorphological questions:

- Why are there such huge moraines in such a small mountain range? One of the surprises in the Vosges is not only the size but also the freshness of the morainic landforms.
- Where are the Maximum Glacial Extent (MGE) and Last Glacial Maximum (LGM) moraines on the eastern side of the Vosges? Is the morphometry of moraines in this area so subdued that they do not stand out in the landscape, or are they covered by younger alluvium? Thus far, it is unknown where the Middle Würmian deposits are.
- The relations between glacial and periglacial processes in the valleys are still to be investigated.

Climatological questions:

- What are the Equilibrium Line Altitude (ELA) relations between Vosges and Alps during Younger Dryas and the Holocene? Mercier *et al.* (1996) show a difference of 1500 m for the ELA calculations between Alps and Vosges during Younger Dryas.

Chronological questions:

- How old are the Husseren-Wesserling (Thur valley) or Kirchberg (Doller valley) moraines? These two moraines appear very similar to the Gschnitz Phase moraines in the Austrian Alps. Does the higher continentality in Austria explain this differences in shape? Is this a climatological or a chronological problem? Did the Maximum Glacial Extent, Last Glacial Maximum (LGM), etc. in the Alps and Vosges occur at the same time?

Glaciological questions:

- Why are the Holocene moraines so frequent and so fresh on both sides of the Vosges? Is this the result of the amplification effect of the local landscape on climate?

References

Baulig, H. (1922). Questions de morphologie vosgienne et rhénane. *Annales de Géographie*, **XXXI**, 132-154.

Becker, J. & Sittler C. (1952). Age de la dénivation dans les niches glaciaires des Vosges gréseuses. *Compte Rendu de l'Académie des Sciences*, **234**, 1191-1192.

Collomb, E. (1846). Sur le terrain erratique des Vosges. *Bulletin de la société géologique de France*, **1 (3)**, 187-197.

Firbas, F., Grunig, G., Weischel, I. & Worzel, G. (1948). Beiträge zur spät- und nacheiszeitlichen Vegetationsgeschichte der Vogesen. *Bibliotheca Botanica*, **121**, 1-76.

Flageollet, J.-C. (1982). Cent cinquante ans de recherches sur le glaciaire vosgien. Historique des idées et état de la question. *Mélanges à A. Journaux*, Caen, 173-192.

Grad, C. (1858). Comparaison chronologique des terrains quaternaires de l'Alsace avec ceux de la vallée du Rhône

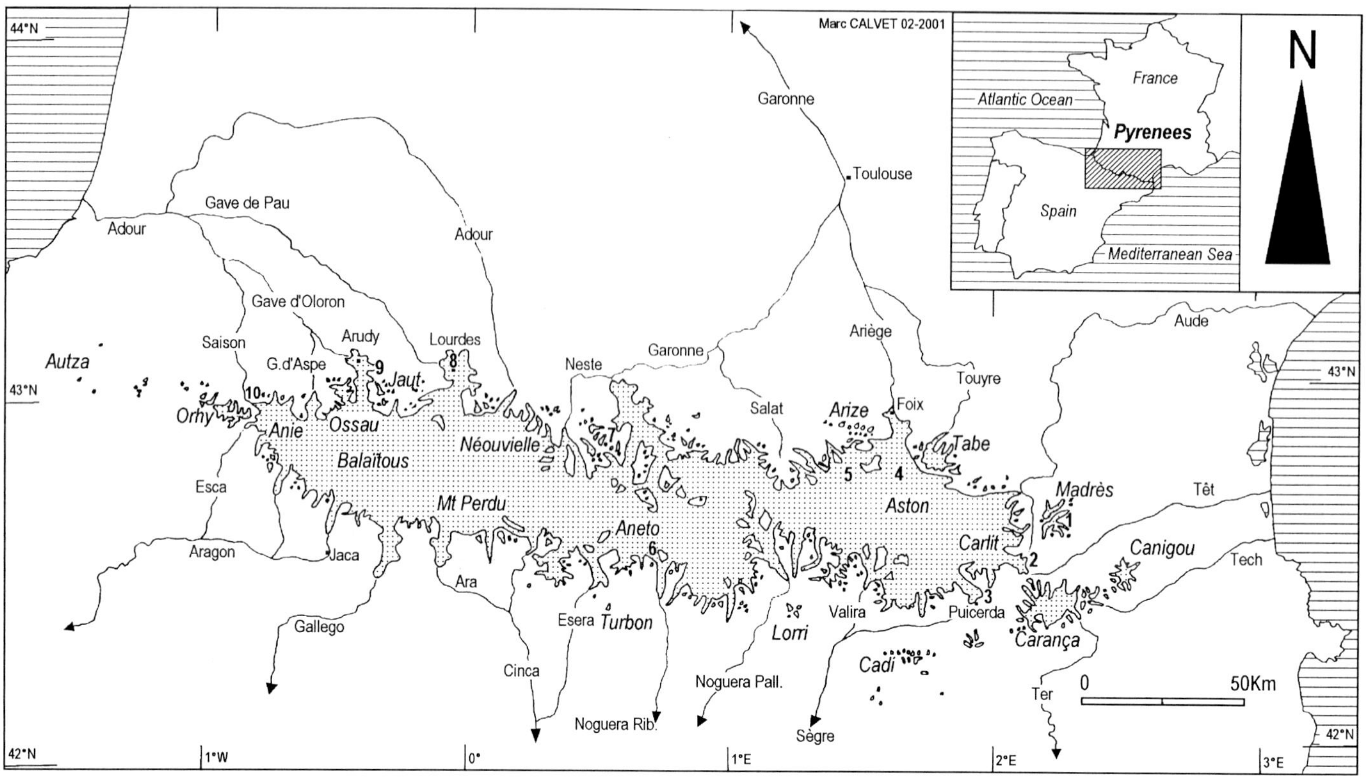

Fig. 1. The Quaternary glaciation in the Pyrenees: location map. Glaciated area is dotted. The numbers 1 to 10 indicate the position of the key sites for the Pyrenean glacial chronology (see after the text).

simple high altitude breaches, others form wide, flared, deep cols namely those (from east to west) of the Puymorens (1920 m), Bonaigue (2070 m), the Pla de Béret (1870 m), the Pourtalet (1794 m) and the Somport (1632 m). An ice-cap stage was never achieved, but plateau glaciers did form at the foot of the highest summits on the remains of Neogene peneplains, including the Carlit, Aston and Arres d'Anie plateaux.

Conversely, downstream the narrowness of the chain, its intense dissection into parallel valleys or the low altitude confluences restricted the formation of powerful compound ice tongues. Branching of tongues were rare e.g. the transfluence from the Pique and Garonne valleys or these of Col de Lers from the Salat and Ariège valleys. The glaciers descended rapidly into the zone of intense ablation resulting in an abundance of lateral and proglacial sediments. The ice spread and thinned in the vast preglacial erosion basins in the marls or the flysch where the ice tongues divided into several diffluent lobes.

The position of the equilibrium line in the Würmian maximum shows, like in the present, a double dissymmetry. It increased in altitude from the west to the east, but much less quickly than it was thought (Barrère, 1954, 1963; Taillefer, 1969a). It is estimated at 1200-1300 m in the small isolated cirques of the Pays-Basque region (Les Aludes, Autza) and at 1400 m a.s.l. in the north-exposed small massifs of the north Pyrenean front. It continued at the same altitude (*c.* 1400 m)from the Saison to the Gave d'Aspe (massifs of Issarbe, Soulaing, Layens - 1625 m) to the east between Garonne and Salat (Paloumère - 1608 m) (Bakalowicz *et al.*, 1980) whereas in the Arize (1716 m) the supply of snow by NW wind even allowed the formation of small cirque glaciers on a south-facing region (Taillefer, 1963). To the east of the Ariège the snowline remained between 1400 and 1500 m, north-west of theTabe massif, and it was at 1600 m in the eastern cirques of the Dourmidou (1843 m) 60 km from the Mediterranean (Fig. 5). In the high axial chain the large north slope glaciers also show an equilibrium line that increased rapidly in altitude from 1500 m in the Aspe valley to 1700 - 1800 m in the Gave d'Ossau, and then stays at the same altitude to the north of the Carlit. The main asymmetry however exists between the north and south slopes of the Pyrenees, with a very quick rising of 600 m. Site conditions and exposure direction play an important role in the south. Glaciation is always very site dependent where the massifs only slightly exceed the snowline altitude. In the eastern Pyrenees this therefore passes over a distance of less than 40 km from 1600 m at the north to 2200 m on the southern slope of the Carlit, in the Canigou and the Puigmal. The limit even climbs to over 2300 m in some sheltered zones at the margins of the Cerdagne. But it locally falls toward 2100 m on the southern slope of the Canigou-Carança, which are exposed to the disturbed Mediterranean airflow.

Regional contrasts in the Pyrenean glaciation

The western Pyrenees were only weakly glaciated because of their very low altitude of only under to 2000 m a.s.l

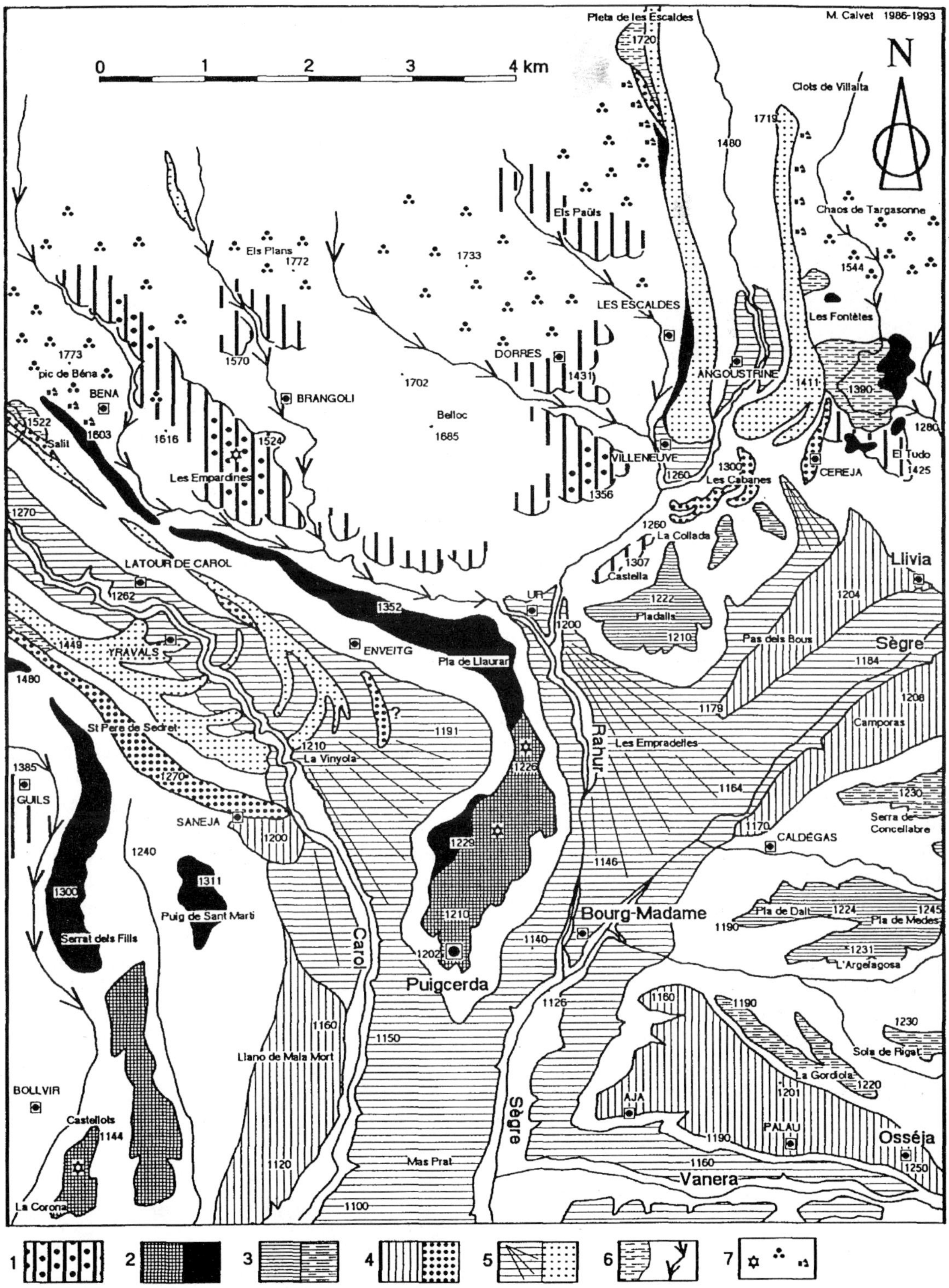

Fig. 2. Map of the Carol and Angoustine glaciofluvial complexes, Cerdagne, eastern Pyrenees. (1) Upper Neogene pediments and associated alluvium. (2) Glaciofluvial unit T4 of early Middle Pleistocene age; weathered old moraines. (3) Glacis and outwash fans T3 of Middle Pleistocene age. The T3 generation of landforms (2 or 3 generations) do not connect to moraines upstream. (4) Glaciofluvial unit T2, weathered moraine (cf. Rissian). (5) Glaciofluvial unit T1, last glacial maximum moraines (Würmian). (6) Glaciolacustrine plain, fluvial gorge in the Hercynian basement and old meltwater channels correlated with the old moraines. (7) Wind-facetted and rusty quartz pebbles like those that occur in the upper terraces T3 to T5 of the Rousillon rivers; granitic tors in situ; tors overridden by the glacier. Nival and glacial meltwater were responsible for excavating the tors from the surrounding thick and very strongly weathered regolith around Targasonne.

Fig. 3. The Maladetta glacier: crests at 3308 m, present equilibrium line at 3000 m and front at 2800 m, largely back from the Little Ice Age frontal moraine, visible on the right and who reaches 2650 m.

(Viers, 1960). The small very localised cirque glaciers of the Basque mountains are found on the north, east and more rarely the south faces (Orzanzurieta, 1570 m), where they were maintained by snow supplied, from massifs that are generally around or under 1400 m high. The north-south assymetry is very strong from the Orhy massif (2017 m) to the Lakhoura massif (1877 m) where only cirque glaciers or rare and very short ice tongues occurred on the southern slopes, whilst small ice tongues 2-4 km long, often compound features, descended towards the Gave de Larrau down to 600 m a.s.l. Only the karstic plateau of Arres d'Anie (1700-2504 m) was glaciated by an ice field of nearly 45 km^2 at this height that supplied several diffluent ice tongues that descended to 600 m in the north but only to 950 m in the Rio Esca valley (Belagua) and 1100 m in the Anso valley in the south.

The northern slope of the chain from the Aspe to the Ariège valleys represents not only three quarters of the glaciated surfaces, but also the most powerful glaciers (Barrère, 1963; Paquereau & Barrère, 1964; Alimen, 1964; Taillefer, 1963, 1966, 1969b, 1977, 1984, 1985; Goron, 1941; Chevalier, 1951, 1954a, b; Bertrand, 1963; Mianes, 1955). Short, several kilometres long tongues existed on the northern flanks of the small pre-Pyrenean massifs where the peaks approach or exceed 2000 m altitude. They include, from west to east, the Mailh Massibé (1973 m), the Jaut (2050 m), the Mont Né (2147 m), the Trois Seigneurs (2199 m) and the Montagne de Tabe (2368 m). At the latter, the Touyre glacier was 8 km long and descended to 650 m. In the main valleys, the glacier tongues mostly descended to 350-400 m a.s.l., but only those that were well-supplied with snow, in the west, formed true piedmont lobes at Lourdes and Arudy (Barrère, 1971; Mardonnes & Jalut, 1983). In the Garonne the terminal lobes were confined in the transverse valleys of the outer range (Taillefer, 1984). On the Ariège the much less fan-shaped glacier barely reached up to the Foix canyon. The Gave de Pau ice tongue reached 50 km in length, that of the Garonne was 70 km long and the Ariège tongue was almost the same length. The thicknesses of these ice masses reached about 1000 m. The Aspe glacier remained within the mountains because of the low altitude of its supply basin and the losses caused by transfluence towards the south, but recent studies propose a broad extrusion to the piedmont (Gangloff *et al.*, 1991). On the Adour, the shortness of the basin slopes and their sheltered position prevented the confluence of Würmian ice-tongues. An ephemeral confluence may have occurred on the Neste basin where protection from the western airflow is still more important and where the glacier reached down to 700 m. If this protection also operated on the high Salat, the narrowness of this mountain barrier and the depth of the valley dissection in the schists (the confluences are nearly all below 700 m a.s.l.), must be invoked to explain the lack of glacial confluences and the shortness of the ice tongues. In contrast the granitic massifs that form the highest topography of any parts of the Salat basin, preserve high basins that favour ice accumulation and explain the major ice tongues of those regions (Riberot, Arget and Garbet). In the Garonne and the Ariège the extensive supply surfaces above 1800 m, the frequent confluences and the almost NW-SE orientation of the valleys, open to the disturbed air flow, certainly compensated for their more continental position.

The glaciation of the southern Pyrenean slope was much more modest (Barrère, 1963; Martí Bono, 1973, 1978; Martí Bono & Garciá Ruiz, 1994; Montserrat Marti, 1992; Garciá Ruiz & Martí Bono, 1994; Peña Monné *et al.*, 1997; Serrano & Martinez de Pisón, 1994). With the exception of the western part where they meet the high chain (from Mount Perdu to Visaurin) the folded pre-Pyrenean mountains were practically not glaciated. With the exception of the Cotiella (2912 m) which was surrounded by a group of radiating small ice-tongues, further east only those massifs that exceed 2000 m, particularly on their northern flank, had several cirque glaciers or very short glacier tongues. They include the Turbon (2500 m), Lorri (2400 m) and the Serra de Cadi (2600 m). The most southerly glacial cirque

Fig. 4. The classic Pyrenean cirques, on the granitic massif of Néouvielle (3091 m), north-central Pyrenees, seen from the North. Note the low height of the headwall and the large embossed floor.

Fig. 5. The lowest cirques of the north-eastern Pyrenees, on the Dourmidou massif (1843 m), floor at 1600-1650 m and terminal moraines at 1300-1400 m. Those cirques are exclusively facing East, in a position of snow overfeeding by western to north-western atmospheric disturbances, as that is still very well illustrated by the present seasonal snow coverage.

in the Pyrenees occurs at 2100 m on the southern summit of the Port de Comte, at a point that favours snow accumulation from the high plateaux of this massif. The ice tongues from the main valleys discharging from the axial zone rarely exceeded 20-30 km in length and were 500 to 800 m thick at most (Fig. 6). In addition, they generally terminated at about 800 m, in the mountain interior, from the Aragon Subordan to the Noguera Pallaresa (Garciá Ruiz, 1989; Garciá Ruiz *et al.*, 1992; Nussbaum, 1956). In Andorra (Prat, 1980; Vilaplana, 1984; Vilaplana & Serrat, 1981), the glacier descended to 950 m. The most powerful ice flows in the west (Barrère, 1966) developed a series of multiple lobes in the flysch basins (Gallego, Ara). They were rarely confluent to form large composite glaciers, except possibly those of the Noguera Pallaresa. Recent research has shown that they descended as far as Rialp, 10 km downvalley from Llavorsi, a termination point that was generally supposed (Serrat *et al.*, 1994).

In the eastern Pyrenees the increased snow line altitude and the fragmentation of the mountain into isolated massifs limit glaciation (Birot, 1937; Nussbaum, 1943; Gómez Ortiz, 1985, 1987; Gielen, 1984; Gómez Ortiz & Salvador i Franch, 1994; Calvet, 1996, 1998). The low dynamics of the ice tongues, which carried a substantial debris load and the weak proglacial drainage system explain the prominent terminal morainic ridges or barriers and those of retreat stages. Only the Carlit (2921 m) because of its climatic position and topography, developed a plateau glacier and several powerful glacier tongues that terminated between 1200 - 1600 m in the high dry basins of the Cerdagne-Capcir (Viers, 1961, 1963, 1968). The exception is the glacier at Donnezan to the north, which descended to 800 m close to the confluence with the Aude. The longest glacier was the Carol which benefitted from the diffluence of the Puymorens, from the Ariège valley, and reached 23 km in length and was 400 m thick. The assymetry is spectacular on the Madres (2469 m) which supported tongues of 6 km that reached to 1200 m on the northern side but only 3 km long glacier tongues in the south which did not reach down beyond 1500 m. Conversely, the glaciation of the Carança (Serrat, 1974, 1977, 1982) and the Canigou (Viers, 1966) was more symmetrical, because of the above-mentioned climatic peculiarities. The Carança (2881 m) had an Alpine-style glaciation that formed well-defined U-shaped valleys. Composite ice-tongues 6 - 8 km long descended lower on the southern than on the northern slopes (1200 m on the Freser). At the Canigou (2784 m) well-defined cirques are found supporting short ice tongues about 3 km long. The lower glacier terminus in the east (Lentilla) reached down to 1350 m a.s.l. in the Würmian.

Glacial chronology in the Pyrenees

The Pyrenees have for a long time represented an exception in the Quaternary stratigraphy of Europe and provided a number of discussions. In contrast to the extreme polyglacialist interpretations of the Pyrenees, that systematically assigned the recessional moraines spaced upstream in the valleys to the four periods of the classic Alpine stratigraphy, the Toulouse school (G. Viers, P. Barrère and to a lesser extent F. Taillefer) defended a monoglacialist interpretation of the Pyrenees' Quaternary features until the 1970s. They were convinced of only one long glacial period in the Upper Pleistocene. This idea was based on an initial report by P. Birot (1937), clearly confirmed by field evidence, that proposed the geomorphological and sedimentological unity of the glacial inherited features. The material was found to be nearly unweathered even in the most outer frontal moraine. Thus only a stage of progression moraine and a series of several retreat phases of the same glaciation could be recognised. Therefore both in the main central Pyrenean valleys, and in the small formerly glaciated valleys of the extreme east or west (Barrère, 1963; Viers, 1960, 1961, 1966; Taillefer, 1963, 1966, 1969a) only one glacial maximum or Pleniglacial phase could be identified. This was marked by an expansion phase, a short period during which erratics and ground moraine was emplaced high on the slopes of the valleys, followed by a prolonged stationary phase during which the external frontal moraines were formed; then a retreat and separation phase. The latter is represented in some places by internal morainic complexes. Finally, a high altitude glaciation phase occurred in the Late- and Neoglacial. It is generally represented by block-covered and rock glaciers. This view was challenged later because of its strong contrast to the Alpine Quaternary stratigraphy. At the 1969 INQUA Congress a compromise interpretation was attempted in which the external moraines were attributed to the Rissian and the internal moraines to the Würmian.

The re-examination of the Pyrenean glaciation since 1980 has benefitted from the work of palynologists and sedimentologists who have examined the karstic deposits and the glaciolacustrine deposits. New results have been

Fig. 6. A great glacial valley of the southern-central Pyrenees: the Esera trough, seen from the frontal lobe of Bisaurri. On the left, the overdeepened terminal basin of Castejón de Sos; on the right, great cultivated benches with lateral moraines; on the background, the Posets massif (3368 m) and the hanging valley of Barbarisa. Note the gullying of the till, on the right valley side.

supported by numerous dates and also by a better understanding of the role of pedogenesis and weathering in the moraines and proglacial terraces. Together these new data have allowed more satisfactory correlations. Speleothems in the karst of the Niaux-Lombrives caves in Ariège contain traces of two major glacial periods between 20,000 and 90,000 B.P. and between 200,000 and 250,000 B.P. and two minor periods between 130,000 and 175,000 B.P. and between 290,000 and 350,000 B.P. (Bakalowicz *et al.*, 1984; Sorriaux, 1981; Taillefer, 1985). Furthermore, in the Western Pyrenees, the La Pierre Saint Martin Caves have provided evidence of at least three glacial stages between 244,000 BP and 579,000 BP, related to MIS 8, 10 and 12, based on U/Th dating (Maire, 1990). In the large external complexes all the well-preserved frontal moraines, composed of almost unweathered sediments, are found to be linked with the low Würmian terraces. Their age is confirmed locally by pollen sequences from ice-dammed

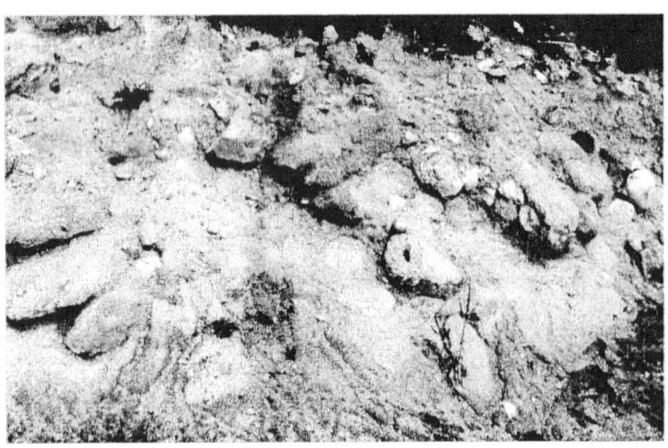

Fig. 7. Very weathered till, ascribed to middle-lower Quaternary, on the outer face of the Angoustrine right lateral moraine, Cerdagne, eastern Pyrenees. Note the metric blocks of granodiorite, entirely disintegrated.

lakes such as that at Sost (Hérail & Jalut, 1986). Conversely the ground moraine of the slopes that dominates the terminal morainic complexes is more altered and therefore appears to indicate an earlier glaciation (Hubschman, 1984) that could be the Rissian. These very discontinuous remains occur in many valleys and indicate a glaciation somewhat more powerful than that during the Würmian. Although it may not have extended further downvalley except locally, the mapping clearly distinguishes the two stages on the Ariège (Taillefer, 1985). However, the mapped glacial limits in the eastern valleys (Gaves de Pau and Ossau), which were formed during the maximum expansion phase (Hazéra, 1975) probably correspond to both the Rissian and the Würmian. The remains of a very old glaciation associated with the Lannemezan formation are indicated north of Arudy (Hétu & Gangloff, 1989). These very ancient vestiges are common in the eastern Pyrenees (Clotet *et al.*, 1985), where they generally extend lower than those of the Rissian and especially the Würmian glaciations, for example in the Canigou to 950 m on the Lentilla (Calvet,

Fig. 8. Forms of Tardiglacial recession in the cirques of the eastern Pyrenees (Carlit massif, cirque of Solana Carnicera, facing ESE). On the background, huge blocks recessionnal moraines, spreaded on the Carlit plateaus, around 2300 m (Les Socarrades). On the centre, two nested arcuate moraines, at the opening of the cirque, that are linked to lateral rock spurs and contain a tongue of a rock-glacier, who is exclusively supplied by the screes of the southern wall of the cirque. View taken from the North, summit at 2878 m.

1996, pp. 897-898). These observations suggest that snow from Mediterranean storm sources had played an important role in the pre-Rissian and even in the Rissian, probably caused by a more intense meridian atmospheric circulation. The Carol complex (Panzer, 1932; Gourinard, 1971; Calvet, 1996, 1998) is exceptional in the Pyrenees (Fig. 2) because three generations of well-preserved moraines and their equivalent alluvial formations (T1, T2 and T4), forming stepped terraces on 70 m. high, can be clearly differentiated on the basis of weathering intensity. The oldest till formation (Fig. 7) appears to date from the early Middle Pleistocene: all the granitic blocks, even the largest ones, up to several meters in diameter, are entirely weathered.

The monoglacialist concept may have been an oddity but in some respects the Pyrenean glacial history differs from that of other regions. These differences are clearly seen in the chronology of the last cold period, the stadials and interstadials of which can be precisely identified (Bordonau, 1992). In the Pyrenees the last glacial maximum was earlier than 45,000 B.P. and the deglaciation began very early (Andrieu *et al.*, 1988) as shown by the terminal complexes at Arudy (Andrieu, 1987) and Lourdes (Mardones & Jalut, 1983; Jalut & Mardones, 1984). In the middle section of the valleys, e.g. the Ariège at Freychinède, the disjunction phase pre-dates 21,000 B.P (Jalut *et al.*, 1982). In the Têt valley, in the eastern Pyrenees, six retreat complexes are preserved, each well separated, upstream from the frontal moraine that stands at 1,600 m. A fossil peat bog beneath the sixth complex at 2150 m altitude has given a ^{14}C age of 16,000 B.P. (Delmas, 1998). The Madrès cirques at 2080 m altitude became ice-free in the Older Dryas (Reille & Lowe, 1993). The dates of the high-altitude glaciation vary depending on the regions. In the western and central Pyrenees a Holocene neoglacial series of moraines were dated at about 5000 B.P. on the Troumouze cirque at 2,100 m altitude (Gellatly *et al.*, 1992). In the same part of the Pyrenees, during the Lateglacial several short ice tongues of 5 to 7 km long were formed in the valleys of the Cauterets region and north of the Balaïtous, with their front around 1,000 to 1,200 m, and named 'Stade d'Estaing' by P. Barrère (Barrère *et al.*, 1980). In the rest of the chain there seems to have been only high-altitude glaciation restricted to the cirques and rock glaciers in the Lateglacial (Taillefer, 1980; Reille & Andrieu, 1993). At Carlit, for example (Delmas, 1998), an ultimate generation of small morainic ridges (Fig. 8) mark an equilibrium line at 2600-2700 m and predate rock glaciers attributed to the Younger Dryas. In this eastern region only the rock glaciers on cirques at the highest altitude of 2600-2700 m suggest reactivation during the Holocene.

Key sites for the Pyrenean glacial chronology

(Location on figure 1)

1. Gorg Nègre (2080 m, Madrès massif, eastern Pyrenees). Pollen sequence of the cirque lake begins in the Older Dryas and indicates a complete and early deglaciation of the Madrès cirques (Reille & Lowe, 1993).

2. La Borde (1650 m, Carlit massif, eastern Pyrenees). Pollen sequence in the infilling of the terminal basin of the Têt glacier began in the Upper Pleniglacial and indicates an earlier glacial retreat (Reille, 1990).

3. The Puigcerda morainic complex (1200 m, Cerdagne). This is the only site that clearly shows three morainic sequences, two pre-Würmian, and their relationship to alluvial terrace deposits. The very large exposure in the glaciofluvial complex at Ur is of early Middle Pleistocene age (Calvet, 1996, 1998).

4. The karstic network of Niaux-Lombrives (600 m, Ariège) which also includes glacial or glaciofluvial sediments that indicate at least three or four glacial phases dated by $^{230}Th/^{234}U$ to the Late to Middle Pleistocene (Sorriaux, 1981).

5. Freychinède (1350 m, Ariège). The lake and peat bog pollen sequence the base of which is dated to 21,000 B.P. indicates the minimum age of the disjunction phase of the Würmian glaciers (Jalut *et al.*, 1982).

6. Llestui (1600 m, Noguera Ribagorzana). A lateral ice dammed lake beside the Llauset glacier moraine indicates a readvance phase following the Würmian glacial maximum. Absolute dates from these lake sediments gave 18,000 - 21,000 B.P. for this readvance phase (Bordonau *et al.*, 1993; Vilaplana & Bordonau, 1989).

7. Sost (650 m, Haute-Garonne). Lateral ice dammed lake at the diffluence of the Garonne glacier and equivalent to the Würmian glacial maximum. The pollen sequence has been attributed to the Early Würmian before 45,000 B.P., by comparison with that at Biscaye (Lourdes) (Hérail & Jalut, 1986).

8. Biscaye peat bog (400 m, Lourdes) in the depression immediately behind the frontal Würmian moraine formed of unweathered material and related to the lowest terrace. It has provided a pollen sequence from the last 45,000 years and demonstrates that the glacial maximum dates from the Early Würmian and that the deglaciation began in the Lourdes basin at 29,500 B.P. (Mardones & Jalut, 1983).

9. Estarrès peat bog (380 m, Arudy), situated in a depression formerly occupied by a lateral lobe of the Ossau glacier and behind Würmian terminal moraines. The deposits show, similar to those at Lourdes, that deglaciation of the Arudy basin began at 24,000 B.P. (Andrieu, 1987).

10. La Pierre St Martin Caves (P. d'Anie), with three glacial stages during the Middle Pleistocene, $^{230}Th/^{234}U$-dated (Maire, 1990).

Acknowledgement

This article was translated by P.L. Gibbard.

References:

Alimen, H. (1964). Le Quaternaire des Pyrénées de la Bigorre. *Mémoire du Service de la Carte Géologique de France*, 394 pp.

Alimen, H. (1967). The Quaternary of France: the Pyrénées. *In:* Rankama, K. (ed.), *The Quaternary*, **2**, 156-177. Wiley & Sons.

Andrieu, V. (1987). Le paléoenvironnement du piémont nord-pyrénéen occidental de 27 000 B.P. au Postglaciaire: la séquence d'Estarrès (Pyrénées-Atlantiques, France). *Compte-rendu à l'Académie des Sciences de Paris*, **304, II, 2**, 103-108.

Andrieu, V., Hubschman, J., Jalut, G. & Hérail, G. (1988). Chronologie de la déglaciation des Pyrénées françaises.

Bulletin de l'Association Française pour l'Etude du Quaternaire, Paris, 2° série, **34-35, (2/3)**, 55-67.

Arenillas, M., Martínez, R., Martínez De Pisón, E. & Navarro, J. (1998). Los glaciares actuales y la nieve en el Pirineo español. *In:* Gómez Ortiz, A. & Pérez Alberi, A. (eds), *Las huellas glaciares de las montañas españolas,* Universidade de Santiago de Compostela, 329-346.

Bakalowicz, M., Debroas, E.J. & Puyoô S. (1980). Nouveaux témoins glaciaires dans le massif d'Arbas. *Bulletin de la Société d'Histoire Naturelle de Toulouse*, **116, 1-2**, 52-56.

Bakalowicz, M., Sorriaux, P. & Ford, D.C. (1984). Quaternary glacial events in the Pyrenees from U. series dating of speleothems in the Niaux-Lombrives-Sabart Caves, Ariège, France. *Norsk Geografisk Tidsskrift*, **38**, 193-197.

Barrère, P. (1954). Equilibre glaciaire actuel et quaternaire dans l'Ouest des Pyrénées centrales. *Revue Géographique des Pyrénées et du Sud-Ouest*, **XXIV, 2**, 116-134.

Barrère, P. (1963). La période glaciaire dans l'Ouest des Pyrénées centrales franco-espagnoles. *Bulletin de la Société Géologique de France*, **(7), V**, 516-526.

Barrère, P. (1966). La morphologie quaternaire dans la région de Biescas et de Sabiñanigo (Haut-Aragon). *Bulletin de l'Association Française pour l'Etude du Quaternaire*, **2**, 83-93.

Barrère, P. (1971). Le complexe morainique d'Arudy (Pyrénées Atlantiques). *Photointerprétation*, **3, (3)**, Paris, éd. Technip, 19-24

Barrère, P., Bois, J.P., Soulé, J.C. & Ternet, Y. (1980). *Notice de la feuille géologique Argelès-Gazost au 1.50000, n° 1070*, Orléans, BRGM édit., 46 pp.

Bertrand, A. (1963). Le modelé glaciaire des versants du val d'Ariège. *Revue Géographique des Pyrénées et du Sud-Ouest*, Toulouse, **34, 3**, 255-272.

Birot, P. (1937). *Recherches sur la morphologie des Pyrénées orientales franco-espagnoles*, Thèse Lettres, Paris, Baillère édit., 318 pp.

Bordonau, J. (1992). *Els Complexos glacio-lacustres relacionats amb el darrer cicle glacial als Pirineus.* Logroño, Geoforma edit., 251 pp.

Bordonau, J., Vilaplana, J.M. & Fontugne, M. (1993). The glacio-lacustrine complex of Llestui (Central Southern Pyrenees) : a key-locality for the chronology of the last glacial cycle in the Pyrenees. *Compte-rendu à l'Académie des Sciences de Paris*, **II**, **316**, 807-813.

Brunet, R. (1956). Un exemple de la régression des glaciers Pyrénéens. *Pirineos*, **39-42**, 261-283.

Calvet, M. (1996). *Morphogenèse d'une montagne méditerranéenne, les Pyrénées orientales*. Documents du BRGM, n° 255, Orléans, BRGM édit., 1177 pp., 3 vols.

Calvet, M. (1998). Los complejos fluvioglaciares de Cerdanya-Capcir (Pirineos orientales) y sus enseñanzas, *In:* Gómez Ortiz, A. & Pérez Alberi, A. (eds), *Las huellas glaciares de las montañas españolas,* Universidade de Santiago de Compostela, 263-290.

Chevalier, M. (1951). Note sur la morphologie glaciaire du haut Salat. *Revue Géographique des Pyrénées et du Sud-Ouest*, **XXII, 2-3**, 207-216.

Chevalier, M. (1954a). Le relief glaciaire des Pyrénées du Couserans. I. Les cirques. *Revue Géographique des Pyrénées et du Sud-Ouest*, **XXV, 2**, 97-124.

Chevalier, M. (1954b). Le relief glaciaire des Pyrénées du Couserans. II. Les vallées. *Revue Géographique des Pyrénées et du Sud-Ouest*, **XXV, 3**, 189-220.

Chueca, J., Peña Monne, J.L., Lampre, F. & Julian, A. (1998). La pequeña edad del hielo en el Pirineo central meridional: inferencias paleoambientales a partir de datos geomorfológicos. *In:* Gómez Ortiz, A. & Pérez Alberi, A. (eds), *Las huellas glaciares de las montañas españolas,* Universidade de Santiago de Compostela, 307-328.

Clotet, N., Gallard, F. & Tricart, J. (1985). Quelques témoins de l'existence d'englacements anté-würmiens dans les Pyrénées catalanes. *Colloque «Héritages géomorphologiques et paléoenvironnements du Quaternaire moyen méditerranéen»*, Paris 16-11-1985, *Physio-Géo*, **14-15**, 87-90.

Copons, R. & Bordonau, J. (1994). La pequeña edad del hielo en el macizo de la Maladeta (alta cuenca del Esera, Pirineos centrales). *In:* Martí Bono, C.E. & García Ruiz, J.M. (eds), *El glaciarismo surpirenaico: nuevas aportaciones*, Logroño, Geoforma édit., 111-124.

Delmas, M. (1998). *Le massif du Carlit, étapes de la déglaciation et dynamiques holocènes*, Mémoire de DEA (inédit), Université de Provence, 144 pp.

Gangloff, P., Courchesne, F., Hétu, B., Jalut, G. & Richard, P. (1991). Découverte d'un paléolac sur le piémont des Pyrénées Atlantiques (France). *Zeitschrift für Geomorphologie N.F.*, **35, 4**, 463-478.

García Ruiz, J.M. (1989). *Sallent (M.T.N. 145), mapa geomorfológico,* Geoforma édit., Cartografía n°1, Logroño, notice 29 pp.

García Ruiz, J.M., Bordonau, J., Martinez De Pison, E. & Vilaplana, J.M. (1992). *Benasque (M.T.N. 180), mapa geomorfológico*, Geoforma édit., Cartografía n° 4, Logroño, notice 39 p.

García Ruiz, J.M. & Martí Bono, C.E. (1994). Rasgos fundamentales del glaciarismo cuaternario en el Pirineo aragonés. *In:* Martí Bono, C.E. & García Ruiz, J.M. (eds), *El glaciarismo surpirenaico: nuevas aportaciones*, Logroño, Geoforma édit., 17-31.

Gellatly, A.F., Grove, J.M. & Switsur, V.R. (1992). Mid-Holocene glacial activity in the Pyrenees. *The Holocene*, **2, 3**, 266-270.

Gielen, P. (1984). Délimitation du domaine englacé sur le socle cristallin dans le massif du Carlit et les bassins du Capcir et de la Cerdagne (Pyrénées orientales). *Colloque «Montagnes-Piémonts», Revue Géographique des Pyrénées et du Sud-Ouest, Travaux*, **I**, 287-298.

Gómez Ortiz, A. (1985). Morfología glacial del macizo de Calmquérdos. Los heleros instalados en su borde meridional. *Notes de Geografia fisica*, **13-14**, 85-96, Universitat Barcelona.

Gómez Ortiz, A. (1987). *Contribució geomorfològica a l'estudi dels espais supraforestals pirinencs. Gènesi, organizació i dinàmica dels modelats glacials i periglacials de la Cerdanya i l'Alt Urgell.* Barcelona, Institut Cartogràfic de Catalunya edit., 161 pp.

Gómez Ortiz, A. & Salvador i Franch, F. (1994). Glaciarismo en el extremo oriental del Pirineo: Baixa Cerdanya y sur de Andorra. *In:* Martí Bono, C.E. & García Ruiz, J.M. (eds), *El glaciarismo surpirenaico: nuevas aportaciones*, Logroño, Geoforma édit., 65-90.

Goron, L. (1941). *Le rôle des glaciations quaternaires dans le modelé des vallées maîtresses des Pré-Pyrénées ariégeoises et garonnaises et leur avant-pays*, Privat édit., Toulouse, 460 pp. Version abrégée*: Revue Géographique des Pyrénées et du Sud-Ouest*, **XII**, 5-61, 147-226, 322-356, 373-430.

Gourinard, Y. (1971). Les moraines de la basse vallée du Carol entre Latour et Puigcerda (Pyrénées orientales franco-espagnoles), *Compte-rendu à l'Académie des Sciences de Paris*, **D 272**, 3112-3115.

Hazéra, J. (1975). Morphologie et chronologie glaciaire dans les Pyrénées atlantiques et centrales. *Revue Géographique des Pyrénées et du Sud-Ouest*, **46, 3**, 338-347.

Hérail, G. & Jalut, G. (1986). L'obturation de Sost (Haute-Garonne) : données nouvelles sur le paléoenvironnement de la phase de progression du glacier würmien dans les Pyrénées centrales. *Compte-rendu à l'Académie des Sciences de Paris*, **II, 303, 8**, 743-748.

Hérail, G., Hubschman, J. & Jalut, G. (1987). Quaternary glaciation in the French Pyrenees, *Quaternary Science Reviews*, **5**, 397-402.

Hétu, B. & Gangloff, P. (1989). Dépôts glaciaires du Pléistocène inférieur sur le piémont nord des Pyrénées Atlantiques. *Zeitschrift für Geomorphologie N.F.*, **33, 4**, 385-403.

Hubschman, J. (1984). Glaciaire ancien et glaciaire récent: analyse comparée de l'altération de moraines terminales nord-pyrénéennes. *Colloque «Montagnes-Piémonts», Revue Géographique des Pyrénées et du Sud-Ouest, Travaux*, **I**, 313-332.

Jalut, G., Delibrias, G., Dagnac, J., Mardones, M. & Bouhours, M. (1982). A palaeoecological approach to the last 21 000 years in the Pyrenees: the peat bog of Freychinède (alt. 1350 m, Ariège, South France). *Palaeogeography, Palaeoclimatology, Palaeoecology*, **40**, 321-359.

Jalut, G. & Mardones, M. (1984). Évolution climatique et dynamique glaciaire en Ariège et sur le piémont de Lourdes, depuis environ 50 000 BP jusqu'à 10 000 BP. *Colloque «Montagnes-Piémonts», Revue Géographique des Pyrénées et du Sud-Ouest, Travaux*, **I**, 313-332.

Maire, R. (1990). *La haute montagne calcaire. Karst, cavités, remplissages, Quaternaire, paléoclimats*. Thèse Doct. État, Karstologia – Mémoires, **3**, 731 pp.

Mardones, M. & Jalut, G. (1983). La tourbière de Biscaye (alt. 409 m, Hautes-Pyrénées): approche paléoécologique des 45 000 dernières années. *Pollen et Spores*, **XXV, 2**, 163-212.

Martí Bono, C.E. (1973). Nota sobre los sedimentos morrénicos del río Aragón. *Pirineos*, **107**, 39-46.

Martí Bono, C.E. (1978). Aspectos de la problemática geomorfológica del alto Aragon occidental. *Estudios Geográficos*, **XXXIX, 153**, 473-493.

Martí Bono, C.E. & García Ruiz, J.M. (eds) (1994). *El Glaciarismo Surpirenaico, nuevas aportaciones*. Logroño, Geoforma édit., 142 pp.

Mianes, A. (1955). Les anciens glaciers des Pyrénées garonnaises. *Pirineos*, **35-38**, 5-71.

Montserrat Marti, J.M. (1992). *Evolución glaciar y postglaciar del clima y la vegetación en la vertiente sur del Pirineo. Estudio palinológico*. Monografías del Instituto Pirenaico de Ecología, **6**, CSIC, 147 pp.

Nussbaum, F. (1943). Orographische und morphologische untersuchungen in den östlichen Pyrenäen. *Jahresbericht der Geographischen Gesellschaft von Bern*, **XXXV, 1942-1943**, 1-148 (1945).

Nussbaum, F. (1956). Observations morphologiques dans la région de la Noguera Pallaresa. *Pirineos*, **39-42**, 57-95.

Nussbaum, F. (1962). Réflexions sur quelques traits morphologiques des Pyrénées. *Actes 2° Congrès international d'études Pyrénéennes, Luchon-Pau* (1954), **5, IV**, 5-11.

Panzer, W. (1932). Die eiszeitlichen Endmoränen von Puigcerda (Ostpyrenäen). *Zeitschrift für Gletscherkunde*, **18**, 411-421.

Paquereau, M. & Barrère, P. (1964). Palynologie et morphologie quaternaires dans les environs d'Arudy. *Actes 4° Congrès international d'études Pyrénéennes, Pau-Lourdes*, **4, IV**, 18-25.

Peña Monné, J.L., Chueca Cia, J. & Lampre Vitaller, F. (1998). Estado actua del conocimiento del glaciarismo pleistoceno de la vertiente meridional del Pirineo central, *in*: Gómez Ortiz, A. & Pérez Alberti, A. (eds.), *Las huellas glaciares de las montañas españolas*, Universidade de Santiago de Compostela, 211-238.

Penck, A. (Traduction L. Braemer) (1885). La période glaciaire dans les Pyrénées. *Bulletin de la Société d'Histoire Naturelle de Toulouse*, **XIX**, 105-200.

Penck, A. (1894). Studien über das Klima Spaniens während der jungeren Tertiärperiode und der Diluvialperiode. *Zeitschrift der Gesellschaft für Erdkunde*, **29**, 109-141.

Prat, M.Cl. (1980). *Montagnes et vallées d'Andorre: étude géomorphologique*, Thèse de 3° cycle inédite, Univ. Bordeaux III, 266 pp.

Reille, M. (1990). La tourbière de La Borde (Pyrénées orientales, France), un site clé pour l'étude du Tardiglaciaire sud européen. *Compte-rendu à l'Académie des Sciences de Paris*, **II, 310**, 823-829.

Reille, M. & Andrieu, V. (1993). Variations de la limite supérieure des forêts dans les Pyrénées (France) pendant le Tardiglaciaire. *Compte-rendu à l'Académie des Sciences de Paris*, **II, 316**, 547-551.

Reille, M. & Lowe, J.J. (1993). A re-evaluation of the vegetation history of the eastern Pyrenees (France) from the end of the last glacial to the Present. *Quaternary Science Reviews*, **12**, 47-77.

René, P. (2001). *Les glaciers des Pyrénées françaises: campagne d'observation de septembre 2000*. Réunion Société Hydrotechnique de France, sect. Glaciologie, Grenoble, 15-16 mars 2001, 16 pp.

Serrano, E. & Martinez De Pisón, E. (1994). Geomorfología y evolución glaciar en el Pirineo aragonés oriental. *In:* Martí Bono, C.E. & García Ruiz, J.M. (eds), *El glaciarismo surpirenaico: nuevas aportaciones*, Logroño, Geoforma édit., 33-64.

Serrat, D. (1974). Nuevos datos sobre glaciarismo en el Pirineo oriental. *Actas 1° reunión nacional del Grupo de trabajo del Cuaternario*, Madrid oct. 1973, *Trabajos sobre Neogeno-Cuaternario*, CSIC, 175-179.

Serrat, D. (1977). *Estudio geomorfológico del Pirineo oriental (Puigmal-Costabona)*. Tesis Doctoral, Univ. Barcelona, 221 pp.

Serrat, D. (1982). Modelado glaciar de la vertiente sur del Pirineo oriental. *Pirineos*, **116**, 13-19.

Serrat, D., Bordonau, J., Bru, J., Furdada, G., Gomez, A., Marti, J., Marti, M., Salvador, F., Ventura, J. & Vilaplana, J.M. (1994). Síntesis cartográfica del glaciarismo surpirenaico oriental. *In:* Martí Bono, C.E. & García Ruiz, J.M. (eds), *El glaciarismo surpirenaico: nuevas aportaciones*, Logroño, Geoforma édit., 9-15.

Sorriaux, P. (1981). Étude et datation de remplissages karstiques: nouvelles données sur la paléogéographie quaternaire de la région de Tarascon (Pyrénées Ariégeoises). *Compte-rendu à l'Académie des Sciences de Paris*, **II, 293**, 703-706.

Taillefer, F. (1957). Le glaciaire pyrénéen: versant nord et versant sud. *Revue Géographique des Pyrénées et du Sud-Ouest*, **XXVIII, 3**, 221-244.

Taillefer, F. (1960). Les terrasses d'obturation glaciaire de la vallée de Saurat (Ariège). *Revue Géographique des Pyrénées et du Sud-Ouest*, **31, 1**, 45-61.

Taillefer, F. (1963). La carte de morphologie glaciaire des Pyrénées au 1.50 000, feuilles de Foix et de Vicdessos. *Revue Géographique des Pyrénées et du Sud-Ouest*, **34, 2**, 5-10.

Taillefer, F. (1966). La feuille de Bagnères de Luchon de la carte du relief glaciaire des Pyrénées. *Pirineos*, **81-82**, 129-142.

Taillefer, F. (1967). The extent of Pleistocene glaciation in the Pyrenees. *In*: Wright, H.E. & Osburn, W.H. (eds), *Arctic and Alpine environments, Proceedings of the VIIth INQUA Congress*, **10**, 255-266, Indianapolis, Indiana University Press.

Taillefer, F. (1969a). Les glaciations des Pyrénées. *Actes du VIII° Congrès international de l'INQUA, supplément au Bulletin de l'Association Française pour l'Etude du Quaternaire*, Paris, 19-32.

Taillefer, F. (1969b). Terrasses de la vallée sous-pyrénéenne de la Neste-Garonne et morphologie glaciaire de la vallée de la Garonne. Le Quaternaire de la vallée de l'Ariège. *Livret-guide excursion A 6, VIII° Congrès INQUA*, Paris, 38-47.

Taillefer, F. (1977). Le glacier de l'Ariège dans le bassin de Tarascon. *Revue Géographique des Pyrénées et du Sud-Ouest*, **48, 3**, 269-286.

Taillefer, F. (1980). les oscillations des glaciers fini- et post-glaciaires des Pyrénées orientales. *Colloque «Oscillation fini et post glaciaire des glaciers»*, Trier, 231-244.

Taillefer, F. (1982). Les conditions locales de la glaciation pyrénéenne. *Pirineos*, **116**, 5-12.

Taillefer, F. (1984). La terminaison du glacier de la Garonne au sud de Montréjeau. *Bulletin de la Section de Géographie, Comité des Travaux Historiques et Scientifiques*, **66**, 271-282.

Taillefer, F. (1985). Idées actuelles sur les glaciations dans les Pyrénées de l'Ariège. *Revue Géographique des Pyrénées et du Sud-Ouest*, **56, 3**, 323-338.

Viers, G. (1960). *Pays Basque français et Barétous, le relief des Pyrénées occidentales et leur piémont*. Toulouse, Privat édit., 604 pp.

Viers, G. (1961). Le glaciaire du massif du Carlit (Pyrénées orientales) et ses enseignements. *Revue Géographique des Pyrénées et du Sud-Ouest*, **32, 1**, 5-33.

Viers, G. (1963). Les moraines externes de la Cerdagne et du Capcir (Pyrénées orientales, France) et leurs rapports avec les terrasses alluviales. *Actes VI° Congrès INQUA, Varsovie 1961*, **III**, 385-393.

Viers, G. (1966). La glaciation quaternaire dans le massif du Canigou (Pyrénées orientales, France). *Pirineos*, **81-82**, 87-94.

Viers, G. (1968). La carte du relief glaciaire des Pyrénées. Feuille Mont-Louis au 1.50 000. *Revue Géographique des Pyrénées et du Sud-Ouest*, **39, 4**, 429-434.

Viers, G. (1971). L'englacement quaternaire des Pyrénées orientales et ses problèmes climatiques. *Colloque «Les milieux naturels supraforestiers des montagnes du bassin occidental de la Méditerranée»*, Perpignan, 57-64.

Vilaplana, J.M. & Serrat, D. (1981). Els deposits d'origen glacial de la cubeta de la Massana-Ordino (Andorra) : significació paleogeográfica. *Acta Geologica Hispanica*, **14 (1979)**, 433-440.

Vilaplana, J.M. (1984). *Estudi del glacialisme de los Valls de la Valira d'Ordino i d'Arinsal (Andorra)*. Institut d'Estudis Catalans, Arxius de la secció de Ciencies, **LXXII**, Barcelona, 84 pp.

Vilaplana, J.M. & Bordonau, J. (1989). Dynamique sédimentaire lacustre de marge glaciaire: le paléo-lac de Llestui (Noguera Ribagorzana, versant sud des Pyrénées). *Bulletin de l'Association Française pour l'Etude du Quaternaire*, **4**, 219-224.

Quaternary Glaciations – Extent and Chronology
Editors J. Ehlers and P.L. Gibbard

Late Pleistocene (Würmian) glaciation of the Caucasus

Ramin Gobejishvili

Vakhushti Bagrationi Institute of Geography, 1, Merab Aleksidze str. Block 8
380093, Tbilisi, Georgia

E-mail: geogr@gw.acnet.ge

Introduction

The shaping of the high mountain relief of the Caucasus is closely connected to the geological and geomorphological activity of ancient glaciations. As on other high mountains, the glaciers left deep traces on the relief of the Caucasus in the form of diverse and variously-sized landforms. Traces of the past glaciations in various states of preservation are found in the river basins. They allow detailed reconstruction of the process of Upper Pleistocene and Holocene glaciation. The investigations of the old glaciations in the Caucasus started in the 19th century. After publication of the works of Penck & Brückner (1909/11), many scientists attempted to correlate the glaciations of the Caucasus with those in the Alps (e.g. Reingard, 1937; Vardanyants, 1937; Tsereteli, 1966; 'Papers of Caucasian expedition', 1960-62, Shcherbakova, 1973). According to these authors, the entire Caucasus was either covered or partially covered by ice, with glaciers emerging from the mountains and moving down to very low altitudes, implying depression of the firn line by some 1100-1300m. Maruashvili (1956) opposed a mechanical application of the Alpine scheme to the Caucasus. He considered that the Late Pleistocene glaciation was of a much more limited character, the depression of the firn line being only about 600-800m. Since then, a number of new investigations, using different methods for the reconstruction of the old glaciations have been published (e.g. Kovalev, 1961; Khazaradze, 1968; Gobejishvili, 1995; Bondarev *et al.*, 1997).

Method of investigation

The Caucasus is of great interest from a theoretical point of view because it allows the comparison of the great difference in regime and dynamics of glaciers formed under a maritime and a continental climate. Because the Caucasus is located mid-way between the Alps and the mountains of Central Asia, they can provide indispensible information for the modelling of natural processes across vast regions.

However, the reconstruction of the extent and morphology of the old glaciations in the Caucasus is difficult. This is because the glaciations have often left only weak traces and there is a widespread occurrence of landforms and sediments resembling glacial deposits. In many regions the original trough forms of formerly glaciated valleys have been greatly changed by weathering, erosion and slope denudation. In these situations, for example, lateral moraines are usually either mostly or completely destroyed. End moraines on the valley floors, which were greatly subjected to glaciofluvial redeposition and mud flows, may be completely removed by streams that redeposited glacial outwash from their original positions to lower altitudinal levels.

In addition, very active exogenetic processes may lead to the formation of assemblages of moraine-like landforms, debris cones and mud-flow deposits. Traditional glacial geomorphological methods do not always allow these features to be differentiated and to assign them to a particular age. Consequently, significant discrepancies exist in the reconstructions of the extent of former glaciations in practically all fthe ormerly glaciated Eurasian mountain regions. Until recently the evidence of modern glaciation have been practically ignored in all investigations into these palaeo-glaciological problems. In more recent studies, however, the relationship between feeding area and length of modern glaciers has been compared to similar parameters from the old glaciations. For example, the length of the Late Pleistocene glaciers and their cirque areas was evaluated by comparison with the data from well-investigated modern glaciated regions. This relationship can be illustrated by the formula

$$Lc/Sc = Ld/Sd$$

Where:
Sc = the snowline altitude of modern glaciers,
Lc = the length of modern glaciers,
Sd = the snowline altitude of Pleistocene glaciers,
Ld = the length of Pleistocene glaciers.

Using this formula the definitions of Sc and Lc were determined by evaluation of topographic maps for all valley glaciers of Georgia and for some large glaciers of the Tien Shan, Spitsbergen, Tibet and the Himalayas. According to the relationship

$$Lc/Sc = K$$

four groups of glaciers could be identified:

- Hanging valley glaciers and simple valley glaciers K= 0,81 (s = 0,96),
- Valley glaciers with multiple source areas K=0,50 (s = 0,95),

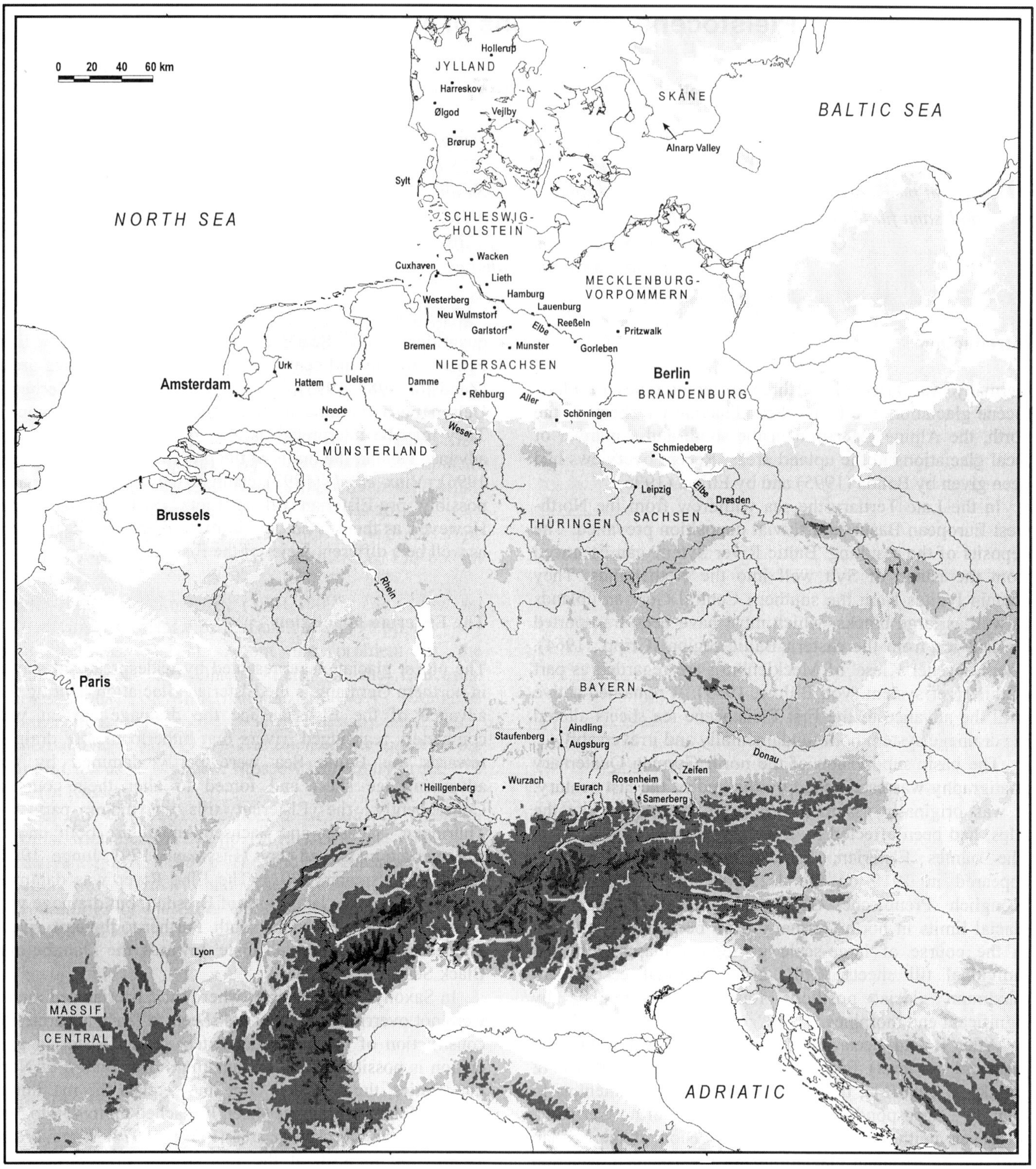

Fig. 1. Location map.

The sedimentary sequence of each advance begins with varved clays, overlain by till and ending with various types of meltwater deposits (Eissmann *et al.*, 1995). The advance velocity of the first Elsterian ice sheet ranged between 600 and 900 m/a (Junge, 1998). Many of the distinct thrust moraines (southern Dübener Heide, Dahlener Heide areas) were probably formed in the Elsterian. Even the 'Muskauer Faltenbogen' (Muskau arch of folds) has recently been re-interpreted as Elsterian in age (Kupetz *et al.*, 1989; Eissmann, 1997).

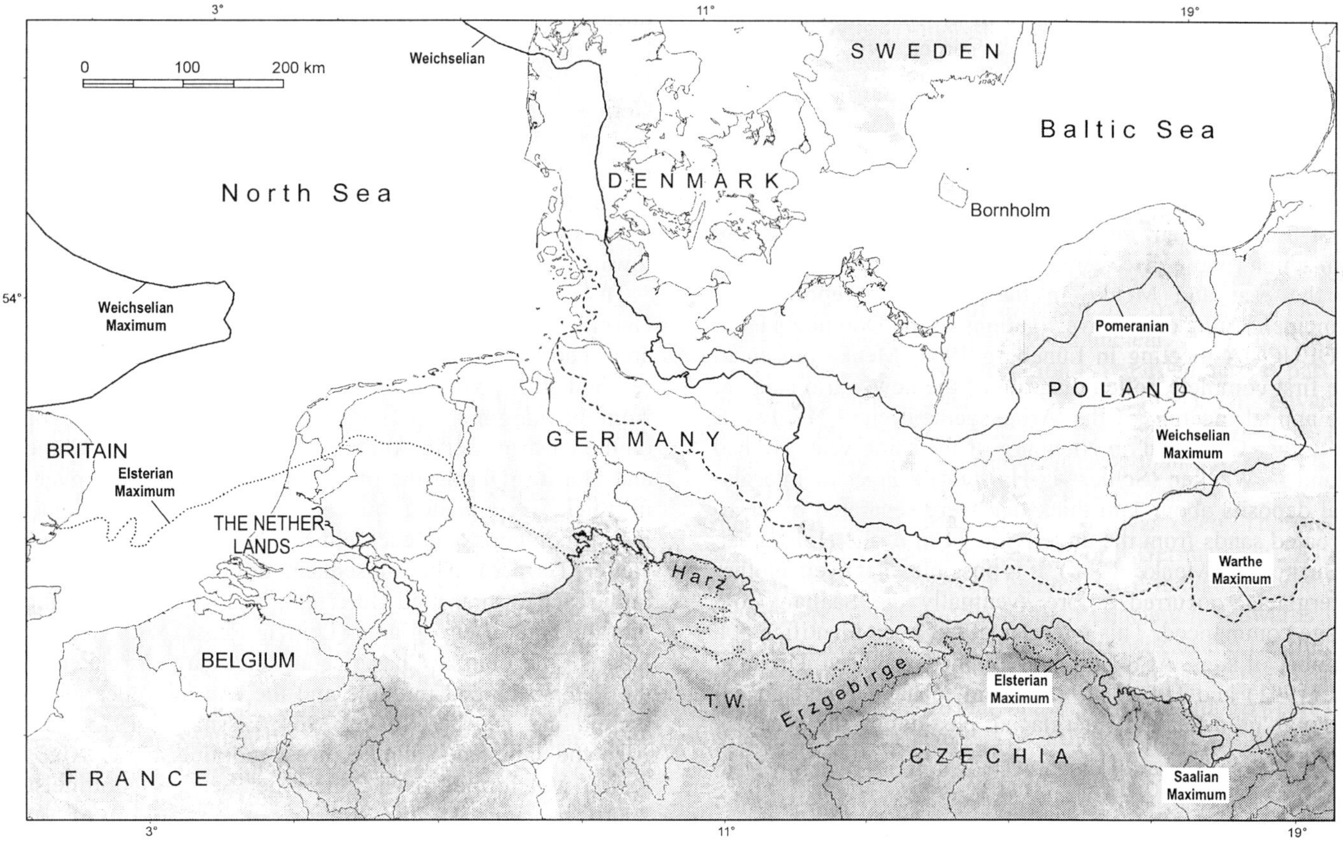

Fig. 2. Glacial limits and end moraines in Germany (converted from digital maps).

During both Elsterian ice advances major erosional zones formed subglacially, partly by meltwater erosion, partly by glacial erosion. The resulting channels connect with the deeply incised buried 'valleys' beneath the North German Lowland (Eissmann, 1967, 1987; Eissmann & Müller, 1979). In Brandenburg and Lower Saxony two separate glaciation phases have been distinguished as well, but the coerrelation with the Elbe-Saale area is tentative only. In the short period of ice retreat between the two advances meltwater sediments were accumulated, devoid of any organic matter.

In the southern parts of Niedersachsen as well as in Nordrhein-Westfalen Elsterian glaciation in most cases can only be inferred from re-deposited Scandinavian erratics in the Middle Terrace. The thrust sequences of the end moraines west of the Weser River contain no Elsterian till (Skupin *et al.* 1993). In contrast to earlier works (Thome, 1980; Klostermann, 1985, 1992) at present most workers maintain that the Elsterian maximum lies north of Osnabrück.

The Holsteinian Interglacial

The term 'Holsteinian' was introduced by Penck (1922), who originally thought that the marine deposits belonged to a post-Eemian, pre-Holocene transgression. This was corrected by Grahle (1936). The Holsteinian Interglacial started with the transition from sub-arctic to boreal climate at the end of the Elsterian and terminated with the deterioration of climate at the onset of the Saalian Complex (Jerz & Linke, 1987).

The sea invaded parts of Jutland and North Germany already at the end of the preceding Elsterian cold stage (Hinsch, 1993, Knudsen, 1993). Marine Late Elsterian deposits in Schleswig-Holstein can be traced as far inland as Kellinghusen, about 50 km from the present North Sea coast. This early transgression is interpreted as a result of isostatic depression of the land surface.

During the Holsteinian Warm Stage Northwest Europe experienced a second, major transgression. For North Germany it was the first marine transgression since the Miocene. The Holsteinian lake deposits show a complete interglacial vegetational succession from subarctic-boreal through temperate to boreal conditions. The interglacial climate was not interrupted by any major climatic fluctuations.

The Saalian Complex

Lower Saalian

The early Saalian cold stage, i.e. the period between the end of the Holsteinian Stage and the first Saalian ice advance, is characterised by extensive valley widening and intensive accumulation of fluvial gravels. These gravel complexes reflect the changing climatic conditions in the Lower each of which could be subdivided into several belts of end moraines. The morphological subdivision of the glaciation, however, is not reflected in the stratigraphy. According to clast-lithological investigations by Cepek (1962), only two Weichselian till units could be distinguished that could be correlated with the Brandenburg and Pomeranian Phases. The end moraines of the so-called 'Frankfurt Stage' thus only represent landforms created by an oscillation during the ice retreat from the maximum Brandenburg ice marginal position (Cepek, 1965). However, minor oscillations of the ice margin have locally deposited more than two till units (e.g. in Berlin; Böse, 1979).

In Mecklenburg, as elsewhere in east Germany, originally only two tills of the Weichselian Glaciation could be distinguished (Cepek, 1972). However, Heerdt (1965) identified a thin, third till unit which he correlated with the 'Rosenthal end moraine'. This till, according to more recent investigations, can be distinguished from the two older Weichselian tills not only stratigraphicly, but also based on its clast composition (Rühberg, 1987). The distribution of this youngest Weichselian till is limited to the area north of the Rosenthal end moraine. It is correlated with a Late Weichselian 'Mecklenburg Advance' (Eiermann, 1984). In cases where the ice sheet encountered steep slopes major push moraines were formed. By geological mapping it could be demonstrated that this ice advanced south to Malchin and Laage. Most eskers in Mecklenburg-Vorpommern were formed during the 'Mecklenburg Advance', whereas major sandurs of this phase are lacking (Rühberg, 1987).

Recently Müller, Rühberg & Krienke (1995) and Rühberg (1999) described another hitherto oldest known Weichselian till in North Germany. Its composition resembles Saalian tills. The correlative ice advance is possibly to be correlated with the early Middle Weichselian.

The three Late Weichselian ice advances into North Germany are relatively poorly dated. The maximum Brandenburg Advance is supposed to have happened about 21-22,000 B.P., but no exact dates are available. According to radiocarbon dates, the Pomeranian Advance occurred around 15,200 B.P. (Liedtke, 1996). The corrected age would be over 17,000 B.P. The Mecklenburg Advance, which must be correlated with the Sehberg Advance of Schleswig-Holstein (Stephan & Menke, 1977, Stephan, 1998), according to Lagerlund & Houmark-Nielsen (1993) is older than 14,200 radiocarbon years, or at least 16,000 calendar years old. Its deposits are overlain by sediments of the Meiendorf Interstadial (Bock *et al.*, 1985).

During the Weichselian Glaciation the lower Elbe Valley between the Havel River mouth and the North Sea served continuously as the main drainage path parallel to the ice margin, so that no changes of river course had to be made. Farther to the east, however, four to five major valley tracts can be distinguished, which as urstromtäler (pradolinas) one after the other drained the southern sector of the Weichselian ice sheet. In the Weichselian the main watershed was placed considerably farther to the east than during the Saalian Glaciation. The easternmost parts of the oldest Weichselian urstromtal, which drained the meltwaters of the Brandenburg Advance, are found at an altitude of about 190 m (Liedtke, 1981). The upper reaches of the most extensive of the four major Weichselian ice-marginal rivers originated in the area around Minsk (Belarus). Farther east the meltwaters drained through the tributaries of the Dniepr towards the east and south.

The Frankfurt ice-marginal position is connected to the Warsaw-Berlin Urstromtal. Liedtke (1957) assumes that during the Pomeranian Advance at first a Torun-Berlin Urstromtal was formed. At that time the drainage in the east via Notec and Warta Rivers and the Oderbruch was already ice-free, whereas in the west the more southerly ice margin forced the meltwaters to drain southwards via the Buckow Gap into the Berlin Urstromtal. This drainage system, however, is not quite safely established. During the Weichselian ice decay phase the drainage was first maintained via the Torun-Eberswalde Urstromtal, which remained in function until the formation of the Rosenthal end moraine. During the formation of the Velgast end moraine, according to Liedtke (1981) the ice had melted back far enough that drainage in the west could possibly be redirected via the Mecklenburger Grenztal into the depression of the western Baltic Sea and farther via the Belt to the North Sea. This youngest ice-marginal stream, the Notec-Randow Urstrom of Liedtke lost its function when finally the Vistula mouth into the Baltic became ice-free.

References:

Andersen, B.G. & Mangerud, J. (1989). The Last Interglacial-Glacial Cycle in Fennoscandia. *Quaternary International*, **3/4**, 21-29.

Benda, L. (ed.) (1995). *Das Quartär Deutschlands*. Berlin, Stuttgart, Gebrüder Bornträger, 408 pp.

Beug, H.J. (1972). Das Riß/Würm-Interglazial von Zeifen, Landkreis Laufen a.d. Salzach. *Bayerische Akademie der Wissenschaften, Mathematisch-Naturwissenschaftliche Klasse N.F.*, **151**, 46-75.

Bock, W., Menke, B., Strehl, E. & Ziemus, H. (1985). Neuere Funde des Weichselspätglazials in Schleswig-Holstein. *Eiszeitalter und Gegenwart*, **35**, 161-180.

Böse, M. (1979). Die geomorphologische Entwicklung im westlichen Berlin nach neueren stratigraphischen Untersuchungen. *Berliner Geographische Abhandlungen*, **28**, 46 pp.

Büchner, L. (1999). Zum geologischen Bau der Schmiedeberger Endmoräne. *Mitteilungen zur Geologie von Sachsen-Anhalt*, **5**, 117-134.

Cepek, A.G. (1962). Zur Grundmoränenstratigraphie in Brandenburg. *Berichte der Geologischen Gesellschaft der DDR*, **7**, 275-278.

Cepek, A.G. (1965). Stratigraphie der quartären Ablagerungen des Norddeutschen Tieflandes. *In:* Gellert, J.F. (Hrsg.), *Die Weichsel-Eiszeit im Gebiet der Deutschen Demokratischen Republik*, 45-65.

Cepek, A.G. (1972). Zum Stand der Stratigraphie der Weichsel-Kaltzeit in der DDR. Wissenschaftliche *Zeitschrift der Ernst-Moritz-Arndt-Universität Greifswald, Mathematisch-Naturwissenschaftliche Reihe*, **21**, 11-21.

Duphorn, K., Stay, B. & Stiller, D. (1979). Deglaciation of the Bara Basin near Malmö, Sweden. *Boreas*, **8**, 141-144.

Eberl, B. (1930). *Die Eiszeitenfolge im nördlichen Alpenvorlande. - Ihr Ablauf, ihre Chronologie auf Grund der Aufnahmen des Lech- und Illergletschers*. Augsburg, Benno Filser, 427 pp.

Ehlers, J. (1978). Die quartäre Morphogenese der Harburger Berge und ihrer Umgebung. *Mitteilungen der Geographischen Gesellschaft in Hamburg*, **68**, 181 pp.

Ehlers, J. (1990a). Untersuchungen zur Morphodynamik der Vereisungen Norddeutschlands unter Berücksichtigung benachbarter Gebiete. *Bremer Beiträge zur Geographie und Raumplanung*, **19**, 166 pp.

Ehlers, J. (1996). *Quaternary and Glacial Geology*. Chichester, Wiley, 578 pp.

Ehlers, J., Meyer, K.-D. & Stephan, H.-J. (1984). Pre-Weichselian glaciations of North-West Europe. *Quaternary Science Reviews*, **3**, 1-40.

Eiermann, J. (1984). Ein zeitliches, räumliches und genetisches Modell zur Erklärung der Sedimente und Reliefformen im Pleistozän gletscherbedeckter Tieflandsedimente - Ein Beitrag zur Methodik der mittelmaßstäbigen naturräumlichen Gliederung. *In:* Richer, H. & Aurada, K. (eds), *Umweltforschung. Zur Analyse und Diagnose der Landschaft*, 169-183. Gotha, Haack.

Eissmann, L. (1967). Glaziäre Destruktionszonen (Rinnen, Becken) im Altmoränengebiet des Norddeutschen Tieflandes. *Geologie*, **16**, 804-833.

Eissmann, L. (1975). Das Quartär der Leipziger Tieflandsbucht und angrenzender Gebiete um Saale und Elbe. Modell einer Landschaftsentwicklung am Rand der europäischen Kontinentalvereisung. *Schriftenreihe für Geologische Wissenschaften*, **2**, 263 pp. + Plates.

Eissmann, L. (1981). Periglaziäre Prozesse und Permafroststrukturen aus sechs Kaltzeiten des Quartärs. Ein Beitrag zur Periglazialgeologie aus der Sicht des Saale-Elbe-Gebietes. *Altenburger Naturwissenschaftliche Forschungen*, **1**, 171 pp.

Eissmann, L. (1986). Quartärgeologie und Geschiebeforschung im Leipziger Land mit einigen Schlußfolgerungen zu Stratigraphie und Vereisungsablauf im Norddeutschen Tiefland. *Altenburger Naturwissenschaftliche Forschungen*, **3**, 105-133.

Eissmann, L. (1987). Lagerungsstörungen im Lockergebirge. Exogene und endogene Tektonik im Lockergebirge des nördlichen Mitteleuropa. *Geophysik und Geologie. Geophysikalische Veröffentlichungen der Karl-Marx-Universität Leipzig*, **III (4)**, 7-77.

Eissmann, L. (1989). Die stratigraphischen und paläogeographischen Hauptbefunde im Saale-Elbe-Raum und ihre Bedeutung für die mitteleuropäische Quartärgeologie. *Ethnographisch-Ärchäologische Zeitschrift*, **30**, 596-635.

Eissmann, L. (1994). Grundzüge der Quartärgeologie Mitteldeutschlands (Sachsen, Sachsen-Anhalt, Südbrandenburg, Thüringen). *Altenburger Naturwissenschaftliche Forschungen*, **7**, 55-135.

Eissmann, L. (1997). Das quartäre Eiszeitalter in Sachsen und Nordostthüringen. *Altenburger Naturwissenschaftliche Forschungen*, **8**, 98 pp.

Eissmann, L., Litt, T. & Wansa, S. (1995). Elsterian and Saalian deposits in their type area in central Germany. *In:* Ehlers, J., Kozarski, S. & Gibbard, P.L. (eds), *Glacial deposits in North-East Europe*, 439-464. Rotterdam, Brookfield, Balkema.

Eissmann, L. & Müller, A. (1979). Leitlinien der Quartärentwicklung im Norddeutschen Tiefland. *Zeitschrift für Geologische Wissenschaften*, **7**, 451-462.

Erd, K. (1987). Die Uecker-Warmzeit von Röpersdorf bei Prenzlau als neuer Interglazialtyp im Saale-Komplex der DDR. *Zeitschrift für Geologische Wissenschaften*, **15**, 297-313.

Felix-Henningsen, P. (1983). Palaeosols and their stratigraphhical interpretation. *In:* Ehlers, J. (ed.), *Glacial Deposits in North-West Europe*, 289-295. Rotterdam, Balkema.

Felix-Henningsen, P. & Urban, B. (1982. Paleoclimatic Interpretation of a thick Intra-Saalian Paleosol, the "Bleached Loam" on the Drenthe Moraines of Northern Germany. *Catena*, **9**, 1-8.

Glapa, H. (1971). Warthezeitliche Eisrandlagen im Gebiet der Letzlinger Heide. *Geologie*, **20**, 1087-1110.

Gottsche, C. (1897a). Die tiefsten Glacialablagerungen der Gegend von Hamburg. Vorläufige Mittheilung. *Mittheilungen der Geographischen Gesellschaft in Hamburg*, **XIII**, 131-140.

Grahle, H.-O. (1936). Die Ablagerungen der Holstein-See (Mar. Interglaz. I), ihre Verbreitung, Fossilführung und Schichtenfolge in Schleswig-Holstein. *Abhandlungen der Preußischen Geologischen Landesanstalt N.F.*, **172**, 1-110.

Graul, H. (1973). Der Stand der Quartärforschung im S der BRD in lithostratigraphischer, pedologischer und geomorphologischer Hinsicht. *Heidelberger Geographische Arbeiten*, **38**, 251-265.

Gripp, K. (1924). Über die äußerste Grenze der letzten Vereisung in Nordwest-Deutschland. *Mitteilungen der Geographischen Gesellschaft in Hamburg*, **36**, 159-245.

Grube, F. (1967). Die Gliederung der Saale-(Riß-)Kaltzeit im Hamburger Raum. *Fundamenta*, **B 2**, 168-195.

Harting, P. (1874). De bodem van het Eemdal. *Verslag Koninklijke Akademie van Wetenschappen*, **N, II, (VIII)**, 282-290.

Heerdt, S. (1965). Zur Stratigraphie des Jungpleistozäns im mittleren N-Mecklenburg. *Geologie*, **14**, 589-609.

Hinsch, W. (1993). Marine Molluskenfaunen in Typusprofilen des Elster-Saale-Interglazials und des Elster-Spätglazials. *Geologisches Jahrbuch*, **A 138**, 9-34.

Höfle, H.-Ch. (1983). Periglacial phenomena. *In:* Ehlers, J. (ed.), *Glacial Deposits in North-West Europe*, 297-298. Rotterdam, Balkema.

Höfle, H.-Ch. (1991). Über die innere Struktur und die stratigraphische Stellung mehrerer Endmoränenwälle im Bereich der Nordheide bis östlich Lüneburg. *Geologisches Jahrbuch*, **A 126**, 151-169.

Höfle, H.-Ch. & Lade, U. (1983). The stratigraphic position of the Lamstedter Moraine within the Younger Drenthe substage (middle Saalian). *In:* Ehlers, J. (ed.), *Glacial Deposits in North-West Europe*, 343-346. Rotterdam, Balkema.

Hövermann, J. (1956). Beiträge zum Problem der saaleeiszeitlichen Eisrandlagen in der Lüneburger Heide. *Abhandlungen der Braunschweigischen Wissenschaftlichen Gesellschaft*, **VIII**, 36-54.

Hoffmann, K, & Meyer, K,-D. (1997). Leitgeschiebezählungen von elster- und saalezeitlichen Ablagerungen aus Sachsen, Sachsen-Anhalt und dem östlichen Niedersachsen. *Leipziger Geowissenschaften*, **5**, 115-128.

Houmark-Nielsen, M. (1994). Late Pleistocene stratigraphy, glaciation chronology and Middle Weichselian environmental history from Klintholm, Møn, Denmark. *Bulletin of the Geological Society of Denmark*, **41**, 181-201.

Jerz, H. & Linke, G. (1987). Arbeitsergebnisse der Subkommission für Europäische Quartärstratigraphie, Typusregion des Holstein-Interglazials (Berichte der SEQS 8). *Eiszeitalter und Gegenwart*, **37**, 145-148.

Junge, F.W. (1998). Die Bändertone Mitteldeutschlands und angrenzender Gebiete. *Altenburger Naturwissenschaftliche Forschungen*, **9**, 210 pp.

Junge, F.W. & Eissmann, L. (2001). Postsedimentäre Deformationsbilder in mitteldeutschen Vorstoßbändertonen - Hinweise auf den Bewegungsmechanismus des quartären Inlandeises. - *Brandenburgische Geowissenschaftliche Beiträge*, **7 (1/2)**, 21-28.

Kaltwang, J. (1992). Die pleistozäne Vereisungsgrenze im südlichen Niedersachsen und im östlichen Westfalen. *Mitteilungen aus dem Geologischen Institut der Universität Hannover*, **33**, 161 pp.

Keilhack, K. (1896). Die Geikiessche Gliederung der nordeuropäischen Glazialablagerungen. *Jahrbuch der Königlich Preußischen Geologischen Landesanstalt für 1895*, 111-124.

Klostermann, J. (1985). Versuch einer Neugliederung des späten Elster- und des Saale-Glazials der Niederrheinischen Bucht. *Geologisches Jahrbuch*, **A 83**, 46 pp.

Klostermann, J. (1992). *Das Quartär der Niederrheinischen Bucht. Ablagerungen der letzten Eiszeit am Niederrhein.* Krefeld, Geologisches Landesamt Nordrhein-Westfalen. 200 pp.

Klostermann, J., Rehagen, H.-W. & Wefels, U. (1988). Hinweise auf eine saalezeitliche Warmzeit am Niederrhein. *Eiszeitalter und Gegenwart*, **38**, 115-127.

Knoth, W. (1964). Zur Kenntnis der pleistozänen Mittelterrassen der Saale und Mulde nördlich von Halle. *Geologie*, **13**, 598-616.

Knoth, W. (1995). Sachsen-Anhalt. *In:* Benda, L. (ed.), *Das Quartär Deutschlands*, 148-170. Berlin, Stuttgart, Gebr. Bornträger.

Knudsen, K.L. (1993). Late Elsterian-Holsteinian Foraminiferal Stratigraphy in Borings of the Lower Elbe Area, NW Germany. *Geologisches Jahrbuch*, **A 138**, 97-119.

Kupetz, M., Schubert, G., Seifert, A. & Wolf, L. (1989). Quartärbasis, pleistozäne Rinnen und Beispiele glazitektonischer Lagerungsstörungen im Niederlausitzer Braunkohlengebiet. *Geoprofil*, **1**, 2-17.

Lagerlund, E. & Houmark-Nielsen, M. (1993). Timing and pattern of the last deglaciation in the Kattegat region. *Boreas*, **22**, 337-347.

Liedtke, H. (1957). Beiträge zur geomorphologischen Entwicklung des Thorn-Eberswalder Urstromtales zwischen Oder und Havel. *Wissenschaftliche Zeitschrift der Humboldt-Universität Berlin, Mathematisch-Naturwissenschaftliche Reihe*, **6**, 3-49.

Liedtke, H. (1981). Die nordischen Vereisungen in Mitteleuropa, 2. Auflage. *Forschungen zur deutschen Landeskunde*, **204**, 307 pp.

Liedtke, H. (1996). Die eiszeitliche Gestaltung des Oderbruches. *In:* Beiträge zur Physiogeographie (Festschrift für Dietrich Barsch). *Heidelberger Geographische Arbeiten*, **104**, 327-351.

Lindner, L. & Marks, L. (1999). New approach to the stratigraphy of palaeolake and glacial sediments of the younger Middle Pleistocene in mid-eastern Poland. *Geological Quarterly*, **43**, 1-8.

Lippstreu, L., Brose, F, & Marcinek, J. (1995). Brandenburg. *In:* Benda, L. (ed.), *Das Quartär Deutschlands*, 116-147. Berlin, Stuttgart, Gebr. Bornträger.

Litt, Th. (1994). Paläoökologie, Paläobotanik und Stratigraphie des Jungquartärs im nordmitteleuropäischen Tiefland unter besonderer Berücksichtigung des Elbe-Saale-Gebietes. *Dissertationes Botanicae*, **227**, 185 S. Berlin, Stuttgart, J. Cramer.

Litt, Th & Turner, Ch. (1993). Arbeitsergebnisse der Subkommission für Europäische Quartärstratigraphie, Die Saalesequenz in der Typusregion. *Eiszeitalter und Gegenwart*, **43**, 125-128.

Litt, Th., Junge, F.W. & Böttger, T. (1996). Climate during the Eemian in north-central Europe - a critical review of the palaeobotanical and stable isotope data from central Germany. *Vegetation History and Archaeobotany*, **5**, 247-256.

Lüttig, G. (1958). Methodische Fragen der Geschiebeforschung. *Geologisches Jahrbuch*, **75**, 361-418.

Lundqvist, J. (1992). Glacial stratigraphy in Sweden. *In:* Kauranne, K. (ed.), Glacial stratigraphy, engineering geology and earth construction. *Geological Survey of Finland, Special Paper*, **15**, 43-59.

Mangerud, J. (1983). The glacial history of Norway. *In:* Ehlers, J. (ed.), *Glacial Deposits in North-West Europe*, 3-9. Rotterdam, Balkema.

Mangerud, J. (1991a). The Last Ice Age in Scandinavia. *Striae*, **34**, 15-30.

Mangerud, J. (1991b). The Scandinavian Ice Sheet through the last interglacial/glacial cycle. *Paläoklimaforschung*, **1**, 307-330.

Menke, B. (1968). Beiträge zur Biostratigraphie des Mittelpleistozäns in Norddeutschland. *Meyniana*, **18**, 35-42.

Menke, B. (1985). Eem-Interglazial und "Treene-Warmzeit" in Husum/Nordfriesland. Mit einem Beitrag von Risto Tynni, unter Mitarbeit von Holger Ziemus. *Geologisches Jahrbuch*, **A 86**, 63-99.

Meyer, K.-D. (1965). Das Quartärprofil am Steilufer der Elbe bei Lauenburg. *Eiszeitalter und Gegenwart*, **16**, 47-60.

Meyer, K.-D. (1983a). Saalian end moraines in Lower Saxony. *In:* Ehlers, J. (ed.), *Glacial Deposits in North-West Europe*, 335-342. Rotterdam, Balkema.

Meyer, K.-D. (1983b). Zur Anlage der Urstromtäler in Niedersachsen. *Zeitschrift für Geomorphologie N.F.*, **27**, 147-160.

Meyer, K.-D. (1987). Ground and end moraines in Lower Saxony. *In:* van der Meer, J.J.M (ed.), *Tills and Glaciotectonics*, 197-204. Rotterdam, Balkema.

Müller, U., Rühberg, N. & Krienke, H.-D. (1995). The Pleistocene sequence in Mecklenburg-Vorpommern. *In:* Ehlers, J., Kozarski, S. & Gibbard, P.L. (eds), *Glacial deposits in North-East Europe*, 501-514. Rotterdam, Brookfield, Balkema.

Penck, A. (1879). Die Geschiebeformation Norddeutschlands. *Zeitschrift der Deutschen Geologischen Gesellschaft*, **31**, 117-203.

Penck, A. (1922). Die Eem-Schwingung. *Verhandelingen van het Geologisch-Mijnbouwkundig Genootschap voor Nederland en Koloniën, Geologische Serie*, **VI**, 91-105.

Picard, K. (1959). Gliederung pleistozäner Ablagerungen mit fossilen Böden bei Husum/Nordsee. *Neues Jahrbuch für Geologie und Paläontologie, Monatshefte*, **(6)**, 259-272.

Präger, F. (1976). Quartäre Bildungen in Ostsachsen. *Abhandlungen des Staatlichen Museums für Mineralogie und Geologie Dresden*, **25**, 125-217.

Różycki, S.Z. (1965). Die stratigraphische Stellung des Warthe-Stadiums in Polen. *Eiszeitalter und Gegenwart*, **16**, 189-201.

Rühberg, N. (1987). Die Grundmoräne des jüngsten Weichselvorstoßes im Gebiet der DDR. *Zeitschrift für Geologische Wissenschaften*, **15**, 759-767.

Rühberg, N. (1999). Über den Wert der Kleingeschiebezählungen (KGZ). *Geschiebekunde aktuell*, **15**, 87-100.

Ruske, R. (1964). Das Pleistozän zwischen Halle (Saale), Bernburg und Dessau. *Geologie*, **13**, 570-597.

Schulz, W. (1962). Gliederung des Pleistozäns in der Umgebung von Halle (Saale). *Geologie, Beiheft*, **36**, 1-69.

Schwarzbach, M. (1942. Das Diluvium Schlesiens. *Neues Jahrbuch für Mineralogie, Geologie und Paläontologie*, **86 B**, 189-246.

Sibrava, V., Bowen, D. & Richmond, G. (1986). Quarternary Glaciations in the Northern Hemisphere. *Quaternary Science Reviews*, **5**, 509 pp.

Sinn, P. (1972). Zur Stratigraphie und Paläogeographie des Präwürm im mittleren und südlichen Illergletscher-Vorland. *Heidelberger Geographische Arbeiten*, **37**, 159 pp.

Skupin, K., Speetzen, E. & Zandstra, J.G. (eds) (1993). *Die Eiszeit in Nordwestdeutschland.* Krefeld, Geologisches Landesamt Nordrhein-Westfalen. 143 pp.

Stephan, H.-J. (1982). Glazialgeologische Untersuchungen im südlichen Geestgebiet Dithmarschens. *Schriften des Naturwissenschaftlichen Vereins für Schleswig-Holstein*, **50**, 1-36.

Stephan, H.-J. (1997). Wie weit reichte die Vergletscherung der letzten Eiszeit in Schleswig-Holstein? *Die Heimat*, **104 (3/4)**, 52-57.

Stephan, H.-J. (1998). Geschiebemergel als stratigraphische Leithorizonte in Schleswig-Holstein, Ein Überblick. *Meyniana*, **50**, 113-135.

Stephan, H.-J. & Kabel, Ch. (1982). The subdivision of the glacial Pleistocene in Schleswig-Holstein and problems of correlations with adjacent areas. *XI INQUA Congress 1982, Moskva, Abstracts*, **II**, 313.

Stephan, H.-J. & Menke, B. (1977). Untersuchungen über den Verlauf der Weichsel-Kaltzeit in Schleswig-Holstein. *Zeitschrift für Geomorphologie N.F., Supplementband*, **27**, 12-28.

Stephan, H.-J., Kabel, Ch. & Schlüter, G. (1983). Stratigraphical problems in the glacial deposits of Schleswig-Holstein. *In:* Ehlers, J. (Hrsg.), *Glacial Deposits in North-West Europe*, 305-320. Rotterdam, Balkema.

Stottmeister, L. (2000). Zur Geologie der Jahrstedt-Steimker Endmoräne (SachsenAnhalt). *Brandenburgische Geowissenschaftliche Beiträge*, **7**, 127-136.

Stremme, H.E. (1960). Bodenbildungen auf Geschiebelehmen verschiedenen Alters in Schleswig-Holstein. *Zeitschrift der Deutschen Geologischen Gesellschaft*, **112**, 299-308.

Stremme, H.E. (1981). Unterscheidung von Moränen durch Bodenbildungen. *Mededelingen Rijks Geologische Dienst*, **34**, 51-56.

Thome, K.N. (1980). Der Vorstoß des nordeuropäischen Inlandeises in das Münsterland in der Elster- und Saale-Kaltzeit - Strukturelle, mechanische und morphologische Zusammenhänge. *Westfälische Geographische Studien*, **36**, 21-40.

Unger, K.-P. (1974). Die Elstervereisung des zentralen Thüringer Keuperbeckens. *Zeitschrift für Geologische Wissenschaften*, **2**, 791-800.

Urban, B. (1995). Palynological evidence of younger Middle Pleistocene Interglacials (Holsteinian, Reinsdorf and Schöningen) in the Schöningen open cast lignite mine (eastern Lower Saxony, Germany). *Mededelingen Rijks Geologische Dienst*, **52**, 175-186.

Urban, B., Lenhard, R., Mania, D. & Albrecht, B. (1991). Mittelpleistozän im Tagebau Schöningen, Ldkrs. Helmstedt. *Zeitschrift der Deutschen Geologischen Gesellschaft*, **142**, 351-372.

van der Vlerk, I.M. & Florschütz, F. (1950). *Nederland in het Ijstijdvak. De geschiedenis van flora, fauna en klimaat, toen aap en mammoet ons land bewoonden.* Utrecht, de Haan, 287 pp.

van Gijssel, K. (1987). A lithostratigraphic and glaciotectonic reconstruction of the Lamstedt Moraine, Lower Saxony (FRG). *In:* van der Meer, J.J.M. (ed.), *Tills and Glaciotectonics*, 145-155. Rotterdam, Boston, Balkema.

van Werveke, L. (1927). Norddeutschland war wenigstens viermal vom Inlandeis bedeckt. *Zeitschrift der Deutschen Geologischen Gesellschaft*, **79**, Monatsberichte, 135.

Vinx, R., Grube, A. & Grube, F. (1997). Vergleichende Lithologie, Geschiebeführung und Geochemie eines Prä-Elster-I-Tills von Lieth bei Elmshorn. *Leipziger Geowissenschaften*, **5**, 83-103.

von Bülow, W. (1969). Altpleistozäne Schotter (Loosener Kiese) in Südwestmecklenburg mit nordischen und südlichen Geröllen. *Geologie*, **18**, 563-589.

von Poblozki, B. (1995). Quaternary geology of the Altmark region. *In:* Ehlers, J., Kozarski, S. & Gibbard, P.L. (eds), *Glacial deposits in North-East Europe*, 473-484. Rotterdam, Brookfield, Balkema.

von Poblozki, B. (1999). Geologische Übersicht - Quartär,- *63. Tagung der Arbeitsgemeinschaft Nordwestdeutscher Geologen, 28. bis 31. Mai 1996 in Helmstedt, Kurzfassung der Vorträge und Exkursionsführer*, 56-74

Wagenbreth, O. (1978). Die Feuersteinlinie in der DDR, ihre Geschichte und Popularisierung. *Schriftenreihe für Geologische Wissenschaften*, **9**, 339-368.

Woldstedt, P. (1925). Die großen Endmoränenzüge Norddeutschlands. *Zeitschrift der Deutschen Geologischen Gesellschaft*, **77**, 172-184.

Woldstedt, P. (1927a). Über die Ausdehnung der letzten Vereisung in Norddeutschland. *Sitzungsberichte der Preußischen Geologischen Landesanstalt*, **2**, 115-119.

Woldstedt, P. (1927b). Die Gliederung des Jüngeren Diluviums in Norddeutschland und seine Parallelisierung mit anderen Glazialgebieten. *Zeitschrift der Deutschen Geologischen Gesellschaft, Monatsberichte*, **1927 (3/4)**, 51-52.

Woldstedt, P. (1928). Die Parallelisierung des nordeuropäischen Diluviums mit dem anderer Vereisungen. *Zeitschrift für Gletscherkunde*, **16**, 230-241.

Woldstedt, P. (1954). Saaleeiszeit, Warthestadium und Weichseleiszeit in Norddeutschland. *Eiszeitalter und Gegenwart*, **4/5**, 34-48.

Worsley, P. (1991). Possible early Devensian glacial deposits in the British Isles. *In:* Ehlers, J., Gibbard, P. & Rose, J. (eds), *Glacial Deposits in Great Britain and Ireland*, 47-51. Rotterdam, Brookfield, Balkema.

Zandstra, J.G. (1983). Fine gravel, heavy mineral and grain-size analyses of Pleistocene, mainly glacigenic deposits in the Netherlands. *In:* Ehlers, J. (ed.), *Glacial Deposits in North-West Europe*, 361-377. Rotterdam Balkema.

Zandstra, J.G. (1993). Nördliche kristalline Leitgeschiebe und Kiese in der Westfälischen Bucht und angrenzenden Gebieten. *In:* Skupin, K., Speetzen, E. & Zandstra, J.G. (eds), *Die Eiszeit in Nordwestdeutschland*, 43-106. Krefeld, Geologisches Landesamt Nordrhein-Westfalen.

Quaternary Glaciations – Extent and Chronology
Editors J. Ehlers and P.L. Gibbard

Pleistocene glaciations of South Germany

Markus Fiebig [1], Susanne J.H. Buiter [2] and Dietrich Ellwanger [3]

[1] *Bayerisches Geologisches Landesamt, Postf. 400346, D-80703 München, Germany*
[2] *Institut für Geologie, Universität Bern, Baltzerstr. 1, CH-3012 Bern, Switzerland*
[3] *Landesamt für Geologie, Rohstoffe und Bergbau, Albertstr. 5, D-79104 Freiburg, Germany*

E-mail: markus.fiebig@gla.bayern.de

Introduction

In 1899 Albrecht Penck presented an outline of the Alpine Quaternary stratigraphy which was based on a simple and convincing train of thought: he had found four periods of gravel accumulation separated from each other by valley incision. Because the gravel spreads were in contact with end moraines, Penck inferred that the gravels were melt-water deposits, laid down during periods of cold climate. On the other hand, the intervening phases of incision seemed to relate to periods when glaciers were absent. From this evidence Penck concluded that warm climate had prevailed during the incision phases.

Between 1901 and 1909 Penck & Brückner presented an overwhelming quantity of field observations concerning their four gravel spreads (1899). They gathered and discussed extensive litho-, pedo-, morpho- and bio-stratigraphic evidence from the whole circum-Alpine area. Based on this foundation Penck & Brückner defined a relative chrono- and climato-stratigraphic system of four glacial units termed Würm, Riß, Mindel and Günz. This model was readily accepted by scientists and amateurs alike, for two particular reasons: 1. Penck & Brückner's definitions could be applied to interdisciplinary evidence since they were valid for most types of investigations (pedo-, morpho-, litho-, bio-, chrono-, terrace-stratigraphic investigations). 2. Consequently the model was well-suited for interdisciplinary correlations.

Subsequently, well known and widely-accepted expansions of Penck & Brückner's terminology were published by Eberl (1930), Schaefer (1956), and Schreiner & Ebel (1981). They added the terms Donau, Biber and Haslach for three additional gravel accumulations again separated from the others by periods of valley incision. The relative chronostratigraphy was, as suggested by Penck & Brückner, reflected in the alphabetical sequence of the inital letters of the terms which are derived from the names of rivers in the northern Alpine Foreland (Fig. 1).

More recently, some authors have avoided labelling the stratigraphical results of their investigations according to the classical terminology (e.g. Ellwanger *et al.*, 1995). According to Ellwanger after deposition of the Mindel gravel spread, in the central Rhine glacier area, a glacial basin was incised and subsequently infilled by at least three lithostratigraphical units, interpreted as independent subglacial events (Fig. 2). This would imply that at least one more glacial event had occurred than the two classical glaciations (Würm and Riß) assumed to follow the Mindel. A revision of the Pleistocene Alpine stratigraphy seems to be neccessary (Doppler & Jerz, 1995, Ellwanger *et al.*, 1995) but it is not yet clear how best to approach this problem.

Stratigraphical results, interpretation and terminology

In most fields of geology, there is a clear distinction between description and interpretation. This also applies to stratigraphy. The standard descriptive stratigraphical category is lithostratigraphy. Lithostratigraphic units are 'characterised on the base of observable lithologic properties' of rocks (Salvador, 1994). In contrast, climate is not a directly observable property of a rock. Salvador's 'International Stratigraphic Guide' contains no separate chapter on climatostratigraphy. However, he points out that "Climatic changes leave a conspicuous imprint on the geologic record in the form of glacial deposits, evaporites, red beds, coal deposits, paleontological changes, and such. Since many climatic changes appear to have been regional or worldwide, their effects on the rocks provide valuable information for chronocorrelation." (p. 96). However, climatostratigraphy must be applied very carefully. Climate changes are not equivalent to colour or lithology but can only be deduced from the interpretation of the red colour or lithology. Thus the resulting chronocorrelations of glacial deposits depend heavily on the quality of the prior interpretation of the sediments and sedimentary sequences with which they are associated.

Stratigraphy is the description of rock bodies and their organisation into mappable units "based on their inherent properties" (Salvador, 1994: 13). On geological maps these mappable units should preferably be lithostratigraphical units bounded by unconformities, in order to exclude interpretation as far as possible. Special attention should be devoted to boundaries and gaps. Unfortunately, some of the terms used in the classic Alpine Quaternary stratigraphy are highly interpretative. Every effort should be taken to avoid those terms as much as possible.

The term 'glaciation' will be used here in a palaeogeographical sense, to refer to glaciers that extended down into the northern foreland of the Alps. The term 'inter-

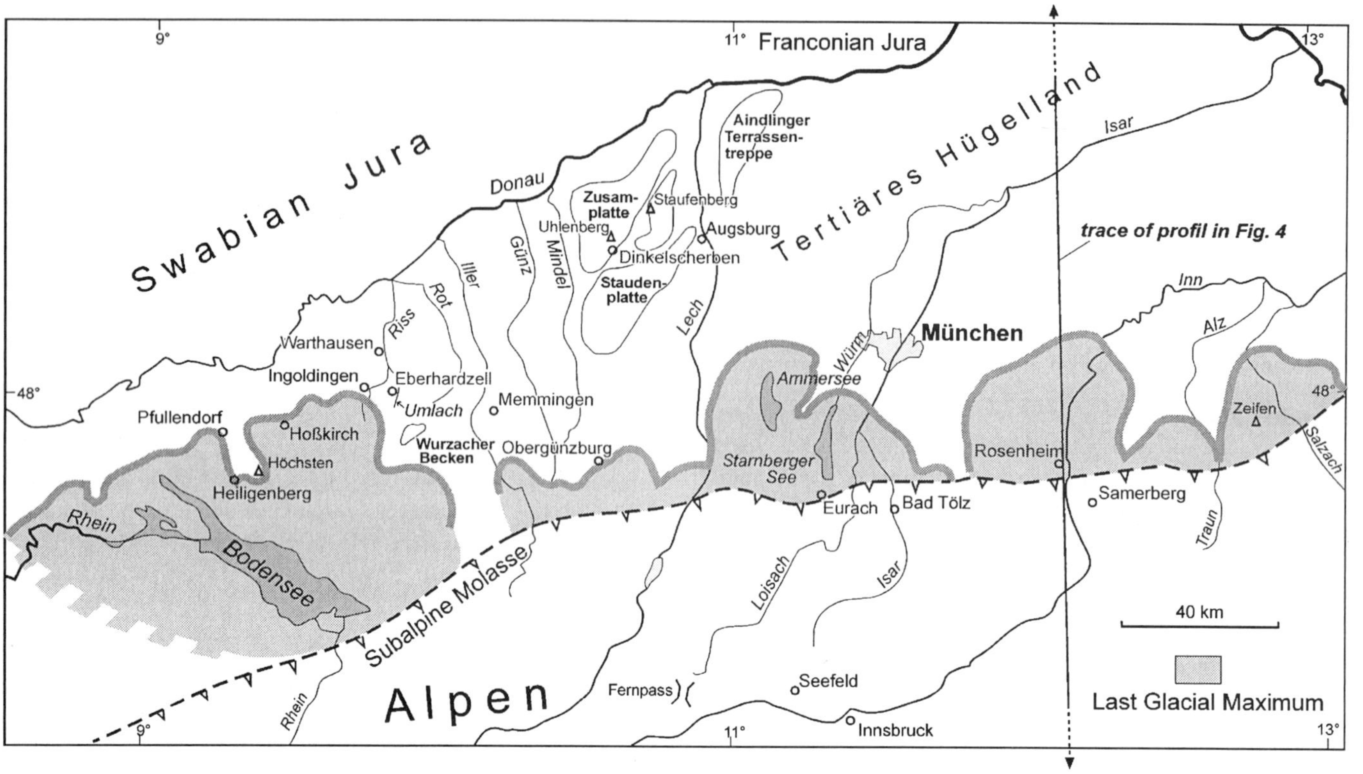

Fig. 1. Overview of localities mentioned in the text.

glacial' refers to a regional climate comparable to that of the Holocene.

Regional description

The southern German Alpine foreland can be roughly subdivided into three large areas (Fig. 3):

1. The west is dominated by the River Rhine and was shaped by the former Rhine Glaciers (Ellwanger *et al.*, 1995). During the Quaternary the Alpine river Rhine evolved from a local contributary of the Danube, to a major stream transporting central Alpine material into the Upper Rhine Graben and further to the North Sea. In the Bodensee (Lake Constance) area, the River Rhine and the Rhine glaciers lowered the land surface by more than 700 m during the Quaternary. Consequently, the landscape changed from a flat Danubian peneplain to the incised present-day relief.
2. The central area between the rivers Riss and Lech is characterised by a post-orogenic rebound of the thickest and deeply-buried stack of Tertiary Upper Freshwater Molasse. Continuous uplift resulted in a landscape consisting of numerous gravel spreads which are preserved at different altitudes and form a kind of tectonostratigraphic sequence. Here the classic model of four gravel spreads (Older Deckenschotter, Younger Deckenschotter, High Terrace and Lower Terrace), related to four Alpine glaciations (Günz, Mindel, Riss and Würm), was established (Penck 1899, Penck & Brückner, 1901/1909). Later, additional gravel spreads were identified on the highest Molasse ridges ('Deckschotter' after Graul) and interpreted as evidence of earlier glaciations (Donau and Biber).
3. In the east, the Danubian character of the landscape was preserved throughout the Quaternary. The Munich gravel plain ('Münchner Schotter-Ebene') was formed by superposition of several gravel accumulations without any further incision (Jerz, 1993). The same type of accumulation occurs between the Ammersee and the Starnberger See lakes and in the Traun-Alz region.

A first attempt to calculate the maximum deflection of the crust in front of the Inn Glacier (Fig. 4) during the last glacial maximum suggests movements of the whole proglacial area up to the Franconian Jura. Thus isostasy might have influenced the development of the circum-Alpine rivers including the Danube which is separated in the eastern Alpine foreland from the formerly glaciated terrain by a wide belt of Tertiary uplands ('Tertiäres Hügelland').

'Deckschotter' and possible older glaciations

In the central south-German Alpine foreland, Eberl (1930) identified ancient, highly-elevated gravels in the Iller-Lech region (Fig. 3). He interpreted these gravels as deposits of a 'Donau Ice Age' dating back to the time before the Older Cover Gravel of Penck & Brückner (1901/09). Schaefer (1953) confirmed the existence of this 'Unterer' (Lower)

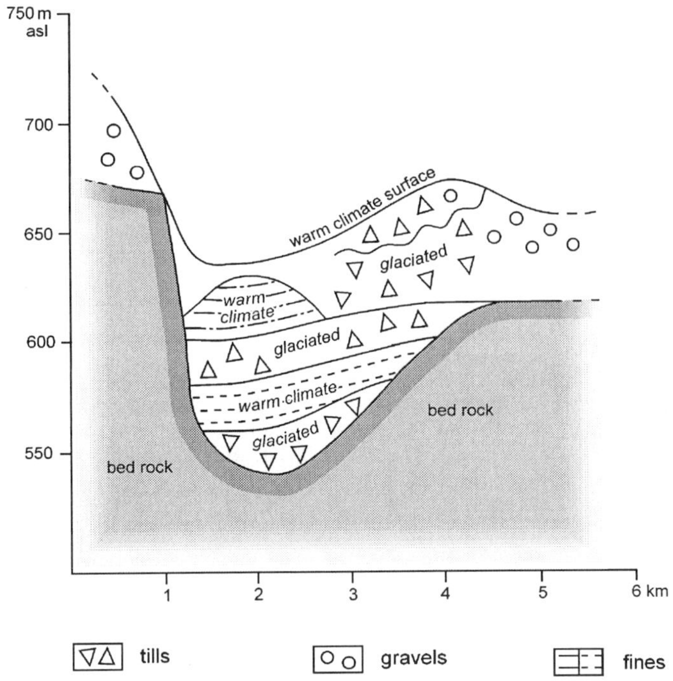

Fig. 2: Section through a glacial basin in the Rhine Glacier area (after Ellwanger et al., 1995). The strata between the tills and on top of the uppermost till contain palynological evidence for warm climate. Thus the basin fill represents at least three glacial and three interglacial periods. This climatostratigraphy comes close to descriptive lithostratigraphy, because evidence for glacials and interglacials are found in a sediment trap in direct superposition. Walther's law of facies applies to stacks like these. In many other cases, Alpine climatostratigraphic interpretation is based on flights of terraces, where proximity of glaciers is inferred for periods of gravel accumulation, and where evidence of interglacials is missing, caused by later erosion. The terrace type of climatostratigraphy is far more interpretative.

'Deckschotter' and later added another, 'Oberer' (Upper) 'Deckschotter', occurring on the Staudenplatte, at Staufenberg and in the Aindlinger flight of terraces east of the River Lech, that predated the Lower 'Deckschotter'. These additional gravel spreads have been generally accepted as products of cold events (cf. for example, Brunnacker, 1986). However, the glaciofluvial origin of these gravels is only implied; the correlative tills are still a matter of discussion.

Chronostratigraphy in sediments overlying the Lower 'Deckschotter' (Uhlenberg site)

Very few sites that have yielded useful bio- and magnetostratigraphic information have been found in the Alpine foothills. Therefore an isolated occurrence of dateable sediments on top of a Lower 'Deckschotter' on the Uhlenberg near Dinkelscherben (west of Augsburg) assumes a key position. The cover sediments start with a 2 m thick fining-upwards sequence containing *Arvicola* molars and molluscs biostratigraphically correlated with the Tiglian (Ellwanger *et al.*, 1994; Rähle, 1995). They are overlain by a peat horizon with an *Alnus-Tsuga-Pinus* pollen spectrum (e.g. Bludau, 1995). On top lies a 3 m thick coarsening-upwards sequence of which some samples show reverse palaeomagnetic polarity (Strattner & Rolf, 1995). This implies that the Lower Deckschotter (and its assumed correlative glaciation) dates back to the Pliocene/Pleisto-cene.

The Older Deckenschotter (Older Cover Gravel) and equivalent glaciation

In the past it had been widely assumed that the tectonostratigraphic unit 'Older Deckenschotter' postdates the Matuyama/Brunhes magnetic epoch boundary. Palaeomagnetic investigation of sites on the Höchsten and near Heiligenberg (north of Lake Constance) have recently revealed (Bibus *et al.*, 1996), that some of the samples from diamictons below a Middle Deckenschotter and from fine lenses at the base of a lithostratigraphically identified Younger Deckenschotter are reverse magnetised. It is still unclear where those deposits belong chronostratigraphically. Consequently, it is still open whether all the deposits formerly assigned to the Older Deckenschotter really represent the same accumulation event.

Grimm *et al.* (1979) and Doppler (1980) reported diamictons overlain by Younger Deckenschotter from the western margin of the Salzach Glacier area. If the sediments below the Younger Deckenschotter are correlatives of the Older Deckenschotter gravels, a glaciation during the accumulation phase of Older Deckenschotter is probable.

Middle Deckenschotter and its equivalent glaciation

The precise number of glaciations that occurred in the Alps is uncertain. In the Württemberg Rottal (south of Ulm) Penck & Brückner's (1901/09) Younger Deckenschotter can be subdivided into a higher (i.e. older) Haslach Gravel and a lower, Tannheim Gravel. They can be distinguished using their clast composition. In the proximal area, the vertical distance between the two units amounts to 10 m. Schreiner & Ebel (1981) assigned the more crystalline-rich Haslach Gravel to a distinctive Middle Deckenschotter and concluded that contemporaneous glaciation had occurred during its accumulation based on the presence of striated pebbles in the proximal part of the gravel. This unit is still not yet generally accepted.

Younger Deckenschotter and equivalent glaciation

The contact of the Younger Deckenschotter with moraines north of Obergünzburg caused Penck to propose that a

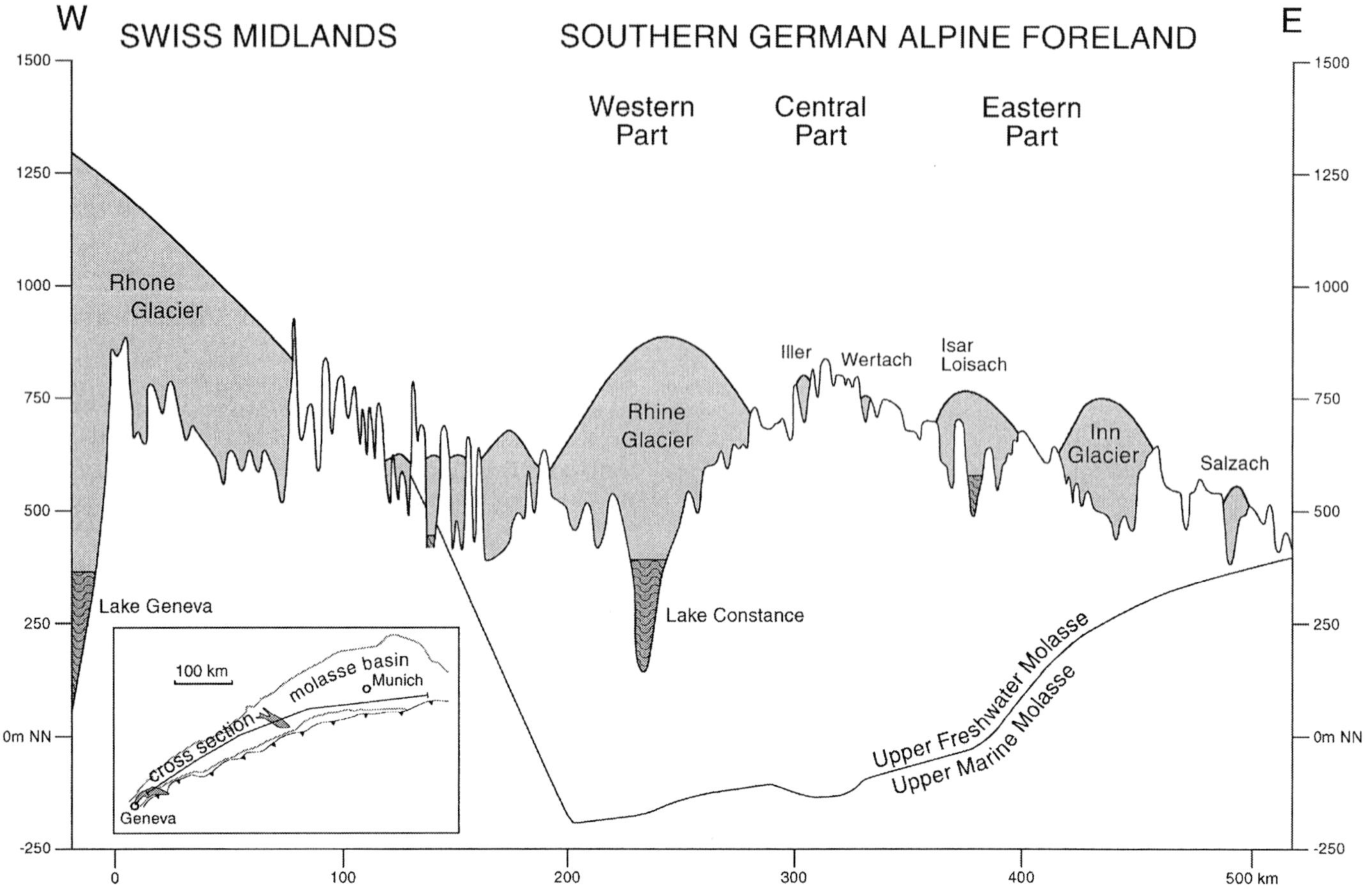

Fig. 3. Cross-section of the northern Alpine foreland with glaciers of the Last Glacial Maximum (LGM).

glaciation had occurred during the gravel accumulation (his Mindel Glaciation). The terminal moraines south of Memmingen, are perfectly shaped. However, they do not continue to the west (Sinn, 1972), but can be followed eastwards as far as Obergünzburg (Penck & Brückner, 1901/1909, Eberl, 1930, Stepp, 1981, Habbe, 1986).

Even further to the east, morphologically-correlated landforms were also identified in the western and northwestern parts of the Salzach Glacier region (Eichler & Sinn, 1974, Grimm *et al.,* 1979). In the eastern Rhine glacier area correlative landforms are pedologically-correlated; they show a weathering depth of 5-8 m. There is no age control for the formation of those landforms.

Incision between Younger Deckenschotter and High Terraces, 'third from last' Glaciation and the Samerberg site

The incision period between the deposition of the Younger Deckenschotter and the High Terrace was traditionally referred to as the 'Great Interglacial'. Today it is thought that this period actually represents a series of two (or even three) interglacials. In the research borehole Samerberg 2 in southern Bavaria, a 'double interglacial' was discovered beneath a thin 'penultimate glacial' diamicton. On the basis of palynological investigations, Grüger (1983) correlated the interglacials with the north German Holsteinian and Wacken/Dömitzian events. Ellwanger *et al.* (1995) think that at least one additional glaciation (the 'third from last' glaciation) occurred in the Hoßkirch basin (Baden-Württemberg) which post-dates deposition of the Younger Deckenschotter and predates the penultimate Glaciation (Riß of the classical Alpine stratigraphy).

The outermost terminal moraine zone in the south German Alpine foreland (see digital map) is interpreted as a product of a 'third from last' glaciation. Whether this glaciation should be correlated with the Younger Deckenschotter or with a later event is still an open question.

The belt of morainic amphitheatres correlated with the penultimate glaciation

Within the Rhine Glacier region, the deposits correlated to the penultimate glaciation are subdivided into three units. The oldest unit is marked pedologically by a greater degree of weathering (Schreiner, 1989, 1992, Schreiner & Haag, 1982). This unit is a diamicton, confined palaeogeographically to the tongue basins in the Riss and neighbouring valleys. The diamicton underlies outwash gravels (the 'High Terrace') which are connected to a very distinct double rampart terminal moraine. The extent of the assumed glacial retreat between deposition of the diamicton bed and accumulation of the outwash gravels is unknown.

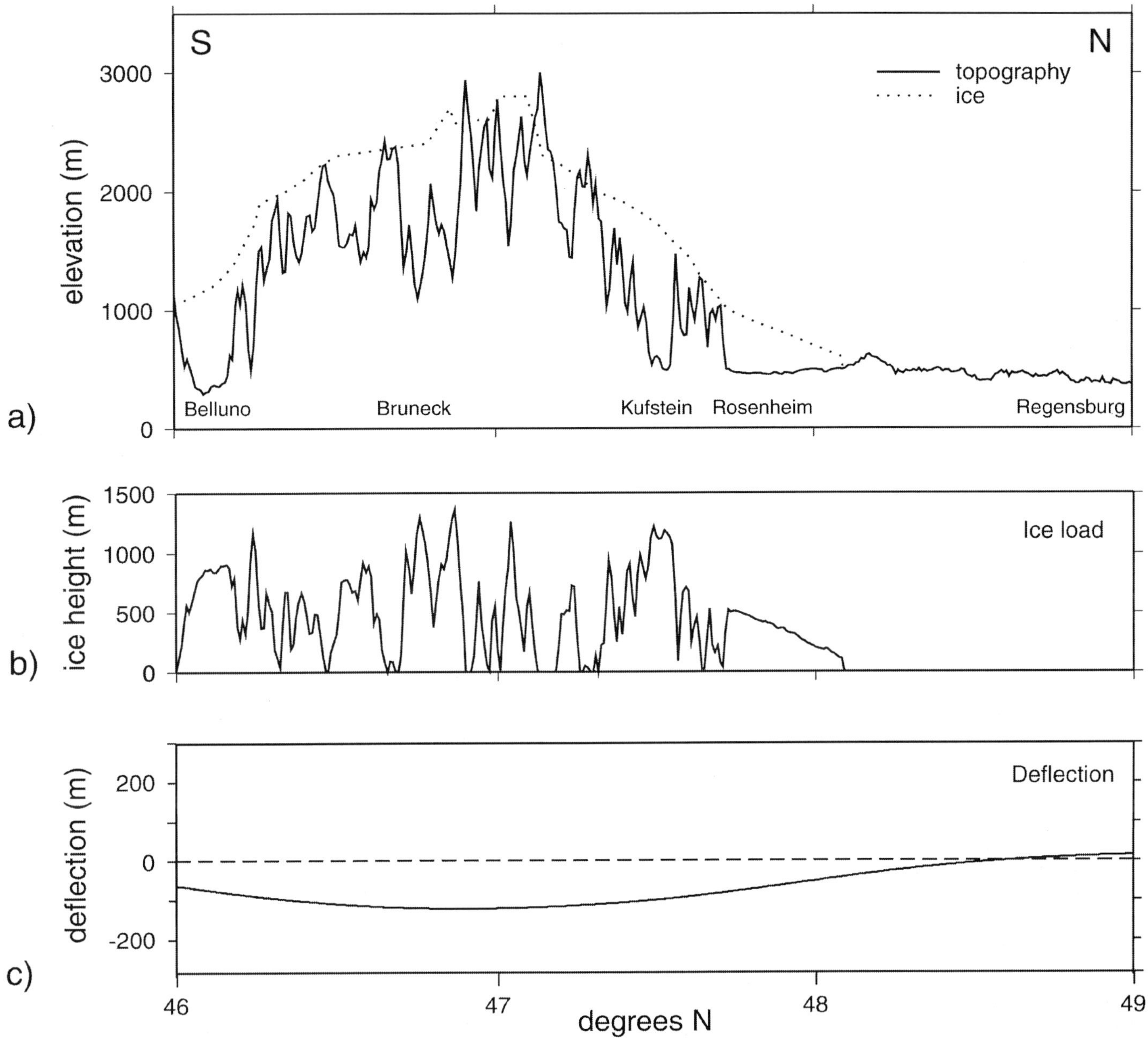

Fig. 4. Three North-South profiles along longitude 12.1 (see Fig. 1) from latitude 46 to 49.
(a) Topography from the GLOBE dataset (www.ngdc.noaa.gov /seg/topo/globe.shtml),
(b) Ice elevation from the map of van Husen (1987).
(c) Deflection under the ice load. The density of the ice is taken as 850 kg/m³. The ice loads an initial horizontal plate with an effective elastic thickness of 30 km (see also Pfiffner et al., accepted for publication in tectonics). Young's modus is 7e10 Pa, Poisson's rate is 0.25. The deflection has been calculated by solving the 1D flexure equation with a finite difference method (Buiter et al., 1998).
A maximum deflection of 135 m results under the ice load. Since the duration of the ice load was very short, we do not assume that the maximum deflection actually occurred, but a substantial deflection is still to be expected. The area of calculated movements reaches the Franconian Jura. This means that the movements could have influenced meltwater streams in front of the ice, including the receiving stream Danube.

In the eastern region of the Rhine Glacier the double rampart terminal moraines ('Doppelwall-Riss') of Schreiner (1989) and Ellwanger (1990) are generally gravelly push moraines 10 to 30 metres in height, formed at a distance of 1 to 3 km from each other. They are interpreted as the morainic amphitheatres of the penultimate glaciation (see digital map).

The occurrence of additional gravel fields in the Riss and Umlach valleys, as far north as Warthausen, at a lower altitude than the High Terrace but higher than the Lower Terrace, suggested that a third stratigraphical unit might belong to the penultimate glaciation (Penck & Brückner, 1901/1909; Schreiner, 1989). Possibly correlative terminal moraines north of Ingoldingen and south of Eberhardzell are hardly recognisable in the field; therefore the existence of this third penultimate glacial event remains doubtful.

Eemian Stage interglacial

In southern Germany Beug (1972) investigated the Eemian interglacial sequences from sites at Zeifen on the Salzach River, and Eurach, near Lake Starnberg (Beug, 1979). The best-investigated pollen section of the Alpine region comprising the last interglacial and the three subsequent interstadials, however, is that at Samerberg in Bavaria, southeast of Rosenheim (Grüger, 1979). Here the vegetational succession corresponds largely to that of the north European Eemian Stage. The Alpine character of the site is expressed through high values of pine pollen. Another locality showing a sequence comparable to the Samerberg sequence has recently been identified in the Wurzacher Becken, a tongue basin of the penultimate glaciation in the Rhine Glacier area (Grüger & Schreiner, 1993).

The Last Glaciation (Würmian)

Cores from the Samerberg and Mondsee (Austria) sites show an undisrupted sequence of lake sediments between undisputed deposits dating from the Last Interglacial and the Last Glacial Maximum (LGM) till (approximately 23,500 cal years B.P.). No evidence is found here of a glaciation intervening between the Eemian and the LGM. The main Würmian ice advance accordingly occurred in the Late Upper Pleistocene, similar to that in northern Europe. Traub & Jerz (1976) obtained an age of 21,650 ±250 (*c.* 25,000 cal years) B.P. from pleniglacial loess molluscs from the margin of the Salzach Glacier; the corresponding strata had been deposited directly before the deposition of the Lower Terrace gravels.

The end moraines do not represent the maximum glacial limit. At various sites (e.g. Ingoldingen, Pfullendorf) diamicton layers are found intercalated into the outwash gravels of the Lower Terrace in front of the terminal moraine zone (see digital map). How far the glaciers retreated between deposition of the diamicton and deposition of the Lower Terrace gravels is again unknown. This is similar to the situation found in the penultimate glaciation (see above).

After the advance to the terminal moraines of the LGM, at least one pronounced readvance is assumed, the 'Innere Jungendmoräne' (see digital map). The significance of numerous minor readvance phases (e.g. Schreiner, 1974, Habbe, 1986, Ellwanger, 1980, Rothpletz, 1917, Knauer, 1929, 1931, Troll, 1924) assigned to the Würmian Glaciation, is still a matter for discussion.

During the LGM, transfluences developed at various points of the ice-stream network. In the Inn valley, around Landeck, for instance, five major tributary glaciers grew together to form one major valley glacier, that advanced eastwards down the Inn Valley. At the same time, around Innsbruck a major tributary glacier from the south had already blocked the Inn valley. Congestion resulted in a rise of the ice surface until finally the threshold to the north was crossed, and ice from the Inn valley flowed over the Fernpass and the Seefelder Senke directly northwards into the Isar and Loisach drainage systems (van Husen, 2000). Such events are reflected in the glacial deposits by far-travelled gravel and major erratics. In the foreland of the Isar-Loisach Glacier, Alpine crystalline rocks constitute up to 35% of the total 20 - 31.5 mm fraction. Further to the east, in the Tölz Glacier area, the contents range from 0 to 1%, because no ice was supplied from the Inn valley (Dreesbach, 1985).

The Alpine 'Late Glacial' comprises the period of recession from the 'Innere Jungendmoräne' until the end of the Younger Dryas Stadial. The basic subdivision of the Late-glacial ice retreat originates from Penck & Brückner (1901/1909) who distinguished three retreat phases: Bühl, Gschnitz, and Daun: Of these, the Bühl represents the last readvance of the still intact ice-stream network probably before 17,000 cal. years B.P. (Patzelt, 2002). Gschnitz is an inner-Alpine local readvance between 16,700 and 16,000 cal. years B.P. The Late Glacial Daun readvance took place before the Egesen readvance which represents the Younger Dryas Stadial between 12,600 and 11,500 cal years B.P. (Patzelt, 2002).

Glaciations of other mountain massifs

Like the Vosges in neighbouring France (Mercier & Jeser, this volume), on the other side of the Upper Rhine graben, the Schwarzwald (Black Forest) was glaciated during the Pleistocene. The first evidence of former glaciation was found by Schimper (1837) in the Titisee region. Later Steinmann (1892) described moraines from the lower Wehra valley on the southern margin of the Schwarzwald. During the last glaciation the Schwarzwald supported a small ice cap, comparable to those in present-day Norway (e.g. Jostedalsbreen, Hardangerjökull). Traces of older glaciations have been identified in a number of places. Those older moraines are summarily attributed to the penultimate glaciation, although concrete evidence is lacking. It had been assumed that during the penultimate glaciation both Alpine and Schwarzwald glaciers might have met (Hantke, 1978). However, according to more recent investigations (Leser, 1979; Wendebourg & Ramshorn, 1987), it is quite clear that this did not happen.

In the Bayrischer Wald (Bavarian Forest) mountains traces of former glaciation have been identified in most areas higher than 1300 m. End moraines of the last glaciation reach down to an altitude of 900 m a.s.l. Traces of older glaciations are also present (Bauberger, 1977; Jerz, 1993).

Traces of former glaciation in the Harz mountains were first identified by Zimmermann (1868). His opinion has been challanged repeatedly, especially considering the relatively low altitude of the mountains; the highest peak, the Brocken, lies only 1,142 m a.s.l. Only after large-scale geological mapping was Duphorn (1968) able to prove beyond doubt that the area had indeed been repeatedly glaciated.

The Riesengebirge (Giant Mountains) further to the east were much less glaciated, as a consequence of the more continental climate. Some 15 former glaciers have been identified, the longest of which extended 5.3 km (Schwarzbach, 1942).

The remaining mountain ranges in Germany, like the Thüringer Wald and the Rhön, were probably not glaciated. It seems doubtful whether the Schwäbische Alb (Hantke, 1974) or Fichtelgebirge had been glaciated (Liedtke, 1981; Jerz, 1993).

Acknowledgements

This text would not have been written without the assistence of Dr. Jürgen Ehlers. Dr. Gerhard Doppler and Dr. Philip L. Gibbard corrected the manuscript. Thanks are also due to S. Tolksdorf and T. Ziegler for drawing Figs 1-3. The work of the first author was partially financed by a grant of the Deutsche Forschungsgemeinschaft (fi693 1-3).

References

Bauberger, W. (1977). *Geologische Karte von Bayern 1: 25.000, Erläuterungen zum Blatt Nr. 7046 Spiegelau und zum Blatt Nr. 7047 Finsterau (Nationalpark Bayerischer Wald).* München, Bayerisches Geologisches Landesamt. 183 pp.

Beug, H.J. (1972). Das Riß/Würm-Interglazial von Zeifen, Landkreis Laufen a.d. Salzach. *Bayerische Akademie der Wissenschaften, Mathematisch-Naturwissenschaftliche Klasse N.F.*, **151**, 46-75.

Beug, H.J. (1979). Vegetationsgeschichtlich-pollenanalytische Untersuchungen am Riß/Würm-Interglazial von Eurach am Starnberger See/Obb. *Geologica Bavarica,* **80**, 91-106.

Bibus, E., Bludau, W., Ellwanger, D., Fromm, K., Kösel, M. & Schreiner, A. (1996). On Pre-Würm glacial and interglacial deposits of the Rhine Glacier (South German Alpine Foreland, Upper Swabia, Baden-Württemberg). *In:* Turner, C. (ed.), *The Early Middle Pleistocene in Europe*, 195-204.

Bludau, W. (1995). Altpleistozäne Warmzeiten im Alpenvorland und im Oberrheingraben.- Ein Beitrag der Palynologie zum „Uhlenberg-Problem". *Geologica Bavarica,* **99**, 119-134.

Brunnacker, K. (1986). Quaternary Stratigraphy in the Lower Rhine Area and Northern Alpine Foothills. *Quarternary Science Reviews*, **5**, 371-379.

Buiter, S. J. H., Wortel, M.J.R. & R. Govers (1998): The role of subduction in the evolution of the Apennines foreland basin. *Tectonophysics*, **296**, 249-268.

Doppler, G. (1980). Das Quartär im Raum Trostberg an der Alz im Vergleich mit dem nordwestlichen Altmoränengebiet des Salzachvorlandgletschers (Südostbayern). Dissertation, München. 198 pp.

Doppler, G. & Jerz, H. (1995). Untersuchungen im Alt- und Ältestpleistozän des bayerischen Alpenvorlands – Geologische Grundlagen und stratigraphische Ergebnisse. *Geologica Bavarica,* **99**, 7-53.

Dreesbach, R. (1985). Sedimentpetrographische Untersuchungen zur Stratigraphie des Würmglazials im Bereich des Isar-Loisachgletschers. Dissertation, München. 176 pp.

Duphorn, K. (1968). Ist der Oberharz im Pleistozän vergletschert gewesen? *Eiszeitalter und Gegenwart,* **9**, 164-174.

Eberl, B. (1930). *Die Eiszeitenfolge im nördlichen Alpenvorlande. - Ihr Ablauf, ihre Chronologie auf Grund der Aufnahmen des Lech- und Illergletschers.* Augsburg, Benno Filser, 427 pp.

Eichler, H. & Sinn, P. (1974). Zur Gliederung der Altmoränen im westlichen Salzachgletschergebiet. *Zeitschrift für Geomorphologie N.F.*, **18**, 132-158.

Ellwanger, D. (1980). Rückzugsphasen des würmzeitlichen Illergletschers. *Arbeiten aus dem Institut für Geologie und Paläontologie der Universität Stuttgart N.F.*, **76**, 93-126.

Ellwanger, D. (1990). Zur Riß-Stratigraphie im Andelsbach-Gebiet (Baden-Württemberg). *Jahreshefte des Geologischen Landesamtes Baden-Württemberg,* **32**, 235-245.

Ellwanger, D., Fejfar, O. & von Koenigswald, W. (1994). Die biostratigraphische Aussage der Arvicolidenfauna vom Uhlenberg bei Dinkelscherben und ihre morpho- und lithostratigraphischen Konsequenzen. *Münchner Geowissenschaftliche Abhandlungen,* **A 26**, 173-191.

Ellwanger, D., Bibus, E., Bludau, W., Kösel, M. & Merkt, J. (1995). Baden-Württemberg. *In:* Benda, L. (ed.), *Das Quartär Deutschlands*, 255-295.

Grimm, W.-D., Bläsig, H., Doppler, G., Fakhrai, M., Goroncek, K., Hintermaier, G., Just, J., Kiechle, W., Lobinger, W.H., Ludewig, H., Muzavor, S., Pakzad, M., Schwarz, U. & Sidiropoulos, T. (1979). Quartärgeologische Untersuchungen im Nordwestteil des Salzach-Vorlandgletschers (Oberbayern). *In:* Schlüchter, Ch. (ed.), *Moraines and Varves - Origin, Genesis, Classification*, 101-119. Rotterdam, Balkema.

Grüger, E. (1979). Spätriß, Riß/Würm und Frühwürm am Samerberg in Oberbayern - ein vegetationsgeschichtlicher Beitrag zur Gliederung des Jungpleistozäns. *Geologica Bavarica*, **80**, 5-64.

Grüger, E. (1983). Untersuchungen zur Gliederung und Vegetationsgeschichte des Mittelpleistozäns am Samerberg in Oberbayern. *Geologica Bavarica*, **84**, 21-40.

Grüger, E. & Schreiner, A. (1993). Riß/Würm- und würmzeitliche Ablagerungen im Wurzacher Becken (Rheingletschergebiet). *Neues Jahrbuch für Geologie und Paläontologie, Abhandlungen,* **189**, 81-117.

Habbe, K.A. (1986). Bemerkungen zum Altpleistozän des Illergletscher-Gebietes. *Eiszeitalter und Gegenwart,* **36**, 121-134.

Hantke, R. (1974). Zur Vergletscherung der Schwäbischen Alb. *Eiszeitalter und Gegenwart,* **25**, 214.

Hantke, R. (1978). Eiszeitalter. *Die jüngste Erdgeschichte der Schweiz und ihrer Nachbargebiete, Band 1.* Thun: Ott. 468 pp.

Jerz, H. (1993). *Das Eiszeitalter in Bayern - Erdgeschichte, Gesteine, Wasser, Boden.* (Geologie von Bayern, Bd. 2.) Stuttgart, Schweizerbart, 243 pp.

Knauer, J. (1929). *Erläuterungen zur Geognostischen Karte von Bayern 1:100 000, Blatt München West (Nr. XXVII), Teilblatt Landsberg,* 47 pp.

Knauer, J. (1931). *Erläuterungen zur Geognostischen Karte von Bayern 1:100 000, Blatt München West (Nr. XXVII), Teilblatt München-Starnberg,* 48 pp..

Leser, H. (1979). *Erläuterungen zur Geomorphologischen Karte der Bundesrepublik Deutschland 1:25.000, GMK 25 Blatt 4, 8313 Wehr,* 60 pp.

Patzelt, G. (2002). Unterlagen für die Exkursionen der 44. Tagung der Hugo Obermaier-Gesellschaft in Innsbruck. Universität Innsbruck, 21 pp.

Penck, A. (1899). Die vierte Eiszeit im Bereich der Alpen. *Schriften der Vereinigung zur Verbreitung naturwissenschaftlicher Kenntnisse,* **39**, 1-20.

Penck, A. & Brückner, E. (1901/1909). *Die Alpen im Eiszeitalter,* 3 Bände: 1199 pp. Leipzig.

Piffner, O.A., Schlunegger, F. & Buiter, S.J.H. (in press): The Swiss Alps and their peripheral foreland basin: stratigraphic response to deep crutal processes. *Tectonics.*

Rähle, W. (1995). Altpleistozäne Molluskenfaunen aus dem Zusamplattenschottern und ihrer Flußmergeldecke vom Uhlenberg und Lauterbrunn (Iller-Lech-Platte, Bayerisch Schwaben). *Geologica Bavarica,* **99**, 103-118.

Rothpletz, A. (1917). Die Osterseen und der Isar-Vorlandgletscher. *Mitteilungen der Geographischen Gesellschaft in München,* **12**, 99-314.

Salvador, A. (1994). *International Stratigraphic Guide.* Boulder, The Geological Society of America, 214 p.

Schaefer, I. (1953). Die donaueiszeitlichen Ablagerungen an Lech und Wertach. *Geologica Bavarica,* **19**, 13-64.

Schaefer, I. (1956). Sur la division du Quaternaire dans l'avant-pays des Alpes en Allemagne. *Actes IV Congres INQUA, Rome/Pise 1953,* **2**, 910-914.

Schimper, K. (1837). Über die Eiszeit. Brief an Prof. Agassiz. *Actes Societ, Helvetique des Sciences Naturelles 22e Session, Neuchâtel,* 38-51

Schreiner, A. (1974). *Erläuterungen zur Geologischen Karte des Landkreises Konstanz und Umgebung 1:50,000. 2., berichtigte Auflage.* Stuttgart, Geologisches Landesamt Baden-Württemberg, 286 pp.

Schreiner, A. (1989). Zur Stratigraphie der Rißeiszeit im östlichen Rheingletschergebiet (Baden-Württemberg). *Jahreshefte des Geologischen Landesamtes Baden-Württemberg,* **31**, 183-196.

Schreiner, A. (1992). *Einführung in die Quartärgeologie.* Stuttgart, Schweizerbart, 257 pp.

Schreiner, A. & Ebel, R. (1981). Quartärgeologische Untersuchungen in der Umgebung von Interglazial-vorkommen im östlichen Rheingletschergebiet (Baden-Württemberg). *Geologisches Jahrbuch,* **A 59**, 3-64.

Schreiner, A. & Haag, Th. (1982). Zur Gliederung der Rißeiszeit im östlichen Rheingletschergebiet (Baden-Württemberg). *Eiszeitalter und Gegenwart,* **32**, 137-161.

Sinn, P. (1972). Zur Stratigraphie und Paläogeographie des Präwürm im mittleren und südlichen Illergletscher-Vorland. *Heidelberger Geographische Arbeiten,* **37**, 159 pp.

Steinmann, G. (1892). Mitteilungen, Über Moränen am Ausgang des Wehratales. *Bericht der 25.Versammlung des Oberrheinischen Geologischen Vereins zu Basel,* 35-39.

Stepp, R. (1981). Das Böhener Feld - Ein Beitrag zum Altquartär im Südwesten der Iller-Lech-Platte. *Mitteilungen der Geographischen Gesellschaft München,* **66**, 43-68.

Strattner, M. & Rolf, C. (1995). Magnetostratigraphische Untersuchungen an pleistozänen Deckschicht-Profilen im bayerischen Alpenvorland. *Geologica Bavarica,* **99**, 55-102.

Traub, F. & Jerz, H. (1976). Ein Lößprofil von Duttendorf (Oberösterreich) gegenüber Burghausen an der Salzach. *Zeitschrift für Gletscherkunde und Glazialgeologie,* **11**, 175-193.

Troll, C. (1924). Der diluviale Inn-Chiemseegletscher - Das geographische Bild eines typischen Alpenvorlandgletschers. *Forschungen zur deutschen Landes- und Volkskunde,* **23**, 1-121.

van Husen, D. (1987). Die Ostalpen in den Eiszeiten. 24 pp. Wien, Geologische Bundesanstalt.

van Husen, D. (2000). Geological Processes during the Quaternary. *Mitteilungen der Österreichischen Geologischen Gesellschaft,* **92**, 135-156.

Wendebourg, J. & Ramshorn, Ch. (1987). Der Verzahnungsbereich alpiner und Südostschwarzwälder Rißvereisung (Baden-Württemberg). *Jahreshefte des Geologischen Landesamtes Baden-Württemberg,* **29**, 255-268.

Zimmermann, K.H. (1868). Über Gletscherspuren im Harz. *Neues Jahrbuch für Mineralogie, Geologie und Paläontologie,* 155-159.

Quaternary Glaciations – Extent and Chronology
Editors J. Ehlers and P.L. Gibbard

Pleistocene glaciation in the mountains of Greece

Jamie C. Woodward,[1] Mark G. Macklin[2] and Graham R. Smith[3]

[1] *School of Geography, University of Leeds, Leeds, LS2 9JT, UK jamie@geog.leeds.ac.uk*
[2] *Institute of Geography and Earth Sciences, University of Wales, Aberystwyth, Dyfed, SY23 3DB, UK*
[3] *Department of Environmental and Geographical Sciences, Manchester Metropolitan University, Manchester, M1 5GD, UK*

Introduction

Much of mainland Greece lies south of 41°N and is predominantly a landscape of upland plateaus and rugged mountains. In his volume *The Mountains of Greece,* Sfikas (1979) indexes over twenty mainland peaks that exceed 2000 m and eight of these (including Crete) are higher than 2400 m. Many of the highest peaks such as Olympus (2917 m), Smolikas (2637 m) and Tymphi (2497 m) are found in Northern Greece, close to 40°N, and these mountains are associated with some of the most extensive and best preserved evidence for the activity of Pleistocene glaciers (Bailey *et al.*, 1990; Lewin *et al.*, 1991; Woodward *et al.*, 1992; 1995; Smith *et al.*, 1997; Boenzi & Palmentola, 1997; Smith *et al.*, 2000). At present, however, field sites in Greece where detailed modern studies of glacial sediments and landforms have been carried out are few in number, and any attempt to review this work is further hampered by the fact that robust age frameworks are not yet available for the glacial sequences. In addition, those studies which have been conducted over the last decade or so have involved contrasting aims and objectives, and a range of approaches has been employed including the lithological analysis of till sediments (Woodward *et al.*, 1992), detailed field-based landform mapping and glacial sedimentology (Smith *et al.*, 1994; 1997), satellite remote sensing for landform recognition and relative age dating (Smith *et al.*, 1998; 2000), and the estimation of Equilibrium Line Altitudes on a local and regional scale (Palmentola *et al.*, 1990; Mastronuzzi *et al.*, 1994; Boenzi & Palmentola, 1997). The aim of this contribution is to briefly review some of the recent literature on glaciation in Greece and to present new data on the age of the glacial deposits on the southern flanks of Mount Tymphi in Northwest Greece. We present the first radiometric (uranium-series) ages obtained for glacial sediments in Greece (Macklin *et al.*, 1998) and discuss the wider implications of this new chronostratigraphic framework. To the authors' knowledge, these are also the first radiometric ages for glacial sediments in the wider Eastern Mediterranean region.

Mountain glaciation in Greece and regional Pleistocene snowlines

In his review of Pleistocene glaciation in the Mediterranean region, Messerli (1967) reported several sites in Greece with evidence for Pleistocene glacial activity and eight of them are listed in Table 1. Despite the absence of detailed morpho- and lithostratigraphic investigations of the glacial sequences at the time, Messerli (1967) assigned them to the Würmian and snowlines were plotted for this period for the entire Mediterranean basin. The principal Greek sites included on Messerli's (1967) map are shown on Fig. 1, yet, for some of these areas, glacial *deposits* have not been recorded on the 1:50,000 geological sheets produced by the Greek Institute for Geological and Mineral Exploration (IGME). This is probably due to the fact that, in some of these localities, the evidence for glacial activity is equivocal (see Pechoux, 1970), or is in the form of small-scale erosional features such as small cirques, nivation hollows and ice-moulded bedrock - with only limited evidence for landforms produced by the movement and deposition of glacial sediments down valley. Boenzi & Palmentola (1997) report that glacial features are only present on those Greek mountains that exceed 2200 m. In fact, the main sites in the country with extensive spreads of glacial sediments and landforms can be identified by examining the 1:500,000 geological map of Greece (IGME, 1983) (Table 1). This map also shows that glacial deposits are important landscape features on some of the highlands in the Southern Pindus Mountains (but to the North of 39.5°), including Mount Peristeri (2295 m) southwest of Metsovo, and further south around Mount Kakarditsa (2429 m) (Sestini, 1933) (Table 1).

Today winters are harsh and snowfall is heavy in the Greek mountains and extensive snow patches are common in the early summer on the highest peaks and ridges. There are marked variations in precipitation across Greece with values in excess of 2000 mm in the northwest and central uplands, falling to <500 mm in the lowland coastal zone of the southeast (Woodward *et al.*, 1995). Boenzi & Palmentola (1997) have calculated Equilibrium Line Altitudes for mountains in Greece, Albania and southern Italy and these are shown in Fig. 1. The ELAs were determined using Kurowski's method and this is based on the mean of the maximum elevation of the source area basin and the minimum elevation of each frontal moraine (Boenzi & Palmentola, 1997 p. 22). These workers have collected data for the mountains (except Olympus) shown on Fig. 1 and these data were "either personally gathered or from bibliographic sources." In common with earlier workers, while they do not provide any radiometric dates for the glacial sediments, Boenzi & Palmentola (1997) have

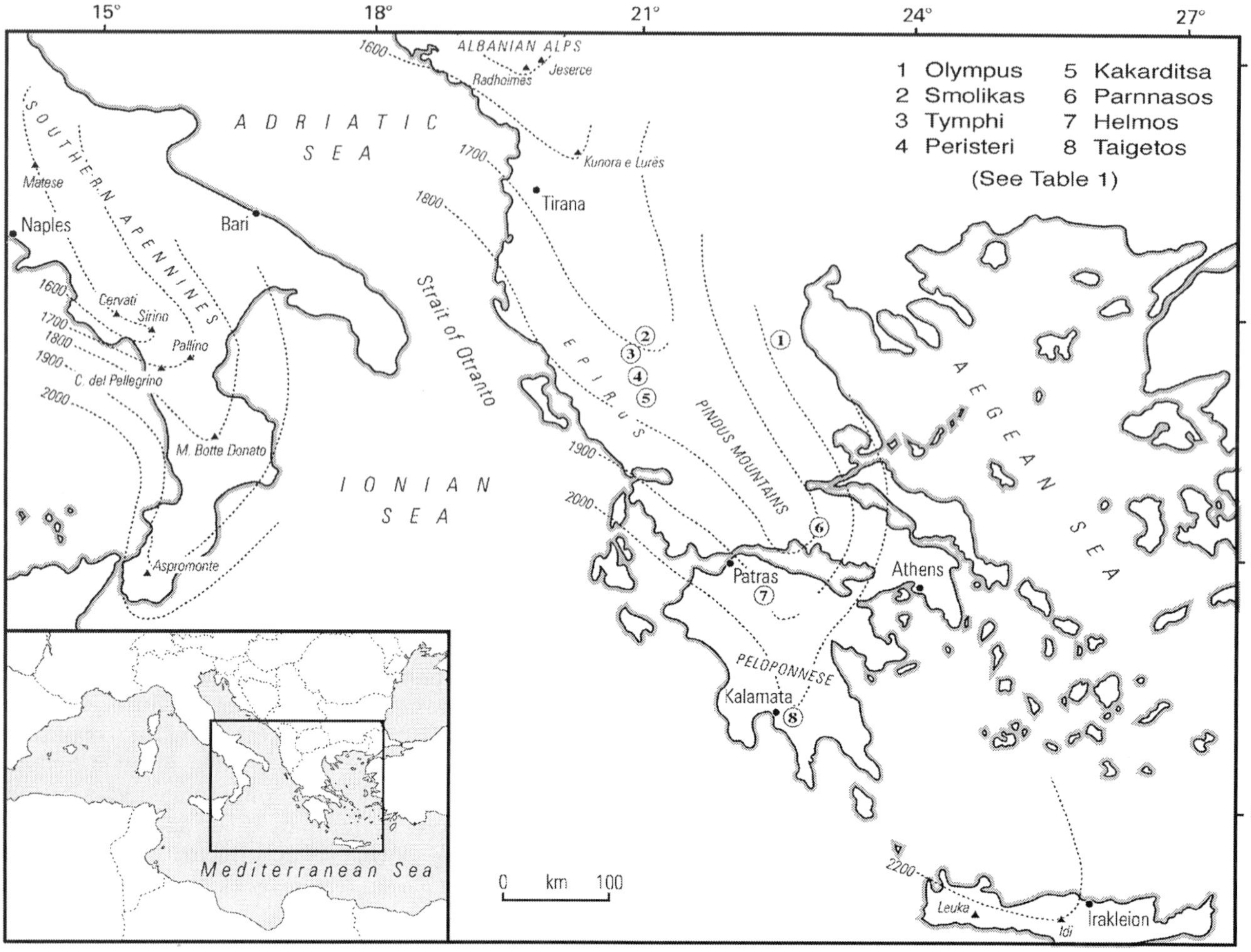

Fig. 1. Map of Greece, Albania and southern Italy showing the sites used in the estimation of regional Equilibrium Line Altitudes by Boenzi & Palmentola (1997). The dotted lines show the ELAs for the maximum extent of glacial activity at each site. The eight numbered sites in Greece are listed in Table 1. Note that dating control is lacking for most of these sites.

assumed that all of the glacial features on these mountains date to the last cold stage:

"On the highest mountains of the Southern Apennines (Italy) and of the region stretching from Albania as far as Crete, many traces of glacial modelling are preserved, all attributable to the last glacial age (Würm)." Boenzi & Palmentola (1997, p. 21).

In fact, for the glacial sequences on Mount Tymphi in Epirus (Fig. 1), they have computed ELA values for four (Würmian) Lateglacial retreat stages – again without any absolute dating control. For this part of the Mediterranean basin, Boenzi & Palmentola (1997) have identified an increase in ELA of about 100 m per degree of latitude moving southwards from Albania across Greece to Crete. From a palaeoclimatic perspective, it is interesting to note that in the Pindus Mountains of Greece glacial landforms are only present on peaks higher than 2200 m whereas in the Southern Appenines of Italy, traces of glacial activity are found on mountains >1900 m. This observation may reflect regional contrasts in Pleistocene climates (humidity and/or temperature) and in the origin of precipitation-bearing air masses.

The glaciation of Mount Olympus

Background

The most detailed work to date on glacial sequences in Greece has been carried out in the Mount Olympus area by an American team led by Geoffrey Smith from the Department of Geological Sciences at Ohio University. Olympus is the highest mountain in Greece (2917 m) and is located to the east of the main Pindus Mountain chain in the northeast corner of peninsular Greece, close to the Aegean coast (Fig. 1). Smith *et al.* (1994; 1997) have argued that previous work in the Mount Olympus area had significantly underestimated the extent of glacial activity since deposits on the eastern piedmont of the massif - that were previously classified as fluvial sediments - are, in fact, glacial diamictons. As Smith *et al.* (1997) have pointed out; the early investigations on Mount Olympus did little to establish the sequence and timing of glacial events in the area (as is also the case for the rest of Greece), although the implication was that glaciation was restricted to the latest Pleistocene or Würm (Messerli, 1967).

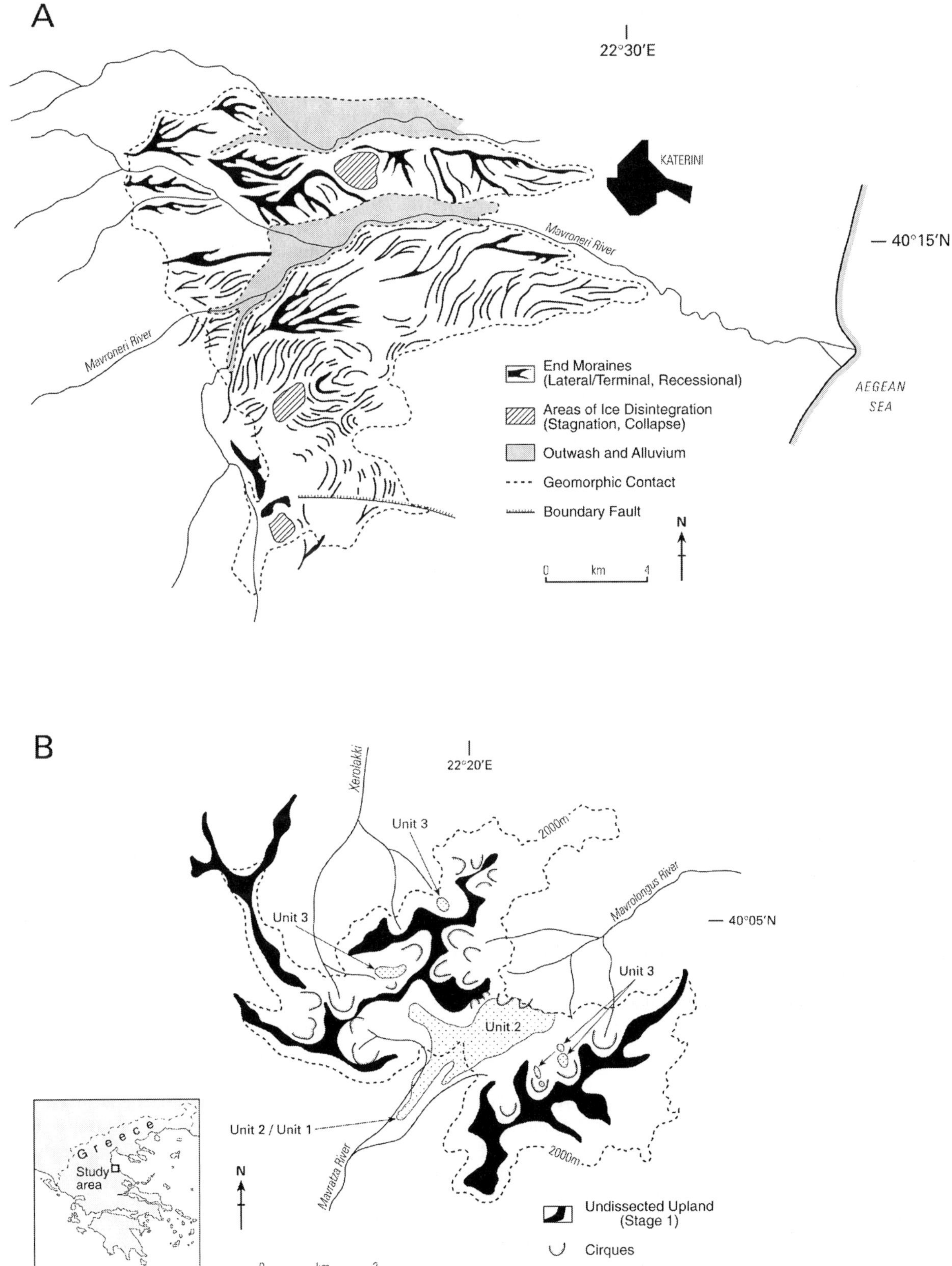

Fig. 2. A) The glacial deposits and landforms of the Mount Olympus upland (Bara Plateau) (after Smith et al., 1997). Much of the terrain shown here is above 2000 m. B) The glacial deposits and landforms in the headwaters of the Xerolakki and Mavroneri rivers on Mount Olympus (after Smith et al., 1997).

Table 1. Mountains in Greece where evidence for significant glacial activity has been reported. The ELA values for sites in the Pindus Mountains are taken from Boenzi & Palmentola (1997) and for Mount Olympus from Smith et al. (1997).

Massif and Region	Peak Elevation	Latitude	ELA*	Reference	Glacial deposits marked on the 1,500,000 IGME map
1. Olympus, Pieria	2917 m	40°05'N	1030 m	Smith *et al.* (1997)	✓
2. Smolikas, Epirus	2637 m	40°05'N	1750 m	Boenzi *et al.* (1992)	✓
3. Tymphi, Epirus	2497 m	40°00'N	1700 m	Woodward *et al.* (1995)	✓
4. Peristeri, Epirus	2295 m	39°50'N	1925 m	Sestini (1933)	✓
5. Kakarditsa, Epirus	2429 m	39°45'N	1925 m	Sestini (1933)	✓
6. Parnnasos, Sterea	2457 m	38°30'N	1800 m	Pechoux (1970)	✗
7. Helmos, Achaia	2341 m	38°00'N	2000 m	Mastronuzzi *et al.* (1994)	✗
8. Taigetos, Messinia	2404 m	36°56'N	1950 m	Mastronuzzi *et al.* (1994)	✗

The glacial sedimentary sequences on Mount Olympus

The stratigraphic record of Pleistocene and Holocene events on Mount Olympus is most clearly preserved on the eastern piedmont within the valleys of the Mavrolongus and Mavroneri rivers which drain to the Aegean coast (Fig. 2). Extensive moraine complexes have been mapped in the Mavroneri valley to the west and southwest of the town of Katerini, with thick proglacial (outwash) sediments present on the valley floor (Fig. 2A). The deposits in this valley represent the convergence of valley ice from Mount Olympus (to the south) and the High Pieria Mountains (to the north). Three discrete sedimentary packages (units 1 to 3) have been identified in the area and each is capped by a distinctive soil. The deposition of these sediments and the subsequent phases of pedogenic weathering reflect periods of glacial and non-glacial activity, respectively, in the region (Smith *et al.*, 1997, p. 809). The deposits of the eastern piedmont comprise a range of glacial, glaciofluvial, fluvial and alluvial fan sediments and each of the three main sedimentary units can be related to a period of glacial activity in the uplands (Smith *et al.*, 1997).

During the glacial phases that led to the deposition of unit 1 and unit 2 sediments, cirque glaciers developed to a size that permitted them to spread from their basins to form a continuous cover of ice on the upland (Smith *et al.*, 1997). The Bara Plateau is part of the Mount Olympus upland (a few kilometres south of the 2917 m Mytikas summit) and its glacial geology is shown in Fig. 2B. The unit 2 sediments are the most extensive in this area and can be traced for almost 2 km from the headwaters of the Mavratza River across the Bara Plateau to the head of the Mavrolongus valley (Smith *et al.*, 1997; Fig. 2B). Thirteen cirques have been identified on the Bara Plateau which have floor elevations ranging from 2180 to 2510 m above present sea level. On the Plateau of the Muses (which includes the Mytikas summit), eleven cirques range in elevation from 2200 to 2660 m above sea level (Smith *et al.*, 1997).

The glacial and proglacial sedimentary sequences in the Mount Olympus area constitute a potentially important record of environmental change to the east of the Pindus Mountains and one that could be usefully compared to the long pollen record at Tenaghi Phillipon *c.* 160 km to the northeast (Wijmstra, 1969). However, the development of a geochronological framework for the glacial sequences has been hampered by the scarcity of suitable samples for radiometric dating. As Smith *et al.* (1997) have stated:

"The lack of radiometric dates on deposits of Mount Olympus and the adjacent piedmont precludes the establishment of a numerical chronology for Pleistocene events in this area. As a result, emphasis has been placed on study of the soils that occur on top of, and within, the Mount Olympus deposits." (Smith *et al.*, 1997, p. 820).

Soil development and relative age dating

Pedogenic maturity indices have been determined for the soil profiles on units 1 to 3 in an attempt to correlate the deposits of the Olympus piedmont with the sequence of soils developed on dated alluvial sediments in the Larissa Basin immediately to the south (Demitrack, 1986; Van Andel *et al.*, 1990). This approach resulted in the tentative correlation put forward by Smith *et al.* (1997), shown in Table 2. This model places the unit 1 sediments before 200 ka within Marine Isotope Stage (MIS) 8 and the unit 2 deposits within MIS 6. On the basis of this age model, the nature of the unit 1 soil suggests that the last major glaciation of the Olympus massif took place at some time during the last cold stage (MIS 2 or 4), but was restricted to valley heads (no upland ice) and glaciers extended to mid-valley positions (Smith *et al.*, 1997).

While Smith *et al.* (1997) present this correlation as tentative, it is important to point out that the chronological control for the alluvial sequence and soils of the Peneios River in the Larissa Plain, studied by Demitrack (1986), is relatively poor by the standards of more recent investigations (see Fuller *et al.*, 1998; Macklin *et al.*, 1997; 2002), and it can be argued that it does not form a reliable yardstick for regional correlation (Table 2). In addition, it is often difficult to make valid inter-site comparisons of soil profile development when sites are located at different altitudes in contrasting geomorphological settings. In such

Table 2. A tentative correlation of the Olympus Piedmont soils with the Pleistocene soils developed in alluvial sediments on the Larissa Plain studied by Demitrack (1986) (after Smith et al., 1997).

Larissa Plain Soil Name	Depositional phase	Olympus Piedmont Soil Name	Marine Isotope Stage
Pinios Group	<200 years	Unit 3 surface soils	1
Deleria soil	Historical		
Girtoni soil	6-7 ka		
Non-calcareous brown soil	10-14 ka		
Gonnoi Group soils	14-30 ka		
Agia Sophia soil	27-42 ka	Unit 3 truncated (buried) soils	3
Rodia Group soils	≥54-125 ka	Unit 2 soil	5
Deep red soil	≤210 ka	Unit 1 soils	7

cases, establishing the long-term constancy of important soil forming factors such as parent material and local climate can be especially problematic (Birkeland, 1984).

Cosmogenic dating of the Olympus moraines

More recently, an attempt to develop an improved temporal framework for the glacial deposits of Mount Olympus has involved the application of cosmogenic dating of unit 1 boulder surfaces using Chlorine 36. Manz (1998) reports cosmogenic dates from two sites within large (unit 1) recessional moraine complexes on the eastern piedmont of Mount Olympus. Limestone boulders at Site 1 (immediately north of the town of Litochoro) have yielded ^{36}Cl ages within the range 32-49 ka. Site 2 is located in the Mavroneri catchment to the west of Katerina (Fig. 2B), and the sampled boulders (igneous, metamorphic and carbonate lithologies) yielded ages ranging from 43-56 ka. One boulder at Site 2 gave an age of 146 ka (MIS 6) and this outlier has been attributed to either previous exposure or evidence of an earlier glacial phase. Manz (1998) argues that the Site 2 ages are to be preferred because silicate rocks are more reliable when using this technique. The tentative age model for units 1 to 3 proposed by Smith *et al.* (1994; 1997) and shown in Table 2 would need to be radically revised if these ages for the unit 1 sediments are accepted. Indeed, it is difficult to reconcile these ages with the extent and thickness of the Pleistocene sediments comprising units 1 to 3 - the ^{36}Cl ages demand much faster rates of soil development and landscape change in this area than those put forward by Smith *et al.* (1997) in Table 2. The deep red soil developed on these [unit 1] sediments is thought to record an extended interval, probably of interglacial duration, during which substantial pedogenesis occurred (Smith *et al.*, 1997 p. 811). Manz (1998) concludes that:

"Although the ages of the sedimentary packages (units 1 to 3) on the eastern Olympus piedmont appear to postdate the last interglacial period, additional ^{36}Cl measurements on deposits on the piedmont and upland areas of the mountain may reveal evidence of earlier glacial episodes." (Manz, 1998 p. 56).

In summary, at present the glacial sediments and landforms of Mount Olympus and the adjacent uplands lack a firm, internally consistent chronological base and, as Smith *et al.* (1997) have pointed out, the sequence of events proposed in Table 2 "can easily be shifted backward or forward in time." Thus, the correlation of glacial phases and periods of soil development on and around Mount Olympus with other sites in Greece and the wider Mediterranean zone is not yet possible.

The glaciation of the southern and southeastern flanks of Mount Tymphi, NW Greece

Background

Mount Tymphi (or Gamila) is located in Epirus in Northwest Greece approximately 15 km to the southwest of Mount Smolikas (Fig. 1). Its several peaks and ridges form part of the high watershed between the catchments of the Voidomatis (384 km^2) and Aoos (665 km^2) rivers. Tymphi has been described as both the most extensive and the most majestic of the Greek mountains (Sfikas, 1979), and includes a series of jagged limestone peaks, the highest of which exceed 2400 m (Fig. 3). The northern side consists of a range of cliffs and cirques and has been dissected by steep ravines. On the southern slopes of Mount Tymphi an extensive tableland is cut by the Vikos Canyon (Figs 3 and 4). A limited area of glacial deposits has also been mapped by IGME (1968) in the western corner of this plateau area (Fig. 3), but these features have not been studied by the authors.

Below the eastern ridges of Mount Tymphi, near the villages of Tsepelovon (39°54'N 20°49'E: *c.* 1100 m a.s.l.) and Skamnelli (39°55'N 20°51'E: *c.* 1200 m a.s.l.), extensive spreads of glacial deposits are marked on the 1:50,000 geological sheet (IGME, 1970; Fig. 3). This area contains an impressive array of glacial landforms and

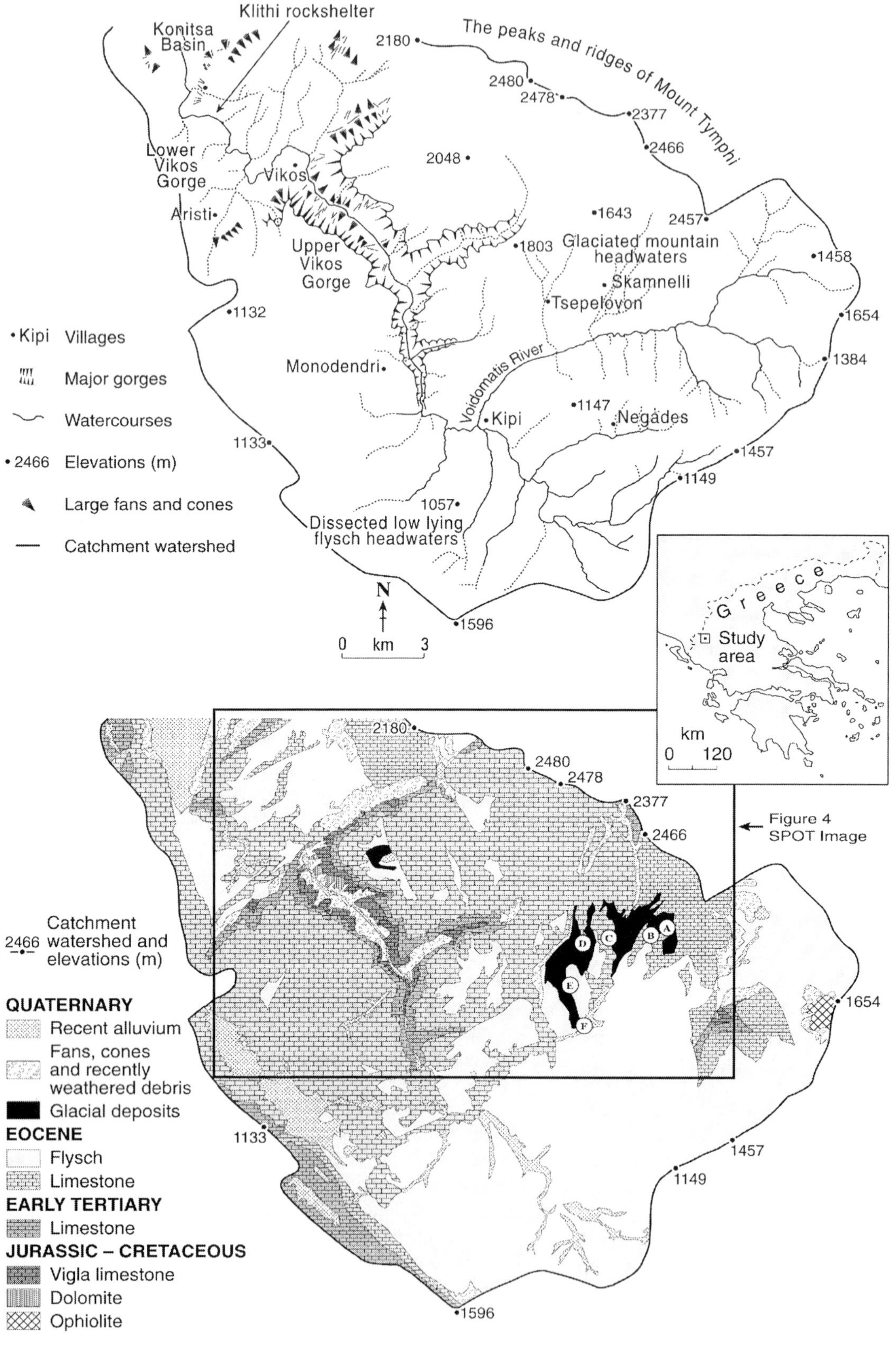

Fig. 3. The drainage network, major physiographic units and solid geology of the Voidomatis River basin (after Lewin et al., 1991 and Hamlin et al. 2000 and based on the geological sheets produced by IGME (1968; 1970). Mount Tymphi forms the highest part of the catchment and its peaks and ridges form the northeastern section of the watershed.

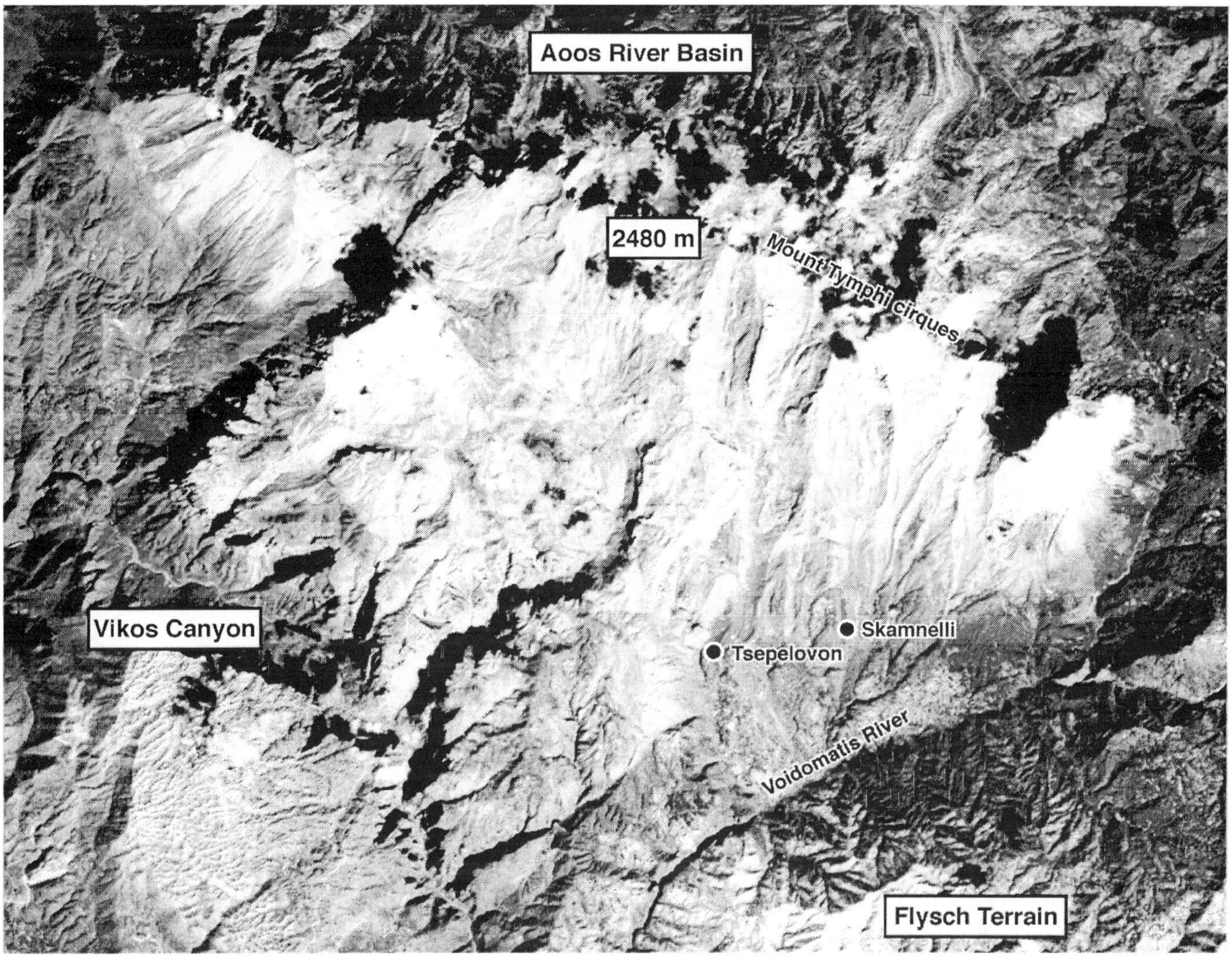

Fig. 4. SPOT satellite image of the glaciated terrain on the southern flanks of Mount Tymphi in the headwaters of the Voidomatis River basin. The Vikos Canyon is also shown. The image covers an area of approximately 25 x 15 km. See Fig. 3 for location.

sediments with perhaps some of the most well preserved glacial terrain on any of the Greek mountains. The cirques and major moraines in this area have been studied by Palmentola *et al.* (1990) as part of their work on central Mediterranean Pleistocene snowlines and their map is shown in Fig. 5. Seven major cirques face to the southwest and a series of 12 small and one large cirque (Manuli) face northwards and northeastwards into the Aoos River basin (Fig. 5).

Reconnaissance field studies of this area were conducted in the mid-1980s as part of the first phase of fieldwork investigating the Pleistocene history of the Voidomatis River and the principal alluvial sediment sources in the basin (Bailey *et al.*, 1990; Lewin *et al.*, 1991; Woodward *et al.*, 1992). This work confirmed that glacial landforms and sediments dominate the terrain on the southeastern flanks of Mount Tymphi and this area is shown on the SPOT satellite image (Fig. 4). This image also shows the two main gorges at the upstream end of the Vikos Canyon where glacial meltwaters and glacially-derived sediments were fed into the Voidomatis River (Woodward, 1990; Woodward *et al.*, 1995; Bailey *et al.*, 1997; Macklin *et al.*, 1997). The glacial deposits south of Tsepelovon reach down to the valley floor of the Voidomatis River to an elevation of between 850 and 900 m above sea level (Fig. 3) (Woodward *et al.*, 1995).

The glacial sediments and landforms

In common with many of the glaciated mountains on the Balkan Peninsula and wider Mediterranean region, hard limestones are the dominant lithology (Menkovic and Markovic, this volume) (Fig. 3). Extensive areas of flysch rocks are present at lower elevations, but these are mainly beyond the limits of glacial erosion (Bailey *et al.*, 1997). Some outcrops of flysch are present within the glaciated area, but all fractions of the till sediments are dominated by limestone (Woodward *et al.*, 1995). The assemblage of glacial landscape features includes cirques of various

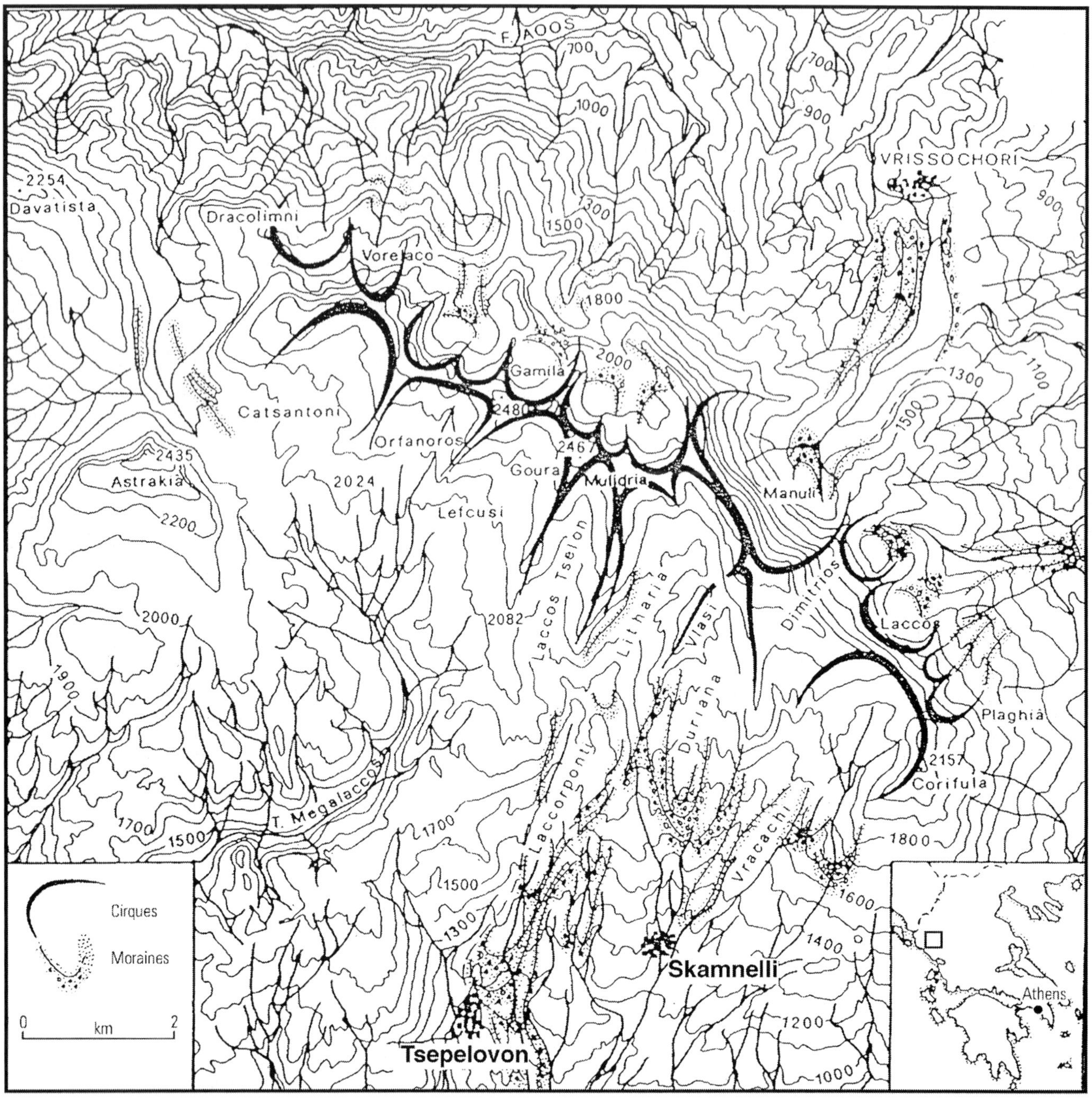

Fig. 5. The cirques and moraines on Mount Tymphi (after Palmentola et al., 1990).

sizes and geometries (Figs 5 and 6A), deep ice-scoured troughs with stepped long profiles (Fig. 6B), ice-scoured limestone bedrock pavements (Fig. 6C), and lateral and terminal moraine complexes (Fig. 6D). Steep limestone cliffs and thick talus deposits are also important landscape features (Fig. 6B). As mentioned above, glacial depositional landform assemblages are especially well preserved on the mountain slopes around the villages of Tsepelovon and Skamnelli (Figs 4, 5, and 7). Palmentola *et al.* (1990) have identified the major moraines in this area. Their map is shown in Fig. 5 and some of their site names are used below.

Remote sensing of glacial landforms and sediments

As limestone is the dominant rock type the spectral characteristics of the glacial sediments and landforms on Mount Tymphi cover only a limited range. This has posed particular problems for attempts to establish the nature and extent of landforms and sediments related to glacial activity using satellite remote sensing because contrasting geomorphological features such as, for example, scree slopes and rocky moraine surfaces, commonly have similar spectral signatures (Smith *et al.*, 1998). In view of the complexity of the glaciated landscape and the low spectral

Fig. 6 A) The cirque close to Tsouka Rossa col on Mount Tymphi ridge. Photograph taken by Graham Smith in June 1997. B) View of the glaciated limestone terrain looking upvalley towards the Mount Tymphi ridge that separates the Voidomatis and Aoos river basins. Photograph taken by Graham Smith in June 1997. C) Glacially-planed limestone pavement with well-developed clints and grikes on the southern flanks of Mount Tymphi. Photograph taken by Jamie Woodward in June 1998. D) A sharp-crested lateral moraine above the village of Skamnelli with an extensive glacio-karst ice-scoured bedrock pavement in the middle background. Unmetalled tracks for scale. Photograph taken by Jamie Woodward in June 1998.

variability within much of the limestone-dominated study area, the effectiveness of three remotely sensed datasets (air photographs, SPOT and TM imagery) and several digital image processing procedures (see Smith *et al.*, 1998; 2000) has been examined. These approaches have been tested for their ability to:

1) determine the spatial extent of landforms on Mount Tymphi and
2) differentiate between glaciated landscape features on the basis of both their genesis and age

One example is given here using Landsat Thematic Mapper (TM) imagery with 30 x 30 m pixels. A supervised image classification of the TM scene (using bands 1 to 5 and 7) was carried out using the Maximum Likelihood Classification (MLC) decision rule (Smith *et al.*, 1998). A portion of the TM image (*c.* 70 km^2) covering much of the glaciated terrain was selected for analysis. Prior to the MLC exercise, the five most important terrain classes were identified from field observations and air photograph analysis as:

1. Sharp-crested ("fresh") moraines
2. Weathered and eroded moraines
3. Scree formations
4. Slopes in flysch bedrock
5. Ice-scoured bedrock and limestone pavement

The results of this procedure are shown in Fig. 8 and Table 3. Given the 30 x 30 m resolution of the TM imagery, the MLC output provides a good representation of the geomorphology of the target area and the overall extent of the glacial deposits. However, some misclassification is evident, perhaps most notably in the overestimation of the extent of weathered and eroded moraines, and several sharp-crested moraines were actually misclassified as limestone bedrock (Smith *et al.*, 1998). Vegetation density and soil development generally increase with decreasing elevation and this is also an important influence on the spectral characteristics of the landscape classes. Nonetheless, the results of the MLC were evaluated in the field and from air photographs and the moraine complex to the north of Skamnelli was correctly classified as being dominated by

Table 3. Supervised Maximum Likelihood Classification results for the portion of the TM image covering the glaciated terrain and adjacent flysch basins shown in Fig. 8. The image covers an area of c. 8.2 x 8.5 km and areal estimates in km^2 of each terrain class are also given.

Terrain Class	% of TM image	km^2
1. Sharp-crested moraines	15.77	10.99
2. Weathered/degraded moraines	19.80	13.80
3. Limestone scree formations	05.21	03.63
4. Slopes on flysch bedrock	32.89	22.92
5. Ice-scoured limestone bedrock and pavement	26.33	18.35
Total	100.00	69.70

sharp crested moraines and the landforms near the village of Tsepelovon were correctly classified as weathered and degraded moraines. This approach is still under development, but it is clear that any relative-age information derived from the MLC can be evaluated against the geochronological data obtained from both the radiometric dating programme (uranium series) and data from soil development indices (e.g. Woodward *et al.*, 1994; Smith *et al.*, 1997). Some of these relative age data from the MLC outputs are discussed further below.

Smith *et al.* (1998; 2000) have shown that satellite imagery offers the potential to improve our understanding of glaciated Mediterranean mountain landscapes and can be a valuable aid in reconnaissance mapping studies. However, for glaciated limestone environments – which are common in the Mediterranean region – it has been shown that the value of such an approach and the accuracy of terrain mapping will increase as a function of both the spatial and spectral resolution of the imagery employed (Smith *et al.*, 1998; 2000).

The impact of Pleistocene glaciation on the long-term behaviour of the Voidomatis River

The middle and lower reaches of the Voidomatis River, some 15 to 30 km downstream of the glaciated terrain described above, contain perhaps the best documented and well-dated Pleistocene alluvial sequence in the Eastern Mediterranean region (Bailey *et al.*, 1990; Lewin *et al.*, 1991; Woodward *et al.*, 1992; Woodward *et al.*, 1994; Macklin *et al.*, 1997; Hamlin *et al.*, 2000; Woodward *et al.*, 2001). The nature of this record will not be repeated in detail here, but it is significant because it demonstrated for the first time in the Mediterranean region that the glaciation of some of the most southerly mountains in mainland Europe exerted a major influence on Pleistocene river behaviour and sediment transfer (Bailey *et al.*, 1990; Lewin *et al.*, 1991; Woodward *et al.*, 1992; Woodward *et al.*, 1995; Macklin *et al.*, 1997). This work involved detailed lithostratigraphic and pedostratigraphic investigations supported by a range of relative and absolute dating methods (see Macklin *et al.*, 1997). The most recent phase of fieldwork in the Voidomatis River basin has shown that uranium series dating of calcretes formed within coarse-grained fluvial gravels can provide excellent age control for Middle and Late Pleistocene sequences which is in good agreement with other methods such as Electron Spin Resonance and Thermoluminescence (Macklin *et al.*, 1998; Hamlin *et al.*, 2000). In conjunction with an extension of the earlier mapping programme, this has allowed the development of a more detailed sub-division of the sequence of Pleistocene coarse-grained alluvial sediments deposited in the lower part of the catchment (Macklin *et al.*, 1998; Hamlin *et al.*, 2000; Hamlin, 2000).

Uranium-series isochron dating of calcretes in the Mount Tymphi tills

Deep sections in the glacial sediments are present where the depositional zone on the lower slopes is traversed by mountain tracks or by the Tsepelovon to Skamnelli road (Fig. 9). Such exposures in the Mount Tymphi tills show that the glacial sediments are commonly matrix supported and rich in sands, silts and clays derived from the comminution of the hard limestone bedrock (Woodward *et al.,* 1995). Sub-rounded boulders, cobbles and gravels are also common and at some locations, the road and track cuttings have exposed cavities within the till matrix where carbonate precipitation has taken place. In some places the fine sediment matrix has been washed away creating interclast voids that have been partially or completely filled with calcite. These calcretes often form a skin on the surface of cobble- and boulder-sized limestone clasts (accreting roughly parallel to the host surface) and, in some cases, they completely fill interclast voids. Calcrete development has been observed within a few metres of the moraine surface and the carbonates are probably derived from corrosion of the fine calcareous till matrix (see Ford & Williams, 1989).

The term calcrete is used here in a broad sense to describe the products of secondary carbonate formation (including a variety of morphologies) in a near surface

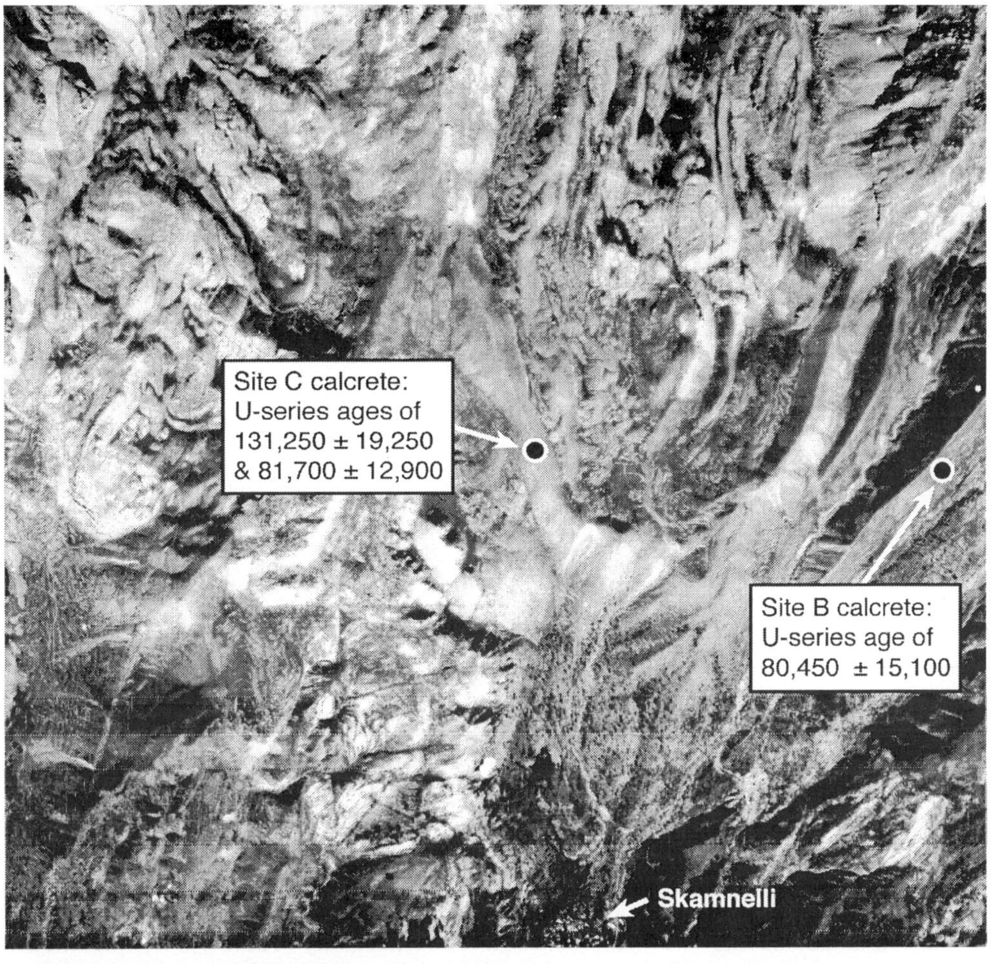

A

B

Fig. 7 A). Air photograph of the moraine complex at an elevation of c. 1600 m (above the village of Skamnelli) on the southern flanks of Mount Tymphi. The locations of calcrete samples B and C listed in Table 4 are shown.
B) Crest of the large end moraine complex on the southern slopes of Mount Tymphi above the village of Skamnelli. Photograph taken by Jamie Woodward in June 1998.

Table 4. Uranium-series ages from calcretes collected from glacial sediments on the southeastern slopes of Mount Tymphi. Sample locations are shown on Fig. 3 and discussed in the text. The errors associated with each U/Th age are given at 2 sigma probability and represent a measure of counting uncertainties in the determination of nuclide concentrations (see Hamlin, 2000). Note that Site F comprised a sequence of glacial and glacio-fluvial sediments exposed on the present valley floor of the Voidomatis River. This scheme will be revised and extended as more dates and field data become available. Marine isotope stages after Martinson et al. (1987). Note that these ages provide only a minimum age for the host sediments (see text for discussion).

Site	Uranium-series age of calcrete sample	Elevation of sample location	Marine Isotope Stage associated with calcrete formation
A.	>350,000	1750 m	11 or earlier
A.	321,500 ± 35,650	1750 m	9
A.	121,400 ± 21,000	1750 m	5e
B.	80,450 ± 15,100	1750 m	5a
C.	131,250 ± 19,250	1800 m	6
C.	81,700 ± 12,900	1800 m	5a
D.	71,100 ± 8,700	1290 m	5a/4
E.	105,730 ± 21,450	1210 m	5c
F.	96,250 ± 15,500	850 m	5c

terrestrial environment (see Dixon, 1994; Kelly *et al.*, 2000). Macklin and Woodward collected calcretes from several exposures on the southern slopes of Mount Tymphi in May 1998 (Figs 3 and 7A) and these materials have been dated by the U/Th isochron method by Dr Stuart Black (see Kelly *et al.*, 2000; Hamlin, 2000). This approach involves the measurement of ^{230}Th ingrown from the decay of authigenic ^{234}U and, indirectly, ^{238}U, the uranium having been co-precipitated from solution with the carbonate (Kelly *et al.*, 2000). This approach employs an isochron method (see Bischoff & Fitzpatrick, 1991) to determine the authigenic radionuclide components in each calcrete sample (see Kelly *et al.*, 2000). Nine U/Th ages from six sites on the southern slopes of Mount Tymphi are listed in Table 4. The errors on the age determinations are quoted at 2 sigma probability and are determined as a measure of counting uncertainties in the radionuclide measurements. It is important to recognise that these U/Th ages provide *minimum* ages for the host glacial sediments as the calcretes were formed sometime after deposition. Each sample site (A to F) is listed in Table 4 and shown on Fig. 3 and described in detail below:

Site A: This sample was collected from an exposure in a body of till sediments that was preserved down-valley of a low bedrock protrusion on the eastern side of the main moraine complexes above the village of Skamnelli (Fig. 10A). In contrast to the other five sites discussed below, the fine sediment matrix (where present) showed some evidence of weathering. The section at Site A revealed sub-rounded limestone glacial clasts with horizons of clast supported material where the fine matrix had been winnowed away by percolating waters allowing calcretes to fill the interclast voids (Fig. 10A). This sample had three clearly different types of calcite present (Fig. 10B) and each has been sub-sampled and dated by uranium series. Several generations of calcite growth were evident (Dr Stuart Black, personal communication) and the oldest of these was beyond the range of uranium-series dating (>350 ka). The two other calcrete samples yielded ages of 321,500 ± 35,650 and 121,400 ± 21,000 and indicate further phases of calcrete development during the interglacial conditions of MIS 9 and MIS 5e respectively (Table 4). The presence of several stages of calcite growth at this site indicates that this section has been stable for an extended period of time (>350 ka). Despite the limited extent of this deposit, the oldest uranium series assay provides the first direct dated evidence of Middle Pleistocene glacial activity in Greece and the wider Mediterranean region before *c.* 350 ka.

Site B: This site was exposed in a section along a vehicle track in the easternmost lateral moraine above the village of Skamnelli. This part of the glaciated terrain (Vracachi complex, see Fig. 5) contains two well-preserved lateral moraines that can be traced up valley to the easternmost cirque shown on Fig. 5. A calcrete sample yielded a uranium series age of 80,450 ± 15,100 years BP (Table 4). This provides a minimum age for this moraine system and indicates that the last major phase of glacial activity in this upland valley took place before MIS 4. This uranium series age is discussed further below in relation to the evidence from Site C (Fig. 7A).

Site C: A calcrete sample was collected from a section exposed in the track on the western side of the major arcuate lateral and terminal moraine complex on the slopes high above the village of Skamnelli. This landform and the sample location are shown in the aerial photograph of this part of the southern slopes of Mount Tymphi (Fig. 7A). This calcrete sample contained evidence of two generations

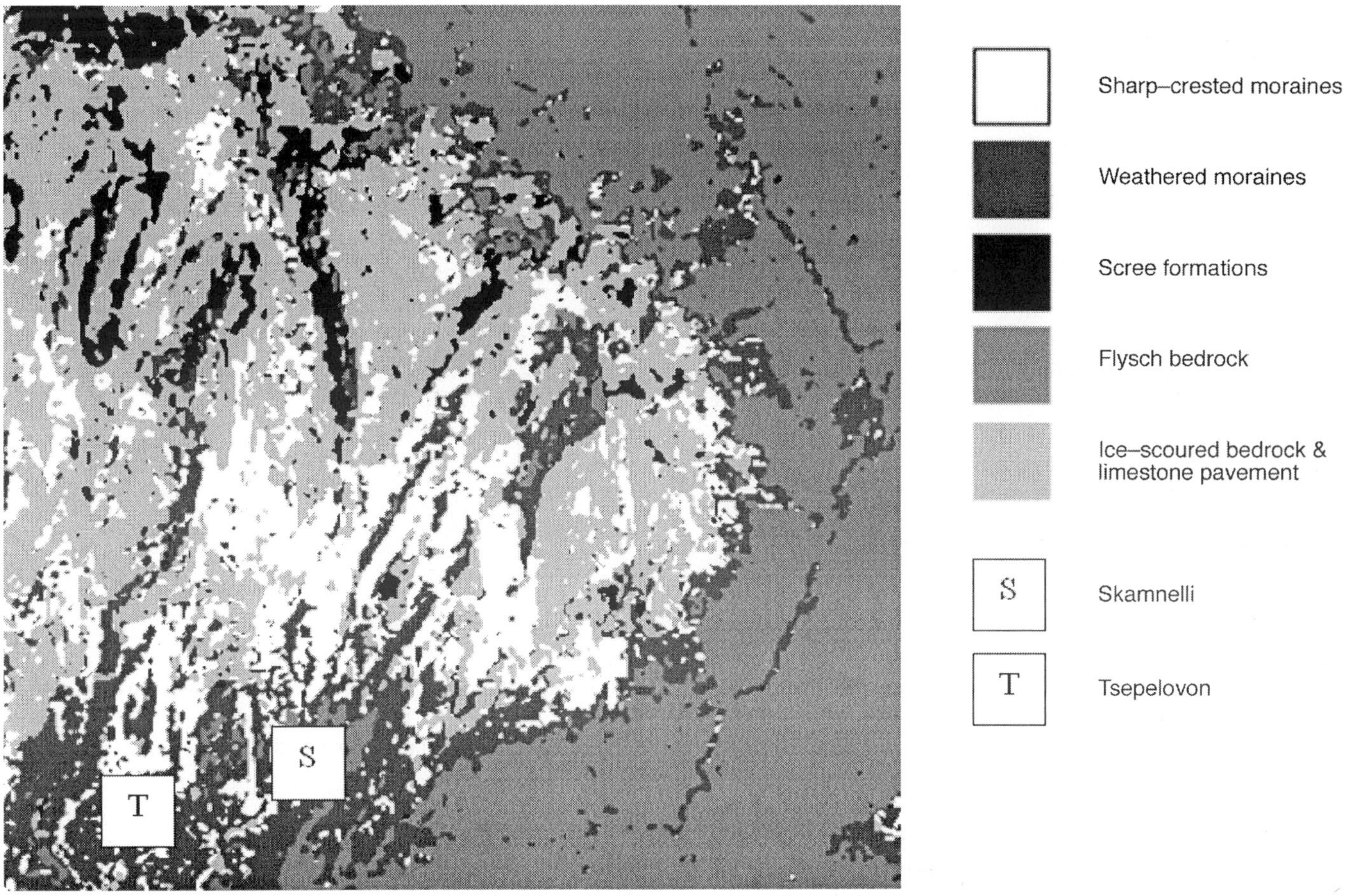

Fig. 8. Supervised Maximum Likelihood Classification of the TM image of the glaciated headwaters of the Voidomatis River (southern slopes of Mount Tymphi). The image covers an area of approximately 8.2 x 8.5 km (after Smith et al., 1988). Elevations: Tsepelovon (c. 1100 m) and Skamnelli (c. 1200 m). Compare with geological map (Fig. 3) and SPOT image (Fig. 4).

of calcite growth and each has been dated giving uranium-series ages of 131,250 ± 19,250 and 81,700 ± 12,900 years BP respectively. The older of the two ages provides a minimum age for this large moraine complex and suggests that the glacial phases to which it relates took place during MIS 6 although an earlier date is possible (Table 4). The similar size, form and degree of preservation between the closely spaced moraines shown in Fig. 7A (associated with Sites B and C) strongly suggests that they relate to the same phase of glacial activity. While the calcrete at Site B has yielded a younger age, the freshness of the topography (sharp-crested moraines shown on the TM imagery and in Fig. 7A) and the uranium series ages from Site C indicate that this moraine complex was probably formed during MIS 6. These ages demonstrate that the glacial sediments and landforms in this part of Mount Tymphi are much older than the global Last Glacial Maximum of MIS 2.

Site D: A calcrete sample was collected from a section revealing limestone-dominated till sediments to the north of Tsepelovon at an elevation of *c.* 1290 m on the margin of the Laccorponti valley system mapped by Palmentola *et al.* (1990) (Fig. 5). This material yielded a uranium series age of 71,000 ± 8,700 years BP. This age falls close to the boundary of MIS 4 and 5a (Martinson *et al.,* 1987). Given the quoted uncertainty associated with this age, the glacial phase could not have taken place later than the early part of MIS 4 and could have taken place during an earlier cold stage.

Site E: The road from Tsepelovon to Skamnelli cuts across several lateral moraines and provides a series of excellent exposures in till sediments that are typical of those found across the southern slopes of Mount Tymphi (Fig. 9). In the first major exposure on this road outside Tsepelovon (see Woodward *et al.*, 1995), cemented till sediments were present at the base of the section just above the level of the road. A calcrete sample from this horizon yielded a uranium-series age of 105,730 ± 21,450 years BP. This moraine is a major landscape feature and is part of a spread of glacial deposits that extend down to the valley floor of the Voidomatis River (Fig. 3) (Woodward *et al.,* 1995). This calcrete formed during MIS 5 (possibly during MIS 5c) and forms a minimum age for the deposition of this feature. This moraine forms part of the "weathered and eroded moraines" category identified on the TM imagery (see Smith *et al.*, 1998 and Fig. 8).

Fig. 9. A road cut section in a large moraine outside the village of Tsepelovon (Photograph taken by John Lewin in August 1986).

In contrast to the sharp-crested moraines at Sites B and C above Skamnelli (whose upper sections are not weathered), exposures in this (Site E) moraine complex have revealed a thick soil with strongly weathered A and B horizons (Woodward and Smith, unpublished) that are similar to those identified on terraced alluvial surfaces in the middle reaches of the Voidomatis River (Woodward *et al.*, 1994). The degraded form of this moraine and the evidence for an extended period of pedogenic weathering indicate that it relates to a phase of glacial activity that took place before the formation of the moraine complex above Skamnelli. The multiple phases of calcrete development evident at Sites A and C demonstrate that calcretes can form a long time after the deposition of the primary glacial sediments and it is important to appreciate that the uranium series age from Site E represents a minimum age for its host glacial sediments. Thus, if the moraines above Skamnelli (Sites B and C) were deposited during MIS 6, the glacial sediments at Site E could relate to an earlier phase of glacial activity during MIS 6 or an earlier (pre-MIS 6) cold stage. It is worth pointing out that Site E is located further down the Laccorponti valley system (than Site D) mapped by Palmentola *et al.* (1990) (Fig. 5) and the above discussion would suggest that the glacial sediments at Site D are much older than the uranium series age of 71,000 ± 8,700 years BP.

Site F: Glacial and glacio-fluvial sediments are exposed on the present valley-floor of the Voidomatis River to the south of Tsepelovon (Fig. 3) at an elevation of approximately 850 m above sea level. A sample of calcrete was collected from a cemented glaciofluvial unit that lay between 5 to 6 m above the modern channel within a much thicker sequence of glacial and glaciofluvial deposits. This sample yielded a uranium series age of 96,250 ± 15,500 (Table 4). In common with the sample from Site E, this age lies within the MIS 5 complex and may have formed during MIS 5c, with the phase of primary coarse sediment deposition taking place during the preceding cold phase (MIS 5d) or earlier.

Towards a new chronostratigraphic framework for glaciation in Greece

While more uranium-series ages and more extensive and detailed field mapping are needed to develop and refine the age model for the Mount Tymphi sequences presented here, it is nonetheless possible to make some important observations about landscape development and glacial history in the mountains of Greece. Furthermore, the geochronological data presented in this paper and the arguments outlined below have significance for our understanding of Pleistocene glacial activity in the wider Mediterranean region. Apart from Site C, the uranium series ages indicate that calcrete development took place during warm periods and, where a suite of ages is available, the oldest ages for a particular deposit can be used to infer glacial activity during a previous cold stage. Additional information from, for example, geomorphological mapping and relative-age dating criteria (e.g. soil development and moraine degradation indices) can be synthesized to develop a chronostratigraphic framework for the Pleistocene glaciation of the Pindus Mountains.

The sediments exposed at Site A contrast with the other exposures discussed above because of the high proportion of clast-supported, matrix free sediment. In addition, unlike the other locations, this body of sediment is not part of an extensive and well-preserved moraine complex and these characteristics are in good agreement with the very old uranium-series ages we have obtained (Table 4). Each of the ages from Sites B to F are from contexts which relate to

A

B

Fig. 10. A) The section exposed at Site A showing the zone of calcrete formation between the two large clasts to the left of Mark Macklin's head and chest.
B) Close up of the sampled calcrete between the boulders showing distinctive layering, well developed columnar calcite crystals and multiple phases of development. The lens cap has a diameter of 6 cm. Photographs taken by Jamie Woodward in June 1998.

distinctive and extensive glacial depositional (and glaciofluvial in the case of Site F) features on the southern flanks of Tymphi, and they range in age from 131,250 ± 19,250 (MIS 6) to 71,000 ± 8,700 (MIS 5a/4).

Work on the cemented alluvial sediments downstream in the Voidomatis gorges and Konitsa basin (Fig. 3) has shown that calcretes can form relatively quickly following deposition of the host gravels. However, the multiple phases of calcrete development at sites A and C demonstrate that it is important to obtain several assays from each suite of glacial sediments and landforms as, under certain conditions, calcrete development can also take place a considerable time *after* the deposition of the parent glacial sediments. Thus, while the ages obtained for Sites E and F (which are located in the "weathered and eroded moraine" category used in the MLC classification of the TM imagery) are younger than the oldest age from Site C, the primary glacial sediments at Sites E and F could be much older. In fact, the presence of deep and strongly weathered soil horizons on the moraine surface down valley of Tsepelovon indicates that the sediments at Site E are considerably older than those at Sites B and C in the "sharp-crested moraine" complex above the village of Skamnelli.

The uranium-series ages discussed above indicate that most of the glacial sediments and landforms in the headwaters of the Voidomatis River basin are much older that the global Last Glacial Maximum (MIS 2) and relate to glacial activity during MIS 6 and earlier cold stages. There is also evidence of glacial activity in the Pindus Mountains before *c.* 350 ka. The evidence for several stages of calcrete growth at two sites (A and C) is highly significant and cautions the development of chronostratigraphic frameworks based on a small number of dates. The multi-phase calcretes at Sites A and C show that these bodies of sediment and their associated landscape features have been stable for an extended period of time and >350 ka in the case of Site A and at least 131,250 ± 19,250 at Site C. These periods span glacial-interglacial cycles and the data from Site C in particular demonstrate that it is unwise to assume that apparently fresh moraines in Mediterranean mountain environments must date to the most recent cold stage (see Palmentola *et al.*, 1990).

It is not yet clear whether ice build up and glacier development took place in Northern Greece during the cooler phases of the MIS 5 complex. The ages from Sites B, D, E and F relate to the Last Interglacial (Table 4), but it is important to remember that each of them represent minimum ages. A larger suite of uranium series ages in conjunction with a detailed mapping programme is required to address this issue. Detailed analysis of the pollen records for these sub-stages would allow a broader picture of the associated regional palaeoenvironments (e.g. humidity and temperature) to be produced.

The bulk of the glacial sediments and major glacial depositional landforms in this area are much older than the global LGM of MIS 2 and this is a very significant outcome of the uranium series dating programme. Glacial deposits of this period do not appear to be extensive, although further mapping and uranium-series dating work is in progress and sediments of this period may be present at higher elevations (Fig. 5). The excellent preservation of the sharp-crested moraines above the village of Skamnelli (Fig. 9A) may be the result of free-draining limestone-rich till sediments and the extensive areas of scree sediments and glacially-scoured limestone pavement that limit surface runoff and erosion by fluvial processes and mass wasting. There is little evidence of soil development on these moraines (Woodward and Smith, unpublished) and it is not yet clear if this is due to very slow rates of pedogenic weathering at this elevation or whether the apparently fresh moraines are eroding slowly and uniformly across their surfaces, thus preventing the development of a marked weathering profile.

As we have mentioned above, the uranium-series method has been used successfully to develop a chronostratigraphic framework for the coarse-grained alluvial units at lower elevations in the Voidomatis River basin (Macklin *et al.*, 1998; Hamlin *et al.*, 2000) providing ages that are in good agreement with other methods such as ESR, TL (Lewin *et al.*, 1991) and radiocarbon (Woodward *et al.*, 2001). It is interesting to note that the Pleistocene alluvial sequence in the Voidomatis River contains evidence for major phases of aggradation (also dated by uranium-series methods) shortly before 113,000 ± 6000 and 80,000 ± 7000 (Macklin *et al.*, 1998; Hamlin *et al.*, 2000). In the southern part of the Konitsa basin, where the river emerges from the confines of the Lower Vikos Gorge (Fig. 3), these terrace surfaces lie approximately 15.5 and 12.75 metres above the present channel (Hamlin *et al.*, 2000). These are the thickest limestone-dominated coarse-grained alluvial units yet identified in the Voidomatis basin and represent major phases of sediment transfer and deposition. The uranium-series ages listed in Table 4 strongly suggest that large glaciers and active moraine complexes were present in the headwaters of this river system during MIS 6 and are in good agreement with the geomorphological implications of the alluvial record. The issue of glacial activity in the Mediterranean during the cold sub-stages of MIS 5 (5d and 5b) is an intriguing one and further work is required to confirm this. Interestingly, in a recent Mediterranean-wide review of Middle and Late Pleistocene river behaviour, Macklin *et al.*. (2002) have identified major periods of valley-floor sedimentation across the Mediterranean region during MIS 6 and the cooler phases of MIS 5 – i.e. during 5d and most notably at the 5b/5a boundary. Most of the river basins considered in their analysis drain mountainous headwater catchments and, during phases of cooler climate, increases in sediment supply and flood frequency promoted widespread aggradation throughout many river valleys. Indeed, as Macklin *et al.* (2002) suggest, these appear to have been important periods of landscape development and sediment transfer in many Mediterranean mountain catchments.

The Mount Tymphi and Mount Olympus records

Smith *et al.* (1994) have argued that the first phase of glaciation on Olympus was the most extensive, with ice lobes covering part of the eastern, northern and western piedmont of the mountain. This period of glaciation was associated with a snowline as low as 1900 m above present sea level. Thus, in common with the available dated evidence from Mount Tymphi, glacial activity was not as extensive during the Last Glacial Maximum (Late Würmian or MIS 2). However, it is worth pointing out that all of the uranium-series ages listed in Table 4 are significantly older than almost all of the ^{36}Cl ages reported by Manz (1998) for the unit 1 moraine complexes on Mount Olympus. As Tymphi and Olympus are both located close to 40°N, one may expect a broad correlation between major phases of ice build up and glacial sediment deposition. If this is the case, it would be difficult to reconcile the extensive pre-Würmian glacial sequence we have identified on Mount Tymphi with the compressed sequence of post-MIS 5 glacial cycles and pedogenic weathering phases required on Mount Olympus if the MIS 3 cosmogenic ages preferred by Manz (1998) are accepted. In fact, notwithstanding the weaknesses discussed above, in the light of the uranium series ages from Mount Tymphi, the timescales associated with the tentative age model presented by Smith *et al.* (1997) would appear to be more realistic (Table 2). It is important to appreciate that, in common with the uranium series data from the calcretes on Mount Tymphi, the cosmogenic ages from Mount Olympus also constitute *minimum* ages. Thus, the cosmogenic ages from Site 2 indicate glacial activity at some time before 43 to 56 ka (and this could have been during MIS 4 or earlier, depending on the geomorphological history of the dated boulders), while the outlier of 146 ka may represent glacial activity during MIS 6. This scenario would be much closer to the age model we have proposed for the glacial sediments and landforms on Mount Tymphi. However, before detailed correlations can be attempted, it is clear that more radiometric dates (ideally using a range of methods) are needed from both regions.

Conclusions and future research needs

Recent work has shown that the mountains of Greece contain evidence for multiple phases of ice build up and decay during the Middle and Late Pleistocene. However, there is still a need for detailed field mapping in many areas to establish the precise spatial extent and style of glacial activity in the Greek mountains. The morphological and sedimentological evidence for glaciation is less extensive and less well preserved in central and southern Greece. At present, the only glaciated terrain in Greece where detailed field-based mapping and lithostratigraphic assessment have been carried out is on Mount Olympus and the surrounding piedmont zone (Smith *et al.*, 1997). However, the absence of a reliable and internally consistent chronology for this work limits its value as a source of palaeoclimatic information and prevents detailed comparison with data from Mount Tymphi and with the long pollen sequences in Greece and other records of proxy climate.

Even though most of the sequences have not been dated, a number of workers have assumed that the glacial sediments and landforms in Greece date to the most recent cold stage (Würmian) of the Pleistocene (Boenzi & Palmentola, 1997) and the global Last Glacial Maximum (MIS 2). The Mediterranean region is dominated by uplifted carbonate terrains and the glacial sediments are often cemented by carbonate materials than can be dated by uranium series methods. The uranium series ages reported here indicate that glaciation was extensive on Mount Tymphi during MIS 6 and earlier cold stages. We also present the first dated evidence for glacial activity in the Mediterranean region before 350 ka. Together, these ages show that many of the glacial landforms have been very well preserved and that ice build up on Mount Tymphi (and perhaps in the rest of the Pindus Mountains) was much less extensive during the Late Würmian. Further work is needed to build on these findings and to address whether glacial activity also took place during the cold sub-stages of MIS 5.

Regional comparisons of Equilibrium Line Altitudes in the Mediterranean should *not* assume that fresh glacial landforms relate to the latter part of the most recent cold stage. Indeed, the timescales involved in the preservation and reworking of glacial landforms and sediments in limestone mountain environments is also a key issue as our data from Mount Tymphi suggest that apparently fresh glacial landforms may be much older than previously thought.

There is little doubt that the glacial sediments and landforms in the mountains of southern Europe form an important record of environmental change. Where appropriate samples are available, uranium series methods may provide the best means of developing good chronological control for sedimentary sequences in glaciated limestone terrains and can provide the necessary age control to develop models of ice build up and decay that may aid palaeoclimatic reconstruction. Future work on the chronology and palaeoclimatic implications of the glacial sequences of the Pindus Mountains will allow detailed comparisons to be made with the region's fluvial records to establish the nature of the interactions between ice build up and decay and coarse and fine sediment inputs to the fluvial system (Woodward *et al.*, 1995; Macklin *et al.*, 1997; Hamlin *et al.*, 2000). Robust chronological frameworks are needed to improve our understanding of long-term landscape development, glacial palaeoclimates, and sediment fluxes in glaciated Mediterranean mountain catchments.

ACKNOWLEDGEMENTS

MGM and JCW acknowledge the support of a grant from the UK Natural Environment Research Council (GR9/2916: *Catastrophic Flooding and Environmental Change in Mediterranean Mountain Environments*) which funded part of the work in the Voidomatis basin reported here. GRS was funded by a PhD studentship from the Faculty of Science and Engineering at Manchester Metropolitan University. We would like to thank the IGME offices in Athens and Preveza for granting permission to undertake fieldwork in Epirus and to export samples from Greece. We also thank David Appleyard (Graphics Unit, School of Geography, University of Leeds) for producing the maps and figures, and Dr Stuart Black (University of Reading) for uranium-series dating of the calcrete samples. We are especially grateful to John Lewin (Aberystwyth), Geoffrey Smith (Ohio) and Phil Gibbard (Cambridge) for reviewing this manuscript. We thank the late Alexis Gouris of Tsepelovon who first welcomed JCW and MGM to Mount Tymphi in 1986 and provided help and hospitality to us and our postgraduate students on many occasions.

Note: A detailed geomorphological mapping programme is currently underway on Mount Tymphi and Mount Smolikas by Philip Hughes as part of his PhD project (2001 to 2004) supervised by Dr P. L. Gibbard

(Cambridge) and Dr J. C. Woodward (Leeds). The U/Th dating programme is also being extended by this team with support from the UK Natural Environment Research Council (NERC).

References

Bailey, G.N., Lewin, J., Macklin, M.G. & Woodward, J.C. (1990) The "Older Fill" of the Voidomatis Valley Northwest Greece and its relationship to the Palaeolithic Archaeology and Glacial History of the Region. *Journal of Archaeological Science,* **17**, 145-150.

Bailey, G.N., Turner, C., Woodward, J.C., Macklin, M.G. & Lewin, J. (1997) The Voidomatis basin: an Introduction. *In:* G.N. Bailey (ed.), *Klithi: Palaeolithic settlement and Quaternary landscapes in northwest Greece. Volume 2: Klithi in its local and regional setting.* Cambridge, McDonald Institute for Archaeological Research, 321-345.

Birkeland, P. W. (1984) *Soils and Geomorphology*. Oxford, Oxford University Press.

Bischoff, J.L. & Fitzpatrick, J.A. (1991) U-series dating of impure carbonates: an isochron technique using total sample dissolution. *Geochimica et Cosmochimica Acta*, **55**, 543-554.

Boenzi, F. & Palmentola, G. (1997) Glacial features and snow-line trends during the last glacial age in the Southern Apennines (Italy) and on Albanian and Greek mountains. *Zeitschrift für Geomorphologie N.F. Supplement-Band,* **41**, 21-29.

Boenzi, F., Palmentola, G., Sanso, P. & Tromba, F. (1992) Le tracce glaciali del Massicio dello Smolikas (Catena del Pindo – Grecia). *Rivista Geografica Italiana*, **99**, 379-393.

Demitrack, A. (1986) *The Late Quaternary Geologic History of the Larissa Plain, Thessaly, Greece: Tectonic, Climatic and Human Impact on the Landscape.* Unpublished PhD Dissertation, Stanford University.

Dixon, J.C. (1994) Duricrusts. *In:* Abrahams, A.D. & Parsons, A.P. (eds), *Geomorphology of Desert Environments*. Chapman and Hall, London: 82-105.

Ford, D.C. & Williams, P.W. (1993) *Karst Geomorphology and Hydrology*. London, Unwin Hyman. 320 pp.

Fuller, I.C., Macklin, M.G., Lewin, J., Passmore, D.G. & Wintle, A.G. (1998) River response to high-frequency climate oscillations in southern Europe over the past 200 k.y. *Geology*, **26**, 275-278.

Hamlin, R.H.B. (2000) *Environmental Change and Catastrophic Flooding in the Voidomatis and Aoos Basins, Northwest Greece.* Unpublished PhD Thesis, University of Leeds, UK, 299 pp.

Hamlin, R.H.B., Woodward, J.C., Black, S. & Macklin, M.G. (2000) Sediment fingerprinting as a tool for interpreting long-term river activity: the Voidomatis basin, Northwest Greece. *In:* Foster, I.D.L. (ed.) *Tracers in Geomorphology,* Chichester, John Wiley and Sons, 473-501.

IGME (1968) *1:50,000 Geological Map of Greece (Doliana Sheet),* Institute of Geological and Mineral Exploration, Athens.

IGME (1970) *1:50,000 Geological Map of Greece (Tsepelovon Sheet),* Institute of Geological and Mineral Exploration, Athens.

IGME (1983) *1:500,000 Geological Map of Greece,* Institute of Geological and Mineral Exploration, Athens.

Kelly, M., Black, S. & Rowan, J.S. (2000) A calcrete-based U/Th chronology for landform evolution in the Sorbas basin, southeast Spain. *Quaternary Science Reviews*, 19, 995-1010.

Lewin, J., Macklin, M.G. & Woodward, J.C. (1991) Late Quaternary fluvial sedimentation in the Voidomatis Basin, Epirus, northwest Greece. *Quaternary Research,* 35, 103-115.

Macklin, M.G., Lewin, J. & Woodward, J.C. (1997) Quaternary river sedimentary sequences of the Voidomatis basin. *In:* G.N. Bailey (ed.), *Klithi: Palaeolithic settlement and Quaternary landscapes in northwest Greece. Volume 2: Klithi in its local and regional setting.* Cambridge, McDonald Institute for Archaeological Research, 347-359.

Macklin, M.G., Woodward, J.C. & Lewin, J. (1998) *Catastrophic Flooding and Environmental Change in Mediterranean Mountain Environments*. Final Report to the UK Natural Environment Research Council (NERC), School of Geography, University of Leeds, Leeds, UK.

Macklin, M.G. Fuller, I.C. Lewin, J. Maas, G.S. Passmore, D.G. Rose, J. Woodward, J.C. Black, S. Hamlin R.H.B. & Rowan J.S. (2002) Correlation of fluvial sequences in the Mediterranean basin over the last 200 ka and their relationship to climate change. *Quaternary Science Reviews*, **21**, 1633-1644.

Manz, L.A. (1998) *Cosmogenic ^{36}Cl chronology for deposits of presumed Pleistocene age on the Eastern Piedmont of Mount Olympus, Pieria, Greece.* Unpublished MSc thesis, Ohio University, USA.

Martinson, D.G., Pisias, N.G., Hays, J.D., Imbrie, J., Moore, T.C. & Shackleton, N.J. (1987) Age dating and the orbital theory of the ice ages: development of a high resolution 0-300,000 year chronostratigraphy. *Quaternary Research,* **27**, 1-29.

Mastronuzzi, G., Sanso, P. & Stamatopoulos, L. (1994) Glacial landforms of the Peloponnisos (Greece), *Rivista Geografica Italiana,* **101**, 77-86.

Menkovic, L. & Markovic, M. (this volume) Glacial morphology of Serbia (Yugoslavia) with comments on the Pleistocene glaciation of Montenegro, Macedonia and Albania.

Messerli, B. (1967) Die eiszeitliche und die gegenwärtige Vergletscherung im Mittelmeerraum. *Geographica Helvetica*, **22**, 105-228.

Palmentola, G., Boenzi, F., Mastronuzzi, G. & Tromba, F. (1990) Osservazioni sulle trace glaciali del M. Timfi, Catena del Pindo (Grecia). *Geografia Fisica e Dinamica Quaternaria*, **13**, 165-170.

Pechoux, P.Y. (1970) Traces d'activité glaciaire dans les Montagnes de la Grece centrale. *Revue de Géographie Alpine*, **58**, 211-224.

Sestini, A. (1933) Tracce glaciali sul Pindo epirota. *Bollettino della Societa Geografica Italiano*, **10 (3)**, 136-156.

Sfikas, G. (1979) *The Mountains of Greece*. Athens, Efstathiadis Group.

Smith, G.W., Nance, R.D. & Genes, A.N. (1994) A re-evaluation of the extent of Pleistocene glaciation in the Mount Olympus region, Greece. *Bulletin of the Geological Society of Greece* (Proceedings of the 7th Congress, Thessaloniki, May 1994), **1**, 293-305.

Smith, G.W., Nance, R.D. & Genes, A.N. (1997) Quaternary glacial history of Mount Olympus, Greece. *Geological Society of America Bulletin,* **109**, 809-824.

Smith, G.R., Woodward, J.C., Heywood, D.I. & Gibbard, P.L. (1998) Mapping glaciated karst terrain in a Mediterranean mountain environment using SPOT and TM data. *In:* Burt, P.J.A., Power, C.H.& Zukowski, P.M. (eds), *RSS98: Developing International Connections.* (Proceedings of the Remote Sensing Society's Annual Meeting), Greenwich, September 1998, 457-463.

Smith, G.R., Woodward, J.C., Heywood, D.I. & Gibbard, P.L. (2000) Interpreting Pleistocene glacial features from SPOT HRV data using fuzzy techniques. *Computers and Geosciences,* **26**, 479-490.

van Andel, T. H., Zangger, E. & Demitrack, A. (1990) Land use and soil erosion in prehistoric Greece. *Journal of Field Archaeology*, **17**, 379-396.

Wijmstra, T.A. (1969) Palynology of the first 30 m of a 120 m deep section in northern Greece. *Acta Botanica Neerlandica*, **18**, 511-527.

Woodward, J.C. (1990) *Late Quaternary Sedimentary Environments in the Voidomatis Basin, Northwest Greece.* Unpublished PhD Thesis, University of Cambridge, Cambrige, UK.

Woodward, J.C., Lewin, J. & Macklin, M.G. (1992) Alluvial sediment sources in a glaciated catchment: the Voidomatis basin, northwest Greece. *Earth Surface Processes and Landforms,* **16**, 205-216.

Woodward, J.C., Macklin, M.G. & Lewin, J. (1994) Pedogenic weathering and relative-age dating of Quaternary alluvial sediments in the Pindus Mountains of northwest Greece. *In:* D.A. Robinson & R.B.G. Williams (eds), *Rock Weathering and Landform Evolution.* Chichester, John Wiley, 259-283.

Woodward, J.C., Lewin, J. & Macklin, M.G. (1995) Glaciation, river behaviour and the Palaeolithic settlement of upland northwest Greece. *In:* J. Lewin, M.G. Macklin & J.C. Woodward (eds), *Mediterranean Quaternary River Environments.* Rotterdam, Balkema, 115-129.

Woodward, J.C., Hamlin, R.H.B., Macklin, M.G., Karkanas, P. & Kotjabopoulou (2001) Quantitative sourcing of slackwater sediments at Boila rockshelter: A record of Lateglacial flooding and Paleolithic settlement in the Pindus Mountains, Northwest Greece. *Geoarchaeology,* **16**, 501-536.

Quaternary Glaciations – Extent and Chronology
Editors J. Ehlers and P.L. Gibbard

Extent and chronology of glaciations in Iceland; a brief overview of the glacial history

Áslaug Geirsdóttir

Deptartment of Geosciences, University of Iceland
E-mail: age@rhi.hi.is

Introduction

The most direct evidence for the existence of glaciers in the past is terrestrial ice contact sediments. However, because only a very small proportion of ice contact sediments is preserved in the Northern Hemisphere, deductions on the extent of glaciers and periodicity of glaciations are in most cases derived from indirect proxies such as microfossils and the distribution of ice-rafted detritus (IRD) in the deep sea sediments. Iceland is exceptional in this regard, where Tertiary and Quaternary terrestrial basal tillites are very common and well preserved. The reason for this unusually well preserved glacial record in Iceland is related to its location within a rifting zone where recurrent volcanic activity with the formation of laterally extensive lava flows provides a shield to the underlying sediments. The geographical position of Iceland in the midst of shifting positions of cold and warm ocean currents and at atmospheric fronts further enhances the island's importance in glacial geological and palaeoclimatic research (Fig. 1). Comparison with the deep-sea record and studies of glacier cyclicity imply that the late Tertiary and early to middle Pleistocene terrestrial record in Iceland does reflect the large-scale palaeoclimatic processes in the North Atlantic region (Fig. 2).

Tertiary to Middle Pleistocene glaciations of Iceland

Geological mapping and sedimentary studies have identified over 20 glacial sediment sequences interbedded with lava flows in the 15 million year old stratigraphy of Iceland (Fig. 2). The oldest indication of the presence of ice is found in the *c.* 5 My old deeply-eroded rock sequence of the southeast, at the margin of the current Vatnajökull glacier. Although no sedimentological analysis has been performed, deposits of supposed glacial origin have been reported from this region (Fridleifsson, 1995). These deposits are associated with hyaloclastites and pillow lavas, the characteristic rock types of subglacial eruptions. Helgason & Duncan (2001) identify a 4.6 to 4.7 My old glacial deposit in the Skaftafell area at the southern margin of Vatnajökull. The Vatnajökull ice cap covers the highest and most humid region of Iceland, with an annual precipitation of 3600 mm. Assuming similar circumstances in the past, local ice caps at the site of the present Vatnajökull may have already existed in the Miocene.

The distribution of glacial deposits and the correlation between rock sequences in Iceland show a fairly distinct trend during the Pliocene-Pleistocene transition indicating a gradually growing ice sheet from the southeast towards the west and north (Geirsdóttir & Eiríksson, 1994, Geirsdóttir & Eiríksson, 1996). The oldest glacial deposit identified based on detailed sedimentological analyses is found imbedded within *c.* 3.8 to 4.0 My old lava flows in Fljótsdalur, eastern Iceland (Fig. 1). Another glacial bed found higher up in the sequence at the same site has an estimated age of 3.4 My. Since it has not been possible to trace these two glacial beds over a larger area, they are thought to represent only local glacier activity. The first indication of a major ice sheet in Iceland dates back to about 2.9 My, based on a correlation between glacial deposits found in Fljótsdalur and Jökuldalur, the two main valleys of eastern Iceland. Intensification in glacier growth in Iceland occurred *c.* 2.7 My ago. Two stratigraphically-separated glacial deposits have been identified both in eastern (Jökuldalur and Fljótsdalur) and western Iceland (Borgarfjörður and Hvalfjörður) interbedded within lava flows dated between 2.7 and 2.5 myr (Geirsdóttir & Eiriksson, 1994). Recent palaeomagnetic study and K-Ar radiometric dating on the stratigraphy of Skaftafell, southeast Iceland, also suggests an increased glaciation at *c.* 2.6 Ma (Helgason & Duncan, 2001). More extensive glaciation is indicated by glacial deposits in all parts of the country dating from 2.4 to 2.5 My ago (Fig. 1). Further periods of glacier expansion and full-scale glacial-interglacial cyclicity appear at 2.2-2.1 My and again at 1.6 My (Fig. 2). From the period between 2.9 and 1.6 My at least 6 glaciations and possibly 8 glaciations are indicated, but the geological record implies a decrease in glacial activity in the period between 1.6 to 1.2 My. (Fig. 2).

Geological studies of the Tjörnes section in northern Iceland (Fig. 1 and 2) suggest an increase in the periodicity of glaciations after 1 My when the 100,000 years ice-volume cycle develops (Eiríksson, 1985). This is in accordance with the findings of Helgason & Duncan (2001) from southeast Iceland. However, the Tjörnes section is the only site known in Iceland with a complete glacial stratigraphy from the earliest regional glaciations to the last glaciation.

The number of glacial deposits identified in different parts of Iceland reflects the growth of the early Icelandic ice sheet from the southeast to the west and north. In southeast Iceland (Skaftafell), Helgason and Duncan (2001) report a 4.6 - 4.7 My old deposit of possible glacial origin. They describe a total of 11 glacial sequences in a

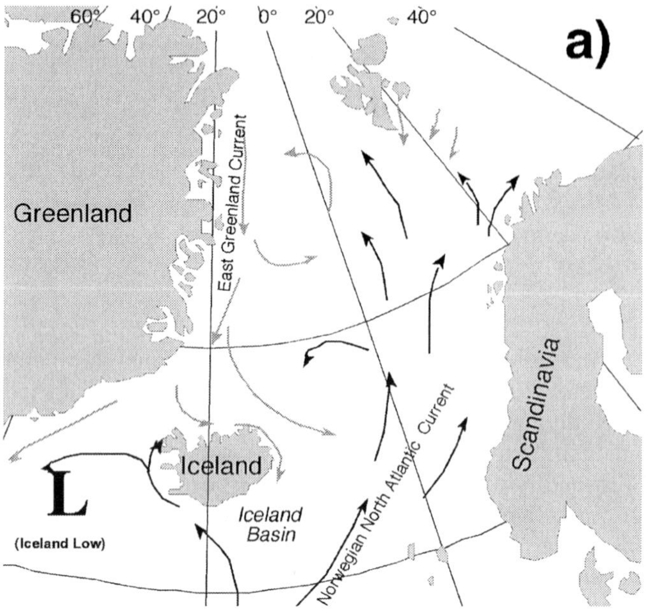

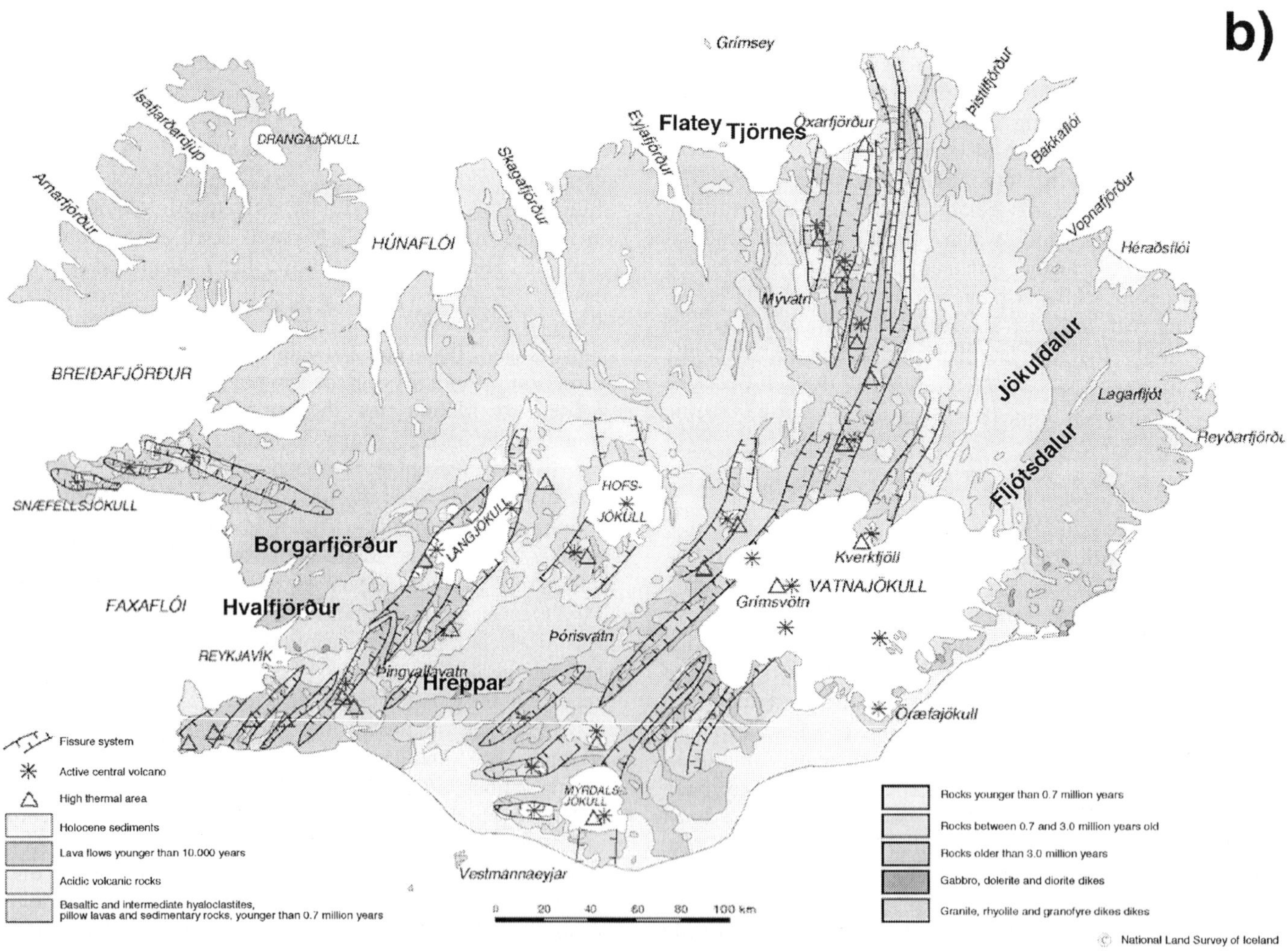

Fig. 1 a. Location of Iceland in the North Atlantic (fronts) b. A geologic map of Iceland. Studied sites containing Tertiary glacial deposits.

stratigraphy, which extends from 4.7 - 0.4 My. In eastern Iceland (Fljótsdalur), a total of 11 glacial units are recorded in a stratigraphic sequence that spans 6.5 - 1.0 My. The oldest glacial deposit in this section is *c.* 4.0 My. old. In northern Iceland, 14 glacial units are identified in a section which extends back to 9.0 My; here the oldest glacial deposit is correlated with rocks aged *c.* 2.5 My. A total of 7 glacial horizons preserved in a sequence from western

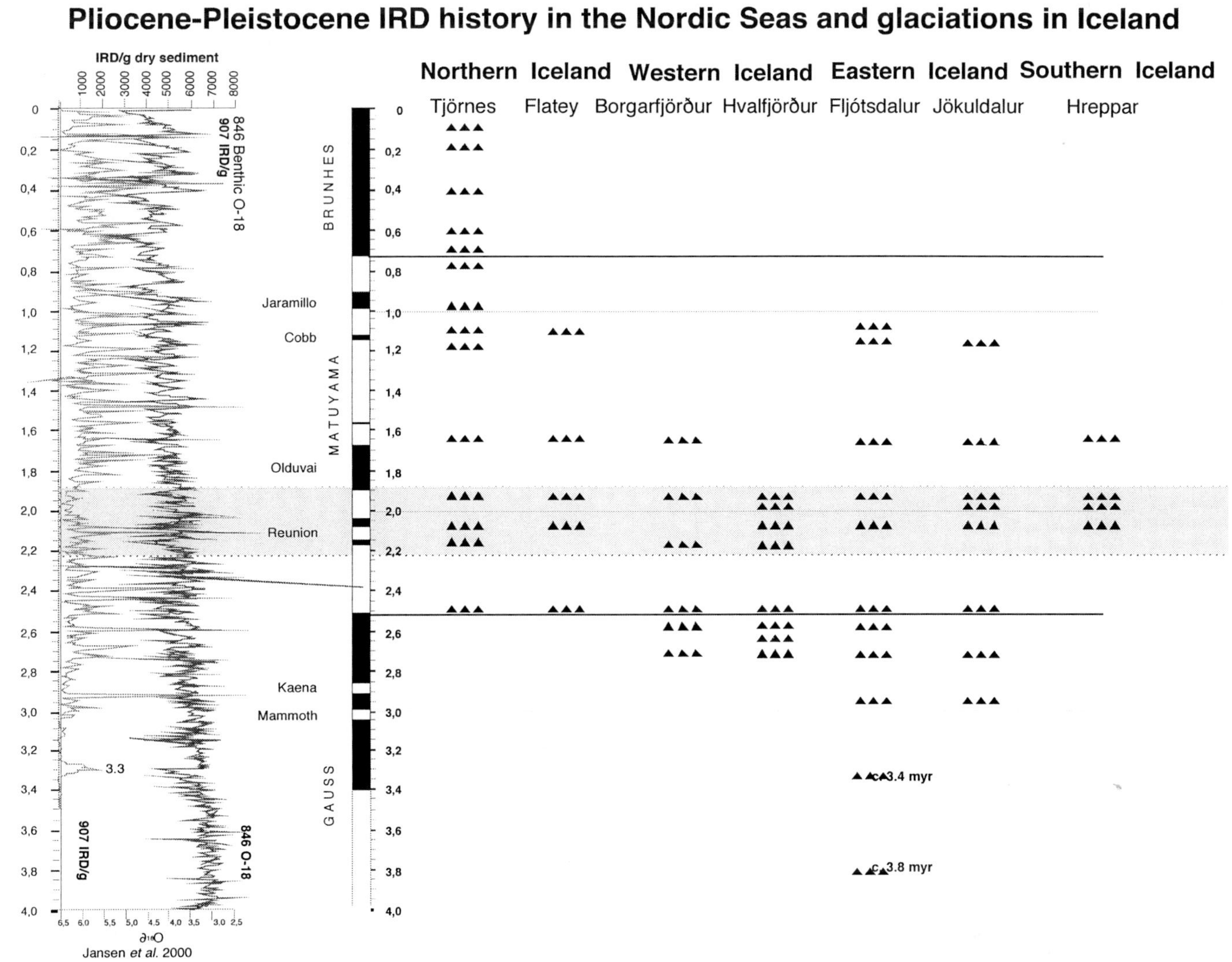

Fig. 2. A tentative correlation of studied glacial deposits in Iceland correlated with the geomagnetic time scale, ‰^{18}O record from the deep sea and IRD record from the deep sea. Modified from Geirsdóttir & Eiríksson (1994).

Iceland (Hvalfjörður and Borgarfjörður) date from 7.0 - 1.8 Ma, with the oldest glacial deposit correlated with 2.6 My old rocks. Whilst in the Hreppar area, southern Iceland, the oldest glacial deposit identified is between 2.5 - 2.2 My. where the oldest rocks are only *c.* 3 million years old (Fig. 3) (Geirsdóttir & Eiríksson, 1994). This figure of glacier distribution through time may reflect the correlation between lava flow formation and preservation potential of glacial deposits where foci of volcanic activity was along the eastern volcanic zone between 6 and 1 million years, but around the Hreppar area not until after 3 My. No Pliocene – Pleistocene glacial deposits have been found within the stratigraphy of the Northwest Peninsula of Iceland, which contains only rocks older than 6 My (Fig. 3).

Middle to Late Pleistocene glaciations in Iceland

The glacial stratigraphy of Iceland from *c.* 1.5 million years to the last deglaciation is very fragmentary, apart from in the Tjörnes section, which contains 9 stratigraphically-separated glacial deposits during this interval (Eiriksson, 1985). The lack of continuous records through the period may be due to a lack of systematic research and radiometric dates since old glacial and interglacial deposits of unknown age have been reported from various sites all around Iceland. Tillites from the Middle Pleistocene have been reported from the north (Jancin *et al.,* 1985), and at Snæfellsnes Peninsula in West Iceland (Einarsson, 1994; Leifsdóttir, 1999). The glacial deposits found on the Snæfellsnes Peninsula were probably formed between 2 and 0.7 m.y based on palaeomagnetic measurements on inter-bedded lava flows.

From the Last Glacial Maximum to the last deglaciation

During the Last Glacial Maximum (LGM), Iceland and the surrounding shelf was presumably ice-covered (Fig.4), although small ice-free areas may have existed along the

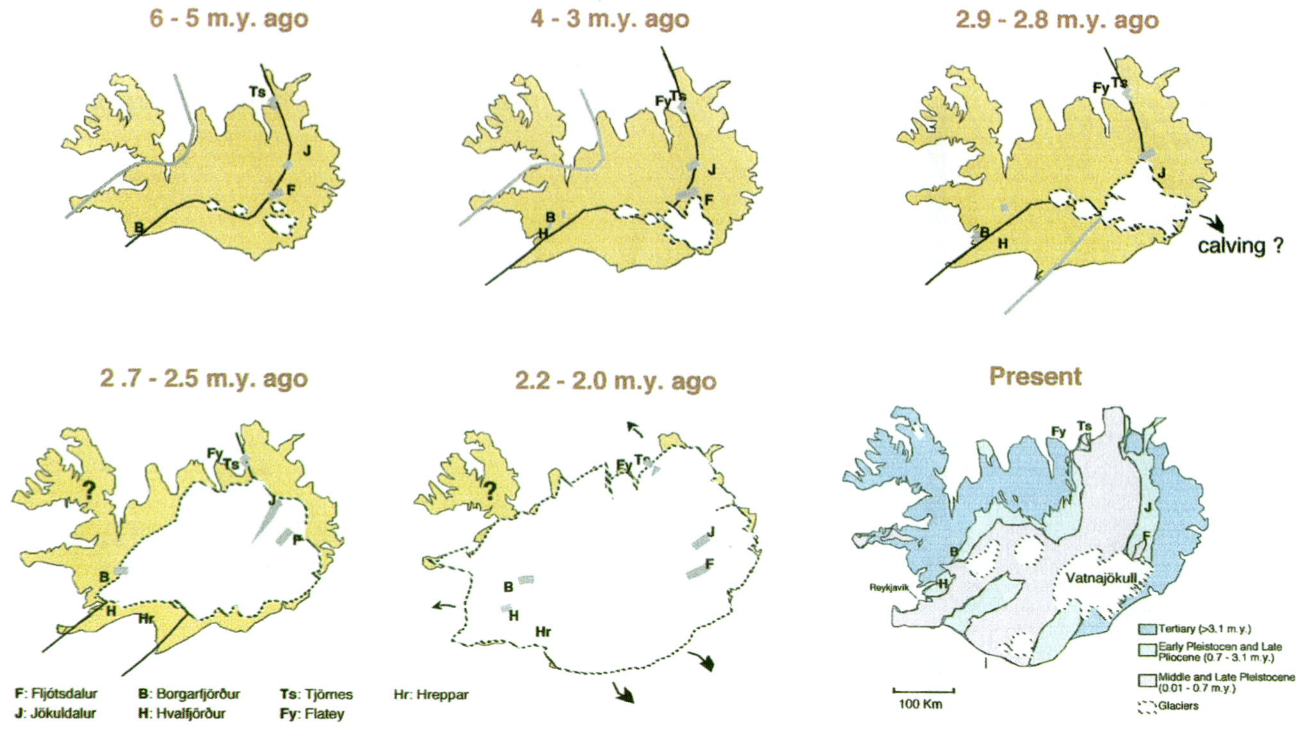

Fig.3. A hypothetical reconstruction of the Tertiary glaciations in Iceland at certain time slices.

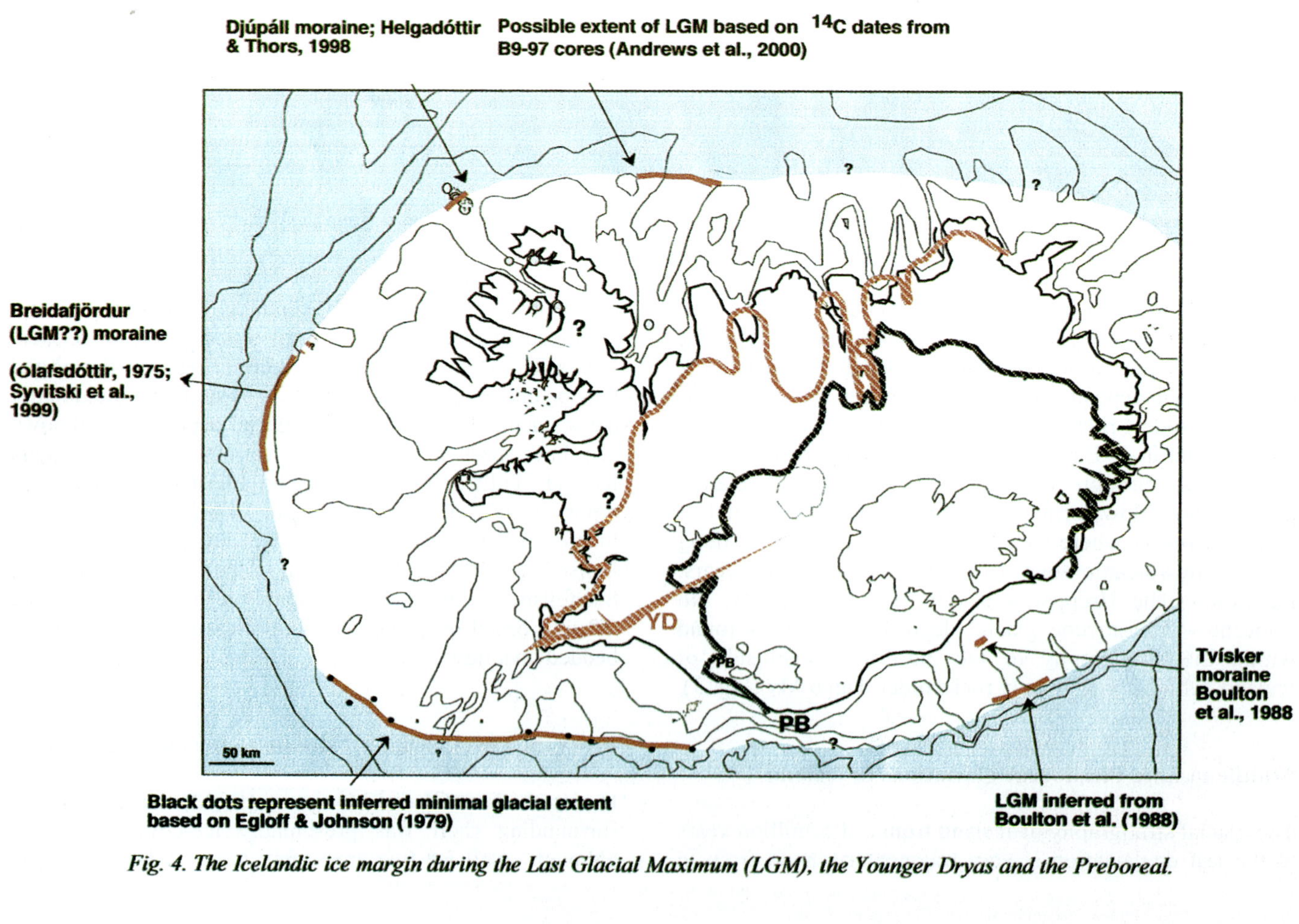

Fig. 4. The Icelandic ice margin during the Last Glacial Maximum (LGM), the Younger Dryas and the Preboreal.

coastal mountains, particularly in the northwest, north and east (e.g. Einarsson & Albertsson, 1988; Ingólfsson, 1991; Norddahl, 1991; Ingólfsson & Norddahl, 1994; Andrews *et al.,* 2000; Rundgren & Ingólfsson, 1999). Although the precise timing of the LGM and the full extent of the ice-sheet is poorly constrained by observations, ice streams and outlet glaciers are thought to have emanated from ice divides in south-central Iceland, terminating at the shelf edge (Fig. 4). The Vestfirdir peninsula in NW Iceland may have supported dynamically independent ice caps and/or valley glaciers with outlet glaciers originating within an ice-divide near the centre of the peninsula. Recent studies, based on sites from all around Iceland, indicate that glacier retreat began by 15 cal ka, (13.000 BP) and that it was rapid, but step-like (e.g. Ingólfsson, 1991; Norddahl, 1991; Ingólfsson & Norddahl, 1994; Ingólfsson *et al.,* 1997; Andrews *et al.,* 2000; Syvitski *et al.,* 1999).

The strongest evidence for step-like retreat (or re-advance) of the main glacier during the last deglaciation is found in the Búði morainic complex in south central Iceland (Geirsdóttir *et al.*, 1997, 2000), and in northern Iceland (Norddahl, 1991). The Búði moraines are 70 km long, extending across the southern lowlands from NW to SE about 25 to 45 km inland from the present coastline. The morainic complex is built up of multiple, discontinuous ridges that parallel the boundary between the interior highlands and the southern lowlands. Most of this morainic complex shows delta characteristics with distinct foreset bedding and sandur accumulation, indicative of a transition between terrestrial and coastal environments (Geirsdóttir *et al.,* 1997, 2000). Shells found both in front and behind the Búði moraines range in age from 11.5 to 10.1 cal ka (10,075-9055 BP), indicating a Preboreal age for the succession. Noting the paucity of ^{14}C dates from Iceland that predate the Preboreal, Hjartarson & Ingólfsson (1988) suggested that the ice extended beyond the present coastline during the Younger Dryas (Fig. 4). The marine limit in the area is 110 m a.s.l. Although not dated directly, Ingólfsson and Norddahl (1994) assume that the transgression maximum occurred during the transition from the Younger Dryas to the Preboreal (between 12.2 and 11 cal ka).

Geirsdóttir *et al.* (1997, 2000) provide an alternative view that implies a much less extensive Younger Dryas ice sheet in south-central Iceland. On the basis of detailed lithofacies analyses of the sedimentary successions in and around the Búði morainic complex they suggest that most of the lithofacies reflect deltaic sedimentation resulting from an interplay of coastal, terrestrial and glacier processes. The sediments in front of the morainic system rest on a stratified diamictite deposited in front of a tidewater glacier prior to 11.1 cal ka (9800 BP) (Hjartarson & Ingólfsson, 1988; Geirsdóttir *et al.*, 1997). The glacier that deposited the diamictite may have been of Younger Dryas age (Geirsdóttir *et al.*, 1997). This interpretation is supported by radiocarbon dates on foraminifera and the occurrence of the Vedde Ash at the base of a discontinuous 25-m-long sediment core from Lake Hestvatn, (15-20 km south of the Búði moraines), covering the last 12 cal ka (Geirsdóttir *et al.*, 2000; Harðardóttir *et al.*, 2001).

Collectively, the Búði moraines and the Lake Hestvatn sediments provide evidence for an ice-free interval in southern Iceland from at least 12.2 (possibly as early as 14) to 11.1 cal ka. A brief glacier readvance towards the Búði moraines took place at around 11.1 cal ka. during the Pre-boreal. Subsequently, the ice margin retreated rapidly to-wards the highlands, followed by rapid isostatic rebound; Hestvatn (50 m a.s.l) was isolated from the sea shortly after 10 cal ka, a drop in relative sea level of almost 60 m in *c.* 1 ka.

One of the most important findings in northern Iceland regarding the deglaciation history, are four separate ice-lake strandlines attributed to four separate periods of glacial isostatic recovery (Norddahl, 1991). The occurrence of the Skógar tephra, correlated with the Vedde Ash in the ice-lake sediments constraints the age of these ice dammed lakes and shows that during the Younger Dryas Chronozone outlet glaciers in northern Iceland extended at least half-way out Eyjafjörður (Fig. 4) (Norddahl & Haflidason, 1990). The reconstructed deglaciation history of North Iceland shows four plausible glacier readvances in the period between 14700-10900 cal. ka (ca. 12,655 and 9,650 BP), where the youngest ice-lakes were formed during the Younger Dryas and the marine limit was reached in late Younger Dryas or early Preboreal time.

Early Holocene thermal maximum and the onset of neoglaciation

By 10 cal ka, the main ice sheet was in rapid retreat across Iceland; pollen studies suggest a climate similar to the present was established by this time (Björck *et al.*, 1992; Hallsdóttir, 1995). Possibly as early as 9 cal ka (although perhaps not until 8 cal ka) the central highlands of Iceland were mostly ice free (Kaldal & Víkingsson, 1991). Gudmundsson (1997) summarizes evidence pertaining to the evolution of Iceland's climate during the Holocene. The expansion of birch shortly after 10 cal ka suggests an early Holocene climate characterized by warm, dry summers; a condition that persisted at least until 7 ka. There is little consensus on the nature of both ice caps and climate between 9 and 5 cal ka, with scattered evidence for glacier activity in the central highlands, but none that is unequivocal. No continuous pollen records span this interval, and the available discontinuous sequences are not conclusive. The pollen record, as summarized by Hallsdóttir (1995), suggests a maximum expansion of birch forests across southern Iceland between about 8 and 5 cal ka, suggesting a possible thermal maximum through this interval.

The onset of Neoglaciation and decline of tree vegetation remains debated: Clear evidence for glacier expansion after 3 cal ka is available, but some authors have suggested glacier growth as early as 5 cal ka BP (*cf.* Guðmundsson, 1997). In almost all cases, the Little Ice Age moraines

(1600-1900 AD) represent the most extensive ice margins since early Holocene deglaciation.

Glacial Limits - Quality of Data, Alternative Interpretation

Tertiary to Middle Pleistocene glaciations

The reconstructed Tertiary glacial history of Iceland is based on the use of multiple sedimentological criteria from two sites in eastern Iceland (Fljótsdalur and Jökuldalur), two sites in northern Iceland (Tjörnes and Flatey), two sites in western Iceland (Borgarfjörður and Hvalfjörður) and one site in southern Iceland (Hreppar) (Geirsdóttir, 1988; Geirsdóttir, 1991, Geirsdóttir *et al.,* 1993; Geirsdóttir *et al.*, 1994; Geirsdóttir & Eiríksson, 1994, 1996). This multi-proxy approach includes a comparison with modern depositional environments, lithofacies analysis, clast fabric measurements, textural studies and rock magnetic measurements, including both remanent magnetization and analysis of anisotropy of magnetic susceptibility. Major emphasis has been placed on detailed mapping and lithofacies analyses where both vertical and lateral changes in facies arrangements could be detected over several kilometres for all sections. The results imply that Iceland has experienced over 20 glaciations since *c.* 4 My ago. This is in reasonable agreement with the number of glaciations retrieved from the ?^{18}O record from the deep sea and suggests that the late Tertiary and early to middle Pleistocene terrestrial record in Iceland may be regarded as an example of global palaeoclimatic processes in the North Atlantic region (Fig. 2).

Map 1 shows the first indication of glacier formation in Iceland. It includes the oldest glacial deposits (ca. 4.0 - 3.8, 3.4 My) mapped in eastern Iceland (Fljótsdalur). Older glacial deposits suggested in southeast Iceland at the margin of the current Vatnajökull glacier are not shown because of a lack of information on the location and description of sites.

Also shown on map 1 is the oldest glacial deposit (2.9 My) that can be correlated between Fljótsdalur and Jökuldalur, eastern Iceland. The deposit rests at both sites on reversed magnetized rocks correlated to the Kaena geomagnetic Event.

Map 2 shows the first indication of regional glaciation in Iceland. Two glacial deposits are identified within the time period from approximately 2.7 My to 2.5 My (i.e. within the upper part of the Gauss geomagnetic Epoch). Glacial deposits from this time period have been mapped in eastern Iceland (Jökuldalur and Fljótsdalur) and in western Iceland (Hvalfjörður and Borgarfjörður)

Map 3 shows the first indication of full glacial-interglacial cyclicity in Iceland at approximately 2.5 My. Glacial deposit lying on normal magnetized lava (presumably Gauss magnetic Event) has been found in eastern (Fljótsdalur, Jökuldalur), western (Hvalfjörður, Borgarfjörður) and northern (Tjörnes) Iceland.

Map 4 shows a clear indication of intense glaciation in Iceland around 2.2 - 2.0 My (Reunion geomagnetic Event). Glacial deposits have been mapped in eastern (Fljótsdalur, Jökuldalur), northern (Tjörnes), western (Hvalfjörður and Borgarfjörður) and southern (Hreppar area) Iceland. These are the oldest glacial deposits in the Hrepppar area in South Iceland.

Middle to Late Pleistocene and early Holocene

Only fragmentary information is available about the Middle to Late Pleistocene glaciations of Iceland. Current work on sedimentary cores from the shelf around Iceland is intended to define the Last Glacial Maximum. The deglaciation history of the last glacial in Iceland is better constrained by radiocarbon age determinations. Figure 4 shows the inferred glacier margin during the Last Glacial Maximum, the Younger Dryas and the Preboreal in Iceland.

Dating of glacial limits - reliability of dates

Tertiary glacial deposits

Correlations between sections/glacial deposits of Tertiary age are still tentative and mainly based on previous palaeomagnetic work that was done on lava flows in order to reconstruct a palaeomagnetic time scale for the Icelandic lava pile (Wensink, 1964; McDougall & Wensink, 1966; McDougal *et al.*, 1976; 1977; Kristjánsson *et al.*, 1980; Eiriksson *et al.*, 1990). Because lava flows chosen for K/Ar dating were not necessarily associated with glacial deposits, some constraints are placed on the reliability of the correlations. Stratigraphical and palaeomagnetical work has further revealed some glacial deposits in southeast Iceland (Helgason and Duncan, 2001). Their K-Ar dates on identified glacial strata confirm previous findings of Geirsdóttir and Eiríksson (1994) on the dating of Tertiary glaciations in Iceland. Although there is insufficient control on individual glacial units, there does appear to be a variation in the frequency of glaciations with time during the last 2.5 My. The author's results show that changes in the ice cover of Iceland correlate reasonably with variations in the deep-sea oxygen isotope records from the North Atlantic (Fig. 2).

Middle to Late Quaternary glacial deposits

C Contentious interpretation of the glacial history of Iceland reflects the lack of continuous chrono- and biostratigraphical records and relatively low age-resolution of the data. However, new studies, particularly on lake sediments, look promising for establishing an improved and more continuous record of the climatic and environmental changes in Iceland for this critical time period (Björck *et al.*, 1992; Rundgren, 1995; Hardardóttir, 1999; Hardardóttir *et al.*, 2001.). The two prevailing conceptual models of the last deglaciation history in southern Iceland agree in regard to the Preboreal age glacier advance, but differ in their interpretation of 1) the extent of the Younger Dryas ice, and 2) the sedimentological history of the Búði morainic complex (Figure 4).

Open questions

It should be noted that some of the Tertiary glacial deposits studied might have a larger lateral extent indicating more extensive regional glaciation than shown on the maps. Old glacial deposits of possible Tertiary age have been reported from various sites all around Iceland, but no dates are available as yet. This makes it impossible to determine precisely the chronological position of the deposits. However, current work is aimed at increasing the number and accuracy of radiometric dates (Ar/Ar and K/Ar) from the terrestrial sections, with emphasis on dating of the lava flows interbedded with the identified glacial deposits. This will give more accurate dates for the actual climatic events.

The absence of Tertiary glacial deposits on the NW peninsula raises questions about the preservation potential of glacial deposits. Does this reflect limited or no glaciation on the peninsula or did subsequent glaciations completely erode the evidence for Tertiary glaciation in the area?

The Tjörnes Peninsula in northern Iceland is the only area in Iceland where we have a complete glacial stratigraphy mapped from about 3.0 My to the Holocene is available. It is however very difficult to draw a map of the extent of the younger glaciations (i.e. younger than *c.* 1.5 My) due to lack of studied sites and radiometric dating of glacial deposits.

In the light of the current knowledge of the deglaciation history of south-central Iceland, there seems to be little doubt about the Preboreal age of some of the Búði moraines. The question that remains is whether some of the sediments in the Búdi morainic complex represent earlier glacial advances. Although much has been written about the Younger Dryas glaciation of Iceland, knowledge of the extent of the Younger Dryas ice sheet in the south is still ambiguous. The main problem with the current hypotheses is that they are based on indirect evidence (i.e. missing sedimentary records). Those who support extensive Younger Dryas glaciation point to the sea for further evidence. Those who support the hypothesis of a more limited extent of the Younger Dryas ice emphasise the importance of the undated basal till in the Búði succession. However, the key to the puzzle is more likely to be the sediment sequence still to be found beneath Lake Hestvatn. Although these problems await future research for their solution, the importance of detailed sedimentological work and facies analyses in glacial geological studies should be emphasized. A comprehensive understanding of the deglaciation history requires a detailed lithofacies analysis in context with good stratigraphic control of the successions. Consequently future research should emphasise detailed sedimentological, stratigraphical and chronological studies of both lacustrine and marine shelf sediments

References:

Andrews, J.T., Harðardóttir, J., Helgadóttir, G., Jennings, A.E., Geirsdóttir, Á., Sveinbjornsdóttir, A.E., Schoolfield, S., Kristjansdóttir, G.B., Smith, L.M., Thors, K. & Syvitski, J.P.M. (2000). The N and W Iceland shelf: insights into LGM Ice extent and deglaciation based on acoustic stratigraphy and basal radiocarbon AMS dates. *Quaternary Science Reviews*, **19**, 619-631.

Björck, S., Ingólfsson, Ó., Haflidason, H., Hallsdóttir, M. & Anderson, N.J. (1992). LakeTorfadalsvatn: a high resolution record of the North Atlantic ash zone I and the last glacial-interglacialenvironmental changes in Iceland. *Boreas*, **21**, 15-22.

Boulton, G.S., Thors, K. & Jarvis, J. (1988). Dispersal of glacially derived sediment over part of the continental shelf of South Iceland and the geometry of the resultant sediment bodies. *Marine Geology*, **83**, 193-223

Egloff, J. & Johnson, G.L. (1979). Erosional and depositional structures of the southwest Iceland insular margin: Thirteen geophysical profiles. *In:* Watkins, J.S., Montadert, L. & Dickerson, P.W. (eds), *Geological and Geophysical Investigations of Continental Margins. American Association of Petroleum Geologists Memoir*, **29**, Tulsa, Oklahoma, 43-63.

Einarsson, Th. & Albertsson, K.J. (1988). The glacial history of Iceland during the past three million years. *Philosophical Transactions of the Royal Society of London*, **Series B 318**, 637-644.

Einarsson, Th. (1994). *Geology of Iceland. Rocks and Landscape*, Mál & Menning, Reykjavík. 309p.

Eiríksson, J. (1985). Facies analysis of the Breidavík Group sediments on Tjörnes. *Acta Naturalia Islandica*, **31**, 56 pp.

Eiríksson, J., Gudmundsson, A.I., Kristjánsson, L. & Gunnarsson, K. (1989). Palaeomagnetism of Pliocene-Pleistocene sediments and lava flows on Tjörnes and Flatey, North Iceland. *Boreas*, **19**, 39-55.

Friðleifsson, G.Ó. (1995). Miocene glaciation in SE-Iceland. *In:* Hróarson, B., Jónsson, D. & Jonsson, S.S. (eds), *Eyjar í Eldhafi*. 77-85.

Geirsdóttir, Á. (1988). *Sedimentologic analysis of diamictites and implications for late Cenozoic glaciation in western Iceland.* Ph.D. Thesis, University of Colorado, Department of Geological Sciences.

Geirsdóttir, Á. (1991). Diamictites of late Pliocene age in western Iceland. *Jökull*, **40**, 3-25.

Geirsdóttir, Á., Thordarson, Th. & Kristjánsson, L. (1993). The interactive impact of climatic deteriorations and volcanic activity on facies assemblages in late Pliocene to early Pleistocene strata of the Hreppar Formation, South Central Iceland. *EOS*, **74 (43)**, p.131

Geirsdóttir, Á., Thórdarson, Th. & Kristjánsson, L. (1994). Áhrif eldvirkni og loftslagsbreytinga á upphledslu jardlaga i Gnúpverjahreppi. Ágrip a Jardfraedaradstefnu íslenska Jardfraedafélagsins, 21. april 1994. (Abstract vol. Geological Society of Iceland).

Geirsdóttir, Á. & Eiríksson, J. (1994). Growth of an intermittent ice sheet in Iceland during the late Pliocene and early Pleistocene. *Quaternary Research*, **42**, 115-130.

Geirsdóttir, Á. & Eiríksson, J. 1996. A review of studies of the earliest glaciations in Iceland. *Terra Nova*, **8**, 400-414.

Geirsdóttir, A., Hardardóttir, J. & Eiriksson, J. (1997). Depositional history of theYounger Dryas (?) - Preboreal sediments, South Central Iceland. *Arctic and Alpine Research*, **29**, 13-23.

Geirsdóttir, A., Hardardóttir, J. & Sveinbjörnsdóttir, A.E. (2000). Glacial extent and catastrophicmeltwater events during the deglaciation of Southern Iceland. *Quaternary Science Reviews*, **19**, 1749-1761.

Gudmundsson, H. (1997). A review of the Holocene environmental history of Iceland. *Quaternary Reviews*, **16,** 81-92.

Hallsdóttir, M. (1995). On the pre-settlement history of Icelandic vegetation. *Búvísindi* (Icelandic Agicultural. Sciences), **9**, 17-29.

Hardardóttir, J. (1999). *Late Weichselian and Holocene environmental history of south and western Iceland as interpreted from studies of lake and terrestrial sediments.* Unpublished Ph.D. Thesis, University of Colorado, Boulder.

Helgadóttir, G. & Thors, K. 1998. Setlog i Isafjardardjupi, Jokulfjordum og Djupal.Vorradstefna 1998. Agrip erinda og veggspjalda. Jardfraedafelag Islands, p. 32.

Hardardóttir, J., Geirsdóttir, A. & Thordarson, Th. (2000). Tephra layers in a sediment core from Lake Hestvatn, S-Iceland: implications for evaluating sedimentation processes andenvironmental impact of tephra fall deposits in the surrounding watershed. *In:* White, J. & Riggs, N. (eds), *Lacustrine Volcaniclastic Sedimentation* International Association of Sedimentologists (IAS) Special Publication. Blackwell, Oxford.

Hardardóttir, J., Geirsdóttir, A. & Sveinbjörnsdóttir, A.E. (2001). Seismostratigraphy and sediment studies of Lake Hestvatn, south Iceland; Implication for the deglacial history of the region. *Journal of Quaternary Sciences*, **16**, 167-179.

Helgason, J. & Duncan, R.A. (2001). Glacial-interglacial history of the Skaftafell region, southeast Iceland, 0-5 Ma. *Geology*, **29**, 179-182.

Hjartarson, Á. & Ingólfsson, Ó. (1988). Preboreal glaciation of Southern Iceland. *Jökull*, **38**, 1-16.

Ingólfsson, Ó. (1991). A review of the late Weichselian and early Holocene glacial andevironmental history of Ice-land. *In:* Maizels, J. & Caseldine, C. (eds), *Environ-mental change in Iceland: Past and present.* Dordrecht, Holland: Kluwer Academic Publishers, 13-29.

Ingólfsson, Ó. & Norddahl, H. (1994). A review of the environmental history of Iceland, 13000-9000 yr BP. *Journal of Quaternary Science*, **9**, 147-150.

Ingólfsson, O., Björck, S., Haflidason, H. & Rundgren, M. (1997). Glacial and climatic events in Iceland reflecting regional North Atlantic climatic shifts during the Pleistocene-Holocene transition. *Quaternary Science Reviews*, **16**, 1135-1144.

Jansen, E., Fronval, T., Rack, F. & Channell, E.T. (2000). Pliocene-Pleistocene ice rafting history and cyclicity in the Nordic Seas during the last 3.5 Myr. *Paleoceanography*, **15**, 709-721.

Kaldal, I. & S. Vikingsson, S. (1991). Early Holocene deglaciation in Central Iceland. *Jökul*, **40**, 51-66.

Kristjánsson, L., Friðleifsson, I.B. & Watkins, N.D. (1980). Stratigraphy and paleomagnetism of the Esja, Eyrarfjall and Akrafjall Mountains, SW-Iceland. *Journal of Geophysics*, **47**, 31-42.

Leifsdóttir, Ó.E. 1999. Ísaldarlög á norðanverðu Snæfellsnesi. Setlög og skeldýrafánur. Unpublished MS thesis, University of Iceland. 101 pp

McDougall, I. & Wensink, H. (1966). Paleomagnetism and geochronology of the Pliocene-Pleistocene lavas in Iceland. *Earth and Planetary Science Letters*, **I**, 232-236.

McDougall, I., Watkins, N.D. & Kristjánsson, L. (1976). Geochronology and paleomagnetism of Miocene-Pliocene lava sequences at Bessastadaá, eastern Iceland. *American Journal of Science*, **276**, 1078-1095.

McDougall, I., Watkins, N.D., Walker, G.P.L. & Kristjánsson, L. (1976). Potassium-Argon and Paleomagnetic Analysis of Icelandic Lava flows: Limits on the Age of Anomaly 5. *Journal of Geophysical Research*, **81**, 1505-1512.

Norddahl, H. (1991). Late Weichselian and early Holocene deglaciation history of Iceland. *Jökull*, **40**, 27-50.

Norddahl, H. & Haflidason, H. (1992). The Skógar Tephra, a Younger Dryas marker in North Iceland. *Boreas*, **21**, 23-41.

Ólafsdóttir, Th. (1975). Jokulgardar a sjavarbotni ut af Breidafirdi. *Náttúrufrædingurinn*, **45**, 31-36.

Jancin, M., Young, K.D. & Voight, B. (1985). Stratigraphy and K&Ar ages across the West flank of the Northeast Iceland axial rift zone, in relation to the 7 Ma volcano-Tectonic reorganization of Iceland. *Journal of Geophysical Research*, **90**, 9961-9985.

Rundgren, M. (1995). Biostratigraphic evidence of the Allerød-Younger Dryas-Preboreal oscillation in northern Iceland. *Quaternary Research*, **44**, 405-416.

Rundgren, M. & Ingólfsson, O. (1999). Plant survival in Iceland during periods of glaciation? *Journal of Biogeography*, **26**, 387-396.

Syvitski, J.P.M., Jennings, A.E. & Andrews, J.T. (1999). High-resolution seismic evidence for multiple glaciation across the southwest Iceland shelf. *Arctic, Antarctic and Alpine Research*, **31**, 50-57.

Wensink, H. (1964). Paleomagnetic stratigraphy of younger basalts and intercalated Plio-Pleistocene tillites in Iceland. *Geologische Rundschau*, **54**, 364-384.

Quaternary Glaciations – Extent and Chronology
Editors J. Ehlers and P.L. Gibbard

Pleistocene glaciations in Ireland

Jasper Knight [1], Peter Coxon [2], A. Marshall McCabe [1] and Stephen G. McCarron [1]

[1]*School of Biological and Environmental Sciences, University of Ulster, Coleraine, Co. Londonderry, BT52 1SA, Northern Ireland, UK*
[2]*Department of Geography, Trinity College, Dublin 2, Ireland*

Abstract

A literature survey and data from recent investigations are used to reconstruct ice limits in Ireland during the last (Midlandian) and penultimate (Munsterian) cold periods which are correlated with Marine Isotope Stages (MIS) 2-5d (Weichselian) and 6-8 (Saalian) respectively. Evidence for Munsterian ice limits and flow directions is equivocal and based mainly on erratic carriage and the presence of striae and subdued glacial landforms found outside well-marked Midlandian end moraines. Ice extent and flow direction is known only from the late Midlandian (MIS 2; 24-10 kyr BP) although ice may well have been present in the early Midlandian (MIS 3-5d; 24-117 kyr BP). Six late Midlandian glacial stages are identified on the basis of morphosedimentary and dating evidence, and patterns of subglacial bedforms including drumlins and Rogen moraines. Previous late Midlandian glacial models are well-established but are generally based on incomplete and/or erroneous datasets, are not age-constrained, and do not consider time-transgressive sedimentation and landform-shaping events.

Recent work shows that repeated ice advance-retreat cycles (oscillations) occurred during the late Midlandian. Oscillations resulted in stratigraphically superimposed, overprinted and cross-cut landform and sediment patterns that record ice activity throughout the glacial cycle. Additionally, subglacial bedforms previously unrecorded in the British Isles, such as flow-transverse ridges (Rogen moraines), are also present. Late Midlandian ice oscillations in Ireland occurred in tempo with millennial-scale changes in North Atlantic climate, suggesting connection to hemispheric shifts of the ice-ocean-atmosphere system.

1. Introduction

Pleistocene glacial deposits in Ireland have been studied extensively since initial observations by Agassiz (1842), but the extent and timing of glaciations is still largely unknown (McCabe, 1987). The significance of warm-climate signatures, including organic sediments and raised beaches, is also under debate (Synge, 1979; Warren, 1979, 1985; Watts, 1985; Coxon, 1993; Gallagher & Thorp, 1997).

Two main glacial events occurred during the Irish Pleistocene (e.g. Mitchell *et al.*, 1973; Bowen *et al.*, 1986; McCabe, 1987; McCabe, 1999). Here, these are correlated with the Marine Isotope Stages (MIS) of Martinson *et al.* (1987). The older (Munsterian) glaciation is correlated with the British Wolstonian and European Saalian glaciations during Marine Isotope Stages 8-6 (~ 302-132 kyr BP) (Mitchell *et al.*, 1973; Coxon, 1993). It is unlikely that ice was always present throughout this time period, but there is no dating evidence to support this (Synge, 1979). The last glaciation occurred within the Midlandian cold stage (~ 117-10 kyr BP), which is correlated with the British Devensian and European Weichselian periods, and with MIS 5d-2 (Mitchell *et al.*, 1973; Bowen *et al.*, 1986; Bowen, 1994). Most glacial landforms and sediments in Ireland stratigraphically postdate middle Midlandian organic sediments (Colhoun *et al.*, 1972) and are therefore assigned to the late Midlandian (~ 25-10 kyr BP). The Munsterian and Midlandian cold periods are separated by an (unnamed) interglacial, assigned to Marine Isotope Substage 5e (130-117 kyr BP; British Ipswichian and European Eemian), for which there is little floristic or dating evidence (Coxon, 1993). Key sections and stratigraphic relationships are discussed in McCabe (1999).

This paper aims to: (1) describe the evidence for ice extent and flow direction during the Munsterian and late Midlandian glaciations; and, (2) in the light of recent research, present an updated glacial model for proposed ice centres, limits and vectors during these glaciations.

2. Munsterian glacial deposits and ice limits

Traditionally, the umbrella term 'older drift' was applied to glacial deposits predating the most recent glaciation (Synge, 1979). 'Older drift' was related to ice activity during the Eastern General or Munsterian glaciation, and distinguished from younger glacial deposits with reference to their subdued relief, strong decalcification and soil development, and depth of cryoturbation (Finch & Synge, 1966). Summaries of this evidence are given by McCabe (1985, 1987), Bowen *et al.* (1986) and Hoare (1991). Stratigraphically, Munsterian deposits are not known to be overlain directly by last-interglacial organic sediments, or by early Midlandian cold-climate deposits, but they underlie components of the late Midlandian-age South of Ireland End Moraine (SIEM) (e.g. Finch & Synge, 1966). It therefore remains possible that Munsterian-age deposits are in fact early Midlandian (Warren, 1985; Coxon, 1993)

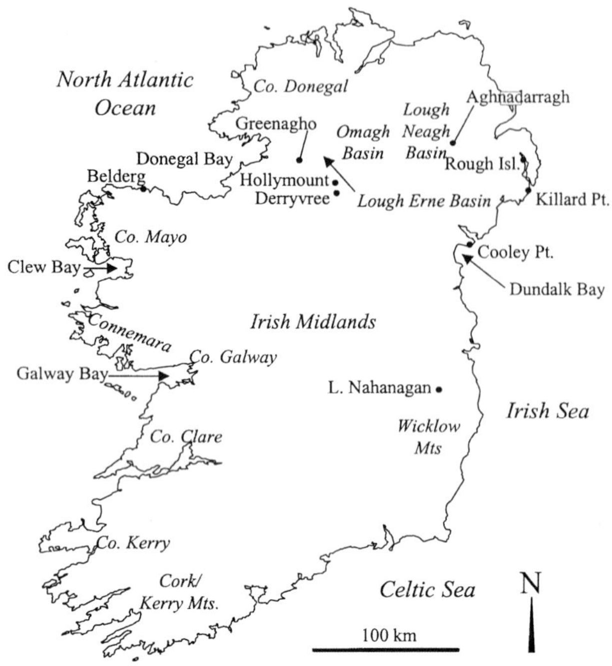

Fig. 1. Map showing the location of places named in the text.

although they are notched by (presumably interstadial) relative sea-level (RSL) highstands along the southeast Ireland coast (Davies, 1960). Additionally, striae south of the SIEM show that pre-late Midlandian ice extended offshore (McCabe, 1998).

Surficial Munsterian deposits and flow indicators are present only outside the SIEM. Inside the SIEM, no sediments have been found that can be correlated lithostratigraphically with those outside. Munsterian sediments and landforms have been described mainly from isolated sites in (1) western, and (2) eastern Ireland (Fig. 1).

(1) In County Kerry (King & Gage, 1961), County Clare (Finch & Synge, 1966), and County Mayo, western Ireland (Synge, 1968), Munsterian landforms include isolated non-aligned hummocks and ridges which generally overlie glacially-stripped and striated bedrock surfaces. Sediments are mainly indifferentiated tills (the Belderg, Erris and Killadoon tills; Synge, 1968). Clasts occasionally form isolated pavements within the till beds, but other primary sedimentary structures have not been described. Sediments are generally heavily cryoturbated to 2-3 m depth (Synge, 1968). Notable erratics include granites derived from Connemara and Galway (e.g. Synge, 1968). Erratic carriage and striae alignment suggest that ice flowed generally to the southwest or west and extended onto the continental shelf (Synge, 1968). There is no equivocal evidence for Munsterian maximal or retreat positions in western Ireland.

(2) In the Wicklow Mountains and adjacent areas in eastern Ireland (Farrington, 1944, 1954; Synge, 1973), Munsterian-age advances of Irish and Irish Sea basin ice are recognised on a lithological basis (Mitchell, 1960; Synge, 1964). Morphologically, terrestrial Irish deposits have little expression (Culleton, 1978), but Irish Sea deposits generally form hummocky and non-aligned sand and gravel landforms, and flat till plains

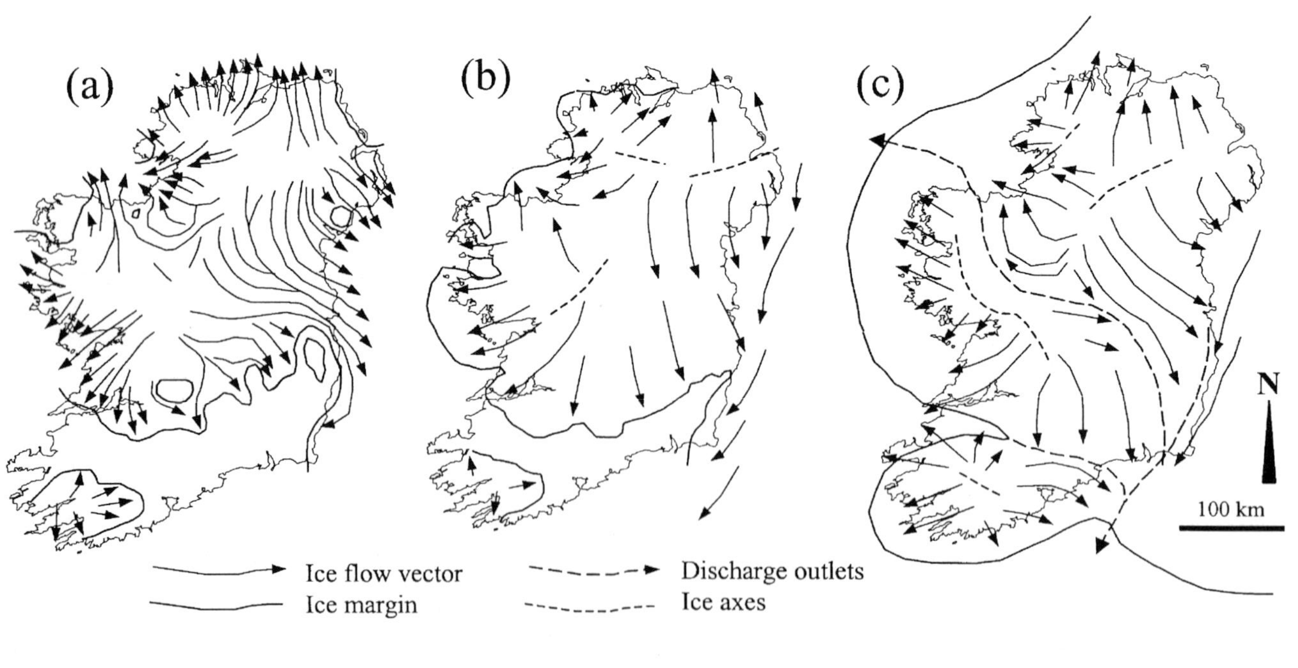

Fig. 2. (i) Munsterian ice flow vectors (redrawn from McCabe, 1985). Note no ice limits are shown: these are unknown. (ii) Late Midlandian ice limits during the LGM, redrawn from (a) Synge & Stephens (1960), (b) McCabe (1985), and (c) Warren & Ashley (1994). Note that ice flow vectors are essentially the same but that the models show different ice-marginal positions. All models may have been essentially correct, but at different times within the LGM period.

Quaternary Glaciations – Extent and Chronology
Editors J. Ehlers and P.L. Gibbard

Glacial history of the southern side of the central Alps, Italy

Alfredo Bini [1] and Luisa Zuccoli [2]

with the co-operation of Claudio Bussolini [1], Daniele Corbari [2], Ornella Da Rold [1], Carla Ferliga [2], Stefano Rossi [2] and Claudio Viviani [1]

[1] *CNR Centro di studio per la Geodinamica Alpina e Quaternaria and Università degli Studi di Milano, Dipartimento di Scienze della Terra, via Mangiagalli 34, 20133 Milano, Italy.*
[2] *Regione Lombardia, Ufficio Progetti Speciali per la Geologia e la Sismica, Milano, Italy.*

1. Introduction

The area of investigation covers the sub-alpine valleys between Lago d'Orta (Cusio), to the east, and Lago d'Iseo (Sebino), to the east and the corresponding morainic amphitheatres of the high Po plain. Geological mapping was carried out at a 1:10,000 scale in order to allow detailed geological reconstruction.

Data from previous studies were not used because preconcieved criteria, poor fieldwork, uncontrollable data and, principally, doubtful location of the outcrops commonly invalidate these works. Therefore the map is limited to areas where new data from the recent survey is available. Moreover, the map does not include any areas which have not been re-investigated by the present authors.

2. Glacial episodes

The term 'glaciation' is rather ambiguous; it is not clear whether it refers to a period of time, a certain climate, or a specific event. Furthermore, it is unclear how much ice is required to justify usage of the term. Therefore for this study Richmond's (1986) definition is adopted: "The term 'glaciation' is applied here to a specific glacial advance and recession, the deposits of which are separable from those of other glaciations by evidence of extensive recession and downwasting of glaciers or by evidence of a warm climate as interpreted from pollen diagrams or weathering profiles. The terms 'warm' and 'cold' are used here in a very general sense to indicate broad differences in the climate of intervals separating glaciations." By application of the above definition only, it is possible, in the authors' opinion, to use the term glaciation, and to recognise different glaciations that take the name of the correspondent alloformation or formation (Bini, 1997b).

The traces of the different glacial episodes are easier to distinguish in the amphitheatres than in the subalpine valleys. In the amphitheatres the morainic deposits are spread over belts several kilometres wide, whereas on the valley flanks only minor differences in elevation of the lateral glacial deposits occur (fig. 1). So, it is only in the valleys that it is possible to separate the deposits belonging to the last glaciation from the other, older deposits that represent many different ice advances.

On the high Po plain the boundaries of many ice advances can be distinguished that moved out of subalpine valleys and reached the plain beyond. According to the terminology of Bini (1997b) glacial episodes, glaciations (G.) and alloformations (Af.) can be distinguished. Accordingly the sediments of the Af. di Cantù were deposited during the Cantù Episode (diachronical unit), which is included in the Late Pleistocene Series (chronostratigraphical unit), and that represent the sedimentary expression of the Cantù Glaciation. The number of glaciations thus distinguished is, obviously, a minimum, because it is possible that sediments related to a minor glacial advance could have been completely reworked by a later, more extensive glacial advance (Bini, 1997a). The different valleys experienced different glacial histories so that the alloformations described from different basins cannot be directly correlated.

A) Verbano Amphitheatre

In the Verbano amphitheatre it is possible to distinguish the following glaciations (beginning with the youngest):

Cantù Glaciation
Daverio Glaciation
Mornago Glaciation
(Montonate) Glaciation
Sumirago Glaciation
Albusciago Glaciation
Golasecca Glaciation
Albizzate Glaciation
Morazzone 2 Glaciation
Morazzone 1 Glaciation
Castronno Glaciation
Immacolata Glaciation
Vivirolo Glaciation.

The two oldest are of Pliocene age (Bini, 1997b; Uggeri *et al.*, 1997), and the Castronno Glaciation could be of Pliocene or Early Pleistocene age (Bini, 1997b; Zuccoli, 1997). The glaciations between the Daverio and Morazzone are, probably, of Middle - Late Pleistocene age. The age of the Cantù Glaciation has been determined by numerous

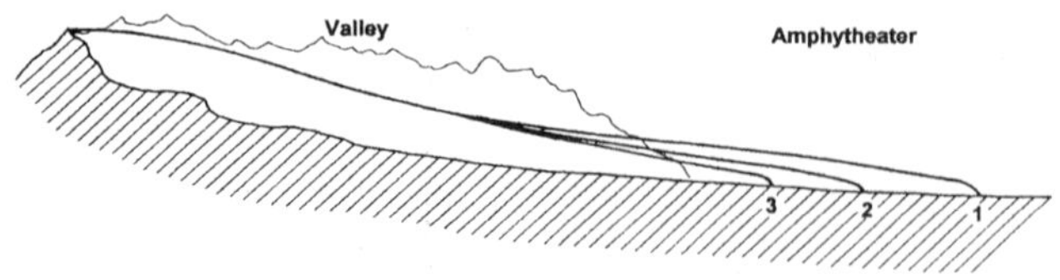

Fig. 1. Schematic longitudinal section of an Alpine valley from the glacial cirque to the plain. The thick lines indicate the glacial limits during three different glaciations, in chronological order (from 1 to 3) but not necessarily consecutive. The extent of the glaciers within the amphitheatre vary greatly. On the valley flanks the glacial limits are closely spaced and in the highest part of the valley only the youngest glaciation can be identified.

radiocarbon dates as Late Pleistocene. It is to be correlated with the Marine Isotope Stage 2. (cf. Bini, 1997b for more details). The following table resume the age of the galcial episodes in the Verbano sector.

Table 1. List of glacial episodes for the Verbano Amphitheatre and their ages.

Glaciation	Age	*On the basis of:*
Cantù	Late Pleistocene	*Radiocarbon dates*
Daverio	Middle Pleistocene	*Stratigraphic position*
Mornago	Middle Pleistocene	*Stratigraphic position*
Montonate	Middle Pleistocene	*Stratigraphic position*
Sumirago	Middle Pleistocene	*Stratigraphic position*
Albusciago	Middle Pleistocene	*Stratigraphic position*
Golasecca	Middle Pleistocene	*Stratigraphic position*
Albizzate	Middle Pleistocene	*Stratigraphic position*
Morazzone 2	Middle Pleistocene	*Stratigraphic position*
Morazzone 1	Middle Pleistocene	*Stratigraphic position*
Castronno	Late Pliocene / Early Pleistocene	*U/Th dates*
Immacolata	Late Pliocene	*Pollen analyses paleomagnetic data stratigraphic correlations*
Vivirolo	Late Pliocene	*Pollen analyses paleomagnetic data stratigraphic position*

B) Lario Amphitheatre

In the Lario amphitheatre, the following glaciations could be distinguished (starting with the youngest):

Cantù Glaciation
Besnate (more than one glaciation) Complex
Binago Glaciation
Specola Glaciation
Cascina Fontana Glaciation
Cascina Ronchi Pella *Glaciation*
San Salvatore Glaciation
Casanova Lanza D Glaciation
Casanova Lanza A/B Glaciation

It is important to observe that the Besnate Allogroup includes more than one glaciation. Thus it is impossible to say, at the present state of knowledge, how many glacial episodes are represented in the Lario amphitheatre. As far as the age of these glaciations is concerned the two oldest (the Casanova Lanza A/B and the Casanova Lanza D) are Pliocene, the San Salvatore Glaciation could be Late Pliocene or Early Pleistocene, the glaciations Cascina Ronchi Pella, Cascina Fontana, Specola, Binago and Besnate are Middle Pleistocene, and the Cantù Glaciation is Late Pleistocene on the basis of radiocarbon dates (Marine Isotope Stage 2).

C) Val Brembana Sector

In this area, the glaciers never advanced beyond the valley, so there is no amphitheatre and unlike for other sectors it is impossible to reconstruct the different phases of glacial advance and retreat. The Maximum Extent of Glaciation (MEG) in the Val Brembana sector and the differentiation of the oldest phases are not yet clear; field surveys are still ongoing. Thus far, several alloformations have been recognized. These alloformations seem to represent a glacier that extended at least as far as S. Pellegrino, with a low tongue ending at a narrow point in the valley.

The Last Glacial Maximum (LGM) in the Val Brembana sector is characterized by reduced and not-coalescent glaciers, following the main branches of the valley. Small feeding basins and a few north-facing slopes seem to be the main reason of this setting. Sedimentary bodies and well-preserved landforms provide good evidence of the extent of the glaciers. Nevertheless it is not possible to link the different alloformations because of a lack of direct relationships. These include the continuity of the moraines and sandurs, or any other clear evidence that the sediments belonging any two alloformations could be regarded as having been deposited during the same depositional event.

D) Val Seriana Sector

In this sector, the survey is also still ongoing, but it is possible to say that no glacier passed beyond the Ponte Costone canyon, whereas the ice spread eastward into the Clusone basin. The petrography of the glacial deposits indicates ice transfluence from the Oglio glacier into the Val Seriana, crossing via Passo della Presolana (1297 m a.s.l.). However, there were also small local ice bodies both during the LGM and during the previous major glacial episodes. Between Clusone and the bottleneck of the valley it is possible to recognize glacial deposits belonging to several ice advances; moraines are still preserved for two of these episodes, whilst the other are represented by buried and differently weathered tills. Changes in the physiography and neotectonic movements fit with the position of the glacial deposits in the middle portion of the valley, but the survey is still ongoing.

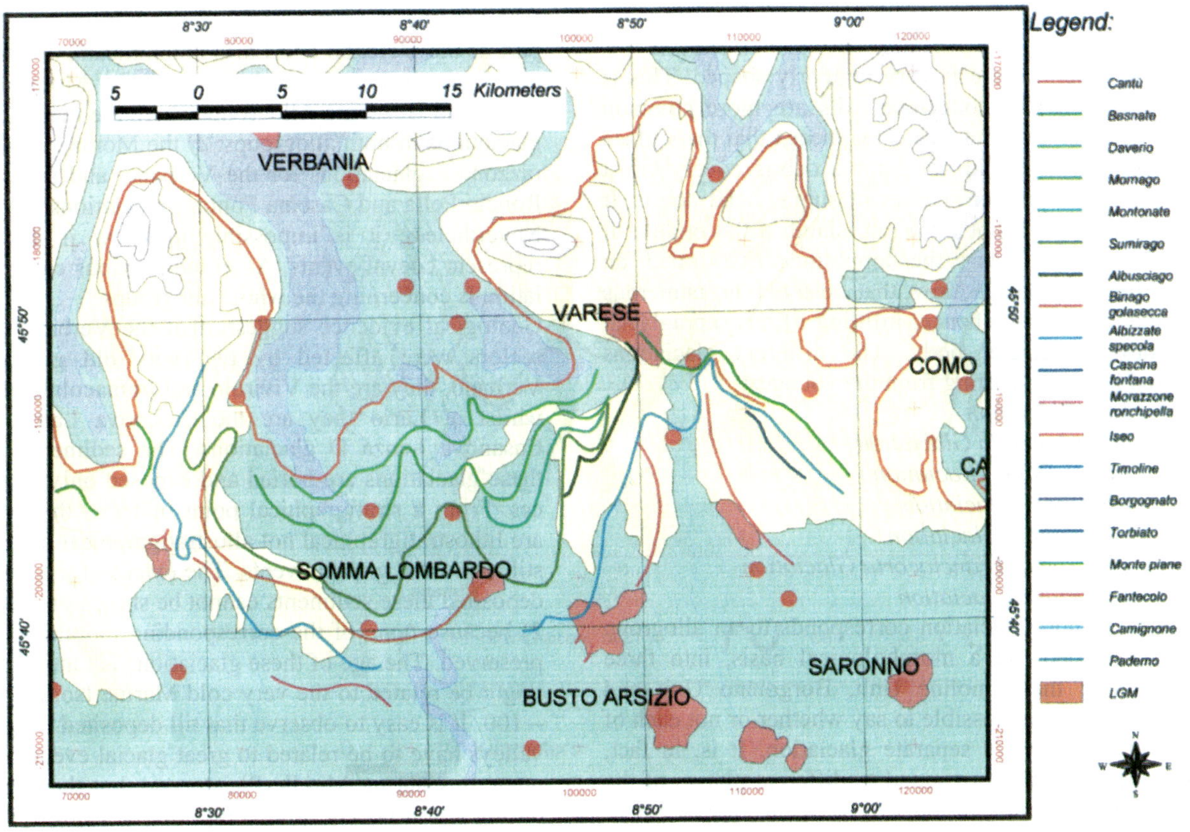

Fig. 2. Schematic limits of the different alloformations in the Verbano – Lario amphitheatre.

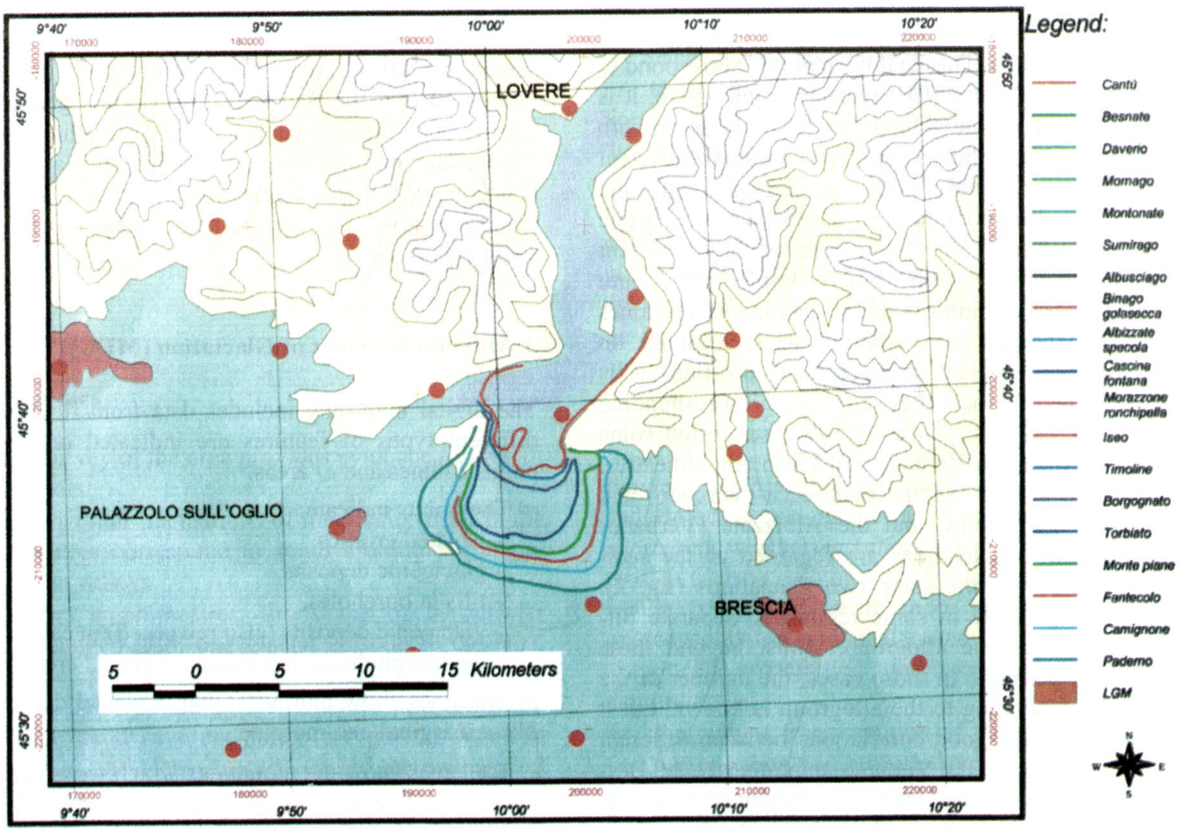

Fig. 3. Schematic limits of the different alloformations in the Sebino amphitheatre.

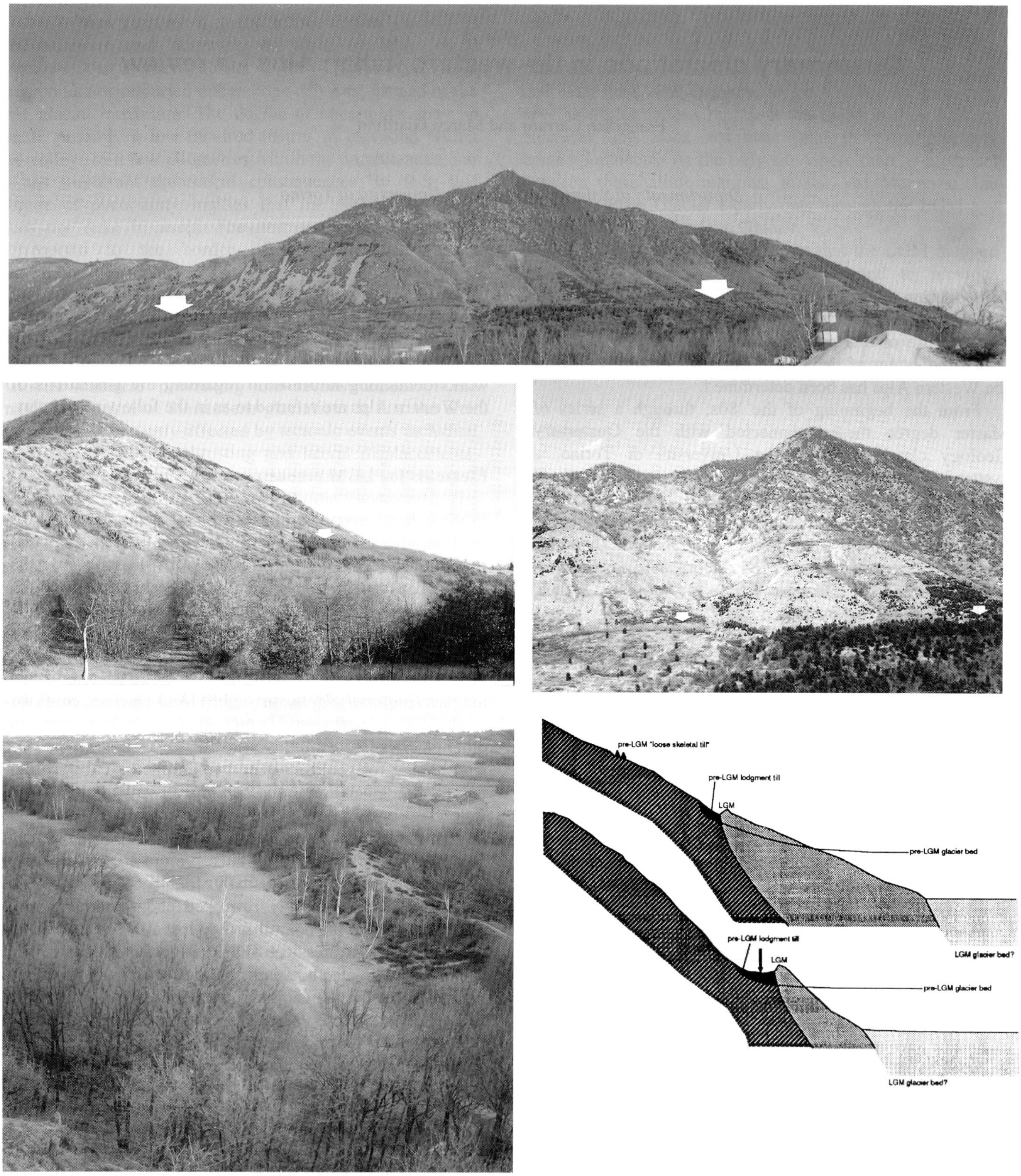

Plate 1. On the left slope of the Susa Valley mouth, traces of two distinct glacial phases are easily recognised along the slope on Mount Musinè: the more recent one is represented by a well-preserved morainic ridge (bold white arrows in pictures a, b and c); the older one is documented by isolated boulders higher up on the slope (thin arrows in picture c). Cosmogenic dating by S. Ivy Ochs is in progress. Photographs and graphics by Carraro & Giardino, 2000. a) Mount Musinè, Lower Susa Valley, and morainic ridge, front view form South. b) Mount Musinè and morainic ridge, profile view form West. c) Detail view of a). d) The elongated depression between the pre-LGM valley floor and the LGM morainic ridge, on the right (black arrow in figure e). e) Geological cross sections through the areas shown in a), b) and c).

Fig. 1. Well-preserved LGM rampart on the right slope of the middle Val Grande di Lanzo (Photograph by Carraro, 1996).

obviously connected to the LGM moraines' role in damming-up meltwater lakes with regard to the following smaller pulsations.

In the morainic amphitheatres located at the mouth of the main Western Alps valleys, we find a succession made up of a number of units, which differs each time. It should be emphasised that the outmost terminal moraine visible in the field does not necessarily correspond to the MGE; for example, in the morainic amphitheatre of Courgnè (Valle Orco; Carraro, 1986), in the stratigraphic log obtained from drillings near Salassa, about 3 km downhill from the outermost terminal moraine visible, *lodgement till* referred to older glacial advances has been found. The same phenomenon is found in front of the Ivrea morainic amphitheatre (see Carraro, in Dal Piaz, 1992, vol. 1: 198). In the same way, in the Rivoli-Avigliana amphitheatre, lodgement till has been found near Rivalta and Bruino, about 2 km beyond the outermost Middle Pleistocene moraines (Fig. 2). This fact is related to the erosion and burying of the outer moraines by the glaciofluvial activities

Fig. 2. An 'inherited' boulder in alluvial deposits near Bruino, some hundred metres beyond the southern margin of the Rivoli-Avigliana Morainic Amphiteatre, testifies that the MGE extent is larger than the outcropping one (Photograph by Anselmo, 2000).

related to younger ice advances. In these cases boulders clearly transported by the glacier and 'inherited' by younger deposits of different nature demonstrate the existence of older glacial deposits. This phenomenon, first described by Feruglio (1920; 1929) for the Tagliamento morainic amphitheatre (Eastern Alps), is a common case. For example, in the Val Pellice opening, the former presence of lodgement till is indicated by boulders, which are today enclosed in fan deposits derived from a northern tributary valley. The form and lithology of the boulders demonstrate that the main valley glacier transported them.

The total obliteration of glacial landforms, both depositional and erosional, is commonly observed. In the middle Val Sesia, a morainic amphitheatre originally existed at the valley opening, where the town of Borgosesia is situated today. Modern fluvial erosion has removed most of this moraine except for a few relict landforms that are aligned transverse to the valley axis. At the junction of the three Lanzo valleys, it is impossible to determine where the LGM glacier terminated because the relating landforms and deposits have been erased. In the Villaretto valley, a left tributary of the Val Chisone, the modern morphology seems exclusively fluvial: only a strip of lodgement till and a large eclogitic erratic, preserved at the valley mouth, represent the previous glacial history (Carraro, 1987). These examples of a very common phenomenon indicate the care that is required in determining the reconstruction of Quaternary geological evolution. These limits are imposed by the degree of preservation of both landforms and deposits that in some cases involve very recent units.

An interpretative model

These recent investigations have aided the improvement of a model for the distribution of glacial landform and deposits in the Western Alps, which was designed and validated in the process. The essential points are:

During the recent surveys, much evidence has been gathered. In most of the glacial valleys, in which the repeated Quaternary glacial pulses have been recognised, more or less marked downcutting seems to have occurred (see Pl. 1). Concerning this problem, Fairbridge, in the *Encyclopaedia of Geomorphology* (1968), defines the term *'Glacial scour, erosion'* (p. 459) as: "*The motion of glacier ice, thanks to its 'arming' by moraine boulders, has long been known to carry out large-scale erosion of the pre-existing landscape. This process is often known as glacial scour. Considerable controversy has ranged about the question of 'how much'.*"

Interpretative models similar to that proposed here are found in the international literature: *cf.* in Owen *et al.* (1977), Smith *et al.* (1997) and Nelson & Shroba (1998). A theoretical analysis of possible glacial erosional deepening is given by Benn & Evans (1998).

This kind of evolution is not common to all glacial valleys but preferably characterises those valleys or valley segments which run transverse to the main Alpine tectonic

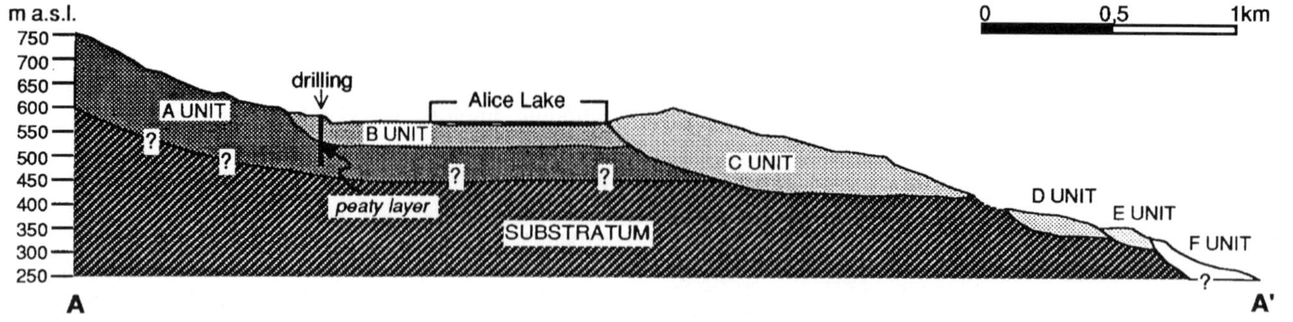

Fig. 3. Schematic cross section through different glacial units of the right lateral sector of the Ivrea Morainic Amphiteatre showing the repeated terracing relationship. Unit A represents the oldest, unit F the youngest glaciation (from Arobba et al., 1997).

structures. In this way they follow the recent uplift gradient, while it is less significant in longitudinal valleys or valley segments. This evidence is found in amphitheatres, at valley mouths, as well as in the valleys.

Hydrogeological investigations in the Rivoli-Avigliana and Ivrea morainic amphitheatres (Arobba *et al.*, 1997) both show the base of the glacial deposits occurs at increasing depths, from the outer and older parts to the inner and younger ones (Fig. 3); so the deposits of all glaciations are in a terraced relationship to one another.

On valley slopes, at some points strips of subhorizontal surfaces are found which constitute the basal support of lodgement till, suddenly cut off toward the valley axis by slopes clearly modelled by glaciers (Fig. 4). In other cases glacial landforms and deposits are found belonging to tributary glaciers downcutting an older main glacial valley floor (Fig. 5).

These situations can be interpreted as relics of original glacial valley floors, dissected by the glacier in a subsequent erosional phase. The first case might be referred to as a glacial terrace. These features are also found in the longitudinal development of valleys, though at a much lower frequency than in fluvial valleys. This gave origin to alignments of landforms on valley's slopes, tending to the valley's opening.

Fig. 4. An example of a 'glacial terrace' at Borgone (Susa Valley): the terraced horizontal surface, covered by lodgement till (interpreted as a remnant of a former glacial valley floor), is cut by a more recent erosional scarp of glacial origin (Photograph by Carraro, 1998).

The reason why glacial terraces in the Alpine valleys are far less frequent than those of fluvial origin is due to the fact that the glacial downcutting responsible for the terracing, during the anaglacial phases, is coupled to the development of the ice mass itself. Upward expansion of the ice inside the valley causes erosion of all the older landforms and deposits (cf. Pl.2). Traces of these are only preserved where the glacier migrated laterally.

The situations described above differ from other types of discontinuities, which are also found on valley slopes. Such discontinuities, in which the passage between strips of subplanar surface and steeper slopes is gradual and the slope breaks themselves are blunted and smoother are known from the literature as glacial shoulders. Glacial shoulders are interpreted either as the product of super-imposition of glacial modelling on a pre-existing fluvial valley, or as the effects of differential erosion on structural discontinuities, or as the net result of the unevenness of the glacial valley erosion itself (Castiglioni, 1979: 270-271). The latter have been explained also as the result of erosional processes caused by glacial confluence from tributary valleys.

The sequence of landforms and deposits, which different authors identified in the formerly glaciated valleys in some cases apart from progressive glacial mass volume reduction, might also be referred to the progressive downcutting. This interpretative model might lead to the conclusion, that the hypothetical ice thickness reconstructed for the LGM glaciers might have been overestimated. In order to verify the effective thickness of the Pleistocene glaciers, the preconsolidation degree of lodgement till is being investigated in the Susa Valley.

Factors controlling glacial valley evolution

The evolution of a glacial valley seems therefore to be based on the interaction of two independent factors: variations in the ice volume controlled by climatic factors,

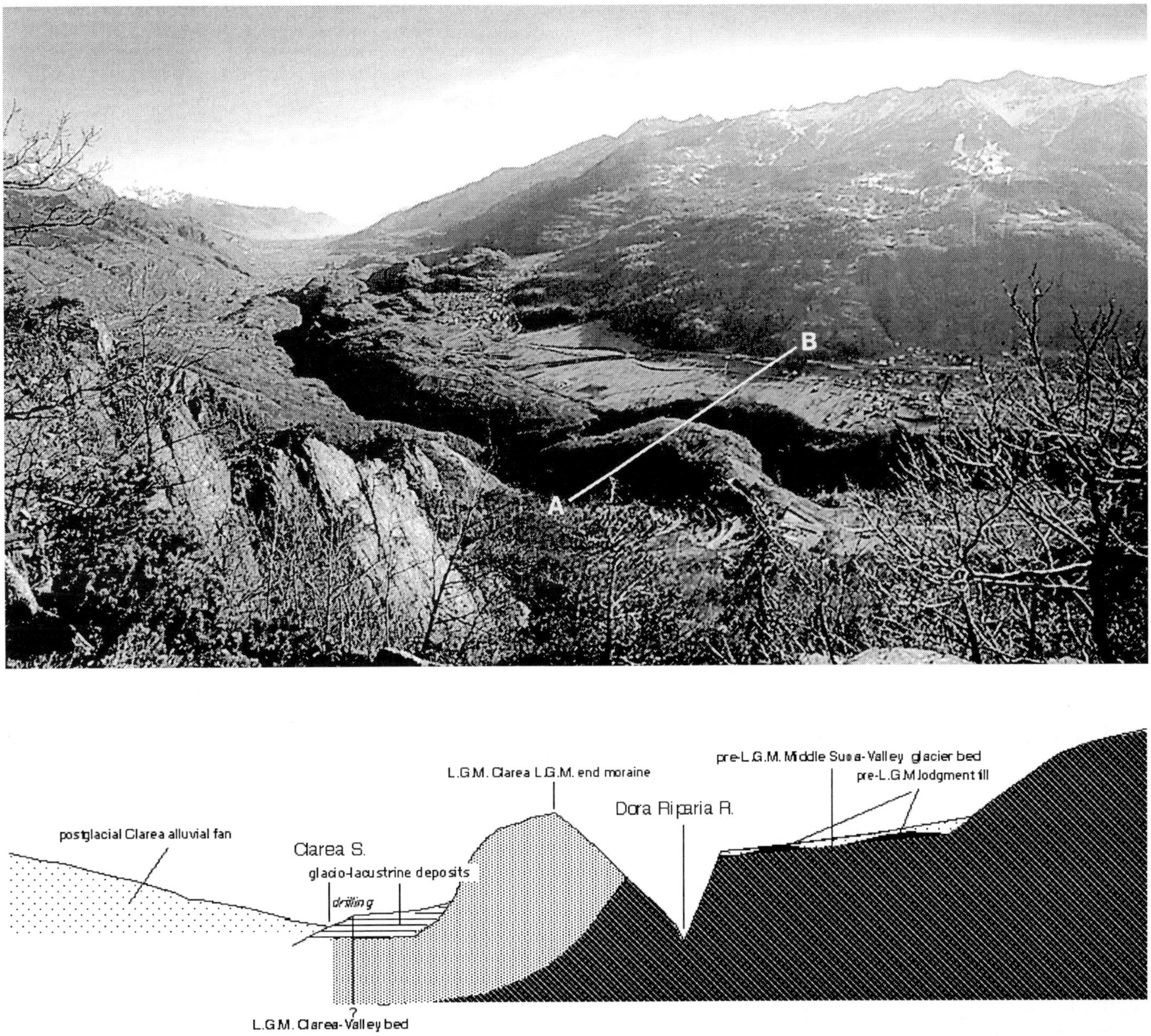

Fig. 5. a. In the middle Susa Valley the late glacial terminal moraine of the Clarea Valley (a left tributary) is clearly entrenched in the valley floor previously modelled by the main glacier.

b. Schematic cross section along the white line shown in a. Photographs and graphics by Carraro & Giardino, 2000.

and erosional downcutting as a response to tectonic uplift prior to and during the morphodynamic activity.

Since these factors are independent, a situation can arise in some valleys in which, at any one period, the ice volume changes repeatedly without any erosional deepening, either because the necessary geodynamic input was lacking, or because climatic variations were too rapid. In these cases, the deposits of successive glacial pulses can lay on the same basal surface (Fig. 1 in Bini & Zuccoli, this volume; Moscariello *et al.*, 1992). In other cases, discontinuous erosion, caused by the modularity of a particularly intense geodynamic signal, can result in the formation of additional sedimentary units (allostratigraphic units, see below) even under constant climatic conditions.

In most cases endogenic and exogenic factors interacted in different ways in any individual basin, which is reflected in variations of ice mass volume and erosional overdeepening. This may explain why the 'cataglacial sequence' is made up of a different number of units in distinct basins: the recognisable and mappable units are not directly related to climatic variations, thus they cannot necessarily be correlated between different basins. Instead they correspond to the influence of different interacting factors. In this respect they resemble units in other (e.g.

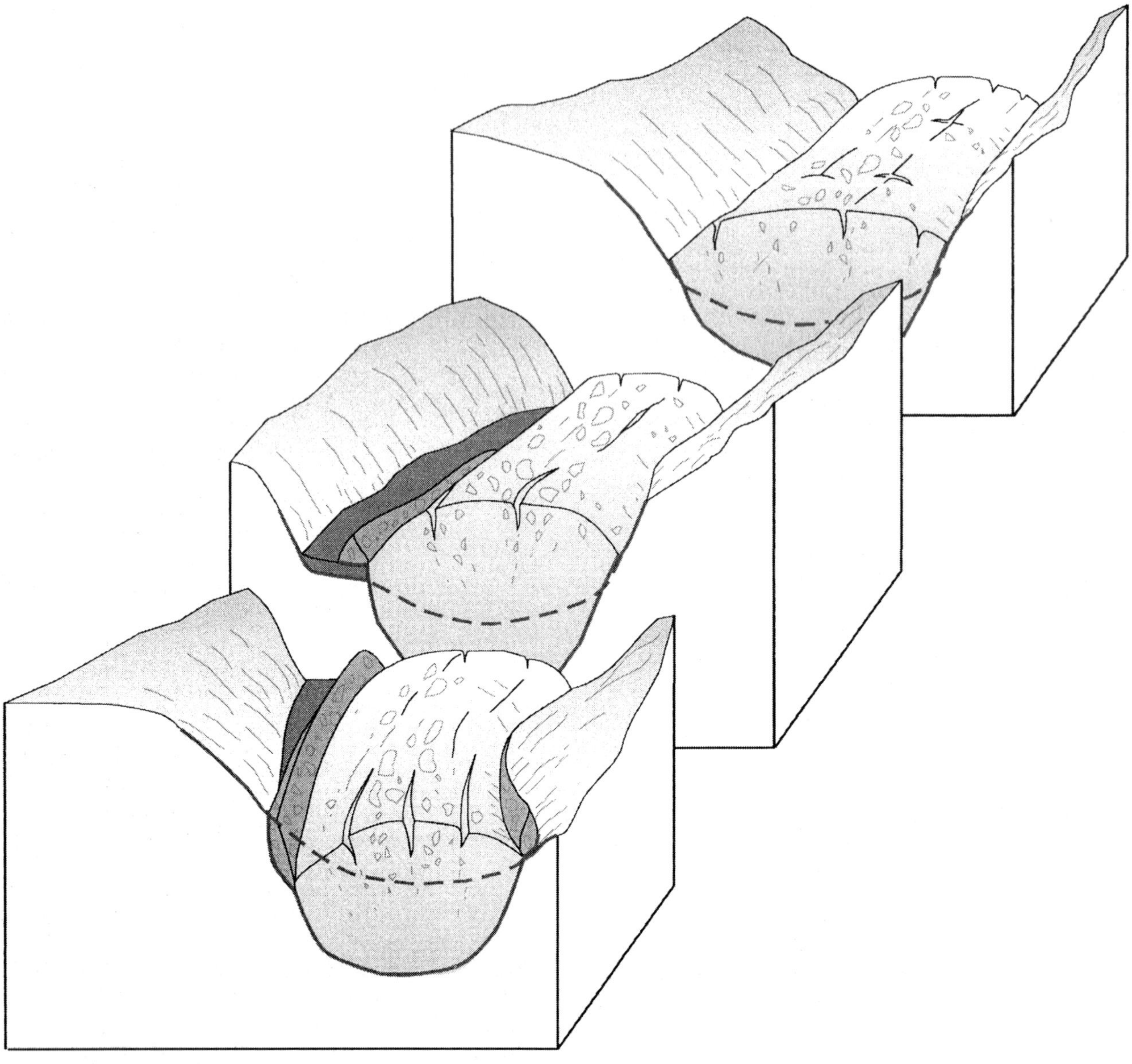

Fig. 6. Block diagrams showing the general offset of landforms and deposits in different parts of a glacial advance by means of bulging and upward expansion of the glacier during successive phases. Note the local relict of the former valley bottom due to lateral migration of the glacial mass.

marine) sedimentary environments. These units fall under the definition of *allostratigraphic unit* (North American Commission on Stratigraphic Nomenclature, 1987) being defined by discontinuities.

Dating

The chance to date the different glacial complexes directly is very rare. There are:

1. Palynological analyses, correlated with radiocarbon dates, carried out by Schneider (1978) in the Rivoli-Avigliana and Ivrea Morainic Amphiteatres. These dates have allowed the reconstruction of the Late- and Post-Glacial phases of the last glaciation.
2. Chronological information derived from the palaeomagnetic analysis of lacustrine deposits linked with the outermost left lateral moraine of the Ivrea Morainic Amphitheatre and pedological evidence from the glacial deposits (Carraro *et al.*, 1991) both indicate a Lower Pleistocene age (*sensu* Richmond, in Šibrava *et al.*, 1986);

3. Palynological analysis and radiocarbon dating of silty-clay and peaty deposits interbedded between two successions of glacial and glaciofluvial deposits in the right lateral sector of the Ivrea Morainic Amphitheatre near Alice Superiore, indicating an early interstadial in the last (Würmian) glaciation (Arobba *et al.*, 1997)

4. Exposure dating of erratics of different units at the Susa Valley mouth (cf. Fig.1) is in progress by S. Ivy Ochs (University of Zürich) using cosmogenic ^{10}Be, ^{26} Al and ^{36}Cl radionuclides.

To compensate for the scarcity of direct dates, soil development has been used as a means of distinguishing different units of glacial deposits in the inner part of the Western Alps. Various authors have demonstrated that pedogenesis is a continuous process and that the rate of soil development is controlled, excluding other factors (parent material, climatic variations, morphology, etc.) by time. In a given region the relative age of a surface can be determined, if only approximately, by the degree of soil evolution.

If the rate of soil evolution is calibrated to the local conditions, the relative age can be determined and utilised for geological dating. However, it must be stressed that the method yields only very approximate results. This approach has been used to identify the deposits and landforms of the LGM in different glacial basins. This approach has been especially successful on glaciofluvial sediments that, because of their morphological expression, better record this evidence.

The results obtained by this method are then compared with the freshness of the landforms. Older deposits have been subjected to stronger remodelling and thus lost part of their original morphological expression (Fig. 6). It is interesting to see that the boundaries between units defined this way are in fact very strong and sharp. The absence of any morphological 'intermediate' features implies that the lapse of time between the deposition of the two units must have been significantly long.

Local Glaciation

Finally, a specific mention must be made about the local glacier problem: in some cases, such as in Val Chisone, the LGM glacier tongue belonging to the main valley was fed only from the tributary valleys. This results from the higher elevation of the tributary valley heads with respect to the main valley (Colle del Sestriere for the Val Chisone).

After the LGM, some of the tributary glaciers remained entirely restricted within their valley, while others stayed active longer than the main glacier. They formed small amphitheatres which are more or less preserved when exiting into the main valley (e.g.: Cogne Valley in the Aosta Valley, Clarea Valley in Susa Valley, etc.) (Fig. 5).

References

Arobba, D., Calderoni, G., Caramiello, R., Carraro, F., Giardino, M. & Quagliolo, P. (1997). Palynological and radiometric evidence of a Last Glacial-Interstadial from peat sediments in the Ivrea Morainic Amphitheatre (NW-Italy). *Geologia Insubrica*, **2/2**, 143-148.

Benn D.L. & Evans D.J.A. (1998). *Glaciers and glaciation.* London, New York, Sydney, Auckland, Arnold, 734 pp.

Carraro, F. (1986). Revisione del Quaternario allo sbocco della Valle Orco'. *In:* Cima M. (ed.), *'Belmonte: alle radici della sua storia.* 9-14, 3ff., 1 carta geol. 1:25 000, Cuorgnè, Ed.CORSAC.

Carraro, F. (1987). Remodelling and reworking as causes of error in distinguishing between glacial and non-glacial deposits and landforms. *Geological Survey of Finland, Special Paper*, **3**, 39-48.

Carraro, F., Lanza, R ., Perotto, A. & Zanella, E. (1991). L'evoluzione morfologica del Biellese occidentale durante il Pleistocene inferiore e medio, in relazione all'inizio della costruzione dell'Anfiteatro Morenico d'Ivrea. *Bollettino. Museo Regionale di Scienze Naturali, Torino*, **9 (1)**, 99-117.

Castiglioni, B. (1940). *L'Italia nell'età quaternaria.* Scala 1:1.200.000, Atlante fisico-economico d'Italia, C.T.I., Roma.

Castiglioni, G.B. (1979). *Geomorfologia.* Torino, U.T.E.T., 436 pp.

Dal Piaz, G.V. (ed.) (1992). *Le Alpi dal Monte Bianco al Lago Maggiore.* Guide geologiche regionali, Societá Geologica Italiana, ed. BE.M.A., Milano, 311 pp.

Fairbridge, R.W. (1968). *The Encyclopedia of Geomorphology.* Stroudsburg, Pennsylvania; Dowden, Hutchinson & Ross, 1295 pp.

Feruglio, E. (1920). I terrazzi della pianura pedemontana friulana. *Pubbliciazioni dell'Ufficio Idrografico del Regio Magistrato alle Acque*, **107**, 1-94.

Feruglio, E. (1929). *Note illustrative della Carta Geologica delle Tre Venezie. F.25, 'Udine'.* Ufficio Idrografico del Regio Magistrato alle Acque Venezia, Padova. 77 pp.

Moscariello, A., Carraro F., & Paganelli A. (1992). Nuovi dati sulla geologia del Quaternario dell'alta Valle di Zoldo (Dolomiti sudorientali). *Il Quaternario - Italian Journal of Quaternary Sciences*, **6 (1)**, 145-168.

Nelson, A.R. & Shroba, R.R. (1998). Soil relative dating of moraine and outwasah-terrace sequences in the upper Arkansas Valley of central Colorado. *Arctic and Alpine Research*, **30 (4)**, 349-361.

North American Commission on Stratigraphic Nomenclature (1987). North American Stratigraphic Code. *American Association of Petroleum Geologists Bulletin*, **67 (3)**, 841-875.

Owen, L.A., Mitchell, W.A., Bailey, R.M., Coxon, P. & Rhodes, E.J. (1977). Style and timing of glaciations in the Lahul Himalaya, Northern India: a framework for reconstructing late Quaternary paleoclimatic changes in Western Himalaya. *Journal of Quaternary Science*, **12 (2)**, 83-109.

Schneider R.E. (1978). Pollenanalytische Untersuchungen zur Kenntnis der Spät- und postglazialen Vegetationsgeschichte am Südrand der Alpen zwischen Turin und Varese (Italien). *Botanisches Jahrbuch der Systematik*, **100 (1)**, 26-109.

Šibrava, V., Bowen, D.Q. & Richmond, G.M. (eds) (1986): Quaternary Glaciations in Northern Hemisphere. *Quaternary Science Reviews*, **5**, 441 pp.

Smith, G.W., Nance R.D. & Geneza, N. (1997). Quaternary glacial history of Mount Olympus (Greece). *Geological Society of America Bulletin*, **109**, 809-824.

Quaternary Glaciations – Extent and Chronology
Editors J. Ehlers and P.L. Gibbard

Quaternary glaciations in the eastern sector of the Italian Alps

Giovanni B. Castiglioni

Dip. di Geografia, via del Santo 26
I-35123 Padova, Italy

1. Introduction

This summary report is intended to give an overview of the current state of knowledge about Quaternary glaciations in the eastern sector of the southern Alps. Some important new results are included, but they are far from a new general interpretation, so that the map does not greatly change the general figure given by older literature. A new survey is now in progress but it will take some years before the results will be available. The survey is part of the CARG National Project (the new Geological Map of Italy, scale 1:50,000), in which special attention will be paid to the Quaternary. At present, surveys in some sectors of the Alpine margin as well as in inner-Alpine basins are in progress.

Where modern studies are lacking, the Last Glacial Maximum (LGM) limit has been drawn according to old geological maps, scale 1:100,000 ('Carta Geologica d'Italia', 'Carta Geologica delle Tre Venezie') and to the map 'L'Italia nell'età quaternaria', published in 1940, scale 1:1,200,000. The latter was a detailed map for that time, the Alpine part of which was compiled by the late Bruno Castiglioni.

For glaciations older than the LGM, all research carried out until 1977 aimed at correlations with the classical Alpine glacial stratigraphy, although in some cases the results were in disagreement with the original works of Penck & Brückner (1909). Unfortunately, the end-moraine complexes ('amphitheatres') bordering the Po-Veneto Plain are separated from each other, and similarities are not clear enough to establish reliable correlations with other south-Alpine or north-Alpine end-moraines. In modern Italian Quaternary research, a broad chronostratigraphical subdivision of the Quaternary into Early, Middle and Late Pleistocene is usually applied. For each area, 'UBSU' (Unconformity Bounded Stratigraphic Units) and/or 'allostratigraphic units' are defined in order to describe the observed sequences in an objective way. Regional correlations are introduced with caution.

2 . A short comment on the map

The glacial limits of the Last Glacial Maximum are sufficiently well-known that it is possible to map their positions in the sub-Alpine and foothill regions, with one major exception. It is difficult to find general agreement about the LGM in the Lake Garda area. In the opinion of the present author, the old reconstruction given by Penck (in Penck & Brückner, 1909, vol. 3, map page 852) comes close to reality (details are discussed in section 3 and in Figs. 1 - 3).

For the older glaciations and their chronology, there are still many open problems. Therefore, in the digital map those boundaries are all drawn in the same colour: solid lines for glacial limits defined by moraine ridges, and dotted lines for uncertain boundaries.

Obviously, there are differences in the freshness of landforms; the older moraines are smoothened and softer in shape, due to their longer exposure to weathering processes. But these morphological differences are less distinct in the foothills of the Italian Alps than in the morainic complexes of the northern Alps, where solifluction phenomena have remodelled the old morainic landscape to a greater extent and where the loess cover reaches greater thickness. Human influence plays a major role in remodelling of the landscape, in particular by terracing and levelling of the terrain for agricultural purposes.

2.1. Last Glacial Maximum

In the west, the first important valley glacier was that of the Chiese Valley, which did not reach the border of the Southern Alps. Its limit is given according to Habbe (1969).

For the Lake Garda Glacier, the limits given follow largely Habbe's Würm moraines (1969) and Cremaschi's 'Solferino Moraine' (1987), although not quite. However, the case of the Lake Garda end-moraines which form a very complex 'amphitheatre', needs further discussion both for the older glaciations and for the LGM. Different interpretations have been given in published studies carried out during the years 1957-1990. This subject is considered in section 3 and Fig. 1. A radiocarbon date obtained from consolidated humic loess deposits covered by the moraines of the eastern side (Valsorda) gives good support to Penck's interpretation as well as to Habbe and Cremaschi.

An ice-dammed lake, the Lago di Ledro, existed on the western side of the tongue of the Lake Garda Glacier, in its Alpine section. A glacier-induced drainage diversion affected the Chiese River, near Salò, in the foothill sector. The course of this river is thought to have gone originally directly towards the Garda basin.

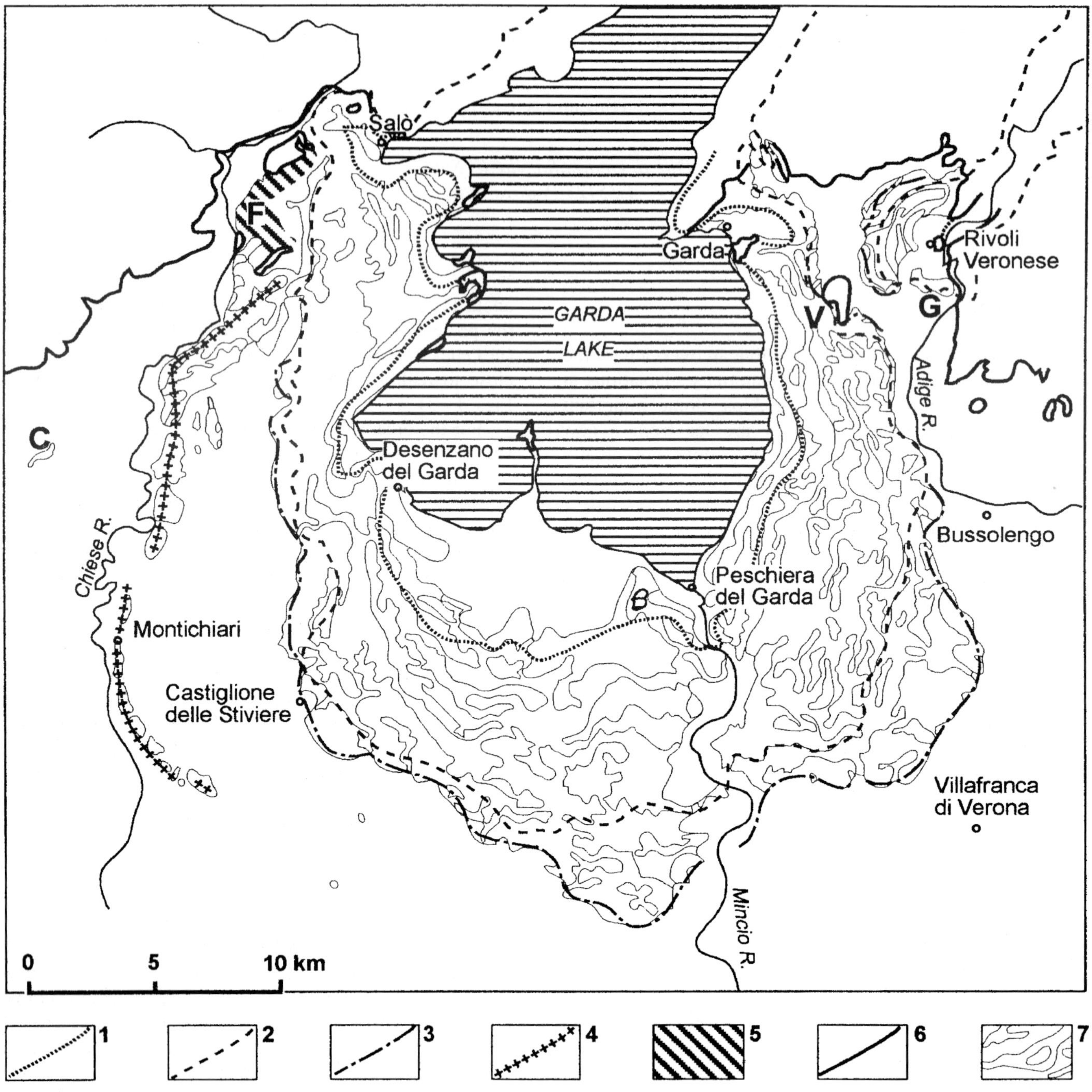

Fig. 1. Sketch-map of Lake Garda moraine complex, with limits of glaciations, according to various authors. Sites of particular interest: C: Ciliverghe hill. F: Monte Faita (see Figs. 2 and 3). G: Gaium quarries. V: Valsorda section.

Legend:

1 - Würm, after Venzo (1969a, 1969b).
2 - Limit of Würm moraines, after Habbe (1969).
3 - Limit of Young Riss moraines according to Habbe (1969) and limit of 'Solferino Moraine' (Late Pleistocene) in the main Garda complex, according to Cremaschi (1987).
4 - Old Riss moraines after Habbe (1969); Middle Pleistocene 'Carpenedolo Moraine' after Cremaschi (1987).
5 - Monte Faita moraine: Young Riss, after Habbe (1968; 1969); Early to early Middle Pleistocene, after Cremaschi (1987).
6 - Margin of outcropping pre-Quaternary relief (Geomorphological Map of Po Plain, 1997).
7 - Morainic ridges and intermorainic depressions.

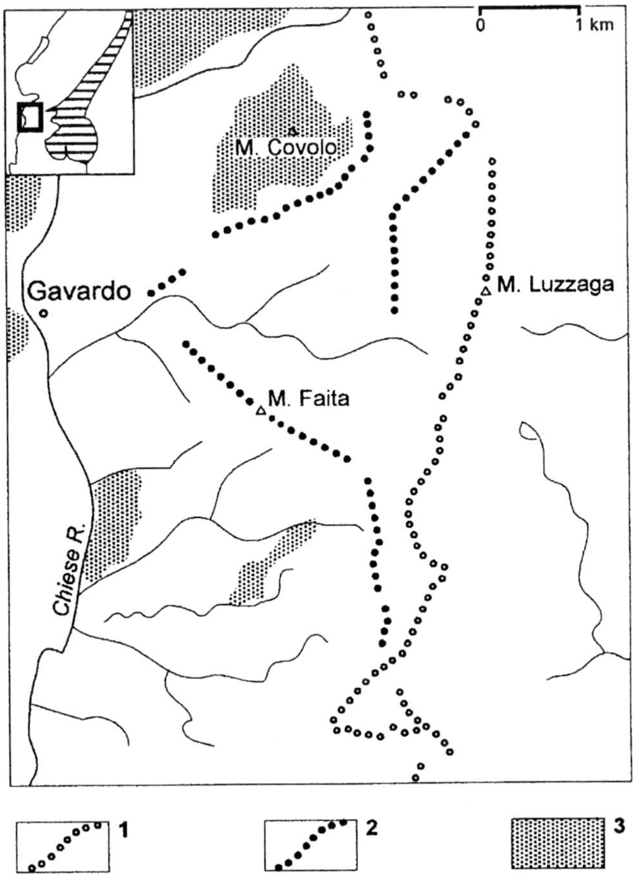

Fig. 2. Riss and Würm end moraines in Monte Faita area, north-western part of Lake Garda end-morainic basin (redrawn and simplified after Habbe, 1968).
1 - Würm end-moraine
2 - Riss II end-moraine
3 - Outcropping hills of pre-Quaternary rocks.

The limits of the Adige Glacier are mapped according to Habbe (1969). They correspond to the ridges of the 'Monte Police Unit' defined by Cremaschi (1987) which is Late Pleistocene in age. This is also supported by a radiocarbon date from local buried colluvial deposits covered by the Adige Glacier end moraine: 40.8 +7.8-3.9 ka B.P. near Rivoli Veronese (Gaium quarries, in Cremaschi, 1990; see also here, Fig. 1, point G).

East of the Adige Valley, the Monti Lessini Plateau was covered by a plateau glacier, surrounded by valley glaciers. The limits are mapped according to Sauro (1973). The mountain massifs between Rovereto and Vicenza also had their mountain glaciers, often in contact with the Adige valley main glacier.

An important valley glacier, fed by transfluences from the Adige Glacier and by outlet glaciers from the surrounding plateaux, was the Astico Glacier: its end-moraines are mapped according to Cucato (in print), who defined the 'Cogollo del Cengio Unit'.

An extensive plateau glacier, together with some valley glaciers are documented, covering the Altopiano dei Sette Comuni (or Asiago Plateau); they were mapped by Trevisan (1939); new details may have to be added depending on the results of studies under the mentioned CARG Project.

The important valley glacier flowing in the deep Brenta River valley is only known from a few lateral moraines, but end moraines are lacking.

The large Piave Glacier was fed by valley glaciers from the Dolomites. It filled most of the Belluno and Alpago intramontane basins. Here, recent surveys were carried out by Pellegrini and others. They determined the altitude of the glacier surface (1,000-1,150 m a.s.l.), obtained important chronological data about the glacial retreat, related to large mass movements from the valley flanks, and they reconstructed the environmental evolution between Late Glacial and Holocene (Pellegrini, 1999, 2000). Further downstream, the Piave Glacier flowed in different directions. The valley glacier tongue following the present-day Piave River ended at the Quero moraine (limit mapped according to Venzo, 1977). Another large valley glacier tongue flowed through the Fadalto Pass and the Valle Lapisina and formed a piedmont lobe in the plain near Vittorio Veneto, together with a minor lobe in the Soligo Valley. The limits are given according to Casadoro *et al.* (1976). South of Vittorio Veneto, the radiocarbon age of a wood sample (a tree embedded in a terminal moraine) was obtained by Bondesan (1999): 17.7 ±0.3 ka B.P., which gives good support for correlation with the LGM. Radiocarbon dates on wood from lacustrine deposits near Revine, connected with the ice retreat, are also known : 14.4 ±115 and 14.8 ±135 ka B.P. (Casadoro *et al.*, 1976).

Further to the east, several valley glaciers formed in the cirques of the Monte Cavallo massif, dominating some karst plateaus and the Veneto-Friuli plain; on its eastern flank, in Val Caltea, a wood sample taken from silt layers covered by till of the LGM was dated at 29.3 ±0.5 ka B.P. (Fuchs, 1969).

The Quaternary deposits linked with Alpine glaciers in the Friuli Region are at present being surveyed within the new CARG Project by a group of earth scientists from the University of Udine, directed by A. Zanferrari. The glacial limits in the digital map are shown only on the basis of older literature: for the Tagliamento Glacier, according to Gortani (1959), and for the Isonzo (Soca) Glacier (mostly in Slovenia) according to Feruglio (1929) and Gortani (1959).

2.2. Moraines older than the LGM

For the moraine complex of the Lake Garda sector only the unsatisfactory data from the literature can be given (for details see section 3). In the main map, the following limits of glaciations older than L.G.M. are marked: 1) 'Sedena Moraine' (Late-Middle Pleistocene according to Cremaschi, 1987); 2) 'Carpenedolo Moraine' (Middle Pleistocene); 3) 'Monte Faita Moraine' (Early Pleistocene to Early Middle Pleistocene); 4) 'Ciliverghe Moraine' (Early Pleistocene);

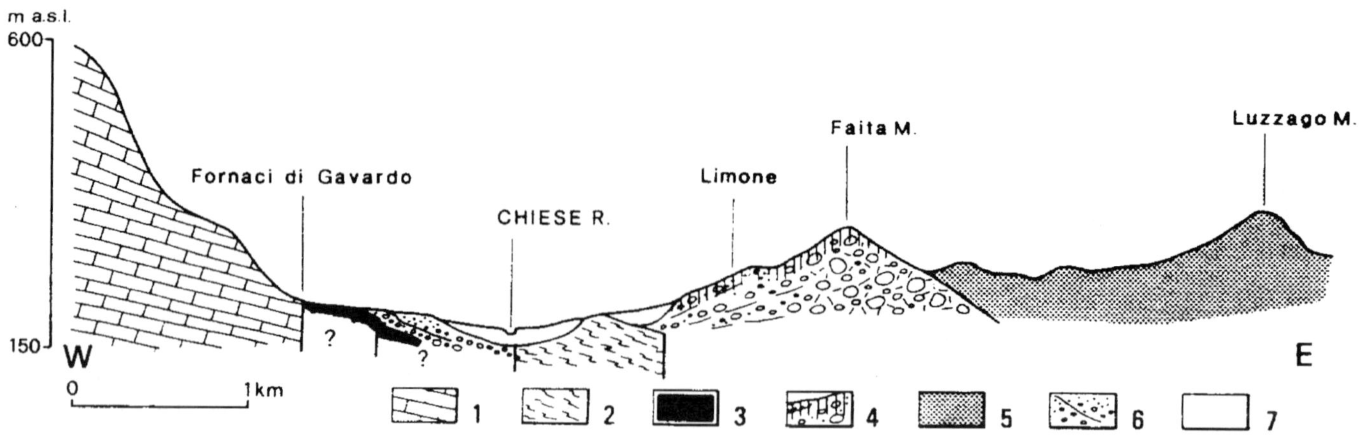

Fig. 3. Cross-section in Monte Faita area, according to Cremaschi (1987).
Key:

1 and 2 - Pre-Quaternary limestones
3 - Strongly weathered clay, more than 5 m thick ('Gavardo 1' stratigraphic unit)
4 - Middle Pleistocene moraine and related fluvioglacial deposits
5 - Late Pleistocene moraine and related fluvioglacial deposits
6 - Gravel, sand, lacustrine sand and silt, loess, colluvium ('Gavardo 4 to 7' stratigraphic units)
7 - Holocene deposits along the Chiese River.

the reported names and chronostratigraphical attributes were given by Cremaschi (1987).

Very instructive is the preservation of a fine distal Middle Pleistocene morainic arc, detached from the more recent moraines, in the western sector of the Garda amphitheatre (Carpenedolo moraine, at Montichiari), now fragmented into a string of small separate hills (Fig. 1). The absence of a similar ridge on the eastern side of this amphitheatre is attributed to a large-scale change in the shape of the glacier terminus between that older glaciation and later glacial events, probably due to tectonic subsidence of the eastern sector, after the old glaciation (Cremaschi, 1987). Similarly, the lack on the eastern side of any trace of the oldest moraine (identified in a stratigraphic section at Ciliverghe hill, SE of Brescia) may be due to the same tilting of this marginal belt of the Po Plain.

In the area in front of the Adige Glacier, the Map shows the following moraines, whose names and chronological attributions are taken from Cremaschi (1990): 'Monte Crivellino Unit' (Middle Pleistocene; corresponds to the Young Riss moraine of Habbe, 1969); 2) 'Caprino Unit' (Early Middle Pleistocene); 3) 'Pesina Unit' (Early Pleistocene). Both units 2 and 3 are strongly eroded, on the right side of the former glacier (dots, north of the morainic amphitheatre of Rivoli Veronese). These are not shown in Fig. 1.

In the area of the Astico Glacier, till of the 'Unità dell'Astico' has been mapped according to new surveys by Cucato (in print).

In the area of the Brenta Glacier, old till is well documented at Bassano del Grappa, according to Bartolomei (1999).

In the area of the Piave Glacier, eastern branch, the limit is reported according to Casadoro *et al.* (1976).

North of Udine, near the LGM end-moraines of the Tagliamento Glacier, old till deposits probably exist, according to Feruglio (1929).

2.3. Maximum limit of Pleistocene glaciation

The only place where a maximum limit of Pleistocene glaciation can be given is on the western side of the Lake Garda glacier, at Ciliverghe hill, as indicated by Baroni & Cremaschi (1987).

3. Discussion on the age of the Lake Garda Glacier end moraines

The most important differences in interpretation are presented in Fig. 1. Note that the limit of the Late Pleistocene moraines according to Cremaschi (1987) is approximately the same as that proposed by Penck (1909).

Great differences exist in the interpretation of the LGM by Venzo (1969a, 1969b) on one hand, and Cremaschi (1987) on the other. Practically, the latter includes the entire main moraine complex of the Lake Garda system, which means all moraines showing a young morainic landscape. It may be added that this interpretation is now supported by new stratigraphic data from the Valsorda section (V, in Fig. 1). There, thick consolidated loess deposits are covered by till corresponding to the external moraine ridge. The radiocarbon age of humic acids extracted from a loess sample is 27.9 ±0.6 ka B.P. (Cremaschi, 1990). In the Valsorda area, the limit of the LGM given by Cremaschi coincides with that proposed 20 years earlier by Habbe.

In other places, differences between Habbe's limit and that proposed by Cremaschi are evident (see limits no. 2 and 3 in Fig. 1). They result from the particular interpretation of the exterior ridge given by Habbe (1968, 1969), who referred it to the Young Riss glaciation. Another point of difference concerns the Monte Faita area on the western side (F, in Fig. 1), which was considered by Habbe as belonging to the Young Riss belt, whereas Cremaschi considered it to be very old (see Figs. 2 and 3).

In a discussion of these and other points, it is worth remembering some aspects which explain the problems highlighted by various authors interested in the Quaternary of the Lake Garda area in the last century.

For several decades until 1960, some Italian geologists expressed disagreement with Penck's view, but their maps and interpretations did not result in any convincing synthesis, especially regarding the extent of the Würm and Riss glaciations. In fact, this problem was the most difficult in the Lake Garda morainic amphitheatre.

In the 1960s, both the Italian Sergio Venzo and the German Karl Albert Habbe were working on systematic new surveys. The geologist Venzo had the support of the well-known pedologist Fiorenzo Mancini, and published his results step by step, in different publications and maps covering parts of the area, in which it is possible to follow the changes in the proposed solutions; finally, he published the map of the central southern sector in agreement with Mancini's view (Venzo, 1965) and, as a synthesis, a map 1:100,000 and 'Illustrative Notes' (Venzo, 1969a; 1969b). The main moraine complex was considered to be Rissian in age, the Würm glaciation being very reduced in its extension, both in the Lake Garda area and at Rivoli Veronese, in the Adige Valley (see Fig. 1).

Habbe's approach, due to his geographical-geomorphological background, considered many components of the landscape evolution. He first mapped the glacial deposits in the mountain valleys, coming finally to the difficult area of the moraine amphitheatre. He discussed the open problems with Venzo at the VII INQUA Congress (U.S.A., 1965), but they were not able to reach a common opinion (see also Habbe, 1968). In his book published in 1969 and in his maps, only the outermost belt of the main end moraine complex is considered as belonging to the Riss glaciation (Young Riss), the Würmian moraines dominating by far (Fig. 1). This distinction is based, in his opinion, on some observations on soils, but also on detailed geomorphological analysis. An illustrative example is the detailed study of the Monte Faita area (Habbe, 1968) on the western side of the amphitheatre, here presented in simple lines in Fig. 2.

It must be said that the general landscape of the morainic hills both of Young Riss and Würmian age according to Habbe, is very similar, due to their 'fresh' morphology. Clear morphological differences may be observed between the Young Riss and Old Riss moraines *sensu* Habbe (4, in Fig. 1). It appears that the pedogenetic processes of the Late Glacial and Holocene were probably underestimated by Venzo and Mancini.

More recently Cremaschi (1987) abandoned the traditional scheme of Alpine glaciations, and only refers to Early / Middle / Late Pleistocene. However, it must be said that the interest of this author was mostly the paleopedology and stratigraphy of a large sector of the Po Plain, and not palaeoglaciology. It is also evident that many points have been chosen for detailed pedological analysis in the large area considered, but that they only partly overlap with the main area of the Lake Garda glacier.

In general terms, the limits given for the Late Pleistocene glaciation approximately fit those of Penck for the Würmian. A geological profile taken from Cremaschi's work (Fig. 3) on the Monte Faita area may easily be compared with Habbe's interpretation (Fig. 2). The very old age of the moraines in that area, according to Cremaschi, is not based on geomorphological evidence, but only on the paleopedologic interpretation of a particular soil profile.

Between 1987 and 1990 Cremaschi participated in two INQUA commissions. In this framework he edited the guide-book for the 1988 excursion, in which the palaeoenvironmental significance of loess and loess-like deposits is evaluated, and the radiocarbon dates are reported, for the Valsorda and Gaium sites, key sections for the history of the Garda and Adige glaciers (Cremaschi, 1990).

The present interpretation of the general extent of the LGM in the eastern part of the Italian Alps is based on these results. The short duration of the LGM has not only been established for the inner part of the Adige basin (Fliri, 1989), but also for the sub-Alpine mountains (Monte Cavallo) and the piedmont area near Vittorio Veneto (Piave Glacier). Recent work in the Garda and Adige catchment areas (CARG Project) has shown that the 'Sintema del Garda' (or 'Alloformazione del Garda') is the stratigraphic unit defining sediments linked to the LGM event. The above-mentioned ice-dammed Lago di Ledro near the upper part of Lake Garda, with its beautiful lateral moraine of the former Garda Glacier, has been chosen as its type area.

References

Baroni, C. & Cremaschi, M. (1987). Geologia e pedostratigrafia della Collina di Ciliverghe (Brescia). *Natura Bresciana*; **23** (1986), 55-78.

Bartolomei, G. (1999). Nuovi elementi su un morenico antico a Bassano del Grappa, Vicenza. *In:* Orombelli, G. (ed.), *Studi geografici e geologici in onore di Severino Belloni*, Università di Milano, G. Brigati, Genova, 9-18.

Bondesan, A. (1999). Una nuova data ^{14}C nell'anfiteatro morenico di Vittorio Veneto e la costruzione del suo sandur (pianura del F. Meschio). 8th Italian Glaciological Meeting, Alpine Glaciers and Climate Change, Bormio, 9-12 Settembre 1999, Abstracts, pp. 87-89. Dip. Scienze dell'Ambiente e Territorio, Università-Bicocca, Milano.

Casadoro, G., Castiglioni, G. B., Corona, E., Massari, F., Moretto, M. G., Paganelli, A., Terenziani, F. & Toniello,

V. (1976). Un deposito tardowürmiano con tronchi subfossili alla Fornaci di Revine (Treviso). *Bollettino del Comitato Glaciologico Italiano, Torino*, **24**, 22-63.

Castiglioni, B. (1940). L'Italia nell'età quaternaria. Plate No. 3, *In:* Dainelli, G. (ed.), *Atlante Fisico-economico d'Italia*. Milano, Consociazione Turistica Italiana.

Cremaschi, M. (1987). Paleosols and vetusols in the central Po Plain (Northern Italy): a study in Quaternary Geology and Soil Development. Doctoral thesis, Amsterdam. Edizioni Unicopli, Milano, 306 pp.

Cremaschi, M. (ed.) (1990). The Loess in Northern and Central Italy: A Loess Basin between the Alps and the Mediterranean Region. INQUA Commission on Loess and INQUA Commission on Paleogeography. *Quaderni di Geodinamica Alpina e Quaternaria*, **1**, Milano, 190 pp.

Cucato, M., (in print). Survey of middle Astico valley (Province of Vicenza): example of application of regulations on geological mapping of continental Quaternary (text in Italian). *Bolletino del Servizio Geologico d'Italia*, Roma.

Feruglio, D. (1929). *Note Illustrative della Carta Geologica delle Tre Venezie, Foglio Udine*. Padova, Ufficio Idrografico del R. Magistrato alle Acque, Sezione geologica, 78 pp.

Fliri, F. (1989). Die Schottergrube von Albeins bei Brixen, eine neue Schlüsselstelle zur Kenntnis der Chronologie der Würmvereisung in Südtirol. *Zeitschrift für Gletscherkunde und Glazialgeologie*, **24**, 137-142.

Fuchs, F. (1969). Eine erste 14-C Datierung für das Paudorf-Interstadial am Alpensüdrand. *Eiszeitalter und Gegenwart*, **20**, 68-71.

Gortani, M. (1959). Carta della glaciazione würmiana in Friuli. *Atti dell' Accademia delle Scienze dell'Istituto di Bologna, Classe di Scienze Fisiche, anno 247, Rendiconti*, **11, t. 6**, 3-11. Bologna.

Habbe, K.A. (1968). The Riss-Würm boundary in northwestern part of the Lake Garda terminal moraine basin of Northern Italy. *University of Colorado Studies, Series in Earth Sciences No. 7, Glaciation of the Alps*, 79-84.

Habbe, K.A. (1969). *Die würmzeitliche Vergletscherung des Gardasee-Gebietes*. Freiburg im Breisgau, H.F. Schulz Verlag, 254 pp.

Pellegrini, G. B. (1999). *Carta Geomorfologica d'Italia: Foglio 063 Belluno, 1:50,000*. Servizio Geologico d'Italia, Regione del Veneto. S.EL.CA., Firenze.

Pellegrini, G. B. (ed.) (2000). *Note Illustrative della Carta Geomorfologica d'Italia alla scala 1:50,000. Foglio 063 Belluno*. Roma, Servizio Geologico d'Italia, Regione del Veneto, Istituto Poligrafico, 141 pp.

Penck, A. & Brückner, E. (1901/1909). *Die Alpen im Eiszeitalter*, 3 vols, 1199 pp. Leipzig.

Sauro, U. (1973). Il paesaggio degli Alti Lessini, studio geomorfologico. *Museo Civico di Scienze Naturali, Verona, Memorie fuori serie*, **6**, 161 pp.

Trevisan, L. (1939). Il glacialismo quaternario nell'Altopiano dei Sette Comuni. *Bollettino del Comitato Glaciologico Italiano, Torino,* **19**, 99-113.

Venzo, S. (1965). Rilevamento geologico dell'anfiteatro morenico frontale del Garda dal Chiese all'Adige. *Memorie della Società Italiana di Scienze Naturali e del Museo Civico di Storia Naturale di Milano*, **14**, 1-82.

Venzo, S. (1969a). *Peschiera del Garda. Foglio 48 della Carta Geologica d'Italia, II ediz., Scale 1:100,000.*

Venzo, S. (1969b). La serie del Quaternario continentale. *In: Note Illustrative della Carta Geologica d'Italia. Foglio 48, Peschiera del Garda*. Servizio Geologico d'Italia, 41-62.

Venzo, S. (1977). I depositi quaternari e del Neogene superiore nella bassa Valle del Piave da Quero al Montello e del Paleopiave nella Valle del Soligo (Treviso). *Memorie Istituti di Geologia e di Mineralogia, Università di Padova*, **30**, 63 pp.

Quaternary Glaciations – Extent and Chronology
Editors J. Ehlers and P.L. Gibbard

The Apennine glaciations in Italy

Carlo Giraudi

ENEA, P.O. Box 2400,
I-00100 Roma A.D., Italy

E-mail: giraudi@casaccia.enea.it

Introduction

The Apennines form the backbone of the Italian Peninsula, thrusting down into the Mediterranean Sea between 38° N and about 44°30' N (Fig. 1). Traces of former glaciation have been known from the Apennines since the 19th century and have been the subject of many investigations. In the present work, however, only the most recent studies will be taken into consideration. The highest peaks in the Apennines are found in the central sector, between about 42°50' N and 41°30' N where they include the Gran Sasso (2912 m), Maiella (2793 m), Velino (2486 m), Sibillini (2476 m) and Laga (2458 m) massifs, but numerous other peaks exceed 2000 m. In contrast, the mountains of the Northern Apennines exceed 2000 m only at Mounts Cusna (2121 m), Cimone (2165 m) and Prato (2054 m), and the mountains of the Southern Apennines exceed 2000 m only in Mounts Sirino (2005 m) and Pollino (2266 m).

At present only a single small glacier which is rapidly melting is found in the Apennines: the Calderone Glacier in the Gran Sasso. Until around 10 – 15 years ago, the study of the Apennines glaciations was limited to examination of the glacial features, the weathering of the till and calculation of the equilibrium line altitude (ELA) of the former glaciers. The chronological framework of the glacial phases relied heavily on correlation between the local ELA and those of the Alps. It is only since 1989 that the use of ^{14}C dates and of tephra layers has made it possible to date the glacial deposits directly, although as yet only very few dates are available (Frezzotti & Giraudi, 1989, Frezzotti & Narcisi, 1989). In the Apennines there are traces of at least two glaciations: the most ancient, while certain, are fairly scanty, whereas traces of the most recent glaciation are very abundant and well preserved. The great majority of the glaciers were present in valleys and glacial cirques orientated towards the north, while only few developed in valleys orientated towards the west and the east, and hardly any occurred in valleys with a southern alignment.

Middle Pleistocene glaciation

It is only recently that the traces of an early glaciation were identified with any certainty (Federici, 1977). Since the glacial limits of the Apennines have been correlated with those of the Alps, the oldest moraines have been attributed to the Rissian glaciation or to a generic pre-Würmian event. Those old moraines have been reported from the Northern Apennines, in the area of Mount Navert (Federici, 1977; Jaurand, 1994), and in the Central Apennines, in the Sibillini Massif (Coltorti & Farabollini, 1995), at Campo Imperatore in the Gran Sasso Massif (Giraudi, 1994; Jaurand, 1994; Giraudi & Frezzotti, 1997; Bisci *et al.*, 1999), in various places of the Velino Massif (Cassoli *et al.*, 1986; Giraudi, 1989; Giraudi, 1998b) and at Mount Greco (Frezzotti & Giraudi, 1989; Cinque *et al.* 1990). According to Damiani & Pannuzi (1991), however, the 'Rissian' moraines at Mount Greco should be attributed to the Early Würmian glaciation. Judging from the distribution of till in the Central Italian massifs, the extension of the glaciers during this period must have exceeded that of the Last Glacial Maximum (LGM) by some 5 - 10%. It is not possible to assess the ELA of the glaciers because the Apennines Chain has been affected by intense neotectonic movements, and it is likely that the characteristics of the catchment areas of the glaciers have changed considerably. Overall the data are so scanty that no glacial limits can be mapped.

In only one very recent study (Giraudi, 1998b) of the Campo Felice area in the Velino Massif (Central Apennines), could the relative age of some of the older moraines be determined. These 'Rissian' glaciation moraines are overlain by aeolian deposits which mainly include volcanic minerals. They contain a palaeosol. In addition, these deposits have also yielded a Mousterian chert artefact. In Central Italy Mousterian cultural remains occur from at least Marine Isotope Substage 5e (i.e. Eemian) and until the end of MIS 4 (i.e. Middle Würmian). Since the moraines predate the aeolian deposits, the palaeosol and the artefacts, it is highly improbable that the ice advance occurred during MIS 4.

According to Kotarba *et al.* (2001), Uranium Series datings have been done on calcite crystals included within moraine deposits at Campo Imperatore: the ages of 121 (+13/-12) and 135 (+10/-9,7) allow to assume a Late-Middle Pleistocene age for the older glaciation phase.

The first major ice advance in the Apennines must therefore date at least from one of the cold stages preceding MIS 5e; it must be Middle Pleistocene. Because the moraines are fairly well preserved, a Late Middle Pleistocene age is most likely (i.e.potentially MIS 6 or 8). According to Jaurand (1994), the Mount Navert moraine, another pre-Würmian feature that was preserved in a palaeolandscape, may be even older. It might indicate glaciation of the Northern Apennines before the Middle Pleistocene.

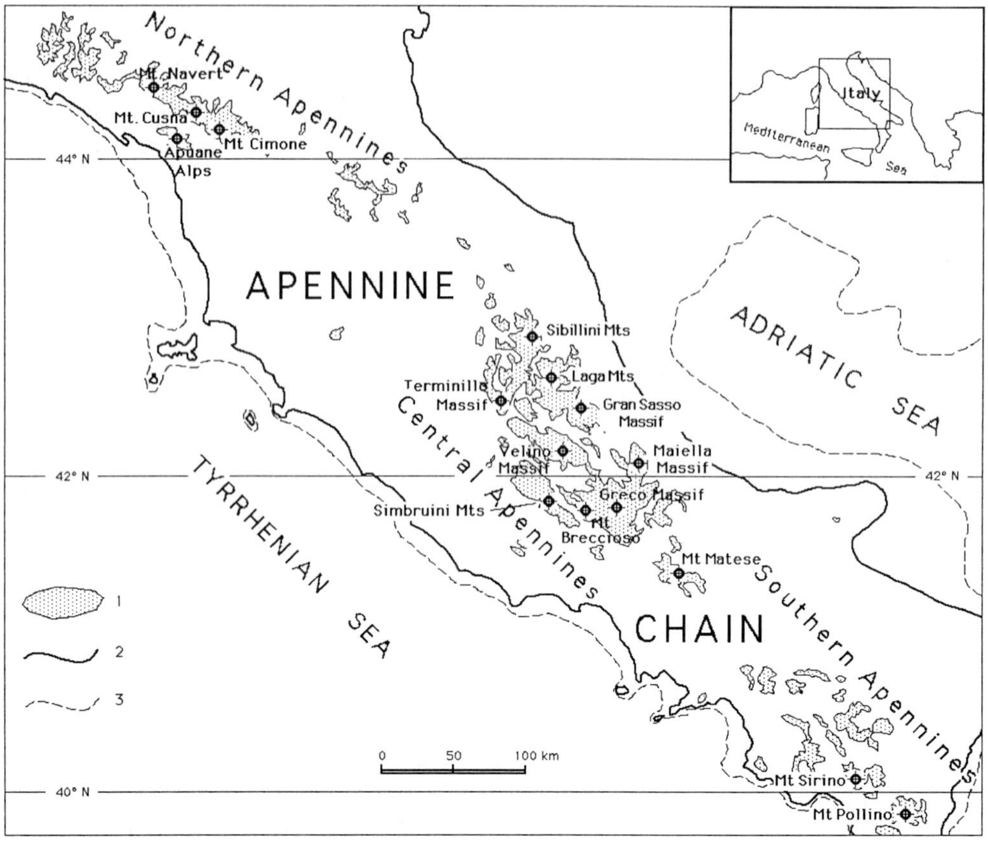

Fig. 1. Location of the Apennines Chain and of the main mountain massifs. Legend: 1 - areas above 1000 m; 2 - present shoreline; 3 - Last Glacial Maximum shoreline.

Upper Pleistocene glaciation, Last Glacial Maximum and its retreat phases

The traces of the Upper Pleistocene Glaciation have been reported in a number of papers, however only the most recent will be considered.

The works on the Northern Apennines glaciers are: Bertolini & Trevisan (1984), Braschi *et al.* (1986), Castaldini *et al.* (1998), Federici (1977, 1978, 1981), Federici & Scala (1966), Federici & Tellini (1983), Gabert (1962), Giglia (1967), Gruppo Ricerca Geomorfologia (1982), Jaurand (1994), Losacco (1982), Nardi (1961), Pelletier (1959), Suter (1950) and Trevisan *et al.* (1971). The works on the Central Apennines glaciers are by Bally (1954), Biasini (1966), Brancaccio *et al.* (1980), Cassoli *et al.* (1986), Castaldo (1970), Catenacci (1974), Cinque *et al.* (1990), Damiani (1975), Damiani & Pannuzi (1976, 1979, 1985/86a, b, 1990, 1991), Demangeot (1965), Dramis & Kotarba (1994), Dramis *et al.* (1987), Dramis *et al.* (1980), Federici (1979, 1980), Frezzotti & Giraudi (1989, 1990, 1992), Frezzotti & Narcisi (1989, 1996), Giraudi (1988, 1989a, b, 1992, 1994, 1997a, b, c, d, 1998a, b, c, d, 2000), Giraudi & Frezzotti (1995, 1997), Jaurand (1994), Mongini (1970), Palmentola (1988), Palmentola *et al.* (1990), Pfeffer (1967), Praturlon (1968), Raffy (1983), Rasse (1994) and Tagliaferro (1982). The works on the Southern Apennines glaciers are by Boenzi & Palmentola (1971, 1972a, b, 1974, 1975a, b), Giraudi (1998f), Lippmann-Provansal (1987), Palmentola & Acquafredda (1983), Palmentola *et al.* (1990) and Palmentola & Pennetta (1979).

Savelli *et al.* (1995) report the presence of glacial remains on Mount Catria (1700 m a.s.l., north-central Apennines). As the authors point out, the glacial traces on Mount Catria are somewhat problematic because they lack age control. The glacier is said to have developed in a south-facing valley and its front must have reached down to an extremely low level (about 580 m a.s.l.), far lower than the levels reached by the major glaciers of the Northern Apennines with a northern exposure. Glacial landforms are absent from the highest peak of the mountain, which is exposed to the west, and is theoretically best suited to support a glacier. Therefore, this glacier cannot be accepted until new evidence becomes available.

Upper Pleistocene glacial morphological features and deposits can almost exclusively be dated to the last glacial maximum and its retreat phases. Older outwash sediments have only been identified (Frezzotti & Giraudi, 1992) from the margins of the Velino Massif (Central Apennines). These sediments predate a fossil soil on volcanic material, dated to 33,140±1700 years BP and developed during MIS 4.

The most recent summaries of data on the late Pleistocene glaciation of the Apennines (Fig. 2), were presented by Federici (1979) for the Central and Northern Apennines, Palmentola *et al.* (1990) for the Southern Apennines and Jaurand (1994) for the whole mountain range.

Federici (1979), using the freshness of the landforms, but without any radiometric dates correlated the moraines of the maximum ice advance of the Central Apennines to the 'Würm III' of the Alps. This expansion was related by

Fig. 2. Dating of Last Glacial Maximum and stadial phases in Federici (1979), Palmentola et al. (1990) and in Jaurand (1994).

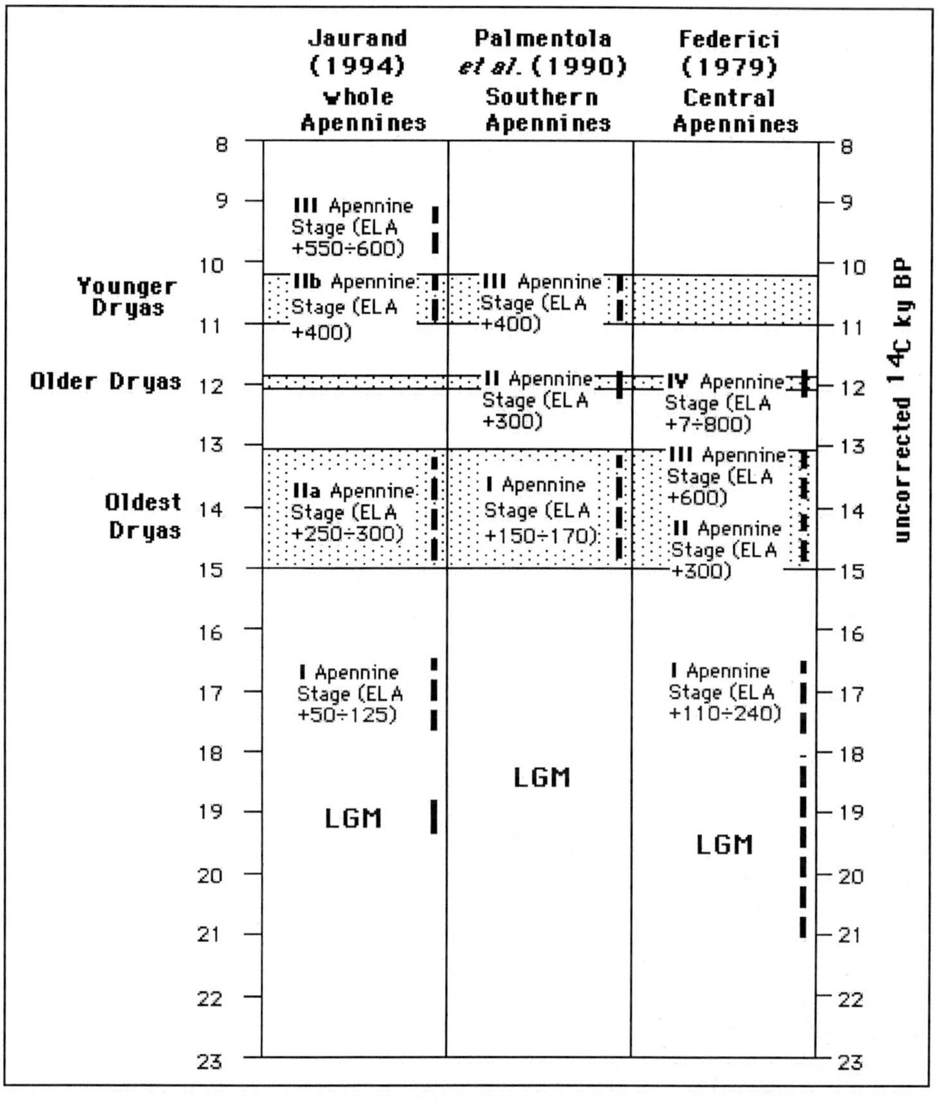

him to between *c.* 21,000 and 18,000 years BP. He also differentiated various successive retreat stages:

> Apennine Stage 1, correlated with the Alpine *Altstadien*, which is marked by a rise of the equilibrium line altitude (ELA) of 110-240 m;
>
> Apennine Stage II, correlated with the Bühl-Schlern Stages of the Alps, with ELA 300 m higher than 'Würm III';
>
> Apennine Stage III, correlated with the Gschnitz of the Alps, with ELA about 600 m higher than 'Würm III';
>
> Apennine Stage IV, correlated tentatively with the Daun of the Alps, with ELA about 700-800 m higher than 'Würm III'.

However, according to Panizza (1985), the Alpine Bühl-Schlern and Gschnitz Stages are included in the Oldest Dryas, while the Daun corresponds to the Older Dryas.

In the Southern Apennines, Palmentola *et al.* (1990), again in the absence of absolute dates, correlate the maximum ice advance to the Alpine Würm III and distinguished three successive retreat stages: Apennine Stage 1, which they correlated with the Oldest Dryas, with an ELA of about 150-170 m higher than Würm III; Apennine Stage II, which they correlated with the Older Dryas, in which an ELA of about 300 m higher than in Würm III was determined and Apennine Stage III, which they correlated with the Younger Dryas, with an ELA about 400 m higher than during Würm III.

Later, Frezzotti & Giraudi (1989, 1992) dated the late-glacial climatic phases in the area of Mount Greco and the outwash sediments of Mount Velino (Central Apennines). For the first time they were able to demonstrate that the Apennine LGM was more recent than a series of depositional and erosional events younger than *c.* 30,000 years BP but older than *c.* 15,000 years BP. Lowe (1992) dated lacustrine sediments behind a number of stadial moraines in the Northern Apennines.

Following study of the glacial events throughout the Apennines, Jaurand (1994) based his synthesis on a number of radiocarbon dates reported in Frezzotti & Giraudi (1989, 1992) and Lowe (1992).

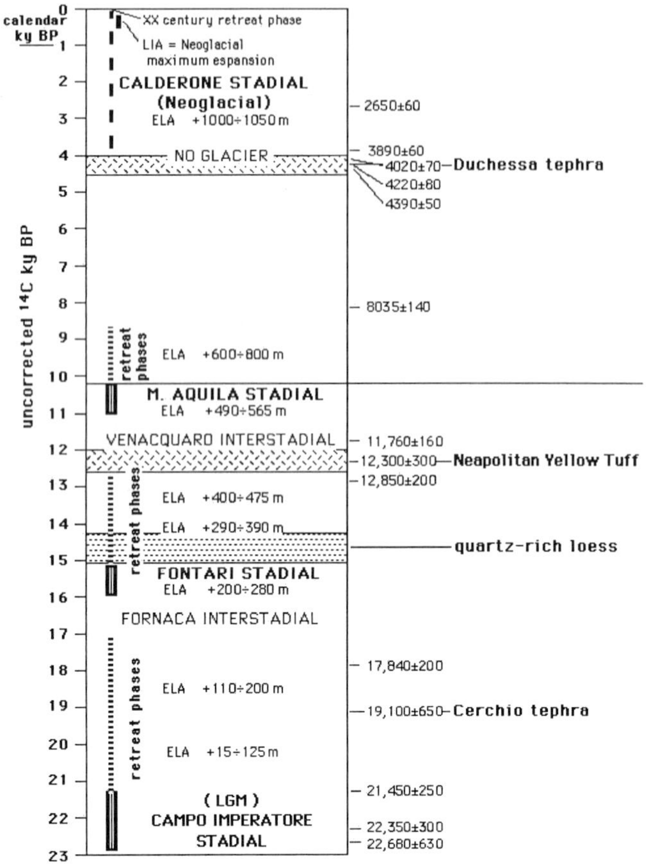

Fig. 3. Chronological framework of the Last Glacial Maximum, late Pleistocene and Holocene Apennines glacial stadials.

Jaurand (1994) attributed the LGM moraines to an ice advance which reached its maximum around 19,000 years BP (Fig. 2). He recognized four successive stages: Apennine Stage I, which corresponded to a rise of the ELA of 50-125 m; Apennine Stage IIa, correlated with the Oldest Dryas, older than 13,000 years BP, with a rise of the ELA of 250-300 m; Apennine Stage IIb, which he correlated with the Younger Dryas, with a rise of the ELA of about 400 m and Apennine Stage III, correlated with the Preboreal, with a rise of the ELA of 550-600 m with respect to that of the Last Glacial Maximum (LGM).

Regarding the Southern Apennines, Jaurand (1994) denies, on the basis of sedimentological and morphological evidence, the glacial origin of deposits in Calabria, south of Mount Pollino formerly postulated by Boenzi & Palmentola (1972b, 1974, 1975b).

More recently, Giraudi (1997a, b, c, d, 1998 a, b, c, d, 2000) and Giraudi & Frezzotti (1997) carried out more detailed investigations on a number of massifs in the Central Apennines (Gran Sasso; Greco, Terminillo, Velino, Maiella, Breccioso, Matese) and in the Southern Apennines (Sirino and Pollino).

On the Gran Sasso massif, thanks to a series of radiocarbon dates and the Neapolitan Yellow Tuff tephra (recognized by Frezzotti & Narcisi, 1989, 1996), Giraudi & Frezzotti (1997) were able to date various phases of the LGM (Campo Imperatore Stadial). The longest glacier in the Apennines had already reached its maximum extent before 22,680±630 years BP. Its glacier tongue blocked a small tributary valley, damming a proglacial lake. Glacial retreat began around 21,450±250 years BP, when thick outwash was deposited. A further phase of intense glacial melting is again dated from outwash deposits to a little less than 17,840±200 years BP. Moreover, many recessional moraines are recognized: all of which, except the highest on Mount Aquila, predate the Neapolitan Yellow Tuff tephra, dated at 12,300±300 years BP (Alessio *et al.*, 1973). The authors were able to distinguish two interstadial periods, the Fornaca Interstadial, at 16-17,000 years BP, and the Venacquaro Interstadial, radiocarbon-dated to 13,000 - 11,000 years BP. Giraudi & Frezzotti (1997) summarize the sequence of events as follows:

> In addition to the Last Glacial Maximum (LGM) moraines of the Campo Imperatore Stadial, on the Gran Sasso there are three recessional moraines that predate the Fornaca Interstadial.
>
> Afterwards these interstadial glaciers readvanced during the Fontari Stadial. A set of three recessional moraines formed prior to the subsequent Venacquaro Interstadial.
>
> This was followed by a minor glacial readvance in the Monte Aquila Stadial, after 11,760±160 years BP and before 8035±140 years BP, which may be correlated with the Younger Dryas.

Despite having no new ^{14}C dates available, Giraudi (1997a, b, c, d, 1998a, b, c, d, 2000) in his studies of other massifs in the Central-Southern Apennines recognized a number of stratigraphic markers (tephra and aeolian sediments) which are fundamental for the dating and correlation of the glacial events on the different massifs. The Cerchio Tephra, a tephra of as yet unknown origin, was identified on the moraines of the first recessional phases after the LGM (with an ELA some 125 m higher than at the LGM) of Mount Breccioso (Central Apennines). The first products of tephra reworking have been dated on the nearby Fucino Plain (Giraudi, 1995) at 19,100±650 years BP.

In all the glaciated areas studied (except for the Gran Sasso and Majella), the most recent moraines (ELA some 400-475 m higher than that of the LGM) are overlain by the Neapolitan Yellow Tuff tephra.

In all glaciated areas, but particularly on Mounts Greco and Matese (Central Apennines) and on Mount Sirino (Southern Apennines), loess was found. This loess consists mainly of quartz. It is the only loess with such a mineralogical composition known in the Central-Southern Apennines in the last 30,000 years BP. Here it overlies moraines which indicate a rise of the ELA of *c.* 200-280 m with respect to the LGM, but is not present on the younger ones. The precise age of this loess, identified for the first

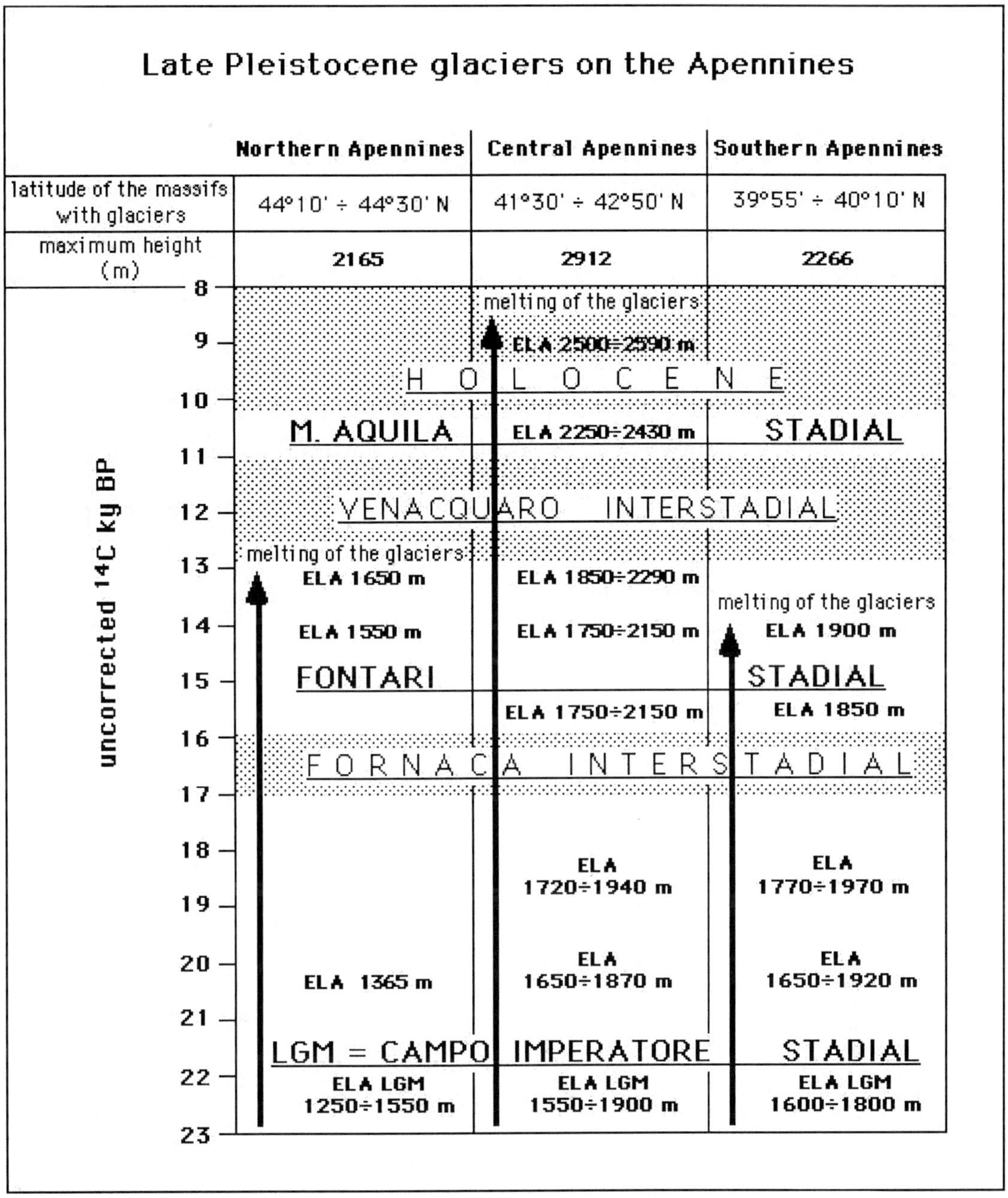

Fig. 4. Glacial stadials and Equilibrium Line Altitude (ELA) variations from Last Glacial Maximum to the early Holocene. The data from the Northern Apennines stadial ELA's are based on Val Parma glaciers facing North (from Jaurand, 1994). Maximum and minimum ELA values reported for the Central and Southern Apennines depend on the different massifs and valley orientations. On the Southern Apennines glaciers melted earlier than on the Northern Apennines despite the maximum elevation being higher. On the Northern Apennines the ELA is clearly lower than that on the Central Apennines, while there is a small difference between Central and Southern Apennines.

time by Frezzotti & Giraudi (1989, 1990), is not known. The date is based on the following criteria (Fig. 3):

> On the Aremogna Plain (Mount Greco - Central Apennines) the loess lies on outwash sediments deposited during the LGM recession. On the basis of the data collected by Giraudi & Frezzotti (1997) these deposits must predate 16,000 years BP. The loess, which has been subjected to soil formation, was redeposited as colluvium and subsequently overlain by peat dated at 12,850±200 years BP (Frezzotti & Giraudi, 1989).
>
> On Mount Matese (Central Apennines) the loess is older than a tephra layer (still under investigation) that predates the Neapolitan Yellow Tuff (Giraudi, 1997a). It may originate from an eruption of the Campi Flegrei or to the Greenish eruption of Mount Vesuvius, datable at *c.* 14,000 years BP.

It is therefore assumed that the loess was deposited between about 15-14,000 years BP. The moraines, which indicate a rise of the ELA of 200-280 m with respect to the LGM, are therefore older than 15,000 years BP.

It is clear (Fig. 3) that the dates obtained generally show an earlier age for the glacial events than assumed by Jauraud (1994). They also differ from the hypotheses proposed by Federici (1979) and Palmentola *et al.* (1990) who lacked radiometric dates.

The difference with Jaurand (1994) regarding the more recent glacial advances derives basically from the data on the Northern Apennines. In the Parmense section of the Apennines, the moraines of Jaurand's (1994) Apennines

Stage IIb (ELA *c.* 400 m higher than at the LGM) are overlain by coarse deposits and then by lacustrine sediments, dated at their base to 10,610±45 years BP (Lowe, 1992). Jaurand (1994) correlates these moraines with the Younger Dryas. He consequently considers the successive moraines of the Apennines Stage III to be Holocene. His dates, however, only indicate that the Stage IIb moraines are older than 10,610±45 years BP; they give no indication of how much older they are.

Holocene Neoglaciation

The presence of glaciers in the Appenines during the Holocene is shown by the Calderone Glacier. This glacier, located on the Gran Sasso Massif, is the southernmost in Europe (Gellatly *et al.*, 1994). It is found at an elevation above *c.* 2670 m in a cirque situated on the northern slope of the Corno Grande (2912 m). It is rapidly melting and almost completely covered by debris, but it is still some 15 m thick in its lower part (Gellatly *et al.*, 1994; Fiucci *et al.*, 1997). The glacier is situated at the head of the Cornacchie Valley. Here new investigations have begun; in the upper part there are some moraines, which were formed by glaciers with an ELA more than 800-1050 m higher than that of the LGM. They must be of Holocene age, because the Younger Dryas ELA was at least 250 m lower.

There are no precise dates for these oldest Holocene moraines: the only chronological indication (Fig. 3) is provided by a sequence of colluvial sediments and soils overlying the till that include a tephra layer. The tephra, of uncertain provenance, may be correlated mineralogically with the Duchessa Tephra. This unit has also been found on other moraines in the Central Apennines, and overlies soils dated to 4390±50 (Beta 117017), 4220±80 (Beta 106450) and 4020±70 (Beta 111004) years BP. The moraines, which indicate a glacier with an ELA about 800 m higher than that of the LGM, may possibly represent the recessional phases of the Mount Aquila Stadial dated to the Younger Dryas Chron.

The moraines situated at the threshold of the Calderone Glacier cirque, with their ELA about 1000-1050 m higher than the LGM, have been attributed to the Little Ice Age (LIA) by various authors (Federici, 1979; Jaurand, 1994; Gellatly *et al.*, 1994). In the course of new observations on these moraines, three different tills have been recognized. The oldest till, almost completely covered by the successive units, has a dark grey matrix of silt rich in organic matter, derived from a soil on volcanic material. Its mineralogical composition is analogous to that of the Duchessa Tephra. The organic matter has been dated to 3890±60 years BP. Consequently, the glacial advance that formed the moraine must be younger than that date. In view of the very steep sides of the glacial cirque, the soil could only be derived from the cirque, which is currently occupied by the glacier. Therefore, starting from about 4300 years BP and until around 3890±60 years BP, the Calderone glacier was absent or was definitely much smaller than at present.

There are no means to date the second till. However, on the basis of its scanty plant cover it must be much more recent. The third till unit, which is best preserved and not covered by vegetation, cuts through the preceding one and is the most extensive, but upstream it connects with the present till. Since part of this till corresponds to the front of the glacier tongue at the end of the 19th century, it may be dated to the LIA. In an exposure a few hundreds of metres from the glacier a soil has developed on the Duchessa Tephra, that is covered by debris. The upper horizon of the soil has been radiocarbon-dated to 2650±60 years BP. The overlying deposits indicate a cold phase and, probably, a glacial readvance.

Thus, the following sequence of events has been deduced: the early Holocene saw the recessional phases of the Mount Aquila Stadial glaciers. These glaciers either melted completely or became reduced to a smaller size than at present. Subsequently they expanded, after 3890±60 years BP, during the Calderone Stadial and reached their maximum extent during the LIA.

Moraines colonized, by only very little vegetation, are also preserved at high altitudes at a number of valley heads in the Maiella Massif (Central Apennines). They have been attributed to the Holocene (Giraudi, 1998e). At the head of the Cannella Valley, an extremely small moraine situated at 2540 m is almost completely devoid of vegetation, comparable to the till of the Calderone Glacier attributed to the LIA above-mentioned. It is therefore probable that in this valley a small glacier also formed during the LIA. Traces of neoglaciation are probably found on only the two highest massifs of the Apennines, since the peaks of the other massifs are below the altitude of the Calderone Stadial moraines.

Conclusions

Knowledge of the Apennine glaciations may be summarized as follows: it is certain that in the Apennines there are traces of at least one glacial expansion older and more extensive than that of the Last Glacial Maximum (LGM). Its age is not known in detail, but it is highly likely that it dates from the Middle Pleistocene, perhaps from the final part. An additional earlier glacial phase might also be represented. The Late Pleistocene glaciation has left few traces relating to Marine Isotope Stage 4, but abundant sediments and features relating to the LGM and to its recessional phases are found. Radiometric dating of these glacial oscillations is still rather scanty. All the main massifs in the Central-Southern Apennines consist of carbonate rocks. Karstic processes, the grain size of the glacial deposits and their rare matrix have meant that except in very rare cases no lakes were formed during the glacial recession. There is thus extremely little organic matter to be dated. Offsetting this, in the last few years tephra layers and some aeolian deposits have been found which, being isochronous, make it possible to obtain a chronological setting and a precise correlation even between the moraines situated at latitudes and elevations very different from one another.

The Upper Pleistocene LGM (Late Würmian) glaciers started to expand earlier than 22,680±630 years BP (Campo Imperatore Stadial), and began to retreat slowly around 21,500 years BP, and then more rapidly until about 16,000 years BP. Shortly after 16,000 years BP there was a readvance. This Fontari Stadial was followed by recessional phases lasting until after 11,760±160 years BP. Another readvance followed, the Mount Aquila Stadial, which may be correlated with the Younger Dryas. The recessional phases of this stadial are probably represented by the earliest Holocene moraines.

Regarding the Holocene glacial oscillations and the Neoglacial, it must be borne in mind that the data are scarce, also because there are extremely few places that lend themselves to investigation. The latter are restricted to small areas of the peaks of the two highest massifs in the Central Apennines. The glaciers present in the initial phase of the Holocene melted or were reduced to a smaller size than at present. After this they reformed or re-expanded in the second half of the Holocene (Calderone Stadial) and reached their maximum extension during the course of the Little Ice Age. The Calderone Glacier, the only one that now exists in the Apennines, has strongly melted back during the 20th century.

References

Alessio, M., Bella, F., Improta, S., Cortesi, C. & Turi, B. (1973). University of Rome carbon-14 dates, IX. *Radiocarbon*, **15**, 165-178.

Bally, A. (1954). *Geologische Untersuchungen in den SE-Abruzzen: Gebiet zwischen der Ebene von Sulmona und der Taler des Pescara und des unteren Sangro*. Zürich, Bühler Buchdruck, 291 pp.

Bertolini, N. & Trevisan, L. (1984). Ghiacciai würmiani sul Monte Molinatico (Appennino settentrionale). *Atti della Società Toscana di Scienze naturali*, **A, 91**, 181-187.

Biasini, A. (1966) Elementi morfotettonici, tratti da un rilievo fotogeologico, al margine dell'altipiano di Ovindoli (Abruzzo). *Geologica Romana*, **5**, 303-312.

Bisci, C., Ciccacci, S., Dramis, F. & Kotarba, A. (1999). The extent of Pleistocene glaciations in the western part of Campo Imperatore (Gran Sasso Massif, Central Italy). *Landform Analysis*, **2**, 37-43.

Boenzi, F. & Palmentola, G. (1971). Tracce della glaciazione würmiana sul Massiccio del Pollino al confine Calabro-Lucano. *Bollettino Società Geologica Italiana*, **90**, 139-150.

Boenzi, F. & Palmentola, G. (1972a). Nuove osservazioni sulle tracce glaciali nell'Appennino lucano. *Bollettino del Comitato Glaciologico Italiano, nuova serie*, **20**, 9-52.

Boenzi, F. & Palmentola, G. (1972b). Alcune considerazioni di neotettonica sull'Appennino calabro-lucano in base al ritrovamento di tracce glaciali sul M. La Mula in Calabria. *Bollettino della Società Naturalisti in Napoli*, **81**, 485-498.

Boenzi, F. & Palmentola, G. (1974). Prima segnalazione di tracce glaciali würmiane nei dintorni di Monte Botte Donato (Sila). *Bollettino Società Geologica Italiana*, **93**, 309-315.

Boenzi, F. & Palmentola, G. (1975a). Osservazioni sulle tracce glaciali della Calabria. *Bollettino Società Geologica Italiana*, **94**, 961-977.

Boenzi, F. & Palmentola, G. (1975b). Effetti di fenomeni glaciali quaternari nella Calabria nord-occidentale. *Bollettino Società Geologica Italiana*, **94**, 871-875.

Brancaccio, L., Sgrosso, I., Cinque, A., Orsi, G., Pece, R. & Rolandi, G. (1980). Lembi residui di sedimenti lacustri pleistocenici sul versante settentrionale del Matese presso S. Massimo. *Bollettino Società Naturalisti in Napoli*, **88**, 275-286.

Braschi, S., del Freo, P. & Trevisan, L. (1986). Ricostruzione degli antichi ghiacciai sulle Alpi Apuane. *Atti Società Toscana di Scienze Naturali*, **A, 93**, 203-219.

Cassoli, A., Corda, L., Lodoli, C., Malatesta, A., Molaroni, M.V. & Ruggeri, A. (1986). Il glacialismo quaternario del gruppo Velino-Ocre-Sirente. *Memorie Società Geologica Italiana*, **35**, 855-867, 5 ff.

Castaldini, D., Caredio, F. & Puccinelli, A. (1998). Geomorfologia delle valli del rio delle Pozze e del torrente Motte (Abetone, Appennino Tosco-Emiliano). *Geografia Fisica Dinamica Quaternaria*, **21**, 177-204.

Catenacci, V. (1974) *Note illustrative della Carta Geologica d'Italia alla scala 1:100,000, Foglio 147 Lanciano*. Roma, Servizio Geologico d'Italia, 87 p.

Cinque, A., Liccardo, C., Palma, B., Pappalardo, L., Rosskopf, C. & Sepe, C. (1990). Le tracce glaciali nel Parco Nazionale d'Abruzzo (Appennino Centrale): nota preli-minare. *Geografia Fisica Dinamica Quaternaria*, **13**, 121-133.

Coltorti, M. & Farabollini, P. (1995). Quaternary evolution of the Castelluccio di Norcia basin (umbro-Marchean Apennines, Central Italy). *Il Quaternario*, **8(l)**, 149-166.

Damiani, A.V. (1975). Aspetti geomorfologici e possibile schema evolutivo dei M. Sibillini (Appennino umbro-marchigiano). *Bollettino Servizio Geologico d'Italia*, **96 (2)**, 231-314.

Damiani, A.V. & Pannuzi, L. (1976). La glaciazione würmiana nell'Appennino laziale-abruzzese (1): il ghiacciaio del Fosso di S. Onofrio nei Monti Sombruini-Ernici. *Bollettino Servizio Geologico d'Italia*, **97**, 85 106.

Damiani, A.V. & Pannuzi, L. (1979). La glaciazione würmiana nell'Appennino laziale-abruzzese (2): I ghiacciai della Valle Granara e della Fiumata (Alta Valle dell'Aniene). *Bollettino Servizio Geologico d'Italia*, **100**, 287-310.

Damiani, A.V. & Pannuzi, L. (1985/86a). La glaciazione würmiana nell'Appennino laziale-abruzzese (3): opportunità di precisazioni terminologiche, metodo-logiche ed introduzione di parametri morfometrici. *Bollettino Servizio Geologico d'Italia*, **105**, 75-96.

Damiani, A.V. & Pannuzi, L. (1985/86b). La glaciazione würmiana nell'Appennino laziale-abruzzese (4): I ghiacciai del gruppo montuoso della Genzana (Scanno Sulmona). La geomorfologia dell'alta Val Parma (Appennino Settentrionale). *Rivista Geografica Italiana*, **90 (3-4)**, 393-428.

Fiucci, A., Gigante, B., Rossi, C., Smiraglia, C. & Veggetti, O. (1997). The Calderone Glacier (Gran Sasso d'Italia). Determination of ice thickness and bedrock morphology by means of radio-echo sounding. *Geografia Fisica Dinamica Quaternaria*, **20**, 305-308.

Frezzotti, M. & Giraudi, C. (1989). Evoluzione geologica tardo-pleistocenica ed olocenica del Piano di Aremogna (Roccaraso Abruzzo): implicazioni climatiche e tettoniche. *Memorie Società Geologica Italiana*, **42**, 5-19.

Frezzotti, M. & Giraudi, C. (1990). Late Glacial and Holocene aeolian deposits and features near Roccaraso (Abruzzo, Central Italy). *Quaternary International*, **5**, 89-95.

Frezzotti, M. & Giraudi, C. (1992). Evoluzione geologica tardo-pleistocenica ed olocenica del conoide complesso di Valle Majelama (Massiccio del Velino Abruzzo). *Il Quaternario*, **5 (1)**, 33-50.

Frezzotti, M. & Narcisi, B. (1989). Identificazione di un andosuolo, possibile livello guida per la cronostratigrafia olocenica dell'Appennino Centrale. *Memorie Società Geologica Italiana*, **42**, 351-358.

Frezzotti, M. & Narcisi, B. (1996). Late Quaternary tephra-derived paleosols in Central Italy's carbonate Apennine range: stratigraphical and paleoclimatological implications. *Quaternary International*, **34-36**, 147-153.

Gabert, P. (1962). Les plaines occidentales du PO et leurs piedmonts. Thèse.Gap, Impr. Louis Jean, 531 pp.

Gellatly, A.F., Smiraglia, C., Grove, J.M. & Latham, R. (1994). Recent variations of Ghiacciaio del Calderone, Abruzzi, Italy. *Journal of Glaciology*, **40**, 486-490.

Giglia, G. (1967). Geologia dell'alta Versilia settentrionale (Tav. M. Altissimo). *Memorie Società Geologica Italiana*, **6**, 67-95.

Giraudi, C. (1988). Segnalazione di scarpate di faglia post-glaciali nel massiccio del Gran Sasso (Abruzzo): implicazioni tettoniche, rapporti tra tettonica recente e morfologia, paleosismicità. *Memorie Società Geologica Italiana*, **41**, 627-635.

Giraudi, C. (1989a). Datazione di un evento sismico preistorico con metodi geologici e radiometrici: Piani di Aremogna e delle Cinque Miglia. *In:* I terremoti prima del 1.000: storia, archeologia, sismologia. Istituto Nazionale di Geofisica, 53-65. Bologna, Ed. S.G.A.

Giraudi, C. (1989b). Datazione con metodi geologici delle scarpate di faglia post-glaciali di Ovindoli Piano di Pezza (Abruzzo Italia Centrale): implicazioni. *Memorie Società Geologica Italiana*, **42**, 29-39.

Giraudi, C. (1992). Segnalazione di scarpate di faglia tardo-pleistoceniche sui Monti della Magnola (Massiccio del Velino Abruzzo). *Il Quaternario*, **5 (1)**, 27-32.

Giraudi, C. (1994). Elementi di geologia del Quaternario della Piana di Campo Imperatore (Massiccio del Gran Sasso Italia Centrale). *Atti Ticinesi Scienze della Terra (Serie speciale)*, **2**, 137-143.

Giraudi, C. (1995). I detriti di versante ai margini della Piana del Fucino (Italia Centrale): significato paleoclimatico ed impatto antropico. *Il Quaternario*, **8 (1)**, 203-210.

Giraudi, C. (1997a). Dating and correlation of glacial deposits using tephra layers and loess: the example of M. Matese (Campania-Molise) Southern Italy. *Studia Geomorphologica Carpatho-Balcanica*, **XXXI**, 45-56.

Giraudi, C. (1997b). Le oscillazioni tardo-pleistoceniche ed oloceniche del lago effimero del Piano di Pezza (Abruzzo Italia Centrale). *Il Quaternario*, **10 (2)**, 191-200.

Giraudi, C. (1997c). I depositi post-glaciali della bassa Valle Peschio di Iorio (Pescasseroli, Parco Nazionale d'Abruz-zo). Datazione ed interpretazione paleoclimatica. *Il Quaternario*, **10 (2)**, 207-212.

Giraudi, C. (1997d). Prima segnalazione dell'apparato glaciale del M. Breccioso (Val Roveto Abruzzo): un contributo all'inquadramento cronologico delle fasi glaciali tardo-pleistoceniche. *Il Quaternario*, **10 (2)**, 201-206.

Giraudi, C. (1998a). Alcuni dati per l'inquadramento cronologico delle fasi glaciali tardo-pleistoceniche dei Monti Greco e Serra Chiarano (Abruzzo Italia Centrale). *Il Quaternario*, **11 (1)**, 115-120.

Giraudi, C. (1998b). The Late Quaternary geologic evolution of Campo Felice (Abruzzo Central Italy). *Giornale di Geologia, sez. 3*, **60**, 67-82.

Giraudi, C. (1998c). I laghi effimeri tardopleistocenici ed olocenici di Campo Imperatore e del Massiccio del Gran Sasso d'Italia (Abruzzo Italia Centrale). *Il Quaternario*, **11 (2)**, 1-9.

Giraudi, C. (1998d). Il glacialismo tardo-pleistocenico del Massiccio del Terminillo (Lazio Appennino Centrale). *Il Quaternario*, **11 (1)**, 121-125.

Giraudi, C. (1998e). Nuovi dati sul glacialismo della Montagna della Maiella (Abruzzo-Italia Centrale). *Il Quaternario*, **11 (2)**, 265-271.

Giraudi, C. (1998f). La deglaciazione tardopleistocenica sui M. Sirino e Pollino (Basilicata, Calabria-Italia Meridionale). *Il Quaternario*, **11 (2)**, 247-254.

Giraudi, C. (2000). Le oscillazioni oloceniche del Ghiacciaio del Calderone, Gran Sasso d'Italia (Abruzzo Italia). *Il Quaternario*, **13 (1-2)**, 31-36.

Giraudi, C. & Frezzotti, M. (1995). Palaeoseismicity in the Gran Sasso Massif (Abruzzo, Central Italy). *Quaternary International*, **25**, 81-93.

Giraudi, C. & Frezzotti, M. (1997). Late Pleistocene glacial events in the Central Apennine, Italy. *Quaternary Research*, **48 (3)**, 280-290.

Gruppo Ricerca Geomorfologia C.N.R. (1982). Geomorfologia del territorio di Febbio tra il M. Cusna e il F. Secchia (Appennino emiliano). *Geografia Fisica Dinamica Quaternaria*, **5**, 285-360.

Jaurand, E. (1994). Les heritages glaciaire de l'Apennin. Thèse pour le Doctorat dès Lettres de l'Université de Paris I Panthéon-Sorbonne. 600 pp.

Kotarba, A., Hercman, H. & Dramis, F. (2001). On the age of Campo Imperatore glaciations, Gran Sasso Massif,

Central Italy. *Geografia Fisica e Dinamica Quaternaria*, **24 (1)**, 65-69.

Lippmann-Provansal, M. (1987). L'Apennin campanien méridional (Italie): étude géomorphologique. Thèse inédite. Université d'Aix-Marseille.II, 2t, 770 p.

Lowe, J.J. (1992). Late glacial and early sediments from the northern Apennines, Italy: pollen stratigraphy and radiocarbon dating. *Boreas*, **21 (3)**, 193-208.

Losacco, U. (1982). Gli antichi ghiacciai dell'Appennino settentrionale: studio morfologico e paleogeografico. *Atti Società Naturalisti Matematici Modena*, **113**, 1-224.

Mongini, G.M. (1970). La morfologia del gruppo orografico del Vettore, Monti Sibillini. *Roma, Istituto di Geografia dell'Università di Roma, pubbl. ser. A*, **10**, 43 pp.

Nardi, R. (1961). Geologia della zona tra la Pania della Croce, Gallicano e Castelnuovo Garfagnana (Alpi Apuane). *Bollettino Società Geologica Italiana*, **80 (2)**, 257-334.

Palmentola, G. & Acquafredda, P. (1983). Gli effetti dei ghiacciai quaternari sulla Montagna del Matese, al confine Molisano-Campano. *Geografia Fisica Dinamica Quaternaria*, **6**, 117-130.

Palmentola, G., Acquafredda, P. & Fiore, S. (1990). A new correlation of the glacial moraines in the Southern Appennines, Italy. *Geomorphology*, **3**, 1-8.

Palmentola, G. & Pennetta, L. (1979). Le tracce glaciali quaternarie: un aspetto della montagna lucana. *Atti dell'8 Simposio Nazionale dell'Istituto di Zoologia dell'Università di Bari sulla conservazione della natura, 26-26 Aprile 1979*, 59-69.

Panizza, M. (1985). Schemi cronologici del Quaternario. *Geografia Fisica Dinamica Quaternaria*, **8**, 44-48.

Pelletier, J. (1959). Le relief de l'Appenin septentrional: fast et problèmes généraux de la partie occidentale. *Revue de Géographie de Lyon*, **34 (2)**, 89-111.

Pfeffer, K.H. (1967). Beitrage zur Geomorphologie der Karstbecken im Bereiche des Monte Velino (Zentralapennin). *Frankfurter Geographische Hefte*, **42**, 86p.

Praturlon, A. (1968). *Note illustrative della carta geologica d'Italia alla scala 1:100.000, Foglio 152, Sora.* Roma, Servizio Geologico d'Italia, 76pp.

Raffy, J. (1983). Le versant tyrrheniѐn de l'Apennin central: étude géomorphologique. Thèse, Saint-Chamond, Impr. G. Montet, 442 pp.

Rasse, M. (1994). L'Apennin ombrien: morphogénèse d'une dorsale récente. Thèse ined. Université de Paris IV Sorbonne, 2t, 538pp.

Savelli, D., Nesci, O. & Basili, M. (1995). Evidenze di un apparato glaciale pleistocenico sul Massiccio del Catria (Appennino Marchigiano). *Geografia Fisica Dinamica Quaternaria*, **18**, 331-335.

Suter, K. (1950). Die eiszeitliche Vergletscherung des Nordapennins. *Vierteljahrsschrift der Naturforschenden Gesellschaft in Zürich*, **95**, 224-241.

Tagliaferro, M.C. (1982). Lembo morenico presso Arquata del Tronto. *Bollettino Società Geologica Italiana*, **101 (2)**, 211-218.

Trevisan, L., Dallan, L. & Federici, P.R. (1971). *Note illustrative della Carta Geologica d'Italia alla scala 1:100.000, Foglio 96, Massa.* Roma. Servizio Geologico d'Italia, 57 pp.

Quaternary Glaciations – Extent and Chronology
Editors J. Ehlers and P.L. Gibbard

Deglaciation history of Latvia

Vitālijs Zelčs and Aivars Markots

Faculty of Geographical and Earth Sciences, University of Latvia, Raiņa Boulevard 19

Rīga, LV-1586, Latvia

Historical background

The deglaciation of Latvia has been studied since the early years of the last century, especially by Doss (1910), Hausen (1913), Philipp (1921), Weimarn (1930), Bucharts (1935) and Zāns (1936). Later Āboltiņš (1963, 1970, 1972, 1975, 1989), Āboltiņš *et al.* (1969, 1972a, b, 1975, 1976), Āboltiņš & Zelčs (1988), Danilāns (1963, 1965), Meirons (1972a, 1975a), Meirons *et al.* (1976), Straume (1979, 1984), Veinbergs (1972), Veinbergs *et al.* (1965) and Zelčs (1997, 1998) described glacial landforms and discussed processes of ice decay and glacial retreat. The reconstruction of the principal lobate and interlobate areas and correlation of glacial retreat within the Baltic region were discussed by Āboltiņš *et al.* (1974a), Āboltiņš *et al.* (1977a, b, c), Chebotareva (1965a), Chebotareva *et al.* (1965a), Chebotareva (1972), Chebotareva & Makaricheva (1974), Punning *et al.* (1967, 1968), Raukas *et al.* (1995a, b), Serebryanny & Raukas (1967) and Zarina & Krasnov (1965). Chronostratigraphical correlation of the ice-marginal formations was attempted by Arslanov (1975), Chebotareva (1965b); Chebotareva & Malgina (1965), Gaigalas & Raukas (1989), Punning *et al.* (1967, 1968), Serebryanny & Raukas (1966), Vidgorchik *et al.* (1970) and Voznyachuk (1972).

Various concepts and criteria have been applied to the deglaciation of the last ice sheet in Latvia. Initially, the ice decay was interpreted as frontal retreat of ice lobes affected only by climatic changes (Hausen, 1ė913; Zāns, 1936). Jaunputnin (1956) and Danilāns (1963, 1965) envisioned areal deglaciation and stressed the importance of dead ice for the formation of glacial landforms. However, Āboltiņš (1970), Āboltiņš *et al.* (1969, 1972b), Veinbergs (1972) and Veinbergs *et al.* (1965) applied a different model based on oscillations of the ice front and large areas covered by passive and dead ice during deglaciation.

The earliest deglaciation schemes of Latvia were based entirely on morphological features. The spatial distribution of glacial landforms, with crests transverse to glacier movement combined with radiocarbon dates and biostratigraphical evidence, have provided the principal data for the reconstruction of the deglaciation process from the mid-1960s until the present (e.g. Āboltiņš *et al.*, 1972a, b; Meirons *et al.*, 1976; Straume 1979; Savvaitov & Veinbergs (1996, 1998, 1999). Glacial landforms, striking parallel to the main direction of ice movement, dispersion fans of indicator boulders, clast fabrics and glacial striations, have been frequently used as indicators of the ice-flow direction (Āboltiņš *et al.*, 1969, 1972b; Dreimanis, 1935; Gaigalas *et al.*, 1967; Markots, 1986; Spriņģis *et al.*, 1963; Straume, 1979; Viiding *et al.*, 1971; Zāns, 1935, 1936; Zelčs, 1993; Zelčs & Dreimanis, 1997; Zelčs *et al.*, 1990). Subglacial glaciotectonic deformations and landforms, which are widespread in Latvia, have been used for palaeoglaciological reconstruction insufficiently and with variable success (Āboltiņš, 1989; Dreimanis, 1935, 1999; Zelčs, 1998, 1999; Zelčs & Dreimanis, 1997; Zelčs *et al.*, 1990).

Overview of the surface topography and glacial dynamics

Latvia is located on the slightly undulating northwestern margin of the East European Plain, that is characterised by moderate variations in elevation. The average height is 87 m. The highest point is Gaiziņkalns (312 m a.s.l.) in the Vidzeme Upland (Fig. 1). Almost 75% of Latvia lies below 120 m a.s.l., and elevations higher than 200 m are restricted to less than 3% of the area. The average local relief rarely exceeds 10-25 m with a maximum up to 90 m in glacial uplands and some places along proglacial spillways.

The present-day topography has largely been formed as a result of Pleistocene glaciations, particularly of the last Weichselian event. More than 60% are glacial lowlands, which occupy large-scale depressions of a sub-Quaternary surface. Most glacial uplands are insular-shaped and bedrock-cored. Uplands and lowlands can be considered as the largest scale or macrorelief glacial landforms. Time-transgressive and regressive processes of glacial accumulation, with particular importance of selective glacial erosion, glaciotectonics and proglacial meltwater activity, have resulted in the formation of lowlands, while proglacial and subglacial accumulation and glaciotectonics predominated in the glacial uplands. According to Āboltiņš (1972, 1975) and Zelčs (1993, 1997), three varieties of glacial uplands and three main types of lowlands can be distinguished in Latvia, based on hypsometric position, thickness and structure of the Pleistocene cover, and glacial topography (Fig.1.).

Latvia occurs at the inner margin of the depositional zone of the Fennoscandian ice sheet where the main features of glacial topography were created by subglacial processes (Āboltiņš & Zelčs, 1988). During the last

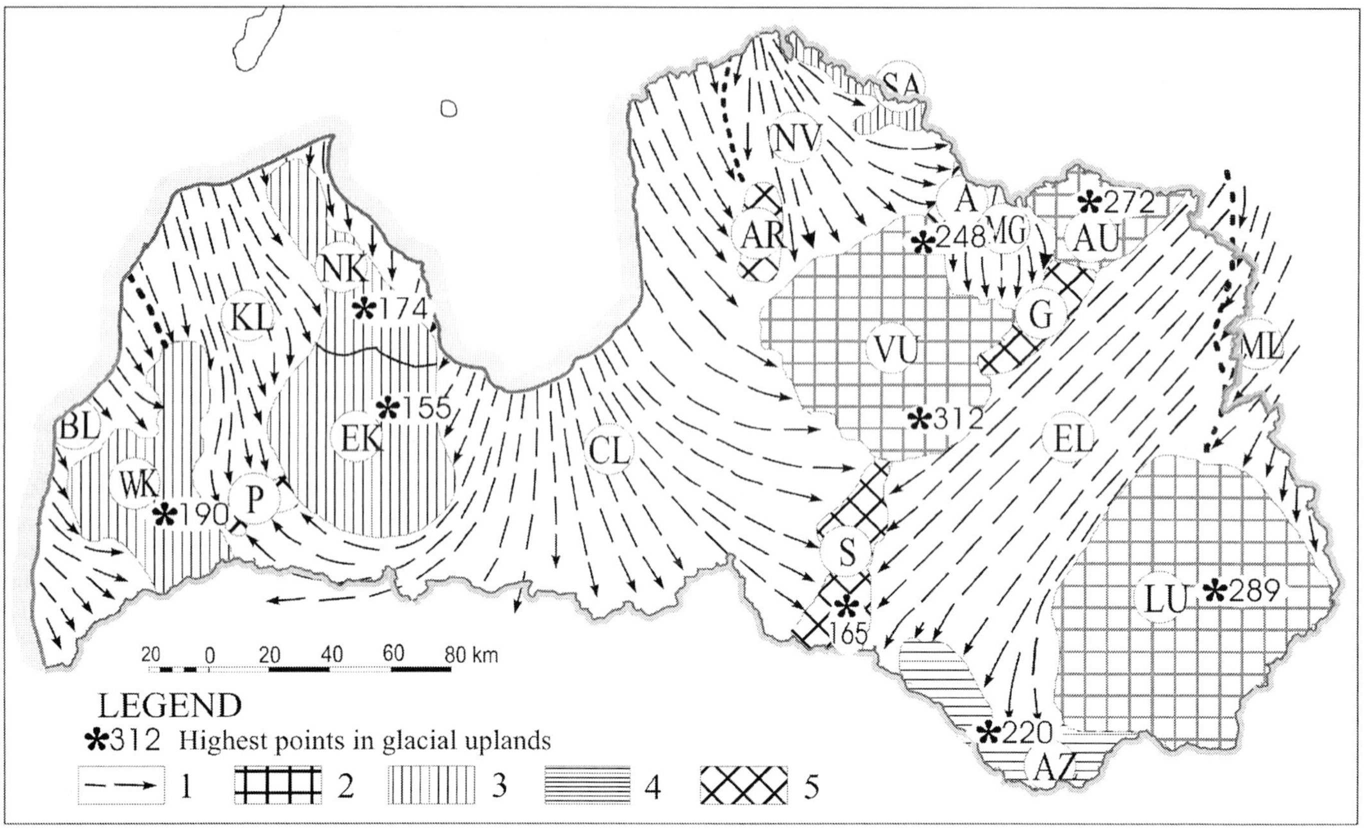

Fig. 1. Glacial geomorphological regions of Latvia.

1 = Glacial lowlands:
Divergent type: BL = Baltic, CL = Central Latvian, NV = North Vidzeme. Convergent type: KL = Kursa, ML = Mudava. Consequent type: EL = Eastern Latvian.
Glacial uplands:
2 = Isometric insular type: VU = Vidzeme, AU = Alūksne, LU = Latgale.
3 = Radial insular type: WK = Western Kursa, EK = Eastern Kursa, NK = Northern Kursa, SA = Sakala.
4 = Marginal upland: AZ = Augšzeme.
5 = Interlobate ridges: A = Aumeistari, AR = Augstroze, G = Gulbene, P = Pampāļii, S = Sēlija.
Arrows indicate main ice flow directions in the lowlands.

glaciation it was affected by the Baltic, Rīga and Peipsijärv (Peipsi Lake) ice streams. The ice streams terminated in lobes and tongues separated by interlobate zones (Fig. 2). The insular uplands and interlobate ridges represent zones of collision of ice lobes and glacier tongues moving in different directions and through separate neighbouring lowlands (Fig. 1).

The formation and location of ice lobes and glacier tongues, and their dynamics were originally controlled by subglacial bedrock (Danilāns 1972a; Raukas *et al.*, 1995a, b; Zāns, 1936; Zelčs, 1993), but during the last glaciation mainly by the pre-Weichselian topography (Straume, 1979). The influence of the subglacial topography increased particularly during deglaciation, as ice thickness decreased. A very complex lobate structure, with many small glacier tongues and sub-tongues, existed at the onset of ice-sheet decay (e.g. the 'insular stage' of Āboltiņš *et al.*, 1972a). Later, during the 'lobate stage' of Āboltiņš *et al.* (1972a), the ice dynamics were simplified and only the largest radial ice lobes and glacier tongues remained active in the lowlands. This reactivation of the lowland ice lobes and glacier tongues was induced not only by climatic and environmental changes but was also caused by the melting and stagnation of the glacier in the adjacent uplands that improved ice mass balance in the lowlands (Zelčs, 1993; Zelčs & Dreimanis, 1997). As a result of deglaciation of stagnant ice, a complex of superimposed glacial landforms in upland areas formed simultaneously with the glacial landforms continuum created by active ice fluctuations in lowlands. Later, related to melting of stagnant ice sedimentation and landform processes occurred in lowlands. The deglaciation of Latvia was controlled by different glacial geological processes interacting both spatially and temporally.

Latvia became ice-free in Daniglacial time, about 16,000-18,000 years BP (Raukas *et al.*, 1995a, b). According to the summary of the IGCP-253 project (Lundqvist & Saarnisto, 1995), the ice sheet finally retreated from Latvia during the Late Weichselian Late-glacial Interstadial, about 14-12 ka BP, but periglacial conditions persisted until the beginning of the Holocene.

Glacial landforms and ice-marginal formations

As emphasised by Hättestrand (1997), the success of using glacial landforms for the reconstruction of former ice sheets depends upon an understanding of the processes of their formation (or preservation). For establishing a deglaciation history, attention must be paid not only to ice-marginal formations but also to other glacial, particularly subglacial landforms.

During recent years, detailed investigations have been undertaken to clarify the relationship between glacial topography and internal composition of the glacial landforms in Latvia (Āboltiņš, 1989; Āboltiņš & Dreimanis, 1995; Strautnieks, 1998; Zelčs, 1993, 1999, 2000; Zelčs & Dreimanis, 1997; Zelčs *et al.*, 1990). These investigations were supplemented by the ongoing Latvian Glaciotectonic Database and Map Project. It has been found that glacial landforms composed of glaciotectonically-deformed glacial and non-glacial strata are predominant throughout Latvia. Most positive glaciotectonic landforms in Latvia are of subglacial origin. They usually contain more stratified beds, including till units (thrusts) of the last glaciation, than the adjoining depressions (Zelčs, 1993). Subglacial glaciotectonic landforms are arranged in fields of radial, transverse and hummocky glaciotectonic land systems. Genetically-related radial, transverse and hummocky glaciotectonic landforms occur in various paragenetic sequences or assemblages separated by ice-marginal formations, including end moraines and active ice-contact slopes. Distinct and specific paragenetic sequences of glaciotectonic landforms occur in both glacial lowlands and uplands.

Four main assemblages of glaciotectonic subglacial landforms can be distinguished in Latvia:

(I) A sequence of end moraines (Linkuva type), Rogen moraines, divergent drumlins, drumlinoids and continuous till plain occurs in glacial lowlands with divergent ice flow patterns.

(II) A sequence of end moraines or active ice-contact slopes, convergent drumlins and continuous till plain is common in glacial lowlands with convergent ice flow.

(III) The assemblage of composite ice-pushed marginal ridge (Trepe moraine) as active ice-marginal formation, fluted moraine and continuous till plain is a typical sequence in glacial lowlands with parallel ice-flow lines.

(IV) The interlobate composite hills, hummocky moraines, continuous till plain and hummocky ice-marginal ridges are present in areas of interlobate insular uplands.

According to Menzies & Shilts (1996), subglacial landforms reflect specific environmental conditions, which favour the development of these landforms within specific regional topographic, sedimentological and glaciodynamic settings. Ultimately, any subglacial landform must be a reflection of the interaction of basal ice-bed stress conditions of a particular mass at a specific location and time (*ibid*). The spatial arrangement and distribution of subglacial glaciotectonic landforms and structures reflects spatial and temporal peculiarities and changes in ice-bed conditions that can be used for palaeogeographical reconstructions. On the basis of the regional distribution of these features, four distinct models of deglaciation can be distinguished (Zelčs, 1993, 1998). Commonly, the formation of basal till occurred only at a distance of 30-40 km (in the insular uplands and convergent lowlands) or up to 60-80 km (in the divergent and consequent lowlands), upglacier from an interlobate moraine or active ice-marginal formation. Between the active ice-marginal formations and till deposition zone, the continuous till cover of the previous advance has been partially eroded by glaciotectonic processes and streamlined subglacial glaciotectonic landforms were formed. In some places, ice-marginal ridges or end moraines, for instance the Linkuva push moraine of the North Lithuanian (Linkuva) advance, overlap transgressively with the continuous till cover and the upglacier ends of drumlins of the preceding Gulbene (Middle Lithuanian) phase (Zelčs *et al.*, 1990).

During deglaciation these assemblages of subglacial landforms were altered to some extent by subsequent subglacial processes and meltwater activity. As a result a complex of superimposed landforms, e.g. ribbed recessive moraines in the areas of hard or impermeable bedrock, eskers in the lowlands with divergent and consequent ice flow, glaciofluvial deltas etc., and other deposits developed (Āboltiņš, 1972; Āboltiņš *et al.*, 1989; Āboltiņš & Zelčs, 1988; Straume, 1979; Zelčs, 1993, 1999). Occasionally these superimposed glacial landforms, particularly ribbed recessive moraines and crevasse fillings, were replaced by ice-marginal formations, for instance in the case of the Plieņi stage in the Central Latvian Lowlands (Āboltiņš, 1970; Āboltiņš *et al.*, 1972a, b; Raukas *et al.*, 1995a, b; Savvaitov & Veinbergs, 1996; 1998). The interlobate ridges (Fig. 1 for location) were considered by some authors as a complex of ice-marginal formations (Veinbergs *et al.* 1965; Āboltiņš *et al.*, 1972a, b; Meirons *et al.*, 1976; Straume, 1979; Savvaitov & Veinbergs, 1996; 1998). Because of these differences in interpretation, the number of deglaciation phases of the last glaciation in Latvia varied from five (Meirons *et al.*, 1976; Straume, 1979) up to six or nine (Āboltiņš, 1970; Āboltiņš *et al.*, 1972a, b; Savvaitov, 2000; Savvaitov & Veinbergs, 1996; 1998). Recent investigations by the present authors suggest that only four phases of the deglaciation can be reliably recognised in Latvia (Zelčs, 1997; 1998). In order of decreasing age, they are the Dagda, Gulbene, Linkuva and Valdemārpils phases.

The topographically highest inner interlobate areas of the isometric glacial uplands originated subglacially due to ice mass convergence during the initial Dagda phase of deglaciation. This interval coincides temporally with the South Lithuanian deglaciation phase (Raukas *et al.*, 1995a, b) or the Indra phase, distinguished by Meirons *et al.* (1976) and Straume (1979). They consist mainly of

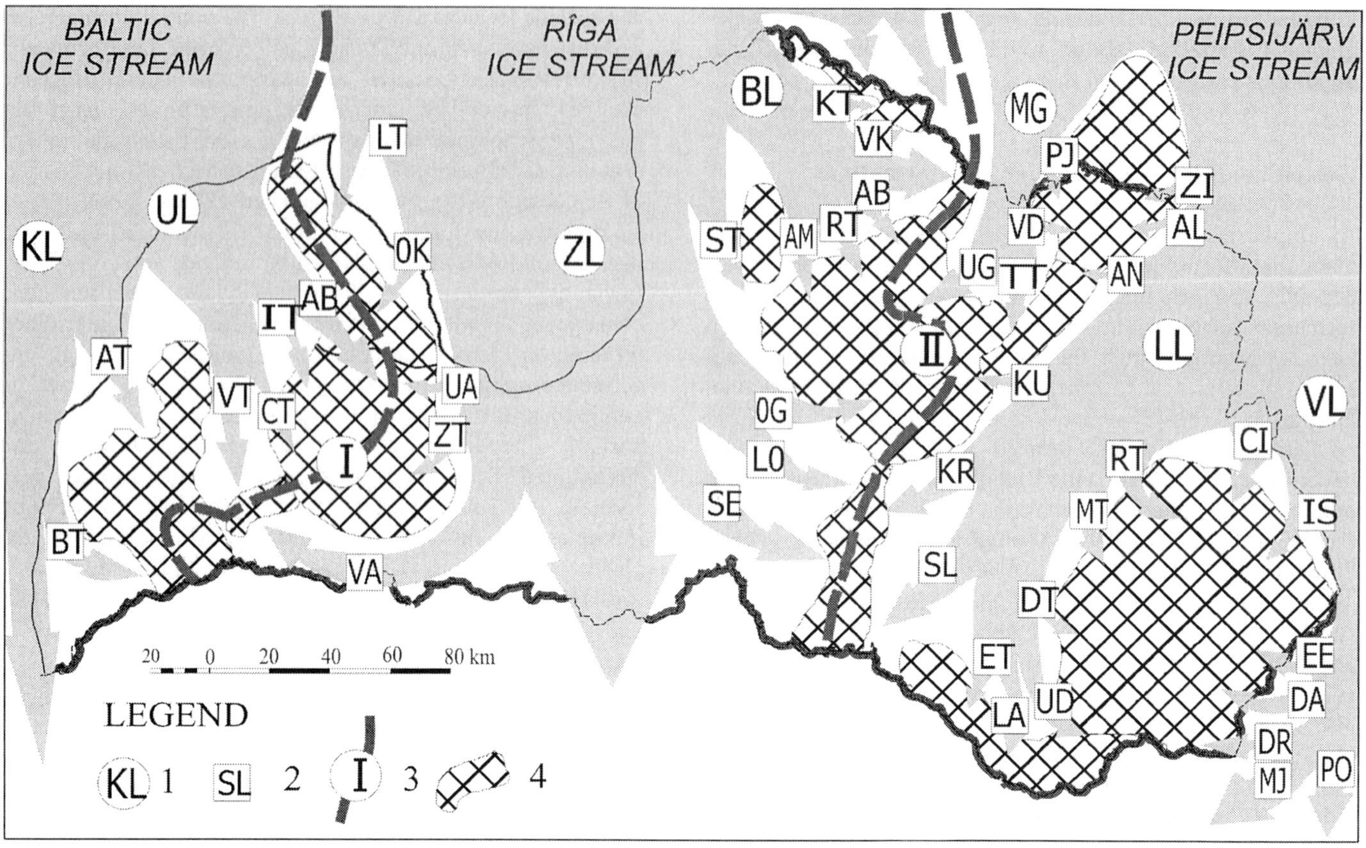

Fig. 2. Lobate structure of the peripheral cover of the Scandinavian ice sheet in Latvia during the last deglaciation.

Ice streams: Baltic; Rīga; Peipsijärv.

1= Ice lobes: KL = Kursian; UL = Usma; ZL = Zemgalian; BL = Burtnieks (North Vidzeme); MG = Middle Gauja; LL = Lubāns; VL = Velikoretski.

2 = Ice tongues: BT = Bārta; AT = Apriķi; VT = Venta; CT = Ciecere; AB = Abava; IT = Imula; LT = Laidze; OT = Okste; UA = Upper Abava; ZT = Zebrus; VA = Vadakste, SE = Selian; LO = Lobe; OG = Ogre; ST = Straupe; AM = Amata; RT = Rauna; AB = Abuls; VK = Valka; KT = Kārķi; UG = Upper Gauja; TT = Tirza; VD = Vaidava, PJ = Perlijõgi; ZI = Ziemeri; AL = Alūksne; AN = Anna; KU = Kuja; KR = Krustpils, SL = Slate; ET = Eglaine; LA = Laucesa; UD = Upper Daugava; DT = Dubna; MT = Malta; RT = Rēzekne; CI = Cirma; IS = Istra; EE = Ežezers; DA = Dagda; DR = Druja; MJ = Mjori;

PO = Polotsk tongue.

3 = Main interlobate zones: I = Baltic - Rīga; II = Rīga - Peipsijärv.

4 = Elevations.

including the glaciotectonic bases of plateau-like hills, morainic and cupola-hills (known in Latvia as *dauguls*).

Hummocky ribbed moraines (hummocky ridges of Āboltiņš *et al.*, 1972a) of the outer or peripheral belt of isometric uplands and ice-marginal formations of the Augšzeme Upland were formed at the contact zone between active and passive ice during the Gulbene phase. Simultaneously, the interlobate ridges and radial uplands, the Vadakste drumlin field of the Central Latvian Lowlands in the Latvian-Lithuanian borderland, and fluted moraine fields superimposed by recessive ribbed moraines in the East Latvian Lowlands originated. Two minor phases of temporary reactivation and stagnation of the active ice margin can be identified after formation of the interlobate ridges of the East Latvian Lowlands and the eastern part of the Central Latvian Lowlands. The Gulbene phase correlates with the Middle Lithuanian phase (Kudaba, 1969; Raukas *et al.*, 1995a, b) and probably with the Sebezha phase of Isachenkov (1972).

Beginning with the Linkuva phase, ice decay began in the Baltic Sea basin, as well as in Northern Vidzeme, Middle Gauja and all of the Central Latvian Lowlands (Fig. 1). The Linkuva phase is the earliest lobate deglaciation phase of Āboltiņš *et al.* (1972a). During the Linkuva, the Burtnieks, Zemgale and Iecava drumlin fields and associated landforms, were formed (Straume, 1979; Zelčs, 1993). The Linkuva line is clearly marked by the push moraine arc in the Central Latvian Lowland (Doss, 1910). It is generally correlated with the North Lithuanian phase in Lithuania (e.g. Āboltiņš, 1970) and with the Haanja phase (Raukas *et al.*, 1995a, b) of Estonia. The continuation of the Linkuva ice-marginal formations along the up-glacier

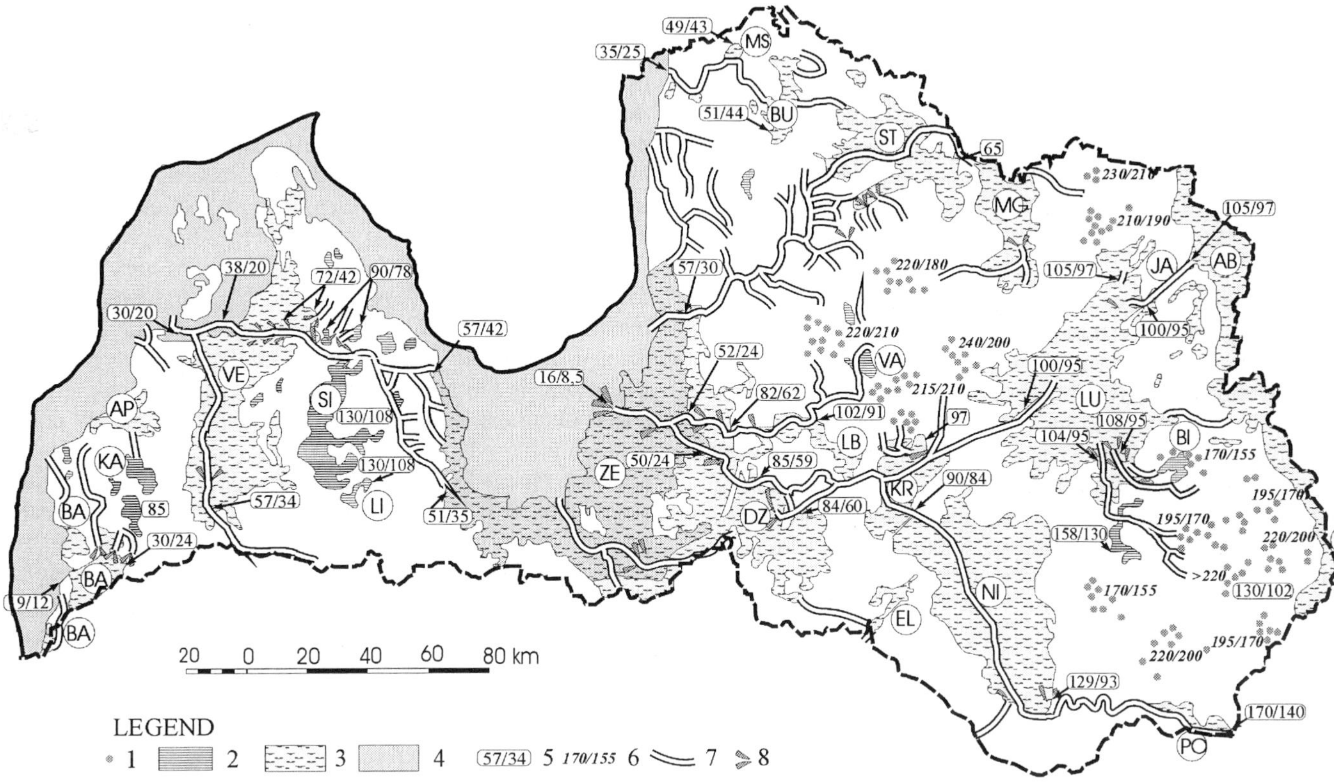

Fig. 3. Distribution of ice-dammed lakes and glaciolacustrine deposition sites in Latvia.

1 = Accumulation of glaciolacustrine sediments on hilltops;
2 = Ice dammed lakes in uplands: KA = Kalvene, SI = Saldus-Imula, LI = Lielauce, VA = Valole, BI = Biži;
3 = Meltwater basins in glacial lowlands: BA = Bārta, AP = Apriķi, VE = Venta, ZE = Zemgale, DZ = Daudzeva, LB = Lobe, EL = Elkšņi, KR = Krustpils,NI = Nīcgale, PO = Polatsk, ME = Mērdzene, LU = Lubāns, JA = Jaunanna, AB = Abrene, MG = Middle Gauja, ST = Strenči, BU = Burtnieks, MS = Middle Salaca;
4 = Baltic Ice Lake;
5 = Maximum/minimum levels of ice-dammed lake;
6 = Maximum/minimal elevations of plateau-like hills;
7 = Proglacial spillways;
8 = Ancient deltas.

margin of the East Latvian Lowlands into Russia is only based on morphological evidence. The Kacēni-Dzērvene ice-marginal composite ridge most probably continues in the Krasnogorodsk marginal ridges of Isachenkov (1972) in Russia, which were formerly correlated with the Gulbene phase of Meirons *et al.* (1976).

The ice-marginal formations of the Valdemārpils phase (North Latvian phase of Raukas & Gaigalas, 1993) mark the latest reactivation of the ice margin in the country. They commonly occupy the lowest hypsometrical level of ice-marginal formations. The maximum extent of this phase is marked by a chain of small push moraine ridges and ramparts in the lowlands, and by active ice-contact slopes along the north-eastern edge of the Northern Kursa and the western edge of the Western Kursa Uplands (Straume, 1979). The lowest parts of the marginal formations were partially eroded and in some places even destroyed by the Baltic Ice Lake or overlain by coastal dune ridges. Controversial views have been expressed concerning the correlation of the Valdemārpils ice-marginal formations with the Estonian deglacial phases. Most frequently the Valdemārpils phase is correlated with the Sakala marginal belt in Estonia (cf. Gaigalas & Raukas, 1989, Meirons *et al.*, 1976; Raukas & Gaigalas, 1993; Raukas *et al.*, 1995a, b; Straume, 1979). Analysis of the Gulf of Rīga sea floor topography, based on large-scale depth charts, reveals a series of subparallel landforms (Juškevičs & Talpas, 1997), which can be interpreted as streamlined features. These forms could have originated subglacially at the beginning of the Valdemārpils reactivation phase.

Ice-dammed lakes and proglacial fluvial systems

The deglaciation history of the last glaciation is to some extent based on the analysis of extra-marginal meltwater formations (Āboltiņš *et al.*, 1974a), particularly the location of ice-dammed lakes, their shorelines and deposits, and proglacial fluvial systems. Due to the flat topography and the inclination of the land surface towards the retreating ice margin, the meltwaters could not drain freely, forming ice-dammed lakes along the ice margin, sometimes covering large areas (Fig. 3). Drainage of these ice-dammed lakes was often a catastrophic process. As a result, in watershed areas deeply-incised (up to 90 m in a depth) proglacial spillways were formed (Āboltiņš, 1971; Eberhards, 1972a; Veinbergs, 1975). Meltwater discharge resulted also in intense deposition of sediments either as deltas in proglacial lakes or glaciofluvial fans in supra-aquatic areas (Danilāns & Lūka, 1969; Āboltiņš, 1971; Eberhards, 1972a, b; Veinbergs, 1975).

The development of the meltwater drainage network has been discussed in a number of publications (Danilāns, 1972b; Āboltiņš *et al.*, 1974c; Straume, 1978, 1979). Regional studies carried out by Āboltiņš (1971), Eberhards (1972a, b, 1973), Meirons (1975b), Veinbergs (1975), Veinbergs & Stelle (1965) have provided data on the spatial and temporal distribution of ice-dammed lakes and correlation of their coastal formations with the terrace levels of proglacial fluvial systems. Extra-regional correlation of the drainage history in the Eastern Baltic was attempted by (Āboltiņš *et al.*, 1974b); Kvasov (1974), Eberhards (1979), Eberhards & Miidel (1984), Raukas (1994), Raukas & Eberhards (1991). The formation of the proglacial fluvial network in the Baltic countries began after the South Lithuanian phase, some 16,000 years ago in southern Lithuania and about 12,600 years ago in northernmost Estonia. In Latvia, a reach of the Daugava River valley, connecting the glacial lakes of Polatsk and Nīcgale, formed during ice retreat, after the Middle Lithaunian stage. Most of the fluvial development, however, was related to the retreat of the Linkuva phase ice sheet about 13.2 ka ago. Commonly, the Late-glacial terraces of the Latvian rivers, except the uppermost terrace fragments of the Daugava and the Jodupe rivers, dip up-glacier or towards the Gulf of Riga and Baltic Sea.

The first meltwater basins that formed in Latvia were supraglacial ice lakes. They are assumed to have appeared in depressions of the ice surface initially as slush ponds. Depressions in ice surface over the composite glaciotectonic hills may have resulted from selective melting of darker, more sediment-rich glacier. These ice-lakes were probably rather small, covering an area from 1-2 up to 15 km^2, with a maximum depth of up to 30-40 m. Local glaciolacustrine deposits overlie the uneven base of the glaciotectonic composite hills in the isometric uplands of the interlobate areas of Eastern Latvia (Fig. 3). They cause the surface of composite hills to appear flat-topped. The elevation of these plateau-like hills usually varies from 180 up to 240 m in the Vidzeme Upland, 190-230 m in the Alūksne Upland and 155-260 m in the Latgale Upland (Āboltiņš *et al.*, 1975, 1976; Meirons, 1975a; Straume, 1979). They rise some 20-40 m above their surroundings. The thickness of the glaciolacustrine deposits reaches on average 6-8 m, with a maximum of 28,5 m in the Gaiziņkalns, the highest point of Latvia rising 60 m above its surroundings (Danilāns, 1972b). Commonly, the plateau-like hills have moderately steep or steep slopes (15-35°), which are terraced in places and crossed by gullies (Āboltinš, 1972; Straume, 1979). The lake sequence begins with massive clays or silt and/or varve-like laminated silt and clay. Coarser sandy material, in places with an admixture and intercalation of gravel, occurs at the margins of these dead ice-dammed lakes. The coarser material was deposited either by supraglacial meltwater streams as small deltas or by slumping of debris. The formation of those lakes occurred at about 16-17 ka and records the end of the insular or initial phases of deglaciation in Latvia (Āboltiņš *et al.*, 1972a; Zelčs, 1997).

Large ice-dammed lakes covering an area of some thousands of km^2 developed in glacial lowlands. The maximum depth of these lakes was up to 40 m. They were formed during the latest episodes of the Gulbene recession. The oldest is the Nīcgale ice-dammed lake, and some small basins along the Sēlija interlobate ridge and at the SW corner of the Vidzeme Upland (Fig. 3). They were probably formed after ice retreat from the Madona-Trepe composite marginal ridge in the East Latvian Lowlands, approximately at the same time as those in the Kursa uplands. The formation of the Lubans and Daudzeva ice-dammed lakes began later when ice retreated from the Rikava-Bērzpils ice-pushed ridge in the East Latvian Lowland and the Taurkalne recessional moraine in the Central Latvian Lowlands. The formation of younger ice-dammed lakes in the lowlands resulted from drainage of the lakes mentioned above, after the North Lithuanian reactivation, and they existed up to the Palivere deglaciation phase in Estonia.

The glaciolacustrine sediments are represented by varved clays and other types of laminated and non-laminated sediments. Gradual transition is rather common from till to the overlying glaciolacustrine sediments. The many attempts to correlate varve sequences of neighbouring ice-dammed lakes have been unsuccessful because of the very complex structure of the varves and the presence of 'drainage' varves. Annual layering can only be distinguished in the lowermost part of the sequences (Kuršs & Stelle, 1964). According to Kuršs & Stinkule (1969), the seasonal lamination is frequently complicated by many fine laminae within an annual cycle. The Zemgale glacial lake is the only basin in Latvia where a direct 'step-by-step' varve correlation has succeeded. Here Kuršs & Stinkule (1966) developed a varve chronology of 46 years, based on varve measurements in 5 clay pits located in intervals of 5 km. According to their calculations, the Linkuva phase ice margin retreated at a rate of *c.* 1000 m per year. Such a rate looks overestimated because in the downglacier and upglacier areas the intervals between individual ridges of the recessive ribbed moraines, the interval is 125-375 m on average with a maximum about 500 m (Zelčs, 1999).

The Baltic Ice Lake covered considerable parts of Latvia (Fig. 3). It existed during the Allerød Interstadial and Younger Dryas Stadial and followed the retreating ice sheet of the Palivere phase in the Baltic depression and adjoining coastal areas. Its waves and currents abraded the offshore glacial plains, including lower parts of marginal formations of the Valdemārpils phase (Veinbergs, 1972). According to Grīnbergs (1957) and Veinbergs (1964), abrasion scarps, bars, spits, accumulative terraces and other coastal landforms allow the reconstruction of several shorelines at

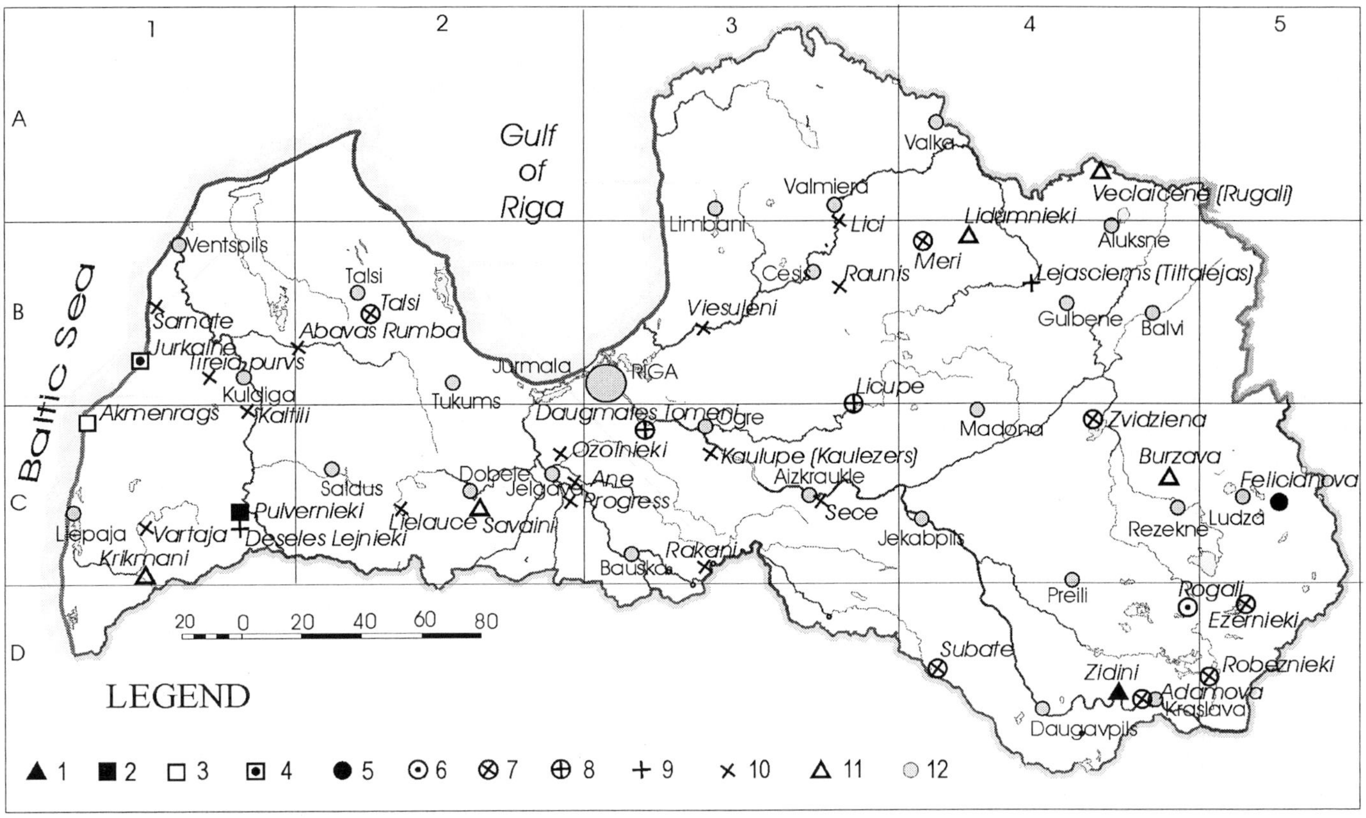

Fig. 4. Location of Pleistocene stratotype sites and dated sites.

PLEISTOCENE STRATOTYPE SITES
1 = Židiņi deposits of Cromerian age
2 = Pulvernieki terrestrial deposits of Holsteinian interglatiation
3 = Akmeņrags marine deposits of Holsteinian interglatiation
4 = Jūrkalne deposits of Early Saalian age
5 = Felicianova deposits of Eemian interglatiation
6 = Rogaļi deposits of Early Weichselian age
7 = TL DATING SITES
8 = ESR DATING SITES
^{14}C DATING SITES
9 = Early and Middle Weichselian age
10 = Late Weichselian age
11 = Questionable dates

12= Major cities and towns

hypsometric levels from 5 to 55 m. All of these Baltic Ice Lake shorelines are tilted northwestwards because of the glacioisostatic uplift. The best developed shorelines are those of Baltic Ice Lake phases BII and $BIII_b$.

Stratigraphical evidence

The stratigraphy of the Pleistocene deposits and history of their investigation have been discussed briefly by Dreimanis & Zelčs (1995, see also references therein). Deposits of four glaciations and three interglacials have been identified in Latvia (Danilāns, 1973). The age of the tills is mainly determined by their stratigraphical position with respect to interglacial, and interstadial for the tills of the last glaciation, deposits. In Latvia, the age of the last glacial and interglacial deposits has been dated using an uncorrected radiocarbon chronology, thermoluminescence (TL) and electron spin resonance (ESR) methods (Table 1, Fig. 4).

According to Meirons & Straume (1979) and Meirons (1986, 1992) litostratigraphically complex sections have been mainly encountered in the eastern Latvian uplands and in buried valleys. The maximum thickness is about 200 m in the Vidzeme Upland and reaches up to 310 m in the Aknīste burried valley, SE Latvia. The glacial lowlands commonly have a relatively thin cover of Pleistocene sediments, except for the Baltic depression in western Latvia.

During the past two decades, detailed investigations (Āboltiņš, 1989; Āboltiņš & Zelčs, 1988; Āboltiņš & Dreimanis, 1995; Zelčs, 1993, 1999; Zelčs & Dreimanis, 1997; Zelčs *et al.*, 1990) have shown that glaciotectonic deformation structures and displacement are very common. Older strata have been reworked and/or emplaced above younger ones as megablocks or as overthrusts, and the lowest part of a till unit may resemble the underlying older ones because of their gross incorporation (Dreimanis & Zelčs, 1995).

Deposits of the last glaciation are present almost throughout Latvia, with the exception of a few places along the Gulf of Rīga and along recent valleys where they have been eroded. They play the dominant role in the geological structure of glacial uplands and lowlands and they form the main glacial landforms (Āboltiņš, 1989; Āboltiņš *et al.*,

Table 1. Age determination of deposits from Pleistocene sites in Latvia.

Site	Location on the map (Fig.3)	ESR age, ka	References
The Early Weichselian age:			
Līčupe:	C3	88.5±7.3 (Tln-205-L1)	A. Molodkov *et al.*, 1998
Portlandia arctica shells from clay lens		97.8±8.2 (Tln-230-086)	
Portlandia arctica shells from clay lens		95.7±8.2 (Tln-231-086)	
Portlandia arctica shells from till			
Daugmales Tomēni:	C3	86.0±7.3 (Tln-205-D1)	A. Molodkov *et al.*, 1998
Portlandia arctica shells from till		105.0±9.2 (Tln-205-D2)	
Portlandia arctica shells from gravel			
Site	**Location on the map (Fig.1)**	**TL age, ka**	**References**
Deposits of the Holsteinian (?) Interglacial:			
Adamova	D4	>161.55 (Tln-TL-48)	Meirons *et al.*, 1981
		106.25 (Tln-TL-49)	Meirons *et al.*, 1981
Till of the Saalian (?) glaciation: Robežnieki	D5		
Intertill sediments between Saalian and Weichselian tills:			
Židiņi	D4	79.15 (Tln-TL-45)	Meirons *et al.*, 1981
Zvidziena	C4	97.00 (Tln-TL-?)	Meirons, 1986
Subate	D4	92.00 (Tln-TL-?)	Meirons, 1986
Ezernieki	D5	40.34 (Tln-TL-?)	Meirons, 1986
Mēri	B4	33.00 (Tln-TL-?)	Meirons, 1986
Glaciofluvial gravel of Early or Middle Weichselian time:			
Talsi	B2	56.056 (Tln-TL-?)	Meirons & Juškevičs, 1984
Site	**Location on the map (Fig. 1)**	**^{14}C age**	**References**
Middle Weichselian age:			
		>34,000 (Mo-318)	Vinogradov *et al.*, 1963
Lejasciems (Tiltalejas)	B4	32,260±730 (LU-159)	Arslanov *et al.*, 1975
		>33,450 (LU-311A)	Arslanov *et al.*, 1975
		34,500±790 (LU-311B)	Arslanov *et al.*, 1975
		$36{,}100\pm\frac{3200}{2300}$ (Tln-483)	Meirons, 1992
The Late Glacial age:			
		13,390±500 (Mo-296)	Vinogradov *et al.*, 1963
		13,250±160 (TA-177)	Punning *et al.*, 1968
Raunis	B3	13,320±250 (Ri-39)	Stelle *et al.*, 1975a, b
		10,780±220 (Ri-5)	Zobens *et al.*, 1969
		10,400±370 (Ri-5A)	
Kaltiķi	C1	11,300±300 (Ri-2)	Zobens *et al.*, 1969
		10,550±180 (Ri-226)	Segliņš *et al.*, 1988
Lielauce	C2	9,840±160 (Tln-239)	
Ozolnieki, moluscs	C2	10,390±105 (TA-128)	Punning *et al.*, 1967
		10,800±280 (Ri-4)	Zobens *et al.*, 1969
Āne (Sarkanais māls), wood	C2	11,950±110 (TA-129)	Punning *et al.*, 1967
Āne, plant remain		11,875±110 (TA-129A)	
Progress	C2	10,590±250 (Ri-24)	Stelle & Veksler, 1975
		10,535±250 (Ri-33)	Stelle *et al.*, 1975b
Sārnate	B1	10,282±250 (Ri-33A)	
Līči	B3	11,270±230 (Ri-105)	Stelle *et al.*, 1975c
		11,114±350 (Ri-74)	
Viesulēni	B3	10,317±230 (Ri-37)	Stelle *et al.*, 1975a
		10,950±250 (Ri-36)	Stelle *et al.*, 1975b
Kaulezers	C3	10,520±190 (Ri-140)	Veksler & Stelle, 1986
Sece	C3	10,410±90 (TA-163)	Danilāns, 1973
Rakaņi	C3	9,870±200 (TA-162)	
Abavas Rumba, wood	B2	10,840±130 (Ri-7A)	Zobens *et al.*, 1969
Abavas Rumba			
Vārtaja	C1	10,180±140 (Ri-305)	Segliņš *et al.*, 1988
Peat			
Tīreļa purvs, peat	B1		

Table 1 (continued). Age determination of deposits from Pleistocene sites in Latvia

Questionable dates:			
Dēseles Lejnieki	C1	>34,000 (Mo-317) >55,000 (TA-199)	Vinogradov *et al.*, 1963 Stelle *et al.*, 1975b
Līčupe	C3	34,300±60 (LU-198A) >41,160 (LU-198B)	Krūkle & Arslanov, 1977
Adamova	D4	21,424±400 (Ri-72)	Stelle *et al.*, 1975b
Sāvaiņi	C2	13,840±350 (Ri-A-I) 13,970±370 (Ri-A-II)	Sakson & Segliņš, 1990
Līdumnieki	B4	13,080±60 (LU-668A) 12,780±100 (LU-668B) 12,830±250 (LU-695)	Arslanov *et al.*, 1981 Arslanov *et al.*, 1981 Arslanov *et al.*, 1981
Veclaicene (Rugāji)	A4	11,300±100 (TA-1865) 12,180±80 (TA-1866)	Segliņš, 1991
Krikmaņi	C1	12,148±100 (Ri-347)	Segliņš, 1988
Burzava	C4	12,080±150 (LU-500A) 12,970±120 (LU-500B) 10,700±190 (LU-501A) 11,200±75 (Tln-7) 10,869±75 (Tln-8) 7,945±200 (Ri-3)	Arslanov & Stelle, 1975

1975, 1976, 1989; Āboltiņš & Dreimanis, 1995; Meirons & Straume, 1979; Straume, 1979, 1984; Zelčs, 1993).

The beginning of the last glaciation is marked by Early Latvian periglacial deposition of non-glacial meltwater sediments (Table 2). These sediments have been encountered in numerous borreholes. They overlie the Felicianova Interglacial deposits, for example at Felicianova, Kaitra, Rogaļi, Sātiķi, Skrudaliena, Subate and Židiņi (Meirons, 1986; Kalnina *et al.*, 1995; Dreimanis *et al.*, 1999). The sediments contain redeposited pollen (Kalniņa & Juškēvičs, 1998a; Kalniņa *et al.*, 1997; Krūkle *et al.*, 1975; Meirons 1972b, 1986) and macroscopic plant remains (Ceriņa 1984, 1999, 2000) derived from the preceding Felicianova (Eemian) Interglacial in various frequencies. The landscape was dominated by treeless vegetation.

At several sites in Latvia the Early Latvian sediments have been TL-dated between 79.15 and 97 ka (Table 1, Fig. 4). The ESR age of redeposited *Portlandia arctica* shells from till and gravel at the Daugmales Tomēni site ranges from 86.0 to 105.0 ka. *Portlandia arctica* shells from a marine clay lens at the Līčupe section have been ESR-dated to between 97.8 and 88.5 ka (Molodkov *et al.*, 1998), indicating that the finite ^{14}C from this site (Table 2) is contaminated (Table 1, Fig. 4). This record also suggests that these Early Weichselian marine deposits are rafted and displaced suglacially from their original site of deposition in the Gulf of Rīga.

The Scandinavian ice sheet seems to have reached NW Latvia first during the final episodes of the first major ice advance of the glaciological model of Holmlund & Fastook (1995, Fig. 4d).

The beginning of the next, Vidzeme, Substage is probably represented by meltwater deposits from the ice recession from NW Latvia. Glaciofluvial gravels at Talsi yielded a TL age of 56,056 ka (Meirons & Juškēvičs, 1984). The Vidzeme Substage correlates with MIS 3 or the 'Long Middle Glacial' of van Andel & Tzedakis (1996). It is unknown whether the territory of Latvia was completely unglaciated or experienced some ice-marginal oscillations during this substage.

According to Meirons (personal communication), a pollen sequence from Ezernieki (Table 2, Fig. 4) indicates a period of warmer climate, TL-dated to 40.34 ka. This might be correlated with the Middle Pleniglacial interstadial complex of Huijzer & Vandenberghe (1998). In NE Latvia non-glacial lacustrine and alluvial deposits of the Vidzeme Substage are known from several sections in the vicinity of Lejasciems, in the Middle Gauja Lowlands (Savvaitov *et al.*, 1964). They formed under severe climatic conditions, possibly in the Middle Pleniglacial cold interval of Huijzer & Vandenberghe (1998), although age control is poor. Gaigalas *et al.* (1987) correlated the sequence at Lejasciems with the Middle Nemunas (Weichselian) lake and river deposits at Rokai in Lithuania.

The major Late Weichselian glaciation of Latvia, probably between 25-12 ka, is represented by the deposits and landforms of Zemgale Substage. The Zemgale Substage correlates with MIS 2 or the Late Pleniglacial and Late-glacial chronostratigraphical units of Huijzer & Vandenberghe (1998). Thus Zemgale Substage includes the Last Glaciation Maximum (LGM) (Table 2). However, TL-dates from glaciofluvial deposits in Mēri, in the Aumeistari interlobate ridge, suggest that the Scandinavian Ice Sheet may have reached northern Latvia as early as 33.0 ka, which is in good agreement with the glaciological model of Holmlund & Fastook (1995, Table 1). The meltwater sediments of the major advance are widespread and are often affected by subsequent glaciotectonic deformation. Commonly they form cores of the glacial hills and streamlined landforms, ribbed moraines and ice-marginal formations (Āboltiņš, 1989; Strautnieks, 1998; Zelčs & Dreimanis, 1997; Zelčs *et al.*, 1990; Zelčs, 1993, 1999, 2000).

Table 2. Comparison of the last glaciation events in Latvia with the Weichselian chronostratigraphy (after Huijzer & Vandenberghe, 1998) in NW Europe and their correlation with oxygen-isotope stratigraphy. In the Latvian interstadial and stadial column the ages refer to the stratigraphic units above them.

NW EUROPE			LATVIA				^{18}O isotope stages	^{14}C age, ka
H O L O C E N E							1	10.2
LATE GLACIAL	LATE	WEICHSELIAN	LATVIAN GLACIAL STAGE	ZEMGALE SUBSTAGE	JELGAVA INTERSTADIAL		2	11.0
					VALDEMĀRPILS OSCILLATION RAUNIS INTERSTADIAL			13.2
LATE PLENIGLACIAL					LINKUVA OSCILLATION GULBENE OSCILLATION DAGDA OSCILLATION MAJOR GLACIER EXPANSION			25
MIDDLE PLENIGLACIAL	MIDDLE			VIDZEME SUBSTAGE	MĒRI INTERVAL Mēri 33.00 ka TL age LEJASCIEMS COOL INTERVAL Lejasciems 32-36 ka ^{14}C age		3	
					EZERNIEKI WARM INTERVAL Ezernieki 40.34 ka TL age			
					Talsi 56.056 ka TL age			59
EARLY PLENIGLACIAL				KURSA SUBSTAGE	TALSI STADIAL		4	74
EARLY GLACIAL	EARLY			EARLY LATVIAN SUBSTAGE	ROGAĻI INTERSTADIAL Židiņi 79.15 ka TL age		5a	85
					Daugmale 86 ka ESR age Subate 92 ka TL age	*Portlandia arctica*	5b	94
					Zvidziena 97 ka TL age			
					Daugmale 105 ka ESR age		5c	105
							5d	117
EEMIAN			FELICIANOVIAN				5e	

Meirons (1986, 1992) distinguished three till beds of the last glaciation, the Augšzeme, Vidzeme and Zemgale, in ascending order. The grey to brownish grey calcareous sandy Augšzeme Till occuring in East Latvian uplands and in western Latvia was considered by Meirons as a formation dating from the end of the Early Latvian substage (the Early Baltija of Meirons). Since the Augšzeme Till closely resembles the underlying Kurzeme Till of the Saalian glaciation, it is possible that it is a reworked and/or redeposited Kurzeme Till (Dreimanis & Zelčs, 1995). The two subsequent Weichselian till units are lithologically very similar. They have a relatively low boulder content and are separated by glaciolacustrine sediments. They occur upglacier from the Linkuva moraine, and were interpreted by Āboltiņš

(1963), Savvaitov & Straume (1963) as deposits of the major and north Lithuanian (Linkuva, Linkava of Āboltiņš, 1963) ice advances. However, according to the drumlin formation model in divergent glacial lowlands (Zelčs, 1993; Zelčs &Dreimanis, 1997), such till structure could have also resulted from subglacial glaciotectonic thrusting. Commonly these tills are reddish-brown, their colour derived mainly from Devonian sandstone. Local variations of colour (up to greyish-brown) result from the incorporation of Upper Devonian dolomite and Permian limestone.

According Zelčs (1993, 1998), all three till beds fall into the depositional sequence of the Zemgale Substage and could be correlated with the main glacier oscillations prior to the Linkuva phase (Table 2). The Augšzeme Till was probably deposited during the initial phases of the major glacier expansion up to the Weichselian maximum. The Vidzeme Till was laid down during the Dagda oscillation but the Zemgale Till was formed during the Gulbene reactivation. The originally continuous till beds were disintegrated by the subsequent oscillations of the active ice margin and subglacial glaciotectonic processes (Zelčs, 1998).

Biological studies in nearby southern Sweden (Berglund *et al.*, 1994) and south Germany (Walker, 1995) suggest that climatic warming prevailed during the Late-glacial Interstadial (13.2-11.0 ^{14}C ka), that contains the time interval formerly termed Bølling and Allerød separated by the Older Dryas. Palynological analyses and investigation of macroscopic plant remains, supplemented by radiocarbon age determinations, provide evidence only for the Late-glacial in Latvia. Most of the Latvian sediments containing organic remains are of lacustrine or fluvial origin (Punning *et al.*, 1968; Danilāns, 1973). The Valdemārpils oscillation, that interrupts the Late-glacial stadial in Latvia, might be correlated with the 'Older Dryas' of Björck & Möller (1987), and Hammarlund & Lemdahl (1994). Such an assumption does not contradict interpretation of the age of the Sakala or Pandivere zone of the ice-marginal formations in Estonia of Pirrus & Raukas (1996), which after corrections based on the Swedish varve chronology, date from 12,250+430 and 12,050+430 varve years, respectively. Thus the problem of the correlation of the Valdemārpils ice-marginal formations with those in Estonia is predominantly morphological rather than chronostratigraphical. Considerable cooling followed during the first half of the Younger Dryas Chron (11-10.2 ^{14}C ka) in Latvia.

About 30 radiocarbon dates for the time interval of 14.0-10.2 ^{14}C ka have been published during the period 1963-1990. Some may be erroneous for a variety of reasons. The intertill organic-bearing deposits occur at a relatively shallow depth in areas of vigorous glaciotectonic deformation. The organic remains may have also been contaminated either by secondary carbonates of pedogenic processes or by calcareous groundwater as emphasised by Veksler & Stelle (1986), Segliņš (1991) and Zelčs *et al.* (1990). Moreover, most of the ^{14}C dates were determined about 30 years ago, when the modern technique of using very small samples was not yet available. Therefore the present authors refer only to those sections where their regional situation, lithostratigraphical and biostratigraphical evidence and ^{14}C dates are in agreement. The only exception is the 'organic-bearing beds' within the Weichselian till sequence. This is because these Raunis deposits are important for the determination of the age of the Luga (Haanja, Linkuva, North Lithuanian) ice advance in the East European Plain (Arslanov, 1975; Arslanov *et al.*, 1981; Arslanov & Stelle, 1975; Chebotareva, 1965a, b; Chebotareva & Malgina, 1965; Chebotareva & Makaricheva, 1974; Chebotareva *et al.*, 1965a, b; Faustova, 1984; Liivrand, 1999; Serebryanny, 1965, 1978; Serebryanny & Raukas, 1966, 1967; Stelle *et al.*, 1975a, b; Vidgorchik *et al.*, 1970; Vonsavičius, 1986; Voznyachuk, 1972; Zarrina & Krasnov, 1965). Other sites with similar or analogous sequences to those at the Raunis site were described by Arslanov *et al.* (1981), Jakubovska *et al.* (1999), Kalniņa & Juškevičs (1998b), Krūkle *et al.* (1963), Meirons (1986, 1992, 1999), Meirons & Straume (1979), Pirrus & Raukas (1996), Punning & Raukas (1985), Raukas *et al.* (1995a, b), Savvaitov (2000), Savvaitov *et al.* (1964), Savvaitov & Veinbergs (1999), Stelle (1999) and Stelle *et al.* (1975a, b; 1999).

Conventionally, the Late-glacial interval in Latvia begins with sedimentation of the Raunis beds (Danilāns, 1961, 1973; Meirons & Straume, 1979; Meirons, 1986, 1992, Stelle *et al.*, 1975a) and ends approximately 10.2 ^{14}C ka ago, as suggested by radiocarbon dates from disseminated organics and macroscopic plant remains from Kaulezers (10,317±230, Ri-37) and Līči (10,282±250, Ri-33A). These dates are in agreement with the last drainage of the Baltic Ice Lake about 10.3 ka BP (Svenson, 1989), that is considered as the beginning of the post-glacial history of the Baltic Sea. According to a ^{14}C date of peat from the Tīreļpurvs bog (Fig. 4, Table 1), the post-glacial autochthonous peat accumulation in the Latvian uplands started at least 10,180±140 (Ri-305) ka BP (Segliņš *et al.* 1988).

Organic-bearing sediments occur between different till beds at many sites in Latvia, e.g. Burzava, Krikmaņi, Līdumnieki, Rugāji (Veclaicene) and probably also at Sāvaiņi. Palynological data from these strata are available, but apparently contain much redeposited pollen that hampers the correlation of these sections. All the dates from these sites (Table 2) are substantially younger than would be expected on the basis of their location with respect to ice-marginal formations and traditional deglaciation chronology. For instance, the Rugāji deposits that were radiocarbon-dated to 11,300±100 (TA-1865) and 12,180±80 (TA-1866) are located in the inner zone of the Alūksne Upland, where the deglaciation began during the South Lithuania phase at least by 16.0 ^{14}C ka BP (Raukas *et al.*, 1995a, b). In addition, the Burzava sections, where the ^{14}C age of organic material ranges from 12,970±120 (LU-500B) to 7,945±200 (Ri-3), relate to the ice-marginal formations of the Gulbene (Middle Lithuanian) phase that dates from 14.5-15.0 ^{14}C ka.

The Raunis section is situated upglacier from the end moraine of the Linkuva phase, in an ice-molded depression at the foot of the Vidzeme Upland. Various data of investigation of the Raunis deposits are presented by Ceriņa (1995), Ceriņa

& Kalniņa (2000), Ceriņa *et al.* (1998a, b), Danilāns (1961, 1973), Jakubovska & Stelle (1996), Punning *et al.* (1968); Savvaitov & Straume (1963), Savvaitov *et al.*, (1964), Zelčs *et al.* (1998) and Vinogradov *et al.* (1963). Initially, Danilāns (1961) suggested that the Raunis organic complex was of pre-Allerød age. However, Savvaitov *et al.* (1964) correlated the Raaunis sediments containing organics with the Burzava Interstadial between the Middle and Upper Valdaian glaciations. They also proposed the name 'Raunis Interstadial' for the interval during which the lower organics were deposited at Raunis. Dreimanis (1966) promoted the use of the term 'Raunis Interstadial' for the region affected by the North European continental glaciation and proposed its correlation with the Susaca Interstadial in SW Europe. Later Danilāns (1973) termed the upper pollen-poor sediments the 'Virsraunis beds', meaning 'overlying the Raunis beds', and suggested that they were formed during the Oldest Dryas time.

The Raunis beds outcrop in the upper part of a 10 m high exposure beside the mouth of a shallow ancient gully on the right bank of the River Raunis. They consist of up to 0.25 m thick fine-grained sand with disseminated organics and irregular lenses of plant remains. The organic-bearing sediments are underlain by a reddish-brown basal till and overlain by a sandy 0.15-0.55 m thick reddish-brown diamicton. Savvaitov & Straume (1963, Fig. 1) interpreted the overlying diamictic sediments as a continuous till sheet representing a glacial readvance. New sections and 4 trenches were excavated and 14 boreholes drilled in the adjoining area during the last 5 years (Ceriņa *et al.*, 1998a, b). These studies suggest that the overlying diamicton only has a distribution of some tens of metres and disappears in the direction of the end moraine. Fabric analyses gave maxima in the SW sector unrelated to the direction of any known glacier movement. It seems likely that the diamicton was produced by slumping or redeposition of till (Danilāns, 1973).

The diamicton is covered by fine-grained material, that contains pieces of wood, an interlayer of lake marl with molluscs, peat and at the top of the section, fine-grained to coarse-grained sand with occasional gravel. The thickness of the overlying deposits decreases from 4.0-5.0 m to 2.5 m in the direction of former ice movement. The ^{14}C dating of molluscs and wood, above the basal till bed, suggest a Pre-Boreal age of 9,230-9,300 yrs BP (Raukas, 1999). The dating of the deformed and partly disrupted peat 0.75-1.5 m is 8,020±70 yrs BP (Beta-70902). However, paleobotanical evidence and the fluorescence of pollen indicate a Pre-Boreal age for the peat (Jakubovska & Stelle, 1996; Stelle *et al.*, 1999). The silt and silty sand material that underlies the peat contains macroscopic and microscopic remains of treeless periglacial vegetation similar to other lower organic-bearing sediments of the Raunis section. As Ceriņa & Kalniņa (2000) have emphasised, palaeobotanical data from the Raunis section suggest a very rapid succession from an open subarctic to boreal vegetation. Thus if the diamicton overlying the Raunis beds is indeed of non-glacial origin, the Raunis section could be considered as a very prominent record of the Late-glacial Interstadial and the transition from Late-glacial to post-glacial conditions in Latvia. The Raunis beds probably correlate with the minimum age of the North Lithuania (Luga) phase in the East European Plain. Consequently, the upper pollen-poor sediments ('Virsraunis beds' of Danilāns, 1973) might have accumulated during a short cooling caused by the Valdemārpils oscillation.

Concluding remarks

A deglaciation history should be based on multiple criteria. Attention should be paid not only to the identification of ice-marginal formations but also complex studies are needed, including the investigation of the spatial distribution and internal composition of subglacial landforms that originated during the reactivation and recession of ice lobes and glacier tongues. The results of geospatial analyses and geological structure of glacial terrain should be supported by palaeoclimatic and environmental reconstructions and age determination by different, particularly up-to-date methods of absolute age determination.

The main problems remaining are the dating of ice-marginal formations and the reconstruction of the environmental change during the transition from the Eemian to Early Weichselian, and during the Middle Weichselian and at the termination of the last glaciation. Most of the key sections need reinterpretation using complex multiproxy climate reconstruction methods, as proposed by Aalbersberg & Litt (1998) for the Eemian and Early Weichselian across northwestern Europe. The geomorphological position and geological structure of these sections are also to be reinvestigated.

Acknowledgements

The technical expenses for this report were provided by a grant from the Latvian Council of Science. We are grateful to Professors Aleksis Dreimanis and Stephen R. Hicock for improving the manuscript by critical comments. Thanks to Professor Ojārs Āboltiņš for discussion of the problems of deglaciation , Laimdota Kalniņa for discussion of the pollen data and Zigurds Meirons for providing useful data on age determination. We wish particularly to thank Jürgen Ehlers and Philip Gibbard for helping us to clarify our presentation and sharpen our focus, also to Professors Jan Mangerud and Anto Raukas for co-operation and support with the latest dating of Raunis.

References

Aalbersberg, G. & Litt, T. (1998). Multiproxy climate reconstructions for the Eemian and Early Weichselian. *Journal of Quaternary Science*, **13,** 367-390.

Āboltiņš, O. (1963). Upper till beds in the Zemgale Plain and problem of formation of the Linkava end moraine. *In*: Temnikova, N. S. (ed.), *Ucheniye zapiski aspirantov Latviyskogo Gosudarstvennogo Universiteta im. P. Stuchki*, **1**, 5-17 (in Russian).

Āboltiņš, O. (1970). Marginal formations of Middle Latvian tilted plain and their correlation to Linkuva (North Lithuanian) end moraine. *In*: Danilāns I. (ed.), *Problems of Quaternary geology*, **V**, 95-107. Rīga, Zinātne (in Russian with English summary).

Āboltiņš, O. (1971). *Formation of the River Gauja Valley*. Rīga, Zinātne, 105 pp (in Russian).

Āboltiņš, O. (1972). On problem of formation of insular-like elevations. *In*: Danilāns I. (ed.), *Lednikoviy morfogenez*, 51-61. Rīga, Zinātne, (in Russian).

Āboltiņš, O. (1975). Glaciodynamic peculiarities of formation of Latvian elevations. *In*: Danilāns I. (ed.), *Problems of Quaternary geology*, **8,** 5-23. Rīga, Zinātne (in Russian with English summary).

Āboltiņš, O. (1989). *Glaciotectonic structure and glacial morphogenesis*. Rīga, Zinātne, 284 pp (in Russian).

Āboltiņš, O. & Dreimanis, A. (1995). Glacigenic deposits in Latvia. *In*: Ehlers J., Kozarski, S. & Gibbard, P. (eds), *Glacial deposits in North-East Europe*, 105-113. Rotterdam / Brookfield, Balkema.

Āboltiņš, O. & Zelčs, V. (1988). Litomorphogenesis of inner zone of ancient glacial region (investigations in Latvia). *In*: Purin, V.R. & Zvejniek, R. (eds), *Ģeogrāfiskās domas attīstība Padomju Latvijā*, 103-126. Latvijas Valsts Universitāte, Rīga (in Russian).

Āboltiņš, O., Veinbergs, I., Savvaitov A., Konshin G. & Stelle, V. (1969). Degradation of the Valdaian glaciation and history of the Baltic and White Sea at the end of glaciation. Latvian SSR. *In*: Gerasimov, I.P., Grichuk, V.P. & Chebotareva N.S. (eds), *Krayevye obrazovaniya materikovikh oledeneniy*, 125-137. Moscow, Nauka (in Russian).

Āboltiņš, O., Veinbergs, I., Danilāns, I., Stelle, V., Straume, J., Eberhards, G., Juškevičs, V. & Jaunputniņš, A. (1972a). Main features of glacial morphogenesis and peculiarities of deglaciation of the last glaciation in the territory of Latvia. *In*: Danilāns, I. & Āboltiņš, O. (eds), *Putyevoditel polevogo simpoziuma III Vsesoyuznogo mezhvedomstvennogo soveschaniya po izucheniyu krayevykh obrazovaniy materikovogo oledeneniya*, 3-16. Rīga, Zinātne (in Russian).

Āboltiņš, O., Veinbergs, I., Stelle, V. & Eberhards, G. (1972b). Main complexes of marginal formations and glacial retreat in the territory of Latvian SSR. *In*: Goretskiy, G.I., Pogul'ayev, D.I. & Shick, S.M. (eds), *Krayevye obrazovaniya materikovikh oledeneniy*, 30-37. Moscow, Nauka (in Russian).

Āboltiņš, O., Mikalauskas, A. & Raukas, A. (1974a). Morphogenetic classification of extramarginal glacio-aquatic formations based on investigation in the Baltic States. *In*: Biske, G. & Mikalauskas, A. (eds), *Predfrontalnye krayevye lednikovye obrazovaniya*, 25-31. Vilnius, Mintis (in Russian).

Āboltiņš, O., Mikalauskas, A. & Raukas, A. (1974b). Peculiarities of the distribution of ice extramarginal formations of the last glaciation in the territory of the eastern Baltic countries and some problems on their formation. *In*: Biske, G. & Mikalauskas A. (eds), *Predfrontalnye krayevye lednikovye obrazovaniya*, 36-43. Vilnius, Mintis (in Russian).

Āboltiņš, O., Veinbergs, I. & Eberhards, G. (1974c). On formation proglacial basins and river valley network during degradation of the last glaciation in the territory of Latvian SSR. *In*: Biske, G. & Mikalauskas A. (eds), *Predfrontalnye krayevye lednikovye obrazovanya* (Proglacial marginal formations), 60-85. Vilnius, Mintis (in Russian).

Āboltiņš, O., Straume, J. & Juškevičs, V. (1975). Relief peculiriaties and main stages of morphogenesis of the Central Vidzeme Elevation. *In*: Danilāns, I. (ed.), *Problems of Quaternary Geology*, **8,** 31-47. Rīga, Zinātne (in Russian with English summary).

Āboltiņš, O., Straume, J. & Juškevičs, V. (1976). The relief and main stages of glacial morphogenesis of the Alūksne Elevation. *In*: Danilāns, I. (ed.), *Problems of Quaternary Geology*, **9,** 74-89. Rīga, Zinātne (in Russian with English summary).

Āboltiņš, O., Danilāns, I., Ilyin, E.A., Isachenkov V.A., Karukapp, R., Raukas, A. & Faustova, M. A. (1977a). The structure of the principal ice shed zones. *In*: Chebotareva, N. S. (ed.), *The structure and dynamics of the last ice sheet of Europe*, 101-112. Moscow, Nauka (in Russian).

Āboltiņš, O., Isachenkov V.A., Faustova, M. A.. & Chebotareva N.S. (1977b). The Chudskoye ice stream. *In*: Chebotareva, N. S. (ed.), *The structure and dynamics of the last ice sheet of Europe*, 44-54. Moscow, Nauka (in Russian).

Āboltiņš, O., Vaitekunas, P., Danilāns, I., Karukapp, R., Klive, G., Raukas, A., Roshko, L., Chebotareva N.S. & Yanke, V. (1977c). The Baltic ice stream. *In*: Chebotareva, N. S. (ed.), *The structure and dynamics of the last ice sheet of Europe*, 17-44. Moscow, Nauka (in Russian).

Āboltiņš, O., Aseyev, A., Vonsavičius, V., Isachenkov, V., Mozhayev, B. & Raukas, A. (1989). Development and use of glacial accumulative insular heights. *Proceedings of the Academy of Sciences of the Estonian SSR, Geology*, **38,** 23-33 (in Russian with English summary).

Arslanov, Kh.A. (1975). Upper Pleistocene geochronology based on radiocarbon dating in the European part of the USSR (glacial and periglacial zone). *Byuleten kommissii po izucheniju chetvertichnogo perioda*, **43,** 3-25 (in Russian).

Arslanov, Kh.A. & Stelle, V. (1975). Radiocarbon datings of intertill deposits at the railway station Burzava in the Latgale Upland. *In*: Savvaitov, A.S. & Veksler, V.S. (eds), *Opyt i metodika izotopno-geokhimicheskih issledovaniy v Pribaltike i Belorussii*, 82-83.

Arslanov, Kh.A., Filonov, B.A., Chernov, S.B. & Makarova, E.V. (1975). Radiocarbon dating of the

Geochronology Laboratory of NIGEI at the University of Leningrad. Report No.2. *Byuleten kommisii po izucheniju chetvertichnogo perioda*, **43,** 212-226 (in Russian).

Arslanov, Kh.A., Kondratienė, O., Krūkle, M., Stelle, V. & Lyadov, V.V. (1981). On age and paleogeography of the Raunis interstadial Late Pleistocene deposits in Latvia. *Vestnik Leningradskogo Universiteta. Geologiya i Geografiya*, **6(1),** 96-102 (in Russian with English summary).

Berglund, B.E., Björck, S., Lemdahl, G., Bergsten, H., Nordberg, K. & Kolstrup, E. (1994). Late Weichselian environmental change in Southern Sweden and Denmark. *Journal of Quaternary Science*, **9 (2),** 127-132.

Björck, S. & Möller, P. (1987). Late Weichselian environmental history in southeastern Sweden during the deglaciation of the Scandinavian ice sheet. *Quaternary Research*, **28,** 1-37.

Bucharts, S. (1935). Quaternary deposits in the region of Talsi. *Talsu novads. Enciklopēdisks rakstu krājums*, **I,** 17-50 (in Latvian).

Ceriņa, A. (1984). Specification of boundaries and content of Felicianovaian formation in Latvia. *In*: Kondratienė, O. & Mikalauskas, A. (eds), *Paleogeografiya i stratigrafiya chetvertichnogo perioda Pribaltiki i sopredel'nikh rayonov*, 149-154. Vilnius.

Ceriņa, A. (1995). Difference between complexes of plant macroscopic remains from outcrops on the Raunis River bank. *In*: *LU 54. zinātniskā konferences Ģeogrāfijas un Zemes zinātņu sekcijas tēzes un programmas*, p. 18. Latvijas Universitāte, Rīga (in Latvian).

Ceriņa, A. (1999). Plant macrofossil assemblages of Eemian and Early Weichselian deposits in Latvia. *In*: Kirsimäe, K. & Ainsaar, L. (eds), *Field Symposium INQUA Commission on Glaciation and Peribaltic Work Group on Pleistocene stratigraphy and glacial geology - Southern Estonia 1999, Abstract volume*, 26-27. University of Tartu.

Ceriņa, A. (2000). The Rogaļi layers and problems of their correlation. *In*: *Latvijas Universitātes 58. zinātniskās konferences Zemes un vides zinātņu sekcijas referātu tēzes*, 27-29. Rīga, Latvijas Universitāte (in Latvian).

Ceriņa, A. & Kalniņa, A. (2000). A new investigation of sections on the right bank of the Raunis River. *In*: *Latvijas Universitātes 58. zinātniskās konferences Zemes un vides zinātņu sekcijas referātu tēzes*, 29-32. Rīga, Latvijas Universitāte (in Latvian).

Ceriņa A., Danilāns I., Dreimanis A., Jakubovska I., Stelle V. & Zelčs V. (1998a). Raunis late glacial deposits southeast from Cēsis. In: *Field Symposium on glacial processes and Quaternary environment in Latvia, 25-31 May 1998. Excursion guide*, 66-73. Rīga, University of Latvia.

Ceriņa, A., Jakubovska, I., Savvaitov, A. & Stelle V. (1998b). Raunis interstadial deposits in Latvia. In: *Field Symposium on glacial processes and Quaternary environment in Latvia, 25-31 May* (*1998. Abstracts of papers and posters*, 8-10. Rīga, University of Latvia.

Chebotareva, N.S. (1965a). General problems of palaeogeography of the last (Valdai) glaciation in the north-western part of the Russian Plain. *In*: Neustadt, M. (ed.), *Upper Pleistocene and Holocene paleogeography and chronology in the light of radiocarbon dating*, 7-22. Moscow, Nauka (in Russian).

Chebotareva, N.S. (1965b). Chronology of the last glaciation in the north-western part of the Russian Plain. *In*: Neustadt, M. (ed.), *Upper Pleistocene and Holocene paleogeography and chronology in the light of radiocarbon dating*, 60-64. Moscow, Nauka (in Russian).

Chebotareva, N.S. (1972). Ice streams of the Valdaian ice sheet. *In*: Goretskiy, G.I., Pogul'ayev, D.I. & Shick, S.M. (eds), *Krayevye obrazovaniya materikovikh oledeneniy*, 69-77. Moscow, Nauka (in Russian).

Chebotareva, N.S. & Makaricheva, I.A. (1974). *Last glaciation and its geochronology in Europe*, 216 pp. Moscow, Nauka (in Russian).

Chebotareva, N.S. & Malgina, E.A. (1965). On the absolute age of some Upper Pleistocene interstadials. *In*: Neustadt, M. (ed.), *Upper Pleistocene and Holocene paleogeography and chronology in the light of radiocarbon dating*, 27-37. Moscow, Nauka (in Russian).

Chebotareva, N., Grichuk, V., Vidgorchik, M., Faustova, M., Biske, G., Vaitekunas, P., Gudelis, V. & Devyatova E. (1965a). Main degradation stages and marginal zones. *In*: Chebotareva, N. (ed.), *Last European glaciation,* 26-44 (in Russian with English summary).

Chebotareva, N., Vidgorchik, M., Grichuk, V. & Faustova, M. (1965b). Stratigraphy of deposits of the Valdai glaciation. *In*: Chebotareva, N. (ed.), *Last European glaciation,* 15-20 (in Russian with English summary).

Danilāns, I. (1961). Problems of the history and paleogeography of the Lateglacial in the territory of Latvia. *In*: Neishtadt, M.I. & Gudelis, V. (eds), *Voprosy holotsena*, 201-209. Vilnius (in Russian).

Danilāns, I. (1963). On history of determination of marginal landforms in the territory of Latvia. *In*: Goretskiy, G.I., Ivanova, I.K. & Orviku, K. (eds), *Trudy kommissii po izucheniyu chetevertichnogo perioda (Krayevye formy relyefa materikovogo oledeneniya na Russkoy ravnine)*, **XXI,** 76-83. AN SSSR, Moscow (in Russian).

Danilāns, I. (1965). Some pecularities of deglaciation and glacial morphogenesis in the territory of Latvia. *In*: Basalikas, A. (ed.), *Krayevye obrazovaniya materikovikh oledeneniy*, 65-71. Vilnius, Mintis (in Russian).

Danilāns, I. (1972a). An influence of subglacial surface on processes deglaciation and glacial morphogenesis. *In*: Goretskiy, G.I., Pogul'ayev, D.I. & Shick, S.M. (eds), *Krayevye obrazovaniya materikovikh oledeneniy*, 19-22. Moscow, Nauka (in Russian).

Danilāns, I. (1972b). On origin and history of glacial lake development in Latvia. *In*: Danilāns I. (ed.), *Problems of Quaternary geology*, **VI,** 59-68. Rīga, Zinātne (in Russian with English summary).

Danilāns, I. (1973). *Quaternary deposits of Latvia*. Rīga, Zinātne, 312 pp (in Russian).

Danilāns, I. & Lūka, K. (1969). The structure and morphology of one of the types of ice meltwater deltas. *In*: Danilāns I. (ed.), *Questions of Quaternary geology*,

IV, 65-81. Rīga, Zinātne (in Russian with English summary).

Doss, B. (1910). Über das Vorkommen von einer Endmoräne sowie von Drumlins, Åsar und Bänderton in nördlichen Lithauen. *Centralblatt für Mineralogie, Geologie und Paläontologie*, **22,** 123-131.

Dreimanis, A. (1935). *The rock deformations, caused by inland ice, on the left bank of Daugava at Dole Island, near Riga in Latvia.* Rīga, Gulbis, 30 pp (in Latvian with English summary).

Dreimanis, A. (1966). The Susacá - interstadial and the subdivision of the late-glacial. Discussion. *Geologie en Mijnbouw,* **45,** 445-448.

Dreimanis, A. (1999). Need for three-dimentional analysis of structural elements in glacial deposits for determination of direction of glacier movement. *In*: Mickelson, D.M. & Attig, J.W. (eds), *Glacial processes past and present*, 59-67. Boulder, Colorado, Geological Society of America Special Paper, **337**.

Dreimanis, A. & Zelčs, V. (1995). Pleistocene stratigraphy of Latvia. *In*: Ehlers, J., Kozarski, S. & Gibbard, P. (eds), *Glacial deposits in North-East Europe*, 105-113. Rotterdam / Brookfield, Balkema.

Dreimanis, A., Danilāns, I. & Zelčs, V. (1999). A comparison of the Last Ice Age events in Latvia with S. Ontario, Canada, and their correlations with oxygen-isotope stratigraphy. *In*: Kirsimäe, K. & Ainsaar, L. (eds), *Field Symposium INQUA Commission on Glaciation and Peribaltic Work Group on Pleistocene stratigraphy and glacial geology - Southern Estonia 1999, Abstract volume*, 7-9. University of Tartu.

Eberhards, G. (1972a). *Structure and development valleys of catchment area of the River Daugava.* Rīga, Zinātne, 132 pp (in Russian).

Eberhards, G. (1972b). Data on the morphology, structure and development of Lubāna plain relief during the late glacial time. *In*: Danilāns, I. (ed.), *Questions of Quaternary Geology*, **IV,** 59-63. Rīga, Zinātne (in Russian with English summary).

Eberhards, G. (1973). Morphogenesis of the Salaca River valley. *In*: Jaunputnin, A. (ed.), *Ucheniye zapiski Latviyskogo gosudarsvennogo universiteta* (1**86 (II),** 21-29. Riga, Latviyskiy gosudarstvenniy universitet, (in Russian).

Eberhards, G. (1979). Main features of the development of river valleys in the territory of last European glaciation. *In*: Biske G.S. (ed.-in-chief), *Paleogeografiya oblastu Skandinavskikh materikovikh oledeneniy*, 10-18. Geograficheskoye obschestvo SSSR, Leningrad (in Russian).

Eberhards, G. & Miidel, A. (1984). Main features of the development of river valleys in the East Baltic. *Proceedings of the Academy of Sciences of the Estonian SSR. Geology*, **33,** 136-145.

Faustova, M.A. (1984). Late Pleistocene glaciation of European SSSR. *In*: Velichko, A.A., Wright, H.E., Jr. & Barnosky, C.W. (eds), *Late Quaternary environments of the Soviet Union*, 3-12. Minneapolis, University of Minnesota Press,.

Gaigalas, A. & Raukas, A. (1989). Correlation and reconstruction of the Pleistocene glaciations in the Baltic region. *In*: Velichko, A.A., Gutovaya, E.E. & Faustova, M.A. (eds), *Paleoklimaty i oledeneniya v pleistocene*, 154-160. Moscow, Nauka (in Russian).

Gaigalas, A., Gudelis, V., Spriņģis, K., Konshin, G., Savvaitov, A., Veinbergs, I. & Raukas, A. (1967). Orientation of pebble long axes in the tills of the last glaciation in the East Baltic region and its relation to ice sheet shrinkage. *Baltica,* **3,** 215-233 (in Russian with English and German summary).

Gaigalas, A., Arslanov, Kh., Banis, Y., Kazartseva, T.I. & Tertychnaya, T.V. (1987). The radiocarbon dating of Late Pleistocene, Holocene and arsheological records on the territory of Lithuanian SSR. *In*: Punning J.M., Ivanova I.K., Kind, N.V. & Chichagova, O.A. (eds), *In new data in Quaternary Geochronology*, 878-97. Moscow, Nauka (in Russian).

Grīnbergs, E. (1957). *The late glacial and postglacial history of the coast of the Latvian SSR.* Rīga, The Publishing House of the Academy of Sciences of the Latvian SSSR, 122 pp (in Russian).

Hammarlund, D. & Lemdahl, G. (1994). A Late Weichselian stable isotope stratigraphy compared with biostratigraphical data: a case study from southern Sweden. *Journal of Quaternary Science*, **9,** 13-31.

Hättestrand, C. (1997). *Ribbed moraines and Fenoscandian Palaeoglaciology. Doctoral dissertation.* 47 pp. Stockholm University.

Hausen, H. (1913). Materialien zur Kenntnis der pleistozanen Bildung in der russischen Ostseelandern. *Fennia*, **34 (2),** 181 pp.

Homlund, P. & Fastook, J. (1995). A time dependent glaciological model of the Weichselian ice sheet. *Quaternary International*, **27,** 53-58.

Huijzer, B. & Vandenberghe, J. (1998). Climatic reconstructions of the Weichselian Pleniglacial in northwestern and central Europe. *Journal of Quaternary Science*, **13,** 391-417.

Isachenkov, V.A. (1972). Marginal formations and some peculiarities of retreating of Valdaian ice sheet from the territory of the Pskov Region. *In*: Basalikas, A. (ed.), *Krayevye obrazovaniya materikovogo oledeneniya*, 211-217, Moscow, Nauka (in Russian).

Jakubovska, I., Savvaitov, A., Veinbergs, I. & Stelle. V. (1999). Stratigraphy of Raunis interstadial in glaciation curve of Late Pleistocene in Latvia. *In*: Kirsimäe, K. & Ainsaar, L. (eds), *Field Symposium INQUA Commission on Glaciation and Peribaltic Work Group on Pleistocene stratigraphy and glacial geology - Southern Estonia 1999, Abstract volume*, 28-29. University of Tartu.

Jakubovska, I. & Stelle. V. (1996). Peculiarities of fluorescence of polen content from the section on the Raunis River bank. *In*: *LU 55. zinātniskā konferences Ģeogrāfijas un Zemes zinātņu sekcijas tēzes un programmas*, p. 21. Rīga, Latvijas Universitāte (in Latvian).

Jaunputnin, A. (1956). On problem of origin of hummocky landforms in Latvian SSR. *In*: *Zinātniskie raksti*, **7**, 25-37. Rīga, Latvijas Valsts Universitāte (in Russian).

Juškevičs, V. & Talpas, A. (1997). Explanatory note to the map (scale 1:200,000). *In*: Segliņš, V. & Kajak, K. (eds), *The map of the Quaternary deposits of the Gulf of Rīga*, 39 pp. Rīga.

Kalniņa, L. & Juškēvičs, V. (1998a). Felicianova (Eemian) interglacial deposits in Latvia. *In*: *Field Symposium on glacial processes and Quaternary environment in Latvia, 25-31 May* (1998. *Abstracts of papers and posters*, 28-29. Rīga, University of Latvia.

Kalniņa, L. & Juškēvičs, V. (1998b). Late Pleistocene lithostratigraphy and palaeoenvironment in Latvia. *Geologija*, **25**, 39-48.

Kalniņa, L., Juškevičs, V. & Mūrniece, S. (1997). Felicianovian interglacial deposits and conditions of their formation. *In*: Kļaviņš, M. (ed.), *Cilvēks, Vide, Resursi, LU 56. konferences Ģeogrāfijas un Zemes zinātņu sekcija*, pp. 44. Rīga, Latvijas Universitāte (in Latvian).

Kalniņa, L., Juškevičs, V. & Segliņš, S.V. (1995). Pleistocene stratigraphy at the Židiņi site, south-eastern Latvia. *In*: Robertsson, A.-M., Hicks, S., Åkerlund, A., Risberg, J. & Hackens, T. (eds), *PACT 50*: *Landscape and Life*, 31-40. PACT Rixensart, Belgium.

Krūkle, M. & Arslanov, Kh.A. (1977). On stratigraphy and geochronology of marine Pleistocene deposits of the Līčupe section. *Baltica*, **6**, 181-192 (in Russian with English summary).

Krūkle, M., Lūsiņa, L. & Stelle, V. (1975). Profile of Pleistocene deposits at Felicianova (Latvian SSR). *In*: Danilāns, I. (ed.), *Questions of Quaternary Geology*, **II**, 7-34. Rīga, Zinātne (in Russian with English summary).

Krūkle, M., Stelle, V. & Veinbergs, I. (1963). Interstadial deposits at the railvay station Burzava, Latgale Upland. *Izvestiye Akademii Nauk Latviyskoi SSR*, **5 (190)**, 77-84 (in Russian).

Kudaba, I. (1969). Ice-marginal formations of the Baltic Ridge and diagnostic of ice margin dynamics. *In*: *Materikovye oledeneniya i lednikoviy morfogenez*, 155-226. Vilnius, Mintis (in Russian).

Kuršs, V. & Stelle, V. (1964). Conditions of formation and age of limnoglacial clays in the Saldus basin. *Izvestiye Akademii Nauk Latviyskoi SSR*, **11 (208)**, 55-64 (in Russian with English summary).

Kuršs, V. & Stinkule, A. (1966). On pecularieties of bedding and conditions of varved clays deposition in the Middle Latvian basin. *Izvestiye Akademii Nauk Latviyskoi SSR*, **10 (231)**, 47-54 (in Russian).

Kuršs, V. & Stinkule, A. (1969). On the varieties of varved bedding in limnoglacial clays of Latvian SSR. *In*: Danilāns, I. (ed.), *Questions of Quaternary Geology*, **IV**, 83-101. Rīga, Zinātne (in Russian with English summary).

Kvasov, D.D. (1974). *Late Quaternary history of large lakes and inland seas of Eastern Europe*, 278 pp. Leningrad, Nauka (in Russian).

Liivrand, E. (1999). Weichselian periglacial deposits and till beds in Estonia and correlations in the Baltic area. *In*: Lukševičs, E., Stinkulis, Ģ. & Kalniņa, L. (eds), *Problems and methods of modern regional stratigraphy. Abstracts of the fourth Baltic stratigraphic conference*, pp.50-52. Rīga, University of Latvia.

Lundqvist, J. & Saarnisto, M. (1995). Summary of project IGCP-253. *Quaternary International*, **28 (1)**, 9-18.

Markots, A. (1986). On the problem of the eastern boundary of the dispersion fan of the Dalarnian porphyry. *In*: Āboltiņš, O., Eberhards, G. & Klane, V. (eds), *Morfogenez relyefa I paleogeografiya Latvii*, 122-129. Riga, Latviyskiy gosudarstvenniy universitet (in Russian).

Meirons, Z. (1972a). Structure of the Quaternary cover and relief formation in the middle part of the North-Kurzeme Elevation. *In*: Danilāns, I. (ed.), *Problems of Quaternary Geology*, **VI**, 45-58. Rīga, Zinātne (in Russian with English summary).

Meirons, Z. (1972b). Section of interglacial deposits at Rogaļi. *In*: Danilāns, I. (ed.), *Problems of Quaternary Geology*, **VI**, 7-13. Rīga, Zinātne (in Russian with English summary).

Meirons, Z. (1975a). The relief peculiriaties of the Latgale Elevation and the bordering areas of the East-Latvian Lowland. *In*: Danilāns, I. (ed.), *Problems of Quaternary Geology*, **8**, 48-81. Rīga, Zinātne (in Russian with English summary).

Meirons, Z. (1975b). Glaciolacustrine basins in North-East Latvia. *In*: Kvasov, D. D. (ed.), *Tezisy dokladov IV Vsesoyuznogo simpoziuma po istorii ozyor*, 2, 82-90. Leningrad (in Russian).

Meirons, Z. (1986). Stratigraphy of Pleistocene deposits in Latvia. *In*: Kondratienė, O. & Mikalauskas, A. (eds), *Issledovaniya lednikovykh obrazovaniy Pribaltiki*, 69-81, Vilnius (in Russian).

Meirons, Z. (1992). Stratigraphic scheme of Pleistocene deposits in Latvia. *In*: Veinbergs, I., Danilāns, I., Sorokin, V. & Ulst, R. (eds), *Paleogeografiya i stratigrafiya fanerozoya Latvii i Balstiyskogo morya*, 84-98 (in Russian).

Meirons, Z. (1999). Some problems of the Upper Pleistocene in Latvia. *In*: Lukševičs, E., Stinkulis, Ģ. & Kalniņa, L. (eds), *Problems and methods of modern regional stratigraphy. Abstracts of fourth Baltic stratigraphic conference*, 69-70. Rīga, University of Latvia.

Meirons, Z. & Juškevičs, V. (1984). Quaternary deposits. *In*: Misāns, J., Brangulis, A. & Straume, J. (eds), *Geologija Latviyskoi SSR*, 89-122. Rīga, Zinātne (Explanatory text with geological maps on scale of 1:500,000, in Russian).

Meirons, Z. & Straume, J. (1979). Cenozoic group. *In*: Misāns, J., Brangulis, A., Danilāns, I. & Kuršs, V. (eds), *Geologicheskoe stroyenie i poleznye iskopayemye Latvii*, 176-268. Rīga, Zinātne (in Russian).

Meirons, Z., Straume, J. & Juškevičs, V. (1976). Main varieties of the marginal formations and retreat of the last glaciation in the territory of Latvian SSR. *In*: Danilāns, I. (ed.), Problems of Quaternary Geology, **9**, 50-73. Rīga,

Academy of Sciences of Latvian SSR (in Russian with English summary).

Meirons, Z., Punning, M.J. & Hütt, G. (1981). Results obtained through the TL dating of Southeast Latvian Pleistocene deposits. *Eesti NSV Teaduste Akademia Toimetised, Geologia*, **1**, 28-33 (in Russian with English and Estonian summaries).

Menzies, J. & Shilts, W. W. (1996). Subglacial environments. *In*: Menzies J. (ed.), *Past glacial environments. Sediments, forms and techniques*, 15-136. Oxford, Butterworth-Heinemann.

Molodkov, A., Dreimanis, A., Āboltiņš, O. & Raukas, A. (1998). The age of *Portlandia arctica* shells from glacial deposits of Central Latvia: an answer to a contraversy on the age and genesis of their enclosing sediments. *Quaternary Geochronology*, **17**, 1077-1094.

Philipp, H. (1921). Beitrag zur Kenntnis des Endmoränenverlaufs im östlichen Balticum. *Neues Jahrbuch für Mineralogie, Geologie und Paläontogie*, **11 (2)**, 9-34. Stuttgart.

Pirrus, R. & Raukas, A. (1996). Late-glacial stratigraphy in Estonia. *Proceedings of the Estonian Academy of Sciences*, **45**, 34-45.

Punning, J.-M., Raukas, A. (1985). Late Pleistocene paleogeography in North Europe. *Paleogeografiya*, **2**, 216.

Punning, J.-M., Raukas, A. & Serebryanny, L.R. (1967). Geochronology of the last glaciation in Russian Plain in the light of new absolute age dating of the relict lacustrine and mire deposits in the Eastern Baltic. *In*: Zavriyev, V.G. (ed.), *Materialy II Simpoziuma po istorii ozyor Severo-Zapada SSSR (23-28 maya 1967)*, 139-147. Minsk (in Russian).

Punning, J.-M., Raukas, A., Serebryanny, L.R. & Stelle, V. (1968). Paleogeographical peculiarities and absolute age of the Luga stage of the Valdaian glaciation on the Russian Plain. *Doklady Akademii nauk SSSR, Geologiya*, **178 (4)**, 916-918 (in Russian).

Raukas, A. (1994). Annually laminated late glacial sediments in the eastern Baltic countries and the evolution of ice-dammed lakes. *In*: Hicks, S., Miller, U. & Saarnisto, M., T. (eds), *PACT 41*: *Laminated sediments*, 45-55. PACT, Rixensart, Belgium.

Raukas, A. (1999). Main outlines of the Estonian Quaternary stratigraphy. *In*: Lukševičs, E., Stinkulis, Ģ. & Kalniņa, L. (eds), *Problems and methods of modern regional stratigraphy. Abstracts of fourth Baltic stratigraphic conference*, 88-89. Rīga, University of Latvia.

Raukas, A. & Eberhards, G. (1991). The problems of palaeohydrological reconstructions in the areas of the Last Glaciation on the example of the Soviet Baltic Countries. *In*: Gullentops, F. (ed.), *Vetalnds in Flanders, Aardkundige Mededelingen*, **6**, 181-183. Leuven University Press.

Raukas, A. & Gaigalas, A. (1993). Pleistocene glacial deposits along the eastern periphery of the Scandinavian ice sheets - an overview. *Boreas*, **22**, 214-222.

Raukas, A., Āboltiņš, O. & Gaigalas, A. (1995a). The Baltic States. Overview. *In*: Shirmer, W. (ed.), *Quaternary field trips in Central Europe*, **1**, 146-151. München, Verlag Dr. Friedrich Pfeil.

Raukas, A., Āboltiņš, O. & Gaigalas, A. (1995b). Current state and new trends in the Quaternary geology of the Baltic States. *Proceedings of the Estonian Academy of Sciences*, **44**, 1-14.

Sakson, M. & Segliņš, V. (1990). The section of the pre-Linkuvaian interstadial deposits from Savaiņi at Dobele. *In*: Miidel, A.. & Molodkov, A.N. (eds), *Chetvertichniy period: metody issledovaniya, stratigrafiya i ekologiya. Tezisy VII Vsesojuznogo soveschaniya po izucheniyu chetvertichogo perioda*, 88-89. Tallinn.

Savvaitov, A. (2000). Late deglaciation stades in the North-western Vidzeme. *In*: *Latvijas Universitātes 58. zinātniskās konferences Zemes un vides zinātņu sekcijas referātu tēzes*, 144-147 (in Latvian).

Savvaitov, A.S. Stelle, V. & Krūkle, M. (1964). On stratigraphic subdivision of deposits of the Valdaian glaciation in the territory of the Latvian SSR. *In*: Danilāns, I. (ed.), Questions of Quaternary Geology, **III (18**3-201. Rīga, Zinātne (in Russian).

Savvaitov, A.S. & Straume, J. (1963). On the problem of two till units of the Valdaian Glaciation in the area between the lower reaches of the rivers Daugava and Gauja. *In*: Danilāns, I. (ed.), Questions of Quaternary Geology, **II**, 71-86. Rīga, Zinātne (in Russian with English summary).

Savvaitov, A. & Veinbergs, I. (1996). Peculiarities of glacier dynamics in Latvia during the last glaciation. *In*: Danilāns, I., Kuršs, V. & Zelčs, V. (eds), *Latvijas devona un kvartāra nogulumu pētījumu materiāli*, 47-57, Rīga, Latvijas Universitāte (in Latvian).

Savvaitov, A. & Veinbergs, I. (1998). The features of deglaciation of the last ice sheet in Latvia. In: *Field Symposium on glacial processes and Quaternary environment in Latvia, 25-31 May* (*1998. Abstracts of papers and posters*, 59-61. Rīga, University of Latvia,.

Savvaitov, A. & Veinbergs, I. (1999). Linkuva stage of the last glaciation in Latvia. *In*: Lukševičs, E., Stinkulis, Ģ. & Kalniņa, L. (eds), *Problems and methods of modern regional stratigraphy. Abstracts of fourth Baltic stratigraphic conference*, pp.99-100. Rīga, University of Latvia.

Segliņš, V. (1988). Late glacial in Western Latvia based on results of investigation of the Krikmaņi section. *Izvestiye Akademii Nauk Estonskoy SSR, Geologiya*, **37**, 89-92.

Segliņš, V. (1991). Condition and estimation of the results of the Pleistocene radiocarbon dating of the Upper Pleistocene separate intervals of Eastern Latvia. *In*: Gaigalas, A. (ed.-in-chief), *Geochronological and isotope geochemical research into Quaternary geology and archaeology*, 79-87. Vilnius, Vilnius University Press (in Russian with English summary).

Segliņš, V., Stelle, V., Veksler, V.S., Lācis, A. & Prēde, E. (1988). Radiocarbon dating of late and postglacial deposits of western Latvia. *In*: Punning, J.M. (ed.-in-chief), *Izotopno geokhimicheskiye isssledovaniye v Pribaltike i Belorussii*, 195-205. Tallinn (in Russian with English summary).

Serebryanny, L.R. (1965). *Progress of radiocarbon dating in Quaternary geology*. Moscow, Nauka, 271 pp (in Russian).

Serebryanny, L.R. (1978). *Dynamics of continental ice sheet and glacio-isostasy in Late Quaternary time*, 270 pp. Nauka, Moscow (in Russian).

Serebryanny, L.R. & Raukas, A. (1966). Transbaltic correlations of the Late Pleistocene marginal ice formations. *In*: Grichuk, V.P., Kind, N.V. & Ravskiy E.I. (ed.), *Verkhniy pleistotsen, Stratigrafiya i absolyutnaya geokhronologiya*, 12-28. Moscow, Nauka (in Russian).

Serebryanny, L. & Raukas, A. (1967). Correlation of Gotiglacial ice-marginal belts in the Baltic Sea depression and the neighbouring countries. *Baltica*, **3**, 235-249 (in Russian with English and French summary).

Spriņģis, K., Konshin, G.I. & Savvaitov, A.S. (1963). Data concerning pebble orientation and changes of the ice movement direction of the Valdaian glaciation. *In*: Danilāns, I. (ed.), *Questions of Quaternary Geology*, **II,** 35-43. Rīga, Zinātne (in Russian with English summary).

Stelle, V. (1999). Palynological criteria for stratigraphical subdivision of the late glacial deposits in Latvia. *In*: Lukševičs, E., Stinkulis, Ģ. & Kalniņa, L. (eds), *Problems and methods of modern regional stratigraphy. Abstracts of fourth Baltic stratigraphic conference*, pp.103. Rīga, University of Latvia.

Stelle, V. & Veksler, V.S. (1975). Radiocarbon dating of deposits from section of the Sārnate bog. *In*: Savvaitov, A.S. & Veksler, V.S. (eds), *Opyt i metodika izotopno-geokhimicheskikh issledovaniy v Pribaltike i Belorussii*, 85-86 (in Russian).

Stelle, V., Savvaitov, A.S. & Veksler, V.S. (1975a). Absolute age of chronostratigraphical stages and boundaries of late-glacial and postglacial in the territory of central Baltic. *In*: Afanasyev, G.D. (ed.-in-chief), *Sostoyaniye metodicheskikh issledovaniy v oblasti absolyutnoy geokhronologii*, 187-191. Moscow, Nauka (in Russian).

Stelle, V., Savvaitov, A.S. & Veksler, V.S. (1975b). Dating of Pleistocene deposits in the territory of Latvia. *In*: Savvaitov, A.S. & Veksler, V.S. (eds), *Opyt i metodika izotopno-geokhimicheskikh issledovaniy v Pribaltike i Belorussii*, 80-81 (in Russian).

Stelle, V., Veksler, V.S. & Āboltiņš, O. (1975c). Radiocarbon dating of alluvial deposits from middle course of the River Gauja. *In*: Savvaitov, A.S. & Veksler, V.S. (eds), *Opyt i metodika izotopno-geokhimicheskikh issledovaniy v Pribaltike i Belorussii*, 87-88 (in Russian).

Stelle, V., Savvaitov, A.S.& Jakubovska, I. (1999). Complex of deposits on the bank of the Raunis River, its structure and age. *In*: Kļaviņš, M. (ed.), *Zeme Daba Cilvēks, LU 57. konferences ģeogrāfijas, ģeoloģijas un vides zinātņu sekcija,* 130-132. Rīga, Latvijas Universitāte (in Latvian).

Straume, J. (1978). Limnoglacial plains and the main late glacial basins in Latvia. *In*: Āboltiņš, O., Klane, V. & Eberhards, G. (eds), *Problemy morfogeneza rel'efa i paleogeografii Latvii,* 45-66. Riga, Latviyskiy gosudarstvenny universitet (in Russian).

Straume, J. (1979). Geomorphology. *In*: Misāns, J., Brangulis, A., Danilāns, I. & Kuršs, V. (eds), *Geologicheskoe stroyenie i poleznye iskopayemye Latvii*, 297-439. Rīga, Zinātne (in Russian).

Straume, J. (1984). Geomorphology. *In*: Misāns, J., Brangulis, A. & Starume, J. (eds), *Geologija Latviyskoi SSR*, 123-153. Rīga, Zinātne (Explanatory text with geological maps on scale of 1:500,000, in Russian).

Strautnieks, I. (1998). *Glacigenic landforms of the Eastern Kursa Upland and their genesis. Dissertation work summary*. 55 pp. Rīga, University of Latvia.

Svenson, N.O. (1989). Late-Weichselian and Holocene shore displacement in the Central Baltic based on stratigraphical and morphological record from Eastern Småland and Gotland. *LUNDQUA thesis*, **25,** Lund, 1-195.

van Andel, T.H. & Tzedakis, P.C. (1996). Palaeolithic landscapes of Europe and environs, 150.000-25.000 years ago. *Quaternary Science Reviews*, **15 (5-6)**, 481-500.

Veinbergs, I. (1964). Coastal morphology and dynamics of the Baltic Ice Lake on the territory of the Latvian SSR. *In*: Danilāns, I. (ed.), *Questions of Quaternary Geology*, **III**, 331-369. Rīga, Zinātne (in Russian with English summary).

Veinbergs, I. (1972). Marginal landforms (Western Latvia is used to illustrate). *In*: Goretskiy, G.I., Pogul'ayev, D.I. & Shick, S.M. (eds), *Krayevye obrazovaniya materikovikh oledeneniy*, 37-45. Moscow, Nauka (in Russian).

Veinbergs, I. (1975). Formation of the Abava-Slocene system of glacial meltwater drainage valleys. *In*: Danilāns, I. (ed.), *Problems of Quaternary Geology*, **8,** 82-102. Rīga, Zinātne (in Russian with English summary).

Veinbergs, I. & Stelle, V. (1965). On proglacial lakes in Kurzeme during decaying of the last glaciation. *In*: Kalesnik, S.V. (ed.), *Istoriya ozyor Severo-Zapada*, 85-87. Leningrad (in Russian).

Veinbergs, I., Konshin, G. & Savvaitov A. (1965). On dynamics of ice retreat during final episodes of the last glaciation in the territory of the Latvian SSR. *Izvestiye Academii Nauk Latviyskoy SSR*, **8,** 87-92 (in Russian).

Veksler, V. & Stelle, J. (1986). Cosmogenic influence on the radiocarbon age of the wood from the Rakaņi section. *In*: Punning. J. M. (ed.), *Izotopno-geokhimicheskiye issledovaniy v Pribaltike i Belorussii*, 36-41. Tallinn (in Russian).

Vidgorchik, M.E., Auslender, V.G., Znamenskaya, O.M. & Dolukhanov, P.M. (1970). New radiocarbon adtings lake deposits in north-west RSFSR and scale of absolute geochronology of the last glaciation. *In*: Kabailienė, M., Kalesnik, S.V., Kvasov, D.D., Klane, V., Klimkaite, I., Korde, N., Kudaba, Č. & Kuderskiy A. (eds), *Istoriya ozyor. Trudy simpoziuma po osnovnym problemama presnovodnykh ozyor (25-29 maya 1970)*, **2,** 5-32. Vilnius.

Viiding, H., Gaigalas, A., Gudelis, V., Raukas, A. & Tarvidas, R. (1971). *Crystalline indicator boulders in the East Baltic area*, 95 pp. Vilnius, Mintis (in Russian).

Vinogradov, A.P., Devirts, A.L., Dobkina, E.I. & Markova, N.G. (1963). Determination of the absolute age according to ^{14}C. Communication 4. *Geochemistry*, **9,** 795-812 (in Russian).

Vonsavičius, V. (1986). Eastern Baltic countries. *In*: Krasnov, I.I. & Zarrina, E.P. (eds). Resheniye 2-go Mezhvedeomstvennogo stratigraficheskogo soveschaniya po chetvertichnoy sisteme Vostochno-Yevropeyskoy platformy, 38-44. VSEGEI, Leningrad (in Russian).

Voznyachuk, L. (1972). Age of the maximum stage of the Valdai glaciation at the north-western part of the USSR and the main phases of glacier degradation. *In*: Danilans, I. (ed.), *Problems of Quaternary Geology*, **VI,** 29-43. Rīga, Zinātne (in Russian with English summary).

Walker, M.J.C. (1995). Climatic changes in Europe during the last glacial/interglacial transition. *Quaternary International*, **28,** 63-76.

Weimarn, C. (1930). Zur physischen Landesnatur Lettgallens. *Zeitschrift für Geomorphologie*, **5,** 167-200.

Zarrina, E.P. & Krasnov, I.I. (1965). Problem of correlation of ice-marginal formations in the north-western European part of SSSR and adjoining foreign areas. *In*: Basalikas, A. (ed.), *Krayevye obrazovaniya materikovogo oledeneniya*, 5-21, Vilnius, Mintis (in Russian).

Zāns, V. (1935). Glacial striae and phenomena of ice-friction on the rocks of Latvia. *In*: *Ģeogrāfiski raksti (Folia Geographica)*, **V,** 63-84 (in Latvian with English summary).

Zāns, V. (1936). Ice age and post glacial in Latvia. *In*: Malta, N. & Galenieks, P. (eds), *Latvijas, zeme, daba un tauta,* **I,** 49-127. Rīga, Valters un Rapa (in Latvian).

Zelčs, V. (1993). *Glaciotectonic landforms of divergent type glaciodepressional lowlands. Dissertation work synthesis,* 105 pp. Rīga, University of Latvia.

Zelčs, V. (1997). Landforms development. *In* Kavacis, G. (ed.-in chief): *Latvijas daba. Enciklopēdija*, **4,** 228-231. Rīga, Preses nams (in Latvian).

Zelčs, V. (1998). Glaciotectonic structures and landforms as indicators of deglaciation in Latvia. *In*: *Guide book of excursion and paper session abstracts of the conference of the INQUA Work Group on Glaciotectonics on Areal versus frontal deglaciation of the vistulian ice sheet. Poznań, Adam Mickiewicz University, 21-25 September 1998*, 94-96.

Zelčs, V. (1999). Ribbed moraines in Latvia. *In*: Kļaviņš, M. (ed.), *Zeme, Daba, Cilvēks, LU 57. konferences Ģeogrāfijas un ģeoloģijas un vides zinātnes sekcija,* 149-163. Rīga, Latvijas Universitāte (in Latvian).

Zelčs, V. (2000). Morphology, internal structure and origin of megascale flute ridges and glacial lineations in East Latvian Lowlands. *Abstracts of International Field Symposium of the Peribaltic Group and the INQUA Commission on Glaciation on Quaternary Geology in Denmark, August 29 - September 3, 2000*, 56-58.

Zelčs, V. & Dreimanis, A. (1997). Morphology, internal structure and genesis of the Burtnieks drumlin field. *Sedimentary Geology*, **111 (1-4),** 73-90.

Zelčs, V., Markots, A. & Strautnieks, I. (1990). Genesis of drumlins in the Middle Latvian lowlands. *In*: Eberhards, G., Zelčs, V. & Vanaga, A. (eds), *Acta Universitatis Latviensis*, **547,** 111-130. Rīga, Latvijas universitāte (in Russian).

Zelčs, V., Markots, A. & Strautnieks, I. (1998). New data on the bedding of the Raunis beds, Central Vidzeme, Latvia. *In*: *Sbornik tezisov K Vserosiyskomy soveschaniyu "Glavneyshie itogi v izucheniyu chetvertichnogo perioda i osnovnye napravleniya issledovaniy v XXI veke",* 158-159. St.Petersburg, VSEGEI (in Russian).

Zobens, V., Putāns, B. & Stelle, V. (1969). The first dating of absolute age in the Riga radiocarbon laboratory. *In*: Danilāns, I. (ed.), *Problems of Quaternary Geology*, **IV,** 141-145. Rīga, Zinātne (in Russian with English summary).

Quaternary Glaciations – Extent and Chronology
Editors J. Ehlers and P.L. Gibbard

A brief outline of the Quaternary of Lithuania and the history of its investigation

Rimantė Guobytė

Geological Survey of Lithuania, Konarskio 35,

LT-2600 Vilnius, Lithuania

Quaternary sediments underlie all of Lithuania. The thickness of the Quaternary sediments varies from 2-10 m in northern Lithuania, up to 250-300 m in the highlands and the buried valleys. The sediments of no less than 6 glacial and 8 interglacial periods have been identified in the country (Gaigalas, 1995; Satkūnas, 1997). The correlation of the Lithuanian Quaternary stratigraphical scheme to that of the European Quaternary Time Scale is shown in Table 1. The scheme is based on comprehensive mapping, on litho-and biostratigraphical investigations and on physical (isotopic) dating of the sediments. All stratotype sections or stratotype areas have been documented (Kondratienė *et al.*, 1993). The scheme presented here is conventional and simplified in the case of the Late Nemunas subdivision. A more detailed stratigraphic subdivision of the Upper Pleistocene was recently proposed by Satkūnas (1999).

Detailed geomorphological investigations of the land surface of Lithuania, carried out by Basalykas in the 1950s, originally served as the basis for the stratigraphical subdivision of the Nemunas (Weichselian) glaciation. His geomorphological map, which has been modified several times, is still useful today (Basalykas, 1959, 1981). The map represents a very detailed analysis of landforms, their origin and relative ages, based on high-precision field observations. However, even though the ice-marginal complexes and their comparative age are shown in detail, these maps include no glacial limits. The first attempt at drawing the positions of glacial limits in the country and to correlate them with the stadials of the Last Glaciation then used in Poland and Germany, was undertaken by Gudelis (1955, 1958). The first Quaternary stratigraphical scheme for Lithuaniaappeared in 1957 (Gudelis, 1957).

By 1962 a Frankfurt (Žiogeliai) Stage of the Last Glaciation had been established and outlined in the southeast (Vaitiekūnas, 1962; Vaitonis & Vaitiekūnas, 1981) (Fig. 1). In the published stratigraphical scheme the Late Nemunas has been subdivided into three stadials, i.e. the Grūda, Žiogeliai and Aukštaitija, which are correlated with the Brandenburg, Frankfurt and Pomeranian stages respectively (Vaitiekūnas, 1962, 1965, 1968, 1969). Furthermore, the Antaviliai and Ūla interstadials, discovered in outcrops at Ūla, Antaviliai, Zervynos and elsewhere, were tentatively placed between the Grūda, Žiogeliai and Aukštaitija Stadials (Vaitiekūnas, 1965, 1969). However, later investigations have forced a revision of the stratigraphical positions of these interstadials, some being younger (Blažauskas *et al.*, 1997) and others being older than the phases mentioned above (Satkūnas & Hütt, 1999). Gaigalas (1988, 1995), however, still considers the Pavytė Interstadial to occur between the Žiogeliai and the Baltija stadials.

Recently, much effort has been spent on the investigation of the interstadials and oscillations of the Late Nemunas glaciation. The best conditions for the accumulation of deposits from these events might have been close to the margin of the last ice sheet, where glaciolacustrine sediments are frequently deposited (Satkūnas, 1993). Over 300 boreholes have been drilled with the purpose of identifying such deposits, but without positive results. However, in the course of those investigations additional Early and Middle Nemunas interstadial sites have been found (Satkūnas *et al.*, 1997) and the stratigraphy of the sites known already has been revised (Satkūnas, 1993; Satkūnas & Grigienė, 1996, 1997a, 1997c).

Following revision of the Quaternary geological and Geomorphological maps at a scale of 1:200,000 (Guobytė, 1998, 2000), the number of Last Glaciation ice-marginal moraine ridges has been reduced because their presence could not be confirmed by field investigations. No geomorphological evidence of the existence of the Žiogeliai (Frankfurt) stadial have been so far found. Furthemore, the boundary of the South Lithuanian phase is not well established in the northeastern and western (coastal area) of Lithuania. Only geomorphologically and geologically well-established marginal ridges are interpreted as Last Glaciation stadial or phase limits.

Last Glaciation stadial and phase limits

The Last Glaciation in Lithuania is referred to as the Nemunas Glaciation (Gudelis, 1957). Its maximum extent was first drawn by Halicki (1934) in southeastern Lithuania. This line was later corrected and presented in small-scale sketches by Gudelis (1955, 1958, 1964), Vaitiekūnas (1962, 1968, 1969), Kudaba (1983), and others. Gudelis correlated the outermost boundary of the Last Glaciation with the Brandenburg stadial, and the onset of the Last Glaciation was estimated at *c.* 30,000 B.P. (Gudelis, 1964).

According to Gaigalas (1988, 1995), the last glacial period in Lithuania began at *c.* 22,000 B.P. More probably,

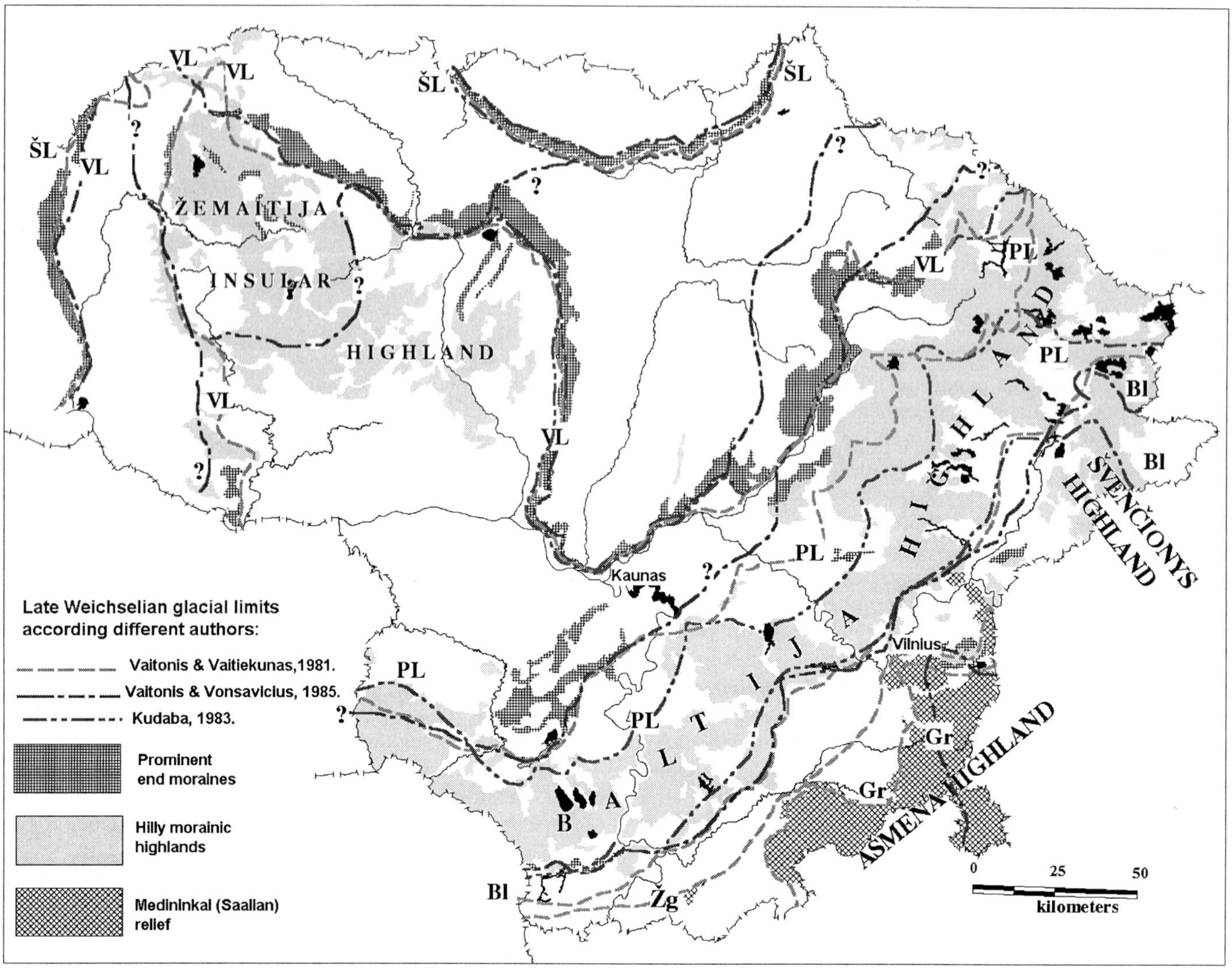

Fig. 1. Late Weichselian ice sheet marginal positions in Lithuania from different published maps. Gr – Grūda Stage, Žg – Žiogeliai Stage, Bl – Baltija Stage, PL – South Lithuanian Phase, VL – Middle Lithuanian Phase, ŠL – North Lithuanian Phase, Pj – Pajūris Phase, ? – the oscillations of the Baltija Stage.

the last ice sheet invaded the Lithuanian landmass at *c.* 25,000 B.P. (Satkūnas, 1993, 1997). The maximum extent of the Last Glaciation corresponds to the Grūda (Brandenburg) stadial.

The existence of the outermost limit of the Last Glaciation in SE Lithuania has been challenged by some researchers who were of the opinion that the whole of Lithuania was covered by the last ice sheet (e.g. Baltrūnas *et al.*, 1984). This opinion was originally presented in the published Quaternary geological maps of Lithuania (Vonsavičius, 1980; Vaitonis & Vonsavičius, 1985).

Eventually the Saalian age of the Medininkai highland had been demonstrated by detailed geological mapping in 1986-1991 (scale 1:50,000) in the vicinity of Vilnius (Satkūnas *et al.* 1991, geological report in preparation; Guobytė, 1996). The maximal position of the Last (Late Nemunas) Glaciation has been established at the foot of the Ašmena highland in southern Lithuania. There is a clear contrast between the 'fresh-looking' landforms of the Late Nemunas Glaciation and the more monotonous 'mature' old morainic landscape beyond its limit. Eemian interglacial and Weichselian interstadial deposits have been identified and dated from several marshy depressions in the Medininkai highland (Kondratienė *et al.*, 1986; Kondratienė & Vonsavičiūtė, 1986; Satkūnas & Kondratienė, 1998). They are not overlain by till.

Baltija Stadial

The name Baltija (Bl) Stage (originally: Baltija Stage) was introduced by Kondratienė *et al.* (1964). According to Gaigalas (1988, 1995), its age is *c.* 17,000 B.P. Originally, a glacial limit running along the outer margin of the Baltija Highland was called the 'Main Pomeranian Stage'. From correlation with the Danish 'Bælthav Stage', Gudelis (1955, 1958) estimated its age to be *c.* 18,500-15,750 B.P. Later this phase was renamed the 'Aukštaičiai Stage', and its age

Fig. 2. Aerial view of the ice contact slope between hummocky morainal feature (A) and outwash fan (B) to the east of Merkinė. The slope margin is traditionelly considered to be a limit of the Baltija Stage of the Last Glaciation. Aerial photograph (1952), approximate scale 1:17 000 (Archives of the Geological Survey of Lithuania).

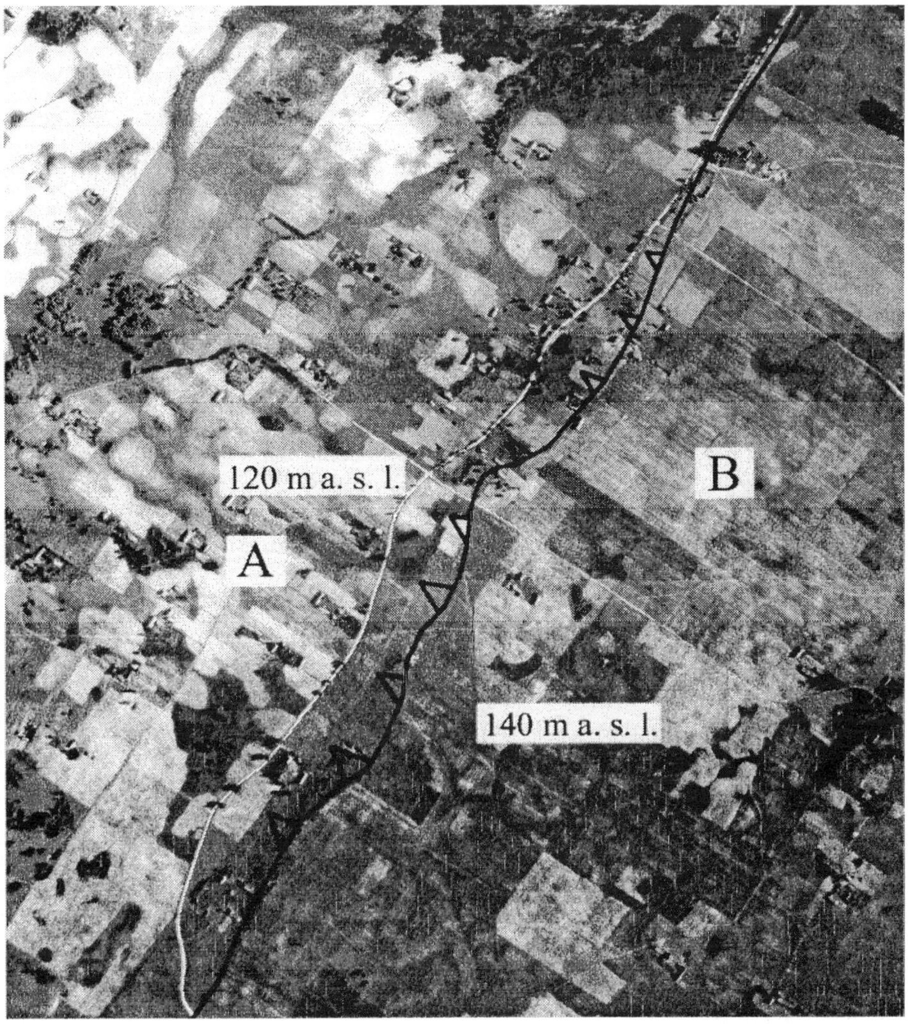

corrected to *c.* 17,340±840 – 11,630±120 B.P. (Vaitiekūnas, 1969). Since 1974, the Baltija glacial limit has also been referred to as the East Lithuanian Phase (Kudaba, 1974).

Mortensen (1924) was the first who described and showed the Baltija Highland's moraine on a map. Detailed geomorphological investigations in the Baltija Highlands were carried out by Kudaba (1974, 1983). Traditionally, the boundary of the Baltija Stage is drawn along the distal (southern) margin of the Baltija Highland (Fig.2).

South Lithuanian (Pietų Lietuvos) Phase

The South Lithuanian (PL) Phase was named by Gudelis (1955) and correlated with the 'Bælthav Phase'. The age of this phase is unknown. A fragment of a ridge, which is assigned to the PL Phase, was first mentioned by Hausen (1913).

The South Lithuanian Phase limit is correlated with some poorly-developed fragments of marginal ridges in southern Lithuania. In the eastern part of the Baltija Highland no continuous marginal ridges are observed. These marginal ridges have been formed more likely by oscillation of the Baltija (Pomeranian) Stage glacier.

Middle Lithuanian (Vidurio Lietuvos) Phase

The Middle Lithuanian (VL) Phase was also named by Gudelis (1955) and is undated. He correlated it with the Danish 'Langeland Stage' (1955, 1958). A relict of the Middle Lithuanian marginal moraine ridge was first described by Mortensen (1924).

Overall, it is a prominent moraine ridge occupying an area between Seda, in the west, to Rokiškis, in the east. The ice-pushed ridge up to 8 km wide consists of small, low hills. Small outwash plains are related to the ridge near Šiauliai and Anykščiai. The Middle Lithuanian phase limit is drawn along the the distal slope of the ridge. The boundary, however, is less-clearly expressed to the northeast of Rokiškis and to the northwest of Seda because of the generally more hilly glacial landscape around the ridge. Some investigators (e.g. Gaigalas & Kazakauskas, 1997) consider the extensive glaciolacustrine basins situated to the south of the ridge as being formed by

Fig. 3. Distal slope of the Linkuva marginal ridge representing the limit of the North Lithuanian Phase. (Photograph: V. Mikulėnas, 1999).

melting of the Middle Lithuanian phase glacier. However, on the basis of the morphology, the basins must have been formed before the VL Phase.

North Lithuanian (Šiaurės Lietuvos) Phase

The North Lithuanian (ŠL) Phase was named by Gudelis (1955). Its age has been determined at *c.* 13,000-14,000 BP (Gudelis, 1955, 1958; Vaitiekūnas, 1969; Gaigalas, 1988). Gudelis (1955, 1958) correlated it with the 'North Rügen Stage'. The ridge was first mentioned by Doss (1910).

The North Lithuanian Phase is represented by an arcuate ridge that bends towards the south, which is about 10 m high and 2 km wide. It is well expressed in the present lanscape and is known as the Linkuva Ridge (Fig. 3). This ridge is considered to represent the limit of the North Lithuanian Phase in maps and publications (Fig. 1).

Pajūris Phase

The Pajūris (Pj) Phase was introduced by Vonsavičius (1980), but has not been dated. The phase is represented by a subdued, low ridge along the Baltic Sea coast.

In the vicinity of Klaipėda, a fragment of the ridge was erroneously described by Wichdorf (1911) as a drumlin landscape. It is a correlative of the Linkuva ridge, and thus belongs to the ŠL Phase, according to some of the Lithuanian investigators (e.g. Gudelis, 1955; Basalykas, 1959; Straume, 1982). Kudaba (1983) assigns it to the VL Phase (Fig. 1). The ridge is shown to be younger than the North Lithuanian Phase in the Quaternary geological map of the Baltic States (scale 1:500,000), and in the revised Quaternary geological and Geomorphological maps of Lithuania at a scale of 1:200,000 (Vonsavičius, 1980; Guobytė, 1998, 2000). It might have been formed during an oscillation of the North Lithuanian Phase of the last glaciation.

Discussion and conclusions

The reconstruction of the Late Weichselian glacial limits is based on geomorphological evidence combined with geological data. The maximum extension of the Last Glaciation has been established in southeastern Lithuania. Here the outermost Late Nemunas glaciation boundary has been traced to the foot of the Medininkai Highland. In general, it corresponds to the boundary between different glacial landscape types. The precise position of the Last Glacial maximum is unknown.

The Late Nemunas ice advance is considered to comprise two parts, the Grūda Stadial (the older) and the Baltija Stadial. However, they can only be distinguished using lithostratigraphic criteria. The Late Nemunas maximum, i.e. the Grūda Stadial, is dated approximately to 25,000- 22,000 B.P. (Gaigalas, 1995; Satkūnas, 1997). The Baltija Stadial is correlated with Pomeranian Stage and its age is *c.* 17,000 B.P. according to Gaigalas (1995). The major part of the present topography of Lithuania was formed during this Baltija Stadial. There are several marginal morainic complexes north-west of the Baltija Highland, which may indicate three or even four additional ice front phases (South-Lithuanian, Mid-Lithuanian, North-Lithuanian and Pajūris) of the retreating Pomeranian ice sheet. These limits have been traced on the basis of various geomorphological criteria. As phase limits, mostly have been interpreted ice-marginal moraine ridges. In Lithuania there are no dates available for these stillstands, represented in recent correlation schemes (Andersen & Borns, 1994; Raukas *et al.*, 1995). No sites with Late Nemunas inter-phase deposits have been discovered in Lithuania.

References:

Andersen, B. & Borns, H. (1994). *The Ice Age World.* Scandinavian University Press. 208 pp.

Baltrūnas, V., Vonsavičius, V., Mikalauskas A., Jurgaitis, A. & Melešytė, M. (1984). O granitse poslednego lednikovogo pokrova v yugo-vostochnoy Litve. *Paleogeografiya i stratigrafiya chetvertichnogo perioda Pri-baltiki i sopredelnykh raiyonov.* Vilnius. 3-18.

Basalykas, A. (1959). Zur Frage der Entwicklung der Oberflächengestaltung Litauens (Lithuanian with German summary). *Moksliniai pranešimai, Geologija, Geografija,* **IX**, 57-86.

Basalykas A. (1981). Geomorfologinis žemėlapis, 1:1,000,000. *Lietuvos TSR atlasas.* Maskva. 50-51.

Doss, B. (1910). Über das Vorkommen einer Endmoräne, sowie Drumlins, Åsar und Bänderton im nördlichen Litthauen. *Centralblatt für Mineralogie, Geologie und Paläontologie.* 723-731.

Gaigalas, A. (1988).Geokhronologicheskaya otsenka razvitya prirodnoy sredy posdnego pleystocena i holocena na territorii Litovskoy SSR. *Isotopno-geoximicheskye issledovanya v Pribaltike i Belorussii.* Tallinn. 23-33.

Gaigalas, A. (1995). Glacial history of Lithuania. *In:* Ehlers, J., Kozarski, S. & Gibbard, P.L. (eds), *Glacial deposits in North-East Europe*, 127-137. Rotterdam, Balkema.

Gaigalas, A. (1997). The Rokai outcrop: the Middle Nemunas (Weichselian) interstadial. *The Late Pleistocene in Eastern Europe: stratigraphy, paleoenvironment and climate. Abstract volume and Excursion Guide of the INQUA-SEQS Symposium.* Vilnius. 82-84.

Gaigalas, A., Arslanov, Kh., Tertychnaja, T., Banys, J., Melešytė, M. & Brazauskas, J. (1986). O vozraste srednenemunskikh (srednevaldayskikh) mezhstadialnykh otlozheniy po dannym izuchenya razreza Rokai na r. Jiesia. *Issledovanya lednikovykh obrazovaniy Pribaltiki.* Vilnius. 82-88.

Gaigalas, A., Arslanov, Kh., Chernov, T., Tertychnaja, T. & Kazarceva, T. (1994a). New radiocarbon dating of the Middle Nemunas (Valday) in the Rokai exposure near Kaunas. *Conference on geochronology and dendrochronology of old towns and radiocarbon dating of archeological findigs, Lithuania, Vilnius October 31-November 4. Abstract and Papers,* 16.

Gaigalas, A. & Hütt, G. (1995). OSL dating of Merkinė (EEM) Interglacial (in Jonionys) and the Nemunas Glaciation (Rokai Section) in Lithuania. *Landscapes and Life. Studies in honour of Urve Miller. PACT,* **50**, 59-69.

Gaigalas, A. & Kazakauskas, V. (1997). Sedimentation sequence of glaciolacustrine clays of Baltija Stage. *The Late Pleistocene in Eastern Europe: stratigraphy, paleoenvironment and climate. Abstract volume and Excursion Guide of the INQUA-SEQS Symposium.* Vilnius. 15-16.

Gaigalas, A., Hütt, G. & Melešytė, M. (1994b). The OSL age of the Merkinė (Mikulino) Interglacial and the Nemunas (Valday) glacial period in Lithuania. *Conference on geochronology and dendrochronology of old towns and radiocarbon dating of archeological findigs, Lithuania, Vilnius October 31-November 4. Abstract and Papers,* 15.

Gaigalas, A., Molodkov, A. & Melešytė, M. (1994c). The first ESR dating results of Būtėnai (Likhvin) and Merkinė (Mikulino) Interglacial deposits in Lithuania. *Conference on geochronology and dendrochronology of old towns and radiocarbon dating of archeological findigs, Lithuania, Vilnius October 31-November 4. Abstract and Papers,* 14.

Gaigalas, A., Serebryanny, L. & Valueva, M. (1992). Middle Valdaian forest environment at Biržai, northern Lithuania. *Boreas,* **21**, 289-293.

Gudelis, V. (1955). Lietuvos TSR Baltijos pajūrio geologinės raidos vėlyvajame glaciale ir postglaciale (Holocene) pagrindiniai etapai. *Vilniaus V. Kapsuko vardo universiteto mokslo darbai,* **VII**, 119-139. Vilnius.

Gudelis, V. (ed.) (1957). The explanatory note to the regional stratigraphical scheme of Quaternary deposits of the Baltic states and the Byellorussian SSR (In Russian). *Scientific reports. Academy of Science of Lithuania,* **IV**, 425-437.

Gudelis, V. (1958). Some problems of stratigraphy and paleogeography of Late-Quaternary in Europe and in North America according to new data. (In Lithuanian with English summary). *Geografinis metraštis,* **I**, 145-174. Vilnius.

Gudelis, V. (1964). *Reljef i chetvertichnye otlozhenya Pribaltiki.* Vilnius. 262.

Guobytė, R. (1996). *Quaternary geological map of vicinities of Vilnius, scale 1:50,000.* Geological Survey of Lithuania, Vilnius.

Guobytė, R. (1998). *Quaternary geological map of Lithuania, scale 1:200,000.* Geological Survey of Lithuania, Vilnius.

Guobytė, R. (2000). *Geomorphological map of Lithuania, scale 1:200,000.* Geological Survey of Lithuania, Vilnius.

Halicki, B. (1934). *Mapa diluwium pn-wsch. Polski. Skala 1:300,000.* Wilno.

Hausen, H. (1913). Über die Entwicklung der Oberflächenformen in den Russischen Ostseeländern und angegrenzenden Gouvernments in der Quartärzeit. *Fennia,* **34, 3**. Helsingfors.

Kondratienė, O. (1965). Stratigraficheskoye raschleneniye pleystocenovykh otlozheniy yugo-vostochnoy chasti Litvy na osnove palinologicheskikh dannykh. *Stratigrafiya chetvertichnykh otlozheniy i paleogeografiya antropogena Yugo-Vostochnoy Litvy.* Vilnius. 189-261.

Kondratienė, O. & Riškienė, M. (1983). Lednikovaya flora v doline reki Jiesia (Litva). *Palinologiya v geologicheskikh issledovaniyakh Pribaltiki I Baltiyskogo moria.* Riga. 54-57.

Kondratienė, O. & Vonsavičiūtė, A. (1986). Novye stratigraficheskye dannye o granitse Nemunskogo lednikovogo pokrova na territorii Litvy. *Issledovanya lednikovykh obrazovaniy Pribaltiki.* Vilnius. 48-59.

Kondratienė, O., Klimasauskas, A., Gaigalas, A. & Cepulyte, V. (1965). Table 59 (Appendix). *In:* Garunkstis, A. (ed.), *Stratigrafiya chetvertichnykh otlozheniy i paleogeografiya antropogena Yugo-Vostochnoy Litvy.* Vilnius.

Kondratienė, O., Šinkūnas, P., Gaigalas, A. & Satkūnas, J. (1993). Stratotypes of Quaternary of Lithuania. (in Russian with English summary). *In:* Kondratienė, O. (ed.), *Catalogue of Quaternary stratotypes of the Baltic region,* 7-30. Vilnius.

Kondratienė, O., Žiedelis, A. & Riškienė, M. (1986). Usloviya zaleganiya i inzhenerno-geologicheskaya kharakteristika Merkinskikh otlozheniy na Medininkskoy vozhvyshennosti. *Issledovanya lednikovykh obrazovaniy Pribaltiki.* Vilnius. 32-47.

Kudaba, Č. (1974). Vzaimosviaz krajevykh lednikovykh i distalnykh vodno-lednikovykh obrazovaniy (na primere Baltiyskikh vysot). *Predfrontalnye krayevye lednikovye obrazovanya.* Vilnius. 44-59.

Kudaba, Č. (1983). *Lietuvos aukštumos.* Vilnius. 186 pp.

Mortensen, H. (1924). Beiträge zur Entwicklung der glacialen Morphologie Litauens. *Geologisches Archiv.* **3, 1-2**, 1-93.

Raukas, A., Aboltinš, O. & Gaigalas, A. (1995). The Baltic States. *In:* Schirmer W. (ed.), *Quaternary field trips in central Europe. XIV INQUA Congress, Berlin*, **1**, 146-151.

Satkūnas, J. (1993). Stratigrafya i litologicheskiy sostav srednenemunskoy (srednevaldaiyskoy) podsvity po dannym kartirovaniya Vilniasko-Nerisskoy glatsiodepressii (Yugo-Vostochnaya Litva). *Chetvertichnye otlozhenya i noveyshaya tektonika lednikovykh oblastey Vostochnoy Evropy*, 39-48. Apatity.

Satkūnas, J. (1997). Outline of Quaternary stratigraphy of Lithuania. *The Late Pleistocene in Eastern Europe: stratigraphy, paleoenvironment and climate. Abstract volume and Excursion Guide of the INQUA-SEQS Symposium*, 65-68. Vilnius.

Satkūnas, J. (1999). The Upper Pleistocene stratigraphy and geochronology in Lithuania. *National scientific programe. Evolution of Earth and forecast of its resources changes in Lithuania. Lithosphere*, **3**, 43-57.

Satkūnas, J. & Grigienė, A. (1996). Correlation of stratigraphic events of Upper Pleistocene in central and peripherial parts of the Last Glaciation. Excursion guide of the international workshop in Lithuania, October 3-6, 1996. *Correlation of stratigraphic events of Upper Pleistocene in central and peripherial parts of the Last Glaciation. Abstract volume and excursion guide of the international workshop in Lithuania, October 3-6, 1996*,41-88. Vilnius.

Satkūnas, J. & Grigienė, A. (1997a). The Jonionys site-sequense of the Eemian interglacial and the Weichselian interstadials. *The Late Pleistocene in Eastern Europe: stratigraphy, paleoenvironment and climate. Abstract volume and Excursion Guide of the INQUA-SEQS Symposium*, 77-81. Vilnius.

Satkūnas, J. & Grigienė, A. (1997b). The Medininkai section - typical Eemian-Weichselian sequense outside the Weichselian glaciation. *The Late Pleistocene in Eastern Europe: stratigraphy, paleoenvironment and climate. Abstract volume and Excursion Guide of the INQUA-SEQS Symposium*, 89-91.Vilnius.

Satkūnas, J. & Grigienė, A. (1997c). The Mickūnai glacial depression and the Eemian-Weichselian sequences inside the glaciated area. *The Late Pleistocene in Eastern Europe: stratigraphy, paleoenvironment and climate. Abstract volume and Excursion Guide of the INQUA-SEQS Symposium.* Vilnius. 92-95.

Satkūnas, J. & Hütt, H. (1999). Stratigraphy of the section Antaviliai, eastern Lithuania, and its implication for the Upper Weichselian climatostratigraphic subdivision. *Geological Quarterly*, **43, 2**, 213-218.

Satkūnas, J. & Kondratienė, O. (1995). Quaternary stratigrafic scheme of Lithuania for nation geological mapping. *XIV International congress, INQUA, Berlin, August 3-10. Abstracts*, 239.

Satkūnas, J. & Kondratienė, O. (1998). Merkinė (Eemian) Interglacial deposits in the Vilnius region and the limit of the Last Glaciation in Southern Lithuania. *Pact*, **54**, 203-210.

Satkūnas, J., Guobytė, R., Versickas, J., Damusyte, A. (1991). Report of integrated geological hydrogeogical mapping at a scale of 1:50 000 of Vilnius, Svencionys district and in Belarus area. (Project leader J. Satkunas). Book 1, 2. (With maps). Vilnius. Geological Survey of Lithuania.

Satkūnas, J., Grigienė, A., Guobytė, R., Marcinkevičius, V. & Račkauskas, V. (1997). Upper Pleistocene stratigraphy based on data from Smalvas and Dysnai localites, Northeastern Lithuania. *Geologija*, **22**, 26-35. Vilnius.

Straume, J. (ed.) (1982). *Geomorphological map of the Soviet Baltic Republics, 1:500,000.* Leningrad.

Vaitiekūnas, P. (1962). Lietuvos teritorijos regioninė padėtis pleistoceninių apledėjimų srityje. *Geologija ir Geografija*, **I**, 183-199. Vilnius.

Vaitiekūnas, P. (1965). Stadyalnost poslednego oledenenya v juzhnoy Pribaltike i strukturno-facialnye osobennosti stadyalnykh I mezhstadyalnykh obrazovaniy. *Krayevye obrazovanya materikovogo oledenenya*, 195-208.Vilnius.

Vaitiekūnas, P. (1968). Degradacya poslednego skandinavskogo lednikovogo pokrova na territorii Pribaltiki. *Geografija ir Geologija*, **V**, 123-135. Vilnius.

Vaitiekūnas, P. (1969). Über stratigraphische Gliederung der jungpleistozänen Bildungen auf dem Inlandvereisungsgebiet (Baltikumsgebiet). *Inlandvereisung und glazigene Morphogenese. Zum VIII. INQUA Kongress, Frankreich, 1969*, 227-268 (Russian with German and French summaries), Vilnius.

Vaitonis, V. & Vaitiekūnas, P. (1981). Kvartero nuogulos, 1:1,000,000. *Lietuvos TSR atlasas.* Moskva. 34-34.

Vaitonis, V. &Vonsavičius, V. (1995). Lietuvos kvartero geologinis žemėlapis, 1:1,000,000. *Lietuvos geologija.* Vilnius. 1995. Appendix.

Vonsavičius, V. (ed.) (1980). *Geological map of the Quaternary deposits of the Soviet Baltic Republics, 1:500,000.* Leningrad.

Vonsavičius, V. & Klimašauskas, A. (1972). Pirmasis tarpmoreninis ežerų mergelio radinys Lietuvoje. *Lietuvos kvarterinių naudingųjų iškasenų geologijos klausimai*, 107-112, Vilnius.

von Wichdorf Hess, H. (1911). Über eine Drumlinlandschaft bei Nimmersatt an der russischen Grenze. *Jahrbuch der Preussischen Geologischen Landesanstalt und Bergakademie zu Berlin,* **32, 1-2**, 407-411.

Quaternary Glaciations – Extent and Chronology
Editors J. Ehlers and P.L. Gibbard

Pleistocene glaciation in The Netherlands

Cees Laban [1] and Jaap J.M. van der Meer [2]

[1] *TNO-NITG, P.O. Box 80015 3508 TA Utrecht, The Netherlands; c.laban@nitg.tno.nl*

[2] *Fysisch Geografisch en Bodemkundig Laboratorium, University of Amsterdam, Nieuwe Prinsengracht 130, 1018 VZ Amsterdam, the Netherlands;*

present address: Dept. of Geography, Queen Mary and Westfield College, Mile End Road, London El 4NS, U.K.; E-mail: j.meer@qmw.ac.uk

Pre-Elsterian glaciations

Introduction

Glacigenic or 'cold' marine and fluvial sediments related to Early Pleistocene glaciations are only present at a few locations in boreholes in the Dutch sector of the North Sea or on the Dutch mainland. In the Dutch sector, the oldest indications of glacial influence probably date from the Tiglian C4c cold Stage. Stiff marine clays recorded in at least one, or possibly two boreholes are thought to date from this early glaciation. Indications of sediments deposited during the Eburonian Stage in a pro-delta and delta front environment are present in a deep borehole and are referred to the Winterton Shoal Formation. Fluvial sediments deposited by the rivers Elbe and Weser during the Menapian Stage are present in only two boreholes. The oldest glacigenic sediments recorded have been found in a borehole near the Dogger Bank and were probably deposited during the Cromerian Glacial A or B (Laban, 1995). 'Cold' marine sediments of Cromerian age are known to be present in a borehole farther east. Correlation of the 'cold' sediments between boreholes is not possible because of the great distance between localities and also the fact that in most boreholes sediments of only one cold stage are recorded.

Tiglian C4c glacial phase and Eburonian Stage

In the Dutch sector of the North Sea the oldest deposits indicating a cold Pleistocene phase are marine deposits probably dating from the Late Tiglian (C4c). They have been found in the well-cored borehole GI 6-22 in which a deposit of stiff clay is present between 260 m and 253 m below MSL (Laban, 1995). Based on the evidence of dinoflagellate cyst analysis the clay probably settled in a shallow marine (prodelta, delta-front to mouth bar environment) low salinity environment during the Late Tiglian and Early Eburonian. The pollen content of the clay is poor, indicating cold climatic conditions (de Jong, 1991). The foraminiferal fauna is poor to very poor, again pointing to arctic conditions with species such as *Elphidium albiumbilicatum, Elphidium excavatum f. clavata* and *Nonion orbiculare* (Neele, 1991b). On top of this clay (between 242.90 m and 198.10 m), sand is present, thought to have been deposited in a fluvial environment during the Eburonian Stage (Burger, 1992). During deposition of the fluvial sand the pollen indicate cold climatic conditions (de Jong, 1991). On land periglacial structures, like frost wedges and involutions (van Straaten, 1956; Maarleveld, 1960) point at widespread cold conditions during different stages of the early Pleistocene.

Menapian Stage

Indications of cold, post-Eburonian and pre-Crornerian conditions have been encountered in several boreholes in the Dutch sector of the North Sea. In borehole F8-6, a stiff marine clay occurs between 244.60 m and 242.50 m below MSL, its foraminiferal content (dominated by *Elphidium excavatum f. clavata*) represents a rich, high arctic, fauna poor in species. Between 242.56 m and 242.50 m a relatively high percentage of *Nonion orbiculare* is present (Neele, 1991a), molluscs are absent (Pouwer, 1991), while pollen indicate cold climatic conditions (Cleveringa; pers. comm.). In borehole LI 0-6 fluvial sediments are present below Cromerian III/IV deposits (Zagwijn, 1977) between 129.35 m and 80.35 m below MSL. They were deposited under cold climatic conditions during the Menapian Stage. The pollen assemblage contains a mixture of Pleistocene and Tertiary pollen which have been derived from glaciofluvial deposits. Near the base of the borehole, at 129.35 m below MSL the pollen assemblage contains a high percentage of herbaceous pollen also suggesting cold conditions. In borehole GI 6-22 fluvial sand is present between 179 m and 149 m below MSL. Again the pollen record indicates the sand accumulated under a cold climate (de Jong, 1991) and is again correlated with the Menapian Stage. In The Netherlands deposition of the Enschede Formation, of central European origin, continues into the Cromerian. At the base of the Enschede Formation, the Hattem Complex/Layer is found in the Netherlands and in adjoining parts of Germany. This lag deposit is taken as evidence of glaciation in northern Europe, because it contains (weathered) Scandinavian erratics that could not have reached the Netherlands by fluvial transport alone (van der Meer, 1987).

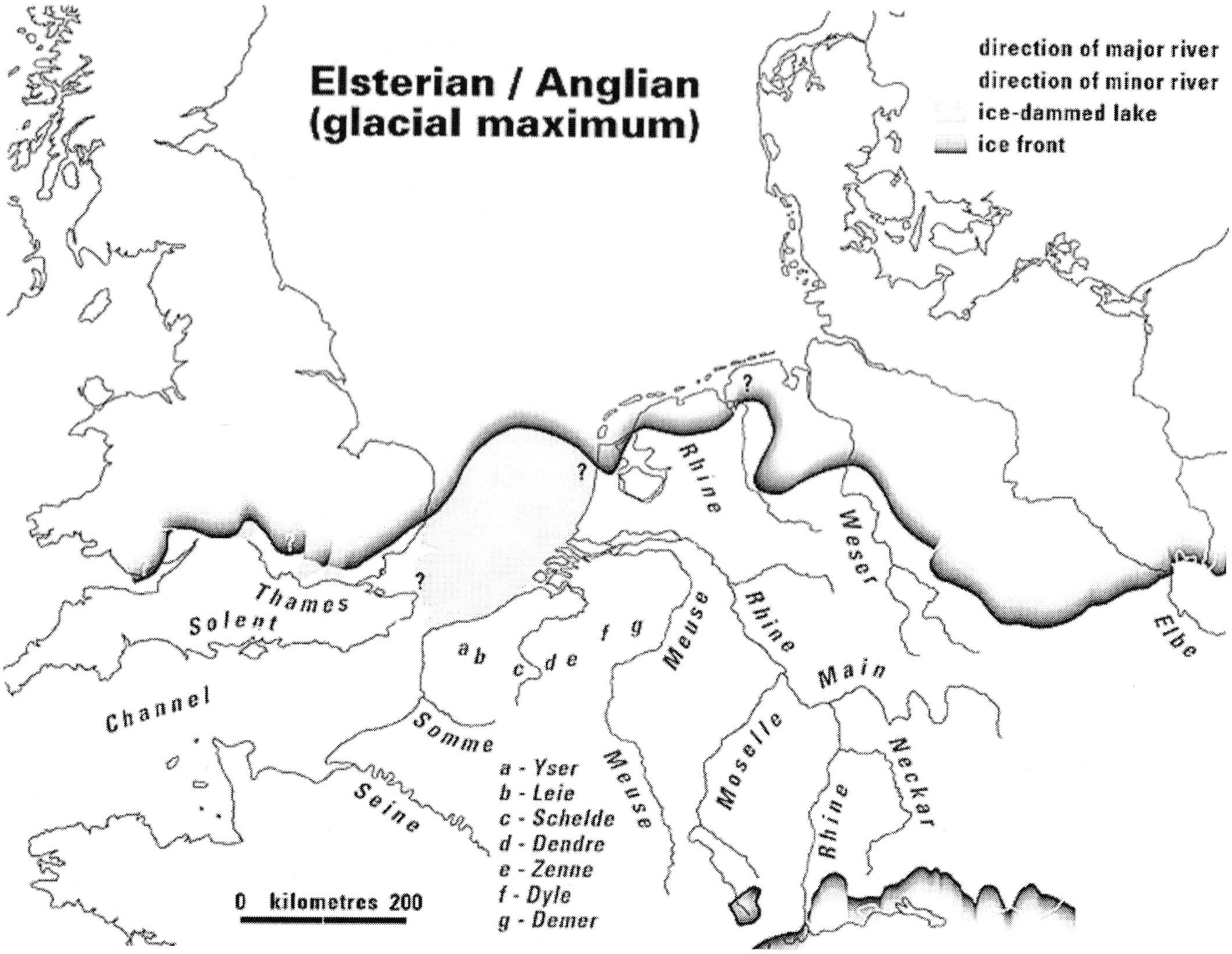

Fig. 1. Elsterian drainage in the southern North Sea region.

Cromerian Stage

Deposits dating from the 'Cromerian Complex' Stage are found in the Dutch sector in borehole E I -10 drilled on the south-east flank of the Dogger Bank. In this borehole very fine, grey, slightly silty sand with sporadic clay laminae and organic matter has been sampled between 153.20 m and 130.30 m. Pollen analysis points to a Scandinavian, non-British glacigenic provenance (de Jong, pers. comm.), while foraminifera or molluscs are absent (Neele, 1986b). Zagwijn (1986) suggests that the age of this bed is Cromerian and probably belongs to Glacial B or A. The sand layer is overlain by a marine bed which Zagwijn (1986) suggested to be probably of Cromerian III age at the base and Glacial C towards the top. Near borehole E I - 10, borehole E8-6 in block E8 was drilled to a depth of 183.50 m below MSL. In this borehole a clay bed is present between 152.50 m and 147.50 m in with a high arctic foraminiferal fauna, dominated by *Elphidium excavatum f clavata, Nonion orbiculare,* and *Elphidium ustulatum* (Neele, 1991b). The pollen record also shows evidence of cool climatic conditions (de Jong, pers. comm.). The glacial, laterally equivalent sand deposit in borehole E I - 10 is, however, non-marine and it is suggested that a coastline lay between these two boreholes at this particular time. Evidence for 'cold' deposits on land during the Cromerian Stage were found in the eastern Netherlands near Emmerschans. Here glaciofluvial sands, the 'Weerdinge Beds', are regarded as glaciofluvial, and deposited close to an ice margin during the Cromerian Glacial C (Ruegg & Zandstra, 1977; de Jong & Maarleveld, 1983) because like the Hattem Layer they contain Scandinavian erratics.

The Elsterian glaciation

The Elsterian glaciation in the Netherlands

The effect of the Elsterian glaciation on the northern Netherlands has always been enigmatic, as a morphologically distinct ice margin is absent, and tills have only been recognised relatively recently. This resulted in two types of reconstruction of the maximum extent of the Elsterian ice sheet. One type showed the ice margin coming from the east to take a wide swing N of the Dutch mainland. The other followed the southern boundary of the Elsterian Peelo Formation.

At present the following Elsterian deposits and landforms are known on land: in the northern Netherlands, Elsterian meltwater deposits occur in depressions varying in depth between *c.* 10 m and over 100 m. They consist of hard, dark brown clay, fine, micaceous, locally laminated, sand and fine- to medium-grained aeolian sands.

In a pit near Peelo, the type locality of the Elsterian Peelo Formation (Zagwijn, 1973; Ruegg, 1975), typical Scandinavian gravel has been found at the top of the Elsterian deposits. In two boreholes, notably Den Burg (913-36) and Tzum (6D-49) in Friesland, rhythmites of Elsterian age have been found. The Tzum borehole was drilled in a depression over 100 m deep. Ter Wee (1983a) and Bosch (1990) describe 20 km to 30 km long channels, 3 km to 5 km wide and locally up to 350 m deep from the northern Netherlands. Measurements on sedimentary structures point to an infill from north to south (Bosch pers. comm., 1994). In the northern Netherlands, Elsterian till has been found in only three boreholes. In borehole 5B-7 in the Waddenzee, Oostmeep, 0.30 m of till has been sampled at a depth of between 51.70 m and 51.40 m at the base of a depression and in a relatively small glacial valley. The gravel in this till is only of eastern provenance, Scandinavian rock fragments are lacking. In the Witmarsum borehole (I OB/191) in Friesland greyish-black, calcareous, very sandy till at a depth of between 226 m and 220 m a was sampled at about 20 m above the floor of a glacial valley. This till is poor in coarser particles and consists of both fluvial material of eastern provenance and a few fragments of flint and Fennoscandinavian granite (Zandstra, 1983). In another borehole on the former island of Wieringen (I 4E/ I 10, Stroe 11), two till layers occur. The deepest till is present between 60 m and 59 m at the base of a glaciolacustrine clay, equated with the Peelo Formation. The upper till between 13.40 m and 1.20 m belongs to the Drente Formation (Saalian glaciation). The gravel content of the lower Elsterian 'till' is of eastern origin and derived from the fluvial deposits of mid-German rivers. The flint, however is of typical Scandinavian provenance (Zandstra, 1986).

The Elsterian glaciation in the Dutch sector of the North Sea

The first glaciation in the North Sea from which abundant glacial sediments and features are preserved is the Elsterian Stage. The most striking feature of this glaciation is the occurrence of a complex system of deeply-eroded anastomosing valleys and isolated oval depressions in the southern North Sea in a broad zone between 53° N and 56° N (Fig. 3). The valleys have been eroded into pre-Elsterian deposits. The zone containing this type of depressions continues towards the west into the British sector (Cameron *et al.*, 1986; Jeffery *et al.*, 1989; Cameron *et al.*, 1992). The valleys in the Dutch sector of the North Sea have been mainly mapped from a regular and dense pattern of north-east/south-west and north-west/south-east seismic lines. The valleys appear to occur within the maximum ice limit of the Elsterian glaciation and mainly trend north-north-west/south-south-east. A difficulty in interpreting the valleys from seismic lines is that the shoulders are not always clearly defined or visible. They occur on the upper part of the profiles where they are often obscured by multiple reflections from the sea bed, even after processing. Most of the valleys have steep slopes in cross section with angles ranging between 5 and 25° in the Dutch sector of the North Sea (Joon, 1987) and up to 55° on land in northern Germany (Kuster & Meyer, 1979). From the longitudinal profile it is clear that the base of the valleys is not flat but shows sub-basins and thresholds. The valleys vary in width between less than 1 km and 23 km, and are generally 100 m to 250 m deep, while exceptionally a depth of about 510 m below MSL is reached. The valleys are most numerous between 53° N and 54° N. South of 53° N comparable valleys are absent. On two seismic lines between 52°50' N and 53° N and *c.* 02°27' E, some valleys up to 50 m deep and several kilometres wide are present incised into the Yarmouth Roads Formation. They are probably associated with the southern end of deep valleys north of the 53° N (Cameron *et al.*, 1984). Between 54° N and 56° N only isolated valleys are present. Based on the presence of these valleys, as well as the presence of glaciotectonic structures, the maximum extent of the Elsterian ice has been reconstructed (Fig. 2). It ran from the Dutch coast at 52°50'00"N in a south-west direction towards Ipswich in East Anglia, near the British east coast. It appears to link up well with what is known from the British side.

Drainage history during the Elsterian glacial maximum

Gibbard (1988) reconstructed the palaeogeography of the Elsterian (Anglian) Stage at its glacial maximum with an ice-dammed lake in front of the ice sheet in the southern North Sea (Fig. 1). Drainage partly took place over the Wealden-Artois anticline and over the north-western French coastal. area. This is supported by the occurrence of fluvial sediments at Wissant in France (Roep *et al.*, 1975) dating from the Saalian or an older glaciation. Sea-level according to Valentin (1952) was 90 m below present level and Fairbridge (1961) calculated a sea-level of about 48 m below present level. Praeg (1994) in his reconstruction projected a lake at least 100 m deep in the southern North Sea dammed by the Wealden-Artois anticline in the south and the ice-margin in the north.

However, the drainage history of meltwater rivers during the Elsterian glaciation is still problematical. Until now, no depositional evidence has been found to support the presence of a lake between the Wealden-Artois anticline

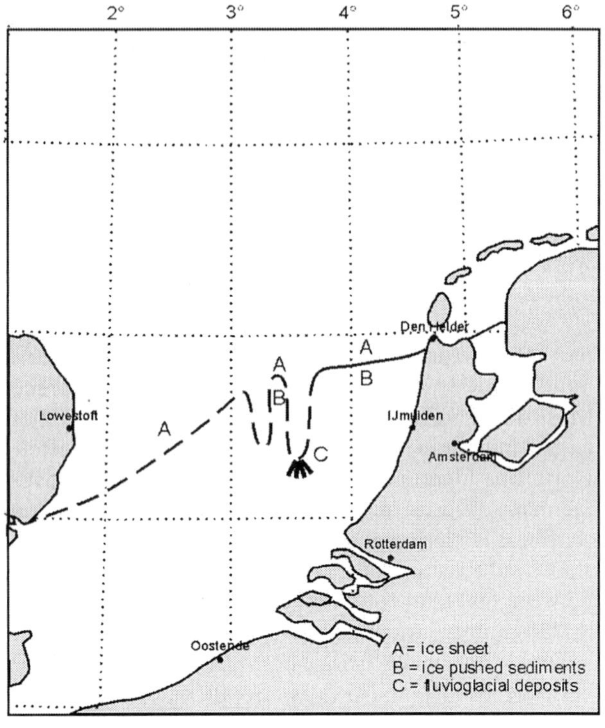

Fig. 2.The maximum extent of the Elsterian ice sheet in the southern North Sea.

in the south, and an ice sheet in the north. In the Dutch sector the glacigenic sediments must have settled on top of fluvial deltaic sediments and in the British sector on top of mainly Lower Pleistocene and Tertiary clay or clayey sand deposits. The possibility exists that there was an overspill of water over the Wealden-Artois anticline. If it is assumed that during the Elsterian glaciation large amounts of sediment became available, then it must have been deposited elsewhere. This leads to the conclusion that no single large lake existed and sediments were transported directly south into the Atlantic Ocean. The present deeps in the southern part of the North Sea are probably the channels along which transport of the material took place. However, as Elsterian sediments are unknown from these channels post-Elsterian erosional processes must have further deepened the channels thereby removing the Elsterian sediments.

The Saalian glaciation in the Netherlands

Introduction

Many of the sediments and features associated with the cold Saalian Stage are much better known than the preceding stage. As a result of this the geometry of the Saalian deposits has been mapped in more detail. From the period immediately preceding the Late Saalian glaciation the only sediments present in the northern part of the Dutch sector of the North Sea are very fine- to fine-grained periglacial sediments. In a valley in the eastern part of the sector, however, marine sediments are present, overlain by Late Saalian till, which may indicate arctic to boreal marine conditions during an Early Saalian Interstadial. On land Saalian deposits mainly consist of coarser fluvial sediments of the Rhine-Meuse system. The Scandinavian ice sheet entered the southern North Sea during the Late Saalian and extended into part of the area. Stiff to very stiff glaciolacustrine clay was deposited in the north and north-west parts of the Dutch sector. Tills are present near the northern coast of The Netherlands and based on their lithology two different facies can be distinguished. In the east, the till contains up to 47 % of gravel, while in the west only up to 3% of gravel has been recorded. At sea fine- to medium-grained glaciofluvial sediments are only found very locally. The extent of the formation was probably greater, but much of it was reworked during the Eemian transgression. On land, glaciofluvial sediments are mainly known from the fringe of the ice-pushed ridges, while it is now known that glaciofluvial deposits are also included in the ice-pushed ridges (Ruegg, 1977). Here the extent may never have been much greater, as this fringe was in touch with the westward diverted Rhine. In the eastern part of the country a continuous esker/entrenched valley has been mapped, feeding into the proglacial meltwater-Rhine system (van den Berg & Beets, 1987). In the North Sea, the occurrence and nature of the Saalian subglacial valleys is

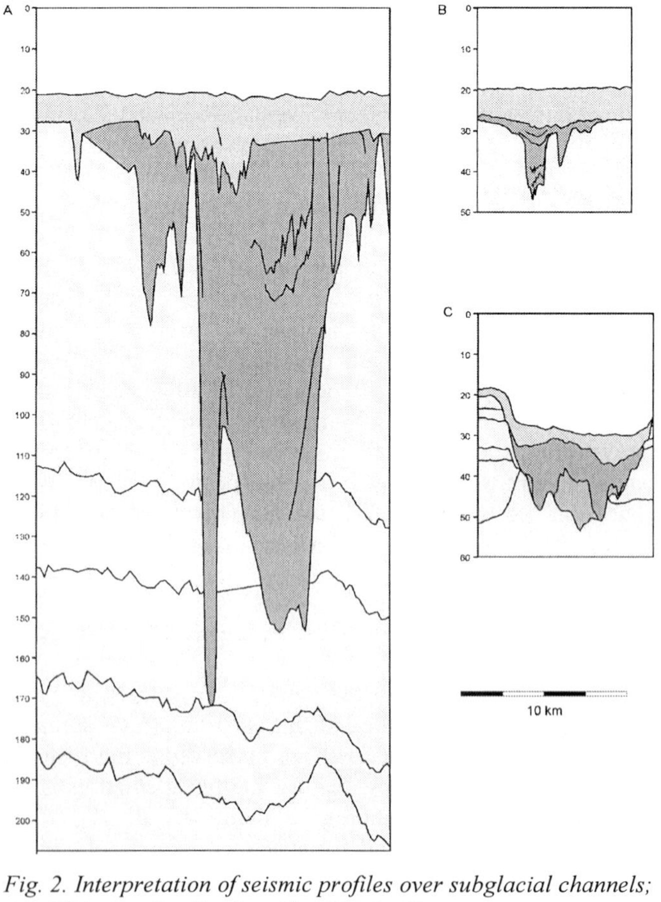

Fig. 2. Interpretation of seismic profiles over subglacial channels; A = Elsterian, B = Saalian, C = Weichselian.

problematical. The dimensions of the valleys are much smaller than those of the Elsterian glaciation and they are partly filled with Eemian marine sediments. Since, on 3.5 kHz seismic records, the base of some of the valleys occurs below the penetration depth, it is difficult to distinguish Saalian valleys from those formed during the Eemian transgression. Tongue-shaped basins (with deformation structures along their flanks) although forming a continuous belt on land (de Gans *et al.,* 1987; van der Wateren, 1992) have only been found at two locations at sea (Laban, 1995).

Saalian glaciation in the Netherlands

The glaciation of the northern half of the country has had a pronounced influence on the geomorphology of The Netherlands. In this relatively short timespan all the relief in this part of the country has been created. The present-day relief has been strongly influenced by younger processes; originally the largest height differences created by the ice must have been in the order of 250 metres. A new map of the glacial geomorphology of The Netherlands was published in 1987 (van den Berg & Beets, 1987); it shows the extent of ice-pushed ridges surrounding five (known) deep glacial basins as well as the extent of glaciofluvial deposits. The ice-pushed ridges are asymmetric in cross section with steep proximal and more gentle distal slopes, the latter grading into glaciofluvial deposits. The ice-pushed ridges consist mainly of older Pleistocene and Tertiary deposits, with a trend towards increasingly younger materials towards the West. In the eastern part of the country, Tertiary deposits are involved, while in the central part the pushed sequence consists of Early and Middle Pleistocene deposits. The westernmost basins are deeply buried under younger deposits and are the least known (de Gans *et al.,* 1987); an overview of the glaciotectonics is given by van der Wateren (1981, 1985, 1992). In the northern part of the country, the Saalian till plateau ocurs, similar to the 'Geest' areas in northern Germany. The most prominent feature on this slightly undulating plateau is a series of streamlined ridges, collectively known as the Hondsrug, and nowadays interpreted as a set of megaflutes (Rappol, 1983; 1984). The strong NW-SE orientation of these ridges, in combination with the till stratigraphy, plays an important role in the reconstruction of events during glaciation. Moreover, the stratigraphy of the Saalian till is quite well known nowadays. The most complicated element is the occurrence of 'red' till floes of completely different composition and derivation (Rappol, 1987; van der Meer, 1987 and references therein). In the Netherlands the interpretation of these red till floes is difficult because the top of the till is always eroded and the original configuration of the floes is unknown (van der Meer & Lagerlund, 1991). Even when the till surface has been preserved by postglacial burial, as exposed in a deep pit near Grouw in the northern Netherlands, the till had been weathered as shown by decalcification and soil formation (unpublished). Till stratigraphy has been studied in a number of deep exposures related to motorway construction. Provenance and directional studies have demonstrated a more complex stratigraphy than hitherto suggested, i.e. it was found that 'red' till occurs at two stratigraphic levels (Rappol *et al.*, 1989).

The studies also demonstrate that on land the last recorded ice-movement was from the NW (Rappol *et al.,* 1989; Kluiving *et al.,* 1991). As the Scandinavian ice was moving upslope and also created its own relief, ice-dammed lakes developed. The most prominent of these must have occurred in the central Veluwe area, around the present-day Leuvenumse Beek valley. From this area mass-flow deposits related to this lake have also been described (Postma *et al.*, 1983; Postma, 1997). During deglaciation lakes developed in the glacial basins, as demonstrated by the deposition of rhythmites (de Gans *et al.,* 1987). The largest of these lakes, in the present-day IJssel valley, was part of the course of the river Rhine and acted as a sediment trap. Because of the sediment depletion the 'Rhine', after leaving the lake, cut a deep channel in the subsoil of the northern Netherlands (van den Berg & Beets, 1987). This channel had previously been interpreted as ice-marginal in origin and formed during the glacial advance.

Saalian glaciation in the Dutch sector of the North Sea

In the south-western and north-eastern parts of the Dutch sector of the North Sea a range of Saalian glacial and periglacial sediments and phenomena is found. Based on these, the outer limit of the Saalian glaciation is drawn in the Dutch sector of the North Sea (Fig. 4). The most obvious feature of this limit is that it curves towards the NE and does not cross the North Sea. This implies that at least in the southern North Sea there was no connection between Scandinavian and British ice. If there was British ice, it has not left recognizable traces in the southern North Sea. In The Netherlands all Saalian glacial sediments are placed in one lithostratigraphic unit, the Drente Formation (Zagwijn, 1961). This formation includes one till sheet, indicating a single ice advance. This contrasts with northern Germany where three Saalian tills occur in certain areas. Correlation with the glacial deposits on land in the northern Netherlands (ter Wee, 1983b; van den Berg & Beets, 1987) and northern Germany (Ehlers, 1990) indicate that the advance of Scandinavian ice in the North Sea took place during the Drenthe Substage.

Reconstruction of the maximum extent of the Saalian ice sheet in the Dutch sector of the North Sea

The reconstruction of the maximum extent of the ice sheet by the present authors is based on a study of its legacy (sediments and landforms). Within the maximum extent of the Saalian ice in the Dutch sector of the North Sea till plateaux and tongue-shaped basins were formed together

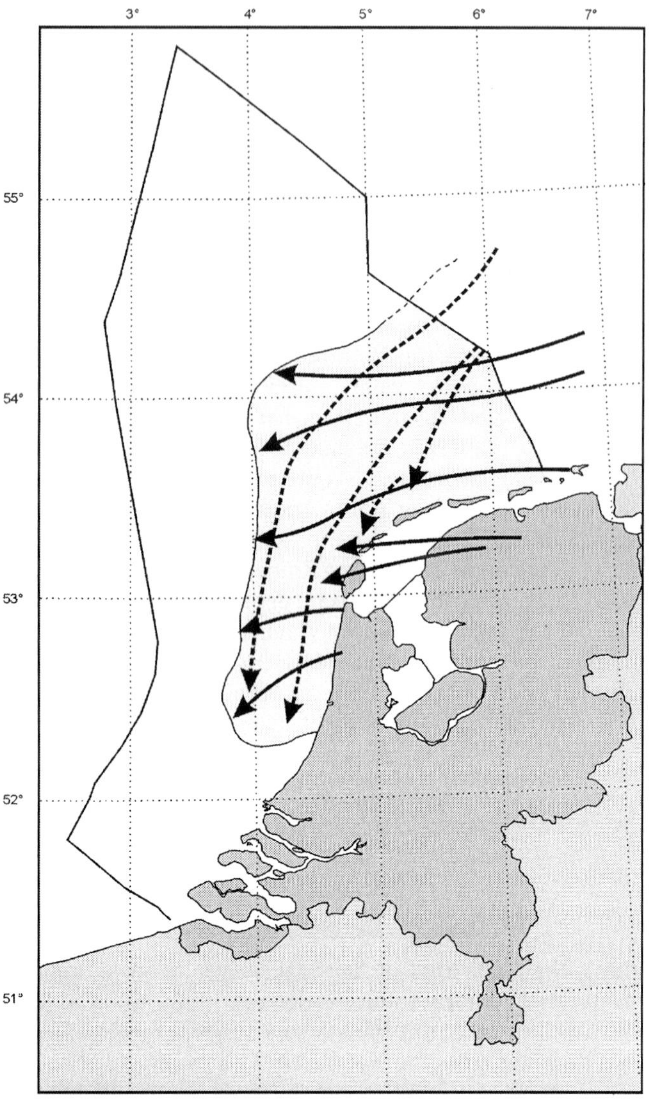

Fig. 4. Maximum extent of the Saalian ice sheet in the Dutch sector of the North Sea, showing suggested flow directions of ice.

with deformation structures along their margins. In addition subglacial channels, esker-like features, and ice-contact gravels were also formed. In the area west and north of the reconstructed ice margin none of the above features have been recorded in boreholes or are evident from seismic data. Glaciolacustrine and glaciofluvial deposits, overlying Saalian periglacial sediments, dominate the area west and north of the ice margin. The ice margin in the North Sea forms a continuation of the north-west/south-east trending margin on land. Only in block PI 5 has the ice moved further south than on land. During the ice advance from the north-east, and in the northeastern and western parts of the Dutch sector, tongue-shaped basins were formed pointing to the formation of a lobate ice margin. The subglacial valleys which are present between 53° N and 54° N latitude indicate a straight ice-front in that area while south of the 53° N a lobate margin again appears to be developed as in the eastern part of The Netherlands. In this area no tills have so far been found. The maximum extent of the ice is based in this area on the presence of the tongue-shaped basins and glaciotectonic structures. The reconstruction of the maximum extent of the ice during the Saalian is also supported by the distribution of Eemian sediments. Postglacial isostatic rebound is reflected in the thickness of Eemian marine sediments in the coastal area of The Netherlands and their progressive thinning in westerly and north-westerly directions outside the Saalian glacial limit.

According to Rappol (pers. comm.), the maximum extent of the ice sheet in the Dutch sector of the North Sea is too close to the coast of the northern Netherlands to allow the ice to change direction from NE-SW to N-S or even NW-SE. Rappol (1987) and van den Berg & Beets (1987) assumed that the ice sheet extended much further north in the North Sea area in order to explain the origin of the north-west/south-east trending Hondsrug in the eastern Netherlands.

Based on data collected by NITG-TNO, the extent of the Saalian ice sheet in the North Sea is far enough to the west to allow ice to enter the northern Netherlands from a north-easterly direction. There is a possibility however that at the end of the glaciation a drawdown system developed in the Norwegian Trench, effectively cutting off the ice flow towards the south-west (Rappol *et al.*, 1989). The effect of cutting off the ice source may have resulted in the development of a local ice dome with a radial flow pattern, which caused a change of direction in the North Sea and the northern Netherlands. In the North Sea the direction changed from SW to NE, while on land it changed from SW to NW. It must be clear that for the system to sustain an ice stream as envisaged by the different authors, this ice dome cannot only have been a slowly melting dead-ice field. Instead it must have acted as an accumulation and dispersal centre of its own. In that it may be similar to marginal ice domes envisaged for the SW Baltic margin of the Weichselian ice sheet (Lagerlund, 1987).

Drainage during the Saale glaciation

During the Saalian glaciation the NW-European rivers drained through the Straits of Dover which opened during this period. The ice barrier in the northern and central Netherlands forced the rivers into a southerly direction. Sea level was lowered to about 130 m ± 10 m below present level (Chappell & Shackleton, 1986; Lambeck pers. comm.). The glacial drainage was directed both north and west, as concluded from the distribution of the meltwater deposits.

Models for the Saale glaciation of the Netherlands

Since the Second World War, a number of glaciation models (based on ice-tectonic stratigraphy, geomorphology and more recently also on till stratigraphy) have been presented. For many years investigators have worked with the Maarleveld/ter Wee model (ter Wee, 1962 and

Maarleveld, 1981) in which five retreat phases were distinguished. Although this model was sometimes criticised (Zonneveld, 1975), it has stood for many years. It was replaced by the Jelgersma & Breeuwer model (1975), which to some extent reshuffled some of the previously-known five phases, but which did not take into account all that was known about the glaciation. Since then new models have quickly emerged: in 1987 the van den Berg & Beets model was presented, which only recognised two phases. The youngest phase is characterised by the existence of an ice stream coming from the North Sea Basin, across the northeastern part of the country. A more recent model by Rappol *et al.* (1989) is similar to the van den Berg & Beets model, but it recognises three phases and the existence of at least two ice streams in The Netherlands and one in the adjoining part of Germany. It also tries to explain ice movement by presenting a model for the behaviour of the southwestern part of the Scandinavian ice sheet. These last two models rely heavily on the existence of coalescing Scandinavian/British ice in the North Sea basin. However, the available evidence does not support this interpretation (Laban, 1995). Unfortunately, the German and Danish sectors of the North Sea have not been mapped in the same detail as the British and the Dutch sectors, which means that the NE continuation of the ice margin is virtually unknown. This leaves much room for speculation about sequences and events of the Saalian Glaciation.

The Weichselian glaciation in the North Sea

Introduction

During the Weichselian, glaciers did not reach the present terrestrial territory of the Netherlands. However, British ice did invade the Dutch sector of the North Sea and consequently this section of the report will only deal with the marine area. The authors would like to stress that common opinion is strongly influenced by the modern distribution of land and sea. As sea level is highly variable in time, this notion of 'wet-and-dry' is often irrelevant to the reconstruction of Quaternary glaciation.

A wide range of fluvial, lacustrine, glacial, glaciofluvial, glaciolacustrine and periglacial sediments were formed during the Weichselian glaciation in the Dutch sector of the North Sea. The oldest Early Weichselian sediments are stiff clays, which were deposited in the Southern Bight of the North Sea in a lacustrine environment. In the central and southern parts of the Southern Bight, medium to coarse-grained fluvial sediments were deposited by the rivers Rhine and Meuse in the form of a south-west prograding delta. During the Late Weichselian, an ice sheet from Britain entered the southern North Sea and extended into the Dutch sector. The Scandinavian ice sheet, however, did not extend as far south and west as the Dutch sector. Fine- to medium-grained glaciofluvial sand was deposited both prior to the advance and during the retreat of the British ice sheet and is recorded locally, while stiff glaciolacustrine clays are widespread. In the north - along the median line between the German and Dutch sectors - a glaciomarine facies is present. Tills are present in the southern part of the glaciated area and also as patches locally in the northern part. Microstructures indicate deposition of both flow and lodgement tills. Studies on the glacial gravels have identified them as originating from Britain. Two different types of subglacial valleys are present. The first type consists of a braided system of partly open valleys with their floors up to 80 m below MSL and occurring near the southern limit of the ice. They are infilled with soft clays and fine-grained sediments. The second type consists of V-shaped valleys. They are found in the north and are infilled with stiff glaciolacustrine clay. Based on the occurrence of subglacial valleys two main ice-flow directions of the British ice have been recognised, one from the west and south-west near the southern limit of the British ice and one from the north-east in the northern part of the Dutch sector. In the Dutch sector only one tongue-shaped basin has been recorded and is present in the Dogger Bank Formation. This indicates ice cover without accompanying deposition of tills. Deformation structures formed by ice-push are rare. Very fine- to fine-grained periglacial sediments are widespread, occur mainly in the eastern half of the Dutch sector and were deposited during the Early, Middle and Late Weichselian. Locally interstadial peat has been sampled and dated by the radiocarbon method. Various ages, including 45,090 + 3750/-2550 BP, 11,280 ±40 BP and 10,945 ±50 BP have been recorded.

Weichselian glaciation in the Dutch sector of the North Sea

Twice during the Weichselian glaciers extended over large areas of Great Britain. The early glaciation probably took place during the Early Weichselian before 46,000 BP (Worsley, 1991). No evidence for this Early Weichselian glaciation has, however, been found so far in the southern North Sea (Balson & Jeffery, 1991; Cameron *et al.*, 1992). Evidence for an Early Weichselian Scandinavian ice sheet in northwestern Europe is still a subject of debate (Houmark-Nielsen, 1987). Recent investigations in the northern North Sea by Sejrup *et al.* (1994) indicate that the maximum Weichselian glaciation in the northern North Sea took place between 29,400 and c. 22,000 BP with a coalescing British and Scandinavian ice sheet. According to Sejrup *et al.* (1994) during a second advance of the Weichselian ice sheet between 18,500 and 15,100 BP the British and Scandinavian sheets did not coalesce.

During the Weichselian glacial maximum, as well as at approximately 18,000 BP, sea level dropped substantially, probably to more than 130 m below present level. According to Oele & Schüttenhelm (1979), in the Dutch sector of the North Sea sea level was more than 50 m below the present level during the Weichselian Stage. This is based on Early Weichselian freshwater deposits sampled at

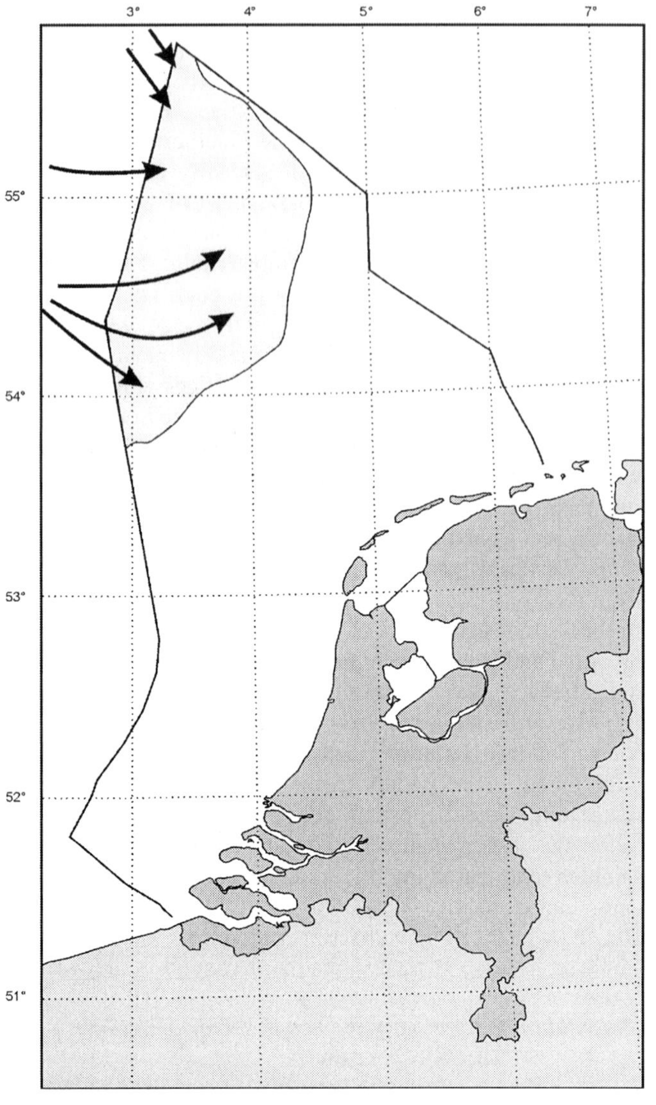

Fig. 5. Maximum extent of the British Weichselian ice sheet in the Dutch sector of the southern North Sea with probable flow directions.

depths up to 48 m below MSL. The freshwater deposits overlie open marine to brackish marine Eemian sediments.

Maximum extent of the Weichselian ice sheet in the southern North Sea.

It is widely accepted that the maximum glaciation occurred around 18,000 BP (Jardine, 1979). Many authors attempted reconstruction of the maximum extent of the Weichselian ice sheets in the North Sea, predominantly by using evidence of the gravels exposed on the sea bed. All authors suggest a broadly similar location of the limits (Laban, 1995). Some assume a connection between the Scandinavian and British ice sheets, whereas others consider such a connection never existed.

The revised map (Fig. 5) shows the extent of the British ice sheet in the Dutch sector, based on the occurrence of tills, deformation structures and subglacial valleys as mentioned above. The arrows indicate the suggested directions of ice flows.

References

Balson, P. S. & Jeffery, D.H. (1991). The glacial sequence of the southern North Sea. *In:* Ehlers, J., Gibbard, P.L. & Rose, J. (eds), *Glacial deposits in Great Britain and Ireland.* Rotterdam, Brookfield, Balkema, 245-253.

Bosch, J.H.A. (1990). Landijs, zee en rivieren als geologische "opbouwwerkers" van het Noorden. *Grondboor & Hamer*, **44**, 90-94.

Burger, A.W. (1992). Zware mineralenonderzoek aan boring F8-6 in de Noordzee. Internal report 958. Geological Survey of The Netherlands.

Cameron, T.D.J., Laban, C. & Schüttenhelm, R.T.E. (1984). *Flemish Bight: sheet 52°N/02°E.* Quaternary Geology, 1:250 000 series British Geological Survey and Geological Survey of The Netherlands.

Cameron, T.D.J., Laban, C. & Schüttenhelm, R.T.E. (1986). *Indefatigable: sheet 53°N/02°E.* Quaternary Geology, 1:250 000 series British Geological Survey and Geological Survey of The Netherlands.

Cameron, T.D.J., Crosby, A., Balson, P.S., Jeffery, D.H., Lott, G.K., Bulat, J. & Harrison, D.J. (1992). *United Kingdom Offshore Regional Report: The Geology of the southern North Sea.* London. HMSO for the British Geological Survey, 150 pp.

Chappell, J., & N.J. Shackelton (1986). Oxygen isotopes and sea level. *Nature*, **324**, 137-140.

de Gans, W., de Groot, Th. & Zwaan, H. (1987). The Amsterdam basin, a case study of a glacial basin in the Netherlands. *In:* van der Meer, J.J.M. (ed.), *Tills and glaciotectonics*. Rotterdam, Balkema, 205-216.

de Jong, J.D. & Maarleveld, G.C. (1983). The glacial history of the Netherlands. In: J. Ehlers (ed). *Glacial deposits in north-west Europe.* Balkema, Rotterdam, 353-356.

de Jong, J. (1991). Palynological investigation of the borehole 89/2 (EEC). Internal report 1112. Geological Survey of The Netherlands.

Ehlers, J. (1990). Untersuchungen zur Morphodynamik der Vereisungen Norddeutschlands unter Berücksichtigung benachbarter Gebiete. *Bremer Beiträge zur Geographie und Raumplanung*, **19**, 166 p.

Fairbridge, R.W. (1961). Eustatic changes in sea level. *Physics and Chemistry of the Earth*, **4**, 99-185.

Gibbard, P.L. (1988). The history of the great northwest European rivers during the past three million years. *Philosophical Transactions of the Royal Society London*, **318B**, 559-602.

Houmark-Nielsen, M. (1987). Pleistocene stratigraphy and glacial history of the central part of Denmark. *Bulletin of the Geological Society of Denmark*, **36**, 1-189.

Jardine, W.G. (1979). The western (United Kingdom) shore of the North Sea in Pleistocene and Holocene times. *In:* Oele, E., Schüttenhelm, R.T.E. & Wiggers, A.J. (eds), *The Quaternary History of the North Sea*. Uppsala, Acta Universitatis Upsaliensis, 159-174.

Jeffery, D.H., Frantsen, P., Laban, C. & Schüttenhelm, R.T.E. (1989). *Silver Well: sheet 54°N/02°E*. Quaternary Geology, 1:250,000 series British Geological Survey and Geological Survey of The Netherlands.

Jelgersma, S. & Breeuwer, J.B. (1975). Toelichting bij de geologische overzichtsprofielen door Nederland. *In:* Zagwijn, W.H. & van Staalduinen, C.J. (eds), *Toelichting bij Geologische Oversichtskaarten van Nederland*, 91-93. Haarlem, Rijks Geologische Dienst.

Joon, B. (1987). The Tertiary History of the Terschelling Basin. Internal Report, KSEPL (unpublished).

Kluiving, Sj., Rappol, M. & van der Wateren, D. (1991). Till stratigraphy and ice movernents in eastern Overijssel, The Netherlands. *Boreas*, **20**, 193-205.

Kuster, H. & Meyer, K.-D. (1979). Glaziäre Rinnen im mittleren und nordöstlichen Niedersachsen. *Eiszeitalter und Gegenwart*, **29**, 135-156.

Laban, C. (1995). The Pleistocene glaciations in the Dutch sector of the North Sea. A synthesis of sedimentary and seismic data. PhD Thesis University of Amsterdam, 194 pp.

Lagerlund, E. (1987). An alternative Weichselian glaciation model, with special reference to the glacial history of Skåne, south Sweden. *Boreas*, **16**, 433-450

Maarleveld, G.C. (1981). The sequence of ice-pushing in the central Netherlands. *Mededelingen Rijks Geologische Dienst*, **34-1/11**, 2-6.

Maarleveld, G.C. (1960). Les phénoménes périglaciaires au Pleistocéne Ancien et Moyen aux Pays-Bas. *Biuletyn Peryglacjalny*, **9**, 135-141.

Neele, N.G. (1986). Micropaleontologisch onderzoek aan het traject 0.00 - 151.13 m van de boring El-10 en correlatie met de boringen E8-4 (Rapport 1516). en E2-3 (rapport 1495). Internal report 1517. Geological Survey of The Netherlands.

Neele, N.G. (1991a). Verslag van micropaleontologisch onderzoek aan de boring Noordzee F8-89/006, traject 0.00 - 195). 10 m onder zeebodem. Internal report 1620. Geological Survey of The Netherlands.

Neele, N.G. (1991b). Micropaleontological investigation of borehole North Sea 89/2 (A). Internal report 1614. Geological Survey of The Netherlands.

Oele, E. & R.T.E. Schüttenhelm (1979). Development of the North Sea after the Saalian glaciation. *In:* Oele, E., Schüttenhelm, R.T.E. & Wiggers, A.J. (eds), *The Quaternary History of the North Sea*. Uppsala, Acta Universitatis Upsaliensis, 191-216.

Postma, G., Roep, Th.B. & Ruegg, G.H.J. (1983). Sandy-gravelly mass-flow deposits in an ice-marginal lake (Saalian, Leuvenumse Beek Valley, Veluwe, the Netherlands). with emphasis on plug-flow deposits. *Sedimentary Geology*, **34**, 59-82.

Postma, G. (1997). De Leuvenumsche Beek-vallei. Een voormalig door landijs gevormd meer met zandige massatransport-afzettingen. *Grondboor & Hamer,* **51**, 117-124.

Pouwer, R. (1991). Molluskenonderzoek van Noordzee-boring F8/006). Internal report 1582, Geological Survey of The Netherlands.

Praeg, D. (1994). Quaternary seismic morphology and stratigraphy - glacial and fluvial channels in the southern North Sea. 2nd Int. Conf. on The Geology of Siliciclastic Shelf Seas 24-28 May, Gent, 96.

Rappol, M. (1983). Glacigenic properties of till. Studies in glacial sedimentology from the Allgäu Alps and the Netherlands. Dissertatie Universiteit van Amsterdam, 225 pp.

Rappol, M. (1984). Till in southeast Drente and the origin of the Hondsrug Complex, the Netherlands. *Eiszeitalter und Gegenwart,* **34**, 7-27.

Rappol, M. (1987). Saalian till in The Netherlands: A review. *In:* van der Meer, J.J.M. (ed.). *Tills and Glaciotectonics*, Balkema, Rotterdam/Boston, 3-22.

Rappol, M., Haldorsen, S., Jørgensen, P., van der Meer, J.J.M. & Stoltenberg, H.M.P. (1989). Composition and origin of petrographically-stratified thick till in the Northern Netherlands and a Saalian glaciation model for the North Sea basin. *Mededelingen van de Werkgroep voor Tertiaire en Kwartaire Geologie*, **26**, 31-64.

Roep, Th.B., Holst, H., Vissers, R.L.M., Pagnier, H. & Postma, D. (1975). Deposits of southward-flowing, Pleistocene rivers in the Channel Region, near Wissant, NW France. *Palaeogeography, Palaeoclimatology, Palaeoecology*, **17**, 289-308.

Ruegg, G.H.J. (1975). Sedimentary structures and depositional environments of Middle and Upper Pleistocene glacial time deposits from an excavation at Peelo, The Netherlands. *Mededelingen Rijks Geologische Dienst*, **N.S. 26**, 17-37.

Ruegg, G.H.J. (1977). Features of middle Pleistocene sandur deposits in the Netherlands. *Geologie en Mijnbouw*, **56**, 5-24.

Ruegg, G.H.J. & J.G. Zandstra (1977). Pliozäne und pleistozäne gestauchte Ablagerungen bei Emmerschans (Drenthe, Niederlande). *Mededelingen Rijks Geologische Dienst*, **(NS) 28**, 65-99.

Sejrup, H.P., Haflidason, H., Aarseth, I., King, E., Forsberg, C.F., Long, D. and Rokoengen, K. (1994). Late Weichselian glaciation history of the northern North Sea. *Boreas*, **23**, 1-13.

ter Wee, M.W. (1962). The Saalian glaciation in the Netherlands. *Mededelingen Geologische Stichting*, **N.S., 15**, 57-76.

ter Wee, M.W. (1983a). The Elsterian Glaciation in the Netherlands. *In:* Ehlers, J. (ed.), *Glacial Deposits in North-West Europe*. Balkema, Rotterdam, 413-415.

ter Wee, M.W. (1983b). The Saalian glaciation in the northern Netherlands. *In:* Ehlers, J. (ed.), *Glacial deposits in North-West Europe*. Rotterdam, Balkema, 405-412.

Valentin, H. (1952). Die Küsten der Erde. *Petermanns Geographische Mitteilungen, Ergänzungsheft*, **246**, 118 pp.

van den Berg, M.W. & Beets, D.J. (1987). Saalian glacial deposits and morphology in The Netherlands. *In:* van der Meer, J.J.M. (ed), *Tills and Glaciotectonics*. Rotterdam, Balkema, 235-251.

van der Meer, J.J.M. (1987). Field trip 'Tills and end moraines in The Netherlands and NW Germany'. *In:* van der Meer, J.J.M. (ed.), *Tills and glaciotectonics*. Rotterdam, Balkema, 261-268.

van der Meer, J.J.M. & Lagerlund, E. (1991). First report of a preserved Weichselian periglacial surface in NW-Europe - the "van der Lijn," geological reserve in The Netherlands. *Eiszeitalter und Gegenwart,* **41**, 100-106.

van der Wateren, F.M. (1981). Glacial tectonics at the Kwintelooijen sandpit, Rhenen, The Netherlands. *Mededelingen Rijks Geologische Dienst*, **35**, 252-268.

van der Wateren, F.M. (1985). A model of glacial tectonics, applied to the ice-pushed ridges in the Central Netherlands. *Bulletin of the geological Society of Denmark*, **34**, 55-74.

van der Wateren, F.M. (1992). Structural geology and sedimentology of push moraines. PhD Thesis University of Amsterdam, 230 pp.

van Straaten, L.M.J.U. (1956). Structural features of the "Papzand" formation at Tegelen (Netherlands). *Geologie en Mijnbouw*, **18**, 416-420.

Worsley, P. (1991). Possible early Devensian glacial deposits in the British Isles. *In:* Ehlers, J., Gibbard, P.L. & Rose, J. (eds), *Glacial deposits in Great Britain and Ireland*. Rotterdam / Brookfield, Balkema, 47-51.

Zagwijn, W.H. (1986). Preliminary interpretation and summary of the micro- and macropalacontological and pollen analysis of the boreholes El-10 and E8-4). Internal report. Geological Survey of The Netherlands.

Zagwijn, W.H. (1961). Vegetation, climate and radiocarbon datings in the Pleistocene of the Netherlands. Part 1: Eemian and Early Weichselian. *Mededelingen Geologische Stichting*, **N.S. 14**, 15-45.

Zagwijn, W.H. (1973). Pollenanalytical studies of Holsteinian and Saalian beds in the northern Netherlands. *Mededelingen Rijks Geologische Dienst,* **N.S. 24**, 139-156.

Zagwijn, W.H. (1977). Stratigrafische interpretatie van boringen tot circa 100 m onder zeeniveau in bet Noordzeegebied (blokken P13, P5, J14, K17, K14, K15, L13, LIO, L7). Internal report 769. Geological Survey of The Netherlands.

Zandstra, J.G. (1983). Fine gravel, heavy mineral and grain size analyses of Pleistocene, mainly glacigenic, deposits in the Netherlands. *In:* Ehlers, J. (ed.), *Glacial Deposits in North-West Europe*. Rotterdam, Balkema, 361-377.

Zandstra, J.G. (1986). Korrelgrootte-onderzoek van glacigene zandige klei op Wieringen. Internal report 865). Geological Survey of The Netherlands.

Zonneveld, J.I.S. (1975). Zijn de noordnederlandse stuwwallen overreden of niet? *Berichten Fysisch Geografische Afdeling Rijks Universiteit Utrecht*, **9**, 3-14.

Quaternary Glaciations – Extent and Chronology
Editors J. Ehlers and P.L. Gibbard

The North Sea basin

Simon J. Carr

Department of Geography, Oxford Brookes University, Gipsy Lane

Oxford, OX3 0BP, UK

E-mail: sjcarr@brookes.ac.uk

Introduction

The North Sea basin originated as a consequence of relaxation of the mid-Miocene Alpine orogenic stress field, with a subsidence rate of approximately 0.35 m/1000 years over the past 10 million years (Ziegler & Louwerens, 1979; Cameron *et al.*, 1992). As a result, there is a thick sequence of Tertiary and Quaternary sediments, with up to 1000 metres of Quaternary sediments in the central graben of the basin (Caston, 1977). This sequence of sediments preserves evidence for repeated glaciation during the Quaternary, with evidence for extensive glaciations during the Elsterian, Saalian and Weichselian, as well as fragmentary evidence for glaciation during the Menapian (Bijlsma, 1981; Gibbard, 1988; Sejrup *et al.*, 1987).

This report synthesises published information from the entire offshore region of the North Sea (Fig. 1), from 52°N to the continental shelf edge, and includes new micromorphological data that adds significant new understanding to the extent and dynamics of the Weichselian ice sheets.

Nature of the evidence

Until the late 1940's, all models suggesting extensive glaciation of the North sea were based on inference from tills identified in coastal sections along the margins of the basin (Wood & Rome, 1868; Geikie, 1874) or from rare seabed samples trawled up in fishing nets (Tesch, 1942). Two critical developments, in seismic analysis, and the recovery of shallow sediment cores, revolutionised the understanding of the glacial history of the North Sea. Acoustic profiling, first applied in the region by Valentin (1957) identified the value of seismic records in identifying geomorphological features in the seabed and the geometry of subsurface sediment units. This enabled seismostratigraphic division of the thick sediment sequence, and also the identification of important glacigenic landforms used to delimit ice sheets during various stages, such as tunnel valleys (Hansen & Nielsen, 1960; Flinn, 1967; Jansen, 1976; Wingfield 1990; Ehlers & Wingfield, 1991) (Fig. 2). Seismic survey can only provide spatial data to trace the geometry of extensive sediment formations and is unreliable in the interpretation of the physical components and small scale structure of such sediments (Cameron *et al.*, 1987). The application of a combined sediment coring and seismic survey approach during the regional mapping programmes of the countries bordering the North Sea has resulted in a large archive of seismic and sedimentological information, much of which has been used to investigate the extent of Quaternary glaciation (Andrews *et al.,* 1990; Balson & Jeffrey, 1991; Cameron *et al.,* 1987, 1989, 1992; Gatliff *et al.,* 1994; Johnson *et al.*, 1993; Long *et al.*, 1988; Peacock, 1995, 1997; Rise & Rokoengen, 1984; Sejrup *et al.,* 1987, 1991, 1994, 1995, 1996, 1998, 1999; Stoker *et al.*, 1985; Wingfield, 1989, 1990).

The limited mass of sample recovered through coring has limited the analytical procedures used for the interpretation of sediments (Fig 3a). This has resulted in widely differing interpretations of the origins of many sediments, with great difficulty in discriminating *in situ* marine or glaciomarine sediments from those that have been tectonised or completely reworked to produce subglacial tills. Van der Meer & Laban (1990) provided an example of the application of thin section micromorphology to the interpretation of such ambiguous sediments, and Carr (1998, 1999) demonstrated how the approach may be applied to a large regional study of the North Sea (Fig. 3b). This approach, combined with data derived from the substantial seismic and sedimentary archive has provided significant new information regarding the genesis of particular formations, and forms much of the evidence used to refine and develop previously published models of ice extent and dynamics (Fig. 4).

Discussion of glacial limits

Maximum extent of pre-Weichselian glaciations

There is scattered evidence for extensive glaciation in Europe including the North Sea during the Menapian stage of the early Quaternary (Gibbard, 1988). The dominant river system feeding the North Sea delta flowed westwards across Poland and northern Germany, draining the entire Baltic region and eastern Europe (Zagwijn, 1979, Gibbard, 1988). The absence of sediments of east European provenance in North Sea sediments younger than the Menapian led Bijlsma (1981) to suggest that the depositional system of Europe had been comprehensively altered by regional glaciation, scouring a 'proto-Baltic' basin. Sejrup *et al.* (1987) provide further support for a regional glaciation of the North Sea during the Menapian from the presence of a

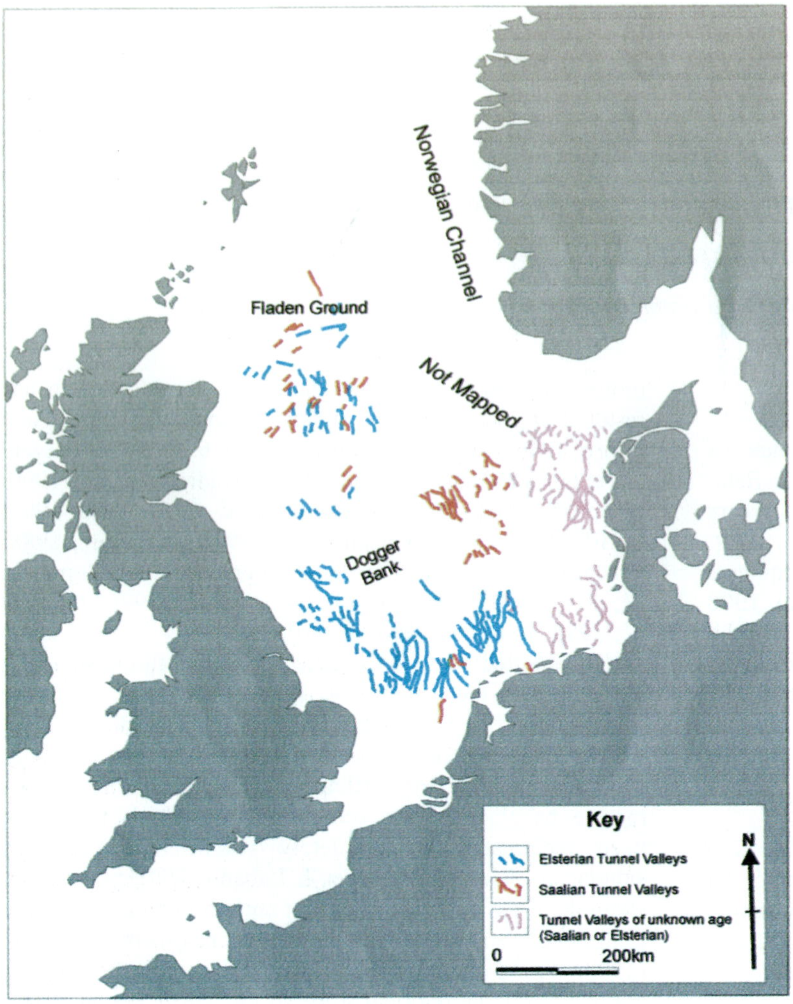

Fig. 1. Location map, North Sea Basin. The approximate pattern and extent of pre-Weichselian tunnel valleys is identified (after Huuse and Lykke-Andersen, 2000). The extent of these tunnel valleys have been one of the major sources of information for the extent of the Elsterian and Saalian ice sheets.

Fig. 2. Interpreted British Geological Survey (BGS) seismic profiles, Fladen area, Central North Sea. It is possibly to trace the subcrop of seismostratigraphic formations across large areas, as well as identifying large scale morphological features such as tunnel valleys (for example the tunnels infilled with the Ling Bank Formation in the diagram). These records also permit the spatial reconciliation of cores and boreholes (identified by their borehole codes) across large areas, essential for reconstruction of regional glacial episodes.

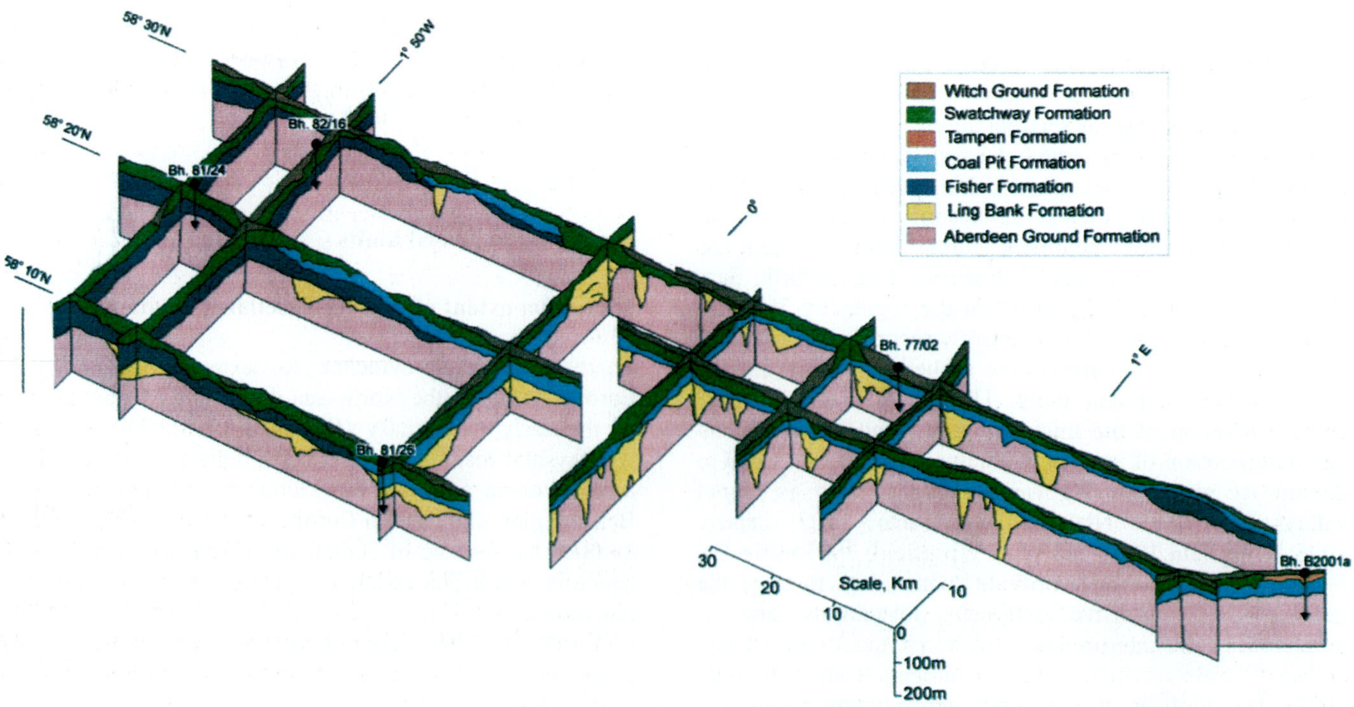

subglacial diamict in borehole 81/26, in the central North Sea, thought to be of similar age.

There is relatively little sedimentary evidence for Elsterian glaciation of the North Sea (Long *et al.*, 1988, Balson & Jeffrey, 1991). Van der Meer (1992) reports the micromorphology of two samples thought to be Elsterian till from the Danish sector of the North Sea, although the stratigraphic position of these sediments is unclear. Balson & Jeffrey (1991) attribute the absence of Elsterian glacial sediments to post-depositional reworking, rather than non-deposition of sediment. There is however considerable seismic evidence of glaciation during the Elsterian, with a significant seismic unconformity interpreted as an Elsterian glacial erosion surface (Cameron *et al.*, 1987; Balson & Jeffrey, 1991). This erosion surface is irregular, with a series of deep, elongate enclosed incisions, interpreted to be tunnel valleys and used to define Elsterian ice limits in the southern North Sea (Cameron *et al.*, 1987; Laban, 1995; Wingfield 1989, 1990, Huuse & Lykke Andersen 2000) (Fig. 1).

Further north, fragments of the Aberdeen Ground Formation have been interpreted as Elsterian till (Stoker & Bent, 1985; Sejrup *et al.*, 1987, 1991), and much of the upper part of the formation is considered to be glaciomarine in nature (Gatliff *et al.*, 1994). The northern and western limit for the Elsterian ice sheet has not been reliably identified, although Elsterian marine and glaciomarine sediments of the Mariner Formation to the north of the Shetland Isles suggest a terminus somewhere near this area (Johnson *et al.*, 1993).

It is possible to reconstruct two stages of Saalian glaciation in the North Sea. The first glacial corresponds to the early glaciation of Rappol *et al.* (1990), with a British ice sheet deflecting the Baltic ice stream in the central North Sea towards the south and north. Subsequent deglaciation occurred, with the deposition of the glaciomarine and marine sediments of the Fisher and Mariner formations. Finally, a second glacial advance occurred, producing an erosion surface across the majority of the North Sea, with a series of tunnel valleys suggesting a southern ice limit at around 56°N and a north western limit at or around the continental shelf edge. All of the evidence for Saalian glaciation of the North Sea basin is suggested to relate to Marine Isotope Stage (MIS) 6 (Ehlers, 1990; Laban, 1995; Rappol *et al.*, 1989), although this may need re-evaluation in the light of the more complex glacial history described by Hamblin *et al.* (2000) from Eastern England.

There are a number of glacigenic sediments interpreted as Saalian in age, resting unconformably upon Holsteinian sediments. This unconformity can be traced across most of the North Sea north of approximately 56°N (Cameron *et al.*, 1987), and has been interpreted as a glacially-eroded surface. A series of tunnel valleys form part of this seismic unconformity (Wingfield, 1989, 1990) (Fig. 1).

Saalian till, similar to that found onshore, forms a continuous sheet up to 100 kilometres offshore from the Netherlands coast, constituting the Borkumriff Formation (Joon *et al.*, 1990) of early Saalian age (Rappol *et al.*, 1989). There is no evidence of till of British origin in the southern North Sea at this time, with non-glacial sedimentation in a terrestrial environment dominating (Cameron *et al.*, 1987; Long *et al.*, 1988; Balson & Jeffrey, 1991). However, the presence of a British ice sheet is required to explain the movement of early Saalian ice south-easterly across the Netherlands (Rappol *et al.*, 1989). This is in contrast to Long *et al.* (1988) who claim the absence of glacigenic sediments of British origin as evidence that no British ice sheet during the Saalian extended far into the North Sea basin.

Evidence of extensive Late Saalian glaciation is found in the central North Sea, within borehole 81/26. Sejrup *et al.* (1987, 1991) identify a till containing fragments of granite, schist, basic igneous and red sandstones probably of Scottish origin. There have been no similar reports of glacial sedimentation in the northern North Sea during the Saalian (Johnson *et al.*, 1993).

Early Weichselian glaciation: The Ferder glacial episode

During MIS 4, possibly at around 70,000^{14}C years BP, substantial glaciation of the North Sea basin occurred, with confluence of the British and Scandinavian ice sheets at the time. Most of the evidence is found as a subglacially-deformed till forming the upper part of the Ferder Formation, in the northern part of the North Sea, and as a number of large, infilled channels of up to 200m amplitude, thought to be tunnel valleys (Carr, 1998; Johnson *et al.*, 1993; Skinner *et al.*, 1986). Sejrup *et al.* (1995) attribute the lower part of a thick diamicton in the Norwegian Channel to glaciation during MIS 4, and may reflect the same episode. The southern limit of the Ferder glacial episode is unclear, although for confluent glaciation of the Scandinavian and Scottish ice sheets, the ice is likely to have extended to the south for a considerable distance, probably to the Dogger Bank region at around 55-56°N.

The age of glaciation is loosely tied to MIS 4, by the fact that the till relating to this event overlies Eemian interglacial and glaciomarine sediments within which the Blake palaeomagnetic Event (0.12 Ma BP) occurs, and underlies middle Weichselian marine and glaciomarine sediments of the Cape Shore Formation (Skinner *et al.*, 1986).

The Last Glacial Maximum: The Cape Shore and Bolders Bank glacial episodes

In this report, a two-phase Last Glacial Maximum (LGM) model is supported, building extensively upon the work of Sejrup *et al.* (1994). Such a model explains conflicting chronologies regarding the timing of extensive onshore glaciation in Britain and Scandinavia, and accounts for the complexity of Late Weichselian glacigenic sediments in the North Sea. Other models of a single stage of glaciation

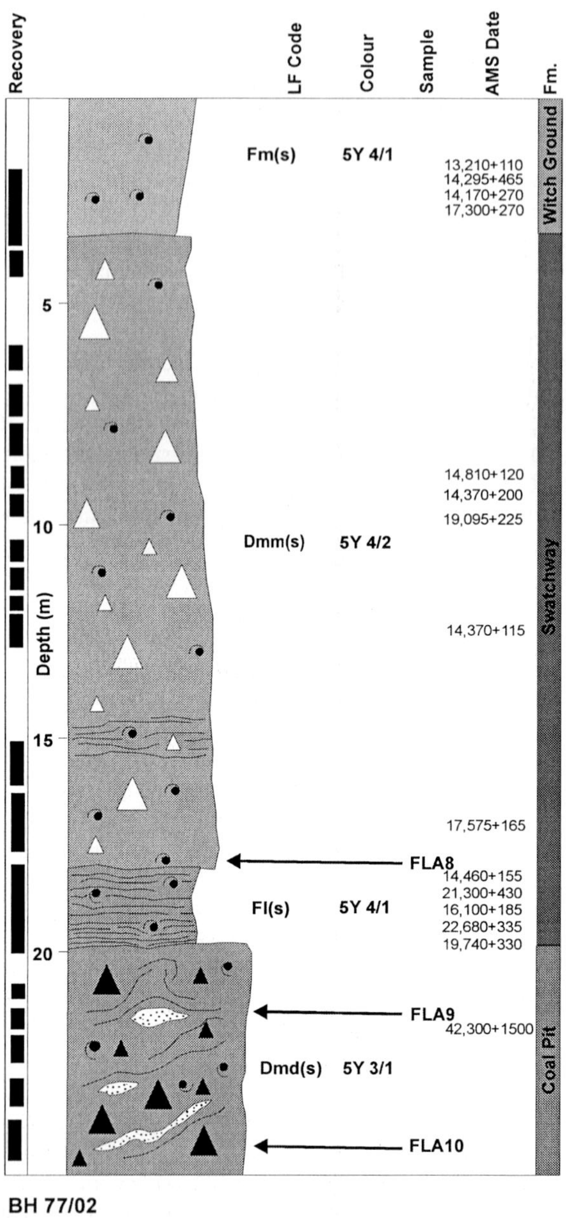

Fig. 3. A (left): Lithofacies description of borehole 77/02, from the Fladen area of the central North Sea. B (right): Summary diagram of micromorphological characteristics used by Carr (1998) in examining the extent and dynamics of Weichselian glaciation in the North Sea. The use of thin section micromorphology has been critical in providing secure sedimentological evidence from small core and borehole samples to support or reject glacigenic sedimentation.

during this time do not fully explain the sedimentological information of multiple stage onshore glaciation (Bent, 1986; Long *et al.*, 1988) or are difficult to reconcile with accepted glaciological modelling (Boulton *et al.*, 1985; Jansen *et al.*, 1979; Ehlers & Wingfield, 1991).

The first phase of the LGM is identified from the Cape Shore and Coal Pit formations in the central and northern North Sea. These formations reflect a transition from glaciomarine to subglacial conditions, with the deposition and subsequent glaciotectonic deformation of a shelly stratified diamicton (Carr, 1998). Bulk and AMS radiocarbon dating of shells and foraminifera in the Cape Shore Formation suggest glaciation took place subsequent to 29,430±390 ^{14}Cyrs BP (Johnson *et al.*, 1993; Milling, 1975; Rise & Rokoengen, 1984). The Cape Shore glacial episode reflects extensive glaciation of confluent British and Scandinavian ice sheets, extending to the continental shelf break.

The southern limit of the Cape Shore ice sheet is placed at the central axis of the Dogger Bank in the southern North Sea. The evidence for this is not secure, because there is poor dating control for the age of the Dogger Bank Formation, and an unclear seismostratigraphic link to the Coal Pit formation to the north. However, the following evidence is used to support a 'Cape Shore' age to the Dogger Bank Formation, and as the southern limit of glaciation during this phase of the LGM;

- Ehlers & Wingfield (1991) identify a series of infilled incisions that are aligned in a belt along the north side of the Dogger Bank (Fig. 5), and are interpreted as a series of intra-formational tunnel valleys formed during the mid- to Late Weichselian. The acceptance of tunnel valleys as an indicator of a former ice margin suggests that the Dogger Bank was a major ice marginal region at some point during the mid- to Late Weichselian.

  In terms of mega-scale geomorphology and sediment content, the Dogger Bank may be closely compared with the structure of Denmark, which is essentially a vast belt of thrust block moraines and associated outwash deposits formed at the Main Stationary Line of the LGM Scandinavian ice sheet. The internal structure of the Dogger Bank and the Dogger Bank Formation is unclear, due to the poor quality of most shallow seismic lines in the area. However, Laban (1995) and Wingfield (*pers. comm.*) suggest that the internal seismic structure of parts of the Dogger Bank Formation seem to reflect thrusting from a northerly direction, associated with an ice sheet in the central North Sea.
- The Dogger Bank Formation overlies a seismic unconformity of mid-Weichselian age (Cameron *et al.*, 1987, 1992; Laban, 1995; Long *et al.*, 1988), and is therefore younger than the early Weichselian 'Ferder' glacial episode.

Sejrup *et al.* (1994) suggest that this first phase of the LGM culminated at around 22 ka BP, with subsequent decoupling of the ice sheets by 20 ka BP.

A second phase of the LGM is preserved as a relatively continuous till sheet that extends up the eastern coast of Britain as the Bolders Bank and Wee Bankie formations (Long *et al.*, 1988). Ice streams, emanating from central and southern Scotland and the north Pennines of England, flowed south-eastwards, piercing the western edge of the Dogger Bank and extending as a lobe across the southern North Sea (Carr, 1999). Further north, British ice was not particularly extensive, only advancing to approximately 60 kilometres offshore (Gatliff *et al.*, 1994). In the Marr Bank

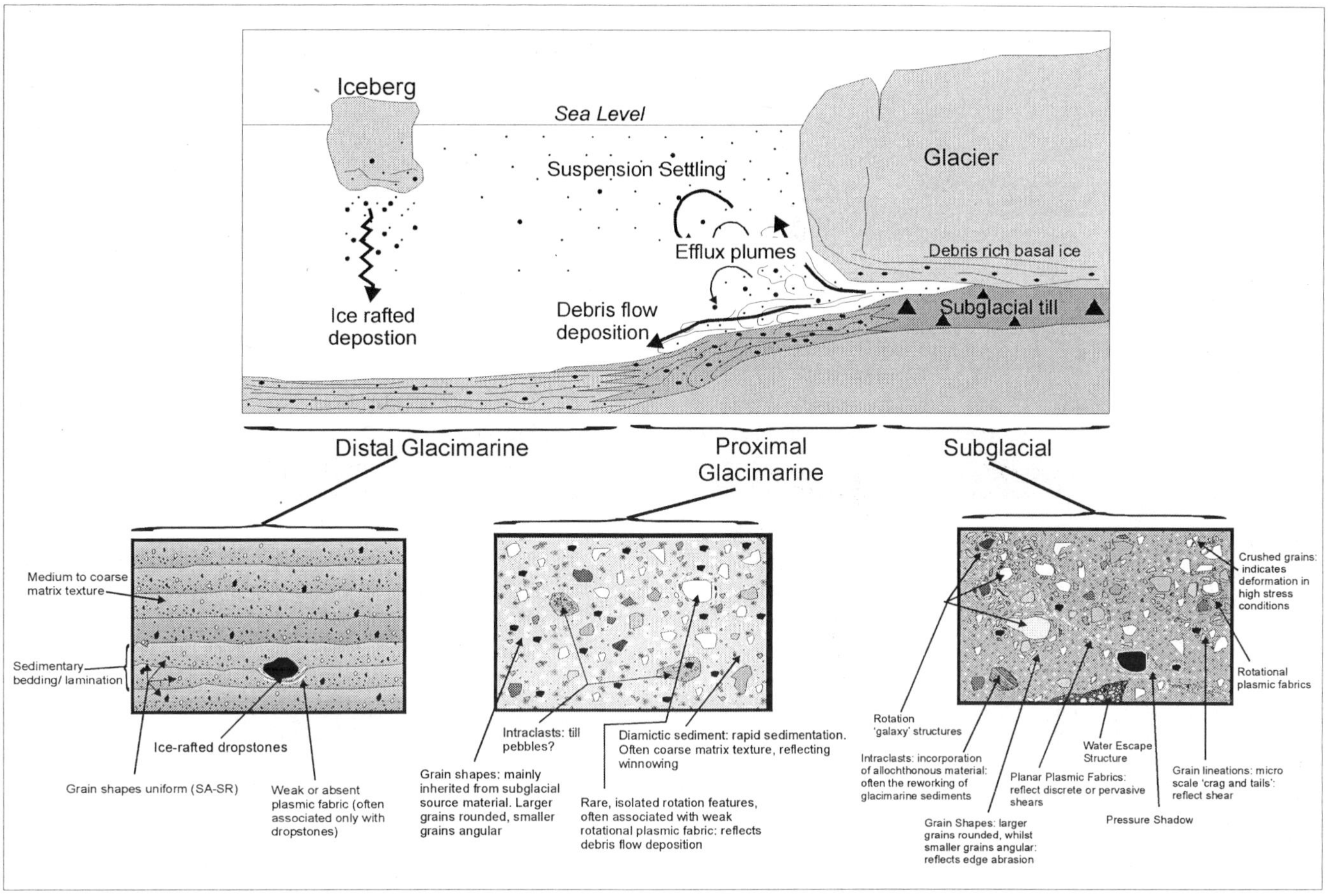

area, the ice terminated as a calving margin into what was probably a shallow lake, whilst across the Southern North Sea, the ice terminated in a terrestrial environment, depositing glacifluvial sediments (Well Ground Formation) (Carr, 1998). There is continuing debate regarding the position of the ice front in the Moray Firth area, with conflict between a minimal model, which accounts for ice free conditions in Caithness, Buchan and the Orkney Isles (Sutherland, 1984, 1991; Sejrup *et al.,* 1987), or a maximal model which implies inundation of the entire onshore region (Hall & Bent, 1986). In this report, the minimal model is adopted, and it is suggested by Carr (1998) that the Bosies Bank end moraine of Bent (1986) and Hall & Bent (1990) probably relates to a stillstand during retreat of the British ice sheet after the Cape Shore maximum. In the northern North Sea, a small, locally-derived ice cap developed over the Shetland Isles, with a limited offshore extent, delimited by the Otter Bank sequence (Stoker & Long, 1985; Sutherland, 1991). Evidence from the east coast of Yorkshire (Rose, 1985) suggests that the Bolders Bank ice advance was a short-lived event, with extensive ice cover in Eastern Britain lasting only 2000 years.

The eastern part of the North Sea preserves evidence of a re-advance of the Scandinavian ice sheet into the central North Sea, with ice extending beyond the Norwegian Channel ice stream, depositing the Tampen Formation (Sejrup *et al.,* 1987, 1994), interpreted by Carr *et al.* (2000) on the basis of micromorphology as a subglacial till.

Post-LGM limited regional ice readvances

Subsequent to the extensive surge of the eastern side of the British ice sheet, widespread ice sheet retreat occurred on both sides of the North Sea. Sejrup *et al.* (1994, 1999) report that the Norwegian Channel was ice free by ca. 15 ka BP, with ice constrained to the western coastal fjords of Norway. In Britain, ice from the Bolders Bank glacial episode was in full retreat by 16 ka BP, with most of the near-shore and lowland areas adjoining the North Sea ice free between 15,000 and 13,000 ^{14}C yrs BP (Boulton *et al.,* 1991). Peacock (1997) presents tentative evidence for a later re-advance of ice from the Moray Firth area into the western North Sea between 15-14 ka BP, supported with evidence suggesting shelf edge glaciation adjacent to the Barra Fan of the Hebridean Isles (Knutz *et al.*, in press). After this event there is no evidence to support large-scale regional glaciation of the North Sea basin.

Discussion

Chronological Framework

Many of the issues resolving the timing and duration of Weichselian glaciation of north-west Europe could be resolved with a secure glacial chronology in the North Sea, enabling a reliable correlation between glacial episodes in

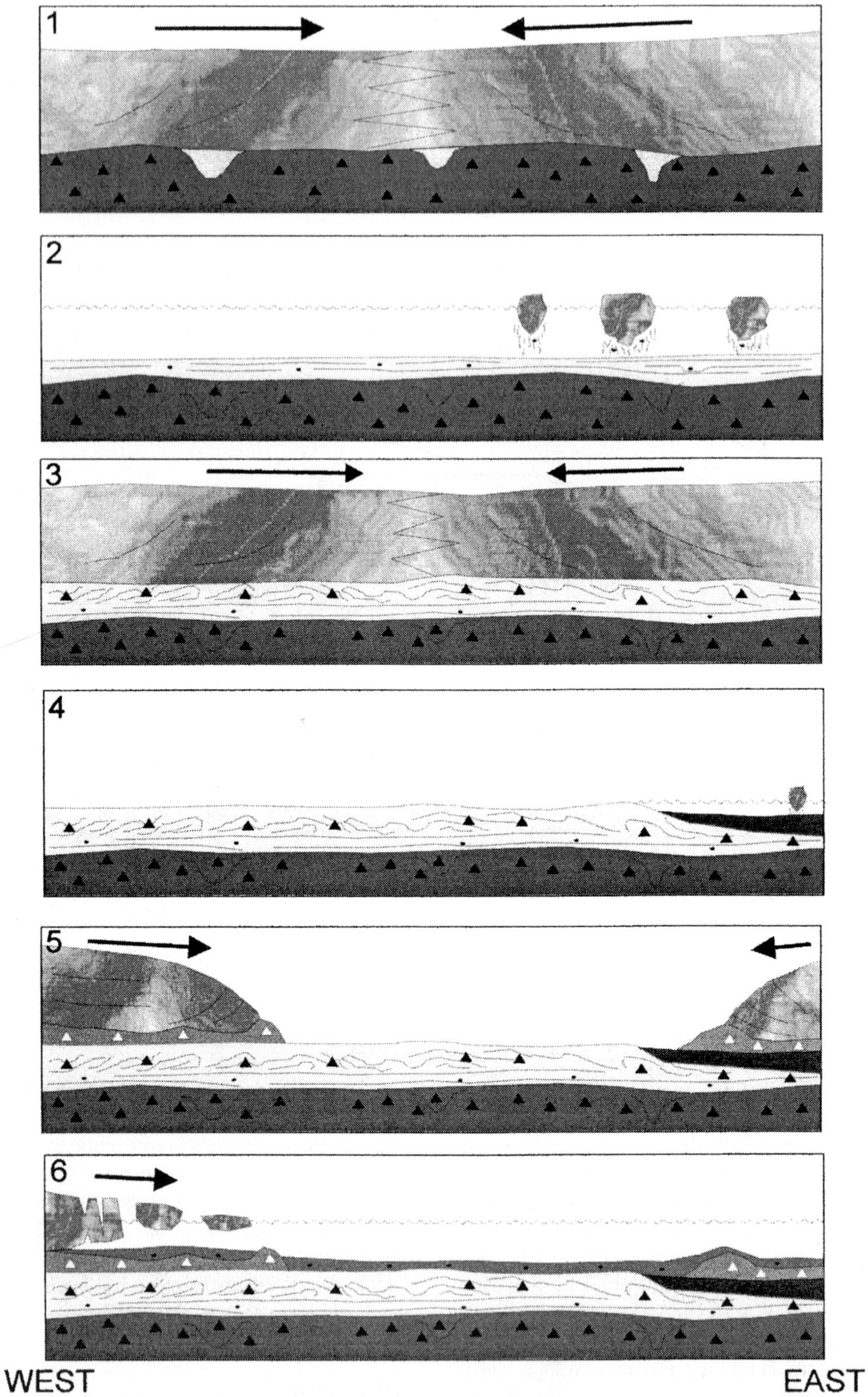

Fig. 4. Example of a schematic diagram illustrating the sequence of events during the Weichselian glaciation between the Shetland Isles and Norwegian Channel. These enable reliable reconstruction of a regional glacial stratigraphy. 1: Extensive glaciation of the North Sea, deposition of Ferder Formation, with associated tunnel valley formation. 2: Marine to glaciomarine conditions with deposition of Cape Shore Formation. 3: Major glaciation of North Sea, tectonizing upper part of Cape Shore Formation. 4: Limited ice sheet retreat results in shallow glaciomarine sedimentation in eastern part of North Sea, and deposition of the Sperus Formation. 5: Still-stand or readvance of separate ice sheets at margins of study area, with the remainder of the North Sea exposed in sub-aerial conditions. Deposition of Otter Bank sequence and Tampen Formation. 6: Retreat of ice in shallow marine conditions with localised glaciomarine sedimentation in west of area as Stormy Bank sequence.

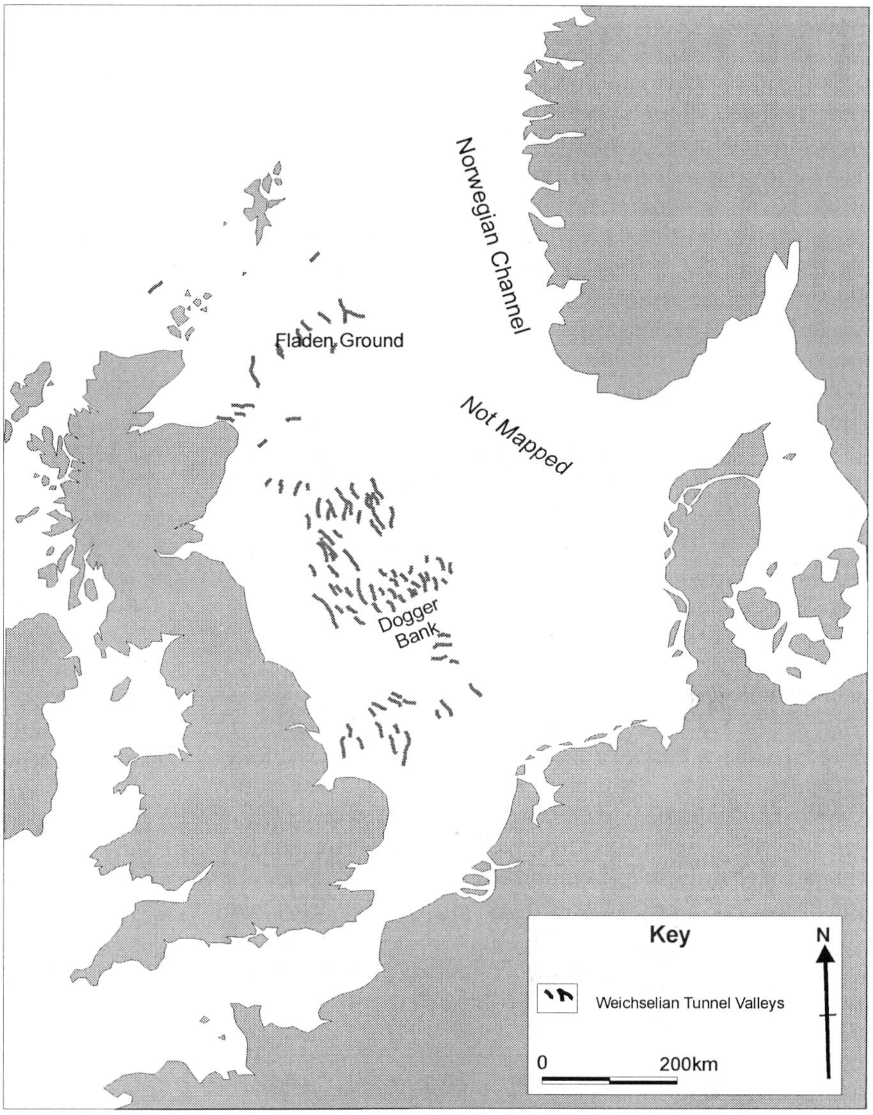

Fig. 5. Extent of Late Weichselian tunnel valleys in the North Sea (after Ehlers & Wingfield, 1990 and Huuse & Lykke-Andersen, 2000). Note the WSW-ENE trending belt of incisions on the northern edge of the Dogger Bank, interpreted as the ice sheet limit during the Cape Shore glacial episode.

the British Isles, Scandinavia and Northern Germany. Probably the main problem in reconstructing the Weichselian glacial history of the North Sea basin is the lack of such a chronological control for many of the seismostratigraphic and sedimentary units in the basin. Critical formations, such as the Coal Pit and Dogger Bank Formations remain as yet insecurely dated, and there is a considerable over-dependence on the few sites that have been radiometrically-dated. In some cases, dating of shells and foraminifera highlight large age ranges, overturned sequences of dates in single sediment facies such as in borehole 77/02 (Sejrup *et al.,* 1987, 1994), and in others, dates have been applied to sediment units without clear knowledge of their genetic origin.

Within the context of this report, the ages of glacial episodes is generalised, and in most cases the dating is supported by other lines of evidence, such as palaeomagnetic and amino-acid chronologies, as well as through following standard approaches to glacial stratigraphy (Rose & Menzies, 1996). The major focus of future research in the North Sea however should involve a fundamental re-evaluation of the existing radiometric dating, and targeting of specific sediment facies for new age determinations.

Rapid ice sheet response

Within the loose chronology presented in this report, it is apparent that on occasions the advance and retreat of the Scandinavian and British ice sheets was dramatic. During the build up to the Cape Shore glacial episode, a period of no more than 8-10,000 years, glaciers advanced up to 1000 km, requiring sustained net ice sheet advance of *at least* 100 ma^{-1} throughout this period. Such a rapid response of large ice sheets to mass balance changes is unprecedented and cannot be easily reconciled with investigation of modern analogues. It is probable that the easily deformable sediments of the North Sea basin would have facilitated rapid ice sheet response to changes in mass balance (*cf.* Boulton & Jones, 1979), but the sheer scale of ice advance and retreat within the chronological framework seems remarkable.

Status of the Bolders Bank glacial episode

Long *et al.* (1988) interpret the ice limit described here as the Bolders Bank glacial episode as reflecting the maximum extent of Late Weichselian glaciation in the North Sea basin. This was first questioned by Sejrup *et al.* (1994) who suggested that this phase of glaciation was simply a readvance subsequent to an earlier, more extensive LGM. This view is supported by Carr (1998) who suggests that the Bolders Bank glacial episode relates to inherent instability within the British ice sheet, and the process of a large surge of the ice sheet down the east coast of England. Such an interpretation is supported by glaciological modelling of the Late Weichselian ice sheets by Boulton *et al.* (1985) who could not model the Bolders Bank ice lobe under steady state conditions. Thus it seems increasingly likely that the LGM in the North Sea basin occurred at around 22 ka BP, following the scheme of Sejrup *et al.* (1994), with retreat and subsequent surging of the British ice sheet at or soon after 18 ka BP.

Conclusions

There has been a substantial volume of research examining glaciation of the North Sea basin, and considerable controversy regarding specific glacial episodes or ice-sheet extents. This report has attempted to summarise the key themes relating to glaciation of the North Sea, and identifies key areas that need further analysis. There are further stratigraphic and chronological issues to address within the North Sea basin, but the models presented in this report address the key lines of evidence currently available.

Acknowledgements

Most of the original material presented is derived from a NERC/ BGS Case studentship (GT 4/94/364/G), supervised by Jim Rose, Jaap van der Meer and Richard Holmes, to whom grateful thanks are offered. The author would like to acknowledge useful discussion and collaboration with Hans Petter Sejrup and Haflidi Haflidason, and the advice of Charles Harris, Doug Peacock and the late Robin Wingfield.

References

Andrews, I.J., Long, D., Richards, P.C., Thomson, A.R., Brown, S., Chesher, J.A. & McCormac, M. (1990). *United Kingdom offshore regional report: the geology of the Moray Firth.* London, HMSO, 96 pp:

Balson, P. S. & Jeffrey, D.H. (1991). The glacial sequence of the southern North Sea. *In:* Ehlers, J., Gibbard, P. & Rose, J. (eds), *Glacial deposits in Great Britain and Ireland.* Rotterdam, Balkema, 245-254.

Bent, A.J.A. (1986). Aspects of Pleistocene glaciomarine sequences in the North Sea. Unpublished Ph.D thesis, University of Edinburgh.

Bijlsma, S. (1981). Fluvial sedimentation from the Fennoscandian area into the North-West European Basin during the Cenozoic. *Geologie and Mijnbouw*, **60**, 337-345.

Boulton, G.S & Jones, A.S. (1979). Stability of ice caps and ice sheets resting on beds of deformable sediments. *Journal of Glaciology*, **24**, 29-43.

Boulton, G.S., Smith, G.D., Jones, A.S. & Newsome, J. (1985). Glacial geology and glaciology of the last mid-latitude ice sheets. *Journal of the Geological Society*, **142**, 447-474.

Cameron, T. Stoker, M. & Long, D. (1987). The history of Quaternary sedimentation in the UK sector of the North Sea basin. *Journal of the Geological Society of London*, **144**, 43-58.

Cameron, T.D.J, Schüttenhelm, R.T.E & Laban, C. (1989). Middle and Upper Pleistocene and Holocene stratigraphy in the southern North Sea. *In:* Henriet, J.P. & de Moor, D (eds), *The Quaternary and Tertiary geology of the Southern Bight, North Sea*, 119-135 Brussels, Belgian Geological Survey.

Cameron, T.D.J, Crosby, A., Balson, P.S, Jeffery, D.H., Lott, G.K., Bulat, J. & Harrison, D.J. (1992). United Kingdom offshore regional report: the geology of the Southern North Sea. London, HMSO, 153 pp.

Carr, S.J. (1998). The last glaciation of the North Sea Basin. Unpublished Ph.D thesis, University of London.

Carr, S.J. (1999). The micromorphology of Last Glacial Maximum sediments in the Southern North Sea. *Catena*, **35**, 123-145.

Carr, S.J., Haflidason, H. & Sejrup, H.P. (2000). Micro-morphological evidence supporting Late Weichselian glaciation of the northern North Sea. *Boreas*, **29**, 315-328.

Caston, V.N.D. (1977). Quaternary sediments in the North Sea. A new isopachyte map of the Quaternary of the North Sea. *Report of the Institute of Geological Sciences*, **77/11**, 1-8.

Catt, J.A. (1991). Late Devensian glacial deposits and glaciations in eastern England and the adjoining offshore region. *In:* Ehlers, J., Gibbard, P & Rose, J. (eds), *Glacial Deposits of Great Britain and Ireland.* Rotterdam, Balkema, 61-68.

Ehlers, J. (1990). Reconstructing the dynamics of the North-West European Pleistocene ice sheets. *Quaternary Science Reviews*, **9**, 71-83.

Ehlers, J. & Wingfield, R. (1991). The extension of the Late Weichselian/Late Devensian ice sheets in the North Sea Basin. *Journal of Quaternary Science*, **6**, 313-326.

Flinn, D. (1967). Ice front in the North Sea. *Nature*, **215**, 1151-1154.

Flinn, D. (1978). The most recent glaciation of the Orkney-Shetland Channel and adjacent areas. *Scottish Journal of Geology*, **19**, 311-320.

Gatliff, R.W., Richards, P.C., Smith, K., Graham, C.C., McCormac, M., Smith, N.J.P., Long, D., Cameron, T.D.J., Evans, D., Stevenson, A.G., Bulat, J. & Ritchie, J.D. (1994). *United Kingdom offshore regional report:*

the geology of the central North Sea. London, HMSO, 118 pp.

Gibbard, P.L. (1988). The history of the great northwest European rivers during the past three million years. *Philosophical Transactions of the Royal Society of London*, **B318**, 559-602.

Geikie, J. (1874). *The Great Ice Age and its relation to the antiquity of man*. London, Isbister & Co., 575 pp..

Hall, A.M. & Bent, A.J.A. (1990). The limits of the last British Ice Sheet in Northern Scotland and the adjacent shelf. *Quaternary Newsletter*, **61**, 2-12.

Hansen, S. & Nielsen, A.V. (1960). *Glacial geology of Southern Denmark: guide to excursions numbers A44 and C39*. 21st Session, International Geological Congress, Copenhagen.

Huuse, M. & Lykke-Andersen, H. (2000). Overdeepened Quaternary valleys in the Eastern Danish North Sea; morphology and origin. *Quaternary Science Reviews*, **19**, 1233-1253.

Jansen, J.H.F. (1976). Late Pleistocene and Holocene history of the northern North Sea, based upon acoustic reflection records. *Netherlands Journal of Sea Research*, **10**, 1-43.

Jansen, J.H.F., van Weering, T.C.E & Eisma, D. (1979). Late Quaternary Sedimentation in the North Sea. In, Oele, E., Schüttenhelm, R.T.E. & Wiggers, A.J. (eds.), The Quaternary History of the North Sea. *Acta Universitatis Upsaliensis: Symposium Universitatis Annum Quingentesium Celebrantis*, **2**, 175-187.

Johnson, H., Richards, P.C., Long, D. & Graham, C.C. (1993). *United Kingdom offshore regional report: the geology of the Northern North Sea*. London, HMSO, 111pp.

Knutz, P.C., Jones, E.J.W., Austin, W.E.N. & van Weering, T.J.C. (in press) Glaciomarine slope sedimentation and formation of contourite drifts on the Barra Fan, U.K. margin. *Marine Geology*.

Laban, C. (1995). The Pleistocene glaciations in the Dutch Sector of the North Sea. Ph.D Thesis, Universiteit van Amsterdam, 200pp.

Long, D., Laban, C., Streif, H., Cameron, T.D.J & Schüttenhelm, R.T.E. (1988). The sedimentary record of climatic variation in the Southern North Sea. *Philosophical Transactions of the Royal Society of London*, **B318**, 523-537.

McCabe, A.M. (1996). Dating and rhythmicity from the last deglacial cycle in the British Isles. *The Journal of the Geological Society*, **153 (4)**, 499-502.

McCabe, A.M. & Clark, P.U. (1998). Ice-sheet variability around the North Atlantic Ocean during the last deglaciation. *Nature*, **392**, 373-377.

van der Meer, J.J.M & Laban, C. (1990). Micromorphology of some North Sea till samples, a pilot study. *Journal of Quaternary Science*, **5**, 95-101.

Milling, M.E. (1975). Geological appraisal of foundation conditions, northern North Sea. *Proceedings of Oceanology International, Brighton*, **1975**, 310-319.

Peacock, J.D. (1995). Late Devensian to early Holocene palaeoenvironmental changes in the Viking Bank area, northern North Sea. *Quaternary Science Reviews*, **14**, 1029-1042.

Peacock, J.D. (1997). Was there a re-advance of the British Ice Sheet into the North Sea between 15 ka and 14 ka BP? *Quaternary Newsletter*, **81**, 1-8.

Rise, L. & Rokoengen, K. (1984). Surficial sediments in the Norwegian sector of the North Sea between 60°30' and 62°N. *Marine Geology*, **58**, 287-317.

Rose, J. (1985). The Dimlington Stadial/Dimlington Chronozone: a proposal for naming the main glacial episode of the Late Devensian in Britain. *Boreas*, **14**, 225-230.

Rose, J. & Menzies, J. (1996). Glacier stratigraphy. *In:* Menzies, J. (ed.), *Glacial Environments: Volume 2: Past Glacial Environments - Sediments, Forms and Techniques*, Butterworth- Heinemann, Oxford, 650pp.

Sejrup, H.P., Aarseth, K.L., Ellingsen, K.L., Reither, E., Jansen, E., Løvlie, R., Bent., A., Brigham-Grette, J., Larsen, E. & Stoker, M.S. (1987). Quaternary Stratigraphy of the Fladen area, central North Sea: a multidisciplinary study. *Journal of Quaternary Science*, **2**, 35-58.

Sejrup, H.P., Aarseth, I. & Haflidason, H. (1991). The Quaternary succession in the northern North Sea. *Marine Geology*, **101**, 103-111.

Sejrup, H.P., Haflidason, H., Aarseth, I., King, E., Forsberg, C.F., Long, D. & Rokoengen, K. (1994). Late Weichselian glaciation history of the northern North Sea. *Boreas*, **23**, 1-13.

Sejrup, H.P., Aarseth, I., Haflidason, H., Løvlie, R., Bratten, A., Tjostheim, G, Forsberg, C.F. & Ellingsen, K.L. (1995). Quaternary of the Norwegian Channel: glaciation history and palaeoceanography. *Norges Geologische Tidsskrift*, **75**, 1-28.

Skinner, A.C., McElvanney, E., Ruckley, N., Rise, L. & Rokoengen, K. (1986). *1:250 000 offshore regional geology maps: Cormorant: Quaternary*. British Geological Survey (HMSO).

Stoker, M.S. & Long, D. (1984). A relict ice-scoured erosion surface in the central North Sea. *Marine Geology*, **61**, 85-93.

Stoker, M.S. & Bent, A.J. (1985). Middle Pleistocene glacial and glaciomarine sedimentation in the west central North Sea. *Boreas*, **14**, 325-332.

Stoker, M.S., Long, D. & Fyfe, A.J. (1985). The Quaternary succession in the central North Sea. *Newsletters in Stratigraphy*, **14 (3)**, 119-128.

Sutherland, D.G. (1991). The glaciation of the Shetland and Orkney Islands. *In:* Ehlers, J., Gibbard, P., Rose, J. (eds), *Glacial Deposits of Britain and Ireland*, Rotterdam, Balkema, 121-128.

Tesch, P. (1942). De Noordzee van historich-geologisch standpunt. *Mededeelingen Rijks Geologische Dienst*, **A 9**, 1-23.

Valentin, H. (1957). Die grenze der letzten Vereisung im Nordseeraum. *Verhandlungen des Deutschen Geographentages*, **30**, 359-366.

Vorren, T.O. & Laberg, J.S. (1997). Trough mouth fans-palaeoclimate and ice sheet monitors. *Quaternary Science Reviews*, **16**, 865-881.

Wingfield, R. (1989). Glacial incisions indicating Middle and Upper Pleistocene Ice limits off Britain. *Terra Nova*, **1**, 538-548.

Wingfield, R. (1990). The origin of Major Incisions within the Pleistocene deposits of the North Sea. *Marine Geology*, **91**, 31-52.

Wood, S.V. & Rome, J.L. (1868). On the glacial and postglacial structure of Lincolnshire and south-east Yorkshire. *Quarterly Journal of the Geological Society of London*, **24**, 146-184.

Ziegler, P.A. & Louwrens, C.J. (1979). Tectonics of the North Sea. In, Oele, E., Schüttenhelm, R.T.E. & Wiggers, A.J. (eds.) The Quaternary History of the North Sea. *Acta Universitatis Upsaliensis: Symposium Universitatis Annum Quingentesium Celebrantis*, **2**, 7-22.

Quaternary Glaciations – Extent and Chronology
Editors J. Ehlers and P.L. Gibbard

Ice sheet limits in Norway and on the Norwegian continental shelf

Jan Mangerud

*University of Bergen, Department of Geology,
Allégt. 41, N-5007 Bergen, Norway
E-mail: Jan.Mangerud@geol.uib.no*

Introduction

Ice sheets and other glaciers have had a spectacular erosional impact on the Norwegian landscape, producing deep fjords, long U-shaped valleys, numerous cirques and thousands of lakes in overdeepened bedrock basins. In the central part of Sognefjorden there has been vertical glacial erosion of at least 1900 m (Andersen & Nesje, 1992; Nesje & Whillans, 1994). Glaciers in the Norwegian and Swedish mountains formed the nucleus from which the Scandinavian Ice Sheet has grown many tens of times during the Quaternary.

In this paper the author will briefly review present knowledge of the glacial history of Norway. The reconstruction of the Late Weichselian maximum and the Younger Dryas ice sheets will be discussed in some detail.

THE OLDEST GLACIAL HISTORY OF NORWAY

The strong glacial erosion by the last large ice sheets has led to scarcity of old Quaternary deposits on land in Norway. Therefore the oldest glacial history of Norway has to be deciphered from deposits in the border zone of the Scandinavian ice sheet to the south, in Denmark-Germany-Netherlands-The North Sea, and to the west, on the Norwegian continental shelf and in the deep Norwegian Sea.

The best record of the oldest glaciations is ice-rafted debris dropped from 'Norwegian ice bergs' into the deep sea beyond the coast. A review and discussion of this record has recently been given in Mangerud *et al.* (1996), who concluded that calving glaciers first appeared along the Norwegian coast at about 11 Ma. When comparing the amount of ice rafting, the global marine oxygen-isotope signal, and the stratigraphy of the Netherlands, these authors concluded that there was a major increase in the maximum size of the Scandinavian Ice Sheet after the onset of the Praetiglian in the Netherlands 2.5-3 Ma (Zagwijn, 1992).

The oldest identified and dated glacial deposits on the Norwegian shelf is the Fedje Till, that is assigned an age of about 1.1. Ma (Sejrup *et al.*, 2000). The most continuous record of the glacial/interglacial history is probably found in the large submarine fans on the continental slope, located in front of troughs where fast-moving ice streams crossed the continental shelf (King *et al.*, 1996; Laberg & Vorren, 1996; Sejrup *et al.*, 2000). However, long cores have not been obtained from these fans off mainland Norway.

On land in Norway, old glacial deposits are found in two areas. Finnmarksvidda in northern Norway (Fig. 1) is a rolling plain with thick Quaternary deposits including several till beds, interglacial deposits and soils dating back to at least Marine Isotope Stages (MIS) 8 or 10 (Olsen, 1998; Olsen *et al.*, 1996). Even thicker Quaternary deposits occur in the Jæren lowlands at the very SW corner of the country. Here glacial and interglacial deposits from MIS 10 and upwards are described from boreholes (Sejrup *et al.*, 2000). Elsewhere in Norway no glacial deposits are proven older than the Saalian (MIS 6), but till of that age has even been found (below Eemian sediments) in the fjord district, where the glacial erosion has been most intense (Mangerud *et al.*, 1981b).

Even older Quaternary formations are found in karst caves. Numerous caves are now known in Norway, many of which contain speleothems (Lauritzen, 1984). The oldest well-dated speleothem yielded non-finite (>350 ka) U series dates by alpha particle counting and had normal polarity and was therefore assumed to have an age between 350-730 ka (Lauritzen *et al.*, 1990). An age of 500 ka (MIS 13) has subsequently been obtained from this speleothem with the mass-spectrometric method (S.E. Lauritzen, written communication, 2000). Even older caves exist (Lauritzen, 1990), but most U-series dates on speleothems have yielded Eemian ages (Lauritzen, 1991; 1995).

The maximum extent of Pleistocene glaciers

As will be apparent from the above discussion, the outermost glacial limit is far beyond the Norwegian territory to the south. To the west glacial deposits are mapped using seismic methods across the entire continental shelf (H. Holtedahl, 1993; King *et al.*, 1987; Sejrup *et al.*, 2000; Vorren *et al.*, 1983). It seems quite obvious that the Scandinavian Ice Sheet would have advanced much further west if shallow water or dry land had extended in that direction. Western expansion was simply limited by the deep water beyond the shelf edge. Where the water depth exceeds some few hundred metres, the ice front generally floats, and further expansion is limited by iceberg calving. In a detailed seismic mapping of Quaternary deposits on the middle Norwegian shelf, King *et al.* (1987) mapped till tongues down to 800-1000 m below present sea level. However, these beds probably have subsided after deposition, as a result of isostatic response of the shelf to sediment loading.

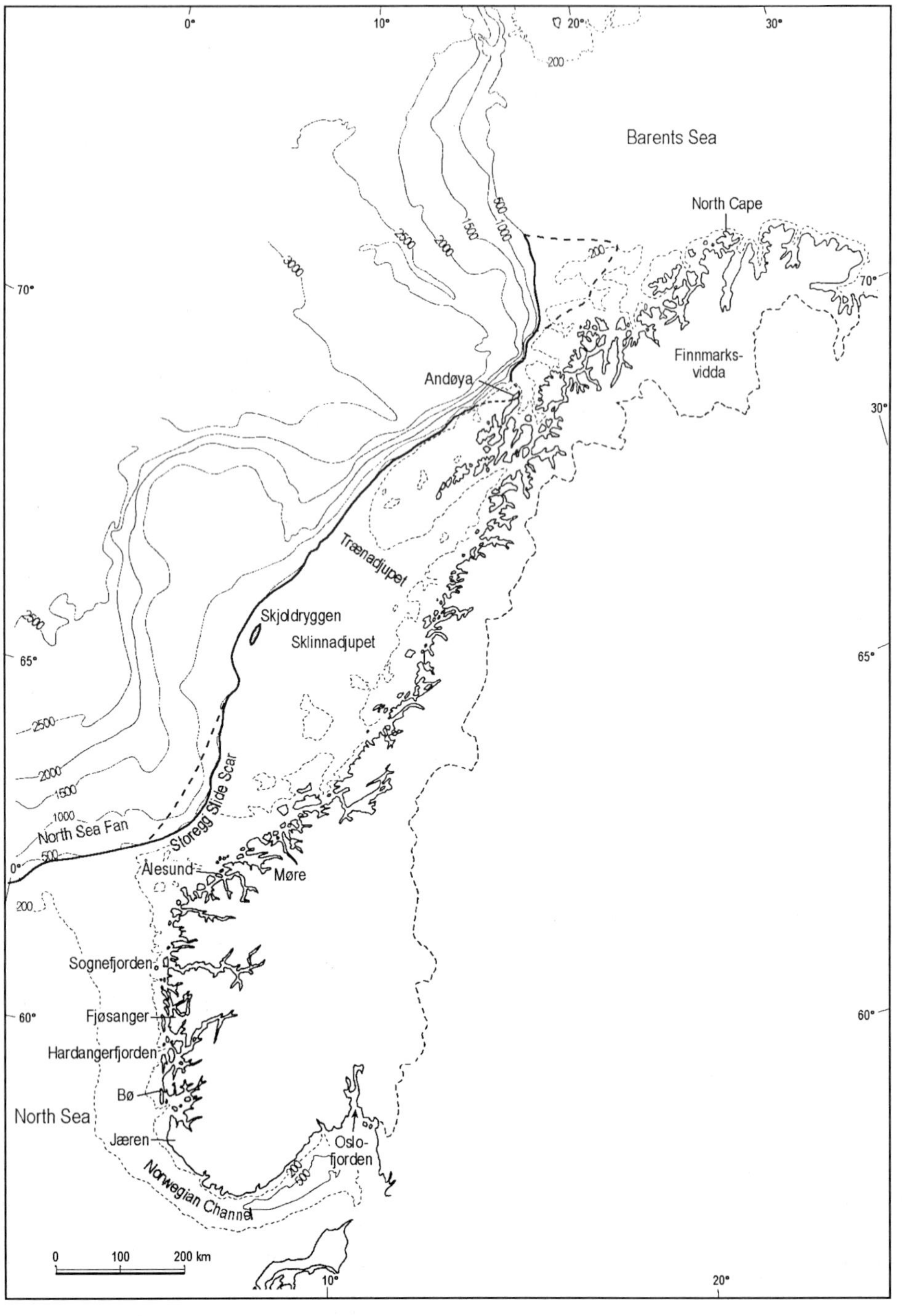

Fig. 1. Map of Norway with the continental shelf. The Late Weichselian glacial limit is marked, in accordance with the digital map on the CD. Geographical names used in the text are given, except some of those related to the discussion of the Younger Dryas which are marked on Figs 12 or 15.

The maximum Pleistocene glacial extent can be considered to be about the same as the Late Weichselian extent, or slightly beyond, as the latter is shown on the digital map.

THE EARLY AND MIDDLE WEICHSELIAN

The glaciation history

The Weichselian is obviously better known than the earlier glaciations. However, the Late Weichselian ice sheet also removed most of the older deposits from this period, so only fragments of the pre-Late Weichselian history are known. Therefore, the interpretation of the older part of the Weichselian is based on observations from very few localities that provide only glimpses into this fascinating history. They are insufficient to allow any accurate mapping of ice sheet limits at different times during the Early and Middle Weichselian.

Dating of events older than the range of the radiocarbon method is also problematic. Different scientists have therefore disagreed on the age of the deposits and the correlation between different sites, and consequently also on the conclusions on the glacial history. Figure 2 shows a glaciation curve for south-western Norway slightly

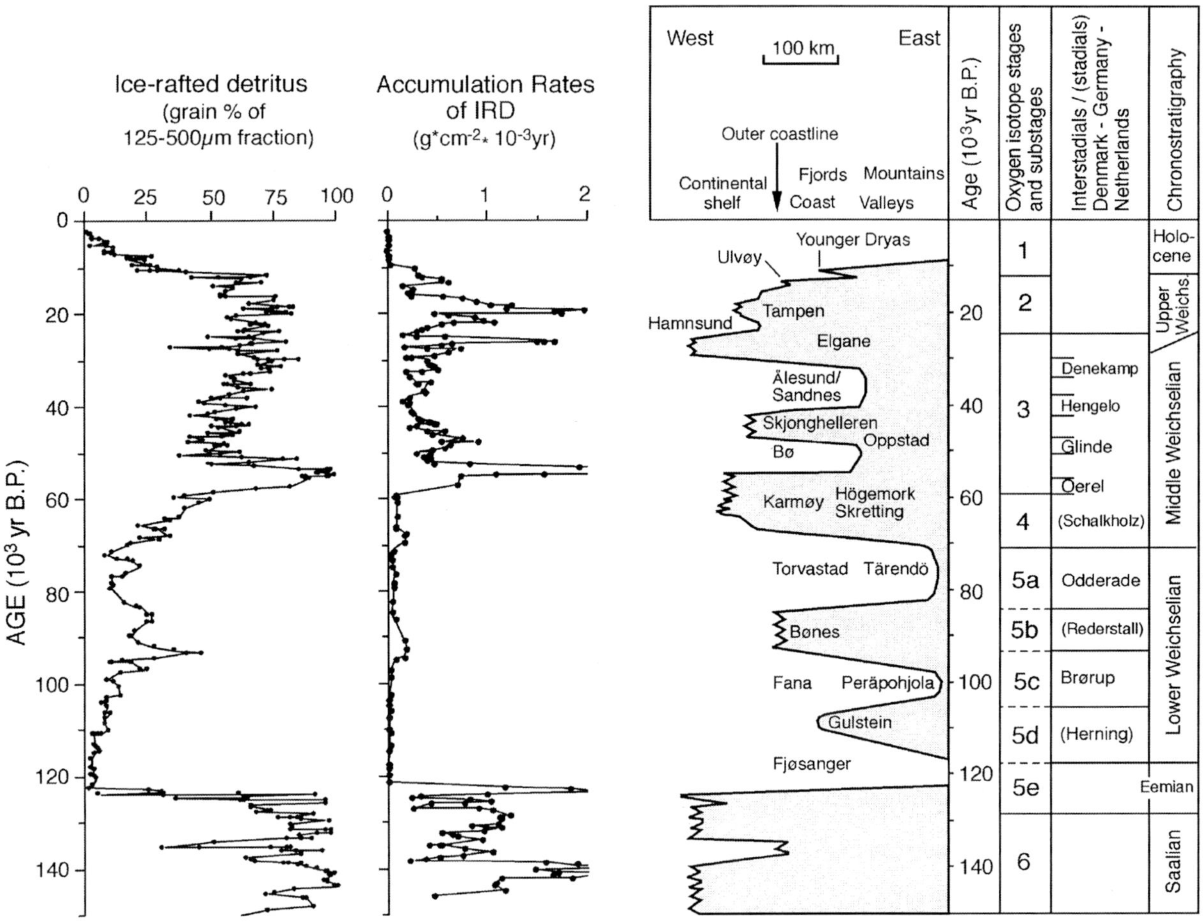

Fig. 2. The curve to the right is a glaciation curve for western Scandinavia constructed from continental data (Mangerud, 1991a), except that the advance during MIS 4 (Karmøy) has been extended into MIS 3, based on the ice-rafted detritus (IRD) curves. Note that the time scale is in calendar years. The curve to the left is a stacked record for grain % of IRD in 5 cores, and the middle curve a stacked record of accumulation rates in 3 cores; all cores collected from the Vøring Plateau west of the Norwegian coast. A three-point moving average smoothing filter was used for the IRD curves. The figure is slightly modified from Baumann et al. (1995). It is postulated that most of the IRD was dropped from icebergs that calved from the margin of the Scandinavian Ice Sheet, off the coast of western Norway. The stratigraphical position of the till beds on Jæren (Elgane, Oppstad, Høgemork/Skretting) are also indicated according to Larsen et al. (2000), except that the Høgemork/Skretting is plotted with an MIS 4-3 age, instead of an MIS 5a age, as suggested by these authors.

modified from Baumann *et al.* (1995). The curve was originally developed from the interpretation of localities on land (Mangerud, 1981; 1991a; 1991b). When comparing the curve with the record of ice-rafted detritus (IRD) off the Norwegian coast, Baumann *et al.* (1995) found a close correlation, supporting the main features of the curve. The only difference was that Mangerud (1991a; 1991b) had assumed that a major glaciation in MIS 4 ended at the transition to stage 3, whereas the IRD record suggested the deglaciation was well into stage 3 (which is adapted in Fig. 2).

Alternative interpretations of the same continental data are presented by Larsen & Sejrup (1990) and Sejrup *et al.* (2000). There is a general agreement with Mangerud (1991a, 1991b) concerning the younger part of the curve. The Skjonghelleren glacial advance is tied to the Laschamp magnetic excursion about 43 ka (in calendar years), and the younger part is based on ^{14}C dates (although note that the scale in Fig. 2 is given in calendar years). However, there are considerable differences in the older part of the curves, which can be explained in the following way: The lowest part of the curve in Fig. 2 is based mainly on the stratigraphy at Fjøsanger (Mangerud *et al.*, 1981b). The correlation of the Fjøsangerian Interglacial with the Eemian is well established. However, the dating of the younger stadials and interstadials has been more problematic, and is partly based on correlation with the western European and deep-sea stratigraphy. The author accepts that the Brørup and Odderade interstadials should be correlated with Marine Isotope Substages 5c and 5a respectively (de Beaulieu & Reille, 1992; Behre, 1989; Lowe & Walker, 1997; Mangerud, 1989) although the boundaries probably were not exactly time parallel. The ages of the MISs are estimated to about 105-93 ka for Substage 5c and thus also for Brørup and 85-74 ka for Substage 5a/Odderade (Martinson *et al.*, 1987). This implies that the younger ^{14}C dates are disregarded (about 61 ka) for the Odderade, which have been duplicated several times (Behre & van der Plicht, 1992).

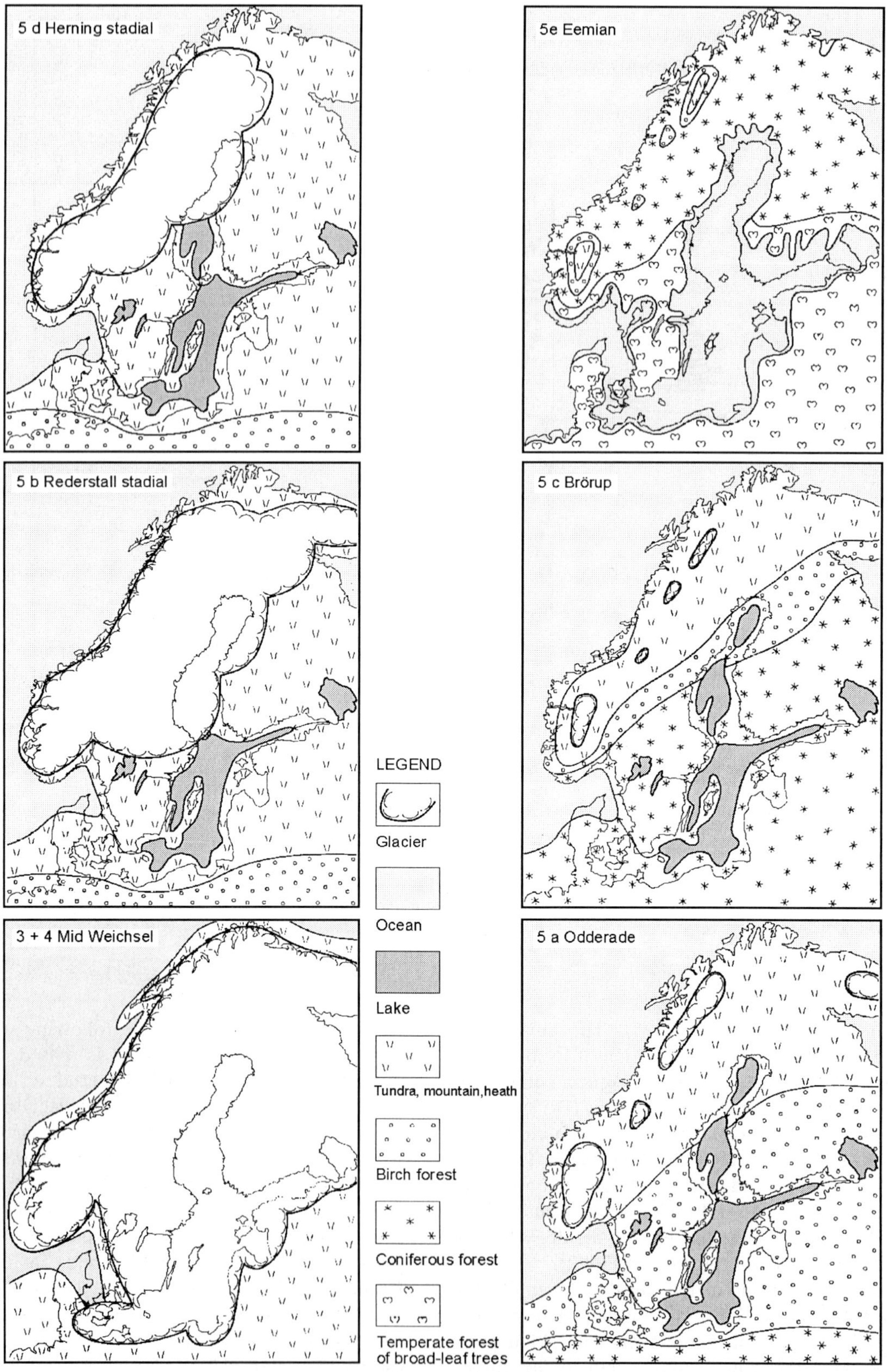

Fig. 3. Sketch maps of the Scandinavian Ice Sheet and surrounding environments during MIS 5e to 3. The data are slightly modified from Lundqvist (1992), who stated that all boundaries were hypothetical and that the maps merely illustrated an interpreted development, although he used the inter-pretations of known sites. Here more extensive glaciation in SW Norway during MIS 5d, 5b and 4 according to Fig. 2 is indicated. In northernmost Norway the ice extent during MIS 5d and 5b are reduced according to Olsen et al. (1996) and L. Olsen (written communication, 2000).

Brørup and Odderade were the last known interstadials with forests in Netherlands-Germany (Behre & van der Plicht, 1992; Zagwijn, 1989), and the author correlates them with two ice-free interstadials in northern Sweden (Lagerbäck & Robertsson, 1988; Lundqvist, 1992). Consequently, the writer postulates that Scandinavia was nearly ice free during the Brørup and Odderade interstadials (Fig. 2). These correlations and interpretations have recently been supported by the results from the Sokli site in northern Finland, which contains a nearly continuous sequence from the Eemian to MIS 3 (Helmens *et al.*, 2000). Sejrup *et al.* (2000), on the other hand, rely more directly on amino-acid dates, and they place for example a major glaciation between 85-70 ka, i.e. during the Odderade interstadial/MIS 5a in my scale, and they propose ice-free conditions in MIS 4. Another difference is that they leave Scandinavia nearly ice-free during Marine Isotope Substages 5b and 5d.

Sejrup *et al.* (2000) *".. note that the new glaciation curve (..), better matches IRD values from the Norwegian Sea (Baumann et al., 1995) than glaciation curves published earlier based on different chronologies (Larsen & Sejrup, 1990; Mangerud 1991c; Baumann et al., 1995)".* The dating and correlation of the individual sites by Mangerud (1991a) and Baumann *et al.* (1995) may of course be wrong. However, if one compares the published glaciation curves with the given IRD curves, the author disagrees with the cited statement from Sejrup *et al.* (2000). They have plotted the Bø Interstadial, when Norway according to their curve was almost ice-free (their Fig. 9), in the period 52-70 ka when the IRD curves (Fig. 2) show the highest peak during the entire Weichselian. In contrast Baumann *et al.* (1995) and the present paper (Fig. 2) postulate that the ice sheet reached the continental shelf during that period (Karmøy Stadial). Sejrup *et al.* (2000), on the other hand, placed this Karmøy glaciation in the period 70-85 ka when the IRD curves show minimum values. The writer will also point out that there is a small increase in IRD during Marine Isotope Substage 5b (about 90 ka), when one assumes the ice front first crossed the coast (Fig. 2), whereas Sejrup *et al.* (2000) postulated no growth of glaciers at that time.

A major new contribution to the Weichselian glacial history of Norway appeared after this paper was written in fact. In the Jæren area (Fig. 1), Larsen *et al.* (2000) demonstrate that the Norwegian Channel ice stream developed in a nearly similar fashion during two earlier Weichselian glaciations as it had during the Late Weichselian. They correlate the youngest advance, represented by the Oppstad Diamicton, with the Skjonghelleren readvance, and the oldest, represented by the Høgemork Diamicton, with the Karmøy readvance (Fig. 2). The author considers that both correlations are probably correct. However, they maintain an age of 80-70 ka for the Karmøy readvance, which the present author finds unlikely. Instead he considers an age corresponding to either MIS 5b or 4 as more probable. When comparing with the IRD curve (Fig. 2), the most probable alternative is MIS 4, extending into stage 3, as discussed above.

Early/Middle Weichselian glacial limits

It is apparent from the above discussion that there is as yet no consensus on the Early/Middle Weichselian glacial history of Norway. However, more important than the disagreement about the interpretation of the few known sites is the scarcity of sites. We simply need more sites to resolve the general stratigraphy, including the timing of the events, and even more sites to be able to map the areal extent of the ice sheets at different times.

In the first attempt to reconstruct the Early (MIS 5d and 5b) and Middle Weichselian (MIS 4) ice sheets of Fennoscandia, the limits along the more than 2000 km long Norwegian coast were based on only four sites/areas (Jæren, Bø, Fjøsanger and Ålesund) (Andersen & Mangerud, 1989). The situation has hardly improved since then. The main new information is the evidence from Jæren discussed above, which shows that in southern Norway the ice sheet twice grew so large that an ice stream developed in the Norwegian Channel (Larsen *et al.*, 2000). The area reconstructions of the ice sheets in Norway still have to be based on observations from these few sites, and on general geological evaluations. The sites are discussed and partly described in the papers cited above (Larsen & Sejrup, 1990; Mangerud, 1991a; 1991b; Sejrup *et al.*, 2000). Primary, and more extensive descriptions for the different sites are for Jæren (Andersen *et al.*, 1987; Larsen *et al.*, 2000; Sejrup *et al.*, 1998), Bø (Karmøy) (Andersen *et al.*, 1983; Sejrup, 1987), Fjøsanger (Mangerud *et al.*, 1981b), Ålesund (Larsen *et al.*, 1987; Mangerud *et al.*, 1981a; Valen *et al.*, 1996).

Figure 3 shows a slightly modified version of the reconstructions by Lundqvist (1992), where the reconstructed ice sheets are not very different from those presented by Andersen & Mangerud (1989). However, Lundqvist included reconstructions for the interstadials, and also the vegetation at different times. Therefore his maps are more instructive and also more provocative for future testing.

A key question is if the ice margin passed the coastline of western Norway during MIS 4-5. According to H. Holtedahl (1993) there is no site that can unambiguously prove whether the ice front reached the continental shelf or not during the Early or Middle Weichselian. In the present author's interpretation (Fig. 2), the ice front passed beyond the coast near Fjøsanger during MIS 5b, and it passed the coast over a wider area during MIS 4 and 3. As discussed above, an ice stream developed in the Norwegian Channel during the two latter events, demonstrating that the ice limit was well outside the coast of southern Norway (Larsen *et al.*, 2000). Within the North Sea Fan, lies a till (M/N) deposited by an ice stream in the Norwegian Channel (Sejrup *et al.*, 2000). These authors conclude that the till is of either Saalian or Early Weichselian age, and they favour the latter interpretation. If correct, it should represent the older of the two ice streams described by Larsen *et al.* (2000), for which the present writer favours a MIS 4 age.

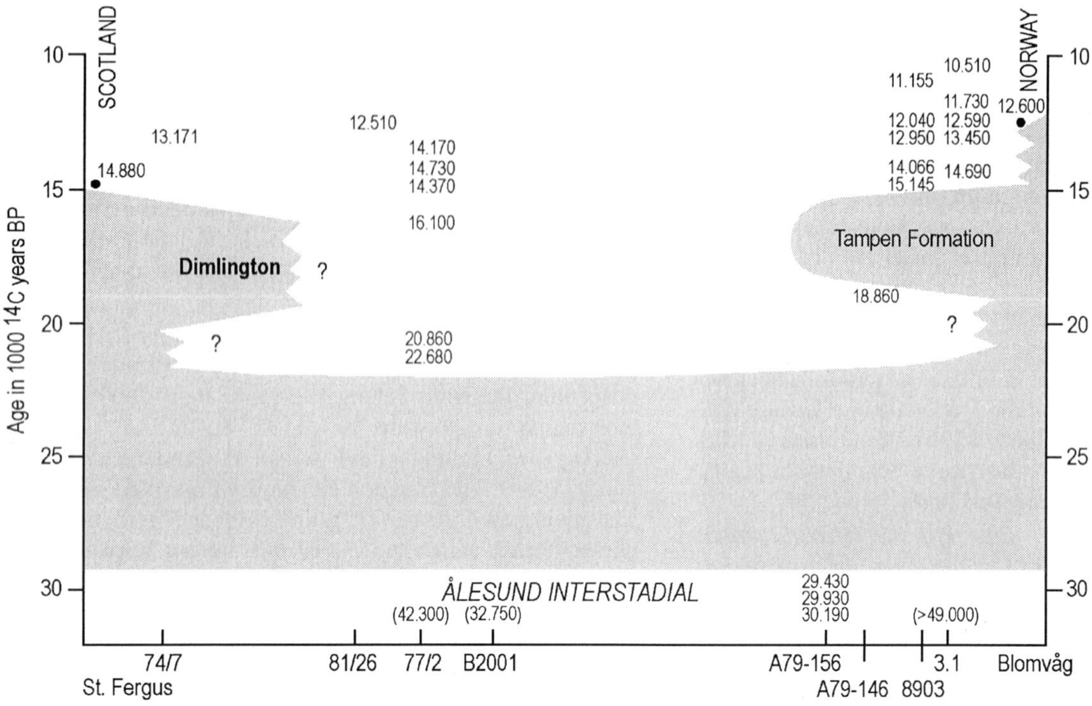

Fig. 4. Time-distance curve for the fluctuations of the ice margins of the British and Scandinavian ice sheets across the North Sea, indicating that the two ice sheets met before about 23 ka. From Sejrup et al. (2000). Note that the time scale is in radiocarbon years, whereas in Fig. 2 it is calibrated to calendar years.

THE LATE WEICHSELIAN GLACIAL MAXIMUM

The ice sheet limit on the Norwegian Shelf

An extensive review and synthesis of the Quaternary geology on the Norwegian shelf is given by H. Holtedahl (1993), to which the author refers for general conclusions and for references to the literature. At present there is an apparent agreement among geologists working on the Norwegian shelf that the Late Weichselian ice sheet reached the shelf edge along its entire length from the mouth of the Norwegian Channel nearly to the North Cape. The only probable exception is a segment west of Andøya (Fig. 1), where the limit according to Vorren & Plassen (2002) (and several earlier authors) was located on land. Thus the main features of the mapping of the Late Weichselian maximum by Andersen (1981) are still valid.

The glacial limit is mapped using three criteria: 1) for most of its length by the western limit of the youngest till sheets, or till tongues, mainly mapped by seismic methods; 2) end-moraine ridges; 3) submarine fans built up mainly by debris flows of glacial sediments. More detailed descriptions of the glacial limit are given in the section 'Comments to the ice sheet limit drawn on the digital map' below.

The age of the glacial limit

There are few radiocarbon dates that can be used to bracket the age of the glacial maximum on the western Norwegian coast or shelf. The best are probably a set from the North Sea (Fig. 4), indicating a radiocarbon age of between 29-22 ka for the maximum and 15-19 ka for a secondary readvance (Sejrup *et al.*, 1994; 2000). Similar dates were obtained from lake sediments on Andøya in Northern Norway (Alm, 1993; Vorren *et al.*, 1988). The stratigraphical sequences in these lakes have been correlated with the stratigraphy in the adjacent fjord and on the shelf (Fig. 5) (Vorren & Plassen, 2002).

Dates related directly to the ice front have been obtained by King *et al.* (1998), dating glacial debris flows deposited from the ice front onto the North Sea Fan (Fig. 6). Several dates bracketing the last main debris flow unit indicate that the ice front rested at the mouth of the Norwegian Channel from before 23 ka until 16 ka. A thinner debris flow unit above indicates that the ice margin remained close to the mouth of the Norwegian Channel until 15 ka (Fig. 6). Further north, about 64° N, a date of 15 ka was obtained from a tongue of till that almost reached the shelf edge (Rokoengen & Frengstad, 1999).

Radiocarbon dates on bones from the Skjonghelleren Cave, including more than 30 AMS dates in the range 29-35 ka, provide a maximum age for a Skjonghelleren ice advance (Fig. 2) (Baumann *et al.*, 1995; Larsen *et al.*, 1987). This advance is also dated by identification of the 28-29 ka old Lake Mungo palaeomagnetic excursion in clay deposited in a lake dammed in the cave by this ice-advance (Larsen *et al.*, 1987). Thus an ice front passed the coast near Skjonghelleren at about 29 ka. These results were supported by many dates and the palaeomagnetic stratigraphy in another cave (Valen *et al.*, 1996). In the latter cave two dates of 24.5 ka were also obtained indicating a withdrawal of the ice-front at that time.

Two radiocarbon dates of 22 and 20 ka have also been obtained on bones of polar bears in a karst cave in Nordland

(Lauritzen *et al.*, 1996; Nesje & Lauritzen, 1996). These dates also support the suggestion that the ice margin in northern Norway withdrew inside the coastline for a period around the time of the Late Weichselian maximum (Fig. 5).

Olsen (1997) has obtained more than 100 AMS dates from terrestrial organic-bearing sediments with low organic content (0.2-1.5%). A description of the individual sites and details of the dated samples will soon be published (L. Olsen, written communication, 2000). The writer is generally sceptical of dates on samples with such a low organic content. However, Olsen's conclusion is that the main glacial maximum occurred 24-21 ka, with an earlier secondary maximum at 30-29 ka and a later one at 14.5-17 ka, more or less similar to the results cited above.

The conclusion from these observations is that the Late Weichselian ice sheet reached its maximum extent off the coast of Norway relatively early, probably at 24-22 ^{14}C ka. There were one or more ice front oscillations up to about 15 ka, when the main ice retreat started. However, the writer emphasises that it is highly unlikely that the ice front reacted in unison along the entire Norwegian coast. The maximum position was probably reached at different times. Many more dates distributed along the position of the ice margin will be necessary to resolve the probable asynchroneity of the glacial maximum.

Ice streams on the shelf

A major improvement in our understanding of glacier dynamics and even ice front deposits on the shelf has come from the discovery and interpretation of submarine fans at the mouth of glacial troughs, the so-called 'trough mouth fans', and the existence of distinct ice streams across the shelf during glaciations. A recent review is given by Vorren *et al.* (1998).

The longest ice stream on the Norwegian shelf was that in the Norwegian Channel (Longva & Thorsnes, 1997; Sejrup *et al.*, 1996; 1998). It was also different from the others in that the channel runs parallel with the coast for some 6-700 km (Fig. 8). This ice stream drained much of the southern part of the Scandinavian Ice Sheet. The huge North Sea Fan was deposited beyond the Norwegian Channel ice stream (King *et al.*, 1996; 1998).

Through detailed morphological mapping of the sea floor, it has been shown that there have been a number of ice streams across the Norwegian shelf (Figs 7 and 8) (Ottesen *et al.*, 2001). There are numerous large, parallel glacial lineations (megaflutes and mega-scale lineations) in the troughs, whereas features diagnostic of glacial flow are absent from the shallow ground. The inference is that there was only slow moving ice in the shallows. However, major trough mouth fans did not form in the front of most of these ice streams. At the mouth of the Sklinnadjupet trough is the morphologically largest end moraine on the shelf, the Skjoldryggen, some 200 m high, 10 km wide and 200 km long (Ottesen *et al.*, 2001).

Comments to the ice sheet limit shown on the digital map

Whether the Norwegian and British ice sheets met in the North Sea and where the ice margins were located have been long-standing questions. It is now documented that they probably met (Carr *et al.*, 2000; Sejrup *et al.*, 1994; 2000). According to the radiocarbon dates mentioned above (Fig. 4) this maximal position was reached before about 22 ka ^{14}C years.

At the mouth of the Norwegian Channel, the limit is placed at the boundary between basal till in the channel and the glacial-debris flows on the North Sea Fan, as identified from seismic signatures by King *et al.* (1998) and King *et al.* (1996). As described above, King *et al.* (1998) dated this limit to 23-16 ^{14}C ka.

The huge, post-glacial Storegga Slide is located along the shelf edge (between about 62° 30' and 64° 40') north of the Norwegian Channel (Bugge, 1980; Bugge *et al.*, 1978; Jansen *et al.*, 1987; H. Haflidason, oral communication, 2000). Its backwall extends for approximately 300 km along the shelf edge. In this area the upper formation on the shelf is a 10-40 m thick till (the Storegga Moraine) which is cut by the slide (Bugge, 1980; Bugge *et al.*, 1978). The till was probably deposited during the Late Weichselian maximum, and the outer limit of the till sheet was therefore removed by the slide. On the map the limit has been drawn as a stippled line across the slide scar.

About 64° N the western limit of the Storegga Till sheet turns eastwards (Bugge, 1980). It is there correlated with till tongue 24 (Rokoengen & Frengstad, 1999) and is underlain by the older till tongue 23. The latter is mapped to a water depth of about 500 m along the shelf edge, and radiocarbon dated to about 15 ka (Rokoengen & Frengstad, 1999). Between 64° and 67° 30'N, the author used the limit of till tongue 23, as mapped by King *et al.* (1987) and Rokoengen & Frengstad (1999). Both till tongues 23 and 24, and thus the Skjoldryggen moraine, were according to these authors deposited during the Late Weichselian glacial maximum. The Skjoldryggen moraine crosses the mouth of the Sklinnadjupet trough and was formed by an ice stream out this trough (Ottesen *et al.*, 2001),

Another major ice stream that reached the shelf edge is mapped in the Trænadjupet trough (67° N) (Ottesen *et al.*, 2001). North of the trough the thickness of Quaternary sediments is less than south of the trough, and the till tongues could not be traced there (King *et al.*, 1987). However, the youngest till sheet generally thickens towards the shelf edge, and probably this records the Late Weichselian ice sheet limit, although the till is not dated (Holtedahl, 1993; Rokoengen *et al.*, 1977). These authors placed the Late Weichselian limit along the shelf edge until the edge turns away from the coast near the southern boundary of the Barents Sea. However, an alternative interpretation is that the north-western part of Andøya, and thus the continental shelf, remained ice-free (Vorren & Plassen, 2002). On the map the author has followed the

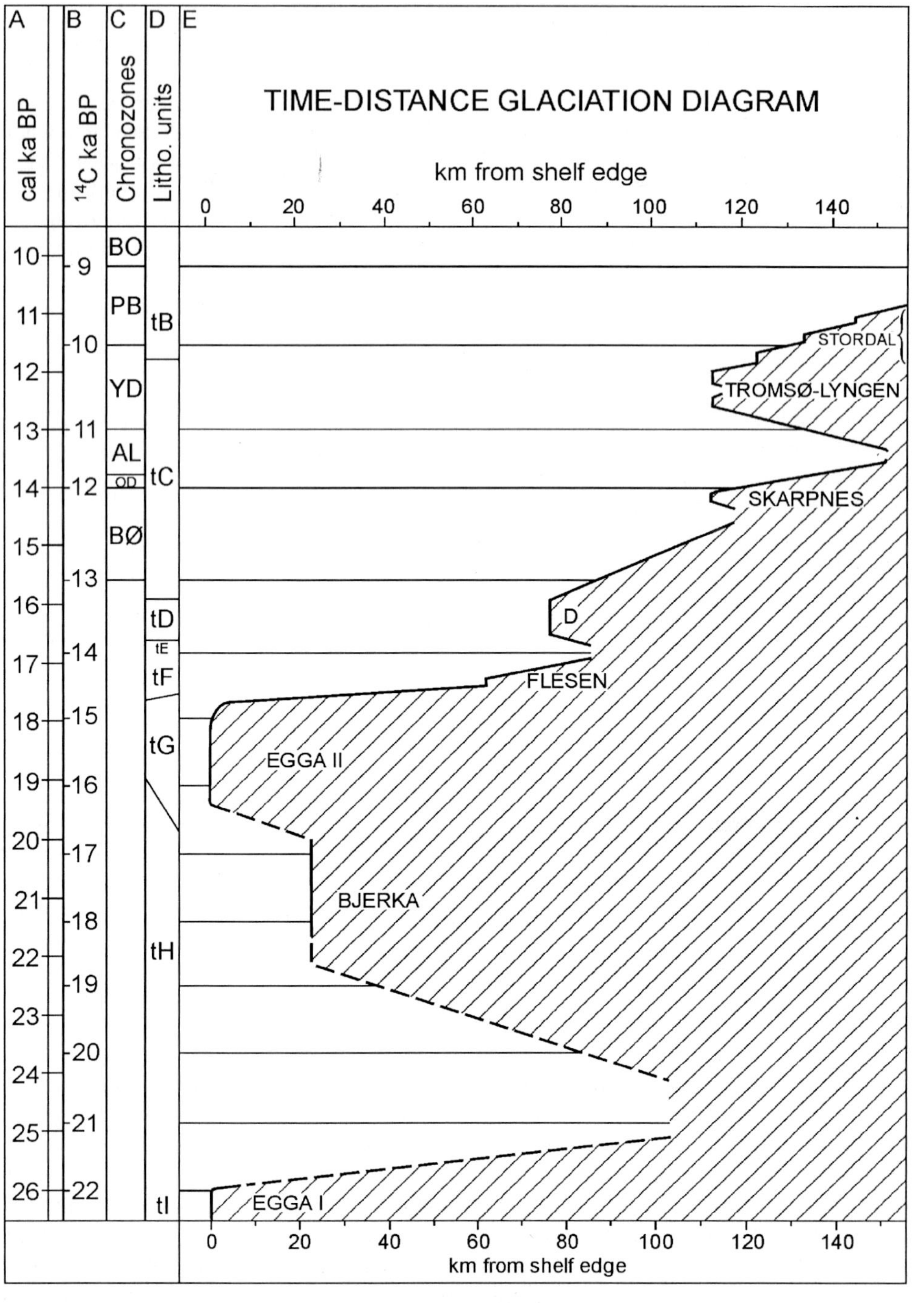

*Fig. 5. Time-distance diagram showing the ice-front variations near Andøya, Northern Norway. From Vorren and Plassen (**2002**).*

latter view, if the former is correct, the limit should simply be drawn along the shelf break.

Vorren & Kristoffersen (1986) mapped a system of end moraines (shown with dashed line on the map) in the south-western corner of the Barents Sea that were considered candidates for the junction between the Scandinavian and Barents ice sheets. However, Sættem (1990) has subsequently mapped some glaciotectonic structures further west, where a minimum age of 13 ka was obtained. On the map the author follows the results of Landvik *et al.* (1998) and draw the limit along the shelf edge, as has also been done in the chapter on the Barents Sea (Svendsen *et al.*, this volume).

Ice thickness – did nunataks exist in Norway?

The surface geometry, and thus the thickness, of the Scandinavian Ice Sheet are much more poorly known than its areal limits. The author will therefore discuss the problems at some length. For more than a hundred years there has been discussion as to whether mountain peaks in Norway protruded as nunataks above the ice surface during the Quaternary glaciations, especially during the Late Weichselian glacial maximum. Recent reviews of the discussion are given in Sollid & Sørbel (1979), Mangerud *et al.* (1979), Nesje *et al.* (1987, 1988) and Nesje & Dahl (1992). Earlier the main focus was whether plants had

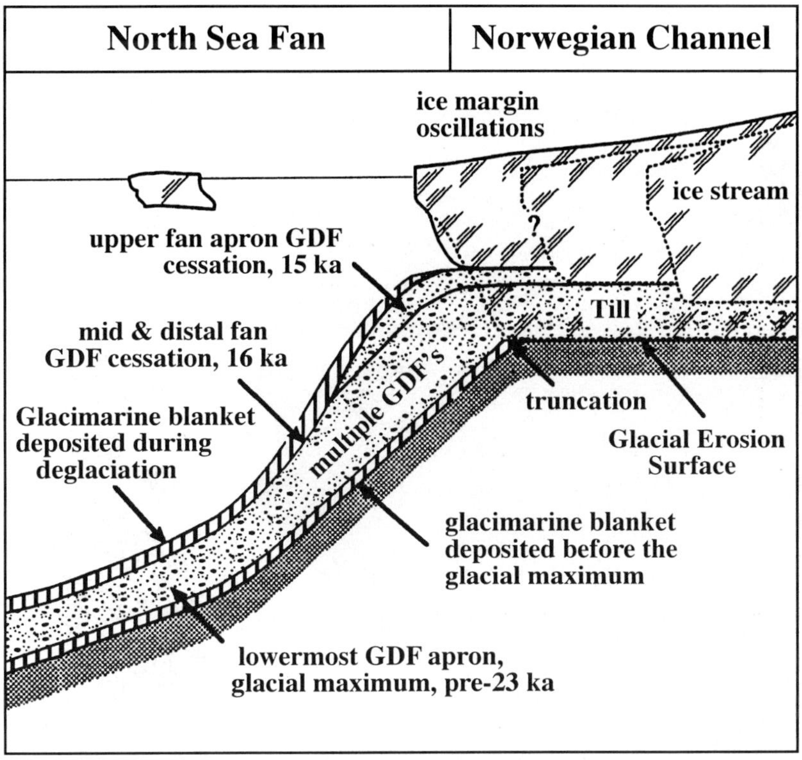

Fig. 6. Schematic diagram showing the development of the Norwegian Channel trough-mouth fan during the Late Weichselian glacial maximum, slightly modified from King et al. (1998). The lowermost glaciomarine blanket was deposited during the glacial advance, after the Ålesund Interstadial (Fig. 2). When the ice stream in the channel reached the slope break, glacigenic debris flows (GDF) travelled down the fan slope, depositing most of the sediments on the fan. The start of this maximum phase is not accurately dated, but was probably about 25-27 ka (in radiocarbon years) whereas the end is dated to 16 ka. An upper GDF indicates that the ice margin had retreated only a short distance at 15 ka.

survived on nunataks or not, whereas here a three-dimensional reconstruction of the Scandinavian Ice Sheet is considered.

Nesje *et al.* (1987) and Nesje & Dahl (1992) argue that there were ice-free summits across much of southern Norway during the Late Weichselian glacial maximum, using the altitudinal distribution of autochthonous block fields as the main argument (Fig. 9). They plotted altitudes of summits with and without block fields and found a geographically consistent altitudinal boundary between the two types. They postulate that the block fields pre-date the Late Weichselian, and that the lower limit represents a glacial erosional boundary showing the maximum elevation of the Late Weichselian ice sheet surface (Fig. 9). They also maintain that their reconstruction is supported by the distribution of alpine pinnacle topography without indications of glacial moulding, and also by the distribution of 'refugial plants'. In the Nordfjord area their interpretation has subsequently been supported by cosmogenic nuclide exposure ages giving 55 ka (^{10}Be age) or 71 ka (^{26}Al age) for bedrock surfaces in the block field, and 21 ka on bedrock below the limit of block fields (Brook *et al.*, 1996). Another area with pinnacles that have traditionally been considered to have remained ice-free is western Andøya in northern Norway (Fig. 1). There this interpretation is supported by mapping of moraines and by the stratigraphy in lakes, as mentioned above (Vorren & Plassen, 2002).

The hypothesis of ice-free nunataks is attractive to explain the cited observations; it is difficult to envisage an eroding glacier overriding the block fields, and for the writer even more so for the tall and narrow pinnacles. Therefore these arguments have been used in favour of the nunatak theory for more than a century. The number of observations and the consistent pattern in the Nordfjord-Møre area in western Norway (Fig. 10) seem to favour that the interpretation might be correct in that area (Nesje *et al.*, 1987; Sollid & Sørbel, 1979). In this area it is also glaciologically reasonable because the deep fjords would efficiently drain the ice flow. However, the interpretation becomes more difficult, or rather impossible, when the limit is extended further across southern Norway, which would imply a maximum ice thickness of less than 800 m, and a maximum altitude of the ice sheet of about 1600 m a.s.l. in the eastern ice divide areas (Nesje & Dahl, 1990; Nesje *et al.*, 1988). A closer look at some of the arguments therefore seems necessary.

Block fields are not unambiguous proof that ice has not overridden the site. On the contrary, it has been demonstrated that both block fields and other unconsolidated sediments have survived below the Scandinavian Ice Sheet and yet show little evidence of glacial overriding (Kleman, 1994; Kleman & Borgström, 1990; Kleman & Hättestrand, 1999; Lagerbäck & Robertsson, 1988). In fact, it is just as difficult to understand how sharp-ridged eskers have survived below glaciers (Lagerbäck & Robertsson, 1988), as it is the pinnacles discussed above. Meltwater channels from the last deglaciation cut through block fields at many localities in the Norwegian and Swedish mountains demonstrating that the ice-sheet surface was above the lower limit of block fields (Borgström, 1999; Sollid & Sørbel, 1994). All of the cited authors argue that the unconsolidated deposits, in most cases, survived beneath

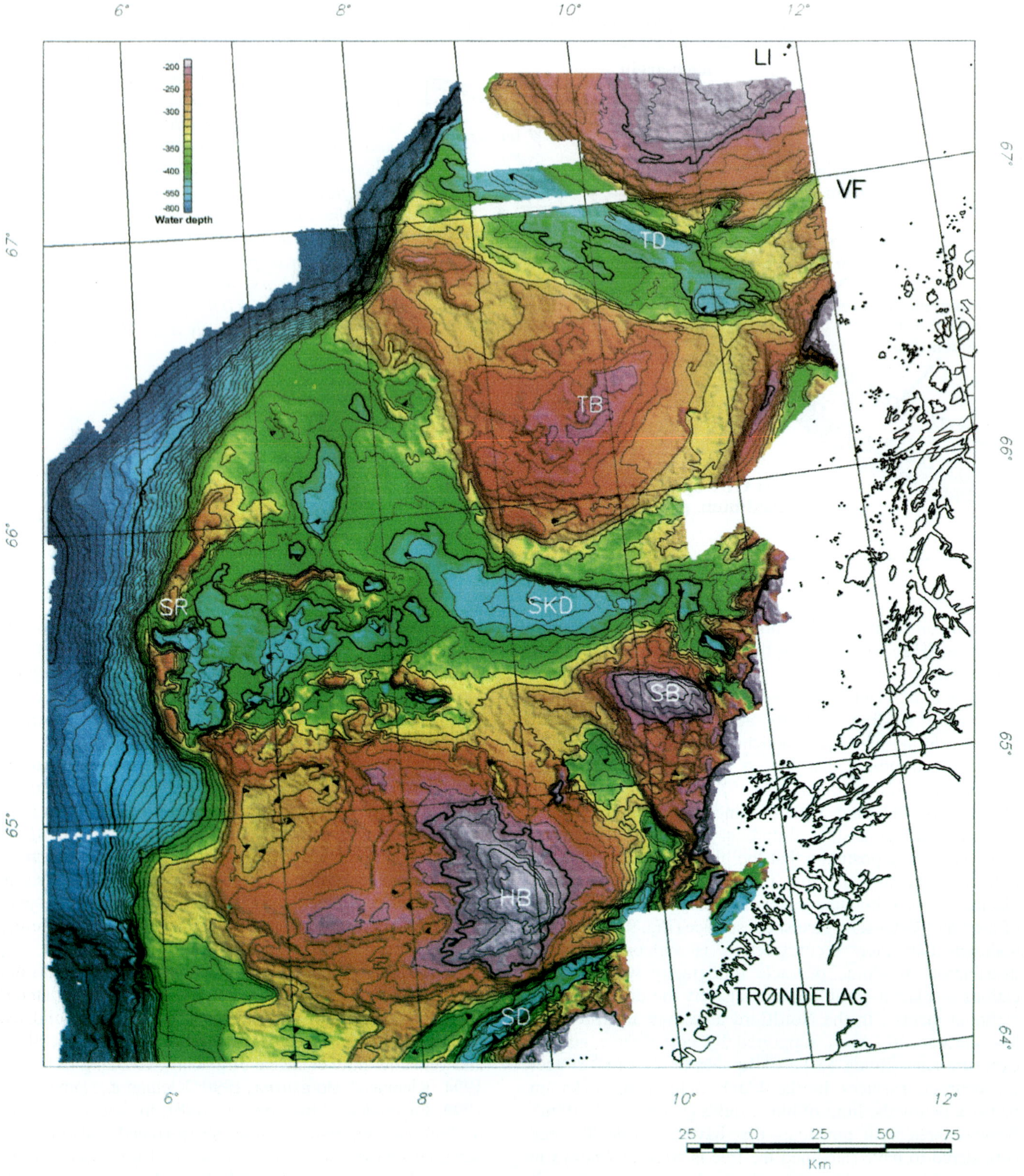

Fig. 7. Relief map from the Norwegian shelf from 64° to 67° N from Ottesen et al. (2001). Contour interval 20 m. LI- Lofoten Islands, VF-Vestfjorden, TD- Trænadjupet, TB- Trænabanken, SKD- Sklinnadjupet, SR- Skjoldryggen, SB- Sklinnabanken, HB- Haltenbanken, SD-Suladjupet.

cold-based ice. In addition the summits with block fields would certainly have lain in the cold-based zone if they were covered by glacial ice. In reply to this Nesje & Dahl (1990) argue that the limit between cold and warm-based ice should not be parallel to the ice surface, and as mentioned above, this could be a valid argument for the

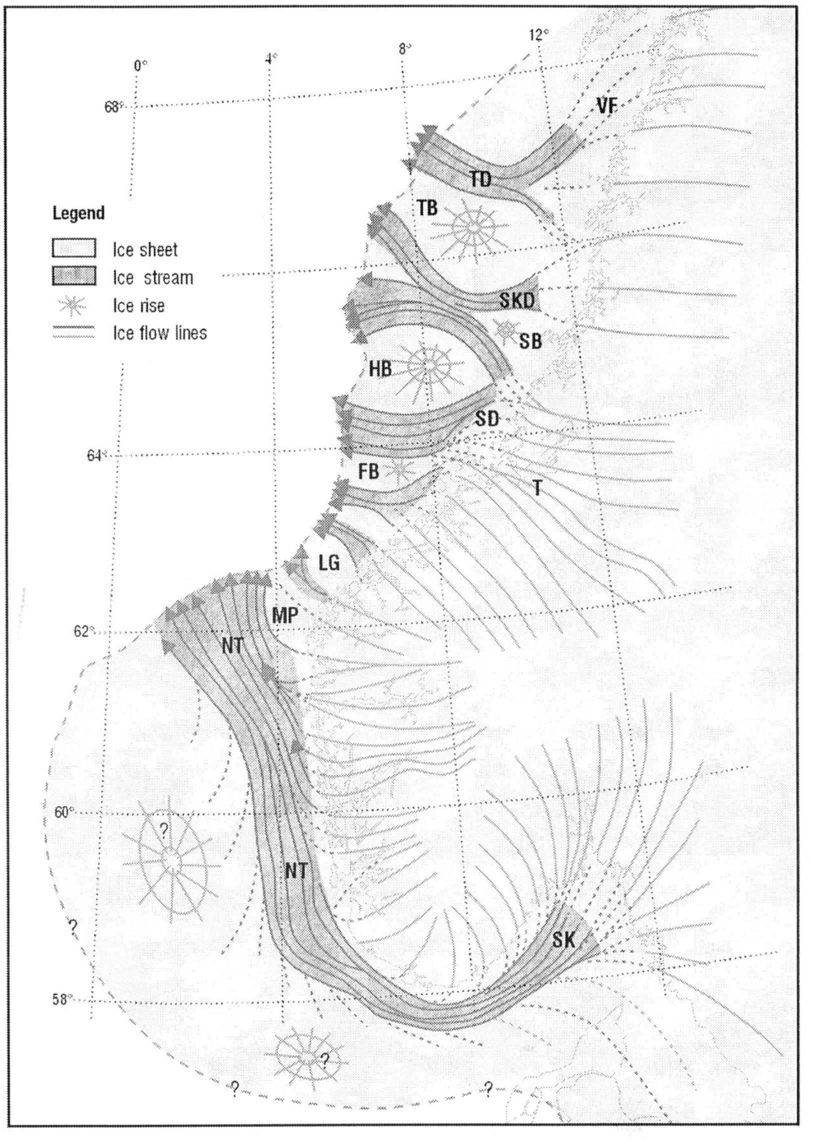

Fig. 8. Interpreted ice-flow model during Late Weichselian glacial maximum from Ottesen et al. (2001).

VF-Vestfjorden,
HB-Haltenbanken,
SKD-Sklinnadjupet,
TD-Trænadjupet,
SB-Sklinnabanken,
SD Suladjupet,
FB-Frøyabanken,
MP-Måløydjupet,
NT-Norwegian Trench (also translated as the Norwegian Channel),
TB-Trænabanken,
LG-Langgrunna,
SK-Skagerrak,
T-Trondheim.

Nordfjord-Møre area where this limit is very consistent. However, an alternative explanation could be that the block fields were covered by cold-based ice, and that the lower limit does not show the boundary to warm-based ice, but a later erosion limit formed at a younger and lower ice-surface.

The botanical argument, first presented around 1890, was in fact the one that started the discussion about ice-free nunataks (Birks, 1993; Mangerud, 1973). The main argument is that some mountain plants, with particular emphasis on endemics and West-Arctic species, have a bi-centric distribution in Norway, which suggests that they survived the glaciation close to these centres. However, it has recently been demonstrated that this distribution pattern can be explained by other factors, and it has no weight as argument for ice-free nunataks (Birks, 1993).

Fjeldskaar (2000) has tested the thin ice sheet interpretation by isostatic modelling. Using the ice sheet of Nesje & Dahl (1990) he found that the predicted tilt of deglacial shorelines was less than 50% of the observed tilt, and he concluded that the results "seem to rule out the thin ice model as a viable option". Lambeck *et al.* (1998; 2000), on the other hand, obtain moderate ice thickness from reverse modelling based on observed sea level curves.

Most glaciological models (e.g. Boulton *et al.*, 1985; Dowdeswell & Siegert, 1999; Holmlund & Fastook, 1993) predict considerably thicker ice than that reconstructed by Nesje & Dahl (1990; 1992), especially in central areas. However, the ice thickness in climate-driven models are very dependent on the amount of precipitation and the duration of ice build up, and both factors are partly dependent on assumptions in the models. All models are also so simple that they cannot be used to reconstruct the regional ice thickness in any detail. Certainly, they cannot be used as an argument against empirical reconstructions in the fjord areas, such as Nesje *et al.*'s reconstruction in Nordfjord.

The writer concludes that the lower limit of block fields is not an unique criterion that can be used to map the ice sheet surface during the Late Weichselian glacial maxi-

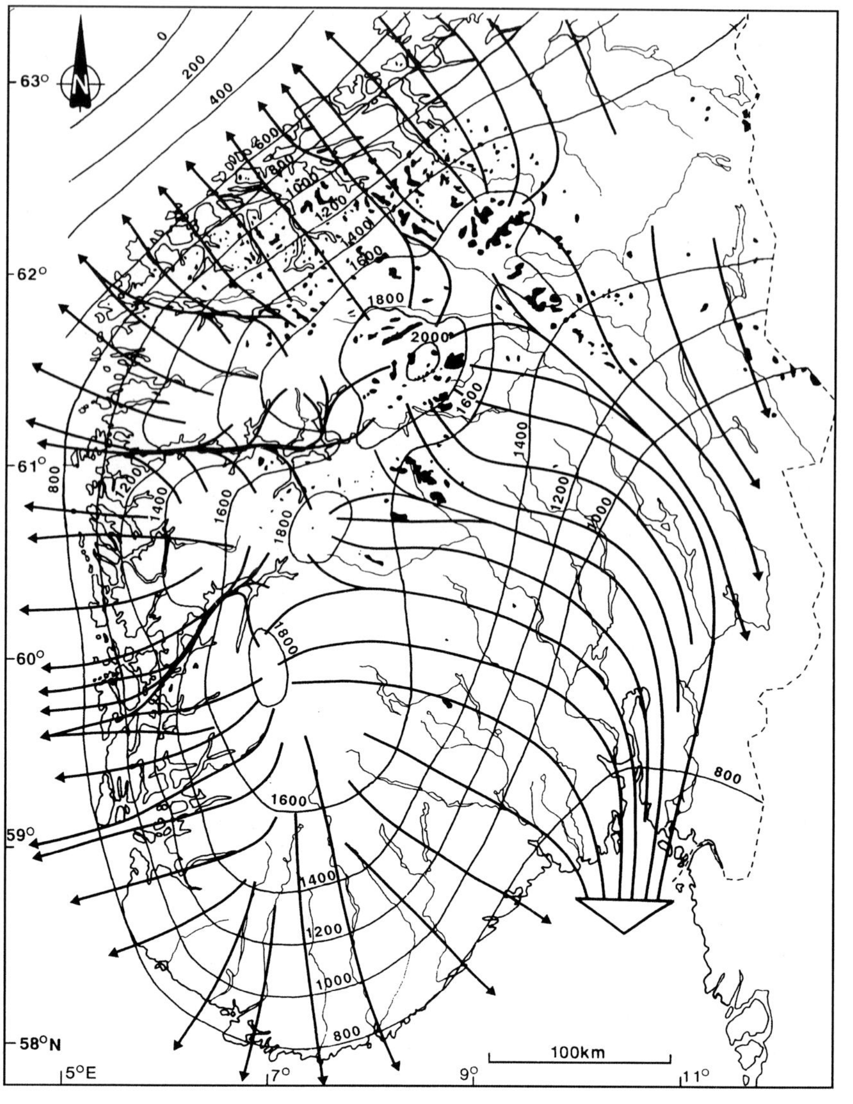

Fig. 9. Map showing block-fields in southern Norway (black spots). The main flow lines and tentatively con-structed contour lines of the ice sheet during the Late Weichselian glacial maximum in southern Norway are also indicated (adapted from Nesje et al., 1988). The present author assumes that the ice sheet was thicker and covered all the block-fields in the eastern part.

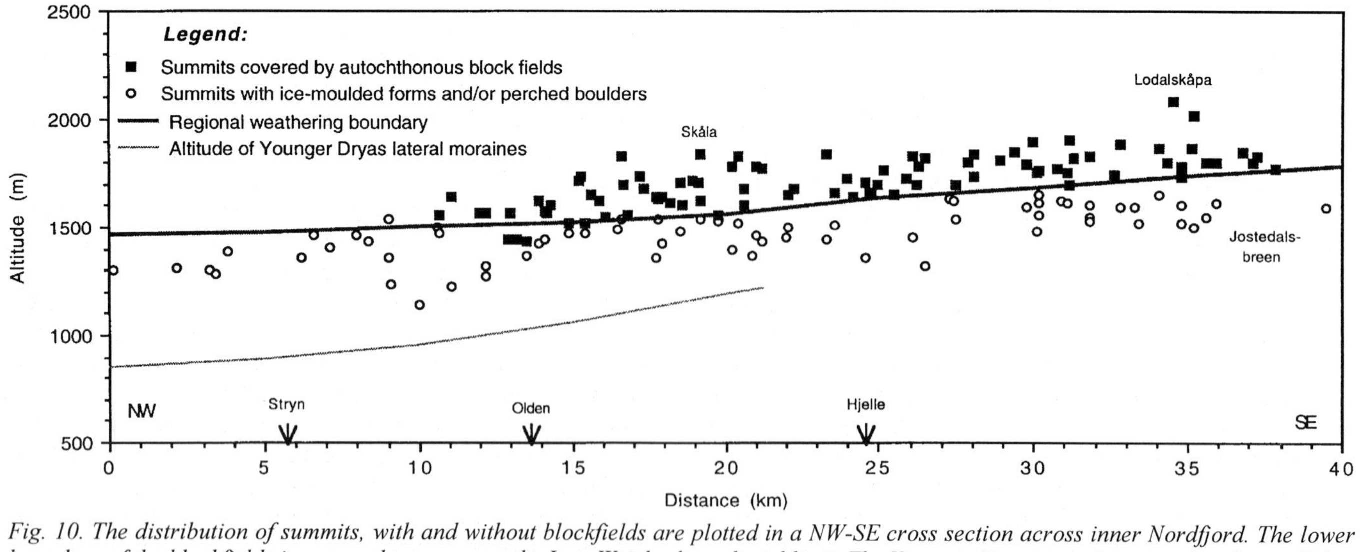

Fig. 10. The distribution of summits, with and without blockfields are plotted in a NW-SE cross section across inner Nordfjord. The lower boundary of the blockfields is assumed to represent the Late Weichselian glacial limit. The Younger Dryas moraines are also shown. Taken from A. Nesje (written communication, 2000), updated from Brook et al. (1996).

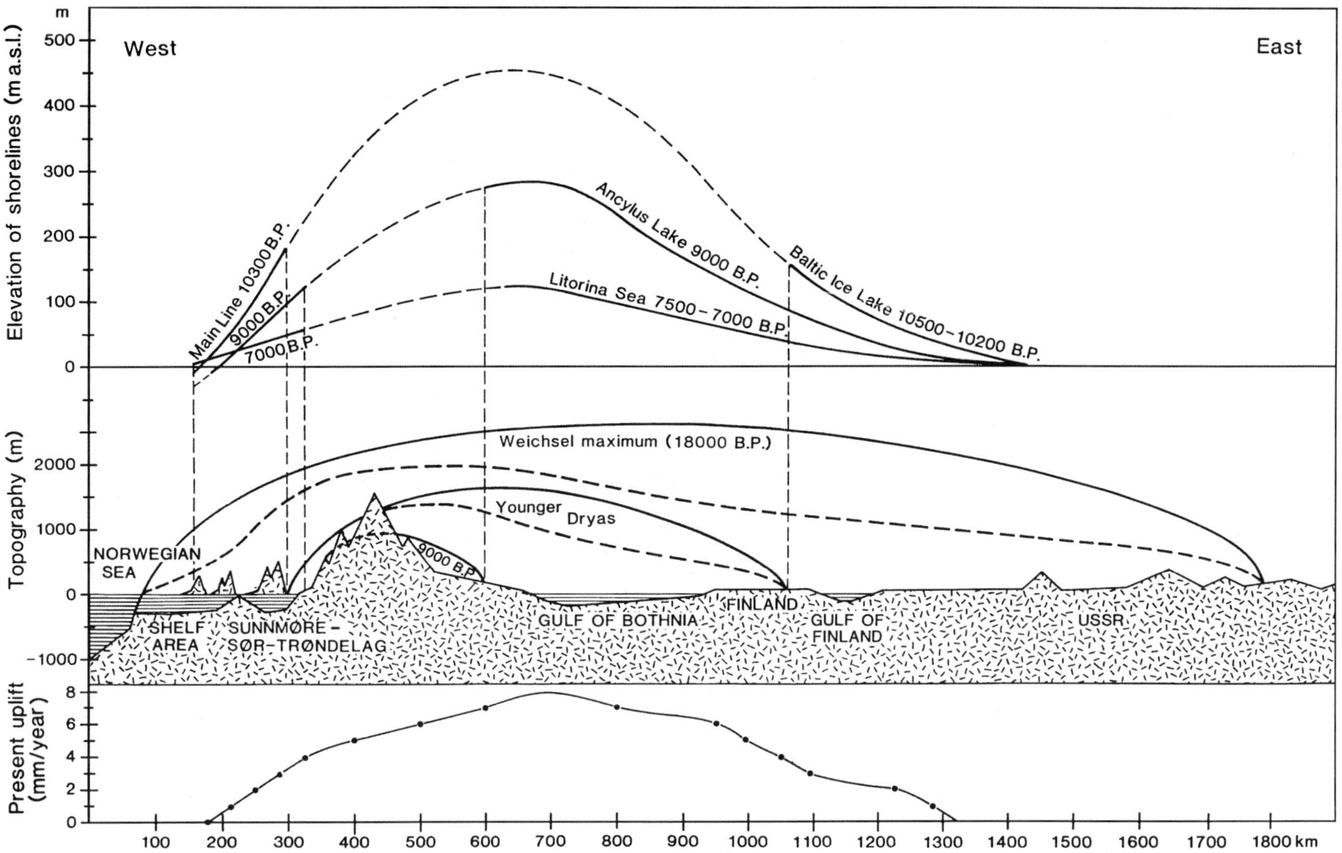

Fig. 11. A profile across Fennoscandia adapted from Svendsen and Mangerud (1987). Upper panel: Shore lines of different ages. Middle panel: Alternative ice-sheet profiles for the Late Weichselian maximum and the Younger Dryas. Full lines show maximum thickness, stippled lines minimum thickness. Lower panel: Present day uplift.

mum, although the interpretation presented for Nordfjord and some other coastal areas may, nevertheless be correct. The conflict with the results based on isostatic modelling has made this problem even more urgent to solve, because isostatic modelling, beside glaciological modelling, is the most used technique to determine past ice sheet thickness world wide (e.g. Dowdeswell & Siegert, 1999; Lambeck *et al.*, 2000; Peltier, 1994).

Here the writer supports and further develops an hypothesis proposed by Longva & Thorsnes (1997). For a century it has been accepted that the first Late Weichselian glacial advance to northern Denmark was from Norway (Sjørring, 1983), an interpretation supported by all recent studies (Houmark-Nielsen, 1999). This conclusion, mainly based upon numerous Norwegian erratics in the north Danish tills, was also supported by till fabric and glaciotectonic deformation features, which showed flow from the north. It is difficult to envisage an ice flow carrying erratics from Norway to Denmark if there was an active ice stream in the Norwegian Channel. In fact it appears impossible if that ice stream was developed as far upstream, as shown in Fig. 8, because flow from Norway would simply be 'cut off'' and drain into the ice stream. Longva & Thorsnes (1997) described three generations of ice flow directions based on the sea-floor morphology in the outer part of Oslofjorden. Deep, diffuse furrows with a direction showing that the ice crossed the Norwegian Channel represent the oldest generation. It is interpreted to show the ice flow to the Late Weichselian ice limit in northern Denmark (Longva & Thorsnes, 1997). Deep furrows, with a more south-westerly direction represent the next generation. This change in direction was likely the result of ice culmination over Central Norway-Sweden moving eastwards. The youngest flutes show a very plastic ice flow along the Norwegian Channel, and are interpreted to represent an ice stream in the channel (Longva & Thorsnes, 1997).

The observations cited from Oslofjorden are compatible with a hypothesis that there first developed a thick Scandinavian ice sheet, for example as the one shown by Kleman & Hättestrand (1999). This ice sheet may have developed without peripheral ice streams, or at least without an ice stream in the Norwegian Channel. Ice from Norway could then move across the Norwegian Channel to Denmark and the North Sea. This ice sheet could have had a steeper surface slope, although it still moved on derformable beds in peripheral areas. In Denmark and the shallow part of the North Sea, there was probably permafrost during the advance, favouring a steeper ice surface (Clark *et al.*, 1999), although that could not have

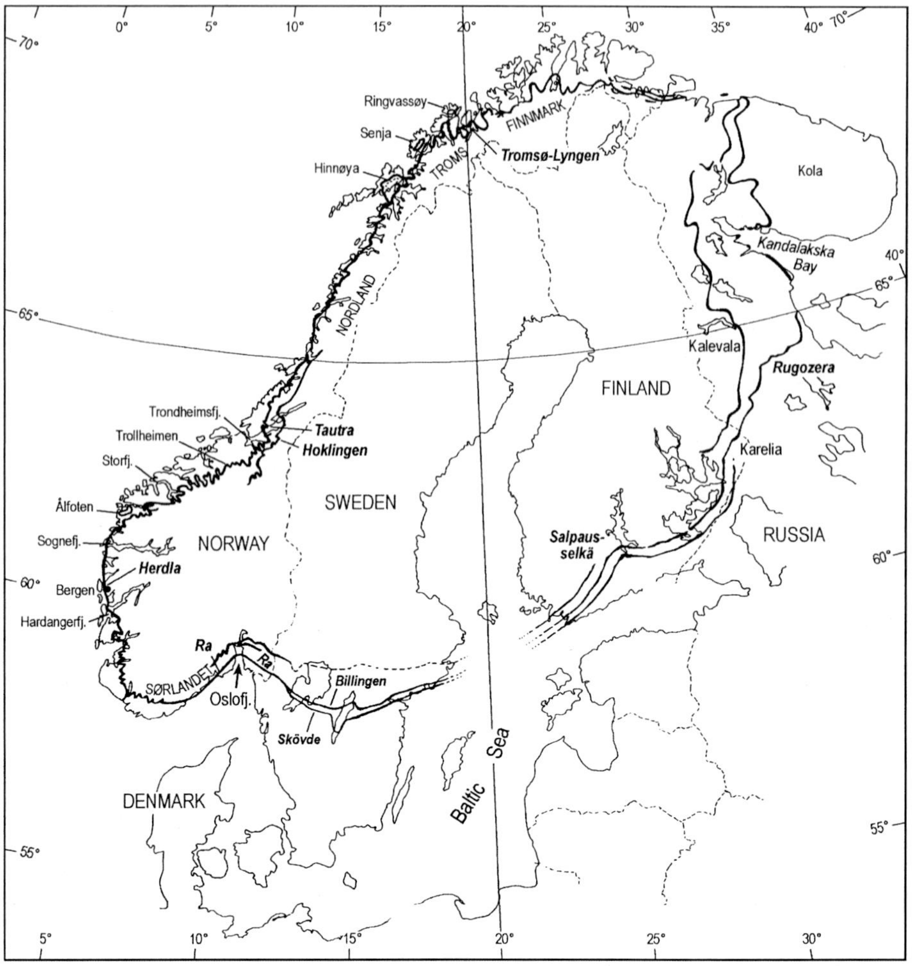

Fig. 12. The Younger Dryas moraines around Fennoscandia slightly modified from Andersen et al. (1995a). The names on the moraines are given with bold letters. Other geographical names used in the discussion of the Younger Dryas moraines are marked with normal text.

been the case in the deep Norwegian Channel. All summits in Central Norway could in this early phase have been covered by frozen-bed ice. Subsequently the ice streams developed, probably from the shelf edge and propagating upstream. That would lead to a major drawdown of the ice-sheet surface, and possibly to a situation similar to that reconstructed by Nesje & Dahl (1990; 1992) (Fig. 9).

Cross-profiles across the Scandinavian Ice Sheet according to Svendsen & Mangerud (1987) are shown in Figure 11. The writer considers that the pattern of the profile of the minimum model for the Late Weichselian maximum probably is correct, showing low surface slopes on deformable beds across the shelf in the west and in eastern areas. However, it is assumed that ice thickness was closer to the maximum model over Scandinavia. The Younger Dryas ice surface profile was probably closer to the shown minimum model.

THE YOUNGER DRYAS

The Younger Dryas moraines constitute the backbone of reconstructions of the deglaciation history of Norway, and indeed of all of Fennoscandia. End moraines from this period have been mapped more or less continuously around the entire former Scandinavian Ice Sheet (Fig. 12) (Andersen *et al.*, 1995a). One of the Younger Dryas moraines in southern Norway was interpreted as an end moraine by the Norwegian geologist Esmark already in 1824 (Andersen *et al.*, 1995b), and that was the first time it was concluded that there had been a major ice sheet over Scandinavia. The large Ra moraines around Oslofjorden were recognised in the middle of the 19th century. However, mapping along the long western and northern coasts was not completed until about a decade ago. As shown below, many problems still remain for the course of the main moraines, but even more so for their accurate dating, the distance of the readvances, and the geometry of the ice surface.

Dating problems

There are three main problems with ^{14}C dates that are relevant for the accurate dating of Younger Dryas glacial events in Norway. First, the calibration scale to calendar years is not well established through the Younger Dryas and Allerød. Secondly, there are plateaux in the radio-carbon scale during the Younger Dryas, including one at the Younger Dryas/Preboreal boundary, so that the

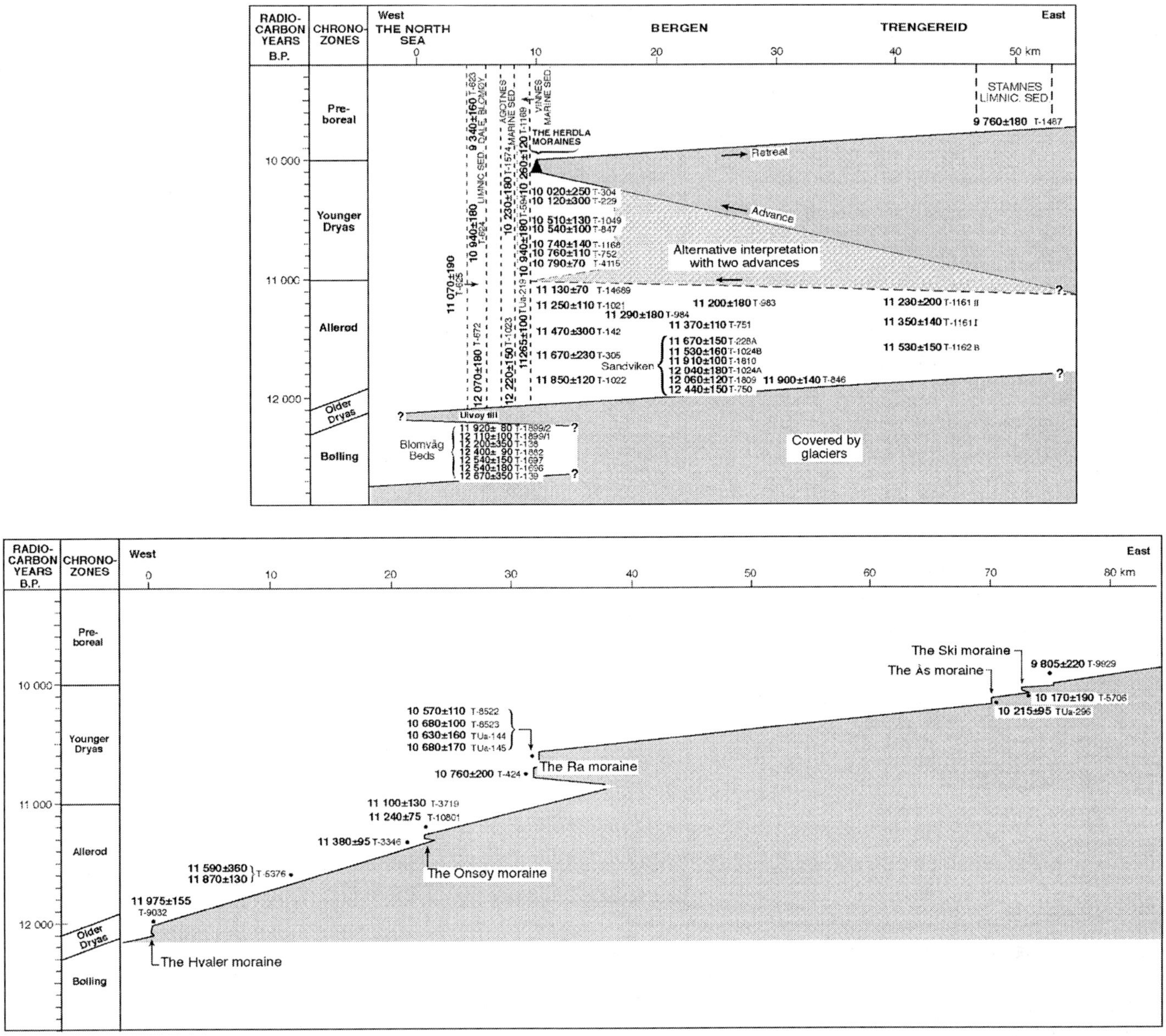

Fig. 13. Upper panel: Ice-front fluctuations in the Bergen district, from Mangerud (2000).Lower panel: Ice-front fluctuations on the east shore of Oslofjorden, redrawn from Andersen et al. (1995b). The location and details about the radiocarbon dates are provided by R. Sørensen (written communication, 2000). Most of the dates are given in Andersen et al. (1995b) and Sørensen (1999). The minimum date for the Hvaler Moraine is from Bergstrøm et al. (1992). The two diagrams are drawn with the same scale and placed so that the main Younger Dryas moraines (Herdla and Ra) are aligned.

boundary is difficult to identify using dates (Gulliksen *et al.*, 1998). Thirdly, and most important, is the uncertainty related to the marine reservoir age, because most ^{14}C dates of the Younger Dryas moraines in Norway are performed on marine molluscs. Uncertainties in the reservoir age hamper precise comparison between dates on marine and terrestrial materials. Conventionally all dates of marine fossils from the Norwegian coast are corrected for a reservoir age of 440 years (Mangerud & Gulliksen, 1975), although it has been known for a long time that the reservoir age could have varied backwards in time (Mangerud, 1972). Until recently, however, the magnitude of this variation has been unknown. For the Allerød, the marine reservoir age is about 380 years for western Norway, identical to the present day reservoir age, if both are calculated the same way (Bondevik *et al.*, 1999). For the Younger Dryas, the reservoir age is so far only estimated for the time of the Vedde Ash fall (Mangerud *et al.*, 1984). For benthic forams from the floor of the Norwegian Channel, Haflidason *et al.* (1995) obtained a reservoir age of 800 years at Vedde Ash time, whereas Bondevik *et al.* (2001) obtained 610 ±55 years for shallow marine shells at the coast. The latter value is considered most relevant for dates on uplifted marine deposits. If this value is correct, it implies that further 170 years should be subtracted from the ^{14}C age of molluscs of mid-Younger

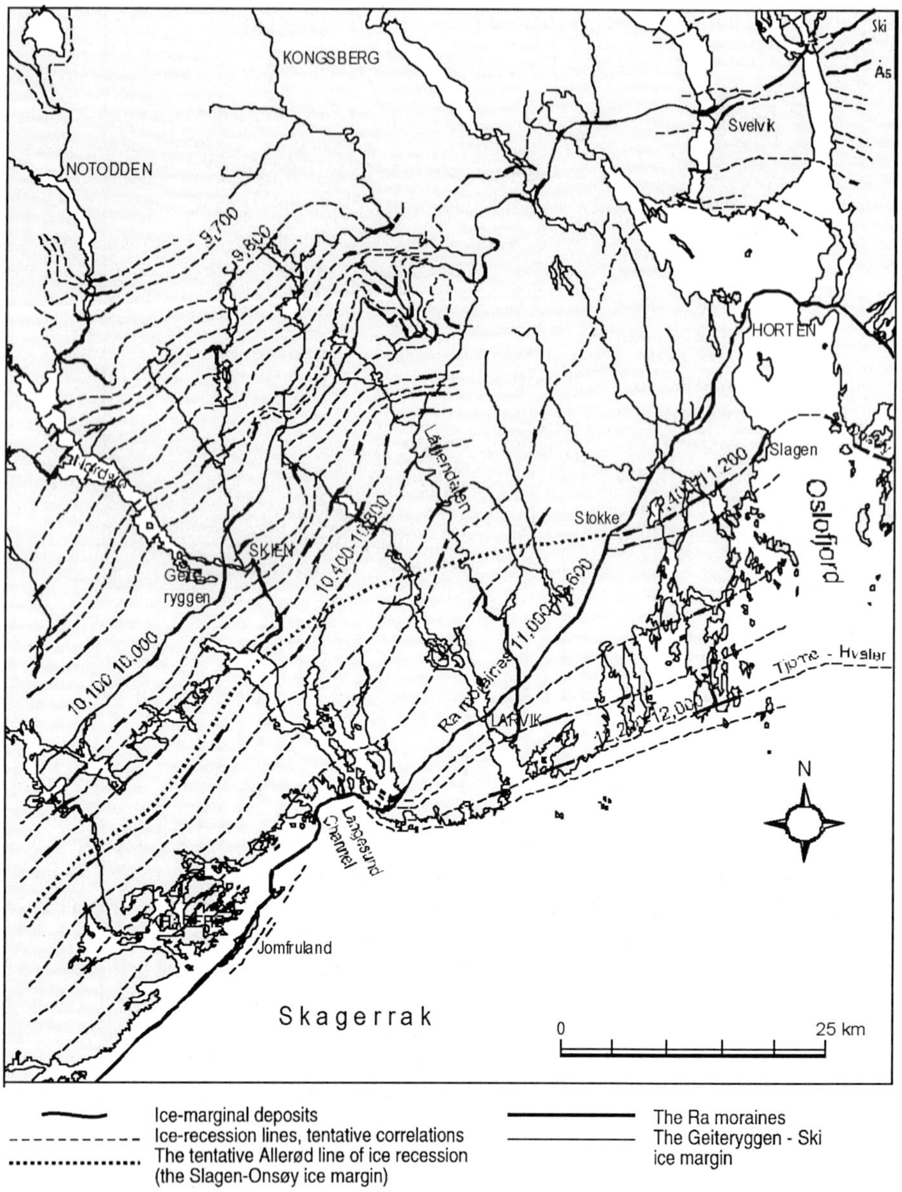

Fig. 14. Map of moraines and constructed retreat lines west of Oslofjorden, slightly modified from Bergstrøm (1999).

Dryas age in order to make them comparable with the ^{14}C age of terrestrial fossils. This is certainly only little, but still enough that some moraines that have during the last decades been considered to be of late Younger Dryas age, based on mollusc dates, might in fact be of early Preboreal age.

Asynchronous moraines and the major re-growth of ice in SW-Norway

In spite of the uncertainties with dates indicated above, it appears quite clear that the outermost and largest Younger Dryas moraines are asynchronous around Fennoscandia (Mangerud, 1980). In Finland, Sweden and eastern Norway the main moraines were formed during the early or middle part of the Younger Dryas. In these eastern areas, the readvances were relatively small and there was a net retreat during the Younger Dryas (Fig. 13, lower panel). In the Bergen area, western Norway, the ice sheet readvanced during the entire Younger Dryas, and formed the Herdla moraines at the very end of the Younger Dryas (Fig. 13, upper panel). Therefore, although the Younger Dryas moraines are mapped morphologically continuously around Scandinavia (Fig. 12), these moraines represent an asynchronous line.

It is apparent from Fig.13 that the asynchroneity of the Younger Dryas moraine in southern Norway mainly results from a long-lasting readvance in the western areas. This problem will be discussed in some detail, partly because it has interesting implications, and partly because the author is familiar with the observations.

The dynamics of the ice front are well documented on the two shores of the Oslofjorden (Andersen *et al.*, 1995b;

Bergstrøm, 1999; Sørensen, 1992) (Fig. 14). On the east shore there was a nearly continuos retreat of the ice front, interrupted only by minor readvances or halts (Fig. 13, lower panel). Between 12 and 10 ka there was a net retreat of more than 70 km, and during the Younger Dryas alone, a net retreat of 30 km. The retreat on the east shore was mirrored on the western shore between 12-11 ka. However, from about 11 ka the ice front on the west shore readvanced with increasing amplitude towards southwest. At Horten, where the Ra moraines cross the Oslofjorden, there was only a minor readvance, whereas at Stokke the Ra crosses the Slagen-Onsøy moraines (Fig. 14). Around the Langesund Channel, the ice front readvanced at least 10 km and at the Ra-time (about 10.6 ka) had the same position as at 12.2-12 ka. This latter reconstruction is similar to the reconstruction in the Bergen area, where the ice front during the Younger Dryas also reached its 12-ka position (Fig. 13, upper panel). However, near Langesund the ice front retreated some 25 km, to the Geiteryggen-Ski moraine, during the later part of the Younger Dryas (from 10.6 to 10.0 ka). In contrast, in the Bergen district, the readvance to the Herdla moraine continued until the very end of the Younger Dryas. If correct, this means that moraines corresponding to the Geiteryggen-Ski moraine somewhere further to the SW had to cross the Ra moraine and connect with the Herdla moraine. This crossing point is yet to be identified.

The readvance in the Bergen area is only documented for a length of about 40 km, although from the fauna at the easternmost site it is thought to have been larger (Andersen *et al.*, 1995b). The readvance occurred along fjords several hundred metres deep, so that the ice reached a thickness of 800-1200 m in fjords that had been ice free during the Allerød (Andersen *et al.*, 1995b). Thus the net accumulation of snow during the Younger Dryas must have been considerable, despite the estimate that winter precipitation was only 60% of the present (Dahl & Nesje, 1992).

The distance of a glacial readvance is difficult to plot, because the mapping requires that some datable sediment survived glacial erosion during the readvance. Therefore our knowledge about the Younger Dryas readvance is certainly incomplete. However, the major readvance which culminated in late Younger Dryas, was apparently confined to the southwestern coast of Norway, from the southern tip to somewhat north of Sognefjorden (Fig. 12). A rise in relative sea level (transgression) of up to 10 m is recorded in an area which coincides with the area of the Late Younger Dryas readvance (Anundsen, 1985). The interpretation has been that the relative sea-level rise was caused by the combined effect of three factors. 1) The ice growth slowed or halted the glacio-isostatic uplift, 2) eustatic sea-level rise and 3) the change in the geoid due to the gravity effects of the advancing ice sheet (Anundsen & Fjeldskaar, 1983; Fjeldskaar & Kanestrøm, 1980). In the Bergen area, relative sea level at the end of the Younger Dryas was as high, or higher, than during the deglaciation about 12 ka (Anundsen, 1985; Krzywinski & Stabell, 1984). At the position of the Younger Dryas ice margin near Bergen, relative sea level was close to 60 m above the present sea level at both 12 and 10 ka (Lohne, 2000). The implication is that the ice load during the Younger Dryas was almost the same as the load during the deglaciation, including the isostatic 'memory' of the glacial maximum load. Certainly, much of the isostatic rebound after the glacial maximum occurred during thinning and retreat of the ice sheet, before the area became ice-free. However, the rising relative sea level during the Younger Dryas in these areas is certainly consistent with considerable ice growth across large mountainous areas in south-western Norway.

It was earlier explained that the large readvance in western Norway, compared to the smaller glacial fluctuations in eastern areas and also in Trondheimsfjorden, mainly as a result of topographic and glaciological effects (Mangerud, 1980). However, here it should be stressed that there must also have been considerably more (snow) precipitation in south-western areas compared to other parts of the ice sheet. This was probably a result of dominant south-westerly winds (Larsen *et al.*, 1984).

Comments to the Younger Dryas moraines on the digital map

A detailed description of the Younger Dryas end moraines in Norway was given in a recent review paper (Andersen *et al.*, 1995b), to which is referred for descriptions and also for more complete references to the literature. In the following the writer mainly comments on new observations and disagreements from the results presented by Andersen *et al.* (1995b), starting in the southeast and following the coast westwards and northwards. The digital map on the CD is much more detailed than the map in Andersen *et al.* (1995b). A number of scientists have therefore examined and improved the map according to published maps and in many cases also unpublished new results. The author has mentioned through the area descriptions below who has examined and contributed to the respective areas, but the writer is certainly responsible for any errors.

The classical view for the Oslofjorden area was that the Ra moraines represented the Younger Dryas, and that the Ås, Ski and younger moraines were of Preboreal age (O. Holtedahl, 1960). New radiocarbon dates subsequently indicated that the Ås and Ski moraines were of (late) Younger Dryas age (Sørensen, 1979), a view that has prevailed until today (Andersen *et al.*, 1995b; Bergstrøm, 1999). However, the moraines are dated mainly by means of marine molluscs and if the reservoir age for the Younger Dryas is larger than 440 years, as discussed above, the Ås-Ski moraines would possibly be of Preboreal age. This will probably not be resolved without an accurate correlation of the moraines, or the corresponding shorelines with well-dated lacustrine sequences. Intuitively, the writer finds it probable that the Ski moraine indeed represents the end of the Younger Dryas, because even with a higher reservoir age, the ice withdrew from the Ra moraine well before the end of the Younger Dryas. The Ski moraine is a relatively

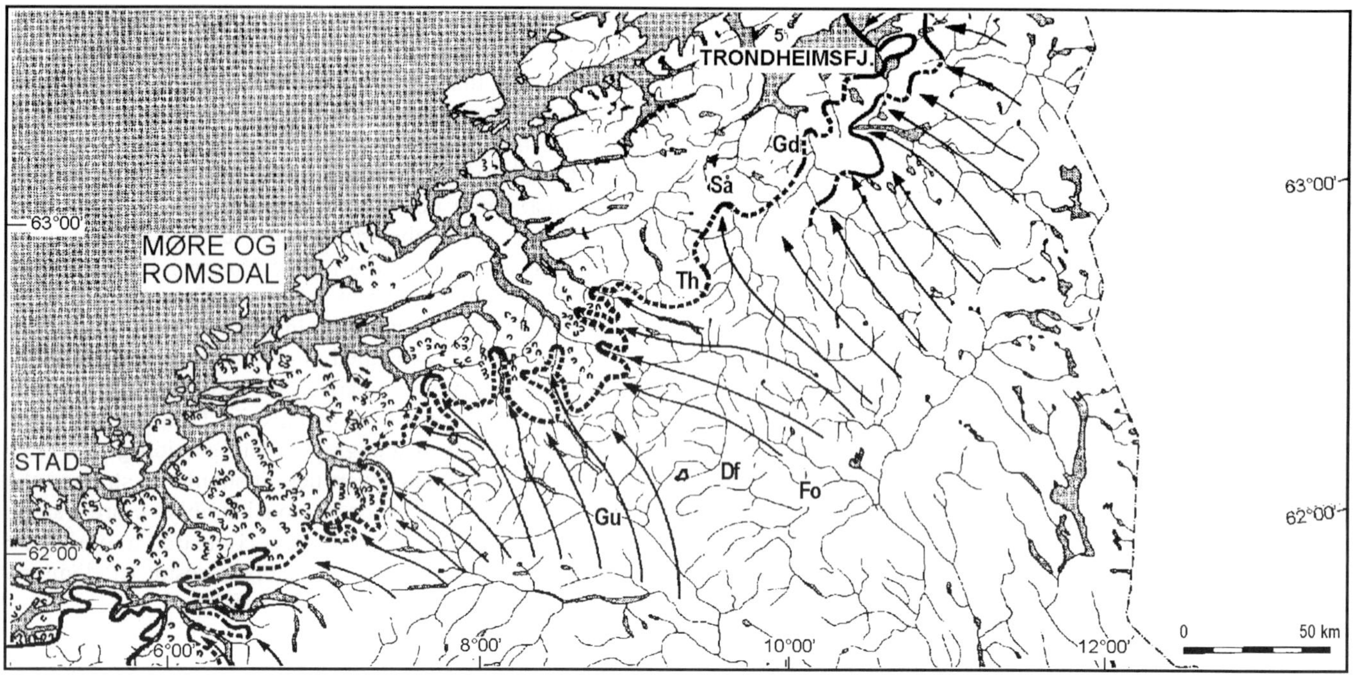

—— Younger Dryas moraine ∩∩∩∩ Younger Dryas cirque moraines Gu=Gudbrandsdal Df = Dovrefjell Fo =Folldal
---- Assumed ice front position ⟶ Younger Dryas ice-flow directions Th = Trollheimen Gd= Gauldal Så = Storås

Fig. 15. Map of the Younger Dryas moraines in the Trollheimen - Trondheimsfjord district, modified from Sollid & Reite (1983).

distinct moraine that probably corresponds in time to the Herdla moraine in western Norway (Fig. 13). On the map, for this area controlled by B. Bergstrøm and R. Sørensen, the Ås and Ski moraines are therefore retained. In Sørlandet, the moraine is drawn according to the detailed map in Andersen (1960).

For western Norway, new descriptions and interpretations have been published for the submarine moraines crossing the fjords (Aarseth, 1997; Aarseth *et al.*, 1997). Several scientists also recently have suggested that Hardangerfjorden remained ice free from the Allerød and throughout the entire Younger Dryas (Bakke *et al.*, 2000; Helle *et al.*, 1997; Helle *et al.*, 2000). This is in contrast to the interpretation given in Andersen *et al.* (1995b), where the Halsnøy Moraine across the mouth of Hardangerfjorden, was mapped as the late Younger Dryas moraine. That problem is discussed in depth in Mangerud (2000), who concluded that the latter interpretation is correct and it is therefore shown on the map. The conclusion that the readvance ended close to 10 ka in the Bergen district (Fig. 13) is mainly based on dates from marine molluscs. Therefore, if the marine reservoir age was larger one should consider if the readvance continued into the very early Preboreal. It is now shown that the readvance certainly ended after deposition of the Vedde Ash, and in fact at the *Betula* increase at the Younger Dryas/Preboreal boundary (Bondevik & Mangerud, 2001; Bondevik & Mangerud, 2002).

Sønstegaard *et al.* (1999) mapped the marginal deposits of two large coalescing plateau glaciers of Younger Dryas age in the Ålfoten area (Fig. 12). These glaciers were located west of, and separate from the Younger Dryas Scandinavian Ice Sheet. It is interesting to note that the plateau glaciers reached their maximum before the deposition of the Vedde Ash, whereas the Scandinavian Ice Sheet reached its maximum after the Vedde Ash in the Bergen district somewhat south of Ålfoten.

There is a distinct boundary for the style of glaciation around the Ålfoten area: south of that line no local glaciers existed outside the limit of the Scandinavian Ice Sheet, whereas along the coast further north there were numerous local glaciers. The change is an effect of topographical differences, in the southern area there are no high mountains beyond the Younger Dryas ice limits, whereas between Ålfoten and Trondheimsfjorden (and also in several areas in northern Norway) there are alpine landscapes with high peaks along the coast (Mangerud, 1980). The local glaciers are not marked on the digital map, but some are seen in Fig. 15. It can be mentioned that it has been demonstrated that some of the local glaciers did not survive during the Allerød, and were formed at the onset of the Younger Dryas (Larsen *et al.*, 1984; Mangerud *et al.*, 1979), whereas others survived the Allerød (Larsen *et al.*, 1998).

The mapped course of the moraine between Hardangerfjorden and Storfjorden (Fig. 12) are controlled and corrected by A.R. Aa, I. Aarseth and E. Sønstegaard.

Conflicting interpretations have recently been presented for the area surrounding the Trollheimen mountain region (Th on Fig. 15) and the Dovrefjell plateau (Df on Fig. 15). The Scandinavian Ice Sheet limit was previously drawn west of Trollheimen, which is a massif located west of the main mountain chain, although continuous marginal moraines have not been found there (Andersen *et al.*, 1995a; 1995b; Sollid & Reite, 1983). Recently Follestad, based on mapping of Quaternary deposits, presented a completely different view (Follestad, 1994a; 1994b; written communication, 2000). He claims that there were only local glaciers in and surrounding Trollheimen. On the digital map the author has partly used the western limits indicated by Follestad to the south of Trollheimen (written communication, 2000), whereas through Trollheimen and towards the north, a limit drawn by A. Reite when he corrected the map is used. Reite emphasises that the limit across Trollheimen is uncertain. In practice, the map does not look very different from the earlier reconstructions south of Trollheimen. However, a major controversy is hidden in this drawing. In Andersen *et al.* (1995b) it was assumed that the moraines were formed by outlet glaciers from the main ice sheet, which is still postulated as correct for most of them. Follestad, on the other hand, postulated that the Trollheimen glaciers were dynamically separated from the Scandinavian Ice Sheet, the only connection being a zone of stagnant ice in northern Gudbrandsdalen (Gu in Fig. 15).

Follestad (written communication, 2000) and Dahl *et al.* (2000; 1997) further propose that mountain plateaux (including Dovrefjell, Df in Fig. 15) in the central part of Norway, and even in the Folldalen valley (presently below 700 m a.s.l.) (Fo in Fig. 15) were ice free during the Younger Dryas. The main arguments are that cirque moraines of postulated Younger Dryas age were mapped down to 1100 m a.s.l., and radiocarbon dates of Allerød and Younger Dryas age were obtained from Grimsmoen in Folldal for example (Dahl *et al.*, 2000). All earlier Younger Dryas ice reconstructions assumed these areas were covered by thick ice flowing towards Trondheimsfjorden, a view supported by geologists currently working in the area (Sveian *et al.*, 2000; A. Reite, oral communication, 2000). They have also mapped the Hoklingen moraine (see below) up to about 1000 m a.s.l. along Gauldalen, which is a large valley to the south of Trondheimsfjorden. That would indicate an ice surface above 1400-1500 m a.s.l. in Folldalen (Sveian *et al.*, 2000). It is concluded that some mountain peaks in Central Norway probably protruded through the ice sheet during the Younger Dryas. However, it is glaciologically very unlikely that the ice surface was below some 1400 m a.s.l., not to speak of 700 m a.s.l, in Central Norway, at a time when there were major glacial readvances out along Trondheimsfjorden and Oslofjorden. It is postulated that the cited radiocarbon samples (Dahl *et al.*, 2000) are from Early Holocene sediments which were contaminated with old carbon (Sveian *et al.*, 2000). On the map the classical view has therefore been retained and the view postulated that Folldalen and Dovrefjell were ice covered.

Over the last decades the area around Trondheimsfjorden has been mapped in detail by the Norwegian Geological Survey and many radiocarbon dates related to the Younger Dryas moraines have been obtained e.g. (Andersen *et al.*, 1995b; Reite, 1994; Sveian & Solli, 1997). The map for this area has been corrected and modified by H. Sveian and A.R. Reite. In the Trondheimsfjorden area there are two major and sub-parallel Younger Dryas moraines. The outer, named the Tautra Moraine, is the larger and is dated to 10.8-10.4 ka; the younger Hoklingen Moraine is dated to 10.3-10.4 ka (Andersen *et al.*, 1995b). Again, most of the dates are from marine shells, and the final results will depend on the marine reservoir age. It is interesting to note that both the glaciation curve (Andersen *et al.*, 1995b; Reite, 1994) for the Trondheimsfjorden area and the picture with two Younger Dryas moraine systems are similar to the Oslofjorden area discussed above. Such distinct double moraines have not been described from elsewhere in southern Norway. Oslofjorden and Trondheimsfjorden are the only fjords in Norway surrounded by wide lowlands, so the similar responses are possibly a result of the similar topography and relation to the ice culminations far inland (Mangerud, 1980). The Tautra Moraine can be mapped nearly continuously towards the north, as shown on the map (Andersen *et al.*, 1995b; Reite, 1994). The Hoklingen Moraine is more distinct than the Tautra Moraine south of the fjord. This moraine is also a result of a readvance and is mapped as a continuous line. On the map it is drawn to the south, as far as it was mapped by Reite (1994). Concerning the age, the situation is again as in the Oslofjorden area: Hoklingen is close in age to the Herdla Moraine in western Norway, therefore its extension has to cross the extension of the Tautra Moraine somewhere between Trondheimsfjorden and Bergen.

In the area from the Trondheimsfjorden to southern Nordland (Fig. 12) B. Bergstrøm and H. Sveian have drawn the results of their recent mapping (Sveian & Solli, 1997). The main difference from the maps in Andersen *et al.* (1995b) is that the Hoklingen Moraine is mapped much further north than previously.

In the southern and northernmost part of Nordland (Fig. 12), the moraine mainly follows Andersen *et al.* (1995b), and the central part it is shown mainly according to Andersen (1975) and Rasmussen (1981). The entire stretch is corrected and somewhat modified by T. Bargel and L. Olsen, according to their recent mapping. On and around Hinnøya Olsen *et al.* (2001) have proposed a re-interpretation. They found that the main ice sheet inundated Hinnøya during an early phase of the Younger Dryas. Subsequently the glaciers in the fjords broke up, and late in the Younger Dryas the ice sheet terminated east of the island, and a small ice cap formed on Hinnøya. Both Olsen *et al.* (2001) and Vorren & Plassen (2002) have concluded that there was a major Younger Dryas readvance of at least some 30 km in this area.

The Tromsø-Lyngen Moraines are indicated mainly according to Andersen *et al.* (1995b), with corrections from B. Bergstrøm, H. Sveian and T. Vorren. In this area, there were numerous local glaciers on the islands outside the ice sheet limit. The two largest of these, Senja (Vorren & Plassen, 2002) and Ringvassøy (Andersen *et al.*, 1995b), are shown on the map.

In Finnmark, the moraines are generally very distinct. They are plotted according to Sollid *et al.* (1973), and the map was corrected by L. Olsen.

Aknowledgements

Øystein Lohne assisted enthusiastically in drawing the map of the Late Weichselian glacial limit, in correctly drawing all the details in the course of the Younger Dryas moraine, finding literature, etc. Terje Finnekås assisted in the first compilation of the Younger Dryas moraine. An earlier draft of the manuscript was sent to a number of geologists. I have received comments to that draft or other scientific information from A.R. Aa, I. Aarseth, B.G. Andersen, T. Bargel, B. Bergstrøm, S. Bondevik, S.-O. Dahl, B. Follestad, H. Haflidason, J. Kleman, E. Larsen, S.E. Lauritzen, L. Olsen, D. Ottesen, A. Nesje, A.J. Reite, K. Rokoengen, J.L. Sollid, E. Sønstegaard, R. Sørensen, H. Sveian and T.O. Vorren. The drawings were made by E. Bjørseth and J. Ellingsen. The English language was corrected by M.R. Talbot. I thank all of these people for their friendly help.

References

Aarseth, I. (1997). Western Norwegian fjord sediments: age, volume, stratigraphy, and role as temporary depository during glacial cycles. *Marine Geology,* **143**, 39-53.

Aarseth, I., Austbø, P.K. & Risnes, H. (1997). Seismic stratigraphy of Younger Dryas ice-marginal deposits in western Norwegian fjords. *Norsk Geologisk Tidsskrift*, **77**, 65-85.

Alm, T. (1993). Øvre Æråsvatn - palynostraatigraphy of a 22,000 to 10,000 BP lacustrine record on Andøya, northern Norway. *Boreas*, **22**, 171-188.

Andersen, B.G. (1960). Sørlandet i sen- og postglasial tid. Norges geologiske undersøkelse 210, 1-142.

Andersen, B.G. (1975). Glacial Geology of Northern Nordland, North Norway. *Norges geologiske undersøkelse*, **320**, 1-74.

Andersen, B.G. (1981). Late Weichselian Ice Sheets in Eurasia and Greenland. *In:* Denton, G.H. & Hughes, T.J. (eds), *The Last Great Ice Sheets,* John Wiley & Sons, New York, pp. 3-65.

Andersen, B.G., Lundqvist, J. & Saarnisto, M. (1995a). The Younger Dryas margin of the Scandinavian Ice Sheet - an introduction. *Quaternary International*, **28**, 145-146.

Andersen, B.G. & Mangerud, J. (1989). The last interglacial-glacial cycle in Fennoscandia. Quaternary International 3/4, 21-29.

Andersen, B.G., Mangerud, J., Sørensen, R., Reite, A., Sveian, H., Thoresen, M. & Bergstrøm, B. (1995b). Younger Dryas ice-marginal deposits in Norway. *Quaternary International*, **28**, 147-169.

Andersen, B.G. & Nesje, A. (1992). Quantification of Late Cenozoic glacial erosion in a fjord landscape. *Sveriges Geologiska Undersökning*, **Ca 81**, 15-20.

Andersen, B.G., Sejrup, H.-P. & Kirkhus, Ø. (1983). Eemian and Weichselian Deposits at Bø on Karmøy, SW Norway: A Preliminary Report. *Norges geologiske undersøkelse*, **380**, 189-201.

Andersen, B.G., Wangen, O.P. & Østmo, S.-R. (1987). Quaternary geology of Jæren and adjacent areas, southwestern Norway. *Norges geologiske undersøkelse, Bulletin*, **411**, 1-55.

Anundsen, K. (1985). Changes in shore-level and ice-front position in Late Weichsel and Holocene, southern Norway. *Norsk Geografisk Tidsskrift*, **39**, 205-225.

Anundsen, K. & Fjeldskaar, W. (1983). Observed and theoretical Late Weichselian shore-level changes related to glacier oscillation at Yrkje, South-West Norway. *In:* Schroeder-Lanz, H. (ed.), *Late- and Postglacial oscillations of glaciers: Glacial and Periglacial forms,* Rotterdam, A.A. Balkema, 133-170.

Bakke, J., Dahl, S.O. & Nesje, A. (2000). Reconstruction of Younger Dryas and Holocene glacier fluctuations and paleoclimate at Folgefonna, southwestern Norway. *Geonytt*, **1, 2000**, 36.

Baumann, K.-H., Lackschewitz, K.S., Mangerud, J., Spielhagen, R.F., Wolf-Welling, T.C.W., Henrich, R. & Kassens, H. (1995). Reflection of Scandinavian Ice Sheet fluctuations in Norwegian Sea sediments during the last 150,000 years. *Quaternary Research*, **43**, 185-197.

Behre, K.-E. (1989). Biostratigraphy of the last glacial period in Europe. *Quaternary Science Reviews*, **8**, 25-44.

Behre, K.-E. & van der Plicht, J. (1992). Towards and absolute chronology for the last glacial period in Europe: radiocarbon dates from Oerel, northern Germany. *Vegetation History Archaeobotany*, **1**, 111-117.

Bergstrøm, B. (1999). Glacial geology, deglaciation chronology and sea-level changes in the southern Telemark and Vestfold counties, southeastern Norway. *Norges geologiske undersøkelse Bulletin*, **435**, 23-42.

Bergstrøm, B., Olsen, K.S. & Sørensen, R. (1992). *Tjøme 1813 II, Kvartærgeologisk kart 1:50 000, med beskrivelse*. Norges geologiske undersøkelse.

Birks, J. (1993). Is the hypothesis of survival on glacial nunataks necessary to explain the present-day distributions of Norwegian mountain plants? *Phytocoenologia*, **23**, 399-426.

Bondevik, S., Birks, H.H., Gulliksen, S. & Mangerud, J. (1999). Late Weichselian marine ^{14}C reservoir ages at the western coast of Norway. *Quaternary Research*, **52**, 104-114.

Bondevik, S. & Mangerud, J. (2001). Tidfesting av innlandsisen si maksimale utbreiing på Vestlandet i yngre dryas. *Geonytt*, **2001**, 33.

Bondevik, S. & Mangerud, J., (2002). A very late Younger Dryas ice sheet maximum in western Norway. *Quaternary Science Reviews* (in press).

Bondevik, S., Mangerud, J. & Gulliksen, S. (2001). The marine ^{14}C age of the Vedde Ash Bed along the west coast of Norway. *Journal of Quaternary Science*, **16**, 3-7.

Borgström, I. (1999). Basal ice temperatures during late Weichselian deglaciation: comparison of landforms assemblages in west-central Sweden. *Annals of Glaciology*, **28**, 9-15.

Boulton, G.S., Smith, G.D., Jones, A.S. & Newsome, J. (1985). Glacial Geology and glaciology of the last mid-latitude ice sheet. *Journal of the Geological Society London*, **142**, 447-474.

Brook, E.J., Nesje, A., Lehman, S.J., Raisbeck, G.M. & Yiou, F. (1996). Cosmogenic nuclide exposure ages along a vertical transect in western Norway: Implications for the height of the Fennoscandian ice sheet. *Geology*, **24**, 207-210.

Bugge, T. (1980). Øvre lags geologi på kontinentalsokkelen utenfor Møre og Trøndelag. *Continental Shelf Insitute Publication*, **104**, 44 pp.

Bugge, T., Lien, R. & Rokoengen, K. (1978). Kartlegging av løsmassene på kontinentalsokkelen utenfor Møre og Trøndelag. *Continental Shelf Insitute Publication*, **99**, 55 pp.

Carr, S., Haflidason, H. & Sejrup, H. (2000). Micromorphological evidence supporting Late Weichselian glaciation of the northern North Sea. *Boreas*, **29**, 315-328.

Clark, P.U., Alley, R. & Pollard, D. (1999). Northern Hemisphere ice-sheet influences on global climate change. *Science*, **286**, 1104-1111.

Dahl, S.O., Follestad, B.A., Nesje, A. & Lie, Ø. (2000). Age and climate when cirque glaciers and ice-dammed lakes existed during the deglaciation in east-central Norway. *Geonytt*, **1 2000**, 54.

Dahl, S.O. & Nesje, A. (1992). Paleoclimatic implications based on equilibrium-line altitude depressions of reconstructed Younger Dryas and Holocene cirque glaciers in inner Nordfjord, western Norway. *Palaeogeography, Palaeoclimatology, Palaeoecology*, **94**, 87-97.

Dahl, S.O., Nesje, A. & Øvstedahl, J. (1997). Cirque glaciers as morphological evidence for a thin Younger Dryas ice sheet in east-central southern Norway. *Boreas*, **26**, 161-180.

de Beaulieu, J.-L. & Reille, M. (1992). The last climatic cycle at La Grande Pile (Vosges, France): A new pollen profile. *Quaternary Science Reviews*, **11**, 431-438.

Dowdeswell, J.A. & Siegert, M.J. (1999). Ice-sheet numerical modelling and marine geophysical measurements of glacier-derived sedimentation on Eurasian Arctic continental margins. *Geological Society of America Bulletin*, **111**, 1080-1097.

Fjeldskaar, W. (2000). An isostatic test of the hypothesis of ice-free mountain areas during the last glaciation. *Norsk Geologisk Tidsskrift*, **80**, 51-56.

Fjeldskaar, W. & Kanestrøm, R. (1980). Younger Dryas geoid-deformation caused by deglaciation in Fennoscandia. *In:* Mörner, N.-A. (ed.), *Earth Rheology, Isostasy and Eustasy,* New York, Wiley, 569-574.

Follestad, B.A. (1994a). *Eresfjord 1320 II. Kvartærgeologisk kart med beskrivelse.* Norges geologiske undersøkelse.

Follestad, B.A. (1994b). *Snota. Kvartærgeologisk kart M 1:50 000 med beskrivelse.* Norges geologiske undersøkelse.

Gulliksen, S., Birks, H.H., Possnert, G. & Mangerud, J. (1998). A calendar age estimate of the Younger Dryas-Holocene boundary at Kråkenes, western Norway. *The Holocene*, **8**, 249-259.

Haflidason, H., Sejrup, H.-P., Klitgaard Kristensen, D. & Johnsen, S. (1995). Coupled response of the late glacial climatic shifts of NW-Europe reflected in Greenland ice cores: Evidence from the northern North Sea. *Geology*, **23**, 1059-1062.

Helle, S.K., Anundsen, K., Aasheim, S. & Haflidason, H. (1997). Indications of a Younger Dryas marine transgression in inner Hardanger, West Norway. *Norsk Geologisk Tidsskrift*, **77**, 101-117.

Helle, S.K., Rye, N. & Stabell, B. (2000). Shoreline displacement and fault activity in the Hardangerfjord, western Norway, after deglaciation. *Geonytt*, **1 2000**, 82.

Helmens, K.F., Räsänen, M.E., Johansson, P.W., Jungner, H. & Korjonen, K. (2000). The Last Interglacial-Glacial cycle in NE Fennoscandia: a nearly continuous record from Sokli (Finnish Lapland). *Quaternary Science Reviews*, **19**, 1605-1623.

Holmlund, P. & Fastook, J. (1993). Numerical modelling provides evidence of a Baltic Ice Stream during the Younger Dryas. *Boreas*, **22**, 77-88.

Holtedahl, H. (1993). Marine geology of the Norwegian continental margin. *Norges geologiske undersøkelse. Special publication*, **6**, 1-150.

Holtedahl, O. (1960). Geology of Norway. *Norges geologiske undersøkelse*, **208**, 1-540.

Houmark-Nielsen, M. (1999). A lithostratigrphy of Weichselian glacial and interstadial deposits in Denmark. *Bulletin of the Geological Society of Denmark*, **46**, 101-114.

Jansen, E., Befring, S., Bugge, T., Edvin, T., Holtedahl, H. & Sejrup, H.-P. (1987). Large submarine slides on the Norwegian Continental Margin: Sediments, transport and timing. *Marine Geology*, **78**, 77-107.

King, E.L., Haflidason, H., Sejrup, H.-P. & Løvlie, R. (1998). Glacigenic debris flows on the North Sea Trough Mouth Fan during ice stream maxima. *Marine Geology*, **152**, 217-246.

King, E.L., Sejrup, H.-P., Haflidason, H., Elverhøi, A. & Aarseth, I. (1996). Quaternary seismic stratigraphy of the North Sea Fan: Glacially fed gravity flow aprons, hemipelagic sediments, and large submarine slides. *Marine Geology*, **130**, 293-315.

King, L.H., Rokoengen, K. & Gunleiksrud, T. (1987). *Quaternary seismostratigraphy of the Mid Norwegian Shelf, 65°-67°30'N - A till tongue stratigraphy.* IKU Sintef Group 114 (ISSN 0332-5288), 1-58 and 7 map enclosures.

Kleman, J. (1994). Preservation of landforms under ice sheets and ice caps. *Geomorphology*, **9**, 19-32.

Kleman, J. & Borgström, I. (1990). The boulder fields of Mt. Fulufjället, west-central Sweden - Late Weichselian boulder blankets and interstadial periglacial phenomena. *Geografiska Annaler*, **72 A**, 63-78.

Kleman, J. & Hättestrand, C. (1999). Frozen-bed Fennoscandian and Laurentide ice sheets during the Last Glacial Maximum. *Nature*, **402**, 63-66.

Krzywinski, K. & Stabell, B. (1984). Late Weichselian sea level changes at Sotra, Hordaland, western Norway. *Boreas*, **13**, 159-202.

Laberg, J.S. & Vorren, T.O. (1996). The Middle and Late Pleistocene evolution of the Bear Island Trough Mouth Fan. *Global and Planetary Change*, **12**, 309-330.

Lagerbäck, R. & Robertsson, A.-M. (1988). Kettle holes - stratigraphical archives for Weichselian geology and palaeoenvironment in northernmost Sweden. *Boreas*, **17**, 439-468.

Lambeck, K., Smither, C. & Johnston, P. (1998). Sea-level change, glacial rebound and mantle viscosity for northern Europe. *Geophysical Journal Int.*, **134**, 102-144.

Lambeck, K., Yokoyama, Y., Johnston, P. & Purcell, A. (2000). Global ice volumes at the Last Glacial Maximum and early Lateglacial. *Earth and Planetary Science Letters*, **18**, 513-527.

Landvik, J.Y., Bondevik, S., Elverhøi, A., Fjeldskaar, W., Mangerud, J., Salvigsen, O., Siegert, M.J., Svendsen, J.I. & Vorren, T.O. (1998). The last glacial maximum of Svalbard and the Barents Sea area: Ice sheet extent and configuration. *Quaternary Science Reviews*, **17**, 43-75.

Larsen, E., Attig, J.W., Aa, A.R. & Sønstegaard, E. (1998). Late-Glacial cirque glaciation in parts of western Norway. *Journal of Quaternary Science*, **13**, 17-27.

Larsen, E., Eide, F., Longva, O. & Mangerud, J. (1984). Allerød - Younger Dryas climatic inferences from cirque glaciers and vegetational development in the Nordfjord area, western Norway. *Arctic and Alpine Research*, **16**, 137-160.

Larsen, E., Gulliksen, S., Lauritzen, S.-E., Lie, R., Løvlie, R. & Mangerud, J. (1987). Cave stratigraphy in western Norway; multiple Weichselian glaciations and interstadial vertebrate fauna. *Boreas*, **16**, 267-292.

Larsen, E. & Sejrup, H.-P. (1990). Weichselian land-sea interaction; western Norway - Norwegian Sea. *Quaternary Science Reviews*, **9**, 85-97.

Larsen, E., Sejrup, H.-P., Janocko, J., Landvik, J.Y., Stalsberg, K. & Steinsund, P.I. (2000). Recurrent interaction between the Norwegian Channel Ice Stream and terrestrial-based ice across southwest Norway. *Boreas*, **29**, 185-203.

Lauritzen, S.-E. (1984). A symposium: arctic and alpine karst. *Norsk Geografisk Tidsskrift*, **38**, 139-143.

Lauritzen, S.-E. (1990. Tertiary caves in Norway: a matter of relief and size. *Cave Science*, **17**, 31-37.

Lauritzen, S.-E. (1991). Uranium series dating of speleothems: A glacial chronology for Nordland, Norway, for the last 600 ka. *Striae*, **34**, 127-133.

Lauritzen, S.-E. (1995). High-resolution paleotemperature proxy record for the Last Interglacial based on Norwegian speleothemes. *Quaternary Research*, **43**, 133-146.

Lauritzen, S.-E., Løvlie, R., Moe, D. & Østbye, E. (1990). Paleoclimate deduced from a multidisciplinary study of a half-million-year-old stalagmite from Rana, Northen Norway. *Quaternary Research*, **34**, 306-316.

Lauritzen, S.-E., Nese, H., Lie, R.W., Lauritsen, Å. & Løvlie, R. (1996). Interstadial/interglacial fauna from Norcemgrotta, Kjøpsvik, north Norway. *In:* Lauritzen, S.-E. (ed.), *Climate change: The karst record,* 89-92. Charles Town, West Virginia, The Karst Waters Institute.

Lohne, Ø. (2000). Sen-weichsel og holosen strandforskyvning i Os-området, Hordaland. Unpulished Cand. scient. (Master) Thesis, Department of Geology, University of Bergen.

Longva, O. & Thorsnes, T. (1997). Skagerak in the past and the present - an integrated study of geology, chemistry, hydrography and microfossil ecology. *Norges geologiske undersøkelse, Special Publication*, **8**, 1-100.

Lowe, J.J. & Walker, M.J.C. (1997). *Reconstructing Quaternary environments*. Harlow, Longman.

Lundqvist, J. (1992). Glacial stratigraphy in Sweden. *Geological Survey of Finland, Special paper*, **15**, 43-59.

Mangerud, J. (1972). Radiocarbon dating of marine shells, including a discussion of apparent age of recent shells from Norway. *Boreas*, **1**, 143-172.

Mangerud, J. (1973). Isfrie refugier i Norge under istidene. *Norges geologiske undersøkelse*, **297**, 1-23.

Mangerud, J. (1980). Ice-front variations of different parts of the Scandinavian Ice Sheet, 13,000-10,000 years BP. *In:* Lowe, J.J., Gray, J.M. & Robinson, J.E. (eds), *Studies in the lateglacial of North-West Europe,* 23-30. Oxford, Pergamon.

Mangerud, J. (1981). The Early and Middle Weichselian in Norway: a review. *Boreas*, **10**, 381-393.

Mangerud, J. (1989). Correlation of the Eemian and the Weichselian with deep sea oxygen isotope stratigraphy. *Quaternary International*, **3/4**, 1-4.

Mangerud, J. (1991a). The Scandinavian Ice Sheet through the last interglacial/glacial cycle. *Paläoklimaforschung*, **1**, 307-330.

Mangerud, J. (1991b). The Last Interglacial/Glacial Cycle in Northern Europe. *In:* Shane, L.C.K. & Cushing, E.J. (eds), *Quaternary Landscapes,* 38-75. Minneapolis, University of Minnesota Press.

Mangerud, J. (2000). Was Hardangerfjorden, western Norway, glaciated during the Younger Dryas? *Norsk Geologisk Tidsskrift*, **80**, 229-234.

Mangerud, J. & Gulliksen, S. (1975). Apparent radiocarbon ages of recent marine shells from Norway, Spitsbergen, and Arctic Canada. *Quaternary Research*, **5**, 263-273.

Mangerud, J., Larsen, E., Longva, O. & Sønstegaard, E. (1979). Glacial history of western Norway 15,000-10,000 B P. *Boreas*, **8**, 179-187.

Mangerud, J., Gulliksen, S., Larsen, E., Longva, O., Miller, G.H., Sejrup, H.-P. & Sønstegaard, E. (1981a). A Middle Weichselian ice-free period in Western Norway: the Ålesund Interstadial. *Boreas*, **10**, 447-462.

Mangerud, J., Sønstegaard, E., Sejrup, H.-P. & Haldorsen, S. (1981b). A continuous Eemian-Early Weichselian sequence containing pollen and marine fossils at Fjøsanger, western Norway. *Boreas*, **10**, 137-208.

Mangerud, J., Lie, S.E., Furnes, H., Kristiansen, I.L. & Lømo, L. (1984). A Younger Dryas ash bed in Western Norway, and its possible correlations with tephra in cores from the Norwegian Sea and the North Atlantic. *Quaternary Research*, **21**, 85-104.

Mangerud, J., Jansen, E. & Landvik, J. (1996). Late Cenozoic history of the Scandinavian and Barents Sea ice sheets. *Global and Planetary Change*, **12**, 11-26.

Martinson, D.G., Pisias, N.G., Hays, J.D., Imbrie, J., Moore, T.C.J. & Shackleton, N.J. (1987). Age dating and the orbital theory of the ice ages: Development of a high-resolution 0 to 300 000 year chronostratigraphy. *Quaternary Research*, **27**, 1-29.

Nese, H. & Lauritzen, S.-E. (1996). Quaternary stratigraphy of the Storsteinshola cave system Kjøpsvik, north Norway. *In:* Lauritzen, S.-E. (ed.), *Climate change: The karst record,* 116-120.Charles Town, West Virginia, The Karst Waters Institute.

Nesje, A., Anda, E., Rye, N., Lien, R., Hole, P.A. & Blikra, L.H. (1987). The vertical extent of the Late Weichselian ice sheet in the Nordfjord-Møre area, western Norway. *Norsk Geologisk Tidsskrift*, **67**, 125-141.

Nesje, A. & Dahl, S.O. (1990). Autochtonous block fields in southern Norway: implications for the geometry, thickness, and isostatic loading of the Late Weichselian Scandinavian ice sheet. *Journal of Quaternary Science*, **5**, 225-234.

Nesje, A. & Dahl, S.O. (1992). Geometry, thickness and isostatic loading of the Late Weichselian Scandinavian ice sheet. *Norsk Geologisk Tidsskrift*, **72**, 271-273.

Nesje, A., Dahl, S.O., Anda, E. & Rye, N. (1988). Block fields in southern Norway: Significance for the Late Weichselian ice sheet. *Norsk Geologisk Tidsskrift*, **68**, 149-169.

Nesje, A. & Whillans, I.M. (1994). Erosion of Sognefjord, Norway. *Geomorphology*, **9**, 33-45.

Olsen, L. (1997. Rapid shifts in glacial extension characterise a new conseptual model for glacial variations during the Mid and Late Weichselian in Norway. *Norges geologiske undersøkelse, Bulletin*, **433**, 54-55.

Olsen, L. (1998). Pleistocene paleosols in Norway: implications for past climate and glacial erosion. *Catena*, **34**, 75-103.

Olsen, L., Bergstrøm, B. & Sveian, H. (2001). New data indicates considerable Allerød-Younger Dryas ice-front fluctuations across Hinnøya and in the adjacent Ofotfjorden-Vestfjorden area, northern Norway. *Geonytt*, **2001**, 88.

Olsen, L., Mejdahl, V. & Selvik, S.F. (1996). Middle and Late Pleistocene stratigraphy, Finnmark, North Norway. *Norges geologiske undersøkelse Bulletin*, **429**, 1-111.

Ottesen, D., Rise, L., Rokoengen, K. & Sættem, J. (2001). Glacial processes and large-scale morphology on the mid-Norwegian continental shelf. *In:* Martinsen, O. & Dreyer, T. (eds), *Sedimentary environments offshore Norway - Palaeozoic to Recent,* Norwegian Petroleum Society.

Peltier, R.W. (1994). Ice Age Paleotopography. *Science*, **265**,195-201.

Rasmussen, A. (1981). The Deglaciation of the coastal area NW of Svartisen, Northern Norway. *Norges geologiske undersøkelse*, **369**, 1-31.

Reite, A.J. (1994). Weichselian and Holocene geology of Sør-Trøndelag and adjacent parts of Nord-Trøndelag county, Central Norway. *Norges geologiske undersøkelse Bulletin*, **426**, 1-30.

Rokoengen, K., Bell, G., Bugge, T., Dekko, T., Gunleiksrud, T., Lien, R., Løfaldli, M. & Vigran, J.O. (1977). Prøvetakning av fjellgrunn og løsmasser utenfor deler av Nord-Norge. *Continental Shelf Insitute (IKU) Publication*, **91**, 67 pp.

Rokoengen, K. & Frengstad, B. (1999). Radiocarbon and seismic evidence of ice-sheet extent and the last deglaciation on the mid-Norwegian continental shelf. *Norsk Geologisk Tidsskrift*, **79**, 129-132.

Sejrup, H.-P. (1987). Molluscan and foraminiferal biostratigraphy of an Eemian-Early Weichselian section on Karmøy, southwestern Norway. *Boreas*, **16**, 27-42.

Sejrup, H.-P., Haflidason, H., Aarseth, I., King, E., Forsberg, C.F., Long, D. & Rokoengen, K. (1994). Late Weichselian glaciation history of the northern North Sea. *Boreas*, **23**, 1-13.

Sejrup, H.-P., King, E.L., Aarseth, I., Haflidason, H. & Elverhøi, A. (1996). Quaternary erosion and depositional processes: western Norwegian fjords, Norwegian Channel and North Sea Fan. *Geological Society of London*, **117**, 187-202.

Sejrup, H.-P., Landvik, J.Y., Larsen, E., Janocko, J., Eiriksson, J. & King, E. (1998). The Jæren area, a border zone of the Norwegian Channel ice stream. *Quaternary Science Reviews*, **17**, 801-812.

Sejrup, H.-P., Larsen, E., Landvik, J., King, E.L., Haflidason, H. & Nesje, A. (2000. Quaternary glaciations in southern Fennoscandia: evidence from southwestern Norway and the North Sea region. *Quaternary Science Reviews*, **19**, 667-685.

Sjørring, S. (1983). The glacial history of Denmark. *In:* Ehlers, J. (ed.), *Glacial deposits of North-West Europe,* 163-179. Rotterdam, A.A. Balkema, pp.

Sollid, J.-L. & Reite, A.J. (1983). The last glaciation and deglaciation of Central Norway. *In:* Ehlers, J. (ed.), *Glacial deposits in north-west Europe,* 41-60. Rotterdam, A.A. Balkema.

Sollid, J.-L. & Sørbel, L. (1979). Deglaciation of western Central Norway. *Boreas*, **8**, 233-239.

Sollid, J.-L. & Sørbel, L. (1994). Distribution of glacial landforms in southern Norway in relation to the thermal regime of the last continental ice sheet. *Geografiska Annaler*, **76A**, 25-35.

Sollid, J.-L., Andersen, S., Hamre, N., Kjeldsen, O., Salvigsen, O., Sturød, S., Tveitå, T. & Wilhelmsen, A. (1973). Deglaciation of Finnmark, North Norway. *Norsk Geografisk Tidsskrift*, **27**, 233-325.

Sveian, H., Reite, A.J. & Olsen, L. (2000). The elevation of the Younger Dryas ice surface in the Røros-Folldal area; two contradiction models. *In:* NORPAST, Past climates of the Norwegian Region, Abstracts, Botanical and Geological Institutes, University of Bergen.

Sveian, H. & Solli, A. (1997). Tid og form - geologisk historie. *In:* Dahl, R., Sveian, H. & Thoresen, M. (eds), *Nord-Trøndelag og Fosen,* 111-130. Trondheim, Norges Geologisk Undersøkelse.

Svendsen, J.I. & Mangerud, J. (1987). Late Weichselian and Holocene sea-level history for a cross-section of western Norway. *Journal of Quaternary Science*, **2**, 113-132.

Sættem, J. (1990). Glaciotectonic forms and structures on the Norwegian continental shelf: observations, processes and implications. *Norsk Geologisk Tidsskrift*, **70**, 81-94.

Sønstegaard, E., Aa, A.R. & Klagegg, O. (1999). Younger Dryas glaciation in the Ålfoten area, western Norway; evidence from lake sediments and marginal moraines. *Norsk Geologisk Tidsskrift*, **79**, 33-45.

Sørensen, R. (1979). Late Weichselian deglaciation in the Oslofjord area, south Norway. *Boreas*, **8**, 241-246.

Sørensen, R. (1992). The physical environment of Late Weichselian deglaciation of the Oslofjord region, south-eastern Norway. *Sveriges Geologiska Undersök-ning*, **Ca 81**, 339-346.

Sørensen, R. (1999). En ^{14}C datert og dendrokronologisk kalibrert strandforskyvningskurve for søndre Østfold, Sørøst-Norge. *AmS-Rapport*, **12A**, 227-242.

Valen, V., Mangerud, J., Larsen, E. & Hufthammer, A.K. (1996). Sedimentology and stratigraphy in the cave Hamnsundhelleren, western Norway. *Journal of Quaternary Science*, **11**, 185-201.

Vorren, T.O., Hald, M., Edvardsen, M. & Lind-Hansen, O.-W. (1983). Glacigenic sediments and sedimentary environments on continental shelves: General principles with a case study from the Norwegian shelf. *In:* Ehlers, J. (ed.), *Glacial deposits in North-West Europe,* 61-73. Rotterdam, A.A. Balkema.

Vorren, T.O. & Kristoffersen, Y. (1986). Late Quaternary glaciation in the south-western Barents Sea. *Boreas*, **15**, 51-59.

Vorren, T.O., Laberg, J.S., Blaume, F., Dowdeswell, J.A., Kenyon, N.H., Mienert, J., Ruhmohr, J. & Werner, F. (1998). The Norwegian-Greenland Sea continental margins: morphology and Late Quaternary sedimentary processes and environment. *Quaternary Science Reviews*, **17**, 273-302.

Vorren, T.O. & Plassen, L. (2002). Deglaciation and palaeoclimate of the Andfjord-Vågsfjord area, North Norway. *Boreas*, **31**, (in press).

Vorren, T.O., Vorren, K.-D., Alm, T., Gulliksen, S. & Løvlie, R. (1988). The last deglaciation (20,000 to 11,000 B.P.) on Andøya, northern Norway. *Boreas*, **17**, 41-77.

Zagwijn, W.H. (1989). Vegetation and climate during warmer intervals in the Late Pleistocene of western and central Europe. *Quaternary International*, **3/4**, 57-67.

Zagwijn, W.H. (1992). The beginning of the ice age in Europe and its major subdivisions. *Quaternary Science Reviews*, **11**, 583-591.

MS submitted 2001

Quaternary Glaciations – Extent and Chronology
Editors J. Ehlers and P.L. Gibbard

Pleistocene glacial limits in Poland

Leszek Marks

Polish Geological Institute, Rakowiecka 4, 00-975 Warszawa, Poland and Institute of Geology, Warsaw University, Żwirki i Wigury 93, 02-089 Warsaw, Poland

E-mail: lmar@pgi.waw.pl

Abstract

The limits of four major ice sheets can be traced in the present landscape of Poland. Glacial deposits and ice-dammed lakes indicate a stream-like pattern of advancing ice bodies, dependent both on ice dynamics in the marginal zones and on the pre-existing landscape in their forefields. The southernmost extent of the Pleistocene glaciers is indicated by the limit of Scandinavian erratics. In most areas this limit was formed by the South Polish Glaciations (Elsterian), partly replaced in the west by the Odranian of the Middle Polish Glaciations (Saalian I). The subsequent Wartanian Glaciation of the Middle Polish Glaciations (Saalian II), and the Vistulian Glaciation (North Polish Glaciation, Weichselian) were limited to areas further to the north. During the Middle Vistulian in Poland, the ice sheet could locally move further to the south than the limit of the Late Vistulian Glaciation.

INTRODUCTION

The landscape of Poland has been considerably open to Pleistocene ice sheets, advancing from Scandinavia. Glacial deposits and ice-dammed lakes in Poland indicate a stream-like pattern of advancing ice bodies, dependent both on their dynamics in marginal zones and landscape in their forefields. Most advancing lobes were of an outlet glacier or surge type, which favoured the development of glaciotectonic phenomena. Individual maximum limits were not always reached precisely synchronously throughout the whole territory; this is obvious for the younger ice sheets (cf. Kozarski, 1986; Marks, 1988, 1991, 2002) and seems also a very reasonable assumption for the older ones.

Regardless of the different stratigraphic subdivisions of the Pleistocene in Poland (Fig. 1), the limits of four main ice sheets can be traced in the present landscape (Fig. 2; Różycki, 1965; Rühle, 1965). In general these limits are well correlated. There are, however, several local discrepancies (Lindner, 1988; Dolecki *et al.*, 1994; Żarski, 1990, 1994; Pożaryski *et al.*, 1995; Lindner & Marks, 1993, 1994, 1995a, b; Marks *et al.*, 1995). In most parts of the country the southernmost glacial limit is attributed to the South Polish Glaciations (Elsterian); only in SW Poland it was overridden locally by the Odranian ice sheet of the Middle Polish Glaciations (Saalian I). The maximum extent of Pleistocene glaciations is indicated by the southern limit of Scandinavian erratics. The southernmost glacial limit is much less controversial in eastern Poland than in the western part of the country. The problem there is that the limits of the South Polish and the Middle Polish Glaciations are close together and overpass one another at many localities (Badura & Przybylski, 1998). Further to the north the limits of the Wartanian of the Middle Polish Glaciations (Saalian II), and of the Vistulian Glaciation (North Polish Glaciation, Weichselian) occur.

Middle Pleistocene Glacial Limits

The ice sheet of the Sanian 2 Glaciation (younger South Polish Glaciation) occupied the largest part of the Polish territory and its outer limit is the most extensive in most of the country. The ice sheet reached the Carpathians and the Sudetes in the south, and entered the Moravian Gate between these two mountain ranges in the upper Oder drainage basin (cf. Lindner & Marks, 1994). The ice sheet margin fragmented into lobes that blocked the mountain valleys from the north. There were several ice-free areas to the north of the maximum ice sheet limit, namely the Polish Jura, the highest ranges of the Holy Cross Mountains and nunatak-like hills in the Sudetian foreland. Valley glaciers developed in the Tatra Mountains and, presumably, also in the Sudetes.

The Odranian Glaciation ice sheet (older Middle Polish Glaciation) reached the northern slopes of the Sudetes, the Polish Jura, the Holy Cross Mountains and the Lublin Upland. It also advanced into the Moravian Gate and its proglacial meltwaters flowed into the Danube drainage basin. The eastern part of the country drained to the east by a sub-Carpathian ice-marginal streamway. In mid-eastern Poland the advancing ice sheet blocked the river valleys of the pre-Vistula, pre-Pilica and pre-Wieprz streams, thus favouring the development of several large ice-dammed lakes (e.g. near Koniecpol and Sandomierz). Valley glaciers developed in the Tatra Mountains and presumably, again in the Sudetes.

The ice sheet of the Wartanian Glaciation (younger Middle Polish Glaciation) occupied Wielkopolska, Mazovia and Podlasie (cf. Baraniecka, 1971; Krupiński & Marks, 1993). The western part of the country was drained by the

Fig. 2. Glacial forms in the Bucura cirque (Retezat Mountains).

only in the Apuseni Mountains, that are hardly higher than 1800 m (1849 m in the Curcubăta Peak, Bihor Mountains), and probably also in the Vlădeasa Mountains.

Lolaia Phase (Rissian)

Analysis of topographic and photogrammetric materials, the use of published geomorphological evidence and geomorphological mapping have facilitated the reconstruction of glacial limits for the maximum Pleistocene glaciation phase for a few representative mountain areas: Făgăraş -Iezer, Retezat, Parâng, Godeanu and Ţarcu.

During this so-called Lolaia Phase (Urdea, 1993), the main glaciers descended to:

- 1035-1100 m a.s.l. in the Retezat,
- 1050-1280 m on the northern slope,
- 1130-1400 m on the southern slope of the Făgăraş and Iezer Mountains,
- 1150-1350 m on the northern slope of Parâng,
- 1420-1450 on the southern slope,
- 1210-1330 m on the northern slope of Godeanu Mountains,
- 1200-1380 m on the southern slope,
- 1290-1460 m on the northern and north-western slope of Ţarcu Mountains,
- 1290-1520 m on the south-eastern slope (Table 1)

The higher positions of the glacier fronts in the Ţarcu and Godeanu Mountains result from the fact that these mountain areas are 200-300 m lower than the Făgăraş and Iezer Mountains. The landscape of the Godeanu, Ţarcu, Şurianu, Cindrel, Lotru, Latoriţei and Căpăţânii Mountains is also dominated by extensive levelled surfaces that belong to the Borăscu sculptural complex (Fig. 1). This geomorphological situation did not favour the formation of well-developed valley glaciers (Table 2). However, plateau glaciers are characteristic here (Table 3). For example, the Pietrele Albe (5.82 km^2) and Ţarcu-Căleanu (4.08 km^2) ice caps in the Ţarcu Mountains represent this type of glaciation.

On the basis of the geomorphological evidence, the most widespread glaciation took place in the Retezat Mountains. This results from their orographic position and preglacial relief. The latter was characterised by a well-developed valley network and drainage basins, that favoured the formation of cirque and valley glaciers, e.g. in the Peleaga-Bucura area (Fig. 2).

During the Lolaia Phase, the Lăpuşnicu Mare Glacier was the largest in the Romanian Carpathians. With an area of 40.1 km^2 and with a maximum length of 18.1 km it represented a real 'river of ice'. It was formed by the confluence of the Bucura (4.2 km length and 8.5 km^2 area) and the Peleaga glaciers (4.4 km length and 5.8 km^2 area), being further fed by the Judele (6.9 km and 11.2 km^2 and then the Paltinu-Galbena glacier (4.6 km and 8.1 km^2) from the Godeanu Mountains. On the northern slope, the large complex Pietrele-Nucşoara glacier included the longest glacier tongue (8.1 km long and a total surface of 27.3 km^2). Its tributaries extending down from the main ridge were:

- the Beagu glacier (5.4 km long and 4.2 km^2),
- the Galeşu glacier (4.4 km long and 7.2 km^2),
- the Valea Rea glacier (3.6 km long and 3.9 km^2) and
- the Stânişoara glacier (5.5 km long and 6.2 km^2).

On the east side, the impressive Râu Bărbat glacier had a length of 7.9 km and covered an area of 15.5 km^2. It was further nourished by ice from some hanging glaciers situated under the Văcăria-Custura-Păpuşa-Porţile Închise Ridge.

The Făgăraş Mountains, - the most massive and highest mountain range in Romania - are characterized by a steeper northern and a comparatively more gentle southern slope. Consequently, there was a difference in the style of the former glaciation. The northern slope was characterised by simple valley glaciers, 3.5-5 km long, separated by sharp-crested and parallel ridges (Fig. 3). The Arpaşul Mare-Podragu Glacier (8.28 km^2, 5 km) was the lowest, reaching down to 1050 m a.s.l. On the northern slope, the largest glacier was the complex Pojorta Glacier of 9.5 km^2 and a length of 6.1 km, with a total length of its branches amounting to 9.9 km. On the southern slope, because of the well-developed preglacial valley network, there were complex valley glaciers, that descended down to 1130 m a.s.l. The aspect and dimension of the prominent Capra Glacier, being comparable with those of alpine glaciers.

The Valea Rea Glacier of 26.77 km^2 and a length of 9 km should be mentioned. Including its branches (Pojarna: 5.68 km^2, 6.6 km, Bândea-Gălăşescu: 8.78 km^2, 8.1 km), it

Table 1. The largest reconstructed Pleistocene glaciers (Lolaia phase = Rissian II) in the main massifs of the Romanian Carpathians.

Mountain area and glaciers	Maximum length (km)	Surface (km^2)	Front altitude (m a.s.l.)	Orientation
Retezat				
Lăpuşnicu Mare	18.1	40.1	1060	S
Pietrele	8.1	27.3	1035	N
Râu Bărbat	7.9	15.5	1230	N
Râu Alb	8.8	5.3	1140	N
Făgăraş				
Valea Rea	26.77	9.0	1160	S
Buda	6.9	23.57	1160	S
Capra	10.0	18.33	1130	S
Pojorta	6.1	9.5	1280	N
Arpaşu Mare-Podragu	5.0	8.28	1050	N
Laita	4.8	6.52	1110	N
Parâng				
Lotru	15,6	23.02	1350	N
Jieţu	8.1	18.5	1150	N
Godeanu				
Lapuşnicu Mic	6.1	16.39	1210	N
Galbena-Paltina	8.74	4.9	1200	N
Cârnea	5.1	5.81	1330	NW
Ţarcu				
Hidegul	4.5	8.11	1290	SW
Şuculeţu	4.2	8.01	1290	NE
Netişul	4.1	6.79	1300	NE
Iezer				
Boarcăşu	5.2	7.5	1360	N
Bătrâna-Piscanu	4	6.06	1300	SE
Bucegi				
Ialomiţa	4.2	8.02	1640	S
Gaura	4	2.47	1360	W
Şurianu				
Cârpa	2	1.1	1490	NE
Cindrel				
Iezeru Mic	4	3.02	1450	N
Iezeru Mare	3.8	1.87	1550	N
Lotrului				
Jipoasa-Cristeşti	2.5	3.17	1450	SE
Ştefleşti	2.8	2.42	1500	N
Rodnei				
Bistricioara-Putredu	7.8	19.8	1100	NE
Repedea-Buhăiescu	6.1	16.92	848	N
Maramureş				
Kvasni	2.8	2.82	1050	N
Groapa Julii	1.8	2.2	1050	NE

Table 2. Medium altitude (m a.s.l.) of the glacier fronts during the maximum Pleistocene glacial phase (Lolaia = Riss II) for some mountain units of the Romanian Carpathians.

Mountain unit	Medium value	Northern slope	Southern slope
Ţarcu	1393	1360	1427
Godeanu	1417	1367	1466
Retezat	1199	1164	1233
Parâng	1387	1327	1447
Făgăraş	1348	1235	1461
Iezer	1395	1371	1420
Bucegi	1500	1360	1640
Southern Carpathians	**1377**	**1312**	**1442**
Rodnei Mountains (Eastern Carpathians)	1232	1063	1400

had a total length of 30.4 km. Another example is the Buda Glacier with a surface of 23.57 km^2 and a maximum length of 6.9 km and a total length of 29.4 km. Finally the Capra Glacier had a surface of 18.33 km^2 and a maximum length of 10 km (26.1 km, when all tributaries are added).

In the Parâng Mountains, the situation was the opposite of that in the Făgăraş Mountains. Here more powerful glaciers developed on the northern slope, as a consequence of the advanced stage of geomorphological preglacial evolution in the Jieţu and Lotru basins. Thus, the Lotru Glacier reached a surface of 23.02 km^2, and a length of 15,6 km, while the Jieţu Glacier had a surface of 18.5 km^2. The latter formed by confluence of the Sliveiul (4.52 km^2 , 4 km), Roşiile (8.86 km^2, 5.5 km) and Coasta lui Rus (3.45 km^2, 3.2 km) glacial tongues.

The largest glacier in the Iezer Mountains was the Boarcăşul Glacier (7.5 km^2, 5.2 km maximum length) on the northern slope. On the southern slope the Bătrâna-Iezeru Mare Glacier (6.06 km^2, 4 km) was developed.

Although the Bucegi Mountains are higher than 2500 m (2505 m in Omu Peak), because of their position in the eastern extremity of the Southern Carpathians, glaciers were less developed than in the west. This is because a more continental climate occurs in this region. Here glaciers descended only to an altitude of 1360 m, e.g. the Gaura Glacier (2.47 km^2, 4.1 km). The largest glacier on the southern slope was the Ialomiţa Glacier (8.02 km^2, 4.2 km).

The largest glaciers in the Ţarcu Mountains were the Hidegul Glacier (8.11 km^2, 4.5 km) on the south-eastern slope and the Şuculeţul Glacier (8.01 km^2, 4.2 km) on the south-western slope of the massif. This area also supported the largest plateau glaciers of the Romanian Carpathians, the Pietrele Albe and Ţarcu-Căleanu.

The largest glaciers in the Godeanu Mountains were also located on the northern slope. They included:

- the Lăpuşnicu Mic Glacier (16.39 km^2, 6.1 km), with its many tributaries,
- the Galbena -Paltina (8.74 km^2, 4.9 km), - tributary of Lăpuşnicu Mare Glacier (Retezat Mountains), the largest of the Romanian Carpathians,
- the Cârnea (5.81km^2, 5.1 km),

Of particular note is the presence of the Borăscu, Paltina Galbena, Micuşa, Moraru and Scărişoara plateau glaciers (Table 3). In the Şurianu, Cindrel, Lotru, Latoriţei, and Leaota Mountains, for the reasons mentioned above, glacier dimensions hardly exceeded a surface of 3 km^2 and a maximum length of 3 km (Table 2).

Major Pleistocene glaciation was limited in the Eastern Carpathians to the Rodnei Mountain, and there only to the northern slope, where glaciers descended down to 848 m, in the case of Repedea-Buhăiescu Glacier (Sawicki, 1911). In this area the largest glaciers of the Eastern Carpathians were present: the Bistricioara-Putredu Glacier (19.8 km^2, 7.8 km maximum length, and 20.6 km total length) and the Repedea-Buhăiescu Glacier (16.92 km^2, 6.1 km, 25.2 km).

The asymmetry of the Pleistocene glaciation in some areas of the Romanian Carpathians has been estimated using a cumulative vector diagram after Evans (1977). The strength of the resultant vector is 23% for the Retezat Mountains, 25.3% for the Godeanu Mountains and 36% for the Ţarcu Mountains. According to Evans' definition, the Retezat and Godeanu Mountains would be areas with weakly-asymmetric glaciation. In the author's opinion, this is due to the major influence of the pre-glacial relief on glacier formation. For the maximum Pleistocene glaciation in the Romanian Carpathians, it is estimated that the snow-line stood at 1640-1700 m a.s.l. in the Southern Carpathians and at 1500-1600 m in the Rodnei Mountains.

Capra-Judele Phase (Würmian)

In the second major glaciation, the Capra-Judele Phase (= Würmian II), the glaciers descended to an altitude of 1330-1360 m in the Southern Carpathians and 1100-1200 m in Rodnei Mountains. Their frontal moraines are located at:

- 1310 m in theCapra Valley (Făgăraş Mountains),
- 1335-1350 in the Judele Valley (Retezat Mountains),
- 1120 m in the Cimpoieşul Valley and
- 1190 m in the Bistricioara Valley (Rodnei Mountains) (Sîrcu, 1978).

Table 3. The main Pleistocene plateau (Lolaia Phase) glaciers of the Southern Carpathians.

Glacier	**Mountain area**	**Surface area (km ^{2})**
Pietrele Albe	Ţarcu Mountains	5.82
Drăgşanu	Retezat Mountains	4.76
Ţarcu-Căleanu	Ţarcu Mountains	4.08
Zănoaga-Zlata	Retezat Mountains	3.55
Borăscu	Godeanu Mountains	3.47
Slăveiu-La Clince	Retezat Mountains	3.11
Berivoiu	Făgăraş Mountains	2.96
Paltina-Galbena	Godeanu Mountains	2.05
Scărişoara	Godeanu Mountains	1.87
Cărbunele	Parâng Mountains	1.57
Ciobanu	Parâng Mountains	1.48
Zârna	Făgăraş Mountains	1.13
Moraru	Godeanu Mountains	0.98
Micuşa	Godeanu Mountains	0.68

These glaciers were still powerful; they occupied the interior parts of the cirques and Rissian glacial valleys, and carved new forms, This is supported by the sculptural benches visible in the main glacial landforms (Fig. 4). The contemporaneous snow-line was situated around 1783 m, and the longest glaciers, Lăpuşnicu Mare and Pietrele in the Retezat Mountains, reached lengths of 9.7 km and 5.5 km respectively. The maximum lengths were 8.5 and 7 km for the Lotru Glacier and the Jieţu Glacier, in the Parâng Mountains and only 6 km for the Capra Glacier, 5.5 km for the Valea Rea Glacier and 4.5 km for the Buda Glacier.

Because of the more arid climate during the Würmian III (Cârciumaru, 1984), glaciation was less extensive and the ice only reached down to 1440 m during the Stânişoara-Pietrele Phase, subsequently getting smaller and smaller. At the end of the Würmian, during the Valea Rea-Lespezi Phase, the glaciers terminated at only 1720-1740 m, the snow-line being situated at 1988-2011 m. They had by then changed from Alpine-type to Pyrenean-type glaciers, with short tongues, less than 2 km in length.

Fig. 3. Alpine crests (gipfelflur) on the northern slope of the Făgăraş Mountains.

Fig. 4. Würmian moraines in the Huluzu Valley (Parâng Mountains).

The Older and Middle Dryas Stadials (Văsiel-Slivei, Roşiile and Ştevia-Arpăşel Phases) are characterised by cirque glaciers in areas higher than 1860 m that were of limited extent, for example, only 8.5% of the Retezat surface.

In the Younger Dryas Stadial (Călţun Phase), glaciation was restricted to cirques situated higher than 2150 m a.s.l. The snow-line was above 2200 m a.s.l. Subsequently, the considerable climate amelioration, reflected in the pollen diagrams (Pop *et al.*, 1971; Farcas *et al.*, 1999), caused final deglaciation of the Retezat Mountains.

Field investigations suggest that during the earlierst Holocene Preboreal (Podu Giurgiului Phase), very small cirque glaciers, niche glaciers and firn and ice masses so-called 'glacierettes' could form only in cirques at altitudes higher than 2200 m, under favourable microclimatic circumstances. They were probably linked to the Vârtopel, Podu Giurgiului, Hârtopu Ursului, Călţun-Râiosu North (Fig. 5, Făgăraş Mountains) and Beagu (Retezat Mountains) cirques.

Table 4. Correlation scheme of the glacial phases in the Southern Carpathians with those in the Tatra Mts and the Alps during the Upper Pleistocene.

a, glacial phase; b, altitude of terminal moraine; c, snow-line; d, stratigraphy of Central and Northern Europe.

Southern Carpathians			Tatra Mountains (Halouzka, 1977)			The Alps (Glückert, 1987)	
a	b	c	a	b	c	a	d
Podu Giurgiu-lui	2250	2306	H	2022	2201	Kromer	Preboreal
Călțun	2150	2203	WH	1838	2075	Egesen	Younger Dryas
Ştevia-Arpăşel	2015	2136	Polana pod Vysokou II	1623	1964	Daun	Middle Dryas
Roşiile	1940-1960	2103-2124	Polana pod Vysokou I	1544	1902	Gschnitz-Clavadel	Older Dryas
Văsiel-Slivei	1840-1860	2035-2058	Rybi potok	1449	1825	Flisur-Steinach	
Lespezi-Valea Rea	1720 1740	1988-2011	Prostredna polana II	1350	1759	Konstanz-Hurden-Bühl	
Genţiana	1640	1948	Prostredna	1310	1740	Zürich	Würmian III
Stâna de Râu	1570	1914	Veza	1213	1683	Stein am Rhein	
Stânişoa-ra- Pietrele	1440	1838	Tatranska-Lomnica	1102	1640	Schaffhausen	
Capra-Judele	1335-1360	1783-1816	Stosy	1094	1631	Aying	Würmian II

Fig. 5. Hanging cirques in the Buda-Râiosu area.

- After the end of the Pleistocene glaciation, a glaciation in the Romanian Carpathians referred to as the 'Carpathian Glaciation' (Velcea, 1973), the mountain area entered a new, postglacial evolution phase. Beginning in the Older Dryas, in the paraglacial environment of the recently deglaciated areas, some relief mesoforms had been formed that are specific to many areas in the Southern Carpathians and in the Rodnei Mountains, i.e. rock glaciers (Urdea, 1992). In many cases the detailed analysis of landforms in the valleys and glacial cirques allows the reconstruction of the sequence of evolution of the latest glaciers in the Southern Carpathians and Rodnei Mountains: ablation complexes → debris-covered glaciers → ice-cored rock glaciers → debris rock glaciers, or secondary rock glaciers (*sensu* Corte, 1976). The latter are found in the
- Pietrele, Galeşu, Valea Rea and Ştevia cirques in the Retazat Mountains,
- Sliveiu, Ghereşiu, Gemănarea, Roşiile, Păsării, Găuri und Gruiu cirques, in the Parâng Mountains,
- Izvorul Grohotişului, Arpăşel, Podrăgel, Căldarea Pietroasă a Doamnei, Căldarea Pietroasă a Arpaşului, Călţun, Boia Mică, Răcorele and Burianului cirques, in the Făgăraş Mountains,
- Cimpoieşul, Iezerelor, Puzdrelor Bila and Lala cirques, in Rodnei Mountains.

Debris-rock glaciers are typically found in glacial cirques and valleys dominated by high crests and peaks, which provide significant quantities of frost-shattered blocks. In other cases the evolution stopped at the stage of debris-covered rock glaciers. That is typically the case for glacial cirques and valleys cut from large levelled surfaces,

Table 5. Altitudinal range of Pleistocene glaciers in the Rodnei Mountains (Eastern Carpathians).

Glacier fronts altitude (m a.s.l.)	Stratigraphy
1930-1950	Younger Dryas
1820-1850	Middle Dryas
1780-1790 1720	Older Dryas
1640 1550 1350-1400 1250	Würmian III
1100-1190	Würmian II
848-890	Rissian II

e.g. the Vlăsia Mică, Bulzul and Mocirliul cirques in the Godeanu Mountains.

In the Holocene the region was situated in the periglacial morphogenetic zone (Urdea, 1997), where only the rock glaciers continued their development.

Conclusions

The results of investigations allow a tentative correlation of glacial phases in the Retezat Mountains with those in the Tatra Mountains (Halouzka, 1977) and the Alps (Glückert, 1987) for the Upper Pleistocene (Urdea, 1993). Starting from the scheme for the Retezat Mountains, a correlation scheme was also proposed for the Southern Carpathians. It is considered that additional field studies are necessary to ascertain the validity of some points of view, thus the present scheme has a provisional status. (Table 4). The fact that the snow limits were lower by about 200 m in Retezat Mountains, as compared to the Tatra Mountains, is due both to the 4° 30' difference of latitude, and the more continental climate, providing lower precipitation for the Romanian Carpathians.

Because the Rodnei Mountains have a particular setting in the Eastern Carpathians, as far as the Pleistocene glaciation is concerned, and because there are differing points of view on this problem -e.g. Sîrcu (1978) *vs.* Sawicki (1912), Kraüthner, (1930) -, a tentative correlation scheme is presented for these mountains (Table 5) This scheme of Pleistocene glaciation phases, is likely to be developed further in the future.

References

Cârciumaru, M. (1980). *Mediul geografic în Pleistocenul superior şi culturile paleolitice din România*, Bucureşti, Edit. Academiei, 268 pp.

Corte, A.E. (1976). Rock glaciers, *Biuletyn Periglacijalny,* **26,** 175-197.

de Martonne, Emm. (1907). Recherches sur l'évolution morphologique des Alpes de Transylvanie (Karpates méridionales), *Rev. de géographie annuelle,* **I** (1906-1907), 286 pp.

Evans, I.S. (1977). World-wide variations in the direction and concentration of cirque and glacier aspects, *Geografiska Annaler* **59 A, 3-4**, 151-175.

Farcas, S., de Beaulieu, J.-L., Reille, M., Coldea, G., Diaconeasa, B., Goeury, C., Goslar, T. & Jull, T. (1999). First ^{14}C datings of Late Glacial and Holocene pollen sequences from Romanian Carpathes, *C.R. Acad. Sci. Paris, Science de la vie,* **322**, 799-807.

Glückert, G. (1987). Zur letzten Eiszeit im alpinen und nordeuropäischen Raum, *Geographica Helvetica,* **42, 2,**

Halouzka, R. (1977). Stratigraphical subdivision of sediments of the last glaciation in the Czechoslovak Carpathians and their correlation with the contemporary alpine and north-european glaciation, *Pr. 73/1/24, IGCP, Quaternary glaciations in the Northern Hemisphere,* Prague, Report nr. **4**, 83-89.

Kraüthner, Th. (1930). Die Spuren der eiszeitlichen Schneegrenze in den Südkarpaten. *Verhandlungen und Mitteilungen des Siebenbürgischen Vereins für Naturkunde zu Hermannstadt*, **LXXIX-LXXX (1929-1930)**, 38-86.

Lehmann, P.W. (1881). Beobachtungen über Tektonik und Gletscherspuren im Fogaraschen Gebirge, *Zeitschrift der Deutschen Geologischen Gesellschaft,* **XXXIII,** 109-117.

Lehmann, P.W. (1885). Die Südkarpaten zwischen Retjezat und Königstein, *Zeitschrift der Gesellschaft für Erdkunde Berlin,* **XX**, 325-336, 346-364.

Niculescu, Gh., Nedelcu, E. & Iancu, S. (1983). Glaciaţia şi relieful glaciar, în *Geografia României, I-Geografia fizică,* Bucureşti, Edit. Academiei, 136-141.

Pop. E., Lupşa, V. & Boşcaiu, N. (1971). Diagrama sporopolinică de la Tăul Zănoguţii (Munţii Retezat). In *Progrese în palinologia românească,* Bucureşti, Edit Academiei, 219-225.

Popescu, N. (1990) *Ţara Făgăraşului. Studiu geomorfologic,* Bucureşti, Edit. Academiei, 180 pp.

Posea, Gr., Popescu, N. & Ielenicz, M. (1974). *Relieful României*, Edit. Ştiinţifică, Bucureşti, cap. „Glaciaţiunea şi relieful glaciar din Carpaţi", p. 186-197 şi cap. „Modelarea periglaciară din ultima parte a Pleistocenului", 198-207.

Puchleitner, S. (1902). Die Eiszeit-Gletscherspuren in den Südkarpathen, *Mitteilungen der k.k. Geographischen Gesellschaft, Wien,* **LII,** 124-139.

Sawicki, L. (1912). Les études glaciaire dans les Karpates. Apercu historique et critique, *Ann. de Géographie,* Paris, vol. **XXI,** p. 230-250.

Sîrcu, I. (1978). *Munţii Rodnei. Studiu morfogeografic,* Bucureşti, Edit. Academiei, 112 pp.

Tietze, K. (1878). Über das Vorkommnis der Eiszeitspuren in den Ostkarpathen, *Verhandlungen der geologischen Reichsanstalt*, Wien, 142-146.

Urdea, P. (1992). Consideraţii asupra dinamicii postglaciare a etajelor morfoclimatice din Munţii Retezat, *Analele Univ. Timişoara, Seria Geografie,* **1,** 25-31.

Urdea, P. (1993). Consideraţii asupra manifestării glaciaţiei cuaternare în Munţii Retezat, *Stud. cerc. de geografie,* **XL**, 65-72.

Urdea, P. (1997). Aspects concerning postglacial and present-day landforms evolution in Southern Carpathians, *Acta Facultatis Rerum Naturalium Universitatis Comenianae, Geographica,* Bratislava, **40,** 107-117.

Urdea, P. (2000). *Muntii Retezat. Studiu geomorfologic.* Bucuresti, Edit. Academiei Române, 272 pp.

Velcea, V. (1973). De l' éxistence d' une glaciation de type carpatique, *Revue de geographie alpine*, **LXI (2)**, 223-230.

Quaternary Glaciations – Extent and Chronology
Editors J. Ehlers and P.L. Gibbard

Pleistocene ice limits in the Russian northern lowlands

Valery Astakhov

St.Petersburg University, Geological Faculty, Universitetskaya 7/9, 199034 St.Petersburg, Russia

E-mail: valery@nb1466.spb.edu

Introduction

This paper is an explanatory note to the digitized map of ice limits compiled by the author for the Russian North, beyond the limits of the Fennoscandian glaciation. The glacial limits presented here are cartographically based on data from the Russian Geological Survey obtained during general geological mapping at standard scales of 1:200,000 and 1:1,000,000. It also incorporates results of specialised studies in critical areas performed by various researchers, the author included. The outermost limit of Pleistocene glaciation is mostly derived from standard geological maps and drilling projects, whereas the Late Pleistocene ice limits are largely products of stratigraphical and photogeological studies during the last two decades. The result is presented in simplified form in Fig. 1.

Although the principal ice limits in Central Russia were already established at the beginning of geological mapping by the Emperor`s Geological Committee in the late 19th century, the size of Pleistocene glaciers in the north and east, excluding the Fennoscandian ice dome, was poorly known until after extensive surveys of these remote areas in Soviet times. Indeed many sheets of the National Geological Map of the USSR appeared in the 1940-50s. This information was summarised in synthetic maps of Quaternary deposits at scales 1:2,500,000 and 1:1,500,000 (Yakovlev, 1950; Zarrina *et al.,* 1961) derived from the first generation geological maps. The second generation of geological maps, that included complementary Quaternary maps, began to appear in the 1960s. The new survey and drilling data were incorporated into more accurate small-scale maps of the Quaternary deposits of European Russia and the entire Soviet Union (Krasnov, 1971; Ganeshin, 1973). Many of the second generation maps were also published in the 1980-90s, albeit at a slower pace. The most inaccessible areas of Central Siberia did not have published Quaternary maps by the end of the 20th century, when the Russian Geological Survey had already commenced compiling the third generation maps in digital form.

The maximum Pleistocene ice limit, related to pre-Eemian glaciations, has only undergone minor changes during the last decades. However, younger ice limits have been under incessant stratigraphic discussion. The number of known ice limits in each given region depend very much both on the logistical accessibility and geographical peculiarities of the area. Several ice limits mapped in European Russia are hardly traceable beyond the Urals, because photogeology, the main tool in the north, does not produce good results in swampy lowlands. Also, topographically expressive ice marginal zones of the Central Siberian ice sheets cannot be directly connected with limits of the Barents-Kara ice sheets in the flat West Siberian Plain. Several morphological boundaries discussed in academic works (e.g., Arkhipov *et al.,* 1980, 1986; Grosswald, 1980, 1993) are not used in the present map, if they are not supported by stratigraphic evidence.

Three main sets of data are presented on the digital map:

1. Ice limits, subdivision of which into certain and uncertain ones is not exactly the same for different glaciations. In the case of Middle Pleistocene glaciations, a certain ice limit in the lowlands is a line distal to the southernmost sections that show glaciotectonised and till-covered interglacial formations. In the uplands of Central Siberia this line is drawn between the last mapped ridge-like diamicton accumulations and heavily dissected fluvial landscape to the south. This line is considered uncertain where no distinct morainic features have been mapped, and only diamicton facies in rare boreholes or erratics on the surface suggest the glacial drift limit. In the case of the Early Weichselian glaciation, the ice limit is shown as certain where it can be drawn between sections of interglacial sediments of Eemian type (mostly marine with warm-water mollusc fauna), overlain by till, on the one hand, and not covered by till, on the other. An uncertain Early Weichselian ice limit is a line interpolated between known stratigraphic sites using such signatures of former glaciation as occurence of buried bodies of stratiform massive ice in the lowlands of West Siberia and/or assemblages of ice-pushed ridges in higher terrains. In many cases marginal features are very spectacular, such as the double Laya-Adzva ridge in the Pechora Basin, which can be seen even from the Moon. Nevertheless the lack of Eemian sites makes this author designate the ice limit there as uncertain. Similarly, the limit of Late Weichselian ice sheet in Central Siberia is drawn as certain where sequences of Weichselian sediments with finite radiocarbon dates have been reported as overlain by till. Where no finite dates are known from sediments beneath till, it is shown as uncertain, even if the limit is clearly expressed by very fresh end moraines. All these distinctions are omitted in the simplified overview map (Fig. 1).

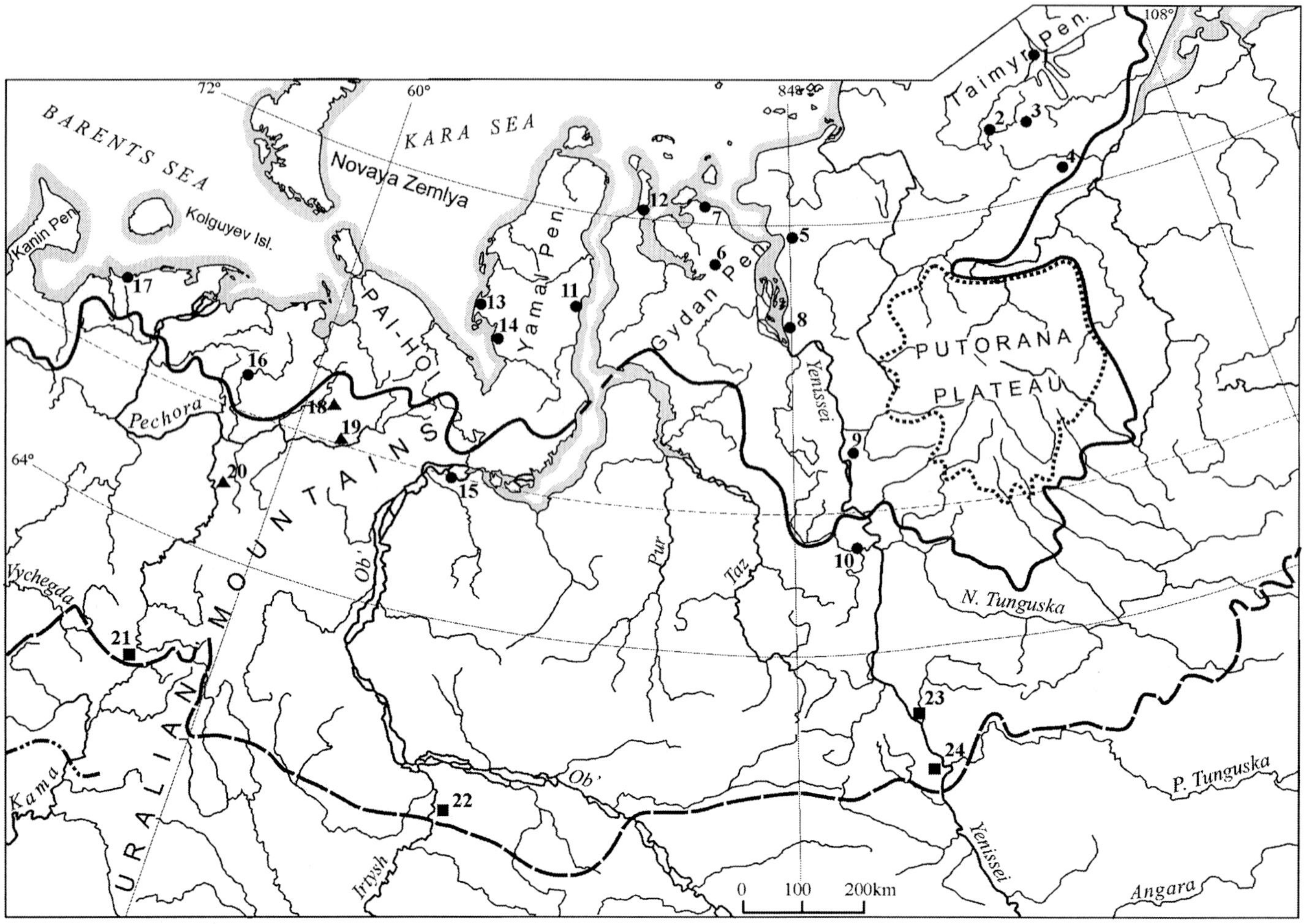

Fig.1. Generalised Pleistocene ice limits of northern Russia beyond the Fennoscandian glaciation.

Dotted line: Late Weichselian glaciation, solid line: Early Weichselian glaciation, dashed line: Saalian glaciation, dash-and-dot line: Don glaciation.

The black circles indicate sites with dated Middle and Late Weichselian sediments not covered by till with their radiocarbon (or OSL) ages:

1- Cape Sabler, 39 to 17 ka (Kind & Leonov, 1982; Möller et al., 1999);
2 - River Logata, 45 to 28 ka (Fisher et al., 1990);
3 - Lake Logata, 45 to 25 ka (Fisher et al., 1990);
4 - Lake Labaz, > 48 to 20 ka (Siegert et al., 1999);
5 - Mokhovaya mammoth, 35.8±2.7 ka (Heintz & Garutt, 1964);
6 - Gyda mammoth, 33.5±1 ka (Heintz & Garutt, 1964);
7 - Leskino mammoth and plants, 30.1±0.3 and 29.7±0.3 (by F. Kaplyanskaya in Astakhov, 1998);
8 - Cape Karginsky, 15.3±0.2 (Kind, 1974);
9 - Igarka permafrost pit, >50 to 35 ka (Kind, 1974);
10 - Farkovo, 42 to 34 ka (Kind, 1974);
11 - Syo-Yakha, 40 to 17 ka (Vasilchuk et al., 2000);
12 - Mongotalyang, 31 to 21 ka (Vasilchuk et al., 1984);
13 - Marresale, 28 to 26 ka (Forman et al., 1999);
14 - Mutny Mys, 30.5 to 27.5 ka (Gataullin et al., 1998);
15 - Salehard moraines covered by interglacial sediments (Astakhov, 2001);
16 - bones in river Kolva terraces, 37 to 26 ka (Mangerud et al., 1999);
17 - Timan Beach, OSL age 52 to 13 ka (Mangerud et al., 1999).

The black triangles are similar sequences at superficial Palaeolithic sites:

18 - Pymva-Shor, 26 to 10 ka (Mangerud et al., 1999);
19 - Mamontovaya Kurya, 37 to 24 ka (Mangerud et al., 1999);
20 - Byzovaya, 33 to 25 ka (Mangerud et al., 1999).

The black quadrangles are interglacial lacustrine and fluvial sediments of Holsteinian type covered by till:

21 - Lake Chusovskoye (Stepanov, 1974);
22 - Semeika (Kaplyanskaya & Tarnogradsky, 1974);
23 - Bakhtinsky Yar (Zubakov, 1972);
24 - Khakhalevsky Yar (Zubakov, 1972).

2. Topographically expressed features of glaciotectonic accretion are shown in the digital map as red arcs the radii of which can be taken as representing former flowlines. There are two kinds of these features dependent on the quality of the substrate. Where substrate is solid bedrock that allowed an easy sliding, these horseshoe-shaped ridges may be positioned close to the outer till limit and thereby considered as end moraines (Fig. 2, a). The internal structure of terminal ridges is poorly known and is normally described as chaotic agglomeration of stony or clayey diamictons distorted together with underlying sand and sometimes massive ice (e.g., Kind & Leonov, 1982; Astakhov *et al.*, 1999). The ridges of sedimentary basins are generally larger and predominantly occur far upglacier irrespective of any marginal formations. In all the sections they invariably demonstrate a regular imbricate structure, with folded slices of various soft rocks divided by listric thrusts and sometimes by clay or ice diapirs (Astakhov, 1979; Astakhov *et al.,* 1996). They are often expressed in the landscape as typical `hill-hole pairs` (Fig. 2, b). In the Middle Pleistocene glaciation area, the imbricate assemblages are mostly constructed of Palaeogene and Cretacious sediments, such as the largest arc between the Urals and the Ob river, the `Malososvinsky Amphitheatre`. The depth of these glacial disturbances may reach 400 m and their width up to 20-25 km. The morainic arcs on the map only show crests of such large structures. These forms, derived from glacial crumpling of perennially frozen clayey formations, must have originated under very thick ice (Astakhov *et al.,* 1996) and therefore cannot be called end moraines. They provide evidence of glacial overriding but not of a quasi-stationary ice margin. In the northern plains where the ice front was fringed by a large proglacial lake and where glacier motion was obstructed neither orographically nor thermally extension ice flow failed to produce any terminal elevations.

3. Named (or numbered) sedimentary sequences (Table 1), constraining temporal and spatial brackets for former ice sheets, are shown mostly for the Late Pleistocene glaciations area, where geochronometric methods can be employed. They are of four kinds: i) interglacial formations of Eemian type which were apparently not overridden by glaciers, ii) glacially disturbed or till-covered Eemian sediments, indicating a Late Pleistocene ice advance, iii) Weichselian sediments with successions of `old` (40 to 15 ka BP) radiocarbon dates, lacking any sign of overriding by a Late Weichselian ice advance, and iv) Weichselian sediments, with finite radiocarbon dates, covered by a till. The most important and well-dated sections are shown in Fig. 1, other can be found on the digital map. For the Middle Pleistocene glaciation area with till-covered interglacial formations of Likvin/ Tobol (i.e. Holsteinian) type only several famous sites have been selected. These formations are known for their typical forest pollen spectra that include exotic taxa, indicating warmer and more humid environments than at present.

The Western European terms Early, Middle and Late Weichselian are rarely applied in the Russian North for specific geological objects, the local names being more popular. However, because of the size of the area considered, with many semi-formalised stratigraphic labels, below the author has used these terms in their geochronological sense common in Russian literature, i.e. as equivalents of the subdivisions of the Wisconsinan.

Pleistocene Glacial Maximum

The spatial resolution of pre-Eemian ice limits, presented here, is mainly in accord with the general maps of Quaternary deposits (Krasnov, 1971; Ganeshin, 1973; Zarrina *et al.,* 1961). However, in many places the configuration of ice margins is corrected using information from later geological maps (Bobkova, 1985; Chumakov *et al.,* 1999; Potapenko, 1985; Rudenko *et al.,* 1981, 1984) and several specialised studies, including Stepanov (1974) for the Pechora-Volga interfluve (21 in Fig. 1), Kaplyanskaya & Tarnogradsky (1974) for river Irtysh (22 in Fig. 1), Astakhov & Fainer (1979) and Zubakov (1972) for Yenissei Siberia (23, 24 in Fig, 1), Arkhipov *et al.* (1976) and Isayeva (1984) for Central Siberia.

The maximum glaciation was traditionally mapped in the USSR as a counterpart of an early Saalian ice advance in Western Europe, which was stratigraphically-traced eastwards across Poland and the Ukraine (Yakovlev, 1956). In the 1980s it was discovered that the till of the southernmost lobe of the Middle Pleistocene glaciation in Central Russia, unlike in the Ukraine, was overlain by interglacial sediments with fauna of the Tiraspol (Cromer) Complex and by 4-5 palaeosols of interglacial character. Accordingly, this so-called Don glaciation, presumably *c.* 0.5-0.6 ma old, could no longer be correlated with either the Saalian in Western Europe, the Dnieper glaciation in the Ukraine, or with the Samarovo glaciation of Siberia (Shik, 1995; Velichko, 1991). In new 1:1,000,000 maps the Pleistocene drift limit is related to the Don glaciation (Marine Isotope Stage 16 or 12, by different estimates), even in the extreme east of the Russian Plain (Chumakov *et al.,* 1999) (Fig. 1).

This age of the maximum glaciation is supported by the coring results around lake Chusovskoye on the Pechora-Volga interfluve (21 in Fig. 1), where two or three more tills occur beneath the superficial pre-Eemian till. Interglacial lacustrine sediments, with pollen of Likhvin (Holsteinian) type, have been found between the two upper tills of this area. The unique find of forest elephant *Palaeoloxodon sp.* is also related to this interglacial sequence (Stepanov, 1974). The uppermost till was previously related to the maximum glaciation of the East European Plain, presumably Saalian (Krasnov, 1971), which is currently split into two different stages - the

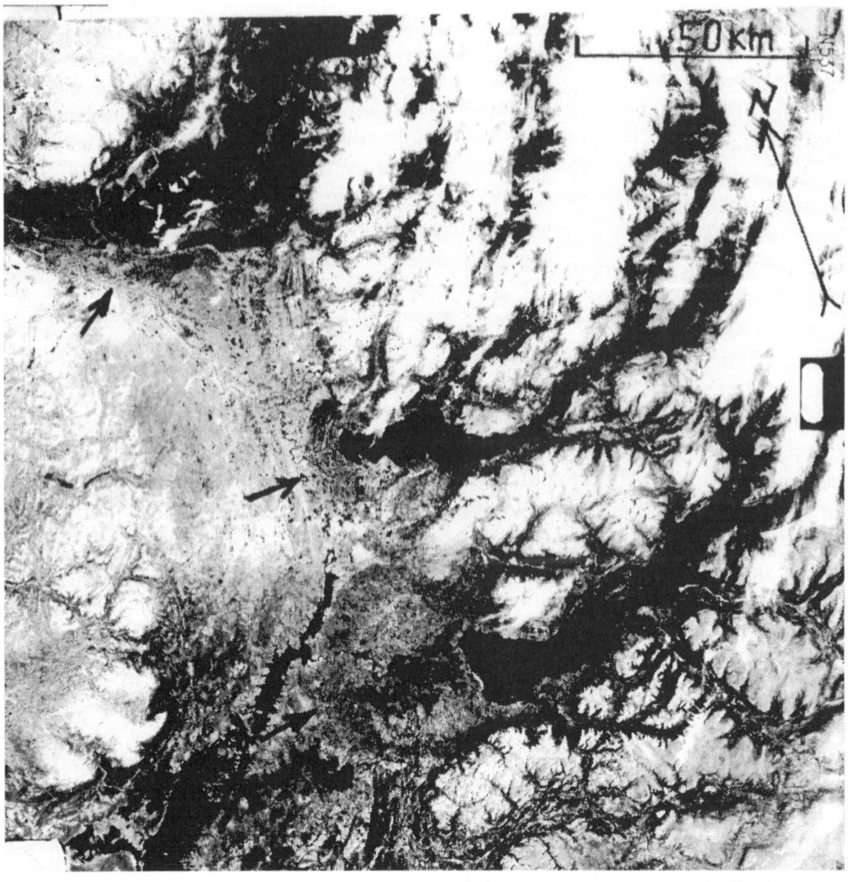

Fig. 2. Ice-pushed ridges in the Yenissei catchment (satellite images).

a. Horseshoe-shaped end moraines (marked by arrows) bounding fjord-like lakes in the western foothills of the Putorana Plateau, Norilsk moraines.

b. Two hill-hole pairs in the sedimentary basin west of the Yenissei, inside the Early Weichselian ice sheet, 60 km upglacier from of its western margin. The northern glaciotectonic ridge, according to Arkhipov et al., 1976 constructed of stony diamicton and cobbly sand, is 130 m above lake Makovskoye which is 65 m deep in the western part.

Dnieper (Saalian) and the Don (Cromerian) (Shik, 1995). It is likely that the new Pleistocene maximum (the Don ice advance) is represented by one of the pre-Holsteinian tills in the lake Chusovskoye area.

The Samarovo glacial maximum of West Siberia is still considered to be of Saalian age. This age is inferred from the underlying Tobol interglacial alluvium (Arkhipov, 1975), which contains forest fauna, shells of central-asiatic fresh-water mollusc *Corbicula tibetensis (fluminalis)*, shows normal magnetic polarity and yields thermoluminescence dates in the range 260 to 390 ka BP. A bone of the elk *Alces latifrons* has been found in sediments beneath till at Bakhtinsky Yar on the Yenissei (23 in Fig. 1). Overlying the Samarovo Till, only one horizon of interglacial soils and peats have been described. Beneath this till two pre-Holsteinian tills can be distinguished in the sedimentary sequences filling buried valleys, overdeepened to 200-300 b.s.l. (Zubakov, 1972; Arkhipov, 1989). These ancient tills have been found proximally to the Samarovo glaciation limit, rather close to it. A pre-Holsteinian glaciation may be the most extensive in the east of Central Siberia, as suggested by Bobkova (1985).

The lingering stratigraphic uncertainty is connected with the relationship between the Moscow till of Central Russia and Dnieper till in the Ukraine. Both are currently related to the Marine Isotope Stage (MIS) 6 by Russian geologists (Shik, 1995), but in the Ukraine the Dnieper till is commonly thought as belonging to an earlier Saalian substage, as suggested by several thermoluminescence dates *c.* 280 ka (Gozhik, 1995).

A similar problem occurs in Siberia where the less extensive Taz till is thought to represent MIS 6, whereas the maximum Samarovo glaciation is related to MIS 8, based on a few thermoluminescence dates (Arkhipov, 1989). Traditionally the Samarovo Stage was correlated with the Dnieper glaciation. Although, unlike in European Russia, both pre-Eemian tills are known in superposition, the independence of the Taz glaciation is questionable, because no unequivocal interglacial formations have ever been described from between this and underlying Samarovo glacial complex. In the present digital map the boundary of the Taz glaciation, commonly drawn along hummocky plateaus and morainic ridges of unknown age (e.g., Arkhipov *et al.*, 1976, 1986), is omitted as unreliable.

Today the only clue for the spatial differentiation betweeen Saalian and pre-Saalian ice sheets is glacial topography which is partly preserved in the Saalian area and totally absent in older glacial landscapes. However, this feeble tool can hardly be employed in the primordially flat lowlands of central West Siberia, where only rare drilling profiles, undertaken for mapping and geotechnical projects, help to interpolate the drift limit between key sections in river valleys. Immediately west of the Urals, the Saalian ice limits are very controversial in different survey maps. Thus the author has assigned the Saalian limit in the Volga-Pechora interfluve using available literature and personal experience in the Pechora Basin. In the extreme northeast of Central Siberia the pre-Eemian ice limits could not so far be defined for want of any reliable data.

Late Pleistocene ice limits

The first cartographic concepts of the last glaciation of northern Russia without the area of the Scandinavian glaciers were put forward by Sachs (1953) and Yakovlev (1956), who authored fundamental syntheses of previous investigations. They established the pattern after which

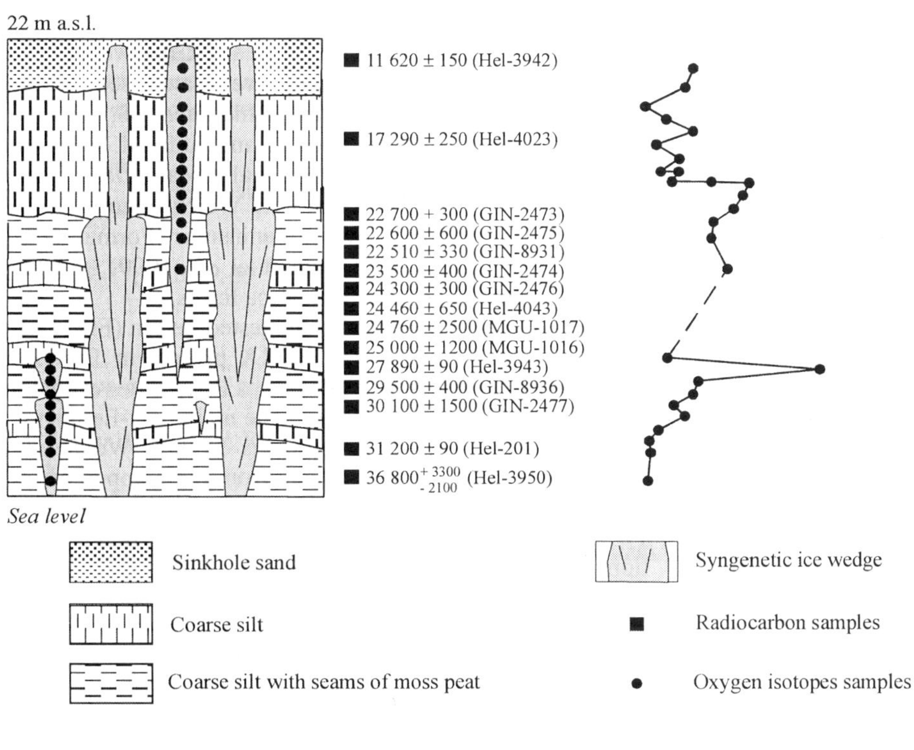

Fig. 3. Syo-Yakha section on the eastern coast of Yamal (11 in Fig. 1), after Vasilchuk et al. (2000).

Note the two generations of telescoped ice wedges that were growing simultaneously with accumulation of organic-rich silts, the latter of probable aeolian origin. No sign of any glacial activity has been found overlying this periglacial sequence. In other sections this formation is underlain by the latest till with blocks of fossil glacial ice.

nology and palaeogeography of the Pleistocene and Holocene of the northern Central Siberian Upland. *In:* Ivanova, I.K. & Kind, N.V. (eds), *Geokhronologia chetvertichnogo perioda*, 198-207, Moscow, Nauka (in Russian).

Bobkova, Z.S. (1985). *National Geological Map of the USSR, scale 1:1,000,000, new series, quadrangle Q-48,49 (Aikhal). Map of Quaternary deposits.* VSEGEI, Leningrad.

Bolikhovsky, V.F. (1987). Yedoma sediments in West Siberia. *In:* Punning, J.-M. (ed.), *Novye dannye po geokhronologii chetvertichnogo perioda*, 128-136, Moscow, Nauka (in Russian).

Chumakov, O.Ye., Shik, S.M. & Kirikov, V.P. (1999). *National Geological Map of the Russian Federation, scale 1:1,000,000, new series, quadrangle O-38,39 (Kirov). Map of Quaternary formations.* St. Petersburg, VSEGEI.

Fisher E.L., Leonov, B.N., Nikolskaya, M.V., Petrov, O.M., Ratsko, A.P., Sulerzhitsky, L.D. & Cherkasova, M.N. (1990). The Late Pleistocene of the central North Siberian Lowland. *Izvestiya, Academy of Sciences of USSR, Seria geograficheskaya,* **6,** 109-118 (in Russian).

Forman, S.L., Ingólfsson, O., Gataullin, V., Manley, W.F. & Lokrantz, H. (1999). Late Quaternary stratigraphy of western Yamal Peninsula, Russia: new constraints on the configuration of the Eurasian ice sheet. *Geology,* **27 (9)**, 807-810.

Ganeshin, G.S., (ed.) (1973). *Map of Quaternary deposits of the USSR, scale 1:2,500*,000. Leningrad, VSEGEI.

Gataullin, V. & Forman, S. (1997). Stratigraphy of Upper Quaternary sediments in western Yamal as a key for last glacial palaeogeography reconstructions. Abstracts of International conference: The problems of Earth cryosphere, Pushchino, 248-249.

Gataulllin, V., Mangerud, J., Svendsen, J.I. (2001). The extent of the Late Weichselian ice sheet in the southeastern Barents Sea. *Global and Planetary Change,* **31 (1-4),** 451-472.

Gozhik, P.F. (1995). Glacial history of the Ukraine. *In:* Ehlers J., Kozarski, S. & Gibbard, P.L. (eds), *Glacial deposits in north-east Europe*. Rotterdam, Balkema, 213-215.

Grosswald, M.G. (1980). Late Weichselian ice sheets of Northern Eurasia. *Quaternary Research,* **13 (1),** 1-32.

Grosswald, M.G. (1993). Extent and melting history of the Late Weichselian ice sheet, the Barents-Kara continental margin. *In:* Peltier, W.R., (ed.), *Ice in the climate system. NATO ASI series,* **112,** 1-20. Berlin, Springer Verlag.

Heintz, A.E. & Garutt, V.Ye. (1964). Determinations of absolute age of fossil mammoth and woolly rhinoceros remains from Siberian permafrost by radiocarbon. *Doklady, Academy of Sciences of USSR,* **154 (6),** 1367-1370 (in Russian).

Houmark-Nielsen, M., Demidov, I., Funder, S., Grøsfjeld, K., Kjær, K., Larsen, E., Lavrova, N., Lyså A. & Nielsen, J.K. (2001). Early and middle Valdaian glaciations, ice dammed lakes and periglacial interstadials in northwest Russia: new evidence from the Pyoza River area. *Global and Planetary Change,* **31 (1-4),** 215-237.

Isayeva, L.L. (1984). Late Pleistocene glaciation of North-Central Siberia. *In:* Velichko, A.A. (ed.), *Late Quaternary Environments of the Soviet Union*, 21-30, Minneapolis, University of Minnesota Press.

Isayeva, L.L., Kind, N.V., Kraush, M.A. & Sulerzhitsky, L.D. (1976). On age and structure of marginal formations in the northern foothills of the Putorana Plateau. *Bulleten Komissii po izucheniyu chetvertichnogo perioda, Academy of Sciences of USSR,* **45,** 117-123 (in Russian).

Kaplanskaya, F.A. & Tarnogradskiy, V.D. (1986). Remnants of the Pleistocene ice sheets in the permafrost zone as an object for palaeoglaciological research. *Polar Geography and Geology,* **10 (4),** 65-72.

Kaplyanskaya, F.A. & Tarnogradsky, V.D. (1974). *The Middle and Lower Pleistocene of the Lower Irtysh area* (Sredny i nizhny pleistotsen nizovyev Irtysha). Leningrad, Nedra, 160 pp. (in Russian).

Kind, N.V. (1974). *Late Quaternary geochronology according to isotopes data* (Geokhronologia pozdniego antropogena po izotopnym dannym). Moscow, Nauka, 255 pp. (in Russian).

Kind, N.V. & Leonov, B.N. (eds) 1982: *The Anthropogene of Taimyr* (Antropogen Taimyra). Moscow, Nauka, 184 pp. (in Russian).

Krasnov, I.I. (ed.) (1971). *Map of the Quaternary of the European USSR and adjacent territories, scale 1:1,500,000.* Leningrad, VSEGEI, 16 sheets.

Lavrov, A.S. (1974). Late Pleistocene ice sheets of the northeast of the European USSR. *Bulleten Komissii po izucheniyu chetvertichnogo perioda, Academy of Sciences of USSR,* **41,** Moscow, 48-55 (in Russian).

Lavrushin, Yu.A., Chistyakova, I.A., Gaidamanchuk, A.S., Golubev, Yu.K. & Vasilyev, V.P. (1989). The structure and composition of the glacial palaeoshelf sediments in Bolshezemelskaya Tundra. *In:* Anonymous, *Litologia kainozoiskikh shelfovykh otlozheniy*, 3-51, Moscow, Geological Institute of Academy Sciences of USSR (in Russian).

Lazukov, G.I. (1970). *The Anthropogene of the northern half of West Siberia, stratigraphy* (Antropogen severnoi poloviny Zapadnoi Sibiri, stratigrafia). Moscow University, 322 pp. (in Russian).

Loseva, E.I. & Duryagina, D.A. (1983). Palaeobotanic evidence for stratigraphy of Cenozoic sediments of the central Pai-Hoi. *Trudy, Institute of Geology, Komi Branch of Academy of Sciences of USSR*, **43,** Syktyvkar, 56-68 (in Russian).

Mangerud, J., Astakhov, V.I., Murray, A. & Svendsen, J.-I. (2001). The chronology of a large ice-dammed lake and the Barents-Kara Ice Sheet advances, Northern Russia. *Global and Planetary Change,* **31 (1-4),** 319-334.

Mangerud, J., Svendsen, J.I. & Astakhov, V.I. (1999). Age and extent of the Barents and Kara Sea ice sheets in Northern Russia. *Boreas,* **28 (1),** 46-80.

Möller, P., Bolshiyanov, D. & Bergsten, H. (1999). Weichselian geology and palaeoenvironmental history of the central Taimyr Peninsula, Siberia, indicating no glaciation during the last global glacial maximum. *Boreas,* **28 (1),** 92-114.

Potapenko, L.M. (1998). *National Geological Map of the Russian Federation, scale 1:1,000,000, new series, quadrangle P-38,39 (Syktyvkar). Map of Quaternary formations.* St.Petersburg, VSEGEI.

Rudenko, T.A., Astakhov, V.I. & Fainer, Yu.B. (1981). The geomorphology. *In:* Kovrigina, Ye.K., (ed.), *National Geological Map of the USSR, scale 1:1,000,000, new series. Explanatory note to quadrangle P-46,47 (Baikit).* Leningrad, VSEGEI (in Russian).

Rudenko, T.A., Faincr, Yu.B. & Fainer T.G. (1984). *National Geological Map of the USSR, scale 1:1,000,000, new series, quadrangle P-48,49 (Vanavara). Map of Quaternary deposits.* Leningrad, VSEGEI.

Sachs, V.N. (1953). *The Quaternary Period in the Soviet* Arctic (Chetvertichny period v Sovietskoi Arktike). Leningrad-Moscow, Institute of Geology of the Arctic, 627 pp. (in Russian).

Shik, S.M. (1995). Pleistocene palaeogeographical events in Europe and problems of their correlation with the events near the Black Sea and in the ocean. *In:* Svitoch, A.A., (ed.), *Korrelatsia palaeogeograficheskikh sobytiy: materik - shelf - okean.* Moscow University, 33-38 (in Russian).

Siegert, C., Derevyagin, A.Y., Shilova, G.N., Hermichen, W.D. & Hiller, A. (1999). Palaeoclimatic indicators from permafrost sequences in the eastern Taymyr Lowland. *In:* Kassens, H., Bauch, H., Dmitrenko, I.A., Eicken, H., Hubberten, H.-W., Melles, M., Thiede, J. & Timokhov, L.A. (eds). *Land-Ocean Systems in the Siberian Arctic.* Berlin-Heidelberg, Springer, 477-499.

Stepanov, A.N. (1974). Stratigraphy and sedimentary environments of the Upper Cenozoic of the Pechora-Kama interfluve. Resume of PhD thesis. Geological Faculty of Moscow University, 34 pp. (in Russian).

Strelkov, S.A. & Troitsky, S.L. (1953). Geological structure and topography of the Bolshaya Heta-Messo interfluve and the Solyonaya catchment. *Trudy, Institute of Geology of the Arctic,* **60,** 5-55 (in Russian).

Sukhorukova, S.S. (1998). The key sequence of marine sediments and climate fluctuations of the Kazantsevo (Eemian) interglacial (northern Siberia). *Geologia i geofizika,* **39 (1),** 74-84, Novosibirsk, Nauka, (in Russian).

Troitsky, S.L. (1966). *Quaternary deposits and landscape of the low coastlands of the Yenissei Estuary and adjacent Byrranga Mountains* (Chetvertichnye otlozheniya i relief ravniinykh poberezhiy Yeniseiskogo zaliva i prilegayushchikh chastei gor Byrranga). Moscow, Nauka, 207 pp. (in Russian).

Troitsky, S.L. (1975). *The Modern Antiglacialism: a critical essay* (Sovremennyi Antiglatsialism). Moscow, Nauka, 164 pp. (in Russian).

Vasilchuk, Yu.K., Serova, A.K. & Trofimov, V.T. (1984). New data on the sedimentary environments of the Karginsky Formation in the north of West Siberia. *Bulleten Komissii po izucheniyu chetvertichnogo perioda, Academy of Sciences of USSR,* **53,** 28-35 (in Russian).

Vasilchuk, Yu.K., van der Plicht, J.,Vasilchuk, A.K., Jungner, H. & Sonninen, E. (2000). First radiocarbon datings on Late Pleistocene syngenetic ice wedges. *Doklady, Russian Academy of Sciences,* **371 (1),** 114-117 (in Russian).

Velichko, A.A. (1991). Loess-palaeosol formation of the Russian Plain. *Quaternary International,* **7/8,** 103-114.

Yakovlev, S.A. (1950). *Map of Quaternary deposits of the European USSR and the adjacent regions, scale 1:2,500,000.* Leningrad, VSEGEI, 10 sheets.

Yakovlev, S.A. (1956). *The fundamentals of the Quaternary geology of the Russian Plain* (Osnovy geologii chetvertichnykh otlozheniy Russkoi ravniny). Moscow, Gosgeoltekhizdat, 314 pp. (in Russian).

Zarrina, Ye.P., Krasnov, I.I. & Tarnogradsky, V.D. (1961). *Map of the Quaternary deposits of the West Siberian Lowland, scale 1:1,500,000.* Moscow, Gostoptekhizdat, 6 sheets.

Zemtsov, A.A. (1976). *Geomorphology of the West Siberian Plain* (Geomorfologia Zapadno-Sibirskoi Ravniny). Tomsk University, 344 pp. (in Russian).

Zubakov, V.A. (1972). *Recent sediments of the West Siberian Lowland* (Noveishie otlozhenia Zapadno-Sibirskoi nizmennosti). Leningrad, Nedra, 312 pp. (in Russian).

Quaternary Glaciations – Extent and Chronology
Editors J. Ehlers and P.L. Gibbard

Valdaian glacial maxima in the Arkhangelsk district of northwestern Russia

Igor N. Demidov [1], Michael Houmark-Nielsen [2], Kurt H. Kjær [3], Eiliv Larsen [5], Astrid Lyså [5], Svend Funder [4], Juha-Pekka Lunkka [6] and Matti Saarnisto [6]

[1] *Russian Academy of Sciences, Karelian Research Centre, Institute of Geology. Pushkinskaya Street 11, Petrozavodsk, Russia, 185610, demidov@krc.karelia.ru*
[2] *Geological Institute, University of Copenhagen, Øster Voldgade 10, DK-1350, Copenhagen K, Denmark.*
[3] *Department of Geology, Lund University, Solvegaten 13, SE-223 62 Lund, Sweden*
[4] *Geological Museum, University of Copenhagen, Østervoldgade 5-7, DK-1350, Copenhagen K, Denmark.*
[5] *Geological Survey of Norway, Leiv Eiriksson vei 39, N-7491 Trondheim, Norway*
[6] *Geological Survey of Finland, Betonimiehenkuja 4, 02150 Espoo, Finland*

Abstract

The marginal configurations and ages of the Valdaian (Weichselian) glacial maxima in northern Russia have hitherto not been well established, thus, numerous versions of the Last Glacial Maximum (LGM) positions of the Scandinavian Ice Sheet are known (Fig.1B). New data on ice sheet growth and decay indicate at least three glacial maxima during the Late Pleistocene each with individual spreading centres and different ages. Results of modern investigations in the Arkhangelsk region, conducted by the authors, are compared with an analysis of previous data on the Late Quaternary geology of Northwest Russia, which comprises a thorough presentation of Russian literature on this subject. This has allowed us to question and revise former models on Valdaian glaciation history and ice-marginal positions in a major part of the Russian North.

An Early-Middle Valdaian (*c.* 70 ka BP) glaciation from the east and southeast, possibly originating on the Timan ridge, crossed the Pyoza River basin and reached the White Sea coast along the Bay of Mezen. The southern terminus of the ice sheet is probably found along parts of the Mezen River (Fig.2 A,B). The presence of an Early-Middle Valdaian Scandinavian glaciation, which covered the Arkhangelsk region and in neighbouring areas of Karelia and Vologda is not supported by geological data. In the Middle Valdaian (*c.* 60 ka BP) an ice sheet from the Barents-Kara Sea flowed from the north, northeast and reached the lower Pyoza River and the south and western shores of Mezen Bay on the White Sea coast. The terminal formations of its maximal stage stretch from west to east just north of Pyoza River and then run marginal to the Timan ridge from the north joining with the Markhida end-moraines on the Pechora Lowland (Fig.2 A,B).

During the Late Valdaian, the Scandinavian Ice Sheet occupied the northwestern part of the Arkhangelsk region around 19-17 ka BP. The limit of this Late Valdaian glacial maximum runs from the White Sea shore of the Kanin Peninsula in the north, along the Kuloi River south of Mezen to the Middle Pinega River, crossing the rivers Severnaya Dvina and Vaga near the villages of Cherevkovo and Ust-Padenga. The glacial boundary bordered the Melovian and Nyandoma high ground and continued southwestwards to Lake Kubenskoe in the Vologda region. The Pyoza, Mezen and Vashka river basins remained ice-free during the Late Valdaian time, this area being covered by fluvial flood plains, with abundant evidence of permafrost and lakes. The latter have yielded pollen evidence indicating an arctic to subarctic environment between 18-10 ka BP (Fig. 2 A,B).

Introduction

During the last decade much new information about the timing and extent of the Late Pleistocene Eurasian ice sheets has been obtained. The age and position of the consecutive Scandinavian ice sheets during the Valdaian (Weichselian) glaciation is fairly well known in north-western Europe compared to that in eastern Europe especially northern Russia (cf. Mangerud *et al.*, 1999). Many questions concerning ice sheet configurations, ages and spreading centres since the Mikulinian (Eemian) interglacial still remains to be solved, however, new achievements on the stratigraphy and glacier-dynamics in northern Russia are now available (Larsen *et al.*, 1999b; Houmark-Nielsen *et al.*, 2001; Kjær *et al.*, 2001; Lyså *et al.*, 2001). Against this background it is important to present an outline of classical and more recent literature on the Valdaian glaciation in northern Russia especially the Arkhangelsk region. Since the vast majority of the literature cited is in the Russian language, which is not easily accessible to the majority of Quaternary geologists in Western Europe, this article contains an extensive review of previous Russian publications.

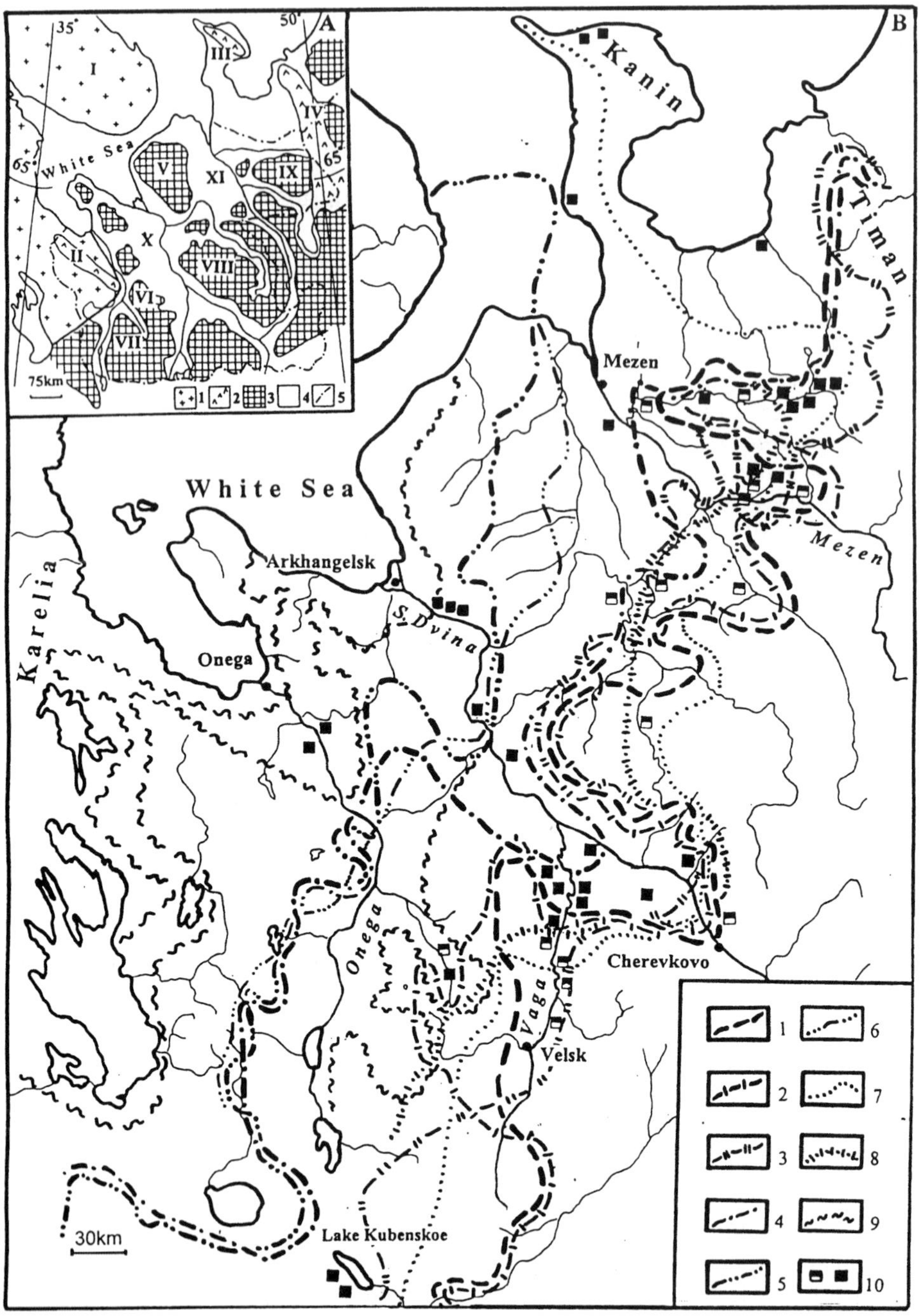

Fig. 1. Main geological-geomorphological setting (A) and main variants of glacial limits (B) in Arkhangelsk district and adjacent areas (compiled by I. Demidov).

A: 1- Archean and Proterozoic largely metamorphosed igneous, volcanic and sedimentary rocks , 2 - lower mountains are composed by Proterozoic or Palaeozoic igneous, volcanic and sedimentary rocks, 3 - water divide plateaux are composed of Palaeozoic sedimentary rocks, 4 - lowlands are composed of Palaeozoic and Mesozoic sedimentary rocks, 5 - administrative boundary of Arkhangelsk district. I - Fennoscandian Shield, II - Vetreny Poyas ridge (Windy Belt), III - Kanin Peninsula, IV - Timan Ridge, V - Belomorian-Kuloi plateau, VI - Melovian uplift, VII - Nyandoma - Konosha uplift, VIII - Pokshenga uplift, IX - Pyoza-Mezen waterdivide, X - Dvina lowland, XI - Mezen lowland.

B: LGM positions .

Early Weichselian:

1 - Krasnov, 1974.

Late Weichselian:

2 - Aseev, 1971,

3 - Yakovlev,1956,

4 - Devyatova,1969,

5 - Ganeshin et al, 1980,

6 - Apukhtin & Krasnov, 1966, Krasnov, 1974, Legkova, 1967,1972,

7 - Stadial (Older Dryas) moraines by Chebotareva,1977.

Key sections of Mikulinian or Early-Middle Valdai interglacial sediments unburied (8) and buried by till (9), 10 – Detail area showed on Fig. 4.

In the Arkhangelsk region, as in other parts of Russia, the outer limits of Pleistocene glaciations have chiefly been defined using geomorphological analyses. Areas of terminal formations - hummocky and end moraines, sandurs and kames – were pieced together into former ice-marginal positions. The development of the lake and river systems was also taken into account. So far, over 30 versions of the LGM position in northern Russia are known (Gey & Malakhovsky, 1998) (Fig. 1.B).

Although glacial deposits formed during the Valdaian have been investigated for over a century, the position and age of the maximum boundaries of ice sheets have not been well documented. This has happened because the areas of interest in North Russia are poorly accessible, and also because the lack of good stratigraphical control prevented reliable conclusions. For example, the geomorphology of older landforms formed during the Moscovian glaciation were often taken as ice-marginal formations formed in the Valdaian glaciation maximum. Moreover, younger lacustrine or fluvial sediments, especially in river valleys, often overprint glacial landscapes. In many areas, the formation of dead-ice landforms during deglaciation has resulted in complex and indistinct marginal zones adding further difficulties to the recognition the maximum position of former ice sheets. It is therefore difficult to trace ice-marginal belts using only aerial photographs and satellite images, even though this has been attempted with variable success (Punkari, 1985; Astakov *et al.,* 1999).

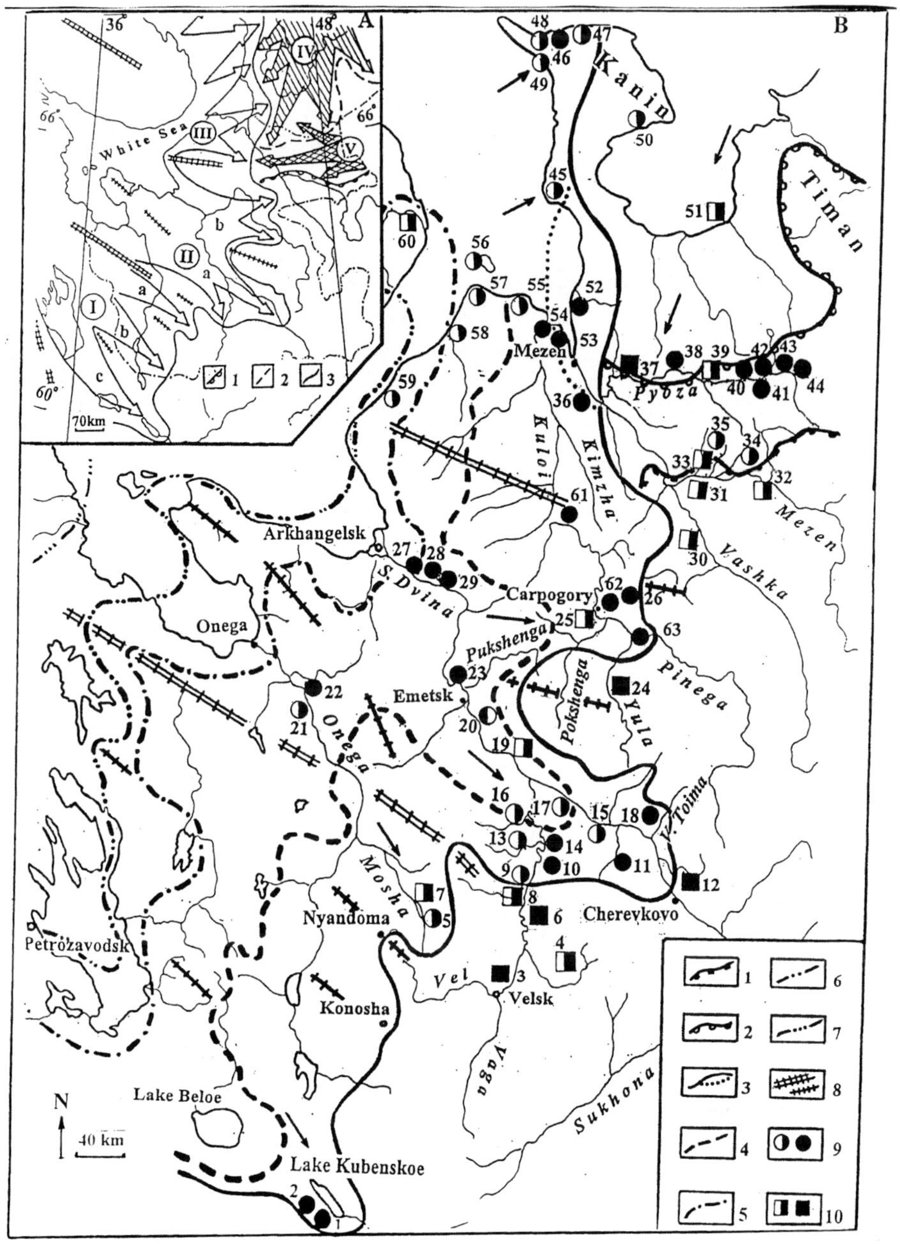

Fig. 2. Main glacial maximum positions of Upper Pleistocene glaciations in the Arkhangelsk district and adjacent areas (A) and current reconstruction of the deglacial stages for the retreat of the Scandinavian Ice Sheet at Late Valdai time (B).

A: Ice flows and tongues of Scandinavian Ice Sheet: I - Karelian ice lobe, a – Mosha ice tongue, b – Lacha ice tongue, c - Lake Onega ice tongue, II - Belomorian (White Sea) ice lobe, a – Dvina ice tongue, b – Pinega ice tongue, III - Kuloi-Mezen ice lobe; IV – Kara/Barents ice sheet, V – Timan (?) ice cap.

B: 1 – assumed maximum position of the southern flank of the Timan ice cap in Early-Middle Valdai time, 2 - maximum position of the southern flank of the Kara/Barents Sea ice sheet in Middle Valdai time, 3 – glacial maximum of the Scandinavian Ice Sheet in Late Valdai time. Deglacial phases of the last Scandinavian ice-sheet: 4 – Vepsovo Phase, 5 – Old Dryas (Luga) Phase, 6 – Middle Dryas (Neva) Phase, 7 - Younger Dryas (Salpausselkä I). 8 - ice-divides, 9 - sections with Mikulino or Middle Valdai interglacial sediments covered by till, 10 - sections with Mikulino or Middle Valdai interglacial sediments unburied by till. Arrows indicate ice flow directions.

Numbers and names of key sections (with ^{14}C, TL or ESR dates):*

1 - Puchka,*
2 - Irkhino,*
3 - Ustya (boring),
4 - Pasva,*
5 - Voiozero,
6 - Koleshka,*
7 - Iksa,
8 - Ust-Padenga,
9 - Osinovskaya,
10 - Raibola,*
11 - Yumizh,*
*12 - Erga**
13 - Led,
14 - Smotrakovka,*
15 - Boltinskaya,
16 - Tarnya,
17 - Konovalovskaya,
18 - Tomasha,*
19 - Krasnaya Gorka,
20 - Lipovic,
21 - Somba,
22 - Verkhnyaya Tyolza,*
23 - Chelmokhta,*
24 - Yula,*
25 - Shilega,
26 - Yezhuga,*
27 - Psaryovo,*
28 - Trepuzovo,*
29 - Bobrovo,*
30 - Olema,
31 - Tseb,yuga I,
32 - Kyma,
33-Tseb,yuga II,
34 - Ona (boring),
35 - Tseb,yuga III,
36 - Zaton,*
37 - Bych,e,*
38 - Viryuga,
39 - Jazevets, Orlovets,
40 - Yolkino,*
41 - Safonovo,
42 - Ust-Varchushka,
43 - Burdui,
44 - Bludnaya,
45 - Konushinskaya Korga,
46 - Krynka,
47 - Madakha,
48 - Tobuev,
49 -Tarkhnov,
50 - Oiva (Zhemchuzhnaya),
51 - Pyosha,
52 - Syomzha,*
53 - Tolstic,*
54 - Kargovsky,*
55 - Abramovsky,
56 - Morzhovets,
57 - Voronov,
58 - Megra,
59 - Tova,
60 - Ponoi,
61 - Kulogora,*
62 - Mezhdurechensky,*
63 - Verkola,*
64 - Kirillov.*

Fig. 3. Huge sections of Upper Pleistocene deposits along the Pyoza River (NE part of Arkhangelsk area). Middle-Early Weichselian till (grey) of the Kara or Timan ice sheet rests on Mikulino and Early Weichselian sediments and is covered by Late Weichselian fluvial sediments (yellow) with traces of cryogenesis. There is no Late Weichselian till present.

As in Western Europe, the coupling of morphology and stratigraphy did not necessarily clarify understanding of the timing and extent of Valdaian glaciations. Stratigraphical examination of numerous cross-sections along the banks of the rivers Severnaya Dvina, Vaga, Pyoza, Onega and their tributaries, and sections along the White Sea and Barents Sea coastlines have been undertaken since the beginning of the early 20th century. Before the availability of reliable dating methods, such as radiocarbon and luminescence, stratigraphical subdivisions of the Pleistocene successions, including interglacial, interstadial and glacial strata were based on a simple 'counting-from-the-top' method. The inaccurate interpretations of diamict deposits and sedimentary successions and palaeo-ecological data has been substantially affected by theoretical concepts on the evolution of the Pleistocene. It has therefore been essential to revise and re-interpret the results of previous studies. Thus, this article presents the history, age and extents of Valdaian glaciations in the Arkhangelsk region, based on a critical review of previously published data and on new results from investigations in northeastern part of the study area.

Fig. 4. Tall cliffs along the Severnaya Dvina River with tills of the Moscovian and Late Weichselian glaciations, separated by Mikulinian and Middle Weichselian marine and fluvial sediments. Section near Bobrovo village.

Diamict beds that rest on marine sediments in the Severnaya Dvina River basin were recognised in the late 19th century by Vollosovich (1900). On the lower Mezen River and on the Kanin Peninsula, Ramsay (1911) recognised three till units separated by marine deposits. Based on the composition and provenance of boulders in these tills, Ramsay concluded that, in addition to the Scandinavian glaciation centre, a second centre existed on Novaya Zemlya. Kalyanov & Androsova (1933), Androsova (1938), Gorbatsky (1932) and Spridonov & Yakovleva (1961) continued research on the Kanin Peninsula. Tolstikhin & Tolstikhina (1935), Barkhatova (1941), Lukoyanov (1941), Devyatova (1961), Evzerov *et al.* (1976) Ekman *et al.* (1974) and Ekman & Iljin (1995) investigated the Onega River basin. The Severnaya Dvina and Vaga basins were studied by Likharev (1933), Zhuze & Poretsky (1937), Pokrovskaya (1937), Lavrova (1937), Legkova & Schukin (1972), Pleshivtseva (1977), Atlasov *et al.* (1978), Ostanin *et al.* (1979), Smirnova (1991), Devyatova & Punning (1976), Devyatova *et al* (1981), Devyatova (1982), Liivrand (1981), Lavrushin & Spiridonova (1995) and Larsen *et al.* (1999b). The Mezen and Pyoza rivers areas were investigated by Korchagin (1937), Malakhov (1934, 1939, 1940), Kalberg (1968), Zekkel (1939), Devyatova & Loseva (1964), Lavrov (1968, 1973, 1974, 1975, 1977) and Filippov & Borodai (1987). The evidence from the Quarternary superficial deposits and the position of different generations of marginal formations recognized by the early 1930s were generalised and published by Yakovlev (1932), Yaunputnin (1934), Gerasimov & Markov (1939) and Sokolov (1946). Yakovlev (1956), Apukhtin & Krasnov (1967), Devyatova (1969), Chebotareva & Makarycheva (1974), Chebotareva (1977), Faustova & Velichko (1992), Grossvald (1980) and Velichko (1997) presented regional summary papers, including the latest information, while Apukhtin & Krasnov (1967), Krasnov *et al.* (1974), Ganeshin *et al.* (1980) and Sergeev *et al.* (1981) contributed to maps showing the Quaternary geology and geomorphology of the USSR, Eurasia and Northwest of the USSR some of which formed part of an international map series under the auspices of INQUA.

Setting

The Arkhangelsk region, in the northern part of the Russian Plain, covers an area of 587 000 km^2, including the polar islands and Nenenian district (Fig. 1A, 2A). Tundra landscapes dominate the subarctic ocean coast grade southwards into north- and middle-taiga forests. According *to The Geology of the USSR* (1963), the north-western part of the region occupies the eastern Fennoscandian Shield, that is composed of highly metamorphosed Precambrian rocks that range up to 120-150 m. a.s.l. On the Vetreny

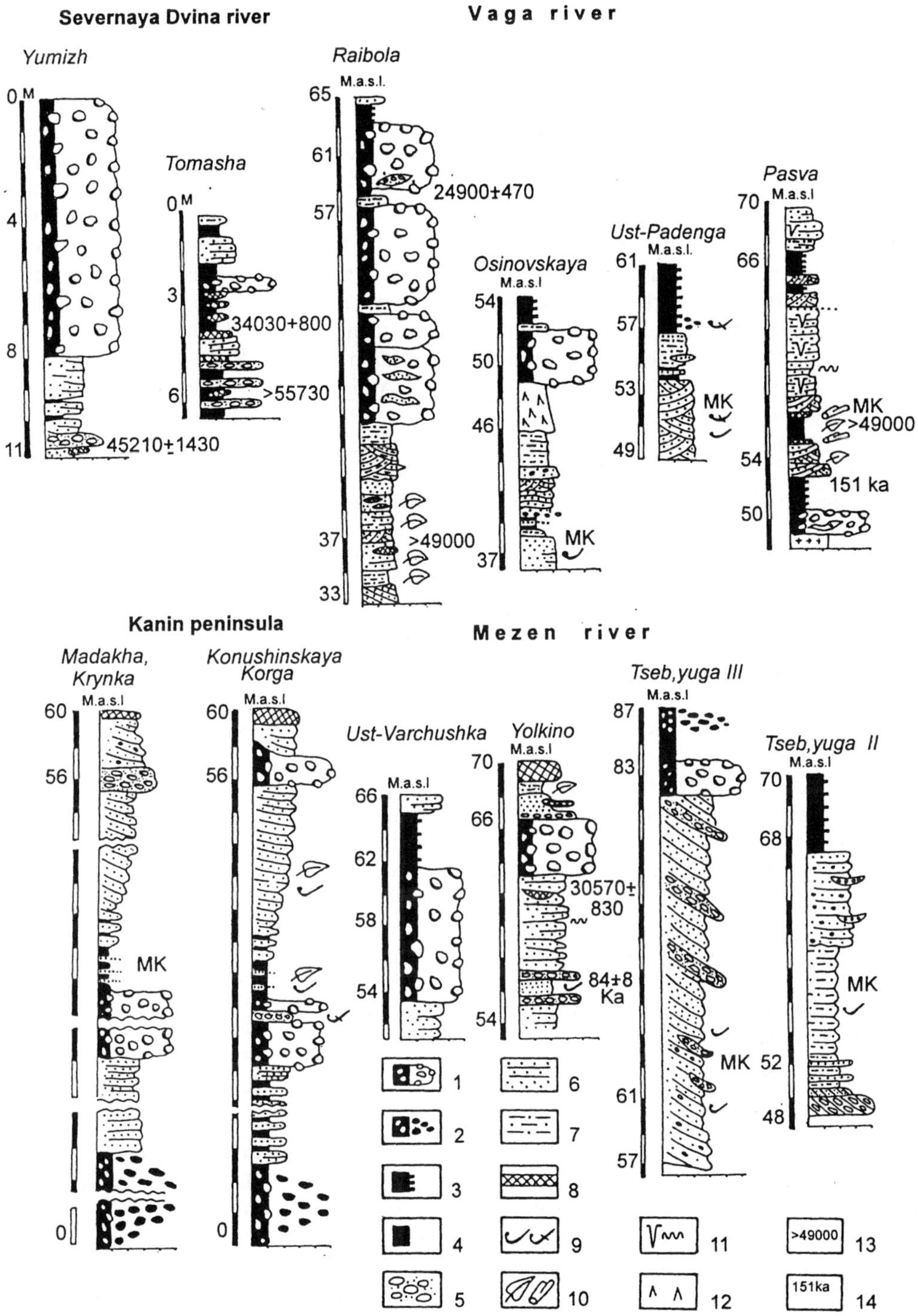

Fig. 5. Key sections (compiled by I. Demidov from the following data: Yumizh and Tomasha - by Arslanov et al. 1984, Krynka, Madakha, Konyshinskaya Korga- by Spiridonov & Yakovleva ,1961, Yolkino - Devyatova, Loseva,1964, Molodkov,1989, Lavrov,1977, Tseb,yuga, Ust-Varchushka - Devyatova, Loseva, 1964, Raibola, Osinovskaya, Ust-Padenga- by Devyatova,1982, Devyatova et al, 1976, 1981):

1-till,

2-mud with clasts (till?),

3-laminated clay,

4-massive clay,

5- gravel,

6- sand,

7-sandy loam, loam,

8-peat,

9- shells and shell fragments,

10- plant remains and twigs,

11- ice wedge and cryoturbations,

12 - colluvial deposits,

13 - ^{14}C age,

14 - TL and OSR age.

MK- Mikulinian sediments.

Poyas (Wind Belt) Ridge, altitudes may reach as high as 320 m. The Timan Ridge is formed of Proterozoic and Palaeozoic magmatic and metamorphic rocks and located in the northeastern part of the region. The maximum altitude here is 250-400 m. The larger part of the region that lies between the Fennoscandian Shield and the Timan Ridge is a rugged, denudation plain consisting of Palaeozoic and, less commonly, Mesozoic sediments (Fig. 1A). Extensive depressions (0-100 m a.s.l.) and large water-divide plateaux (100-250 m a.s.l.) dominate the bedrock relief, just as the valleys of major northward-flowing rivers, such as Severnaya Dvina, Onega and Mezen, are confined to depressions in the present topography.

The Quaternary sequence is dominated by glacigenic sediments deposited mainly during the Moscovian and Valdaian glaciations (Fig. 5), and separated by interglacial Mikulinian and interstadial Middle Valdaian sediments (Apukhtin & Krasnov, 1967; Devyatova, 1969). Mikulinian marine deposits, which are widespread in the region, reach heights of up to 70 m a.s.l. They provide a solid marker bed in stratigraphic sequences. In the southern Arkhangelsk region, drill holes have penetrated strata formed during older, pre-Moscovian glaciations in the River Ustya area (Kotova *et al.* 1977).

Glacial maxima: Early-Middle Valdaian

Scandinavian Ice Sheet

The presence of a Scandinavian Ice Sheet in the Arkhangelsk region and in adjacent areas of Karelia and Vologda during the Early or Middle Valdaian is not supported by geological and geochronological evidence. Only one till bed is found overlying the Mikulinian sediments in Severnaya Dvina (Devyatova, 1982; Atlasov *et al.*, 1978; Ostanin *et al.*, 1979; Arslanov *et al.*, 1984; Larsen *et al.*, 1999b) and the Onega river basins (Barkhatova, 1941; Devyatova, 1961). A few boreholes have penetrated two diamict beds upon the Mikulinian sediments in the lower Vaga area. They are, however, situated inside a terminal belt of the Vepsovo stage (Figs 1B, 2B-4), and may represent readvance till beds of the Late Valdaian ice front (Atlasov *et al.*, 1978; Ostanin *et al.*, 1979). According to Devyatova (1961, 1982), the Mikulinian marine deposits are conformably overlain by Early Valdai lacustrine sediments in boreholes at Shegovary on the lower Vaga River, about 30 km north of Shenkursk (Fig. 6). The pollen analyses by Devyatova (1961, 1982) indicate that Early Valdaian glaciolacustrine laminated clays were deposited in the River Pinega area. However, their age and possible relation to a Scandinavian or Barents-Kara Sea ice sheet is not clearly understood.

It has been suggested (Apukhtin & Krasnov, 1967; Legkova, 1967) that that the Scandinavian Ice Sheet was larger during the Early Valdaian than in Late Valdaian and that it extended eastwards to the rivers Mezen and Pyoza and southeast to the Sukhona River (Fig. 2). These reconstructions were based on geomorphological observations alone and by delineating areas of hummocky moraine landscape and terraces of glaciolacustrine basins in river valleys. A critical examination of Devyatova's (1972) and Lukashov's (1982) evidence of the presence of a till unit between the Mikulinian marine strata and the Middle Valdaian lacustrine sediments in the Petrozavodsk borehole procludes this possibility. To sum up, no sections or boreholes containing Early Valdaian till of Scandinavian origin are known from northwest Russia. Thus, many authors have interpreted end moraines of the Moscovian glaciation as ice-marginal formations of Late Valdaian age (Apukhtin & Krasnov, 1967; Legkova, 1967).

Kara-Barents ice sheet and Timan ice cap

The stratigraphical position of tills in relation to interglacial strata and the age of inter-till deposits interbedded with till sequences are important for determination of LGM (Last Glacial Maximum) position. However, the last ice-sheet maximum position and the Pleistocene palaeogeography of the Mezen-Pyoza River area are still not adequately investigated. According to Malakhov (1934) and Rudovitz (1947), deposits formed during two marine transgressions, separated by Moscow Glaciation till have been recognised in the area (Spiridonov & Yakovleva, 1961; Devyatova & Loseva, 1964; Legkova, 1967). Also, in a section near the village of Yolkino on the River Pyoza, sandy-gravelly marine sediments with shells were interpreted as representing the maximum shoreline of the pre-Late Moscovian Northern transgression (Odintsovo interglacial) (Fig. 5). Further up the River Pyoza, Odintsovo lacustrine and fluvial deposits underlie Moscovian till, whereas Mikulinian marine sediments are common downstream from Yolkino. Bylinsky (1980), who has revised his earlier palaeogeographic concepts of the Mezen basin (Bylinsky, 1962), believes that the sediments were deposited during two glaciations (Moscovian and Valdaian). He also concludes that they were separated by a two-phase Mikulinian transgression. A geochronological study of the Mikulinian key sections at Bychie and Zaton, together with the 'Odintsovo' section Yolkino using electron spin resonance (ESR), support deposition during the Upper Pleisticene, some 100-80 ka ago (Molodkov, 1989) (Fig. 2B-3). Mikulinian marine sediments, identified by their fauna and flora, are present in the banks of the rivers Irasa and Tsebyuga (Devyatova & Loseva, 1964) and proved by a borehole in the River Ona valley (Kalberg, 1968). These sediments are overlain by till deposits in the upper Tsebyuga River valley and locally by glaciolacustrine clay in its middle reaches (Fig. 5).

Three till units, separated by interglacial sediments, are known from the northern and western Kanin Peninsula coasts. The upper till contains boulders derived from the Kola Peninsula, whereas the lower and the middle till contain clasts from the Kara Sea area (Fig. 2B-3). On the eastern Kanin Peninsula, the boulders are only of Kara Sea provenance (Ramsay, 1911; Gorbatsky, 1932; Spiridonov & Yakovleva, 1961).

North of the River Pyoza, a topographically-distinct marginal moraine belt extends roughly from west to east (Fig. 2B). In the Upper Pyoza River, this belt bends northwards and continues across Lakes Varshskie and along the western slope of Timan. Furthermore, a few minor end moraine ridges are situated north of this terminal belt (Korchagin, 1937; Lavrov, 1991). Hilly, morainic topography in the Mezen-Pyoza interfluve occupies large areas in the Upper Tsema River, along the River Ona, and in the Middle Kyma River valley (Korchagin, 1937; Malakhov, 1939, 1940; Devyatova & Loseva, 1964). Fresh glacial landscape is present on the left bank of Mezen at Yuroma village. In addition, a wide, hilly morainic zone extends north-south from the Pyoza River mouth to Yezhuga, a tributary of Pinega River (Lavrov, 1991). These areas, with dense hummocky moraines and the general lack of stratigraphical information, have provided the source for many suggestions on the position of the last glaciation ice maximum limit (Figs 1B, 2B).

Based on this evidence data from the Mezen and Pyoza rivers and on the results of comprehensive litho- and biostratigraphical studies supplemented by a modern geochronology, a revised regional stratigraphy has developed in the past few years. The marine sediments in the Yolkino section are now interpreted as of Mikulinian interglacial age and the overlying diamict is assigned to the last glaciation. Therefore an older ^{14}C date from this site must be regarded at infinite (Fig. 5). Clast fabric studies and the petrography of clasts in tills overlying the Mikulinian sediments along the River Pyoza have shown, that the ice was moving from the east and south-east by the end of the Early Valdaian and from the north and north-east at the beginning of the Middle Valdaian (Houmark-Nielsen *et al.,* 2001, Kjær *et al.,* 2001) (Fig. 3). The two till units, separated by glaciofluvial sediments, occur above Mikulinian deposits in a section near the Viryuga River confluence with the lower Pyoza River. Thus, two separate ice sheets deposited the tills during the Early and Middle Valdaian and it is suggested here that the older ice sheet possibly originated on the Timan Ridge and the younger with more confidence originated in the Barents-Kara Sea.

Three till units, separated by marine and glaciolacustrine sediments, have been recognised in a section on Cape Tolstik on the southern shore of Mezen bay (Fig. 2B). Here clast orientation and the petrographic composition in the lower (Early Valdaian) till, which is enriched by Mikulinian marine shells, indicate ice movement from the east and southeast. Thus, the preliminary results suggest that during the Early Valdaian an ice sheet, possibly from the Timan Ridge, extended westwards over large parts of the northern half of the Arkhangelsk region. The position of the ice margin and the direction of ice movement along the Pyoza-Mezen water divide during the Early Valdaian is unclear, but a tentative position of the LGM is sketched in Fig. 2B-1). Its southern boundary probably stretched along the topographically-distinct hummocky land in the Upper Bludnaya River catchment, intersects the River Kyma about 20 km northwards from its mouth, and extends westwards along the River Ona, where a borehole encountered Mikulinian marine sediments underlying the till (Kalberg, 1968) (Fig. 2B). The boundary might continue westward across the Upper Tsebyuga River, where deposits formed during the Boreal Transgression (Mikulinian) are overlain by till (Devyatova & Loseva, 1964), and intersects the Mezen River near Yuroma village in a hilly-morainic zone. In the Upper Kimzha River catchment (Fig 2B), fragments of marginal formations are truncated by younger Scandinavian Ice Sheet end moraines. On the Ona-Mezen interfluve, near Tsenogora, the ice probably dammed the River Mezen. The terraces of the ice-dammed lake in the Upper Mezen River lie at around 145 m a.s.l., and terraces in the Middle Mezen and Vashka River valley are 15 m lower (Lavrov, 1968, 1975). However, there is no reliable evidence of their age. The southern limit of the Early Valdaian glaciation was probably controlled by Mezen-Vashka-Pinega water divide where the ground reaches to around 250 m a.s.l.

The early Middle Valdaian Barents-Kara Sea ice sheet must have extended westwards at least as far as the Kuloi Plateau, which is over 100 m. a.s.l. high. Otherwise, ice-dammed lakes situated in front of the Barents-Kara Sea ice sheet, and which covered extensive areas in rivers Mezen and Pyoza basins, would have lower run-off thresholds along the Mezen and Kuloi valleys. The Barents-Kara ice sheet probably also blocked the White Sea strait.

Considering the mode of occurrence of Middle Valdai deposits in the Severnaya Dvina basin, Devyatova (1982) assumes that the base level was at least 10 m higher during the Middle Valdaian mainly because the White Sea was isolated from the Sea by glaciation. It is known that the Middle Valdaian Barents-Kara Sea ice sheet covered at least the southern part of Mezen Bay and the lower reaches of the Pyoza River. Here till of Kara Sea provenance rests on Mikulinian and Early Valdai deposits, the ages of interglacial sediments have been determined using ESR and OSL (Molodkov, 1989; Houmark-Nielsen *et al.,* 2001). Maine, glaciolacustrine and fluvial sediments overlying and underlying the Kara till along the River Pyoza and along the Mezen Bay coast give luminescence ages of 10-18 ka and 44 - 67 ka respectively. The huge terminal moraine belt consisting of the Loban-Viryuga, Pyoza and Varshsky end-moraine ridges extends along the Pyoza River and the western slope of the Timan ridge, and marks the maximal southerly advance of Kara-Barents ice sheet (Fig. 2B-2). Several end-moraine ridges located on the Pyoza – Chyoshskaya Bay watershed record recessional stages of the Barents-Kara Sea ice sheet. Glaciolacustrine sandy and clayey sediments, overlying the Barents-Kara Sea ice sheet till are widespread along the Mezen and Chyoshskay bays.

Glacial maxima: Late Valdaian

There is no record of Barents-Kara Sea ice sheets on the Russian mainland in the Late Valdaian (Svendsen *et al.,* 1999, Houmark-Nielsen *et al.,* 2001; Kjær *et al.,* 2001). However, new information on the ages of the last, Scandinavian Ice Sheet expansion and decay in northwestern Russia, is now available (Astakhov *et al.,* 1999; Mangerud *et al.,* 1999; Larsen *et al.,* 1999a). Consistent ages for the last ice sheet maximum have been reported in Russia from only three areas; the Zapadnaya Dvina River basin near the Russian-Belorussian border (^{14}C 17-18 ka BP: Arslanov *et al.,* 1971), near Lake Kubenskoe in the Vologda region (^{14}C 21 ka BP: Arslanov *et al.,* 1970) and in the Arkhangelsk region, where the River Vaga join the River Padenga (17 ka BP, Larsen *et al.,* 1999b). An outline of the Late Valdaian evolution of the Scandinavian Ice Sheet for the Arkhangelsk region is given below.

Scandinavian Ice Sheet

In the Arkhangelsk region, the last, Scandinavian Ice Sheet formed distinct lobes that occupied pre-Valdaian

1984; Fig. 2B-3). On the River Yumizh, a left tributary of Dvina, in an outcrop located 18.4 km from its mouth and 3.2 km south of the village Nikolayevskoye, a peat lens from sand beneath till gives a ^{14}C age of 45 210±1430 (LU-1206). A lens of gyttja from deposits underlying till in the Tomasha River section is of 34030±810 years BP (LU-1257). Beyond the maximum ice-front position, in the Yerga River and in a section at the River Yula, peat layers underlying Late Valdaian lacustrine clay were dated at 43440 ±1460 (LU-1053) and 45000 ±1150 years BP (LU-1262). Here there is no Upper Valdaian till (Arslanov *et al.*, 1984).

Where absolute age estimates are obtained they all point to a Late Valdaian age of the last glacial maximum. For instance, the absolute dating (TL and ^{14}C) from the beds situated beneath Valdaian varved clay in the Koleshka and Pasva sections in the River Vaga yield ages from 31 to 62 ka BP (Devyatova *et al.*, 1976, 1981), and ages for sediments underlying till in river sections on the Yumizh and Tomasha (tributaries of the Dvina River) range from 34 to 45 ka BP (Arslanov *et al.*, 1984) (Fig. 2B-3). A radiocarbon date of 24900 ±470 BP (Vib-40) was obtained from the Raibola section, 10 km north of Shenkursk (Atlasov *et al.*, 1978, Ostanin *et al.*, 1979). New geochronological evidence from key sections on the rivers Vaga (Pasva, Koleshka, Raibola, Smotrakovka) and Severnaya Dvina (Bobrovo, Trepuzovo) from deposits above and below the till or proglacial varved clay have shown that glacial maximum in the Severnaya Dvina basin occurred about 17 ka BP and deglaciation may have started close to 15 ka BP (Larsen *et al.*, 1999b).

The area glaciated in Valdaian time is of 'fresh' glacial form with morainic hills, 10- 20 m high and up to 400 m wide and kames some 25-50 m high and 1-4 km in diameter. Subglacially-formed gullies are up to 5 km long (Atlasov *et al.*, 1978, Arslanov *et al.*, 1984). The maximum ice front position in the Vaga River valley and on the Vaga-Onega watershed has a well-defined festooned pattern (Figs 2 and 6). The Dvina ice tongue reached the Melovian upland (266 m a.s.l.) in the east and the Onega-Mosha tongue in the west, and occupied bedrock depressions near the Mosha River. The relative altitude of the Melovian upland above the surrounding plains is 160 m. In the western part of the Vaga depression a zone of glaciofluvial and glaciolacustrine sand-gravel and less abundant clay deposits *c.* 20 km wide, was formed distal to the ice-front position (Fig. 6). The topography in this zone is characterised by valley outwash trains that alternate with glaciolacustrine plains dissected by fluvial erosion. These sediments rest on Moskovian till or occasionally on Mikulinian deposits (Pasva and Koleshka sections) (Figs 5 and 6). Outside this zone of proglacial deposits lies a territory composed solely of Moscovian till with a characteristic relief subdued by erosion.

In the Vaga and Severnaya Dvina river catchments, the terminal formations of the Dvina tongue are rather indistinct. The end moraines are partially eroded and buried under lake sediments, deposited up to120-130 m. a.s.l.in front of the glacier. Erosional remnants of ice-marginal landforms are seen in aerial photographs as ridges and hills separated by sinkholes (Atlasov *et al.*, 1978, Ostanin *et al.*, 1979; Fig. 6). Terraces in this basin were formed at altitudes of 130 and 110 m a.s.l. on the distal side of the Severnaya Dvina ice-marginal formation. In the southern part of the Vaga basin, the absolute altitudes of terraces vary from 130 to 75 m, indicating a drop in the water level during the decay of the ice tongue.

In the Dvina and Vychegda river valleys, the position of two terminal belts has been mapped (Arslanov *et al.*, 1984). The outermost glacier position is represented in the lower levels of the Severnaya Dvina tributaries of Yumizh, Soiga, Verkhnyaya and Nizhnyaya Toima at altitudes of 60-90 m by a system of end-moraine ridges in the glaciolacustrine plain. In the central part of the Severnaya Dvina depression, end-moraine ridges form a series of convex arcs towards the east, and near the flanks of the depression, they are elongate resembling lateral moraines (Fig. 6). The ridges are 3-12 km long, 0.5 - 2 km wide and 5-10 m above the surroundings. On the Dvina-Pinega interfluve, the external terminal belt stretches NW and approximately N-S at 140-170 m, where they are 1-6 km long and 0.2-1.0 km wide. Inside the maximum position, a terminal belt formed by moraine and sand-gravel ridges in the Pukshenga and Pokshenga River basins and in the Severnaya Dvina-Pinega watershed is recognised at 50-80 m a.s.l. This represents a younger stage of deglaciation (Vepsian?) (Atlasov *et al.*, 1978). In the Vaga basin and on the left bank of Dvina, this stage occur as end moraines, radial and marginal eskers and kames towering 10 - 25 m above surrounding glacio-lacustrine plain. This terminal belt extends northwards as far as Yemetsk, where it merges with marginal formations, studied by Legkova & Schukin (1972). Two Valdaian till beds were revealed here by drilling in the basins of the Mekhrenga, Puya and other rivers, inside this possible Vepsian deglaciation position (Legkova, 1967; Atlasov *et al.*, 1978; Ostanin *et al.*, 1979).

Petrographic study of the Valdaian and Moscovian tills indicate some differences in their composition and supports the Late Valdaian maximum position determined earlier. For example, the Valdaian till typically contains a high percentage of Scandinavian crystalline rock clasts (50-95%) i.e. granites, gneisses, quartzites and 12-45% of local sedimentary rock fragments in the Vaga River valley. In the Moscovian tills, local Palaeozoic limestones, sandstones and dolomites represent 65-96% of the total whereas Scandinavian boulders represent 11-35% (Ostanin *et al.*, 1979).

Belomorian ice lobe (II) -Pinega glacier tongue

The maximal position of the Pinega tongue has still not been adequately studied. Most investigators believe that the Pinega tongue reached the Pinega-Dvina watershed

(Pokshenga Massif) marked by a zone of hummocky relief (Figs 1A, B, 2B). Pre-existing evidence has allowed data its boundary to be extended along the upper part of Berezovitsa River (a left tributary of the Okhtoma River) and continued north of Karpogory. This reconstruction is chiefly based on the fact that Mikulinian marine deposits are not overlain by till in both the upper and middle Beryozovitsa River valley area (Apukhtin & Krasnov, 1967; Chebotareva, 1977; personal communication T.A. Kuznetsova in: Chebotareva, 1977). At a site 20 km west of Karpogory, a 2 m thick Mikulino marine sand, gravel and shell layers unit was found at the base of the sequence (Devyatova, 1982) that was overlain by a periglacial 0.5 m thick sand interbedded with gravel and small boulders (possibly a till). According to the authors' observations, Mikulinian marine sediments, described by Filippov & Borodai (1987) from the River Ezhuga, 50 km north-eastward of Karpogory, are overlain by Scandinavian sandy outwashed till about 1 m thick. In this area the Quaternary sequence is only 1-5 m thick and the ice-dammed lakes in Vashka, Pinega and Severnaya Dvina valleys located at 130 m. (Lavrov, 1968, 1975) should intrude to the area. Thus, thin till might have been eroded both in the Karpogory and in the Ezhuga River areas.

Preliminary luminescence dating by the present authors shows that sand underlying a 1-2m thick till unit and a 1-4 m thick glaciolacustrine unit in the Verkola section and that at Mezhdurechensky was deposited between 100 –200 ka BP. This evidence leads to the assumption that the southern boundary of the Pinega ice tongue reached Verkola and probably overrode the Ezhuga River basin. At that time the ice front extended northwards as far as the watershed of the River Kimzha, where it merged with marginal formations of the Kuloi-Mezen ice lobe (Figs 1 and 2). Thus, the Pinega ice tongue was controlled by the Pokshenga Massif to the west, where ^{14}C-dated Middle Valdai deposits are not overlain by till on the River Yula (Arslanov *et al.*, 1984), and by the Pinega-Vashka watershed (250 m a.s.l) to the east.

Kuloi-Mezen ice lobe (III)

The Kuloi Plateau probably acted as a strong barrier to the flow of Scandinavian Ice Sheet into the Mezen basin (Fig. 1A, 2A). The thinness of the ice sheet is indicated by large areas practically devoid of a Quaternary sediment on the Kuloi Plateau. Scandinavian rock boulders are also encountered very seldom on the banks of the rivers Mezen and Vashka. The authors' data suggest that Mikulinian marine sand in the Olema section on the River Vashka are not overlain by till. Thus, there is no positive evidence of any Scandinavian Late Valdaian till along Mezen River. Consequently there is no stratigraphical evidence for the movement of Scandinavian ice into the Pyoza-Mezen watershed and further south along the Vashka River valley as far as Chuprovo and farther east along the Mezen across the Kyma River confluence, as some researchers have assumed (Krasnov, 1974; Lavrov, 1991) (Fig. 1B).

The zone of marginal formations, which extends approximately N-S across the central part of the Kuloi Plateau, was until recently regarded as the Late Valdai glaciation maximum limit (Legkova, 1967; Legkova & Schukin, 1972; Krasnov 1974; Figs 1B, 2B). However, the writers postulate that these formations probably formed in the younger Vepsian stage, which is represented by topographically-distinct marginal landforms in the Arkhangelsk and Vologda regions. The Valdaian glaciation limit was erroneously drawn along the tectonic scarp which delimits the Kuloi Plateau in the east (Yakovlev, 1956) (Fig. 1B).

Late Valdaian tills are absent along the Pyoza River valley, but practically all the Pyoza sections are overlain by thick, Late Valdaian periglacial fluvial sand beds with traces of cryogenesis (Fig. 3). Eleven OSL dates from these sands gave ages between 18 and 10 ka (Houmark-Nielsen *et al.*, 2001). A ^{14}C date from a thin peat unit in these sands gave an age of 10 150 ±100 BP (lU-556A) (Chebotareva, 1977). The occurrence of these sands up to 80 m a.s.l. (i.e. 18 m higher than modern water level in the Pyoza River) indicates that the Scandinavian ice dammed the Pyoza River near its mouth or somewhat to the north in the Mezen bay during Late Valdaian time (Fig. 3).

Boulders of Scandinavian origin (mainly granites, basic rocks and red sandstones) are widespread on the western Kanin coast. They occur in clayey diamicton and in probably glaciofluvial sediments capping sections near Cape Konushinskaya Korga and on the Tarkhanov River at 25-30 m a.s.l. Following the stratigraphical concept of Ramsay (1911) they are interpreted as 'upper till'. 'Fresh' glaciofluvial ridges up to 18 m high and with very clearly defined slopes were also found in the Shomokhovskie Sopki (Hills) area near Cape Kronushinskaya Korga. However, no Scandinavian boulders were found on the beach or in the sections near the mouth of the River Oiva on the eastern shore of Kanin (Fig. 2B).

Considering these data, it is suggested here that the maximum position of the Scandinavian ice-sheet advocated by some researchers is fairly accurate (Ramsay, 1911; Rudovits, 1947; Yakovlev, 1932; Aseev, 1974). The ice margin extended approximately N-S from the north shore of the Kanin Peninsula across the hilly-morainic landscape of the Shomokhovskie Sopki in the Konushinskaya Korga area and further south across a hummocky zone near the River Nes (Devyatova & Loseva, 1964), towards the lower part of the Mezen River. The margin then presumably intersected the Mezen River near the Pyoza confluence and extended along the Kimzha River, where it formed a wide zone of morainic hills (Lavrov, 1991). This is connected with the marginal formations of the Pinega ice tongue on the River Yezhuga (a tributary of the Pinega) and River Kimzha interfluve. Alternatively, the ice margin extended from the western shore of the Kanin Peninsula, from the hummocky terrain near Konyshinskay Korga, over the

Mezen Bay to the mouth of the Kuloi River and then to the south. This suggestion is based on the absence of Scandinavian till in the Cape Tolstik section, on the left bank of Mezen Bay, and the occurence of Scandinavian till on Cape Kargovsky near the Kuloi River mouth (Fig. 2.B). In this case, an ice-dammed lake must have existed in the southern part of the Mezen Bay and in the lower Pyoza River during Late Valdaian time. This finds some support from the fact that glaciolacustrine sediments are widespread capping the sections on Mezen Bay and along the lower part of the Mezen River.

Conclusions

The Scandinavian ice sheet, the Barents-Kara Sea ice sheet and possibly a Timan ice sheet occupied different parts of the Arkhangelsk region during the Late, the Middle and the Early Valdaian glaciations. Their tills, with a characteristic pattern of directional properties and clast provenances, are separated by marine, fluvial and lacustrine sediments, which have been dated by radiocarbon and luminescence. This means that since the Mikulian interglacial, ice sheets developed independently in different parts of the European North. In the Late Valdaian, the maximum of the last Scandinavian glaciation was asynchronous in various regions of Russia. Possibly, this is controlled by the distance from the centre of glaciation in the Gulf of Bothnia, but less dependent on peculiarities in the topography beneath different ice flows and the regional palaeoclimate.

At present there are three major uncertainties on the Valdaian glacial history in the Arkhangelsk area. One deals with the configuration and glaciation centre of the so-called 'Timan ice sheet' from the Early Valdaian. Another question is the position of the western boundary of the Barents-Kara Sea glaciation during the Middle Valdaian and whether or not it was confluent with a Scandinavian Ice Sheet west of the Arkhangelsk region. A third option is to map the distribution of deposits belonging to the Late Valdaian Scandinavian glaciation in the Kuloi area and possibly trace these deposits northwards to the Kanin Peninsula.

Acknowledgements

We thank K. Grøsfjeld, Geological Survey of Norway; J.K. Nielsen, University of Copenhagen for putting their data at our disposal. We also thank V. Astakhov, VNIIKAM, St.Petersburg; V. Gey and V. Kiseleva, Petersburg Geological Survey; E. Devyatova and V. Kopylova for helpful advice and consultation. D. Nikolaev and many other inhabitants of Arkhangelsk district and Karelia are thanked for practical assistance during long field trips along the Rivers Svernaya Dvina, Mezen and Pyoza.

The field work was supported by the ESF programmes 'Quaternary Environments of the Eurasian North' and 'Eurasian Ice Sheets'. The Danish contribution was funded by grants from the Danish Research Council to the 'Catlina' and 'Client' projects. The Norwegian contribution was funded through the European Community project 'Eurasian ice sheets', contract ENV4-CT97 and through the Norwegian Research Council.

References

Androsova, V.P. (1938). Geomorphologicheskie raboty po dostochnomu i severnomu poberezhiyu Kaninskogo poluostrova letom 1932 g. (Geomorphological works on eastern and northern shores of Kanin Peninsula in the summer of 1932. *In: Raboty po geologii* morya (The works on the geology of the Sea), *Transcations of the Institute of Marine Fishers and Oceanography of the USSR*. **V**, 165-189, Moscow.

Apukhtin, N.I. & Krasnov, I.I. (eds) (1967). *Geologia chetvertichnykh otlozheniy Severo-Zapada Evpopeiskoi chasti SSSR.* Geology of the north-western European USSR (with Map of Quaternary deposits of the north-western European part of the USSR, scale 1: 2,500,000, 1966, Leningrad, Nedra, 344 pp.

Arslanov, Kh., Auslender, V., Gromova, L., Zubkov, A. & Khomutova, V. (1970). Palaeogeograficheskie osobennosti i absolyutnyi vozrast maksimalnoi stadii Valdaiskogo oledenenia v raione Kubenskogo ozera. (Palaeogeography and absolute age of the Valdaian glacial maximum in the Lake Kubenskoe area). *Doklady Academii Nauk SSSR,* **195, 6**, 1395-1398.

Arslanov, Kh.A., Voznyachuk, L.N., Velichkevich, F.Yu, Kurerova, L.V. & Petrov, G.S. (1971). Vozrast maksimalnoi stadii poslednego oledenenia na mezhdureche Zapadnoi Dviny i Dnepra. (Age of the maximal stage of the last glaciation on the watershed of the Zapadnaya Dvina and Dnepr rivers). *In:* Bondarchuk, V.G. (ed.), *Doklady Academii Nauk SSSR,* **196, 1**, 161-164.

Arslanov, Kh.A., Levina, N.B., Ostanin, V.E., Baranova, V.N., Smirnova, V.M. & Tertychnaya, T.V. (1984). O geokhronologii i stratigrafii pozdnego pleistotsena i rannego golotsena basseina r.Severnoi Dviny (On geochronology and stratigraphy of the Late Pleistocene and Early Holocene in the basin of the Severnaya Dvina river). *Vestnik Leningradskogo Gosudarstvennogo Universiteta, Geologia, Geografia,* **12**, 57-66.

Aseev, A.A. (1974). *Drevnie materikovye oledenenia Evropy*. (Ancient continental glaciations of Europe), 319pp. Moscow. Nauka.

Astakhov, V.I., Swendsen, J.I., Matiochkov, A., Mangerud, J., Maslenikova & O., Tveranger, J. (1999). Marginal formations of the last Kara and Barents ice sheets in northern European Russia, *Boreas,* **28**, 23-45.

Atlasov, R.R., Bukreev, V.A., Levina, N.B. & Ostanin, V.E. (1978). Osobennosti reljefa kraevoi zony valdaiskogo oledeneniya na Onego-Vazhskom mezhdurechje i v doline r. Vagi. (Relief peculiarities of terminal zone of Valdai glaciation on Onega - Vaga waterdivide and in r. Vaga valley). *In: Kraevye obrazovania materikovych oledeneniy* (Terminal formation of continental glaciations), 30-38. Kiev, Naukova Dumka.

Barkhatova, V.P. (1941). *K geologii basseinov yugo-vostochnogo poberezhia Onezhskogo ozera i verkhoviev r. Onega.* (On geology of basins of Onego lake south-eastern shores and Onaga river headwaters) *Trudy Severnogo geologicheskogo upravlenia,* **9**, 116 pp. Moscow-Leningrad, Gosgeolizdat.

Bylinsky, E.N. (1962). Novye dannye po stratigraphii chetvertichnykh otlozhenii i palaeogeografii basseina r. Mezeni. (New data about Quaternary stratigraphy and palaeogeography of the River Mezen basin) *Doklady Akademii Nauk SSSR,* **147**, **6**,1421-1424.

Bylinsky, E.N. (1980). Prichiny morskih pleistosenovyh transgressii severa Evrazii. (The causes of marine transgressions in northern Eurasia). *Byulleten komissii po izucheniyu chetvertichnogo perioda,* **50**, 35-56.

Chebotareva, N.C. & Makarycheva, I.A. (1974). *Poslednee oledenenie Evropy i ego geokhronologia.* (Last glaciation of Europe and its geochronology), 216 pp. Moscow, Nauka

Chebotareva N.S. (ed.) (1977). *Struktura i dinamika poslednego lednikovogo pokrova Evropy* (Structure and dynamics of the last ice sheet of Europe), 142 pp. Moscow, Nauka.

Devyatova E.I (1961). *Stratigraphia chetvertichnych otlozneniy i palaeogeographia chetvertichnogo perioda v basseine r. Onegi* (Stratigraphy of Quaternary deposits and the palaeogeography of Quaternary period in River Onega basin), 89pp. Moscow, Academiya Nauk SSSR.

Devyatova, E.I. (1969). Degradatsia Valdaiskogo oledenenia i pozdnelednikovaia istoria Baltiiskogo i Belogo morei. Arhangelskaia oblast (Degradation of Valdai glaciation and lateglacial history of Baltic and White Sea) *In: Poslednii lednikovyi pokrov na severo-zapade evropeiskoi chasti SSSR,* 231-241. Moscow, Nauka

Devyatova, E.I. (1972). Palinologicheskaya kharakteristika verkhnechetvetichnykh otlozheniy Karelii (Palynological characteristics of Upper Quaternary sediments of Karelia). *In:* Biske, G.S. (ed), *Chetvertichnaya geologia i geomorphologia vostochnoi chasti Baltiiskogo schita,* 59-96. Leningrad, Nauka.

Devyatova, E.I. (1982). *Prirodnaya sreda pozdnego pleistotcena i yeyo vliyanie na rasselenie cheloveka v Severodvinskom basseine i v Karelii* (Late Pleistocene environments as related to human migrations in the Severnaya Dvina basin and in Karelia), 156 pp. Petrozavodsk, Karelia.

Devyatova, E.I & Loseva, E.I. (1964). *Stratigraphia i palaeogeographia chetvertichnogo perioda v basseine r. Mezeni.* (Stratigraphy and palaeogeography of Quarternary period in r.Mezen basin, 105 pp. Leningrad, Nauka.

Devyatova, E.I. & Punning, J-M.K. (1976). Verkhnepleistotsenovyi rasres Koleshka (Arhangelskaya oblast) i ego stratigraficheskoe znachenie [Upper Pleistocene deposits in the Koleshka profile (Arkhangelsk region) and their stratigraphical interpretation]. *Izvestiya Akademii Nauk ESSR,* **25**, *himia, geologia*, **2**, 152-160.

Devyatova, E. I., Raukas, A.V., Rajamae & R.A., Hyutt G.I., (1981). *Verkhnepleistotsenovyi rasres Pasva (r. Vaga, Arhangelskaya oblast) i ego stratigraficheskoe znachenie.* [Upper Pleistocene deposits in the Pasva profile (River Vaga, Arkhangelsk region) and their stratigraphical interpretation]. *Byulleten komissii po izucheniyu chetvertichnogo perioda,* **51**, 38-50.

Ekman., I. M., Devyatova, E. I. & Ilyin, V. A. (1974). Predfrontalnye kraevye obrazovania yugo-vostochnoi okrainy Baltiiskogo schita (Frontal terminal glacial formations of south-eastern part of the Baltic shield). *In:* Biske G. & Mikalauskas A. (eds), *Predfrontalnye kraevye lednikovye obrazovania,* 86-98. Vilnius, Mintis.

Ekman., I. & Iljin.,V. (1995). Deglaciation, the Yonger Dryas end moraines and their correlation in Russian Karelia and adjacent areas. *In:* Ehlers, J. Kozarski, S. & Gibbard, P.L. (eds), *Glacial deposits in North-East Europe*, 195-209. Rotterdam/Brookfield, A.A. Balkema.

Evzerov, V.Ya., Gydina, V.I. & Pervuninskaia, N.A. (1976). K palaeogeographii i stratigrafii verkhnego pleistosena v basseine r. Onega. (On the palaeogeography and stratigraphy of the Upper Pleistocene in the River Onega basin, *In:* Biske, G. (ed.), *Priroda i hozyaistvo Severa,* 5-11. Leningrad, Geographicheskoe obschestvo SSSR.

Faustova, M.A. & Velichko, A.A. (1992). Dynamics of the last glaciation in northern Eurasia. *Sveriges Geologiska Undrsökning*, **Ca 81**, 113-118.

Filippov, V.V. & Borodai, L.V. (1987). Opornye razrezy mikulinskih otlozhenii na severe Mezenskoi vpadiny (Stratotype sections of the Mikulino marine deposits in the north of the Mezen hollow). *Vestnik Leningradskogo Gosudarstvennogo Universiteta, Geologia, Geografia,* **21**, **3**, 72-74.

Ganeshin, G.S., Krasnov, I.I., Marinov, N.A. & Murzaeva, V.E. (eds) (1980). *Karta chetvertichnyh otlozhenii Evrazii* (Map of Quaternary deposits of Eurasia. Scale 1:5mln). Ministerstvo geologii SSSR, VNII zarubezh-geologia.

Geologia SSSR. (1963). *Arkhangelskaia i Vologodskaia oblasti i Komi ASSR. Tom2, chast 2* (Geology of USSR. Vol 2. Part 2. Arkhangelsk and Vologda districts and Komi ASSR), 1079 pp. Moscow, Nedra.

Gerasimov, I.P & Markov, K.K. (1939). *Lednikovyi period na territorii SSSR.* (The glacial period in the territory of the USSR), 462 pp., Moscow-Leningrad, Akademia Nauk SSSR.

Gey, V.P. & Malakhovsky, D.B. (1998). O vozraste i rasprostranenii maksimalnogo verkhnepleistotsenovogo lednikovogo nadviga v zapadnoi chasti Vologodskoi oblasti (On the age and expansion of maximal late Pleistocene glacial thrust in western part of the Vologda area). *Izvestiia Russkogo geograficheskogo obschestva.* **1**, 43-53.

Gey, V.P., Auslender, V.G., Demidov I.N., Kiselyeva, V.B., Lunkka, J-P., Pleshivtseva, E.S. & Saarnisto, M. (2000). *Problemy stratigrafii chetvertichnykh otlozhenii i kraevye lednikovye obrazovania Vologodskogo regiona*

(Severo-Zapad Rossii) (Problems of Quaternary stratigraphy and glacial terminal formations of Vologda region. North-West of Russia), 100 pp. Moscow, GEOS.

Gorbatsky, G.V. (1932). Postpliotsenovye otlozhenia i relief yugo-vostochnogo poberezhia poluostrova Kanin (The post-Pliocene and relief of the South-East coast of the Kanin Peninsula). *Izvestia geograficheskogo obschestva.* **LXIV, 6**. 448-479.

Grosswald, M.G. (1980). Late Weichselian Ice Sheets of Northern Eurasia. *Quaternary Research,* **13**, 1-32.

Houmark-Nielsen, M., Demidov, I., Funder, S., Grosfjeld, K., Kjær, K.H., Larsen, E., Lavrova, N., Lyså, A. & Nielsen, J.K. (2001). Early and Middle Valdaian glaciations, ice-dammed lakes and periglacial interstadials in northwest Russia: new evidence from the Pyoza River area. *Global and Planetary Change*, **31**, 215-237.

Juze, A.P, Poretskiy, V.S. (1937). Diatomovye mezhlednikovykh otlozheniy po r.Vage (Diatoms of interglacial sediments of the River Vaga). *Trudy Sovetskoi sektsii INQUA,* **1**, 178-205.

Kjær, K.H., Demidov, I., Houmark-Nielsen, M. & Larsen, E. (2001). Discrimination between easterly- and westerly-flowing Valdaian ice streams in Arkhangelsk region, Northwest Russia. *Global and Planetary Change*, **31**, 201-214.

Kalber, E.A. (1940). Geoloicheskoe opisanie Onezhskogo poluostrova. (Geological description of the Onega Peninsula). *In: Trudy severnogo geologicheskogo upravlenia. Vyp.5.* 178-205

Kalberg E.A. (1968). *Gosudarstvennaya karta chetvertichnyh otlozhenii. List Q-38-XXIX-XXX. Tsenogora (State map of Quaternary deposits. Sheet Q-38-XXIX-XXX. Tsenogora. Scale 1:200000)*

Kalyanov, V.P. & Androsova, V.P. (1933). Geograficheskie nablyudenia na Kanine. (Geographical observation on Kanin.). *Zemlevedenie,* **25, 1**. 16-25.

Korchagin, A.A., (1937). O konechno-morennych landshaftah basseina r.Mezeni. (On end-moraine land-scapes of the River Mezen basin. *Izvestia geograficheskogo obschestva*, **69, 4**, 623-631.

Kotova, S.F., Nukzarova, V.V., Sanina, G.N. (1977). Novye dannye po stratigrafii nizhne- i srednechetvertichnykh otlozheniy yga Arhangelskoi oblasti (New data on stratigraphy of Lower and Middle Pleistocene on the south of Arkhangel area). *In:* Biske, G.S. (ed.), *Stratigrafia i palaeogeografia chetvertichnogo perioda severa Evropeiskoi chasti SSSR* (Stratigraphy and palaeogeography of Quaternary on the north of European USSR), 63-71, Petrozavodsk, Karelsky filial Akademia Nauk SSSR.

Krasnov, I.I. (ed.) (1974). *Karta chetvertichnyh otlozhenii Evropeiskoi chasti SSSR.* (Map of Quaternary deposits of the European part of the USSR scale 1:1.5 mln). Ministerstvo geologii SSSR.

Larsen, E., Funder, S. & Thiede, J. (1999a). Late Quarternary history of northern Russia and adjacent shelves – a synopsis. *Boreas,* **28**, 6-11.

Larsen, E., Lyså, A., Demidov, I., Funder, S., Houmark-Nielsen M. & Kjær, K. (1999b). Age and extent of the Scandinavian Ice Sheet in north-west Russia. *Boreas,* **28,** 115-132.

Lavrov, A.S. (1968). Verhnepleistotsenovye dolinnye ozera v basseinah Pechry, Vychegdy i Mezeni.(Upper Pleistocene valley lakes in Pechora, Vychegda and Mezen basins), *Izvestija VGO,* **100, 2**, 146-151.

Lavrov, A.S. (1973). Drevnie oledeneniya severo-vostoka Russkoi ravniny.(Ancient glaciations of North-East of the Russian Plain). *Izvestia AN SSSR. Seria geograficheskaia,* **6**, 29-37.

Lavrov, A.S. (1974). Pozdnepleistosenovey lednikovye pokrovy severo-vostoka Evropeiskoi chasti SSSR.(Late Pleistocene ice sheets of the North-East of the European part of the USSR). *Byulleten komissii po izucheniyu chetvertichnogo perioda,* **N 41**, 48-55.

Lavrov, A.S. (1975). Pleistosenovye podprudnye ozera severo-vostoka Russkoi ravniny. (Pleistocene ice-dammed lakes of the North-East of the Russian Plain. *Istoria ozer v pleistosene,* **II**, 136-141. Leningrad, Nauka

Lavrov, A.S. (1977). Kolsko-Mezenskiy lednikovyi potok. (Kola-Mezen ice flow). *In:* Chebotareva, N.S. (ed.), *Struktyra i dinamika poslednego lednikovogo pokrova Evropy* (Structure and dynamics of the last ice sheet of Europe), 83-89, Moscow, Nauka.

Lavrov, A.S. (1991). *Karta chetvertichnych otlozenii. List Mezen. Scale 1:1 mln.* (Map of Quaternary deposits. Sheet Mezen). Leningrad, VSEGEI.

Lavrova, M.A. (1937). O srtatigraphii chetvertichnykh otlozheniy Severnoi Dviny ot ystja Vagi do Konetsgorja (On the Quaternary stratigraphy of the Severnaya Dvina from the mouth of the River Vaga to Konetsgorje). *Trudy Sovetskoi sektsii INQUA,* **1**, 152-177.

Lavrushin, Yu.A. & Spiridonova, E.A. (1995). The Late Pleistocene events in the North of European Russia: geology and palaeoenvironments. *In:* Velichko, A.A. (ed.), *Climate and Environment Changes of East Europe during Holocene and Late-middle Pleistocene*, 63-68, Moscow, Nauka.

Legkova, V.G. (1967). Stratigraphia. Severo-zapad Arkhangelskoi oblasti. (Stratigraphy. The north-west of Arkhangel region). *In:* Apukhtin, N.I. & Krasnov, I.I. (eds), *Geologia chetvertichnykh otlozheniy Severo-Zapada Evropeiskoi chasti SSSR* (Geology of Quaternary deposits of the north-west European part of USSR), 172-191, Leningrad, Nedra.

Legkova, V.G. & Schukin, L.A. (1972). Poiasa kraevyh obrazovanii v severo-zapadnoi chasti Arhangelskoi oblasti. (Belts of terminal formations in north-western part of Arkhangelsk district). *In:* Goretsy, G.I., Pogulyaev, D.I. & Shik S.M. (eds), *Kraevye obrazovania materikovyh oledenenii,* 202-205, Moscow, Nauka.

Likharev, B.P. (1933). *Obschaya geologicheskaya karta Evropeiskoi chasti SSSR. List 69. Shenkursk - Velsk* (Geological map of European USSR. Shenkursk-Velsk), 101 pp. Leningrad - Moscow. Trudy VGRO, **240**.

Liivrand, E.D. (1981). O vozraste i correlyatsii sloev razrezov Pasva i Koleshki v Arhangelskoi oblasti po geologicheskim i palinologicheskim dannym (The age and correlation of Pasva and Koleshka strata based on geological and palynological data). *In:* Yevzerov, V.Ya. (ed.), *Geologia pleistotsena Severo-Zapada SSSR,* 72-86, Apatity, Academy of Sciences of the USSR. Kola Science Center, Geological Institute.

Lyså, A., Demidov, I., Houmark-Nielsen, M. & Larsen, E. (2001). Late Pleistocene stratigraphy and sedimentary environment of the Arkhangelsk area, northwest Russia. *Global and Planetary Change*, **31**, 179-199.

Lukashov, A.D. (1982). *Guidebook for excursion A-4, C-4. Karelia. INQUA. XI congress*. 57 pp. Moscow, VINITI.

Lukoyanov, S.M. (1941). O borealnoi transgressii v raione r.Moshi. (On the boreal transgression in the Mosha River area). *Izvestia geograficheskogo obschestva*, **73, 3**, 468-470.

Lunkka, J-P., Saarnisto, M., Gey, V., Demidov, I. & Kiseleva, V. (2001). The area and timing of the Last Glacial maximum in the Valdaian (Weichselian) cold stage in Vologda and adjacent areas of the NW Russian Plain. *Global and Planetary Change*, **31**, 407-426.

Malakhov, A.A. (1934). O stratigrafii chetvertichnykh otlozhenii srednego techenia Mezeni i Pyozy (On the stratigraphy of the Quaternary deposits of the middle Mezen and Pyoza basins). *Izvestia geograficheskogo obschestva,* **66, 3**, 469-478.

Malakhov, A.A. (1939). Chetvertichnye otlozhenia i geomorfologia basseina r.Mezeni (Quaternary deposits and geomorphology of the River Mezen basin. *Trudy Sovetskoi sektsii INQUA*, **4**, 43-44.

Malakhov, A A. (1940). Geologia srednego Timana i zapadnogo Pritimania (Geology of middle Timan and western Pritimania). *Trudy Severnogo geologicheskogo upravlenia*, **6**, 112 pp.

Mangerud, J., Svendsen, J.I. & Astakhov, V.I. (1999). Age and extent of the Barents and Kara ice sheets in Northern Russia. *Boreas,* **28**, 46-80.

Molodkov, A.N. (1989). Opyt ispolzovania metoda ESR-datiovania rakovin molluskov chetvertichnykh otlozenii (An experience of ESR-dating of mollusca shells applied for stratigraphical schemes correlation of Quaternary deposits). *In:* Janshin A.L. (ed), *Chetvertich-nyi period. Stratigraphia*, 204-213, Moscow, Nauka.

Ostanin, V.E., Atlasov, R.R., Bucreev, V.A. & Levina, N.B. (1979). Kraevye obrazovania i granitsa valdaiskogo oledenenia v basseine r. Vagi (Terminal formations and the Valdai glaciation boundary in the basin of the River Vaga). *Geomorphologia,* **1**, 72-76

Pleshivtseva, E.S. (1977). Izmenenie palaeogeograficheskih uslovii severo-Dvinskoi vpadiny v pozdne-poslelednikovoe vremia (The changes of palaeogeographical conditions in the Severnaya Dvina lowland). *Priroda i hoziaistvo Severa*, **6**, 39-47, Petrozavodsk.

Pokrovskaya, I.M. (1937). Necotorye dannye micropalaeobotanicheskogo izuchenia mezhlednikovykh otlozheniy Severnoi Dviny i Vagi (Some data of micropalaeobotanical investigations of interglacial sediments of the Severnaya Dvina and Vaga rivers). *Trudy Sovetskoi sektsii INQUA,* **1**, 206-220.

Punkari, M. (1985). Glacial geomorphology and dynamics in Soviet Karelia interpreted by means of satellite imagery. *Fennia,* **163**, 113-153.

Ramsay, W. (1911). Beiträge zur Geologie der Halbinsel Kanin. *Fennia*, **21 (7)**, 66 pp.

Rudovits, Yu.L. (1947). O kolichestve oledenenii, borealnyk transgressii i granitsa poslednego oledenenia v svyazi s novymi issledovaniiami na srednem Timane. (Quantity of glaciation of boreal transgressions and border of the last glaciation in connection with new research in the middle Timan region) *Byulleten komissii po izycheniyu chetvertichnogo perioda,* **9**, 15-22.

Sergeev, E.M. & Babak, V.I. (eds) (1981). *Karta geomorfologo-nentectonicheskogo raionirovania nechernozemnoi zony SSSR* (Map of geomorphological-neotectonic regions of the chernozem-free zone of the Russian Federation. Scale 1:1.5 mln.). Minisyerstvo geologii SSSR, Moskovsky Universitet.

Smirnova, .V.M. (1991). Arcticheskoe pozdnemoskovskoe more v basseine srednego techenia reki Severnoi Dviny (Arctic Late Moscow sea in the basin of the middle Severnaya Dvina River). *In:* V.Ya.Yevzerov (ed.), *Geologia pleistocena Severo-Zapada SSSR*, 87-97, Apatity, Geologichesky Institut.

Sokolov, N.N. (1946). O polozhenii granits oledenenii v evropeiskoi chasti SSSR (On the position of glacial limits in the European part of the USSR). *Trudy Instituta geografii AN SSSR*, **37**. 83-99.

Spiridonov, M.A. & Jakovleva, S.V. (1961). Chetvertichnye otlozhenia poberezhia poluostrova Kanin i basseina r. Pyoza (Quaternary deposits of the Kanin Peninsula shore and River Pyoza basin). *Materialy po chetvertichnoi geologii i geomorfologii SSSR*, **3**, 75-89.

Svendsen, J.I., Astakhov, V.I., Bolshiyaniv, D.Yu., Demido, I.N., Dowrswell, J.A., Gataulin, V., Hjort, Ch., Hubberten H.W., Larsen, E., Mangerud, J., Melles, M., Möller, P., Saarnisto, M. & Siegert, M. (1999). Maximum extent of the Eurasian ice sheets in the Barents and Kara Sea region during the Weichselian. *Boreas,* **28**, 234-242

Tolstikhin, N.I. & Tolstikina, M.M. (1935). Materialy k geomorphologii Onego-Dvinskogo mezhdurechia (On geomorphology of the Onega - Dvina interfluve). *Izvestia geograficheskogo obschestva,* **67, 3**, 287-317

Velichko, A.A., Kononov, Y.M. & Faustova, M.A. (1997). The Last Glaciation on Earth: size and volume of ice sheets. *Quaternary International,* **41/42**, 43-51.

Vollosovitch, K.A. (1900). Zametki o postpliotsene v nizhnem techenii r. Severnaya Dvina (Notes about post-Pliocene of the lower Severnaya Dvina River). *Materialy dlya geologii Rossii,* **XX. S. Pb**, 251-261.

Yakovlev, S.A. (1932). O karte otlozhenii chetvertichnoi sistemy Evropeiskoi chasti SSSR i soprdelnykh s nei territorii v masshtabe 1: 2,500,000 (On the Quaternary map of the European part of the USSR and adjacent areas, scale 1: 2,500,000). *Trudy II mezhdunarodnoi konferentsii po izucheniyu chetvertichnogo perioda*

Evropy. **1**. (Reports of the II International Quaternary Research Conference of Europe), 91-103. Leningrad-Moscow, GNTGRI.

Yakovlev, S.A. (1956). Karta chetvertichnoi sistemy Evropeiskoi chasti SSSR. 1: 5,000,000 (The map of the Quaternary system of the European part of the USSR. 1: 5,000,000). *In:* Yakovlev S.A. (ed.), *Osnovy geologii chetvertichnykh otlozheniy Russkoi ravniny. Stratigraphia.* (The fundamentals of the Quaternary geology of the Russian Plain. Stratigraphy). 314 pp. Moscow, GOS-GEOLTEH-IZDAT.

Yaunputnin, A.I. (1934). K voprosu ob usloviiah otstupania poslednego lednikovogo pokrova na severo-zapadnoi okraine Russkoi ravniny (On the conditions of last ice-sheet retreat in the north-western part of the Russian plain). *Izvestia geograficheskogo obschestva.* **66, 3**, 447-457.

Zekkel, Ya.D. (1939). Geologicheskie issledovania v raione Zimnego berega Belogo moria, Basseina Kuloia i nizoviev Mezeni (Geological investigations in the areas of the White Sea Zimnii coast, River Kuloi basin and lower River Mezen). *Trudy Severnogo geologicheskogo upravlenia,* **3**, 96 pp.

Quaternary Glaciations – Extent and Chronology
Editors J. Ehlers and P.L. Gibbard

Glaciations of the East European Plain - distribution and chronology

A.A. Velichko, M.A. Faustova, Yu.N. Gribchenko, V.V. Pisareva and N.G. Sudakova

Laboratory of Evolutionary Geography, Institute of Geography, Russian Academy of Sciences, Staromonetny 29 Moscow, 109017 RUSSIA

E-mail: palaeo@glasnet.ru

The East European Plain is one of the regions which were repeatedly subjected to the expansion of large continental ice sheets. A notion of three ice sheets expanding successively into European Russia developed in the 1930s. They are known as the Oka, Dnieper and Valdai glaciations and were tentatively correlated with Mindel, Riss, and Würm of the Alpine region. Since then views on the number of glaciations and on the periodicity of glacial events has undergone a considerable revision (Fig. 1). Biostratigraphical and lithostratigraphical evidence suggest that first strong cooling which might have resulted in glaciation occurred as early as 2.4 to 1.8 million years ago (Frenzel, 1967; Grichuk, 1981), although no glacial deposits dating from this interval have been found. Vast ice sheets expanded into the European part of Russia for the first time beginning with the Brunhes Epoch. They are

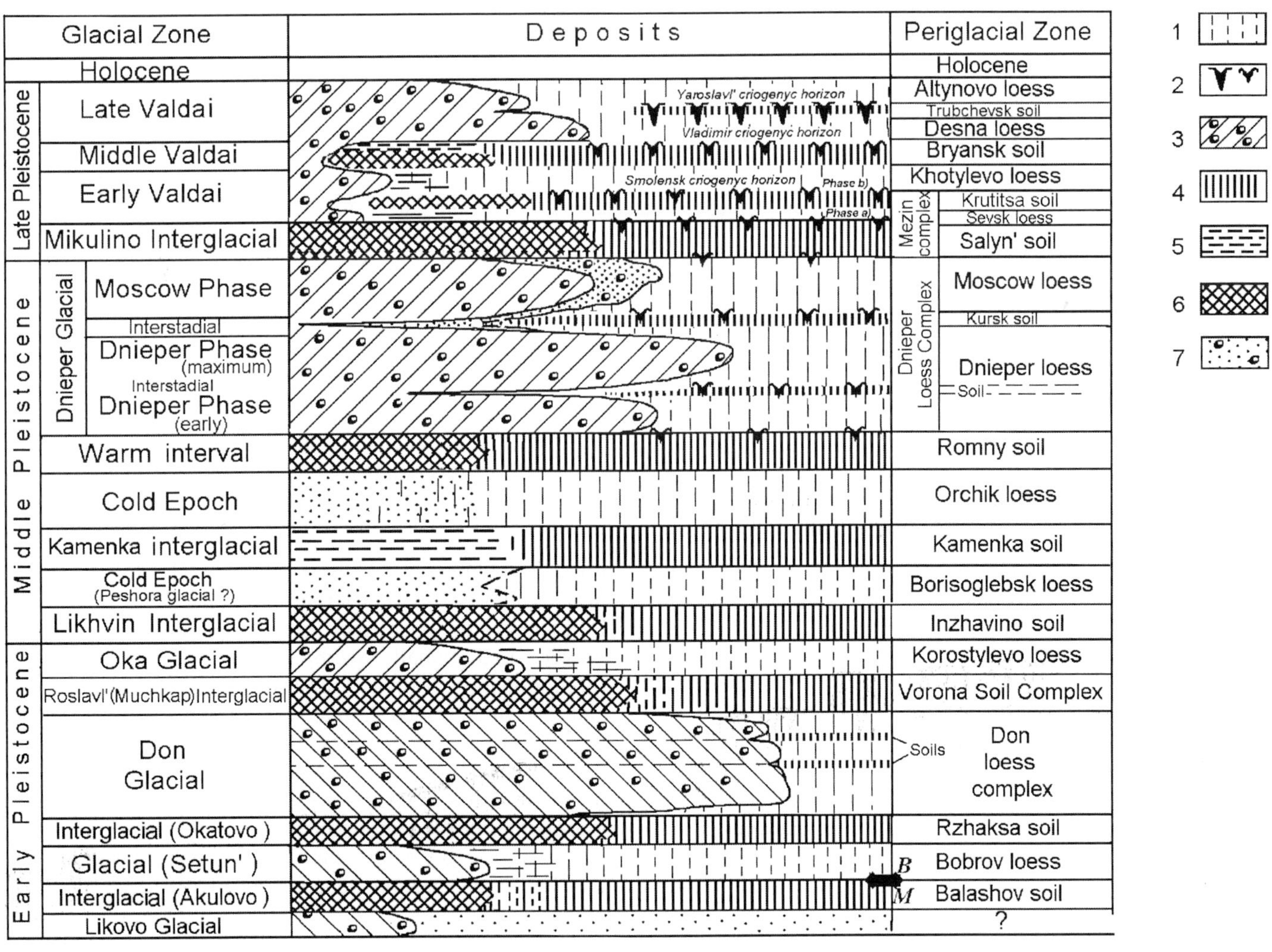

Fig. 1. Correlation between glacial and periglacial stratigraphy.

1 – loess, 2 – frost deformations, 3 – till, 4 – fossil soil, 5 – lacustrine deposits, 6 – peat deposits, 7 – glaciofluvial deposits.

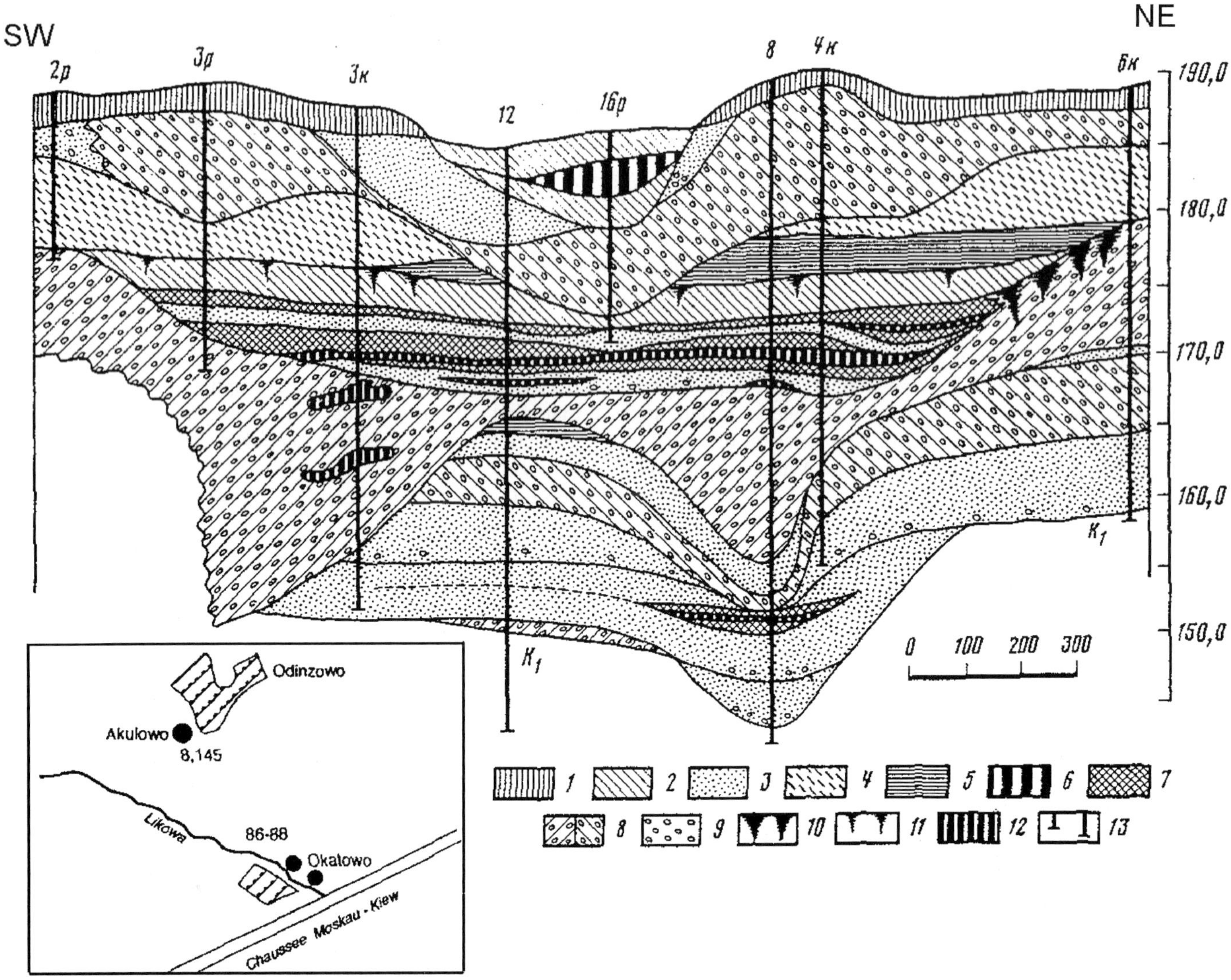

Fig. 2. Longtitudinal section of the Akulovo site.
1 - cover loam, 2 - lacustrine loam, 3 - sand, 4 - sandy loam, 5 - clay, 6 - peat, 7 - gyttja, 8 - diamicton (till), 9 - gravel and pebble, 10 - fossil soils, 11 - weakly comminuted fossil soils, 12 - detached blocks, 13 - boreholes.

associated with the Tiraspolian (faunal) period that corresponds to the 'Cromerian Complex' Stage in the Western European chronostratigraphy.

Likovo Glaciation

The oldest glaciation is correlated conventionally with 'glacial a' of the 'Cromerian Complex'; it expanded as far south as the latitude of Moscow. The so-called 'Likovo' till, attributed to this ice sheet, was only discovered in the 1980s, after members of the Geological Survey of the Central Regions of the East European Plain had thoroughly examined many boreholes in the Moscow region (Fig. 2). It has not yet been found elsewhere (Maudina *et al.*, 1985). Boreholes west of Moscow, near the western margin of the town of Odintsovo (Akulovo village) penetrated the till near the base of the Quaternary sequence. It is represented by black or greenish, dark grey loam including abundant gravel and small pebbles of sedimentary rocks (predominantly flint and limestone). Crystalline (augitite and rosé fine-grained granite with a grain size less than 1-2 mm) gravel is only occasionally found. No rocks of certain Scandinavian origin are recorded in any of the sections. The heavy-mineral composition is typically garnet-disthene-tourmaline. In the Don drainage basin, this cryochron (cold stage) is probably represented by the loess unit exposed in the Troitskoye section beneath the oldest soil. The fossil fauna of the latter is dominated by the ground squirrel *Citellus* (Krasnenkov *et al.*, 1997). The Troitskoye fauna's antiquity is suggested by presence of *Prolagurus pannonicus* KORM. and the abundance of European vole *Pitymis hintoni* KRETZOI (Agadjanian, 1992). A similar

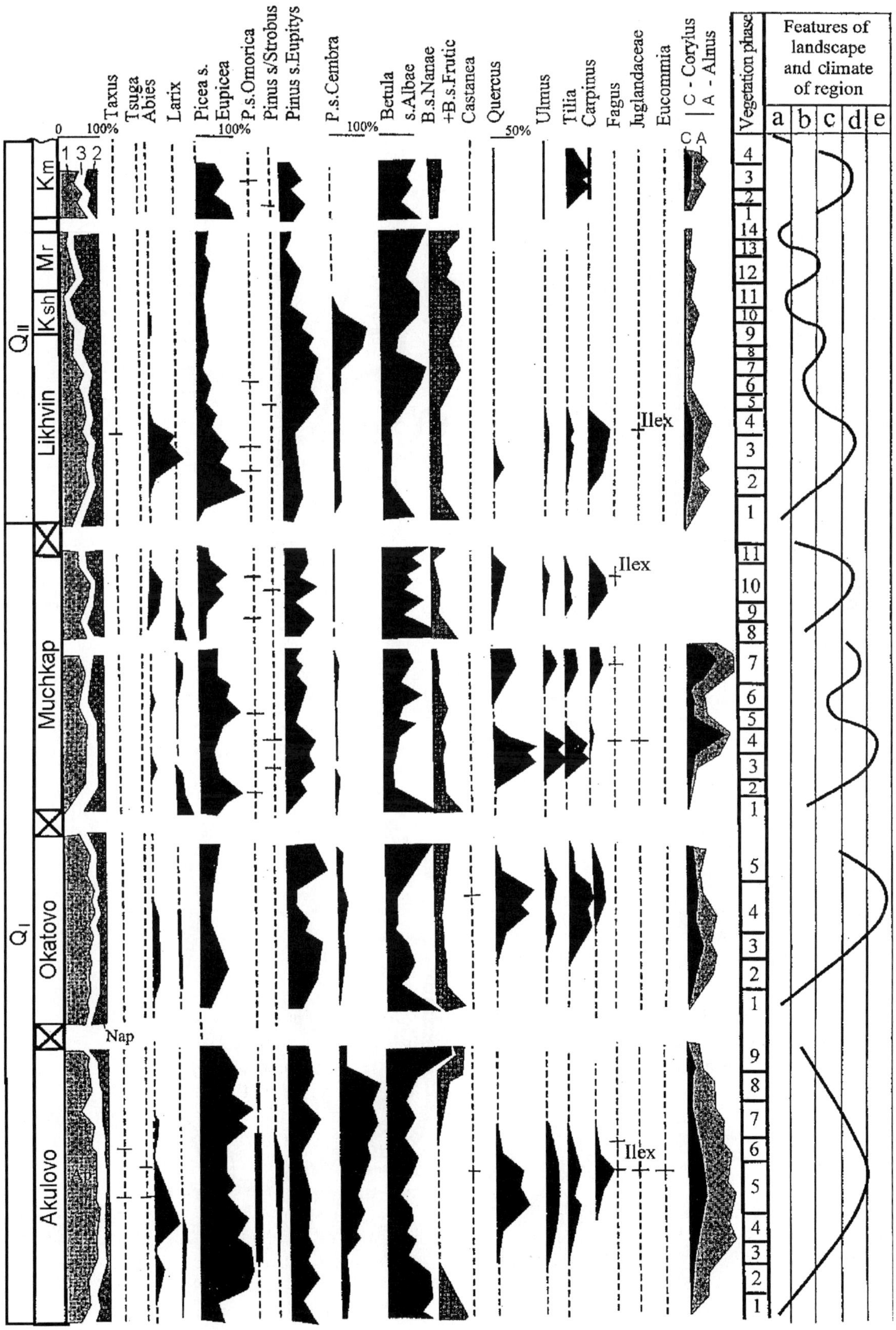

Fig. 3. Composite pollen diagram of the Lower- and Middle Pleistocene deposits in the northern and western Moscow region. General composition: 1 – AP, 2 – NAP, 3 – Spores; a - periglacial forest-steppe, b - tundra and forest-tundra, c - coniferous and deciduous forests, d - coniferous-broad-leaved forests, e - broad-leaved forests. Indices: Ksh – Kosha Interstadial, Mr - Maryno Interstadial, Km – Kamenka Interglacial.

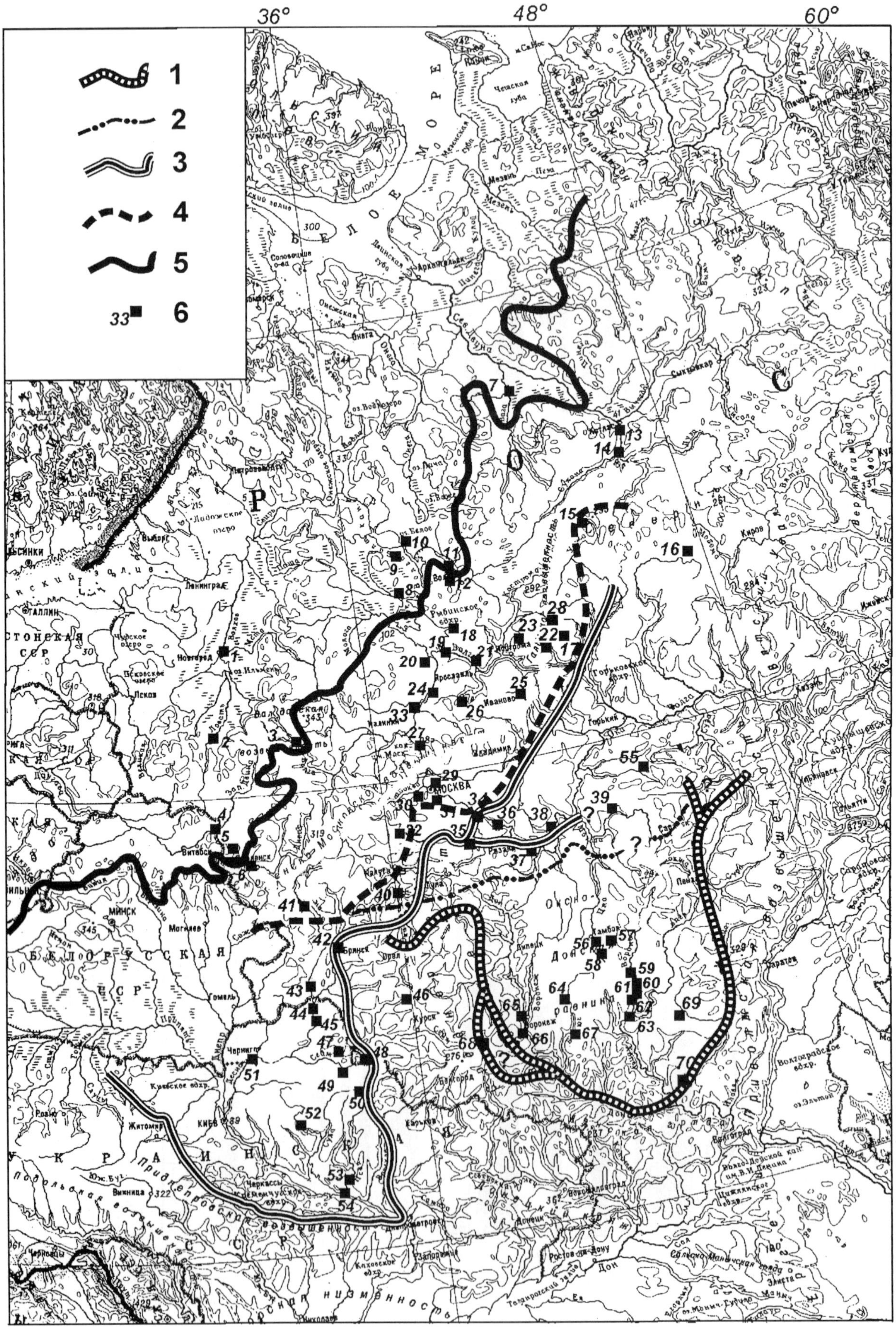
36°
48°
60°
64°
56°
48°
1
2
3
4
5
6
33
МОСКВА
Новгород
Псков
Витебск
МИНСК
Могилев
Гомель
Брянск
Орел
Курск
Белгород
Харьков
Воронеж
Липецк
Тамбов
Рязань
Калуга
Ярославль
Иваново
Кострома
Владимир
Горький
Казань
Киров
Сыктывкар
Саратов
Волгоград
Киев
Житомир
Винница
Чернигов
Черкассы
Днепропетровск
Запорожье
Донецк
Ростов-на-Дону
Ленинград
Таллин
Хельсинки
Архангельск
Онега
Мезень
Печора
Ухта
Пенза
Ижевск
Тольятти
Рыбинское вдхр.
Горьковское вдхр.
Кременчугское вдхр.
Киевское вдхр.
Каховское вдхр.
Волгоградское вдхр.
Цимлянское вдхр.
Куйбышевское вдхр.
Саратовское вдхр.
Чудское озеро
Псковское озеро
оз. Ильмень
Ладожское озеро
оз. Белое
Чешская губа
Онежская губа
Двинская губа
Кандалакшский залив
БЕЛОЕ МОРЕ
БЕЛОРУССКАЯ ССР
ЭСТОНСКАЯ ССР
УКРАИНСКАЯ ССР
Приднепровская возвышенность
Донская равнина
Окско-Донская
Среднерусская возвышенность
Дон
Волга
Ока
Десна
Днепр
1
2
3
4
5
6
7
8
9
10
11
12
13
14
15
16
17
18
19
20
21
22
23
24
25
26
27
28
29
30
31
32
33
34
35
36
37
38
39
40
41
42
43
44
45
46
47
48
49
50
51
52
53
54
55
56
57
58
59
60
61
62
63
64
65
66
67
68
69
70
?

Fig. 4. Glacial limits and key sections in the East European Plain.(opposite page)

The boundaries of the ice sheets:
1 - Don,
2 - Oka,
3 - Dnieper,
4 - Moscow (Warta),
5 - Valdai.

Key sections:
1 - Savino (dn, ms, vd),
2 - Lyakhovo (dn, ms, vd),
3 - Bulatovo, Tyaglitsy, (ok, lh, ms),
4 - Drichaluki, Shapurovo (ms, vd),
5 - Mikulino (mk, vd),
6 - Smolensk, Kuchino (dn, ms, mk),
7 - Shenkursk (vd),
8 - Cherepovets (dn, ms, mk),
9 - Moeksa (dns, dn, vd),
10 - Ferapontovo (dns, dn, vd),
11 - Puchka (dn, ms, vd),
12 - Molochnoe (dn, ms),
13 - Kotlas (dns, dn, ms),
14 - Veliky Ustyug, Dymkovo (dns, dn, ms),
15 - Anyuk (dn, mc),
16 - Vavilyata (pre-dns, dns),
17 - Pepelovo (dns, mč),
18 - Yakovlevskoe, (ok, lh, dn, ms),
19 - Rybinsk, Chermenino (pre-dns, dns, lh, dn, ms),
20 - Shestikhino (ms, vd?),
21 - Tutaev, Dolgopolka (ms, mk,),
22 - Chelsma (dns, mč, dn-ms),
23 - Zakharyino(dns, mč , dn, ms),
24 - Verkhniye Ploski, Altynovo (dns, dn, ms),
25 - Bibirevo(dns, mč, ms),
26 - Cheremoshnik (ms, mk),
27 - Dmitrov (ms, mk),
28 - Gorki (dns, mč, ms),
29 - Balashikha (dns, mč ms),
30 - Akulovo (pre-dns, dns, mč, ms),
31 - Okatovo (pre-dns, dns, ok, ms),
32 - Borovsk, Satino (pre-dn, dn, ms),
33 - Alkhimkovo (pre-dns, dns, ok, lh, dn - ms),
34 - Golutvin, Gololobovo (dn, ms),
35 - Zaraisk, Solomovo (dns, dn, mk),
36 - Alpatyevo (dns, pre-dn, mk),
37 - Troitsa, Fatyanovka (dns, ok, pre-dn, dn),
38 - Elatma, Kasimov (pre-dn, dn),
39 - Narovatovo (dns, ok, lh),
40 - Chekalin, Bryankovo (dns, ok, lh, dn),
41 - Roslavl, Konakhovka, Malakhovka, Sergeevka (pre-dns, dns, mč, ok, lh, ms),
42 - Bryansk (dn, mk),
43 - Pogar (dn, mk),
44 - Pushkari (dn, mk),
45 - Arapovichi, Mezin (dn, mk).
46 - Zheleznogorsk (pre-dn, mk),
47 - Banichi (dn, mk),
48 - Markovo (dn, mk),
49 - Igorevka (lh, dn, mk),
50 - Gadyach (dn, mk),
51 - Novoselki (dn, mk),
52 - Priluki (pre-dn, dn, mk),
53 - Ostapye, Gunki (lh, dn, mk),
54 - Gradizhsk (dn, mk),
55 - Lukoyanov (pre-dns. dns), .
56 - Tambov (dns, mč - dn),
57 - Rasskazovo (dns, mč, lh - dn),
58 - Kotovsk (dns, pre-dn),
59 - Inzhavino (dns, mč, lh, pre-dn),
60 - Muchkapsky, Korostelevo (dns, mč, lh, pre-dn, mk),
61 - Bolshaya Rzhaksa, Perevoz, Posevkino (pre-dns, dns, lh, pre-dn),
62 - Borisoglebsk (dns, mč, pre-dn, mk),
63 - Novokhopersk (pre-dns, dns),
64 - Polnoye Lapino (dns, mč),
65 - Bogdanovka, Verkhnyaya Emancha (dns, mč, lh, pre-dn),
66 - Uryv (pre-dns, dns),
67 - Klepki (pre-dns, dns),
68 - Stolinsky (pre-dns, dns),
69 - Nizhne - Dolgovsky (dns, mč),
70 - Mikhailovka (pre-dns. dns, pre-dn, mk).

faunal assemblage was recovered from ancient fluvial sediments near the Karai-Dubina (Dnieper drainage basin) by Markova (1982) who attributed it to the final stage of the Matuyama Epoch.

Akulovo Interglacial

The following interglacial, the Akulovo, was identified in the Akulovo section mentioned above and at another site at Krasikovo (south-eastern Tver region, near Kimry). It is characterized by the oldest seed flora (Maudina *et al.*, 1985), including over 30% of local and regional exotic taxa and more than 11% of extinct species. The antiquity of the Akulovo flora which bears similarities to undoubtedly Pliocene assemblages is suggested by the diversity of the coniferous pollen present with high percentages of *Pinus sec. Cembra* (up to 30-50%), *P. sec. Strobus* (up to 5-8%), as well as *P. sec. Mirabilis, P. sec. Omorica, Tsuga* and *Taxus*. Of the broad-leaved taxa (which comprise 35-50% at the climatic optimum) *Quercus, Ulmus, Carpinus, Zeltis, Zelkova, Fagus, Pterocarya, Juglans, Castanea, Ilex, Morus, Eucommia, Vitis, Ligustrum, Corylus, Ostrya* and *Myrica* are all represented (Fig. 3).

The subsequent cooling restored a tundra-like landscape to the Moscow region.

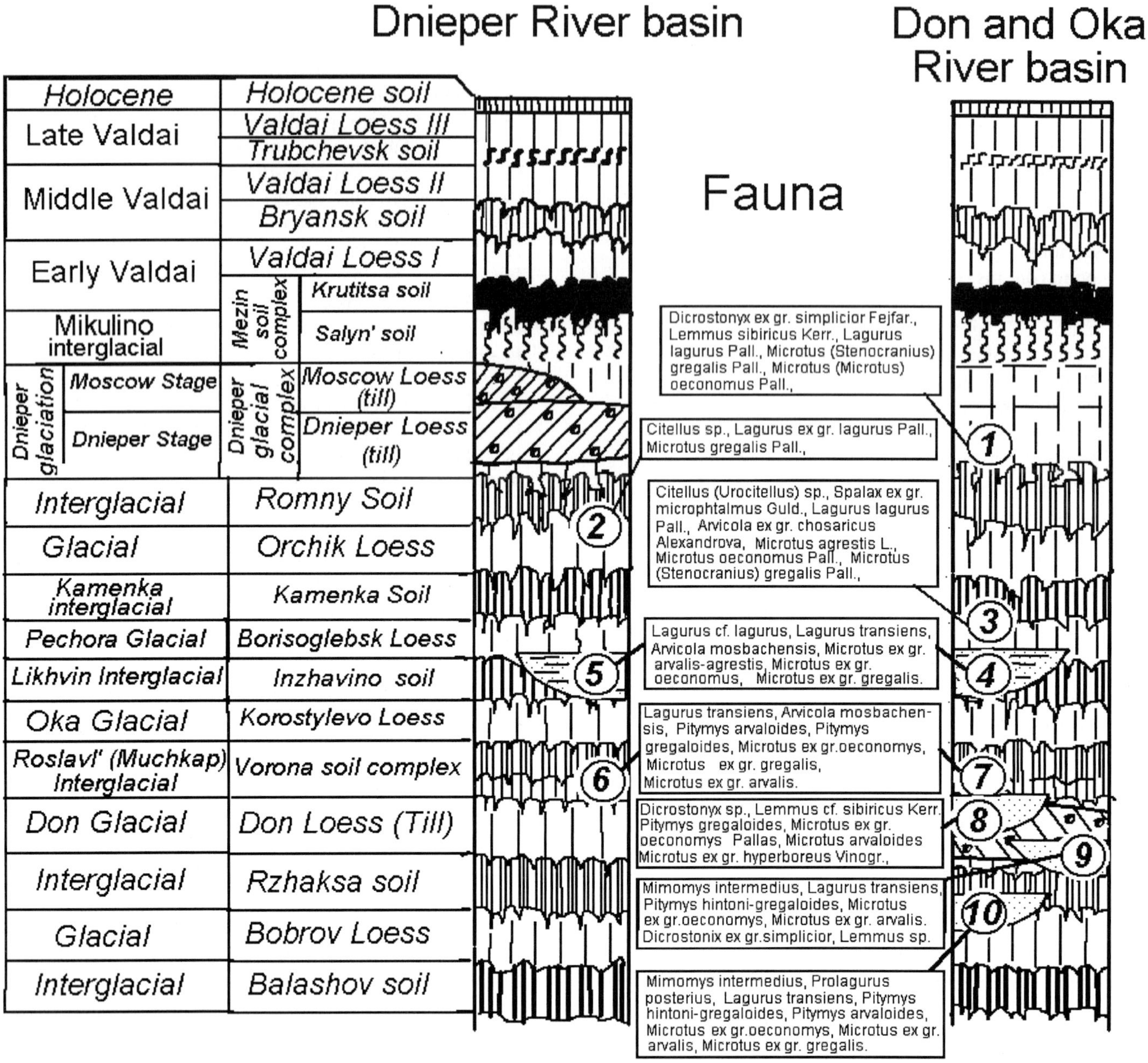

Fig. 5. Correlation between the glacial and interglacial units and stratigraphical position of tills in the Dnieper and Don river basins. Key sections: 1-Alpatjevo, Troitsa, 2-Priluki, 3-Rasskasovo, 4-Verkhnyaya Emancha, 5-Gunki, 6-Roslavl, 7-Posevkino, Perevoz, Muchkap, 8-Troitsa, 9-Bogdanovka, 10-Klepki (by Velichko et al., 1977).

Setun' Glaciation

The so-called Setun' ice sheet (named after the Setun' River in the Moscow region) expanded as far south as the northern margin of the Tula region. The Setun' Till (described by the Geological Survey team) is represented by massive brownish and greenish dark-grey loams and sandy loams. Magmatic rocks of Scandinavian provenance comprise about 40 to 60% of the clasts. The till is characterised by an amphibole-epidote-garnet-disthene mineral association.

The interglacial that followed the Setun' Glaciation has been studied in detail, however, much of its sediments were destroyed by the subsequent (and by far the largest) ice sheet. Fragmentary data on flora and fauna have so far only been obtained for the middle Volga drainage basin, in the vicinity of Nizhny Novgorod (Pisareva, 1992).

Okatovo Interglacial

The subsequent interglacial is known as the 'Okatovo' warm Stage. Its deposits were drilled west of Moscow, 4 km east of Vnukovo station, near Okatovo (Fursikova *et al.*, 1992) and near Skhodnya station, at Dubrovka. This younger flora is poorer than the Akulovo in every respect. Unlike the previously described interglacial, its climatic optimum featured polydominant broad-leaved forests (with several - 5 to 10 on the average - dominant tree species), primarily composed of oak, elm, and lime trees, and later with hornbeam and other species characteristic of a mild temperate climate (Fig. 3).

That the Akulovo and Okatovo warm stages are really separate interglacials is attested by the fact that each of them was preceded by an essential cooling when open birch forests with some cryoxerophyllic plant communities expanded into the region.

Don Glaciation

The subsequent expansion of the ice sheet known as the Don Glaciation corresponds to the maximum ice extent on the East European Plain. Its age is determined primarily from its relation to palaeontological evidence from over- and underlying sediments. Until the late 1970s it was generally assumed that the maximum glaciation on the East European Plain was the late Middle Pleistocene Dnieper (Saalian) Glaciation, and the oldest known ice sheet, the Oka ice sheet was correlated to the Elsterian Stage in western, central Europe.

New multidisciplinary studies and analysis of the relations between tills belonging to the two largest ice lobes of Russia (the Dnieper and the Don lobes) on the one hand, and loess-palaeosol periglacial sequences on the other, revealed that the tills of the two lobes differed substantially in age (Gerasimov & Velichko, 1980). Studies of loess and soil horizons exposed in many sections and analysis of palaeontological evidence allowed the identification of several glacial stages in which active loess accumulation occurred (Fig. 5). In contrast to the glacial sequence of the Dnieper lobe that was formed by the largest Middle Pleistocene ice sheet, the Don Till was deposited by the largest of all the Pleistocene ice sheets dated to the Tiraspolian. The limits of the Don ice sheet are only clearly defined east of the Central Russian Uplands, while to the west their position is indistinct. The maximum glaciation of the East European Plain, the Don Glaciation, expanded far south, into the drainage basin of the middle and lower Don (fig. 4). Unfortunately, thermoluminescence (TL) dates obtained for the till exposed in sections west of Moscow and in the Kostroma region show a wide scatter - from 365±90 to 595±150 ka BP (Fursikova *et al.*, 1992).

The correlation of the Don Glaciation to the 'Cromerian Complex' Stage has been substantiated by investigations of glacial and loess-palaeosol sequences in the Don and Dnieper drainage basins. Palaeontological evidence has confirmed beyond any doubt the late Tiraspolian age of fossil rodent remains recovered from numerous sections both above and below the Don Till (Gerasimov & Velichko, 1980; Krasnenkov *et al.*, 1997). At most sites (Vol'naya Vershina, Korotoyak-4, Korostelevo-2, Kuznetsovka and others in the Don basin, as well as in the vicinity of Roslavl', at Konakhovka, Podrudnyansky and Sergeevka in the Dnieper basin) the late Tiraspolian small mammal remains recovered from the beds immediately underlying the till are somewhat older than those found in the overlying sediments. Thus, a fauna dominated by *Lemmus* in combination with *Mimomys, Pitimys* and *Microtus oeconomus* has been identified in the overlying sediments. These assemblages compare closely to the Tiraspolian faunas in Moldavia, Czechoslovakia, Hungary, and France (Aleksandrova, 1982; Agadjanian, 1992). The investigations have provided corroborative evidence for the different age of the ice lobes extending southward into the Dnieper and Don basins, as well as for the different age of their tills. Consequently, the Cromerian palaeogeography of the East European Plain has had to be revised (Fig. 5). The Don Formation glacial sequence as a rule is between 3-5 and 15-20 m thick and consists of several till units which can be traced all over the Don lobe area. Individual strata differ slightly in both composition and clast orientation (Grishchenko, 1976; Shick, 1984; Sudakova & Faustova, 1995). The most characteristic feature of the Don Till is its abundance of small gravel, while large pebbles and boulders are only occasionally found. Most large particles are derived from local sedimentary rocks varying in composition throughout the area, whilst far-travelled clasts are chiefly granites, metamorphic rocks, and quartzite-like sandstone (Fig. 6). The maximum content of the 2 to 5 cm debris (up to 15-20%) has been found in the western portion of the lobe; in the central and eastern parts debris concentration decreases to 4-7% (Gribchenko, 1980; Maudina *et al.*, 1985).

With regard to crystalline indicator rocks, the Don Glaciation tills differ markedly from all the other tills in the central, northern and northwestern Russian Plain (Fig. 6). The fact that the Don Till lacks the main index rocks which Chirvinsky (1914), Yakovlev (1939), Viiding *et al.* (1971) have identified in the Middle and Late Pleistocene tills, suggests a more northeasterly position of the Don Glaciation centre. This is supported by a southwestern orientation of clasts in many sections (e.g. Gerasimov & Velichko, 1980). The mineralogy of the Don Till matrix varies considerably regionally. In contrast to the younger tills, minerals of local and medium distance provenance dominate, whilst the actual percentages of exotic minerals, such as hornblende, amphibole, pyroxene and biotite, do not exceed 6-10%. Amphiboles are more frequent in the western sector, whilst epidote, tourmaline, ilmenite and rutile are most typical of the central and eastern sectors. The most conspicuous result of relief-forming activities of the Don ice sheet are numerous linear depressions eroded

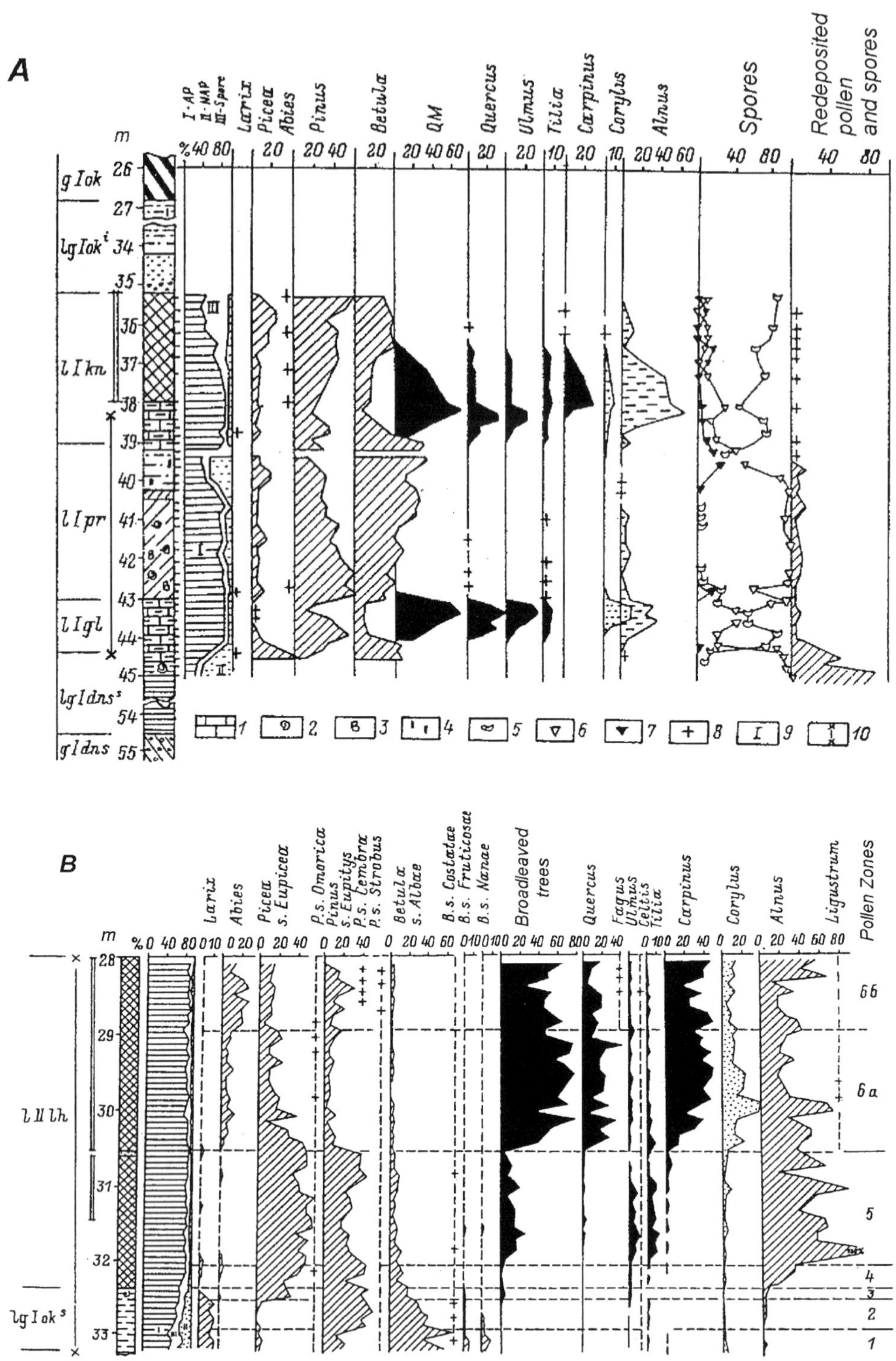

Fig. 8. Pollen diagrams of the deposits in the Roslavl stratotype area (Rumyantsev, 1998)

A-Muchkap Interglacial (Konakhovka site), B-Likhvin Interglacial (Malakhovka village)

1-marl, 2-shells of fresh-water molluscs, 3-vivianite, 4-peat inclusions, 5-Polypodiaceae, 6-Bryales, 7-Sphagnales, 8-rare grains, 9-deposits, where fossil plants were studied, 10-deposits, where the diatoms were studied.

Indices: 1- gIdns-Don till, 2-lgIdns^{s}-glacio-lacustrine clays dated to the Don ice retreat, 3-lIgl-Glazov optimum lacustrine deposits, 4-lIpr- Podrudnyanski cooling lacustrine deposits, 5-lIkn-Konakhovka optimum lacustrine deposits, 6-lgIok^{i} glaciofluvial deposits associated with the advance of the Oka ice sheet, 7-gIok-detached block of the Oka Till, 8-lgIok^{s}-glaciolacustrine deposits of the time of Oka ice retreat, 9-lIIlh-Likhvin Interglacial lacustrine deposits.

of the landscape zone dynamics (Grichuk, 1989; Pisareva, 1998). The whole area under consideration, including the southernmost regions, was positioned within the forest zone, although the forest differed in composition. South of the latitude of Moscow, a mixed polydominant coniferous-broad-leaved forest grew, in which first oak and hornbeam and later hornbeam and fir assemblages prevailed. Increased rainfall enabled the forests to expand into the Don basin. The temperate, slightly oceanic climate favoured the continued presence of relict plants, such as *Taxus, Ilex, Castanea, Buxus, Pterocarya* and *Fagus*; their pollen have been recovered from the Likhvin stratotype section at Chekalin (Grichuk, 1989). Many of the taxa only occurred occasionally at the latitude of Moscow and were completely absent farther north. There composite spruce forests with *Quercus, Carpinus* and *Abies* occurred.

A sequence of alternating warm and cold phases between the Likhvin Interglacial and Dnieper Glaciation

The results of recent investigations have shed new light on the sequence of Middle Pleistocene cold and warm stages. Additional large-scale climatic fluctuations have been recognised within the interval between the Likhvin Interglacial and the Dnieper (Saalian) Glaciation. Detailed investigations into the relationship between glacial and loess-palaeosol series in the periglacial areas have revealed that the Likhvin Interglacial is represented in these areas by the Inzhavino soil (Velichko *et al.*, 1997a). Until the late 1960s the Dnieper Glaciation was assumed to have occurred immediately after the Likhvin Interglacial. However, Moskvitin (1967) pointed out that there were palaeosols between the Likhvin and Dnieper layers. Recently, studies of loess-palaeosol sequences have confirmed the existence of an additional interglacial between the Likhvin Warm Stage and the Dnieper Glaciation which corresponds to another palaeosol, the Kamenka soil (Velichko, *et al.*, 1992, 1997a). As follows from palaeobotanic data, no less than two interstadials occurred subsequent to the Likhvin Interglacial (Fig. 3) The Kosha Interstadial is exposed in sections at the Bolshaya Kosha river near Bulatovo and Tyaglitsy in the Tver' region Grichuk, 1989; Rumyantsev, 1998). The Maryino Interstadial is known from a section in Maryino village north of Moscow (Pisareva, 1998). In addition to the interstadials, at least one warming of interglacial rank has also been recognised in these sections and in those at Lipna (in the Klyazma drainage basin) and Bibirevo (in the Yaroslavl' region) (Zarrina, 1991; Rumyantsev, 1998). At its optimum the central regions of the Russian Plain were covered with mixed coniferous broadleaved forests of somewhat limited floral diversity.

This interglacial was preceded by a deep cooling known from the central East European Plain under the name of Kaluga (Markov, 1977). It is not inconceivable to associate it with the lower Middle Pleistocene till which will be discussed below (the till is not shown in the maps nor in Fig. 4 as its spatial limits have not been ascertained as yet).

The Middle Pleistocene sequence of the northern East European Plain suggests at least two large glacial phases; two or even three till units can be distinguished in the composite geological transects across the East European Plain from north-west to south-east. The units differ considerably in their characteristics.

A distinctive feature of the lowest Middle Pleistocene till, (named Pechora after the drainage basin of the Pechora river) in both the central and the north-eastern regions, is a rather high content of local rocks and minerals. In the north-eastern regions there are sedimentary and metamorphic rocks derived from the Urals and Timan found in the till, including agate-bearing basalts. Among the minerals there are glauconite, sulphides, siderite and other minerals of local and transit provenance.

An ilmenite-garnet heavy mineral association predominates and is characterised by a remarkable presence of epidote and a relatively low content of hornblende (less then 20%). The clast orientation in the till suggests that the ice moved southwards (in the western regions) and south-eastwards (in the central and south-eastern regions An extensive ice sheet which formed three large ice streams (Belomorsky, Cheshsko-Mezensky and Pomorsky) moved into the East European Plain from the Scandinavian and the Ural - Novaya Zemlya glacial centres. Judging from the indicator-clast composition, the ice streams were related to different source areas. The Pomorsky stream flowed from the Pai-Khoi and Novaya Zemlya provinces (indicators: Silurian bituminous limestones and dolomites with coral fauna, as well as Permian and Triassic polymict sandstones). The source area of the Cheshsko-Mezensky stream was primarily in Novaya Zemlya, as indicated by boulders of rose marble-like crinoid-bryozoan Ordovician limestones. As for chronological correlation of the Pechora till, there is still disagreement among specialists. Some researchers (e.g., Andreicheva *et al.*, 1997) correlate it with the Dnieper (Saale) one, while others object against such a far-fetched correlations.

It is not inconceivable that the Pechora Till may be older than the Dnieper Till of the central regions, considering the possibility of heterochroneity in the development of the Scandinavian and Kara-Novaya Zemlya glacial centres. Therefore, the spatial correlations of the lower tills in the central and Pechora regions must be considered as tentative. The Middle Pleistocene age of the latter was demonstrated by the find of remains of *Dicrostonyx* genera both in underlying and overlying sediments.

In the northeast of the East European Plain the Pechora till is overlain with the Rodionov Interglacial deposits; stratigraphically the latter lie above the Likhvin layers. Pollen spectra present suggest that during the Rodionov Interglacial, the Pechora basin was covered by Pinus and Betula forests with Picea, Abies and some broad-leaved species, such as Ulmus and Tilia (e.g. Guslitser, 1981; Duryagina & Konovalenko, 1993). As for the central regions of the East European Plain, the Rodionov Interglacial presumably corresponds to a warming (Fig. 3) when the Kamenka soil developed in the Dnieper basin and the Chekalin one in the Oka basin (Likhvin stratotype).

Dnieper Stage

In the west (in the Middle Dnieper drainage basin, the Ukraine and Belarus) there is one Middle Pleistocene glacial complex. In the Dnieper basin, where it consists of several strata, its stratigraphical position is everywhere fixed by the overlying Mezin soil complex, the earlier part of which is related to the Mikulino Interglacial. This Dnieper Till is closest to the surface and can be traced northeastwards into the Upper Dnieper and Upper Oka drainage basins. In the Likhvin stratotype section the till rests directly on lacustrine silts dating from the early Dnieper Substage which contain a lemming fauna (Markov,

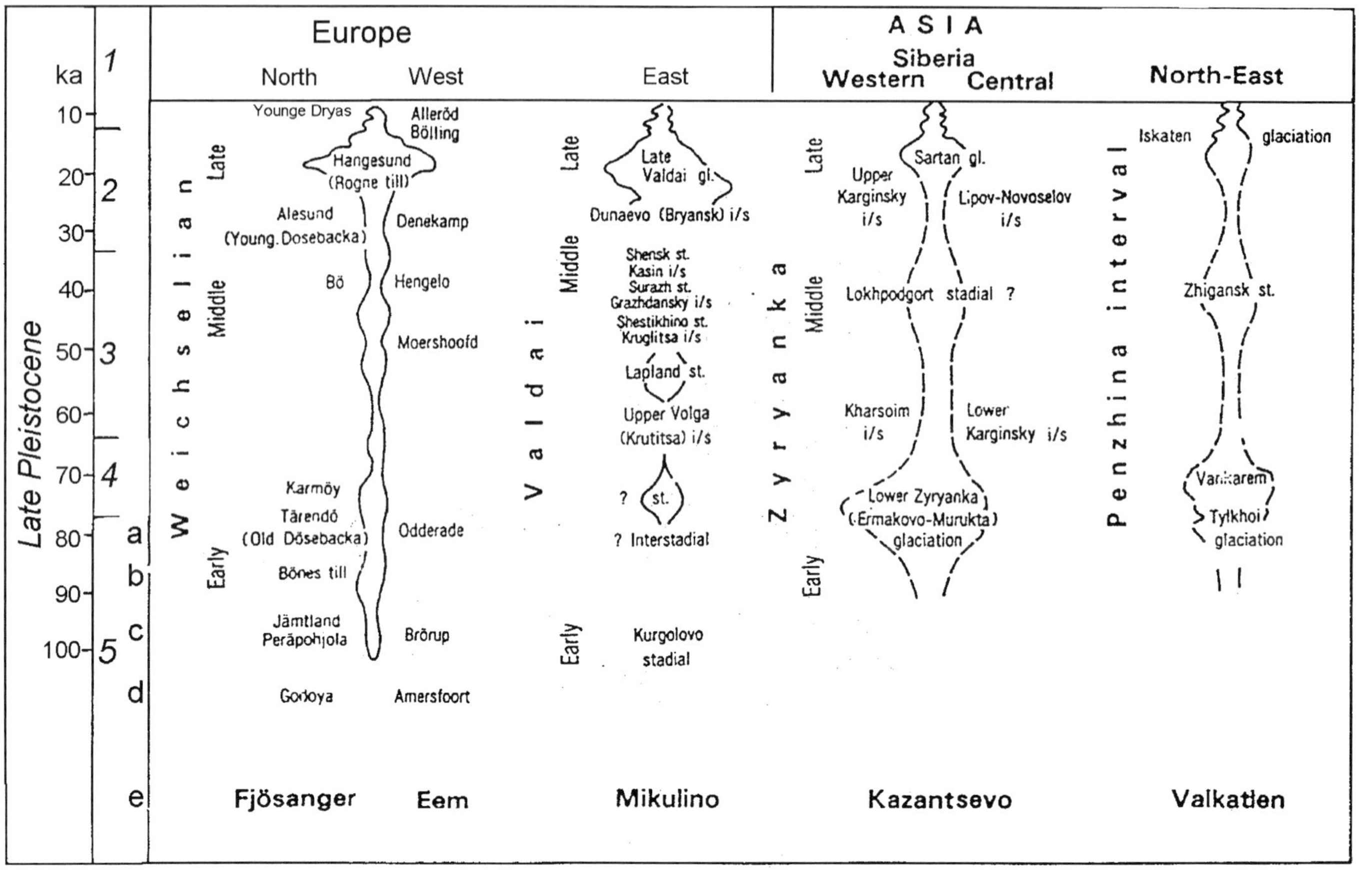

Fig. 10. Temporal and spatial asymmetry of glacial systems in the Late Pleistocene.

also present.A change in the environment about 50 ka BP marked the beginning of a new - Middle Valdai - substage charac-terized by alternating warmer and cooler phases against a background of a moderately cold and comparatively wet climate. Both warming and cooling were less prolonged, and coolings were less severe than in the Early Valdai (Gerasimov & Velichko, 1982; Arslanov *et al.*, 1981; Borisova & Faustova, 1994). In the north-west and central regions of the East European Plain climate was more severe than today. During warmer intervals a taiga-like vegetation dominated by dark coniferous species occurred, while the cold intervals were marked by an open northern taiga in combination with shrub tundra. The last warming of interstadial rank during the Middle Valdai, the so-called Dunaevo Interstadial (31 to 25 ka BP) can be broadly correlated with that recorded in the periglacial zone (the Bryansk fossil soil) (Gerasimov & Velichko, 1982; Borisova & Faustova, 1994). The subsequent pre-glacial cooling (24 to 17 ka BP) was marked by the expansion of the Late Valdai ice sheet into an open tundra-steppe land-scape (Velichko *et al.*, 1987; Nazarov, 1984; and others).

Typically the Late Valdai till includes regular changes in erratic clast composition over the area related to the flow pattern of the Scandinavian ice sheet. In comparison to older tills, these deposits include boulders and pebbles derived from more westerly source areas (the Baltic and Ladoga-Onega provinces). In addition, the Late Valdai till lithology depends to a large extent on the composition of underlying older tills, especially within the ice-marginal formation zones (Faustova & Gribchenko, 1995).

The reconstruction of the Late Valdai glacial system in northern Eurasia is based on the results of integrated international and national research programmes; important evidence has been obtained by drilling on the sea floor and on land, comprehensive studies of sections, seismo-acoustic profiling, micropalaeontological studies, and analysis of the glacial isostasy isobases. All these data suggest rather limited glaciation on the shelf and islands of the Russian Arctic (Velichko & Faustova, 1989; Velichko *et al.* 2000). During the Valdai Glaciation, the ice moved into the East European Plain from two glacial centres: Scandinavia in the NW and Pai-Khoi - Polar Urals - Novaya Zemlya in the NE (Fig. 11). In the late Valdai, the activities of the latter were restricted to the islands proper. Estimates based on geological, geomorphological and palaeobotanical infor-mation suggest that the ice sheet was less than the present land area in the archipelago (Krasnozhen *et al.*, 1987; Malyasova & Serebryanny, 1993). No fresh morainic ridges have been found south of the islands, and there is no evidence of contact between the Scandinavian and Novaya Zemlya ice sheets (moraines or glacial dislocations) (Pav-lidis, 1992).

The question of the age of glacial landforms in the northeastern part of the East European Plain is closely

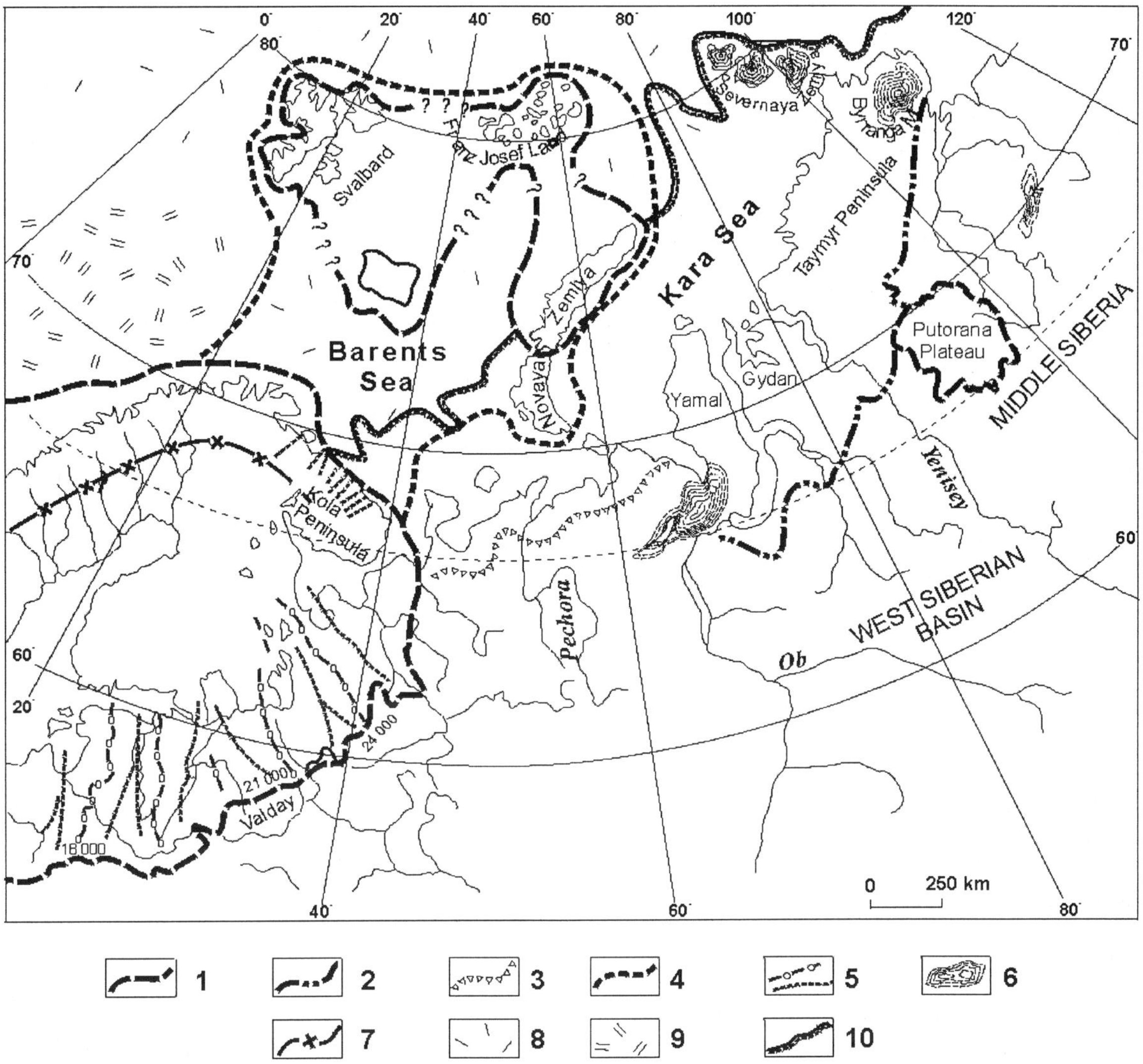

Fig. 11. Glaciation of northern Eurasia during the last glacial maximum. (by A. Velichko, M.Faustova and Yu. Kononov)
1 - glaciated areas during the last glacial maximum
2 - boundary of the Kara ice sheet after Astakhov (1979),
3 - boundary of the Novaya Zemlya ice sheet after Astakhov et al.(1998),
4 - glacial limit during the last glacial maximum according to Svendsen et al.(1999),
5- ice divides between the large ice streams and ice lobes,
6- areas of reticulated and mountain-valley glaciation during the last glacial maximum and ice cover on the shelf,
7- main ice divide of the Scandinavian ice sheet,
8 - perennial sea ice,
9 - seasonal sea ice,
10 - coastline at the last glacial maximum.

related to estimates of the Ural-Novaya Zemlya ice sheet size. Radiocarbon dates obtained for deposits underlying till near the presumed limit of the Valdai Glaciation in the north-eastern region (the Pechora drainage basin) are much older than those obtained for identical deposits near the Scandinavian ice sheet boundary (Velichko & Faustova, 1989). The dates suggest an age of 33 to 45 ka BP or older. This was originally interpreted as evidence of a heterochronous expansion of the ice sheets from the Scandinavian and northeastern glacial centres, so that the NE of the East European Plain was glaciated well before the so-called 'LGM' (Last Glacial Maximum - 20 to 18 ka BP). This interpretation disagreed with another concept, according to which the Pechora basin had been invaded by ice both during the LGM and the early Holocene - Preboreal periods (Lavrov & Potapenko, 1989). This was based on preliminary investigations of the Markhida section (Pechora River basin). However, the results of recent geological-geomorphological studies of the section, together with newly-obtained radiocarbon dates (Mangerud *et al.*, 1996) have revealed that the submorainic sediments predated by far the LGM, thus confirming the concept of an older (in all probability - Early Valdai) age of the ice sheet in the Pechora basin. In more recent publications the cited authors (Svendsen *et al.*, 1999) consider the glacial relief age in NE European Russia to be older than Late Valdaian. The youngest till in the extreme northeast of the East European Plain contains material of Ural - Pai-Khoi - Novaya Zemlya provenance (Andreicheva, 1992). Farther west, in the northwest of the Pechora lowland, Fennoscandian material predominates, with Palaeozoic dark grey and black limestones and boulders of rosé crinoid-bryozoan limestones indicative of the northeastern glacial centre.

The most extensive glacial advance during the Valdai was from the Scandinavian centre. At its maximum, this ice sheet approached the upper reaches of Nieman, Dnieper and Zapadnaya Dvina rivers, invaded the Mologa-Sheksna lowland and extended into the Onega basin.

The Scandinavian ice sheet advanced onto the East European Plain forming several large ice streams - Baltic, Chudsky, Ladoga, and Onega - Karelian ones; the relief-forming activities of the streams were discussed in detail in a number of monographs and papers (e.g. Chebotareva, 1977; Gerasimov & Velichko, 1982). Three major phases of deglaciation have been distinguished; they differ in degree of ice-marginal mobility. The initial phase (before 16 ka BP) is noted for its gradual and synchronous retreat of the ice margin (a regressive deglaciation), the integrity of the ice sheet being preserved, together with the outlines of the ice lobes. Well-pronounced marginal deposition left linear or festoon-like (large festoons) landform assemblages. The ice margin retreated spasmodically, phases of rapid ice melt apparently alternated with short, cool intervals of stagnation. The receding ice-marginal zones can be easily correlated, thus enabling the reconstruction of the ice-margin configuration. In the outermost about 30 km wide zone, the ice thickness presumably did not exceed 100-150 m. The dominant landforms are flat-topped pre-Valdai elevations covered by a thin mantle of clayey Valdai Till in association with the diverse landforms related to dead-ice found on slopes of higher areas and in meltwater valleys. The area north of the peripheral zone was modelled by thicker ice which became detached from the thin, stagnant peripheral ice in the process of retreat. The retreating ice tongues, still highly active, formed 2 to 4 belts of hills and ridges, often composed of sand and gravel. At the contacts of adjacent ice lobes broad zones of radial deposition developed, typified by isolated uplands and hilly massifs, while zones of ice divergence are marked by angular massifs, often associated with deep linear depressions. The proglacial zone included vast ice-dammed basins and smaller glaciofluvial plains.

The type of deglaciation changed after a short but pronounced glacial readvance along the whole ice front; this Vepsovo (= Pomeranian) advance occurred at about 15.5 ka BP. It was preceded by a cold interval. During the advance, the ice margin became twice as dissected as before and oscillated incessantly. The well-defined end moraines of this phase form a continuous belt from the Baltic ridge, in the west, to the Onega drainage basin in the east. Numerous minor radial landforms were formed, such as 'inter-tongue' angular massifs. Twice during the retreat the ice margin underwent reactivation, as shown by the Luga and Neva ice-marginal formations. As ice thickness decreased, all major oscillations ceased. This accounts for the narrower ice-marginal zones and thinner glacial deposits. The ice itself molded to the underlying topography, and transgressive-regressive lobate deglaciation gave way to individual decaying glacier tongues. This second stage of deglaciation was characterised by a rapid transition from active to passive ice. The most important of these glacial 'litho-morpho-complexes' are various in-version landforms related to dead-ice (zvontsy, for example) which are closely associated genetically to active-ice landforms, such as terminal moraines and hummocky topography. At the second stage of deglaciation substantial proglacial lakes appeared which rapidly changed their margins as the ice retreated; some smaller lakes came into being as a result of thawing of isolated ice blocks, together with ice-dammed lakes in the marginal zones. The proglacial zone featured various levels of valley trains and multiple-cone deltas. At the end of that stage, in the Bølling, the East European Plain became completely ice-free (Faustova, 1994).

The last stage of deglaciation (predominantly areal decay) began about 11 ka BP. Its marginal formations are found in the adjacent Baltic region. In Russia, the last advance of the ice front in the Younger Dryas only affected Karelia and the Kola Peninsula (Velichko, 1993).

References

Agadjanian, A.K. (1992). Stages of small mammals evolition in the central regions of the Russian Plain. *In: Quaternary stratigraphy and palaeogeography of Eastern Europe*. Moscow, Institute of Geography, Russian Academy of Sciences, 37-49 (in Russian).

Aleksandrova, L.P. (1982). A new species of Dicrostonyx okaensis and its significance for Oka till dating in the Likhvin stratotypical section. *In: Stratigraphy and palaeogeography of the Anthropogene*. Moscow, Nauka Publishers, 17-21 (in Russian).

Andreicheva, L.N. (1992). *Tills of the European North-East of Russia and their lithostratigraphic significance*. St.-Petersburg, Nauka Publishers, 125 pp (in Russian).

Andreicheva, L.N., Nemtsova, G.M. & Sudakova, N.G. (1997). *Middle Pleistocene tills of the North and Centre of the Russian Plain*. Ekaterinburg, Uralian Branch of the Komi Research Center, Russian Academy of Sciences. 83 pp (in Russian).

Arslanov, Kh.A., Breslav, S.L., Zarrina, E.P., Znamenskaya O.M., Krasnov I.I., Spiridonova E.A. (1981). Climastratigraphy and chronology of the Middle Valdai in the northwest and centre of the Russian Plain. *In:* Velichko, A.A. & Faustova, M.A. (eds), *Pleistocene glaciations of the East European Plain*. Moscow, Nauka Publishers, 12-27 (in Russian).

Astakhov V.I. (1979). New data on the latest activity of Kara-shelf glaciers in West Siberia. *Project 73/1/24 Quaternary Glaciations in the Northern Hemisphere. Report*, **5**, Prague, 22-31.

Astakhov V.I. (1998). The last ice sheet of the Kara Sea: terrestrial constraints on its age. *Quaternary International* **45/46**, 19-28

Biryukov, I.P., Agadjanian, A.K., Valueva, M.N., Velichkevich, F.Yu. & Shik, S.M. (1992). Quaternary deposits of the Roslavl stratotype region. *In:* Velichko, A.A. & Shik, S.M. (eds), *Quaternary stratigraphy and palaeogeography of Eastern Europe*. Moscow, Institute of Geography Russian Academy of Sciences, 152-180 (in Russian).

Bolikhovskaya, N.S. (1995). *Evolution of the loess-soil formation of Northern Eurasia*. Moscow University Press. 270 pp. (in Russian).

Borisova, O.K. & Faustova, M.A. (1994). Sequence of natural phases of the Valdai ice age in European Russia. *In:* Velichko, A.A. & Starkel, L. (eds), *Paleogeographical basis of the modern landscapes*. Moscow, Nauka Publishers, 17-25 (in Russian).

Chebotareva, N.S. (ed.) (1977). *Structure and dynamics of the last ice sheet of Europe*. Moscow, Nauka Publishers, 143 pp (in Russian).

Chirvinsky, V.N. (1914). Materials to understanding of chemical and petrographic composition of glacial deposits in the south-west of Russia in connection with problem of ice sheet motion. *Zapiski Kiev. obshchestva esetestvoispytateley*, **24**, 316 pp. (in Russian).

Duryagina, D.A. & Konovalenko, L.A. (1993). *The Pleistocene palynology of the north-east of the European Russia*. St.-Petersburg, Nauka Publishers, 124 pp (in Russian).

Faustova, M.A. (1994). Deglaciation and types of glacial relief in the European part of Russia. *In:* Velichko, A.A. & Starkel, L. (eds), *Paleogeographical basis of the modern landscapes*. Moscow, Nauka Publishers, 30-40 (in Russian).

Faustova, M.A. & Gribchenko, Yu.N. (1995). Lithology of glacial deposits of the last Pleistocene glaciation. *In:* Ehlers, J., Kozarski, S. & Gibbard, P.L. (eds), *Glacial deposits in North-East Europe*, Rotterdam, A.A.Balkema, 183-188.

Frenzel, B. (1967). *Die Klimaschwankungen des Eiszeitalters*. Braunschweig, Vieweg, 296 pp.

Fursikova, I.V., Pisareva, V.V., Yakubovskaya, T.V., Vlasov V.K., Kulikov O.A. & Semenenko L.T. (1992). Key section of the Pleistocene near Okatovo Village west of Moscow. *In:* Shik, S.M. (ed.), *Phanerozoic stratigraphy of the centre of the East European platform*. Moscow, Tsentrgeologiya Publishers, 59-82 (in Russian).

Gerasimov, I.P. & Velichko, A.A. (eds) (1980). *Age and distribution of the maximum glaciation in Eastern Europe*. Moscow, Nauka Publishers, 212 pp. (in Russian).

Gerasimov, I.P. & Velichko, A.A. (eds) (1982). *Palaeogeography of Europe during the last 100 000 years (atlas-monograph)*. Moscow, Nauka Publishers, 156 pp. (in Russian).

Goretskiy, G.I., Chebotareva, N.S. & Shick, S.M. (eds) (1982). *Moscow ice sheet of Eastern Europe*. Moscow, Nauka Publishers, 240 pp. (in Russian).

Gribchenko, Yu.N. (1980). Petrographic composition of tills of Dnieper and Don glacial lobes. *In:* Gerasimov, I.P. & Velichko, A.A. (eds), *Age and distribution of the maximum glaciation in Eastern Europe*. Moscow, Nauka Publishers, 73-88 (in Russian).

Grichuk, V.P. (1981). The oldest ice sheet in Europe - its indicators and stratigraphic position. In: Problems of palaeogeography of the Pleistocene in glacial and periglacial regions. Moscow, Nauka Publishers, 7-35 (in Russian).

Grichuk, V.P. (1989). *History of flora and vegetation of the Russian Plain in the Pleistocene*. Moscow, Nauka Publishers, 175 pp (in Russian).

Grishchenko, M.N. (1976). *Pleistocene and Holocene of the Upper Don drainage basin*. Moscow, Nauka Publishers, 228 pp (in Russian).

Guslitser, B.I. (1981). Comparison of Pleistocene sections in the Pechora and Vychegda drainage basins on the basis of fossil rodent remains. *In:* Yakhimovich, V.L. (ed.), *Pliocene and Pleistocene of the Volga-Ural region*. Moscow, Nauka Publishers, 28-37 (in Russian).

Janczyk-Kopikowa, Z. (1975). Flora interglacialu Mazowieckiego w Ferdynandowie. *Z. badan czwartorzedu w Polsce*, **17**, 5-70.

Krasnenkov, R.V., Iosifova, Yu.I. & Semenov, V.V. (1997). The Upper Don drainage basin - an important stratoregion for climatic stratigraphy of the lower Middle Pleistocene (Lower Neopleistocene). of Russia. *In:* Alekseev, M.N. & Khoreva, I.M. (eds), *Quaternary geology and palaeogeography*. Moscow, Geos, 82-96 (in Russian).

Krasnozhen, A.S., Baranovskaya, A.F., Zarkhidze, V.S., Malyasova, E.S. & Lev, O.M. (1987). Stages of evolution and the Cenozoic mantle structure of the Novaya Zemlya archipelago. *In: Cenozoic deposits of the European North-East and placer geology*. Syktyvkar, Proceedings of the Uralian Branch of the Komi Research Center, the USSR Academy of Sciences, 34-36 (in Russian).

Lavrov, A.S. & Potapenko, L.M. (1989). Comparative description of the Late Pleistocene glacial formation and terraces of the North Pechora and West Siberian lowlands. *In:* Velichko, A.A., Gurtovaya, E.E. & Faustova, M.A. (eds), *Palaeoclimates and glaciations during the Pleistocene*. Moscow, Nauka Publishers, 204-211 (in Russian).

Malyasova, E.S. & Serebryanny, L.R. (1993). Natural history of the Novaya Zemlya. *Novaya Zemlya,* **2**, 10-22 (in Russian).

Mangerud, J., Svendsen, J.I., Astakhov, V.I. & Tveranger, J. (1996). Glaciations and palaeoenvironments in the European Russian Arctic: the Pechora basin. *In: Quaternary Environment of the European North (QUEEN)*. Oral contributions - 1 Annual Workshop, Strasbourg, France, Nov. 29-Dec. 2, 1996. 81-82.

Markov, K.K. (ed.). (1977). *Sections of glacial deposits in the central Russian Plain.* Moscow University Press. 198 pp. (in Russian).

Markova, A.K. (1982). *Pleistocene rodents of the Russian Plain.* Moscow, Nauka Publishers, 185 pp. (in Russian).

Maudina, M.I., Pisareva, V.V. & Velichkevich, F.Yu. (1985). The Odintsovo stratotype in the light of new data. *Doklady Akademia Nauk SSSR*, **284 (5)**, 1195-1199 (in Russian).

Moskvitin, A.I. (1967). *Pleistocene stratigraphy of the European part of the USSR.* Moscow, Nauka. 236 pp. (in Russian).

Nazarov, V.I. (1984). *Anthropogene insects of the north-eastern Byelorussia and adjacent regions.* Minsk, Nauka i Technika, 115 pp. (in Russian).

Pavlidis, Yu.A., 1992. *The World Ocean shelf during the Late Quaternary.* Moscow, Nauka Publishers, 272 pp. (in Russian).

Pisareva, V.V. (1992). Interglacial deposits of the Lukoyanov city. *In:* Shik, S.M. (ed.), *Phanerozoic stratigraphy of the centre of the East European platform.* Moscow, Tsentrgeologiya Publishers, 82-96 (in Russian).

Pisareva, V.V. (1997). Flora and vegetation of the Early and Middle Pleistocene of the Russian Plain. *In:* Alekseev, M.N. & Khoreva, I.M. (eds), *Quaternary geology and palaeogeography*, Moscow, Geos, 124-133 (in Russian).

Pisareva, V.V. (1998). Palaeoclimates of the Pleistocene of the East European Plain. *In:* Palaeoclimates and evolution of palaeoenvironments in the geological history of the Earth. *Abstracts of papers of International Symposium, August 27-31, 1998, Petrozavodsk, Russia*, 71-72 (in Russian).

Robertsson, A.-M., Ringberg, B., Miller, U. & Brunnberg, L (eds). (1992). Quaternary stratigraphy, glacial morphology and environmental changes. *Sveriges Geologiska Undersökning*, **Ca 81**, 380 pp.

Rumyantsev V.A. (ed). (1998). *History of the Pleistocene lakes of the East European Plain.* St.-Petersburg, Nauka Publishers, 404 pp. (in Russian).

Shick, S.M. (ed.) (1984). *Key sections of the Lower Pleistocene of the Upper Don drainage basin.* Voronezh University Press. 213 pp. (in Russian).

Sudakova, N.G. (1990). *Palaeogeographic regularities of glacial lithogenesis.* Moscow University Press. 159 pp. (in Russian).

Sudakova, N.G. & Faustova, M.A. (1995). Ice sheets of the Middle and Late Pleistocene on the East European Plain. *In: Climate and environment changes of East Europe during Holocene and Late-Middle Pleistocene.* Moscow. August 14-18, 69-75 (in Russian).

Svendsen, J.I., Astakhov, V.I., Bolshiyanov, D.Yu., Demidov, I., Dowdeswell, J.A., Gataullin, V., Hjort, Ch., Hubberten, H.W., Larsen, E., Mangerud, J., Melles, M., Möller, P., Saarnisto, M. & Siegert, M.J. (1999). Maximum extent of the Eurasian ice sheets in the Barents and Kara Sea region during the Weichselian. *Boreas*, **28**, 234-242

Velichko, A.A. (ed.) (1993). *Evolution of landscapes and climates of Northern Eurasia.* Moscow, Nauka, 102 pp.

Velichko, A.A. & Faustova, M.A. (1989). Reconstruction of the last Late Pleistocene glaciation of the Northern Hemisphere (18-20 ka BP). *Doklady Akademia Nauk SSSR*, **309 (6)**, 1465-1468 (in Russian).

Velichko, A.A. & Shik, S.M. (eds) (1992). *Quaternary stratigraphy and paleogeography of Eastern Europe.* Moscow, Institute of Geography, Russian Academy of Sciences, 245 pp.

Velichko, A.A., Isaeva, L.L. & Faustova, M.A. (eds) (1987). *Quaternary glaciations on the territory of the USSR.* Moscow, Nauka Publishers, 128 pp (in Russian).

Velichko, A.A., Gribchenko, Yu.N., Gubonina, Z.P., Morozova, T.D., Nechaev, V.P., Sycheva, S.A., Timireva, S.N., Udartsev, V.P., Khalcheva, T.A., Tsatskin, A.I. & Chikolini, N.I. (1997a). *Loess-soil formation of the East European Plain. Palaeogeography and stratigraphy.* Moscow, Institute of Geography, Russian Academy of Sciences, 140 pp. (in Russian).

Velichko, A.A., Kononov, Yu.M. & Faustova, M.A. (1997b). The last glaciation of the Earth: size and volume of ice-sheets. *Quaternary International*, **41/42**, 43-51.

Velichko, A.A., Kononov, Yu.M. & Faustova, M.A. (2000). Geochronology, distribution, and ice volume on the Earth during the Last Glacial Maximum: Inferences from new data. *Stratigraphy and Geological Correlation*, **6, (1)**, 1-12.

Velichko, A.A., Morozova, T.D., Nechaev, V.P., Udartsev, V.P. & Tsatskin, A.I. (1992). Problems of chronostratigraphy and correlation of the loess-soil formation of the Russian Plain. *In:* Velichko, A.A. & Shik, S.M. (eds), *Quaternary stratigraphy and palaeo-geography of Eastern Europe.* Moscow, Institute of Geography, Russian Academy of Sciences, 115-140 (in Russian).

Velichko, A.A., Udartsev, V.P., Markova, A.K., Morozova, T.D., Pevzner, M.A., Gribchenko, Yu.N. & Sycheva, S.A. (1977). New notions of the age of the Dnieper and Don lobes on the Russian Plain. *Izvestia Akademia Nauk SSSR, Seria Geograficheskaya*, **6**, 25-36.

Viiding, H., Gaigalas, A., Gudelis, V., Raukas, A. & Tarvidas, R. (1971). *Crystalline index boulders of the Baltic region.* Vilnius, Mintis Publishers, 95 pp. (in Russian).

Yakovlev, S.A. (1939). Index boulders, tills and limits of the Novaya Zemlya ice sheet on the Russian Plain. *Bulletin of the Commission on Quaternary, USSR Academy of Sciences*, **5**, 18-26 (in Russian).

Yevzerov, V.Ya. & Koshechkin, B.N. (1980). *Paleogeography of the western Kola Peninsula during the Pleistocene.* Leningrad, Nauka, 170 pp (in Russian).

Zarrina, E.P. (1991). *Quaternary deposits of north-western and central regions of the European part of the USSR.* Leningrad, Nedra, 187 pp (in Russian).

Zarrina, E.P. & Shik, S.M. (eds) (2000). *Problems of Quaternary stratigraphy and marginal glacial formations of the Vologda Region (NW Russia). Materials of International Symposium.* Moscow, GEOS, 100 pp.

Quaternary Glaciations – Extent and Chronology
Editors J. Ehlers and P.L. Gibbard

On the age and extent of the maximum Late Pleistocene ice advance along the Baltic-Caspian watershed

V.P. Gey [1], V.V. Kozlov [2] and D.B. Malakhovsky [3]

[1] *Geological Expedition, Odoyevskogo St., 24/1, 199155 St. Petersburg, Russia*
[2] *Smolensk State Pedagogical University, Przevalskogo St., 4, 214000 Smolensk, Russia*
[3] *St. Petersburg State University, Faculty of Geography and Geoecology, 10th Line, 33-35, 199178 St. Petersburg, Russia.*

In northwestern European Russia glacial limits are determined using both geomorphological (e.g., marginal features, drainage network density) and stratigraphical (the presence or absence of till overlying interglacial and interstadial sequences) criteria. Since Markov's work (1940) the stratigraphic approach has increasingly been gaining prevalence. Other parameters, such as distribution of periglacial loess-like silts, lithology of erratics, sub-till topography, proglacial reservoirs, soil thickness and lake density have also been taken into account. The area of the last Scandinavian ice sheet has traditionally been associated with the lake country fringed from the southeast by the wide belt of diverse glacial accumulations on the interfluvial Valdai Upland running between St. Petersburg and Moscow.

Until the late 1940s there was general agreement that a single Valdaian glaciation had followed the single Late Pleistocene interglacial (the Mikulinian). However, in 1950 Moskvitin suggested that a Mologa-Sheksna interglacial divided the large, Valdaian ice age into two glaciations – the Kalinin and Ostashkov (Moskvitin, 1950). This concept was adopted by Yakovlev (1956), Apukhtin & Krasnov (1967), Punning *et al.* (1969), Breslav (1971) and Zarrina (1991). Other investigators maintained that the Valdai glaciation was interrupted only by climatic ameliorations of inter-stadial rank. According to Malakhovsky and Markov (1969), Chebotareva (1974) and Arslanov (1987) the longest, Middle Valdaian interstadial is the equivalent of the so-called `Mologa-Sheksna Interglacial`.

Another long-standing controversy concerns the extent of the two ice advances, namely, which ice sheet, Early or Late Valdaian, was larger and where the major ice limit was located. Vozniachuk (1973) and Chebotareva & Makarycheva (1974) defended an ice-free Early Valdaian, whilst Velichko (1993) suggested the time-transgressive Late Valdai ice limit. In various publications the Valdaian glacial maximum was attributed either to the time around 20 ka BP, or to a period beyond 50 ka ago, and its position in the Upper Volga catchment area varied by 200-300 km (cf. maps by Krasnov, 1971 and Aseyev, 1981).

These disputes largely stemmed from the controversy concerning the origin of the sediments overlying the Mikulino (Eemian) interglacial sequences. Authors advocating an Early Valdaian glacial maximum tend to interpret all superficial diamictons as basal tills. For instance, Sudakova *et al.*, (1997) maintain that the 1-2 m thick diamicton, in places overlying Mikulino interglacial sediments distal to the main belt of terminal moraines, is an Early Valdaian-age basal till. This interpretation implies that the Late Pleistocene ice reached as far as only 20 km north of Moscow, i.e. 230 km southeast from the Late Valdai limit shown in the present digital map. However, Kozlov (1972) and other geologists have repeatedly demonstrated that the known interglacial sequences south of the main belt of Late Valdai end moraines are often overlain by a thin, soft, weathered diamicton instead of the normal glacigenic sedimentary complex. This diamictic layer pinches out upslope and has weak, slope-dependent bedding and fabrics, i.e. it bears signs of redeposition by slopewash or solifluction. Such sites are interspersed with interglacial sequences that lack any Pleistocene sediment cover.

In standard maps of the Geological Survey the thin discontinuous diamictic mantle is usually not regarded as a reliable signature of an ice advance (Auslender, 1989), the attitude shared by the present authors. In the digital map such sequences of Mikulino and Valdai interstadial sediments are also shown as not overlain by till. Accordingly, the limit of the maximum Late Pleistocene ice advance is located distally but close to the main marginal belt comprising diverse hummocky glacigenic accumulations (Gey & Malakhovsky, 1998). However, the geomorphological evidence is not straightforward everywhere. The most well-expressed push moraines, such as the Vepsa and Krestsy stadial formations, are often situated proximally to the maximum ice limit (Chebotareva, 1977). Moreover, in the flat lowland along the Suda river, in the Babayevo-Vesyegonsk area, the till is obscured by glaciolacustrine sediments. Two options for the ice limit in this area are shown on the digital map. Recent data indicate that the proximal option is more correct (Lunkka *et al.*, 2001).

Recently, the stratigraphic knowledge has improved significantly. The `Mologa-Sheksna interglacial` in its type

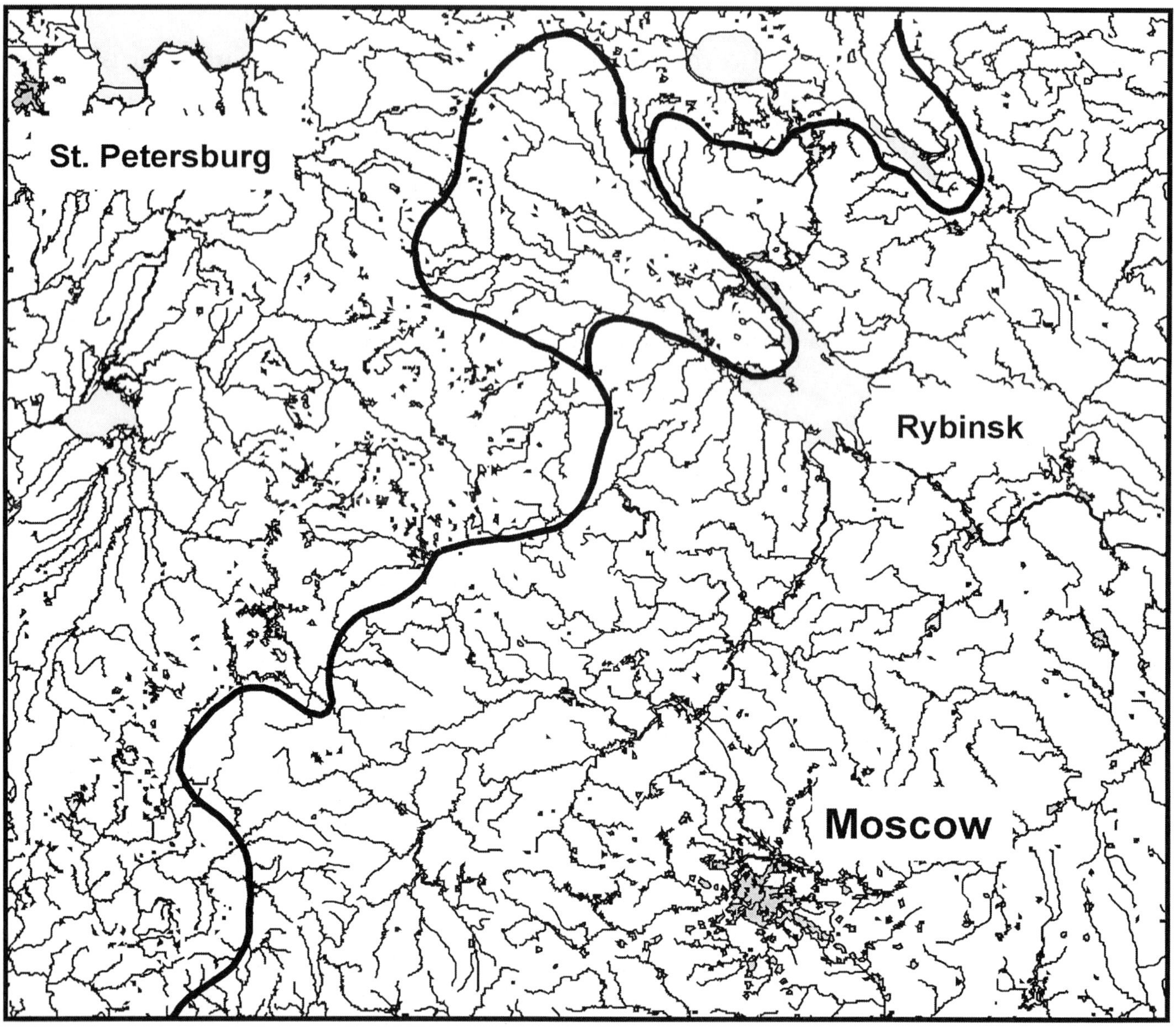

Fig. 1. Position of the maximum LGM in the Baltic-Caspian watershed region.

sections proved actually to represent the Mikulinian (Eemian) interglacial stage and was therefore deleted from the regional stratigraphical scale. Instead an interstadial sequence with the palaeoclimatic characteristics colder than those at present and radiocarbon dates ranging from 25 to 50 ka BP, was chosen as stratotypic for the Middle Valdai (Krasnov & Zarrina, 1986; Velichko & Faustova, 1986). The main climatic phases of the Mikulinian and Valdaian Stages are now more reliably distinguished on the basis of pollen analysis. Therefore, not only sites with interglacial sediments, but also those including Late, Middle and Early Valdaian interstadial sequences, can be used for locating the ice margin, depending on whether they are covered or not by Valdaian-age tills.

The most complete Upper Pleistocene stratigraphical sequence has been described from a borehole which penetrated 90 m of lacustrine sediments overlying Middle Pleistocene tills near the city of Dmitrov at the 1st May Factory, 56°N. The coldest and driest phase, indicated by pollen and diatom assemblages, is found in the interval between 30 and 12 ^{14}C ky BP. However, no Valdaian tills occur in this succession (Semenenko *et al.*, 1981). As to the Early Valdaian time, pollen assemblages in sections from the periglacial zone (e.g., 1st May Factory, Rogachovo, Shestikhino) show unmistakable signs of cooling suggesting a possible glaciation elsewhere. How-ever, basal till lying between the Mikulinian and Middle Valdaian formations has only been found in the very far north of the region (Podporozhye, Petrozavodsk and other sites north of 61°N).

The post-Eemian glacial maximum in northwestern European Russia, according to many works (Vigdorchik *et al.*, 1971; Auslender, 1989; Faustova, 1995; Gey & Malakhovsky, 1998;. Gey *et al.*, 2001), occurred 18-20 ^{14}C ky BP. This estimate, based on conventional radiocarbon dating, has lately been refined by the Russio-Finnish team who applied combined modern techniques of optically stimulated luminescence and AMS radiocarbon methods. They found that the maximum ice advance in the Vologda Region occurred *c.* 18 cal ky BP (Lunkka *et al.*, 2001).

Younger standstills of the retreating ice margin based largely on geomorphological evidence, are more controversial (Vigdorchik *et al.*, 1971). The age brackets for the main retreat stadials the Vepsa-Krestsy, Luga and Neva are determined mainly from evidence obtained from the adjacent Baltic states (Velichko & Faustova, 1986). Data available in the region under discussion suggest that these short-lived readvances occurred during the interval 17 - 13 ka BP. Recently, Saarnisto & Saarinen (2001), using varve chronology, palaeomagnetic events and AMS radiocarbon dating, concluded that the Luga Stadial, in the south of Lake Onega, took place slightly earlier than 14,250 cal yr BP, and the Neva Stadial occurred *c.* 13,300 cal yr BP, i.e. just before the Allerød interstadial.

References

Apukhtin, N.I. and Krasnov, I.I. (eds). (1967). *Geology of the Quaternary deposits of the northwestern European USSR. (*Geologia chetvertichnykh otlozheniy Severo-Zapada Yevropeiskoi chasti SSSR). Leningrad, Nedra, 265 pp. (in Russian).

Arslanov, Kh.A. (1987). *Radiocarbon: geochemistry and geochronology* (Radiouglerod: geokhimia i geokhronologia). Leningrad University, 297 pp. (in Russian).

Aseyev, A.A.(ed.) (1981). *Geomorphological Map of the USSR, scale 1:2,500,00.* Moscow.

Auslender, V.G. (1989). *National Geological Map of the USSR, scale 1:1,000,000, new series. Sheet O-35,36, Quaternary deposits.* Leningrad, VSEGEI.

Breslav, S.L. (1971). *Quaternary deposits. In: Geologia SSSR,* **IV.** *Tsentr Yevropeiskoi chasti SSSR, geologicheskoye opisaniye.* Moscow, Nauka, 74-95 (in Russian).

Chebotareva, N.S. & Makarycheva, I.A.(1974). *The last glaciation of Europe and its geochronology.* (Posledneye oledeneniye Yevropy i yego geokhronologia). Moscow, Nauka, 157 pp. (in Russian).

Chebotareva, N.S., (ed.) (1977). *The structure and dynamics of the last ice sheet of Europe* (Struktura i dinamika poslednego lednikovogo pokrova Evropy*).* Nauka, Moscow, 143 pp. (in Russian, English summary).

Faustova, M. (1995). Glacial stratigraphy of the Late Pleistocene in the northwestern part of the Russian Plain. *In:* Ehlers J., Kozarski, S. and Gibbard, Ph. (eds).*Glacial deposits in north-east Europe.* Rotterdam, Balkema, 179–182.

Gey, V.P.& Malakhovsky, D.B. (1998). On age and extent of the Late Pleistocene ice advance in the western Vologda Region. *Izvestia, Russian Geographical Society,* **1**, 43–53 (in Russian).

Gey, V. Saarnisto, M., Lunkka, J.P. & Demidov, I. (2001). Mikulino and Valdai palaeoenvironments in the Vologda area, NW Russia. *Global and Planetary Change,* **31,** 347–366.

Kozlov, V.B. (1972). On location of the limit of the last glaciation in the catchment area of the Upper Volga and Upper Dnieper. *In:* Goretsky, G.I., Pogulyayev, D.I. & Shik, S.M. (eds). *Krayevye obrazovania materikovykh oledeneniy.* Moscow, Nauka, 73–80 (in Russian).

Krasnov, I.I. (ed.) (1971). *Map of Quaternary deposits of the European USSR and adjacent areas, scale 1:1,500,000.* Leningrad, VSEGEI, 16 sheets.

Krasnov, I.I.& Zarrina, Ye.P. (eds) (1986). *Resolution of the 2nd Inter-Departmental Stratigraphic Conference on the Quaternary of the East European Platform* (Resheniye 2-go Mezhvedomstvennogo stratigraficheskogo soveshchaniya po chetvertichnoy sisteme Vostochno-Yevropeiskoi platformy). Leningrad, VSEGEI,157 pp. (in Russian).

Lunkka, J.P., Saarnisto, M., Gey, V., Demidov, I. & Kiselova, V. (2001). Extent and age of the Last Glacial Maximum in the southeastern sector of the Scandinavian Ice Sheet. *Global and Planetary Change,* 31, 407–425.

Malakhovsky, D.B. and Markov, K.K. (eds) (1969). *The geomorphology and Quaternary deposits of the northwestern European USSR (*Geomorfologia i chetvetichnye otlozhenia Severo-Zapada Yevropeiskoi chasti SSSR). Leningrad, Nauka, 257 pp. (in Russian).

Markov, K.K. (1940). Materials to the Quaternary stratigraphy of the Upper Volga catchment area. *Trudy Verkhnevolzhskoi expeditsii,* **1,** Leningrad, Geografo-ekonomichesky institut*,* 69 pp. (in Russian).

Moskvitin, A.I. (1950). *The Würm epoch (Neopleistocene) in the European USSR.* (Vurmskaya epokha – neopleistotsen- v Yevropeiskoi chasti SSSR). Moscow, Academy of Sciences of USSR, 365 pp. (in Russian).

Punning, J.-M., Raukas, A.V. & Serebryanny, L.R. (1969). The Karakula interglacial deposits of the Russian Plain (stratigraphy and geochronology). *Izvestia, Academy of Sciences of USSR, Geology,* **10,** 46–54 (in Russian).

Semenenko, L.T., Alyoshinskaya, Z.V., Arslanov, Kh.A., Valuyeva, M.N. & Krasnovskaya, F.I. (1981). The key section of the Upper Pleistocene at the 1st of May Factory, Dmitrov District of the Moscow Region (the deposits of the ancient Lake Tatishchevo). *In:* Shik, S.M. (ed.), *Novye dannye po stratigrafii i paleogeografii verkhniego pliotsena i pleistotsena Tsentralnykh rayonov Yevropeiskoi chasti SSSR,* Moscow, 121–136 (in Russian).

Saarnisto, M. and Saarinen, T. (2001). Deglaciation chronology of the Scandinavian Ice Sheet from the Lake Onega to the Salpausselka End Moraines. *Global and Planetary Change,* **31**, 387–405.

Sudakova, N.G., Vvedenskaya, A.I., Voskovskaya, L.T. & Pisareva, V.V. (1997). Stratigraphical problems of Pleistocene on the Klin-Dmitrov elevation. *In:* Alexeyev, M.N. & Khoreva, I.M. (eds) *Chetvertichnaya geologia i*

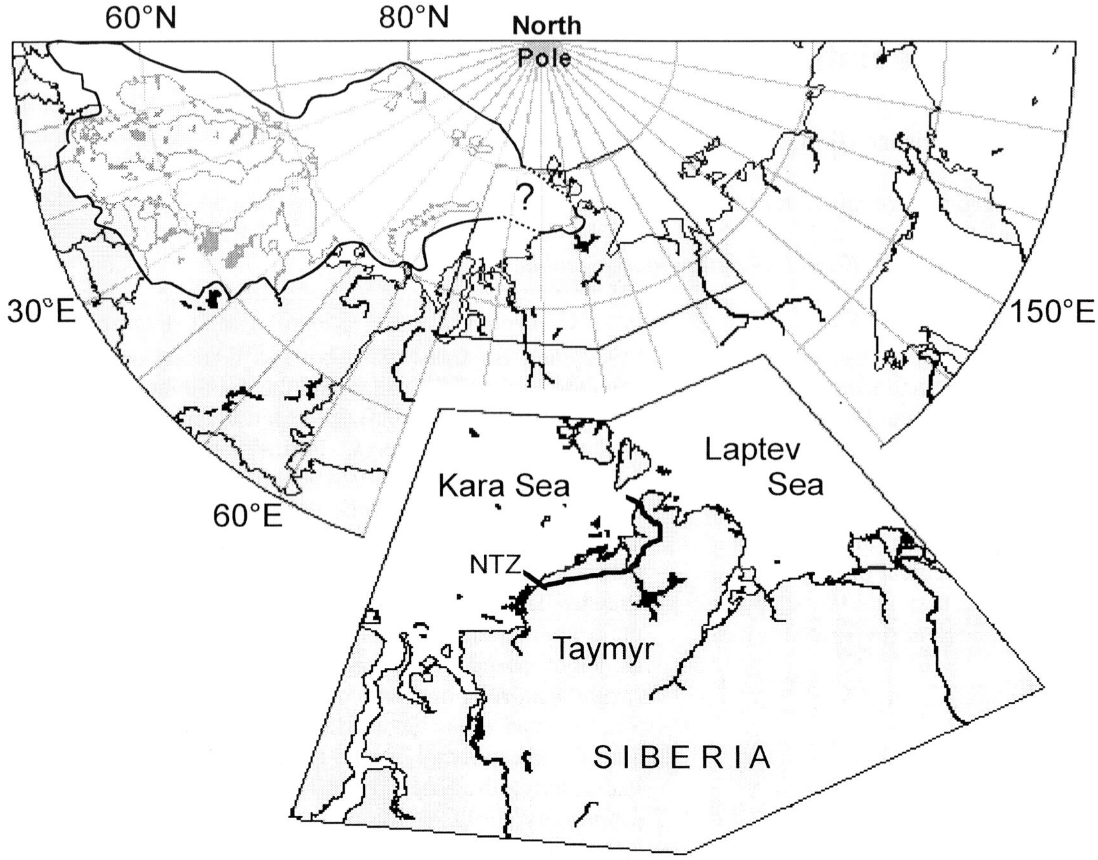

Fig. 1. Map of northern Eurasia and the Taymyr Peninsula (inset), showing the North Taymyr ice-marginal zone (NTZ), and the extent of the LGM glaciation based on the synthesis by Svendsen et al. (1999).

tain valleys and into the marine basin. This meltwater must have emanated from an ice front, which at that time stood along the northern slopes of the mountains (Fig. 3).

This marine event is probably the equivalent of the Karginsk Marine Transgression of Andreeva & Kind (1982), in its earliest phase radiocarbon-dated to 50-39 ka BP. However, these ^{14}C dates, from mollusc shells, were conventional and may well prove to be infinite if the AMS technique is used.

The marine basin, shown in Fig. 3, is a palaeogeographical reconstruction of the situation during the time of the 'Ledyanaya Gravel Event', constructed using shorelines of 100 m a.s.l. along the Byrranga Mountains (the authors' data) and 50 m at Khatanga Bay (Andreeva *et al.* 1982). The inundated areas to the south in this reconstruction are somewhat smaller than might be expected from the known distribution of subaquatic deglaciation sediments (*e.g.* Andreeva & Isaeva 1982; Siegert *et al.* 1999). This is because a substantial part of the regression must have already taken place when the ice front had retreated to the Byrrangas.

There are no signs of any glaciation reaching south of the Byrranga Mountains since this Early Weichselian deglaciation (*e.g.* Möller *et al.* 1999a). These mountains and the Taymyr Lake basin have thus been continuously ice free ever since, as illustrated by the sediment sequence (lacustrine/bog/aeolian sediments) at Cape Sabler, where the base has been radiocarbon-dated to >40 ka BP and continuous deposition to the Holocene is present (Kind & Leonov 1982; Pavlidis *et al.* 1997; Derevyagin *et al.* 1999; Möller *et al.* 1999a). Lake sediment successions from both the Taymyr Lake itself and from the adjacent Levinson-Lessing Lake (Ebel *et al.* 1999; Hahne & Melles 1999; Niessen *et al.* 1999) also indicate ice-free conditions since at least the Middle Weichselian.

Deglaciation of the Chelyuskin Peninsula

The Early Weichselian Kara Sea ice sheet also inundated the Chelyuskin Peninsula on northernmost Taymyr, where it flowed eastwards over the 350 m high Astrup- and Sverdrup Mountains. This conclusion is based on till and glaciotectonic deformations overlying and affecting Eemian interglacial marine sediments, and from the occurrence of Kara Sea crystalline erratics found on the hilltops (Möller *et al.* unpublished). In addition, the glacio-isostatic depression in this area resulted in post-glacial marine inundation, ESR dated to *c.* 95-80 ka BP and thus largely contemporaneous with that in the Taymyr Lake basin. The marine limit from this time lies between 65-80 m on Chelyuskin. Both the Eemian and Early Weichselian marine sediments here con-

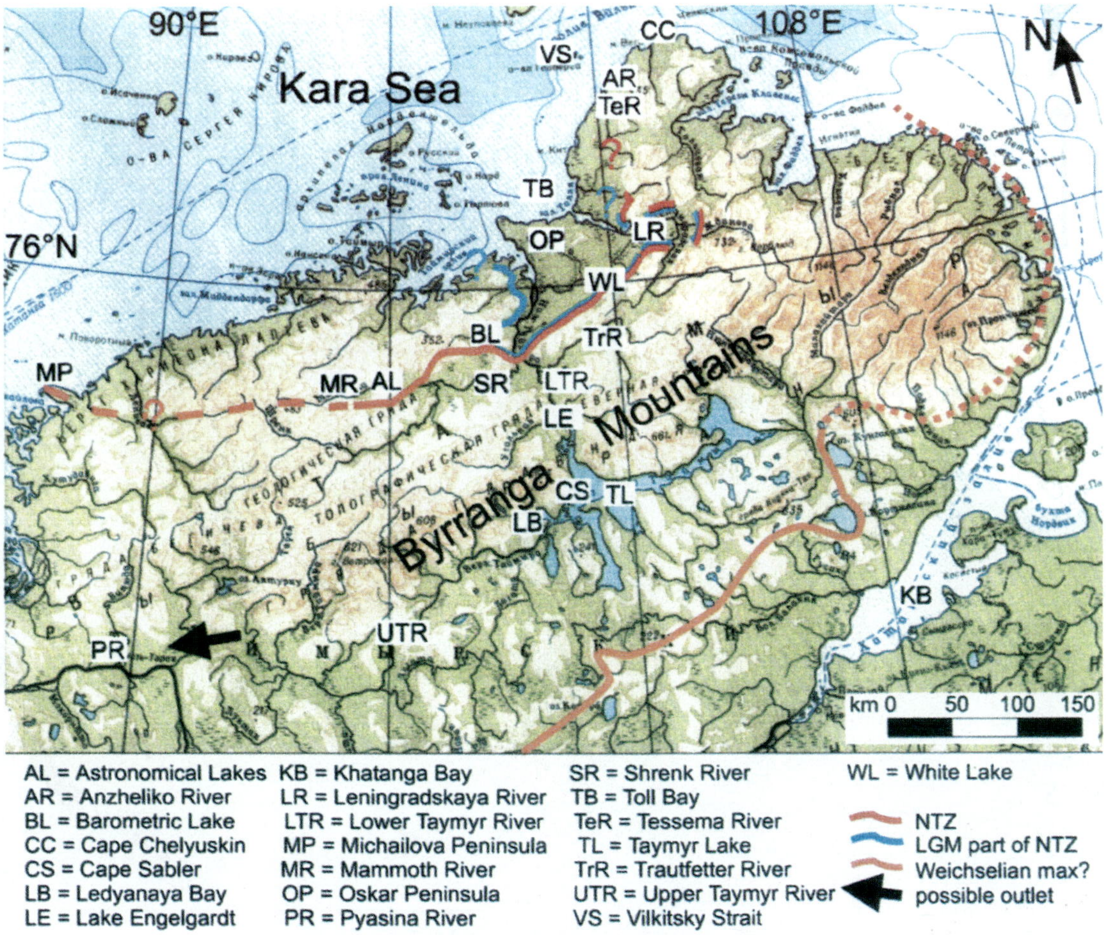

Fig. 2. The northern Taymyr Peninsula with geographical names mentioned in the text, together with the current authors' proposed outlines of the North Taymyr ice-marginal zone (NTZ) and the main ice-front position at the time of the Weichselian glaciation maximum. The Astrup- and Sverdrup mountains lie east of the Angelika River on the Chelyuskin Peninsula. The base map is 'Россия и сопредельные государства' Geodetic and Cartographic Federal Office of Russia, Moscow 1996, original scale 1:4,000,000.

tain shells of *Chlamys islandica,* which indicates considerably warmer water temperatures during the earlier parts of the Weichselian than later or even today.

The North Taymyr ice-marginal zone

The northwestwards retreat of the Early Weichselian ice front from the Byrranga Mountains seems to have proceeded largely by calving into a glacial lake filling the Shrenk-, Trautfetter- and part of the lower Taymyr river valleys and dammed towards the northwest by the ice itself. A new grounding line was reached on the northwestern sides of the Shrenk- and Trautfetter valleys, causing a temporary still-stand of the ice front which resulted in the formation of the North Taymyr ice-marginal zone, the NTZ (Figs. 2 and 4).

The NTZ is a complex of glacial, glaciofluvial and glaciolacustrine deposits, containing large amounts of re-deposited Quaternary marine sediments and also glacially-displaced, coal-bearing Cretaceous sands. It has now been dated for the first time and described in some detail by Alexanderson *et al.* (2001, 2002), but had already been broadly mapped and discussed by Kind & Leonov (1982). When the Kara Sea ice-sheet front stood at this ice-marginal zone it seems to have crossed the present coastline at the Michailova Peninsula at *c.* 75°N. The NTZ can then be followed first eastwards and then northwards for 700-750 km, mostly 80-100 km inland, and seems to re-cross the present coastline south of the Tessema River, around 77°N. It is best developed in its central parts, *c.* 100 km northeast and southwest, respectively, of where it is today cut through by the Taymyr River. The base of the NTZ (Alexanderson *et al.* 2001, 2002) is a series of ridges up to 100 m high and 2 km wide, mainly consisting of, or possibly only covered by, re-deposited marine silts. They are still ice-cored, but in most parts of the zone the present active layer only rarely reaches the ice-surface. Smaller ridges of both till and glaciofluvial material are superimposed onto the main ridges. They often surround lakes that originated as minor over-deepened glacial basins. Associated with the NTZ are deltas, abrasion terraces and shorelines corresponding to two generations of ice-dammed lakes, with shore-levels at between 140-120 m and at *c.* 80 m a.s.l. (Fig. 5). These lakes drained southwards into the Taymyr Lake basin, as recorded by current directions in fluvial sediment sequences along the Taymyr River valley

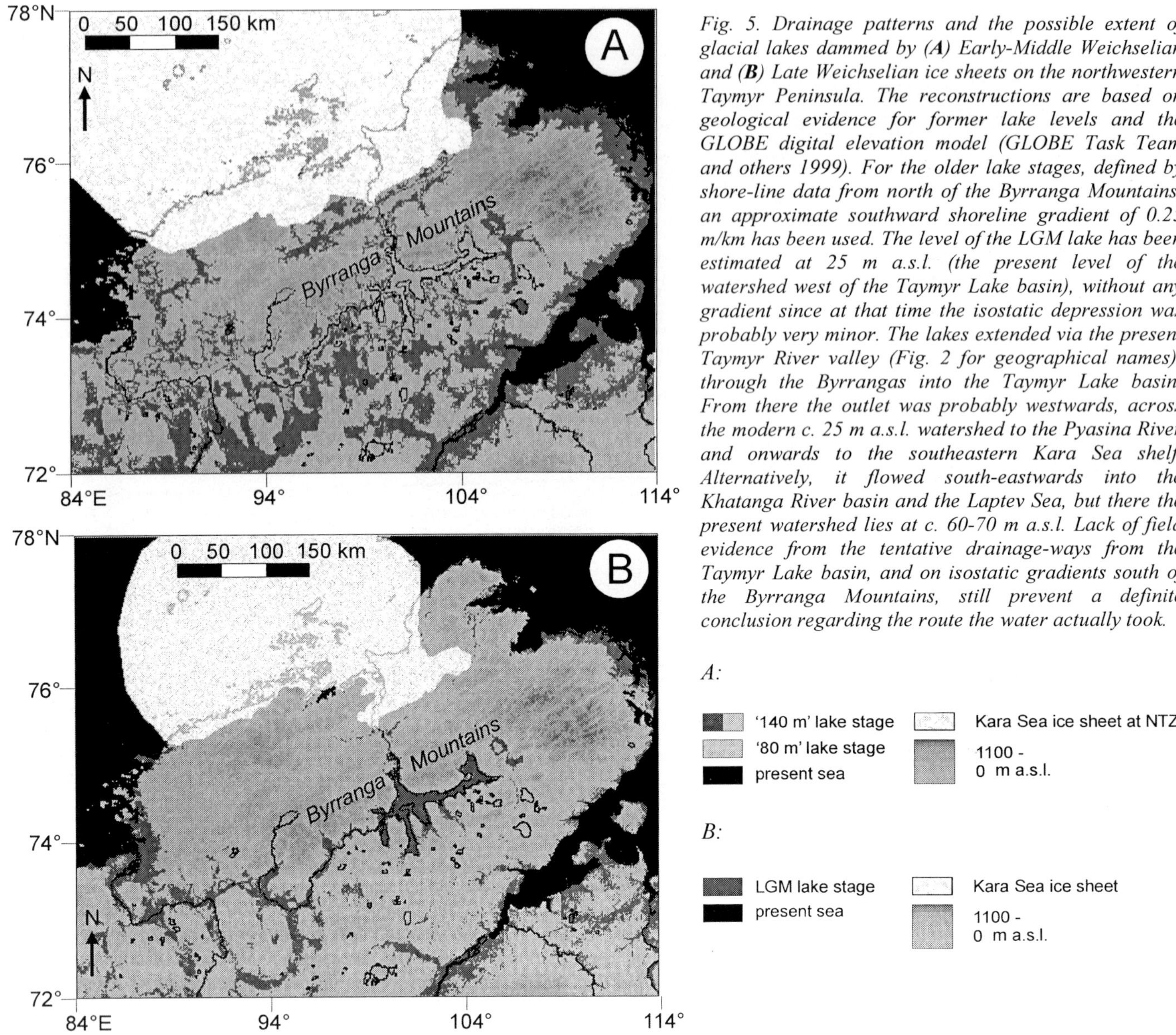

*Fig. 5. Drainage patterns and the possible extent of glacial lakes dammed by (**A**) Early-Middle Weichselian and (**B**) Late Weichselian ice sheets on the northwestern Taymyr Peninsula. The reconstructions are based on geological evidence for former lake levels and the GLOBE digital elevation model (GLOBE Task Team and others 1999). For the older lake stages, defined by shore-line data from north of the Byrranga Mountains, an approximate southward shoreline gradient of 0.25 m/km has been used. The level of the LGM lake has been estimated at 25 m a.s.l. (the present level of the watershed west of the Taymyr Lake basin), without any gradient since at that time the isostatic depression was probably very minor. The lakes extended via the present Taymyr River valley (Fig. 2 for geographical names), through the Byrrangas into the Taymyr Lake basin. From there the outlet was probably westwards, across the modern c. 25 m a.s.l. watershed to the Pyasina River and onwards to the southeastern Kara Sea shelf. Alternatively, it flowed south-eastwards into the Khatanga River basin and the Laptev Sea, but there the present watershed lies at c. 60-70 m a.s.l. Lack of field evidence from the tentative drainage-ways from the Taymyr Lake basin, and on isostatic gradients south of the Byrranga Mountains, still prevent a definite conclusion regarding the route the water actually took.*

Fig. 6. Remnant Late Weichselian (LGM) glacier ice under a melt-out till cover of only c. 0.5 m., near White Lake. Photograph by C. Hjort.

sediments'; Funder *et al.* 1999). They indicate damming between the present land and an ice front receding onto the Kara Sea shelf, after retreating from the NTZ. The OSL dates of *c.* 70 - 80 ka BP obtained seem to relate much of this sedimentation to the deglaciation after the Early Weichselian NTZ event (Funder *et al.* unpublished).

The Late Weichselian NTZ stage

During the third NTZ generation the Kara Sea ice sheet was much thinner than previously and inundated a much smaller area (Alexanderson *et al.*, 2001, 2002). Because it did not cross a 300-500 m high range of coastal hills (e.g. Fig. 4), which were overridden during the two previous stages, its thickness near the present coastline could not have been more than 500 m. Nonetheless, it penetrated 100 km inland, on a 150 km broad front centred along the lower Taymyr River valley and terminated at altitudes below 150 m a.s.l. Northeast of the valley the front abutted a system of bedrock cuestas, whilst to the southwest it was in contact with the pre-existing NTZ moraine and, in one case, formed an independent lobate moraine (Fig. 4). The area overridden by this ice sheet, the most recent to inundate the Taymyr Peninsula, is to a large extent covered by dislocated marine sediments, identifiable on satellite images by their dendritic erosional pattern. In a 5-10 km wide zone behind the former ice front, where the ice contained most debris, the landscape is patterned by a multitude of shallow slides, exposing remnant glacier ice under a melt-out till cover of only about 0.5 m. (Fig. 6). Further northwest (up-ice), there are fewer indications of the former overriding, probably the consequence of a cleaner and more rapidly melting ice. However, a boulder-lag on top of the glacio-lacustrine sediments at the Kara Sea coast (Funder *et al.* 1999) may date from this glacial event.

This youngest ice sheet advance is pre-dated by two radiocarbon dates of mollusc shells *(Hiatella arctica, Astarte sp.)* of *c.* 20 ka BP, from glacially re-deposited marine silt sampled *c.* 2 km behind the former ice front position near White Lake (Alexanderson *et al.*, 2001, 2002). It is post-dated by a radiocarbon date of *c.* 12 ka BP from *in situ* terrestrial material from just inside the present coast on Oskar Peninsula (Bolshiyanov *et al.* 2000), and possibly also by organic material retrieved from the sea bottom off the coast in the Toll Bay. Here a radiocarbon date (a conventional bulk sample) has given *c.* 16 ka BP (Bolshiyanov *et al.* 1998). The glaciation thus dates from the Weichselian Last Glacial Maximum (LGM). This brief advance (8000 years or less) of a thin ice sheet onto the present land may have been of surge character, utilizing an easily deformable substratum. It is not yet clear whether it emanated from the growth of a regional ice cap on the very shallow, and for global eustatic reasons at that time, mostly dry shelf in the northeastern corner of the Kara Sea, or if it was connected to the west to ice centred near Novaya Zemlya.

No evidence of any glacial lake dammed by this LGM ice sheet have been found north of the Byrrangas and it is therefore thought that meltwater from this thin ice sheet mainly drained southwards via the Taymyr River valley into the Taymyr Lake basin (Fig. 5B, and Alexanderson *et al.*, 2001, 2002). Indications of an increasing sedimentation rate in the lake around 19 ka BP (Möller *et al.* 1999a, Möller *et al.* unpublished) suggest a causal connection with the meltwater input.

No raised marine shorelines dated to the LGM or thereafter have been found on Taymyr which is not surprising considering the thin, short-lived and thus isostatically insignificant ice and the extremely low eustatic sea level at the time, persisting into the Holocene.

Summary of results

The main results of this study of the glacial and marine history of the Taymyr Peninsula, summarized in Fig. 7, are as follows:

Three main phases of Weichselian glaciation of successively decreasing amplitude have been mapped (Figs. 2, 3 and 5) and dated. The most extensive glaciation dates from the Early Weichselian (culminating ≥100 ka BP), and a Middle Weichselian event of intermediate extent dates from *c.* 65 ka. The last and least extensive glaciation, contemporaneous with the Last Glacial Maximum (LGM), was short, lasted only 8000 years or less, culminated between 18 - 16 ka BP, and had largely disappeared from present onshore areas by 12 ka BP.

The ice sheets that covered the Taymyr Peninsula on all three occasions during the Weichselian emanated from the Kara Sea continental shelf, from which they advanced generally southeastwards across the land. At most, the ice-front reached some 400 km from the coast, leaving a series of more or less distinct zones of ice-marginal features south of the Taymyr Lake basin. In the south and east the ice-front reached the Laptev Sea drainage basin. No sign of any local Weichselian glaciation has been encountered and the only recorded glaciation that affected most of the Byrranga Mountains has been that of overriding Kara Sea ice. However, the higher easternmost part of the Byrrangas, not studied by the authors, may at times have acted as a local centre of glaciation. Today it still supports some minor cirque glaciers.

The Kara Sea ice sheets dammed large glacial lakes, filling the lake- and river basins both north and south of the Byrranga Mountains (Fig. 5) and, during the final stages of the different deglaciations, also lowland areas along the coast. The water from north of the mountains drained southwards along the Taymyr River valley (where today the water flows northwards) into the Taymyr Lake basin, and in most cases thereafter probably westwards to the Kara Sea shelf.

The glacio-isostatic depression of the Earth's crust, arising from the ice-load, led to substantial marine submergence of present land areas on Taymyr. Most

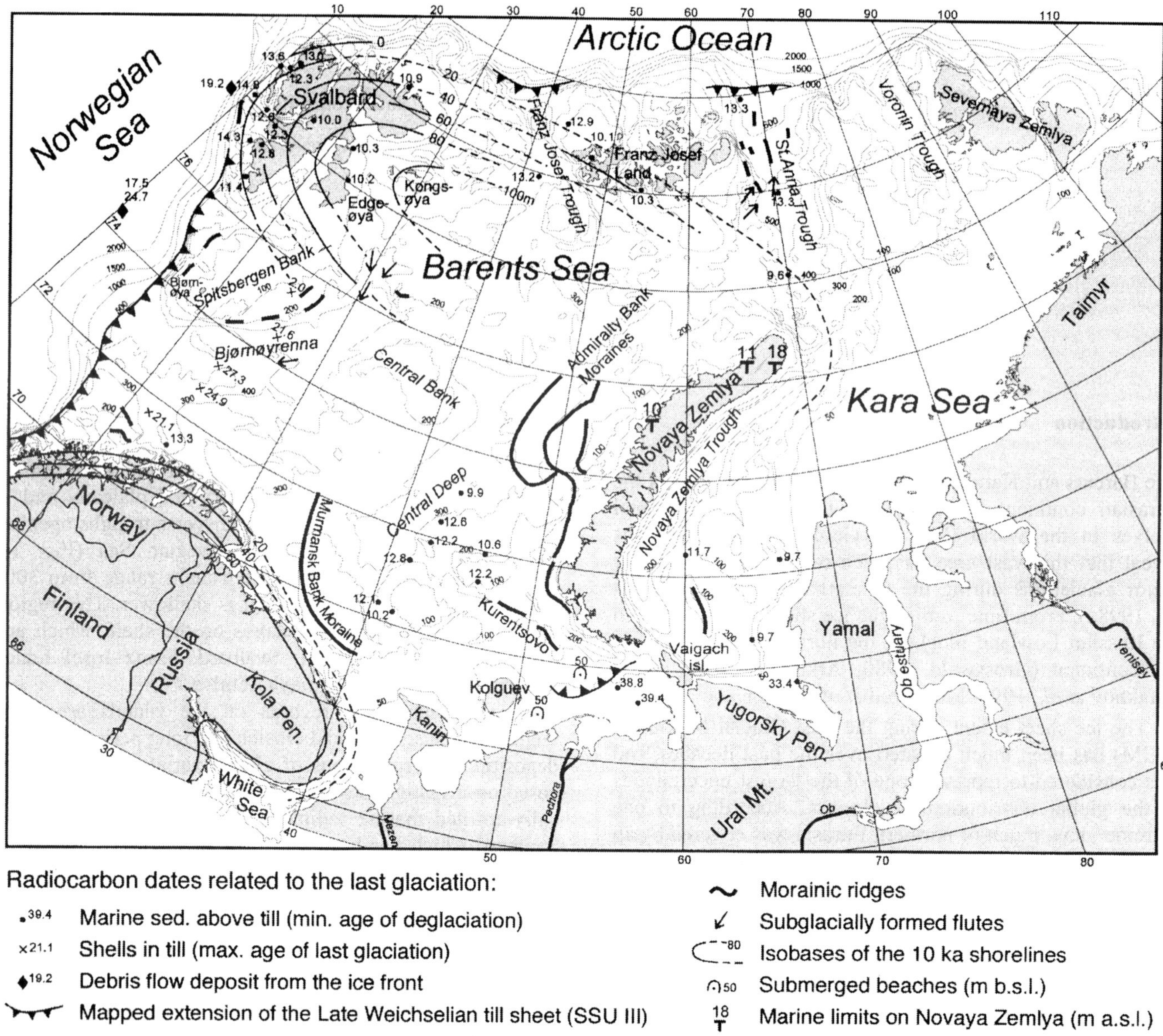

Fig. 1. Index map of the Barents- and Kara Sea Region showing some of the evidence used for reconstructing the Late Weichselian ice sheet extent (Fig. 2). Most of the data from the western part of the study area were presented in a compilation by Landvik et al. (1998). Glacial features and dates from the eastern Barents Sea and from the Kara Sea are discussed in the text. The reconstructed isobases of the10 ka shoreline in the northern Barents Sea region and along the Norwegian coast are slightly modified from Landvik et al. (1998). The corresponding isobases around the Kola Peninsula in Russia are simplified from Koshechkin (1978). Marine limits on Novaya Zemlya are taken from Forman et al. (1999).

that incipient glaciations occurred on adjacent land areas some 12-14 Ma ago (Thiede *et al.*, 1998). The increased supply of IRD at approximately 3.2-2.7 Ma reflected a strengthening of northern hemisphere large-scale glaciations (Jansen & Sjøholm, 1991; Mangerud *et al.*, 1996). In the Barents Sea region, the initial glacial growth is seen in a pronounced increase in the general sedimentation rate along the Svalbard margin from around 2.3-2.5 Ma (Faleide *et al.*, 1996; Butt *et al.* 2000). The volume of thick wedges of glacigenic sediments along the western continental slope suggest that there has been as much as 1000-1500 m of erosion over the Barents Sea region since the onset of glaciations (cf. Solheim *et al.*, 1998). During the oldest glaciations, the physiography of the Barents Sea region was very different from the present and the shelf area was probably well above sea level.

Sedimentological data from the ODP site 986 drilled off western Svalbard reveal low IRD values during the interval 2.3-1.6 Ma, suggesting that glaciers in the Barents Sea region did not reach the sea (Butt *et al.*, 2000). Butt *et al.* (2000) conclude that further glacial expansions took place during 1.5-1.3 Ma and that ice sheets expanded to the shelf break several times after 1.3 Ma. From the seismic stratigraphy it has previously been concluded that the ice

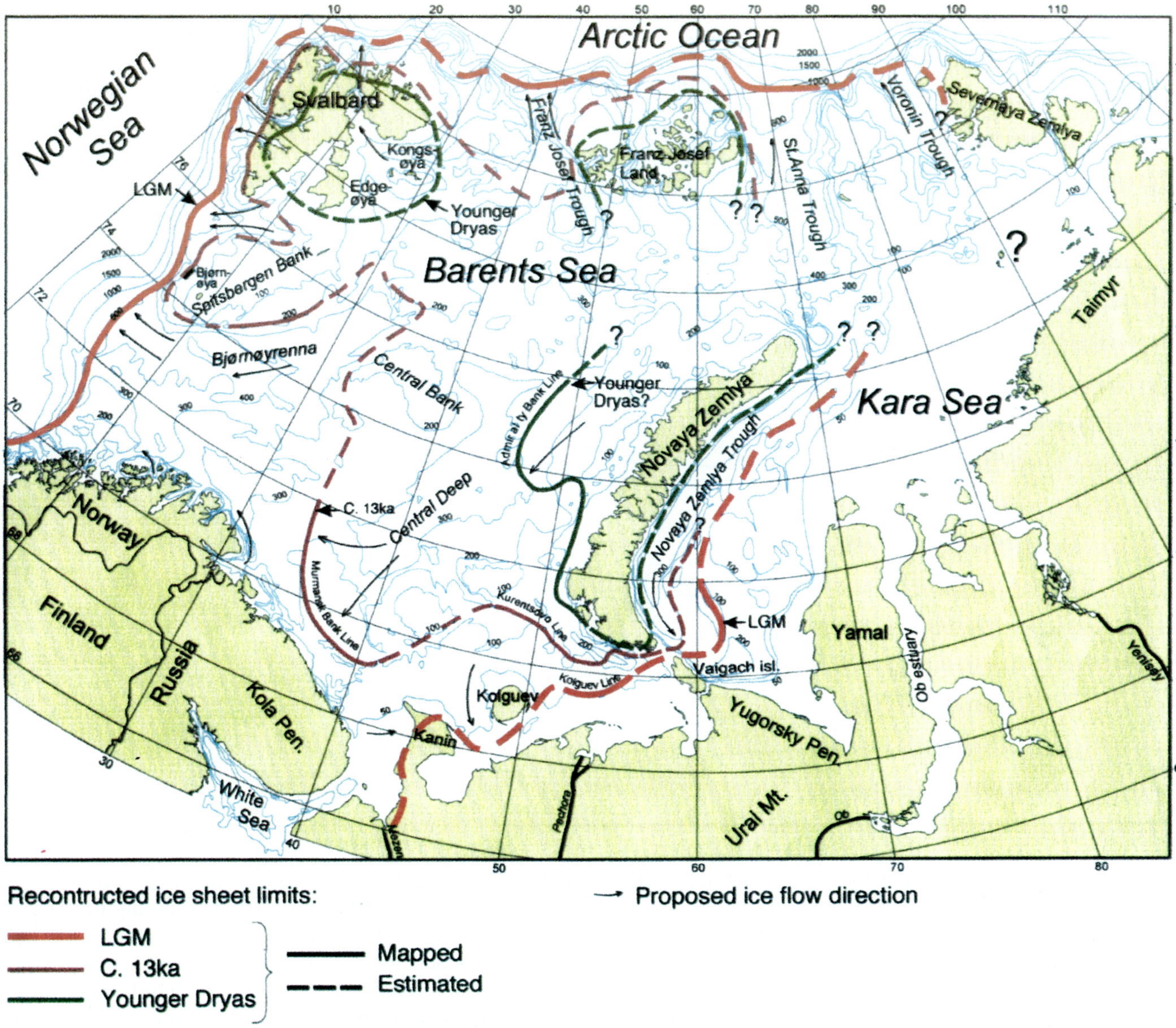

Fig. 2. Map of the Barents and Kara Sea Region showing the interpreted ice sheet limit during the Last Glacial Maximum (21-18 ka). Judging from the distribution of radiocarbon dates and mapped morainic ridges on the shelf (Fig 1) tentative ice-recessional limits have also been drawn at around 13 ka and during the Younger Dryas Chron.

sheet in the southwestern Barents Sea expanded to the outer shelf at least eight times (Laberg & Vorren, 1996). The last period of glacial deposition appears to be dominated by aggradation of glacial sediments on the shelf margin rather then progradation. This may reflect a different style of glaciation as the shelf was glacially eroded and eventually became submerged. Accordingly, Solheim *et al.* (1998) suggest that thick eroding ice sheets existed during the oldest glaciations whereas the younger ice sheets were characterised by thinner and more fast-flowing ice streams at the margins.

The maximum extent of Quaternary glaciation

Tills and other glacial features show that the shelf-centered ice sheets expanded far onto the Russian mainland several times (Yakovlev,1956; Arkhipov *et al.*, 1995; Astakhov *et al.*, 1999 and this volume; Hjort *et al.*, this volume). In West Siberia the Quaternary glaciation limit is located as much as *c.* 1400 km to the south of the Arctic coastline, testifying to the existence of at least one enormous ice sheet (Krasnov, 1971; Astakhov, this volume). Some observations suggest that the ice sheet at some point was fringed by a thick marine ice shelf that grew into the Arctic Basin. Ploughmarks on the Yermak Plateau reflect grounded ice down to more than 850 m below present sea level (Vogt *et al.*, 1994); an interpretation that is supported by coring results (Myhre *et al.*, 1995). More surprisingly, seismic profiles and side-scan records from the Lomonosov Ridge in the Arctic Basin show not only ploughmarks of several kilometres in length, but also a pronounced regional erosional truncation to 1000 m below present sea level,

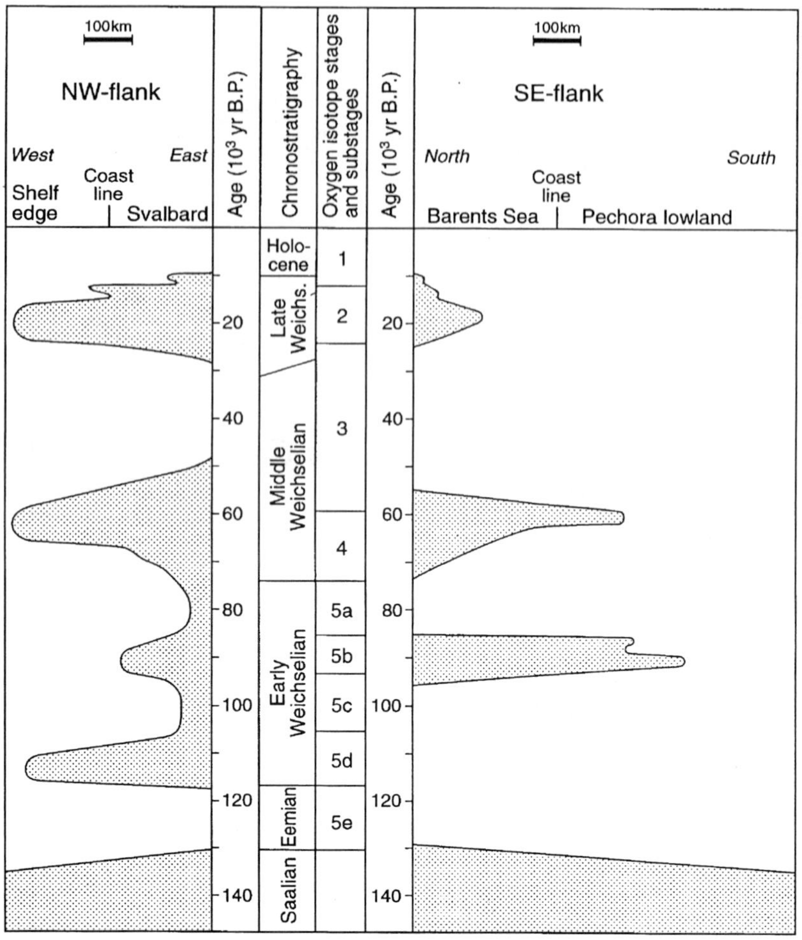

Fig. 3. Time-distance diagrams showing recorded advances of the northwestern (Svalbard) and southeastern (northern Russia) flank of the Barents-Kara Ice Sheet during the last 150 ka (modified from Mangerud et al., 1998; 2001).

interpreted to have been eroded by a grounded ice shelf (Jakobsson, 1999; Polyak *et al.,* 2001).

The age of the Quaternary glacial maximum when the Lomonosov Ridge was eroded is uncertain. One of the most extensive glaciations occurred during the Late Saalian (Marine Isotope Stage 6) when the entire shelf in the Barents and Kara seas was ice covered (Mangerud *et al.*,1998; Knies *et al*., 2001). According to Jakobsson *et al.* (2001) the possible ice grounding on the crest of the Lomonosov Ridge took place at this time. However, it should be noted that this chronology is at variance with the age model suggested by Spielhagen *et al.* (1997) who date the correlative horizon with MIS 16 (625-690 ka). This latter chronology is also supported by Flower (1997) who concludes that the last ice grounding on the Yermak Plateau occurred over 660 ka ago.

The ice sheet history during the Early/Middle Weichselian Substages

A synthesis of the Late Pleistocene glacial fluctuations in the northwestern Barents Sea have been given by Mangerud *et al.* (1998), who correlate the onshore record of Svalbard with IRD in cores from the ocean to the west (Fig. 3). They identified three major glacial advances during the Weichselian which, according to their chronology terminated on the shelf west of Svalbard during the Marine Isotope Substage 5 d (110 ka), MIS 4 (60 ka) and 2 (20 ka). The large glaciations were probably separated by long-lasting periods when glaciers were not significantly larger than today. During the Middle Weichselian (50-30 ka), Mangerud *et al.* (1998) consider that the shelf area east of Svalbard was ice free. However, it remains unclear if the shelf between Svalbard and mainland Norway was also affected by three separate glaciations during the Weichselian. Thus far only one till unit has been identified overlying the last interglacial (Eemian) in the outer part of the Bjørnøya Trough (Sættem *et al.*, 1992; Laberg & Vorren, 1995). This may imply that this part of the shelf remained ice free during the Early and Middle Weichselian glaciations. However, more investigations are necessary to resolve the Weichselian glacial history of this area.

During the Early and Middle Weichselian the Barents-Kara Ice Sheet expanded much further southeastwards, inundating the northern rim of the Russian mainland (Astakhov *et al.*, 1999; Forman *et al.*, 1999; Alexandersson *et al.,* 2001; Hjort *et al.,* this volume). Large ice dammed lakes formed and flooded the Russian lowland in front of the advancing Early Weichselian ice sheet. A series of optically stimulated luminescence (OSL) dates indicate that the maximum ice sheet extent was attained at around 90 ka

(Marine Isotope Substage 5 b) (Mangerud *et al.*, 2001; Alexandersson *et al.,* 2001). A regrowth of the ice sheet probably occurred during MIS 4 leading to another advance that culminated at around 60 ka (Henriksen *et al.*, 2001; Houmark-Nielsen *et al.*, 2001) (Fig. 3). Along the Pechora River, in the European Russian Arctic, the ice front reached as far south as the preceding ice advance, but the ice extent was more restricted further to the east than during the glacial maximum at 90 ka. Sediment cores from the Pechora Sea indicate that the deglaciation after the last advance occurred before 35 ka (Polyak *et al.*, 2000a). Moreover, radiocarbon dates from raised shorelines on Severnaya Zemlya have yielded ages ranging from 21 ka to more than 50 ka (Bolshiyanov & Makeyev, 1995). These shorelines are thought to reflect a significant rebound of the crust following comprehensive Middle Weichselian deglaciation. Investigations of sediment cores from the continental slope to the northeast of Severnaya Zemlya show a pronounced peak in IRD content around 60 ka, with a smaller peak around 90 ka (Knies *et al.*, 2000). Knies *et al.* (2000) maintain that the youngest till on the shelf margin to the east of Severnaya Zemlya is of Middle Weichselian age, reflecting an ice sheet with a grounding line at least 340 m below the present sea level.

As will be apparent from the above review, the LGM was clearly preceded by at least two major Weichselian glaciations. However, the proposed age for the Early Weichselian glacial maximum is not quite the same in the different sectors. The Svalbard records suggest that the first ice advance occurred during Marine Isotope Substage 5d (110 ka), whereas in northern Russia the maximum ice sheet position has been ascribed to MI Substage 5b (90 ka). This age difference may be real, but it is possibly also an artifact of dating uncertainties.

The geological evidence for the reconstructed ice-sheet advances compare fairly well with a recent model simulation for the repeated growth and decay of the Eurasian ice sheets during the Weichselian (Siegert *et al.,* 2001). It is also worth noting that the model, which is forced by global sea level and solar insolation changes, creates three major glaciations in the Barents and Kara Sea region that culminated at around 90 ka, 60 ka and the LGM. Moreover, in accordance with the empirical reconstructions, the Siegert *et al.* (2001) model predicts that each ice-sheet advance was followed by an almost total deglaciation of the shelf.

The Late Weichselian glacial maximum (LGM)

A major ice sheet formed over the Barents Sea shelf during the Late Weichselian Substage (e.g. Landvik *et al.*,1998). Till overlain by glaciomarine sediments that date from the Late Weichselian has been mapped on the sea floor throughout most of the Barents Sea and is also found on many islands (Fig. 1). One of the most convincing lines of evidence for a grounded ice sheet in the Barents Sea is the occurrence of subglacially-formed flute bedforms. Such features have been discovered in the Bjørnøy Trough, near the central part of the Barents Sea (Solheim *et al.*, 1990) and in the St.Anna Trough to the east of Franz Josef Land (Polyak *et al.*, 1997). The pattern of raised shorelines provide further evidence in supporting the existence of a major Late Weichselian ice sheet (e.g. Salvigsen., 1981) (Fig. 1). The reconstructed shoreline isobases define a centre of Holocene glacio-isostatic uplift in the north-central Barents Sea, resulting from a former ice dome in this area (Forman *et al.,* 1995).

According to Landvik *et al.*'s (1998) reconstruction, the ice sheet overrode all islands on Franz Josef Land and Svalbard during the LGM, when the ice front reached the outer shelf margins with a grounding line at 500-600 m below present sea level. Based on radiocarbon dates of shells and foraminifers from sediment cores, the last major ice sheet advance terminated at the western shelf break in the Barents Sea, between 19 and 15 ka (Landvik *et al.*, 1998). In the northern Barents Sea, the Late Weichselian till sheet has been traced to the shelf break at the mouth of the Franz Victoria Trough (Lubinski *et al.* 1996; Kleiber *et al.* 2000). Here the occurrence of debris-flow sediments and glaciomarine diamictons on the continental slope led Kleiber *et al.* (2000) to conclude that the ice sheet reached the shelf break as early as 23 ka (Kleiber *et al.* 2000). It has also been demonstrated that the ice sheet filled the St. Anna Trough to the east of Franz Josef Land (Polyak *et al.*,1997). Here the grounding line probably occurred close to shelf break along the entire shelf margin from Svalbard and eastwards to the St.Anna Trough (Fig. 2). However, more investigations are necessary to define the eastern ice-sheet limit on the northern Kara Sea shelf.

Until recently, it was a common view that the south-eastern margin of the Barents-Kara Ice Sheet expanded well onto the Russian mainland during the LGM (e.g. Landvik *et al.*, 1998). However, new land-based evidence from Russia contradicts this view (Astakhov *et al.,* 1999; Forman *et al.*, 1999; Mangerud *et al.*, 1999). On the basis of marine geological and geophysical data this ice limit has now been identified on the sea floor off the mainland (Figs 1 and 2) (Svendsen *et al.*, 1999; Polyak *et al.*, 2000a; Gataullin *et al.*, 2001). This boundary in the Pechora Sea, termed the Kolguev Line, marks the southern limit of the uppermost till sheet (seismo-stratigraphic unit III) in the Barents Sea. It is assumed that the southern flank of the ice sheet coalesced with the Scandinavian Ice Sheet near the northern tip of the Kanin Peninsula, but the exact boundary remains to be defined in this confluence area. To the south of the proposed ice-sheet limit, and distal to the mouth of the Pechora River, there is a one hundred metre thick wedge of Middle to Late Weichselian prodeltaic marine sediments that is not overlain by till (Polyak *et al.*, 2000a; Gataullin *et al.*, 2001). The relative sea level in the southern Pechora Sea was apparently 15-20 m below the present during the LGM, reflecting a glacioisostatic depression of *c.* 100 m near the ice limit (Gataullin *et al.*, 2001). The glacioisostatic uplift caused the relative sea level to fall after the ice front receded from its maximal position and a minimum level of 50-60 m b.s.l. was established at the very end of the Late Weichselian (10-12 ka).

The maximum ice sheet limit during the LGM in the southern Kara Sea probably occurred at a well-defined morainic ridge SE of the southern end of the Novaya Zemlya Trough (Svendsen *et al.*, 1999; Polyak *et al.*, 2000b). On the floor of this trough there is only a 4 - 5 m thick veneer of marine sediments upon the till, whereas a much thicker ac-cumulation (up to 100 m) has been recorded closer to the morainic ridge. Further north, we assume that the LGM limit has occurred along the eastern margin of the Novaya Zemlya Trough (Fig. 2), although a more extensive ice cover in this part of the Kara Sea cannot be excluded (Polyak *et al.*, 2000b).

The ice-sheet configuration in the northern and eastern Kara Sea is more uncertain and therefore no ice limits have been drawn in this area. Land-based investigations show that the central part of the Taymyr Peninsula was ice free during the LGM (Möller *et al.,* 1999). Two AMS dates of molluscs that were incorporated in glacier ice on the Taymyr Peninsula have yielded ages of around 20 ka suggesting that a younger glacier advance from the north affected the NW Taymyr coast during the LGM (Alexandersson *et al.*, 2001; Hjort *et al.*, this volume). Possibly this was a lobe of the Barents-Kara Ice Sheet that reached Taymyr. However, no observations confirm that there was an ice-dammed lake along the northern rim of the Russian Lowland at this time, suggesting that the northbound drainage on the continent was not blocked by a coherent ice sheet. Accordingly, the glacier advance that reached the NW coast on Taymyr may have been unconnected with the Barents-Kara Ice Sheet, i.e. there was an ice-free corridor on the shelf.

Based on marine geological investigations it is concluded that the continental shelf to the east of Severnaya Zemlya was ice free during the LGM (Knies *et al.,* 2000). Furthermore, the lack of IRD in the marine sediments from this period suggests that that there was no calving ice front on this side of the archipelago. This assumption is supported by radiocarbon dates from mammoth remains on Severnaya Zemlya (Makeyev *et al.*, 1979; Bolshiyanov & Makeyev, 1995) suggesting that the local glaciers on these islands were not significantly larger than today.

The early deglaciation of the shelf

A step-like deglaciation during the period 15-10 ka is thought to have occurred within most of the Barents Sea (e.g. Polyak *et al.*, 1995; Kleiber *et al.* 2000). At the mouth of the Franz Victoria Trough in the northern Barents Sea the initial disintegration of the ice sheet is indicated by a pronounced increase in the flux of IRD to the slope around 15 ka and a subsequent isotopically-depleted meltwater signal (Kleiber *et al.* 2000). Radiocarbon dates obtained from glaciomarine sediments above the upper till bed led Kleiber *et al.* (2000) to conclude that the grounding line started to retreat from its maximum position shortly before 13.4 ka. Moreover, on the continental slope to the west of Svalbard a distinct meltwater event is reflected in the oxygen isotope record at 15 ka (Elverhøi *et al.*, 1995). Sediment cores retrieved from the outer shelf in this region indicate that the ice sheet began receding before 14.8 ka (Svendsen *et al.*, 1996), or perhaps as early as 16.4 ka (Cadman,1996). A very similar age for the onset of deglaciation was proposed by Vorren *et al.* (1988) for the confluence area between the Scandinavian and Barents-Kara ice sheets. The inference that the postglacial uplift in northern Norway and the Kola Peninsula in Russia is little influenced by the former Barents-Kara Ice Sheet has been explained by a significant reduction of the ice load prior to 15 ka (Elverhøi *et al.*,1993). However, there is no positive evidence to suggest that the ice front retreated far inside the shelf margin at this time, and it is possible that this early stage of deglaciation mainly occurred as a thinning of the ice sheet.

There are a number of morainic ridges on the floor of the Barents Sea which, according to the authors' correlations may represent a contemporaneous glacial event (Fig.1). In the southeastern Barents Sea this event is marked by the Kurentsovo Line, a series of long ice-pushed ridges that are located 50-100 km north of the apparent LGM limit (Gataullin *et al.*, 2001). Accumulations of glaciomarine sediments up to 100 m thick are found on the southern side of these ridges, whereas less than 10-20 m occur on the northern side. The Kurentsovo Line can be traced to the southern end of Novaya Zemlya and further into the SW Kara Sea between Novaya Zemlya and the LGM limit at this time. The western continuation of this line is presumably marked by the Murmansk Bank Moraines, a chain of morainic ridges, *c.* 400 km long, north of the Kola Peninsula. It is also possible that a series of 20-50 m high ridges at 300 m water depth along the western and southern margins of the Spitsbergen Bank are of the same age (Elverhøi & Solheim, 1983). These ridges are correlated with thick accumulations of ice-proximal marine sediments on the inner part of Bjørnøyrenna and at the mouth of Storfjordrenna (Elverhøi *et al.*, 1990; 1993).

Investigations of exposed marine sediments indicate that the ice limit at around 13 ka was located near the extreme western coastline of Svalbard (Fig. 2) (Landvik *et al.*, 1998). At this time the Franz Josef Trough and St. Anna Trough were largely ice free, but a major ice dome may have still covered the central part of the Barents Sea shelf. It is likely that the end moraines in the southern and western Barents Sea mentioned above are of a similar or slightly older age. The sediment sequences in cores from the Central Deep suggest that the ice front receded from this line shortly after13 ka and much of the shelf was ice free a few hundred years later (Polyak *et al.*, 1995). Along the western margin of Svalbard, the deglaciation was interrupted by a short-lived glacier readvance that culminated on the inner part of the shelf shortly after 12.4 ka (Svendsen *et al.*, 1996). In spite of a rapid global transgression, the relative sea level was falling beyond the receding ice front, reflecting strong glacio-isostatic uplift.

The Younger Dryas and the final deglaciation

The large Admiralty Bank moraines are located west of Novaya Zemlya (Fig. 1) (Gataullin & Polyak, 1997). Judging from radiocarbon dates from the Central Deep (Polyak *et al.*, 1995), this ice margin is younger than ~12.5-13 ka. This led Gataullin *et al.* (2001) to hypothesise that the moraines were formed during the Younger Dryas Chron. However, more dates are necessary to confirm this assumption. A thickening of the glaciomarine sediments on the distal side of this line may suggest that the ice front halted at this position for some time. The configuration of the ice sheet that deposited the Admiralty Bank Moraines is not clear, but the orientation of the ridge system points towards an ice dispersal centre that was positioned over the northernmost part of the Novaya Zemlya area or even further to the N-NW. A coherent ice cover may have occurred between Novaya Zemlya and Franz Josef Land (Fig. 2). The few radiocarbon-dated sediment sequences on Franz Josef Land suggest that this area was deglaciated at the end of the Younger Dryas and/or very early in the Holocene (Forman *et al.*, 1997).

During the Younger Dryas, local glaciers on the west coast of Svalbard were of the same size as those today or even smaller (Mangerud & Svendsen, 1990). In contrast, a major ice sheet was centred on eastern Svalbard, with long outlet glaciers in the fjords. An almost stable position of the relative sea level along the west coast of Svalbard during the Younger Dryas implies the occurrence of a stable or growing ice load in the east (Landvik *et al.*, 1998). Even though the Younger Dryas limits around Svalbard have not been mapped, the available radiocarbon dates from uplifted marine sediments suggest that eastern Svalbard and the adjacent shelf was still covered by a sizeable ice cap at this time, and that it may have been larger than shown in this reconstruction.

It seems that the last remains of the Barents-Kara Ice sheet occurred over the northern part of the Barents Sea and that the ice caps remained stable or expanded as a response to the Younger Dryas cooling. This pattern of deglaciation is consistent with the 10 ka shoreline isobase that reflects an uplift dome over the northern Barents Sea. The ice caps melted rapidly at around 10 ka as a result of the abrupt Holocene climatic warming and as much as ~ 90 % of glacio-isostatic rebound was completed by *c.* 6 ka (Forman *et al.*, 1997). During the early and middle Holocene the glaciers on Svalbard and Franz Josef Land were smaller than today (Svendsen & Mangerud, 1997; Lubinski *et al.*, 1999).

Acknowledgments

The contribution of Polyak and, partially, Gataullin was supported by USA National Science Foundation grant OPP-9818247. Mangerud and Svendsen were supported through the projects *Paleo-Environment and Climate History of the Russian Arctic (PECHORA)* and *Ice Sheets and Climate in the Eurasian Arctic at the Last Glacial Maximum (Eurasian Ice Sheets)*, funded by the Research Council of Norway and the European Union (grant ENV4-CT97-0563) respectively. Both projects form part of the European Science Foundation Programme *Quaternary Environment of the Eurasian North (QUEEN)*. Eva Bjørseth, University of Bergen, assisted with the figures. Phil Gibbard corrected the English language. We thank all the institutions and persons involved for supporting this work.

References:

Alexandersson, H., Hjort, C., Möller, P., Antonov, O. & Pavlov, M. (2001). The North Taymyr ice-marginal zone, Arctic Siberia – a preliminary overview and dating. *Global and Planetary Change*, **31**, 427-445.

Arkhipov, S.A., Ehlers, J., Johnson, R.G. & Wright, H.E., jr. (1995). Glacial drainage towards the Mediterranian during the Middle and Late Pleistocene. *Boreas,* **24** (196-206.

Astakhov V.I. (1998. The last Kara Sea ice sheet: terrestrial constraints on its age. *Quaternary International,* **45/46**, 19-28.

Astakhov, V.I., Svendsen, J.I., Matiouchkov, A., Mangerud, J., Maslenikova, O. & Tveranger, J. (1999). Marginal formations of the last Kara and Barents ice shelves in northern European Russia. *Boreas,* **28**, 23-45.

Astakhov, V.I., this volume. Pleistocene ice limits in Russian northern lowlands. *In:* Ehlers, J. & Gibbard, P.L. (eds), *Quaternary Glaciations - Extent and Chronology*, Vol. 1. Europe. Amsterdam, Elsevier.

Bolshiyanov, D.Yu. & Makeyev, V.M. (1995). *Archipelago of Severnaya Zemlya: glaciation, environmental history.* St. Petersburg, Gidrometeoizdat, 216 pp.

Butt, F.A., Elverhøi, A., Solheim, A. & Forsberg, C.F. (2000). Deciphering Late Cenozoic development of the western Svalbard Margin from ODP Site 986 results. *Marine Geology,* **169**, 373-390.

Cadman, V. (1996). Glaciomarine sedimentation and environments during the Late Weichselian and Holocene in the Bellsund Trough and Van Keulenfjorden, Svalbard. Unpublished thesis, University of Cambridge.

Dowdeswell, J.A., Elverhøi, A. & Solheim, A. (1998). Glaciomarine Sedimentary Processes and Facies on the Polar North Atlantic Margins. *Quaternary Science Reviews,* **17**, 243-272.

Elverhøi, A. & Solheim, A. (1983). Physical environments in the western Barents Sea, 1: 500,000 surface sediment distribution. *Norsk Polarinstitutt Skrifter*, **179 A**, 1-23.

Elverhøi, A., Nyland-Berg, M., Russwurm, L. & Solheim, A. (1990). Late Weichselian ice recession in the central Barents Sea. *In:* Bleil, U. & Thiede, J. (eds), *Geological History of the Polar Oceans: Arctic versus Antarctic*, Dordrecht, Kluwer, 289-307.

Elverhøi, A., Fjeldskaar, W., Solheim, A., Nydal Berg, M. & Russwurm, L. (1993). The Barents Sea ice sheet – a model for its growth and decay during the last ice maximum. *Quaternary Science Reviews,* **12**, 863-873.

Elverhøi, A., Andersen, E.S., Dokken, T., Hebbeln, D., Spielhagen, R., Svendsen, J.I., Sørflaten, M., Rørnes, A., Hald, M. & Forsberg, C.F. (1995). The growth and decay of the Late Weichselian ice sheet in western Svalbard and adjacent areas based on provenance studies of marine sediments. *Quaternary Research*, **44**, 303-316.

Elverhøi, A., Hook, R.L.B. & Solheim, A. (1998a). Late Cenozoic Erosion and Sediment Yield from the Svalbard-Barents Sea Region: Implications for Understanding Erosion of Glacierized Basins. *Quaternary Research*, **44** (209-241.

Elverhøi, A., Dowdeswell, J., Funder, S., Mangerud, J. & Stein, R. (1998b). Glacial and Oceanic History of the Polar North Atlantic Margins: An Overview. *Quaternary Science Reviews,* **17**, 1-10.

Faleide, J.I., Solheim, A., Fiedler, A., Hjelstuen, B.O., Andersen, E.S., Vannetse, K. (1996). Late Cemozoic evolution of the western Barents Sea – Svalbard continental margin. *Global and Planetray Change,* **12 (1-4)**, 53-74.

Forman, S.L., Lubiniski, D., Miller, G.H., Snyder, J., Matishov, G., Korun, S. & Myslivets, V. (1995). Postglacial emergence and distribution of late Weichselian ice-sheet loads in the northern Barents and Kara seas, Russia. *Geology,* **23**, 113-116.

Forman, S.L., Weihe, R., Lubiniski, D., Tarasov, G., Korsun, S. & Matishov, G. (1997). Holocene relative sea-level history of Franz Josef Land, Russia. *Geological Society of America Bulletin,* **109**, 1116-1133.

Forman, S.L., Ingolfsson, O., Gataullin, V., Manley, W. F. & Lokrantz, H. (1999). Late Quaternary stratigraphy of western Yamal Peninsula, Russia: New constraints on the configuration of the Eurasian ice sheet. *Geology,* **27**, 807-810.

Flower, B.P. (1997). Overconsolidated section on the Yermak Plateau, Arctic Ocean ice sheet grounding prior to 660 ka ? *Geology,* **25**, 147-150.

Gataullin, V. & Polyak, L. (1997). Morainic Ridge Complex, Eastern Barents Sea. *In:* Davies, T. A., Bell, T., Cooper, A.K., Josenhans, H., Polyak, L., Solheim, A., Stoker, M.S. & Stravecs, J.A. (eds), *Glaciated Continental Margins: An Atlas of Acoustic Images,* London, Chapman & Hall, 82-83.

Gataullin, V., Mangerud, J. & Svendsen, J.I. (2001). The extent of the Late Weichselian ice sheet in the southeastern Barents Sea. *Global and Planetary Change,* **31**, 447-451.

Grosswald, M.G. (1980). Late Weichselian Ice Sheets of Northern Eurasia. *Quaternary Science Reviews,* **13**, 1-32.

Grosswald, M.G. (1993). Extent and melting history of the Late Weichselian ice sheet, the Barents-Kara continental margin. *In*: Peltier, W.R (ed.), *Ice in the climate system*, NATO ASI series I: Global Environmental Change. Berlin, Springer-Verlag, 1-20.

Grosswald, M.G. (1998). Late-Weichselian ice sheets in Arctic and Pacific Siberia. *Quaternary International,* **45/46**, 3-18.

Henriksen, M., Mangerud, J., Maslenikova, O., Matiouchkov, A. & Tveranger, J. (2001). Weichselian stratigraphy and glaciotectonic deformation along the lower Pechora River, Arctic Russia. *Global and Planetary Change*, **31**, 297-319.

Hjort, Chr., Möller, P. & Alexanderson, H., this volume. Weichselian glaciation of the Taymyr Peninsula, Siberia. *In*: Ehlers, J. & Gibbard, P. (eds), *Quaternary Glaciations - Extent and Chronology.* Vol. 1. Europe. Elsevier (Amsterdam).

Houmark-Nielsen, M., Demidov, I., Funder, S., Grøsfjeld, K., Kjær, K.H., Larsen, E., Lavrova, N., Lyså, A. & Nielsen, J.K. (2001). Early and Middle Valdaian glaciations, ice-dammed lakes and periglacial interstadials in north west Russia: new evidence from the Pyoza River Area. *Global and Planetary Change*, **31**, 215-237.

Jakobsson, M. (1999). First high-resolution chirp sonar profiles from the central Arctic Ocean reveal erosion of Lomonosov Ridge sediments. *Marine Geology*, **158**, 111-123.

Jakobsson, M., Løvlie, R., Arnold, E., Backman, J., Polyak, L., Knutsen, J.-O. & Musatov, E. (2001). Pleistocene stratigraphy and paleoenvironmental variation from Lomonosov Ridge sediments, central Arctic Ocean. *Global and Planetary Change*, **31**, 1-22.

Jansen, E. & Sjøholm, J. (1991). Reconstruction of glaciation over the past 6 Ma based on Norwegian Sea records. *Nature,* **349**, 600-603.

Kleiber, H.P., Knies, J., Niessen, F. (2000). The Late Weichselian glaciation of the Franz Victoria Trough, northern Barents Sea: ice sheet extent and timing. *Marine Geology*, **168**, 25-44.

Knies, J., Nowaczyk, N., Müller, C., Vogt, C. & Stein, R. (2000). A multiproxy approach to reconstruct the environmental changes along the Eurasian continental margin over the last 160.000 years. *Marine Geology* **163**, 317-344.

Knies, J., Kleiber, H.-P., Matthiessen, J., Müller, C. & Nowaczyk, N. (2001). Marine ice-rafted debris records constrain maximum extent of Saalian and Weichselian ice-sheets along the northern Eurasian margin. *Global and Planetary Change*, **31**, 45-64.

Koshechkin, B.I. (1978). *Golotsenovaya tektonika vostochnoi chasti Baltiyskogo schita.* (Holocene tectonics of the eastern part of the Baltic shield). Leningrad, 176 pp. (in Russian).

Krasnov, I.I. (ed.) (1971). Map of the Quaternary of the European USSR and adjacent territories, scale 1:1 500 000, VSEGEI, Leningrad, 16 sheets

Laberg, J.S. & Vorren, T.O. (1995). Late Weichselian submarine debris flow deposits on the Bear Island Trough Mouth Fan. *Marine Geology*, **127**, 45-72.

Laberg, J.S. & Vorren, T.O. (1996). The Middle and Late Pleistocene evolution of the Bear Island Trough Mouth Fan. *Global and Planetary Change,* **12**, 309-330.

Landvik, J.I., Bondevik, S., Elverhøi, A., Fjeldskaar, W., Mangerud, J., Salvigsen, O., Siegert, M.I., Svendsen, J.I.

& Vorren, T.O. (1998). The Last Glacial Maximum of Svaldbard and the Barents Sea area: ice sheet extent and configuration. *Quaternary Science Review,* **17**, 43-75.

Lubinski, D., Korsun, S., Polyak, L., Forman, S.L., Lehman, S.J., Herlihy, F.A., Miller, G.H. (1996). The last deglaciation of the Franz Victoria trough, northern Barents Sea. *Boreas,* **25**, 89-100.

Lubinski, D.J., Forman, S.L. & Miller, G.H. (1999). Holocene glacier and climate fluctuations on Franz Josef Land, Arctic Russia, 80°N. *Quaternary Science Reviews,* **18**, 85-108.

Makeyev, V.M., Arslanov, K.A. & Garutt, V.Ye. (1979). Age of Severnaya Zemlja mammoths and some issues of Late Pleistocene palaeogeography. *Doklady Academia Nauk USSR,* **242 (2)**, 421-424.

Mangerud, J. & Svendsen, J. I. (1990). Deglaciation chronology inferred from marine sediments in a proglacial lake basin, western Spitsbergen, Svalbard. *Boreas,* **19**, 249-272.

Mangerud, J., Jansen, E. & Landvik, J. (1996). Late Cenozoic history of the Scandinavian and Barents Sea ice sheets. *Global and Planetary Change,* **12**, 11-26.

Mangerud, J., Dokken, T., Hebbeln, D., Heggen, B., Ingolfson, O., Landvik, J.Y., Mejdahl, V., Svendsen, J.I. & Vorren, T. (1998). Fluctuations of the Svalbard-Barents Sea ice sheet during the last 150 000 years. *Quaternary Science Reviews,* **17**, 11-42.

Mangerud, J., Svendsen, J. I. & Astakhov, V.I. (1999). Age and extent of the Barents and Kara ice sheets in Northern Russia. *Boreas,* **28**, 46-80.

Mangerud, J., Astakhov, V.I., Murray, A. & Svendsen, J.I. (2001). The chronology of a large ice-dammed lake and the Barents-Kara Ice Sheet advances, Northern Russia. *Global and Planetary Change*, **31**, 321-336.

Myhre, A.M., Thiede, J., Firth, J.V. *et al.* (1995). Initial Reports, Ocean Drilling Program, Leg 151: College Station, Texas. Ocean Drilling Program, 926 pp.

Möller, P., Bolshiyanov, D.Y. & Bergsten, H. (1999). Weichselian geology and paleoenvironmental history of the central Taymye Peninsula, Siberia, indicationg no glaciation during the last glacial maximum. *Boreas,* **28**, 92-114.

Polyak, L., Lehman, S.J., Gataullin, V. & Jull, A.J.T. (1995). Two step deglaciation of the southeastern Barents Sea. *Geology,* **23**, 567-571.

Polyak, L., Forman, S.L., Herlihy, F.A., Ivanov, G. & Krinitsky, P. (1997). Late Weichselian deglaciation history of Svyataya (Saint) Anna Trough, northern Kara Sea, Arctic Russia. *Marine Geology,* **143**, 169-188.

Polyak, L., Gataullin, V., Okuneva, O. & Stelle, V. (2000a). New constraints on the limits of the Barents-Kara ice sheet during the Last Glacial Maximum based on bore-hole stratigraphy from the Pechora Sea. *Geology,* **28**, 611-614.

Polyak, L., Levitan, M., Gataullin, V., Khusid, T., Mikhailov, V. & Mukhina, V. (2000b). The impact of glaciation, river discharge, and sea-level change on Late Quaternary environments in the southwestern Kara Sea. *International Journal of Earth Sciences,* **89**, 660-562.

Polyak, L., Edwards, M.H., Coakley, B.J. & Jakobsson, M. (2001). Ice shelves in the Pleistocene Arctic Ocean inferred from glaciogenic deep-sea bedforms. *Nature,* **410**, 453-457.

Salvigsen, O. (1981). Radiocarbon dated raised beaches in Kong Karls Land, Svalbard, and their consequences for the glacial history in the Barents Sea. *Geografiska Annaler,* **63A**, 280-291.

Siegert, M.J., Dowdeswell, J.A., Hald, M. & Svendsen, J.I. (2001). Modelling of the Eurasian Ice Sheet through a full (Weichselian) glacial cycle. *Global and Planetary Change*, **31**, 367-385.

Solheim, A. & Kristoffersen, Y. (1984). The physical environment in the Western Barents Sea, 1: 5,000,000. Sediments above the upper regional unconformity: thickness, seismic stratigraphy and outline of the glacial history. *Norsk Polarinstitutt Skrifter,* **179B**, 1-26.

Solheim, A., Russworm, L., Elverhøi, A. & Nydal Berg, M. (1990). Glacial geomorphic features: direct evidence for grounded ice in the northern Barents Sea and implication for the pattern of deglaciation and late glacial sedimentation. *Geological Society of London Special Publication,* **53**, 253-268.

Solheim, A., Faleide, J.I., Andersen, E.S., Elverhøi, A., Forsberg, C.F., Vanneste, K. & Uenzelmann-Neben, G. (1998). Late Cenozoic Seismic Stratigraphy and Glacial Geological Development of the East Greenland and Svalbard-Barents Sea Continental Margins. *Quaternary Science Reviews,* **17**, 155-184.

Spielhagen, R.F., Bonani, G., Eisenhauser, A., Frank, M., Fredrichs, T., Kassens, H., Kubik, P.W., Mangini, A.W., Norgaard-Pedersen, N., Nowaczyk, N.R., Schper, S., Stein, R., Thiede, J., Tidemann, R. & Washner, M. (1997). Arctic Ocean evidence for late Quaternary initiation of northern Eurasian ice sheets. *Geology,* **25**, 783-786.

Svendsen, J.I. & Mangerud, J. (1997). Holocene glacial and climatic variations on Spitsbergen, Svalbard. *The Holocene*, **7 (1)**, 45-57.

Svendsen, J.I., Elverhøi, A. & Mangerud, J. (1996). The retreat of the Barents Sea Ice Sheet on the western Svalbard margin. *Boreas*, **25**, 244-256.

Svendsen, J.I., Astakhov, V.I., Bolshiyanov, D.Yu., Demidov, I., Dowdeswell, J.A., Gataullin, V., Hjort, Ch., Hubberten, H.W., Larsen E., Mangerud, J., Melles, M., Möller, P., Saarnisto, M. & Siegert, M. J. (1999). Maximum extent of the Eurasian ice sheet in the Barents and Kara Sea region during the Weichselian. *Boreas,* **28**, 234-242.

Sættem, J., Poole, D.A.R., Sejrup, H.P. & Ellingsen, K.L. (1992). Glacial geology of outer Bjørnøyrenna, western Barents Sea. *Marine Geology,* **103**, 15-51.

Thiede, J., Winkler, A., Wolf-Welling, T., Eldholm, O., Myhre, A.M., Braumann, K.-H., Henrich, R. & Stein, R. (1998). Late Cenozoic History of the Polar North Atlantic: Results from Ocean Drilling. *Quaternary Science Reviews,* **17**, 185-208.

Velichko, A.A., Kononov, Yu.M. & Faustova, M.A. (1997). The last glaciation of Earth: size and volume of ice-sheets. *Quaternary International,* **41/42**, 43-51.

Vogt, P.R., Crane, K. & Sundvor, E. (1994). Deep Pleistoce Ice-berg ploughmarks on the Yermak Plateau: sidescan and 3.5 kHz evidence for thick calving ice fronts and a possible marine ice sheet in the Arctic Ocean. *Geology,* **22**, 403-406.

Vorren, T.O., Vorren, K.D., Alm, T., Gulliksen, S. & Løvlie, R. (1988). The last deglaciation (20,000 to 11,000 B.P.) on Andøya, northern Norway. *Boreas*, **17**, 41-77.

Vorren, T.O., Lebesbye, E., Andreassen, K. & Larsen, K.B. (1989). Glacigenic sediments on a passive continental margin as exemplified by the Barents Sea. *Marine Geology,* **85**, 251-272.

Vorren, T.O., Lebesbye, E. & Larsen, K. (1990). Geometry and genesis of the glacigenic sediments in the southern Barents Sea. *In:* Scource, J.D. & Dowdeswell, J.A., (eds), *Glacimarine environments: processes and sediments,* London, Geological Society, 309-328.

Vorren. T.O., Laberg, J.S., Blaumme, F., Dowdeswell, J.A., Kenyon, N.H., Mienert, J., Rumohr, J. & Werner, F. (1998). The Norwegian-Greenland Sea Continental Margins: Morphology and Late Quaternary Sedimentary Processes and Environment. *Quaternary Science Reviews,* **17**, 273-202.

Yakovlev, S.A. (1956). Osnovy geologii chetvertichnykh otlozheniy Russkoi ravniny (The fundamentals of the Quaternary geology of the Russian Plain) *Trudy VSEGEI*, **17**, 314 pp. Moscow (in Russian).

Quaternary Glaciations – Extent and Chronology
Editors J. Ehlers and P.L. Gibbard

Glacial morphology of Serbia, with comments on the Pleistocene Glaciation of Monte Negro, Macedonia and Albania

Ljubomir Menkovic [1], Miroslav Markovic [2], Tomas Cupkovic [2], Radmila Pavlovic [2], Branislav Trivic [2] and Nenad Banjac [2]

[1] *Geographical Institute of Serbian Academy of Sciences and Arts, Jaksiceva 7/III, 11000 Belgrade, Yugoslavia*
[2] *Faculty of Mining and Geology, University of Belgrade, Djusina 7, 11000 Belgrade, Yugoslavia*

Abstract

Cvijic was the first to collect evidence on the glacial morphology of the Balkans, at the end of the 19th century. He reported the existence of glacial features on the three highest mountains of Serbia, Prokletije, Sara and Koritnik. The most recent investigations have been carried out using remote sensing techniques supplemented by field observations. This approach has led to the discovery of numerous cirques, glacial troughs, moraines and other characteristic features produced by Pleistocene glaciers in these mountains. Analysis of the positions and mutual relationships between the glacial features of these three mountain areas has enabled determination of the Pleistocene snowline altitude. Results indicate that it occurred at 1900 m elevation on the northern faces and 2200 m on the southern faces. These results indicate that the development of the Pleistocene, more precisely the Würmian glaciation, was restricted to the highest elevations within the areas investigated. The type of glacier, which was formed, was dependent on the snowline position (i.e. on the topographic surface above it). Most glaciers developed in cirques, as hanging glaciers, and troughs, as valley glaciers. On Sara Mountain, where a vast plain existed at 2200-2400 m altitude, plateau-type glaciers developed.

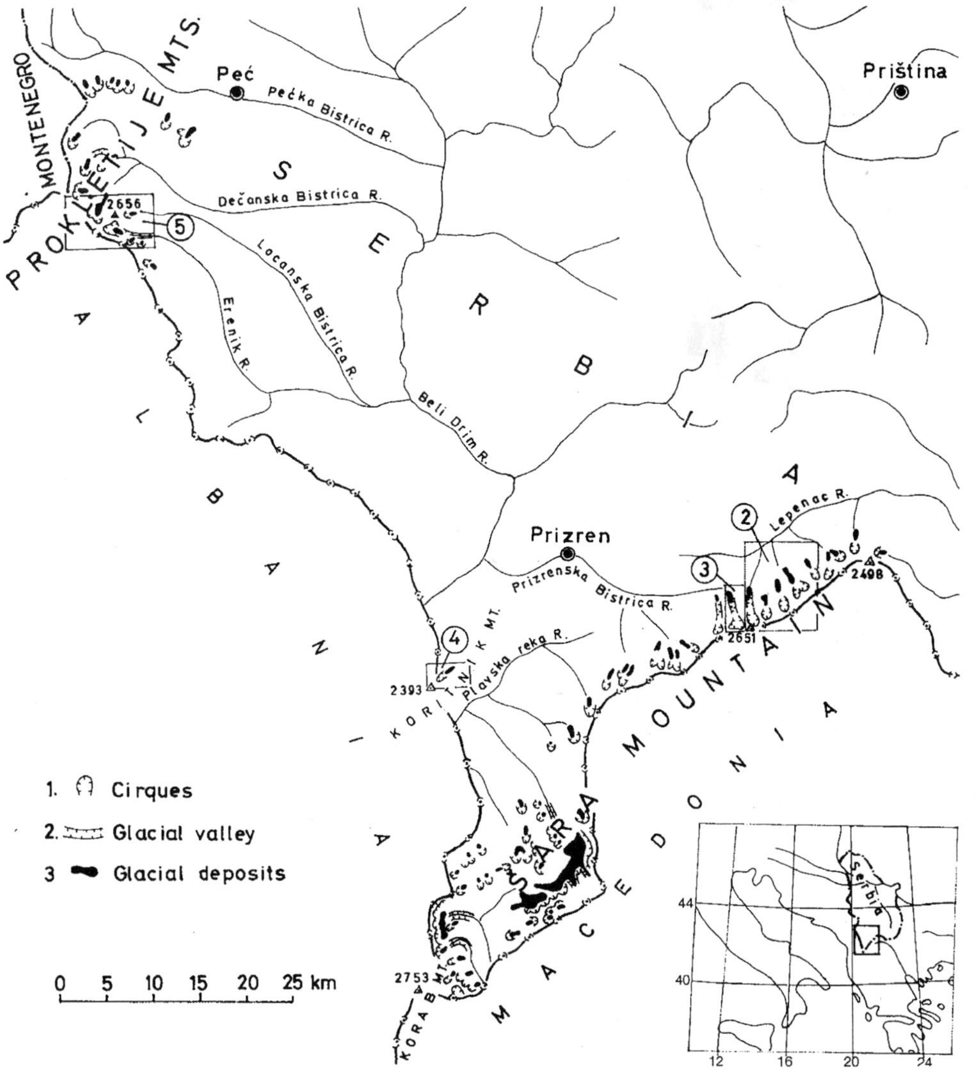

Fig. 1. Location map of the formerly glaciated areas of Serbia. The circled numbers (2, 3, 4 and 5) refer to the sketch maps in Figs. 3, 4, 12 and 9 respectively

Fig. 2. End moraines on the western side of the Orjen Mountain in Monte Negro.

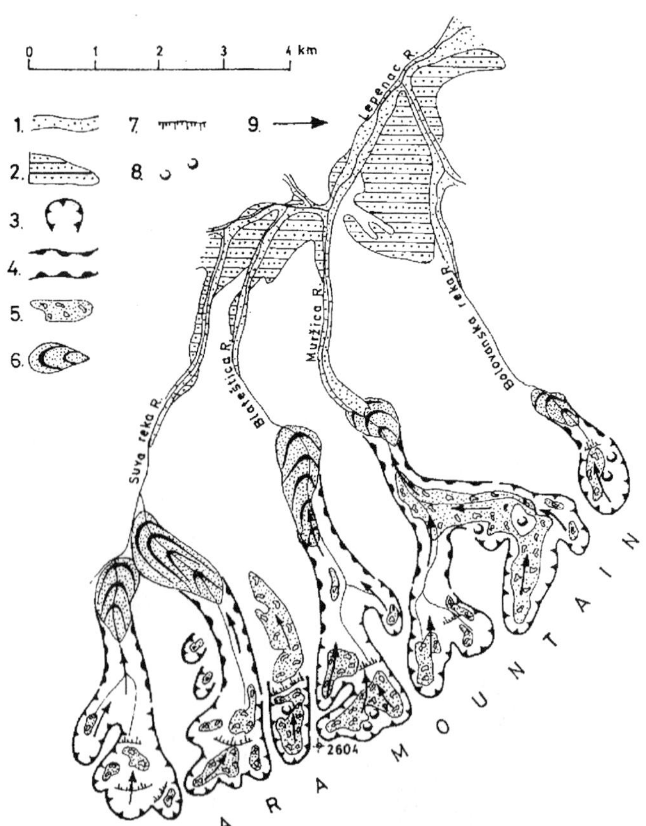

Fig. 3. Sara Mountain, right-bank tributaries of the Lepenac River. Detailed sketch of the glacial morphology. Legend: 1. Alluvial plain; 2. Glacio-fluvial terraces; 3. Cirque; 4. Glacial trough; 5. Moraines; 6. Terminal moraines; 7. Icefall; 8. Roches moutonnées; 9. Direction of ice movement.

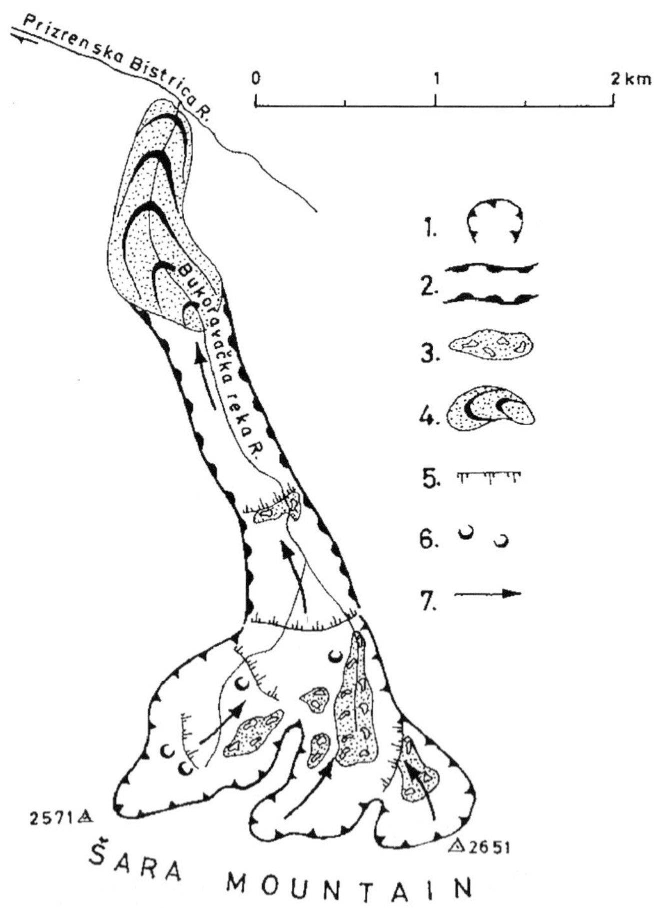

Fig. 4. Sara Mountain, Bukorovacka River area, left-bank tributary of Prizrenska Bistrica River. Detailed sketch of the glacial forms. Legend: 1. Cirque; 2. Glacial trough; 3. Moraines; 4. Terminal moraines; 5. Icefall; 6. Roches moutonnées; 7. Direction of ice movement.

Fig. 5: The Jezerska Cuka cirque, Sara Mountain.

INTRODUCTION

During the Pleistocene, particularly in the Würmian Stage, the highest parts of the Balkans were covered by so-called 'eternal snow'. As a consequence of strong glacial activity, characteristic morphological features developed. Many of these are still recognisable in the present-day topography, in spite of the fact that the glacial processes were long ago replaced by fluvial and slope processes.

The earliest basic evidence of Balkan glaciation were presented by Cvijic (1897, 1899, 1903a, 1903b, 1911, 1913). He identified the main glaciation centres in the high Balkan mountains, and defined the Würmian age of the glaciation. However, his maps showing the evidence for glacial limits have required correcting. Reliable topographic maps were not available to Cvijic. Modern corrections have been made mostly using remote sensing methods and high quality large-scale photogrammetric topographic maps. In addition, present-day investigations were carried out using remote sensing techniques, supplemented by field observations. This approach has led to the discovery of numerous cirques, troughs, moraines and other characteristic features produced by Pleistocene glaciers in these mountains, (Figure 1).

For Yugoslavia (Serbia and partly Monte Negro), Cvijic's original evidence was also used, with the necessary corrections, together with the results of more recent investigations by Marovic and Markovic (1972), Markovic (1973) and Menkovic (1977/78, 1985, 1994). However, for Macedonia and Albania, no published data were available. For these areas information is based on satellite images and small-scale topographic and geological maps. For this reason no detailed sketches or explanatory texts could be given for these areas.

Glaciation in the Balkan area investigated was developed as Alpine-type, characterised mostly by hanging cirques, now occupied by small lakes. The glaciers were short, the most common being 2-3 km in length. The snowline was generally depressed to an altitude of 1900-2200 metres; its height decreased southwards, towards the Adriatic Sea. Longer glaciers were only developed in the Prokletije Mountains area, where end moraines formed terminal basins now occupied by lakes.

MAIN AREAS OF GLACIATION

1. Serbia

Cvijic (1898, 1911, 1913) recorded the first evidence of the glacial morphology in the Balkans. He observed traces of glacial features on Serbia's highest mountains, including Prokletije and Sara. Glacial features were also reported on the Korab Mountain by Nikolic (1912) and on the Koritnik Mountain by Menkovic (1985). All of these occurrences are located in the extreme southwest of Serbia, along its border with Albania, Macedonia and Monte Negro.

Sara Mountain area

Sara Mountain is situated on the border of Serbia and Macedonia. Its ridge trends northeast, at an average elevation of 2000-2200 m. Some peaks rise to over 2500 m. Detailed geomorphological investigations were carried out by Menkovic (1977/78). On the mountain's northwestern (Serbian) side, some thirty cirques, varying in diameter from 100 m to 1 km, have been recognised. All of them are located

Fig. 6. The Durlov Potok cirque, Sara Mountain.

Fig. 7. The Livadicki cirque and lake, Sara Mountain.

Fig. 8. Karanikolica cirque and lake, Macedonian side of the Sara Mountain.

immediately below the main mountain ridge and the highest peaks. Springs in these cirques form the headwaters of the Lepenac, Prizrenska, Bistrica and Plavska rivers.

The cirques are mainly exposed on the northern slopes at elevations of 1900-2400m. From the development, location and density of the cirques, it is concluded that the most severe glaciation during the Pleistocene took place around the watershed (2651m) between the Lepenac and Prizrenska Bistrica rivers, as well as in the extreme southwestern part of Sara Mountain. The glacial valleys originating from the cirques are 2-3 km long. Later they were replaced by the valleys of right-bank tributaries of the Lepenac River. A similar situation occurs in the southwestern part of Sara Mountain.

Glacial valleys end, as a rule, with terminal moraines. However, in the Sara Mountain area they have been partially or completely eroded. The best-preserved examples, with heights varying from 50-200 m, were found in the upper courses of the Lepenac and Prizrenska Bistrica rivers. The

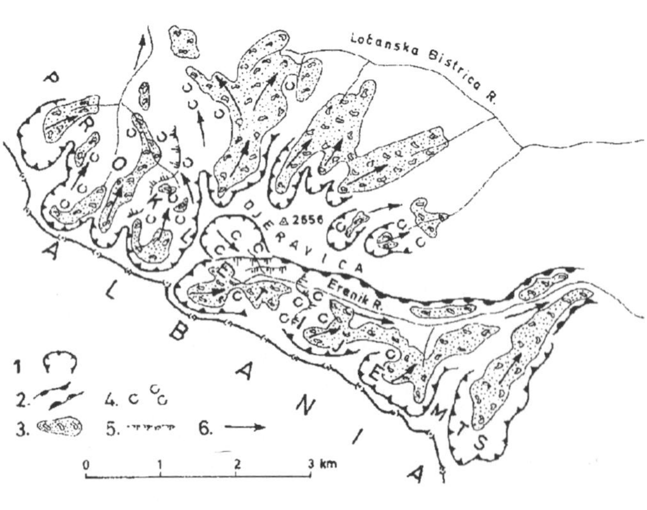

Fig. 9. Prokletije Mountains. Detailed geomorphological sketch map. Legend: 1. Cirque; 2. Glacial trough; 3. Moraines; 4. Roches moutonnées; 5. Icefall; 6. Direction of ice movement.

lowest of them occur at elevations of 1250-1300 m (Figures 3 and 4).

During the glacial retreat, the moraines were left in their cirques and valleys. The valley moraines, however, were subsequently strongly eroded, whilst the cirque moraines are well preserved. Glacial activity also produced other forms and features, including *roches moutonnées* (well developed on the high plain of Sara Mountain, especially in its southwestern part), icefall and glacial lakes (e.g. the well-known Livadica and Jazinac lakes, in the Lepenac River drainage area).

Fig. 10. Prokletije Mountains, Djeravica cirque and lake.

Prokletije Mountains area

The Prokletije Mountains form Serbia's border with Albania and Monte Negro. The eastern part, including the highest peak Djeravica (2656 m), lie within Serbia. The relief here is highly dissected. The first information on glacial features of the area was presented by Cvijic (1913). Detailed modern geomorphological investigations were carried out by Menkovic (1994).

Here cirques formed just below the highest peaks and mountain ridges. The springs which now form the headwaters of the tributaries of the Pecka Bistrica, Decanska Bistrica, Locanska Bistrica and Erenik rivers originate from here. The cirques have diameters of 100-2000 m, and occur at elevations of 2000-2200 m. The cirques continue into short glacial valleys, which are the valleys of the present-day

Fig. 11. Glacial valley of the Erenik river in the Prokletije Mts.

Erenik and Decanska Bistrica rivers, with lengths of some 4-5 km. The moraines are well preserved only in the cirques; at lower altitudes they have been mostly eroded. The moraines reach a minimum recorded altitude of approximately 1500 m a.s.l.

Based on the presence of numerous cirques and moraines, it is concluded that Pleistocene glaciation was most intensive in the Djeravica area. Starting from Djeravica, glaciers spread towards the Erenik, Locanska Bistrica and Peck Bistrica rivers (Figure 9).

In the Prokletije Mountains, particularly in the Djeravica Peak area, *roches moutonnées* and traces of an icefall have been identified. These were produced by selective glacial erosion. The glacial lakes, located in cirques, are of the same origin. The latter include the well known Djeravica Lake, formed in a cirque just below the peak of Djeravica.

Korab Mountain area

Korab Mountain is situated in the extreme southern part of Serbia, on its border with Albania and Macedonia. Only the part of the mountain that lies north of the peak (2753m) belongs to Serbia. On the basis of the glacial landforms recorded, including cirques and moraines, it is concluded that

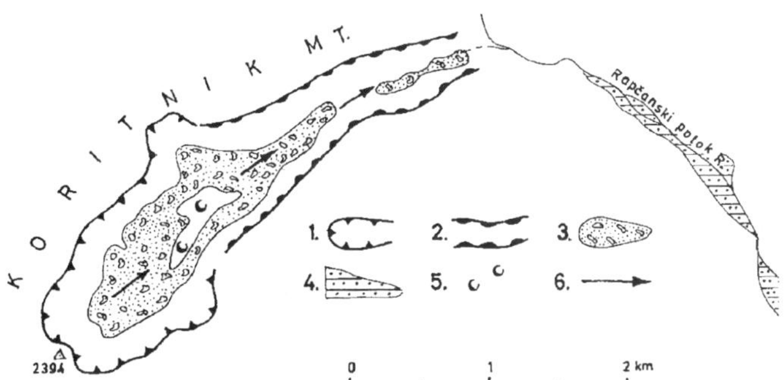

Fig. 12. Koritnik Mountain, detailed sketch map of the glacial morphology,. Legend: 1. Cirque; 2. Glacial trough; 3. Moraines; 4. Glaciofluvial terraces; 5. Roches moutonnées; 6. Direction of ice movement.

Pleistocene glaciers have significantly affected the morphology of this mountain. Numerous semi-circular cirques, 100-1000 m in diameter, occur immediately below the ridge separating Serbia and Albania. All of them occur at elevations above 2000m, and each is covered by moraines. Some of these cirques continue into glacial valleys, generally ending with terminal moraines. The remnants of terminal moraines were found at elevations of 1500m.

Koritnik Mountain area

Koritnik Mountain forms the border between Serbia and Albania. It is clearly separated from the other mountains. During the Würmian maximum, the highest parts of this mountain (at elevations above 2000 m) were under permanent snow cover. A single valley glacier was formed due to the existence of a pre-glacial valley, deeply-incised into the limestone rocks. The topographic position of this valley on the northwestern side of the mountain also favoured snow accumulation and glacier formation, since it is in shadow most of the time.

A cirque, with a diameter of 1200 m and depth of 100-200 m, was also developed here at an elevations of 2000-2200 m. It gradually narrows towards the northeast and continues down into a glacial trough with a length of c. 1 km. The bottoms of both the cirque and the glacial trough are covered by moraines and *roches moutonnées*, composed exclusively of limestone (Figure 12). The moraines can be traced down to elevations of 1400 m.

2. Monte Negro

Two glacial centres are considered in more detail here: the Durmitor and Orjen mountains. In the Durmitor mountain area, more widespread glaciation occurred, both of Alpine and plateau-glacier types. Alpine-type valley glaciers extended down from its highest summits to the high plateau, limited by the deep canyons incised by the Piva and Tara rivers. A large plateau-type glacier was formed on this plateau. Starting from this ice field, additional Alpine-type glaciers descended into the Piva river valley. Their troughs, hanging now c. 600 m above the present level of the deeply-incised Piva river, can be easily recognised, both in the field and on aerial photographs (Marovic and Markovic, 1972).

The Orjen Mountain, rising above the Adriatic coast close to the town of Herceg Novi, was characterised by karstic glaciation. Glacial cirques and valleys were developed that exploited older karstic forms, such as major sinkholes and uvalas. Because of the strongly-dissected initial relief glaciers were often separated into individual valleys. The biggest glacier on the eastern side of the mountain was c. 2 km wide and 3 km long, and its thickness was estimated as 300 m. Near Crkvice the glacier separated into two branches. One was directed northward, and left a huge amount of glaciofluvial and glaciolacustrine deposits in the karstic polje of Grahovo. The second branch of the glacier advanced towards the east, where it terminated on the cliff edge of the Boka Kotorska gulf. Proximity of the sea and increased moisture caused considerable lowering of the snowline altitude, which was here located approximately at 1300 m above sea level. High precipitation was the main reason that the Orjen Mountain, with an altitude barely exceeding 1800 m, was so strongly glaciated. The melting of glaciers at the end of the Pleistocene produced a significant quantity of cold water, so that the karstic processes were reinforced during the early Holocene. The Orjen Mountain belongs to the area with the most intensive karstification in the whole Balkans (Markovic, 1973).

CONCLUSIONS

Based on the occurrence and distribution of glacial features, it is concluded that the Pleistocene (or more precisely, the Würmian) glaciation affected only the highest parts of Serbia's mountains, at elevations above 2000m. Nothing can be said about earlier glaciations since all traces were removed subsequently by the strong erosion. The analysis of the positions and mutual relationships between the glacial features of the mountain areas has enabled the Pleistocene snowline altitude to be determined. The results indicate that it occurred at 1900 m a.s.l. on northern faces and 2200 m on southern faces.

The type of glacier depended on the pre-existing topography and on the snowline altitude. The most favourable sites for glacier formation were the spring areas of pre-glacial rivers, situated above the snowline. These allowed the accumulation of significant amounts of snow. In the course of time these snow patches developed into cirque glaciers. The formation of particular glacier types depended on cirque dimensions (i.e. on the surface catchment area of the former spring). Downstream of the larger cirques, Alpine-type valley glaciers developed. The smaller cirques, however, are characterised by the development of cirque or hanging glaciers. Because of the lack of large surfaces above the snowline, the second type was more frequent than the first.

On Sara Mountain, where a vast plain exists at 2200-2400m a.s.l., a plateau-type glacier developed, the area of which is estimated at 30-35km^2, whilst on the Korab Mountain ridge, facing Albania, a saddle-type glacier developed. This divided into two valley glaciers as it descended: one tongue descended westwards into Albania and the other northeastwards into Serbia.

References

Cvijic J. (1897). Traces of old glaciers in the Rila Mt. (in Serbian). *Glas Srpske Kraljevske Akademije Nauka,* **LIV**, 1-103. Belgrade.

Cvijic, J. (1899). Glacial and morphological studies of the mountains in Bosnia, Herzegovina and Monte Negro (in Serbian). *Glas Srpske Kraljevske Akademije Nauka*, **LVII**, 1-196. Belgrade.

Cvijic, J. (1903a). New results on the glacial epoch in the Balkan Peninsula (in Serbian). *Glas Srpske Kraljevske Akademije Nauka*, **LXV**, 185-333. Belgrade.

Cvijic, J. (1903b). Balkan, Alpine and Carpathian glaciation (in Serbian). *Glas Srpske Kraljevske Akademije Nauka*, **LXVII**, 219-277. Belgrade.

Cvijic, J. (1911). Base for the geography and geology of Macedonia and Old Serbia (in Serbian). *Posebno izdanje Srpske Kraljevske Akademije Nauka*, **III**, 1074-1094. Belgrade.

Cvijic, J. (1913). The ice age in the Prokletije and surrounding mountains (in Serbian). *Glas Srpske Kraljevske Akademije Nauka*, **XCI**, 1-149. Belgrade.

Markovic, M. (1973). Geomorphological evolution and neotectonics of the Orjen Mt. area (in Serbian). D.Sc. thesis, pp. 1-285. Faculty of Mining and Geology, University of Belgrade.

Marovic, M., Markovic, M. (1972). Glacial morphology of the Durmitor Mt. wider area (in Serbian). *Geoloski anali Balkanskog poluostrva*, **XXXVII**, 37-48. Belgrade.

Menkovic, Lj. (1977/78). Glacial and nivation relief of the north-west side of Sara Mt. (in Serbian). *Vesnik Zavoda za geoloska i geofizicka istrazivanja*, **A 35/36**, 99-115. Belgrade.

Menkovic, Lj. (1985). Glacial morphology of the Koritnik Mt. (In Serbian). *Priroda Kosova*, **6**, 31-39. Pristina, Pokrajinski zavod za zastitu prirode.

Menkovic, Lj. (1994). Glacial traces in the Djeravica area - Prokletije Mountains (in Serbian). *Geografski godisnjak*, **30**, 139-146. Kragujevac.

Nikolic, R. (1912). Glaciation of the Sara and Korab Mountains (in Serbian). *Glas Srpske Kraljevske Akademije Nauka*, **LXXXVII**, 49-97. Belgrade.

Quaternary Claciations – Extent and Chronology
Editors J. Ehlers and P.L. Gibbard

The extent of Quaternary glaciations in Slovenia

Miloš Bavec [1] and Tomaž Verbič [2]

[1] *Geological Survey of Slovenia, Dimičeva 14,*
1000 Ljubljana, Slovenia
E-mail: milos.bavec@geo-zs.si)

[2] *Cesta dveh cesarjev 15A,*
1000 Ljubljana, Slovenia
E-mail: tomaz.verbic@guest.arnes.si)

Introduction

Slovenia is positioned at the southeasternmost tip of the Alps, a situation that represents a marginal zone with regard to the Alpine Quaternary glaciations. The basis for the reconstruction of the glacial limits in Slovenia are still largely consistent with the statements of Penck & Brückner (1909). In general, this means that most of the Slovenian part of the Alps was glaciated during the Pleistocene, excluding the foreland, with the exception of rare valley glaciers that reached beyond the limits of the mountainous area. Continuous tectonic activity, that produced deep and narrow river valleys and steep alpine topography reduced the preservation potential of glacigenic sediments to the point that only patches of (mostly redeposited) sediments are preserved. This makes the evaluation of ice-extent dependant upon interpretation rather than on reconstruction based on firm geological evidence.

Geological Setting

The Slovenian part of the Alps extending from the west to the east can be divided into three major units: the Julian Alps, the Karavanke Mountains and the Kamnik – Savinja Alps, respectively (Fig. 1). Elevations here vary from around 200 m above sea level (a.s.l.) in intramontane basins

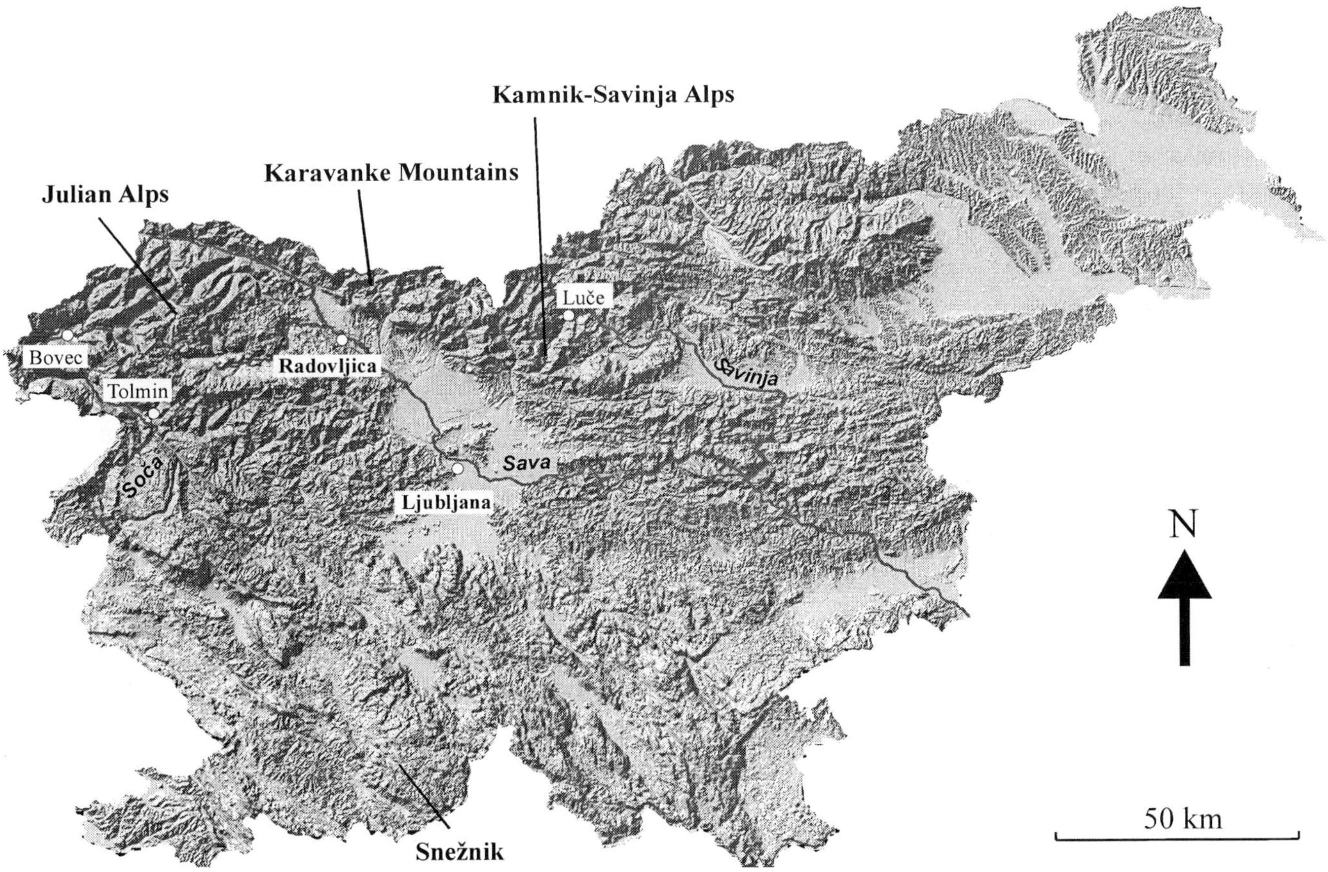

Fig. 1. Map of Slovenia showing the geographic units connected with the Quaternary glaciation.

Fig. 2. Late 1960's mining operation in Srpenica clay pit. Photo courtesy of Geological Survey of Slovenia.

in the west to 2864 m a.s.l. at the highest peak – the Triglav. Most of the area is drained through the Sava and Savinja Rivers towards the Black Sea, with the exception of the western part of the Julian Alps which is drained through the Soča River towards the Mediterranean (Adriatic) Sea. The coincidence between the modern drainage network and the Quaternary ice-flow pattern has been used to name the glaciers extending into the Alpine foreland during the Quaternary, which are referred to as the Soča, Sava, and Savinja glaciers (Penck & Brückner, 1909).

Reconstruction of the maximum ice extent

Snow accumulated in that part of the Julian Alps south of the Black Sea/Mediterranean divide formed the *Soča Glacier* extending along the valley of modern Soča River to the town of Tolmin (Penck & Brückner, 1909; Kuščer *et al.*, 1974, Kunaver, 1975). This interpretation is based on the occurrence of glacially reworked boulders in the area and a patch of glacigenic sediments found about 300 m above the modern Soča River, 3 km north of Tolmin, with the lithology indicating an upstream source. According to recent observations the authors interpret that only the Middle and/or Early Pleistocene glaciers may have been capable of extending so far into the foreland (Bavec & Tulaczyk, 2002). Namely the characteristics of Late Pleistocene sediments in the Bovec Basin (MIS 5 and younger; based on infrared stimulated luminescence, and radiocarbon dating of lake sediments) indicate that all sedimentation and resedimentation took place under paraglacial conditions (*cf.* Church & Ryder, 1972). This excludes the possibility of glacier presence in the basin itself. An extensive lake filled the area around Bovec towards the end of the Pleistocene (Fig. 2), in which more than 200 m of laminated fine-grained lacustrine sediments were deposited. The age of the upper 34 m that have been excavated for commercial purposes in the past (Fig. 3) was dated to 12.490±70 ^{14}C years B.P. at the bottom (Šercelj, 1970) and to 5885±60 ^{14}C years B.P. at the top while the age and sedimentary characteristics of the lower 170 and more meters is yet to be determined.

Snow accumulated in the part of the Julian Alps north of the divide and in the Karavanke Mountains formed the *Sava Glacier*. Its maximum extent was determined by ter-minal moraines in the vicinity of Radovljica (Kuščer, 1955, 1990; Žlebnik, 1971). Four generations of topographically relatively well-expressed terminal moraines - all in a distance of less then 4 km apart - have been identified and correlated with four stages of fluvial valley fill tentatively referred to as 'Late Würmian, Würmian, Rissian and Mindelian' (Žlebnik, 1971). The chronology of the deposits is based on litho- and morphostratigraphical correlations, partly supported by palynological analyses (Šercelj, 1970). According to this chronology, the most distal moraine is the oldest, while the terminal moraines of the subsequent glacial events follow in a proximal direction.

The Kamnik-Savinja Alps represent the accumulation area for the Pleistocene *Savinja Glacier*, with the exception of the southernmost slopes that drained towards the Sava Valley. All sediments deposited by the Savinja Glacier are interpreted as Last Glacial or Late Glacial, respectively (Meze, 1966; Mioč, 1983). Fluvial erosion has

Fig. 3: Late Pleistocene lake extent in the area of Bovec. Arrow is indicating the location of the former clay pit.

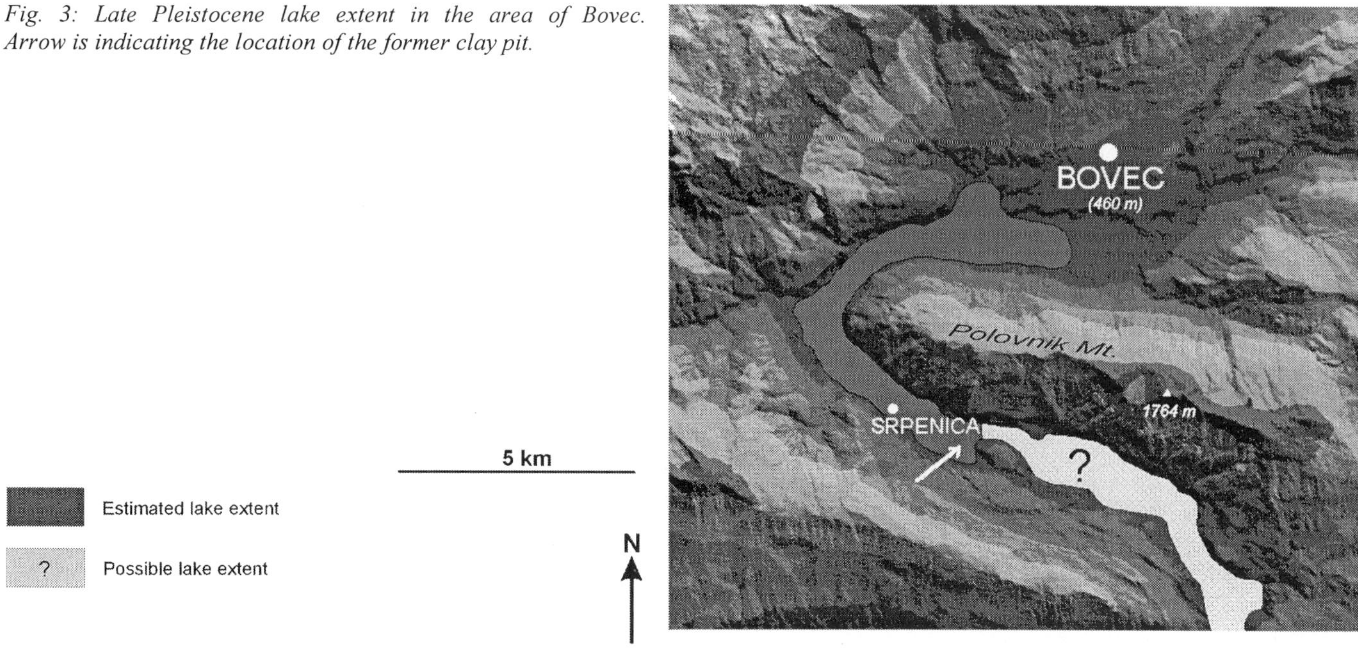

removed all traces of glaciation in the valley's lower reaches, so the interpretation of the glacial maximum extent as far as the town of Luče is based solely on the shape of the valley (Mioč, 1980).

Apart from the Alps and the valleys mentioned, the *Snežnik Mountain* in southern Slovenia (1796 m a.s.l.; Fig. 1) also supported an ice-cap during the Pleistocene. The glacier reached down to approximately 900 m a.s.l. where the most distal terminal moraines have been recognized (Šikič & Pleničar, 1972).

Discussion

High erosion rates in the Slovenian part of the Alps have caused most glacigenic sediments, together with depositional, as well as erosional geomorphological features, to be eroded or redeposited. The stratigraphical record has consequently been damaged to the point that it is rarely possible to put age-constraint on the glacial events identified. The maximum extent of Pleistocene glaciers is therefore interpreted according to the position of their most distal evidence of either glacial erosion or deposition (e.g. striations, moraines), in most cases without assigning an age attribute. Another problem is that, apart from some recent work in the Julian Alps, most of the other intensive research on the glaciogenic sediments in Slovenia was made decades ago using outdated terminology and lacking modern research techniques.

References

Bavec, M. (2000): Clast fabric analyses; an additional tool used for genetic interpretation of diamicts in the Bovec Basin, Log pod Mangartom and Stože, NW Slovenia. *Rudarsko metalurški zbornik,* **47**, 235-243 (in Slovene with English summary).

Bavec, M. & Tulaczyk, S.M. (2002): New simple mathematical model to help evaluating the extent of the Late-Quaternary valley glacier in the Upper Soča region. *Geologija*, **45** (in press).

Church, M. & Ryder, J.M. (1972) Paraglacial sedimentation: A Consideration of Fluvial Processes Conditioned by Glaciation. *Geological Society of America Bulletin*, **83**, 3059-3072.

Kunaver, J. (1975) On the geomorphological development of the Basin of Bovec during the Pleistocene epoch. *Geografski vestnik,* **47**, 11-39 (in Slovene with English summary).

Kuščer, D (1955) Beitrag zur Pleistozängeologie des Beckens von Radovljica. *Geologija,* **3**, 136-150 (in Slovene with German summary).

Kuščer, D. (1974) Geology of the Soča Valley between Bovec and Kobarid. *Geologija*, **17**, 425-476 (in Slovene with English summary).

Kuščer, D. (1990) The Quaternary valley fills of the Sava River and neotectonics. *Geologija*, **33**, 299-314 (in Slovene with English summary).

Meze, D. (1966) La vallée suprérieurre de la Savinja nouvelles constatations dans le développement géomorphologique de la region. Dela IV razreda Slovenske akademije znanosti in umetnosti 20. Inštitut za geogra-fijo, Ljubljana, 199 pp. (in Slovene with French summary).

Mioč, P. (1983) Tolmač lista Ravne, Osnovna geološka karta SFRJ 1:100 000. Zvezni Geološki Zavod, Beograd, 69 pp.

Penck, A. & Brückner, E. (1909) Die Eiszeiten in den Südalpen und im Bereich der Ostabdachung der Alpen. *Die Alpen im Eiszeitalter*, **3**, 718-1199.

Šercelj (1970) Würmeiszeitliche Vegetation und Klima in Slowenien. *Razprave IV razreda Slovenske akademije znanosti in umetnosti,* **13**, 209-249 (in Slovene with German summary).

Šikič, D. & Pleničar, M. (1972) Osnovna geološka karta SFRJ 1:100 000, list Ilirska Bistrica. Zvezni geološki Zavod Beograd.

Žlebnik (1971) Pleistocene deposits of the Kranj, Sora, and Ljubljana Fields. *Geologija*, **14**, 5-51 (in Slovene with English summary).

Quaternary Glaciations – Extent and Chronology
Editors J. Ehlers and P.L. Gibbard

Pleistocene glaciation in Spain

Augusto Pérez Alberti, Marcos Valcárcel Díaz and Ramón Blanco Chao

Departamento de Xeografía. Universidade de Santiago de Compostela (USC)
Plaza de la Universidad 1, 15703 Santiago de Compostela, SPAIN

E-mail: xepalber@usc.es, marcosvd@usc.es, rblanco@usc.es

1. Introduction

The Iberian Peninsula is characterized by a relatively high average altitude and by the predominance of extensive plains. However, mountain ranges occur in the north (the Pyrenees, with summits over 3000 m high, Cordillera Cantábrica, between 2000 and 2500 m), south (Sierra Nevada, 3482 m maximum height) and east (Sistema Ibérico, 2315 m maximum height). Furthermore, in the centre the Sistema Central runs from W to E from the Serra da Estrela, in Portugal, which rises to about 2000 m, to the Peñalara and Ayllón (north of Madrid), which are over 2400 m high.

Pleistocene glaciation was centred primarily in the higher mountain ranges and extended downvalley as glacier tongues, several kilometres long. The largest valley glaciers extended up to 52 km in the Central Pyrenees, Noguera Pallaresa basin (Nussbaum, 1956) and up to 28 km in the Galaico-Leonés massif (Bibei glacier) (Pérez Alberti & Valcárcel Díaz, 1998).

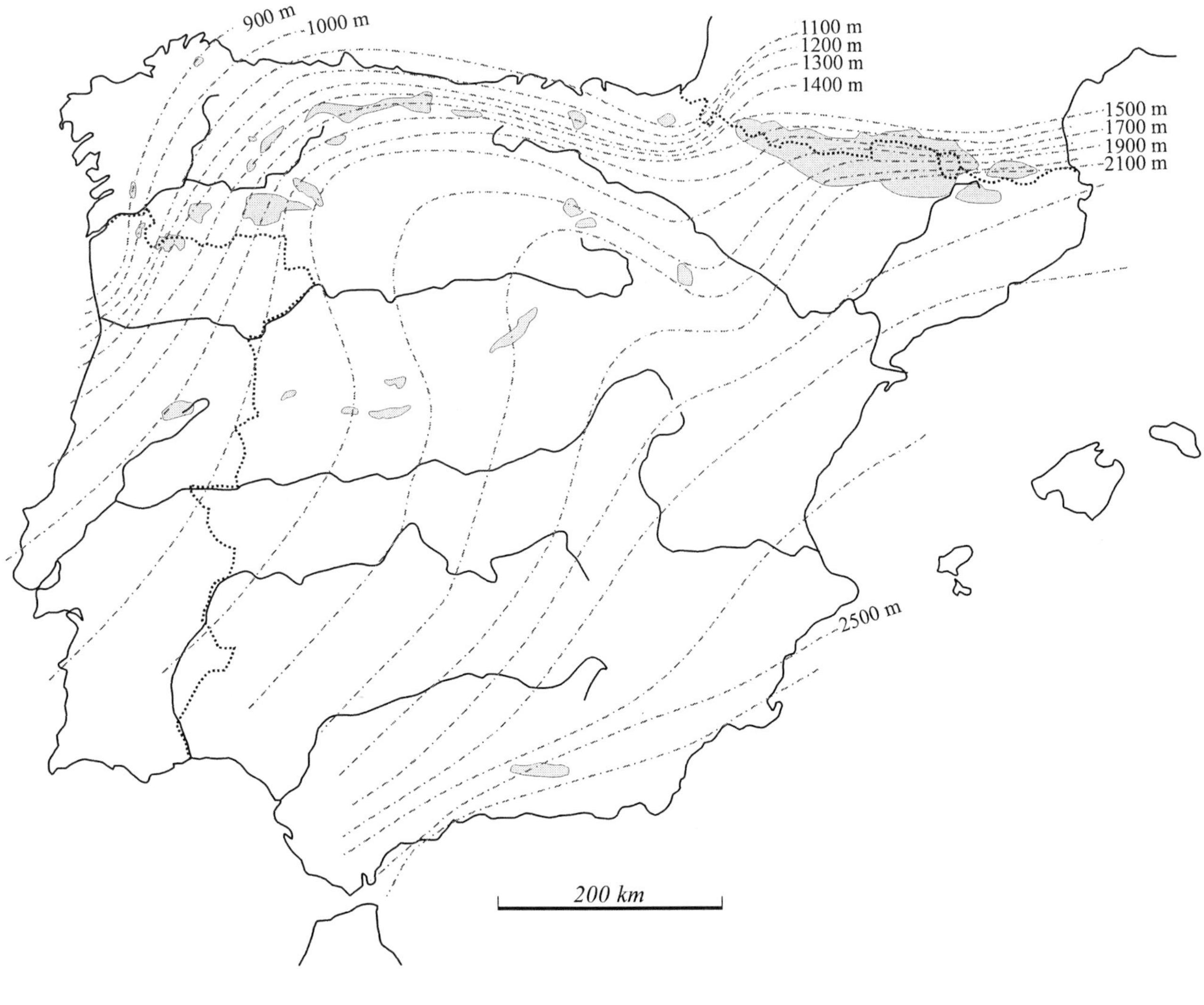

Fig. 1. Altitude of ELA in the Iberian Peninsule during the last glacial maximum (modified from Schmitz, 1969).

2. Studies of Pleistocene glaciation in Spain

The first papers on the Pleistocene glacier dynamics in the Spanish mountains were published in the middle of the 19th century (Prado, 1862; Baysselance, 1883; Penck, 1884). However, it was in the first half of the 20th century that the presence of glacial evidence in numerous mountainous massifs was established. This phase was followed by another that began in the 1970s, when more precise analyses were made of the size and complexity of the different former ice masses. Doctoral theses and other papers from that period deal with:

- the southern Pyrenees (Serrat, 1977; Vilaplana, 1979, 1981, 1983; Gómez Ortiz, 1980; Brú, 1985; Serrano, 1991, Bordonau, 1992; Montserrat Martí, 1992; Chueca, 1992; Peña Monné *et al.*, 1993; Turú, 1994; Martí Bono, 1996),
- the Cordillera Cantábrica (Frochoso, 1980; Castañón, 1989; Frochoso and Castañón, 1992; Alonso, 1992; Jiménez, 1994),
- the Galician mountains (Hernández Pacheco, 1957; Schmitz, 1969; Pérez Albertí, 1991; Pérez Albertí *et al.*, 1993; Vidal Romaní *et al.*, 1995; Kossel, 1996; Valcárcel Díaz, 1998),
- the Sistema Central (Martínez de Pisón and Muñoz Jiménez, 1972; Acaso, 1983; Pedraza, 1993; Marcos and Palacios, 1995), the Cordillera Ibérica (García Ruiz, 1979; Pellicer, 1980; Ortigosa, 1986) and
- the Sierra Nevada (Gómez Ortiz, 1987; Gómez Ortiz and Salvador Franch, 1991).

Based on the results of these studies, monographic syntheses were produced, not only for individual mountain ranges (Martí Bono and García Ruiz, 1994; Chueca *et al.*, 1998) but also for Spain as a whole (Gómez Ortíz and Pérez Albertí, 1998). Nevertheless, the delimitation of the former glacial areas, their extent and the number and age of the glacial phases in each massif are still to be determined in many cases, because the available data are not of consistent quality.

3. Importance of the glacial phenomenon

In order to measure its intensity, glaciation must be linked to the height attained by the ELA (Equilibrium Line Altitude) in each massif at the time of the maximum glacial advance recognized in each sector (MAG), and the area available above it in order to produce ice accumulation. As a whole, a general rise in ELA has occurred in the Iberian Peninsula in a NW-SE direction (Fig 1). Thus, while evidence of low-altitude glaciation was found in the northwestern mountains, with ELAs progressively increasing from 850-900 m in the low mountains closest to the coast (Faro de Avión, 1158 m maximum height; Schmitz, 1969; Serra do Xistral, 1034 m maximum height; Pérez Alberti *et al.*, 1993), to the 1400-1500 m of the mountains in the Galaico-Leonés massif (Trevinca 2114 m maximum height; Schmitz, 1969; Valcárcel Díaz *et al.*, 1998). According to Schmitz (1969), in the Sistema Central, the ELA ranges from 1600 m in its western sector (Serra da Estrela) and above 1800 m in its central sector (Sierra de Gredos), to above 2000 m in its eastern sector (Macizo de Peñalara). In the first sector, the Manteigas glacier reached a length of 13 km (Daveau, 1971), despite the fact that the mountain range reaches less than 2000 m. On the other hand, the glaciers that formed in Peñalara, with a height of 2430 m, were largely confined to the limits of the cirques and rarely reached a maximum length of 1 km. The extreme case in the Iberian Peninsula is the glaciation of the Sierra Nevada, where the ELA was above 2500 m, allowing extensive glaciation of higher parts to the summits of the Mulhacén (3482 m) and the Veleta (3398 m).

Fig 2. Erratic boulder in the Bibei valley (NW Spain).

The different extent of glaciation on the Iberian Peninsula is related to two factors. One is the available area above the ELA at the time of the MAG that progressively increased towards the interior of the Iberian Peninsula. The other factor is associated with the relief and aspect of the mountain ranges, which in some cases provided a far richer supply of snow and glacial accumulation than in others

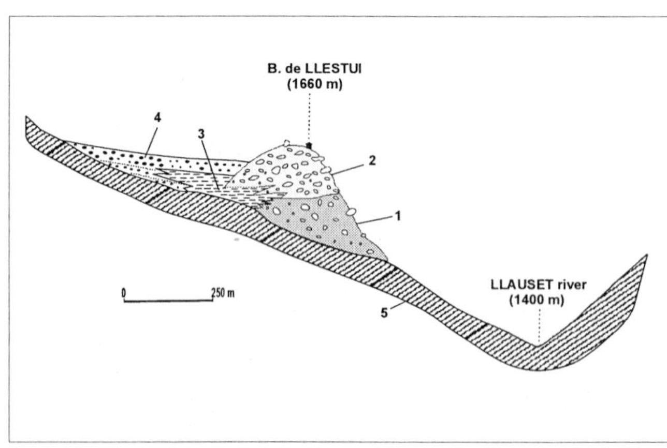

Fig 3. Section of a sedimentary formation in Llestui (South Pirineo). 1.- Subglacial till; 2.- Supraglacial till; 3.- Lacustrine deposits; 4.- Alluvian fan deposits. 5.- Paleozoic bedrock (Vilaplana, 1983).

(Martínez de Pisón and Arenillas, 1979; Martínez de Pisón and Alonso Otero, 1992).

The variation of glaciation at the Iberian Peninsula is related to two factors. One is the available area above the ELA at the time of the MGA that progressively increased towards the interior of the Iberian Peninsula. The other factor is associated with the relief and aspect of the massifs, which in some cases facilitated oversupply of sonw and glacial accumulation (Martínez de Pisón & Arenillas, 1979; Martínez de Pisón & Alonso Otero, 1992).

4. Key sections

The studies about Pleistocene glacial deposits are rare in Spain. Of particular interest are the studies centered in Llestui (Llauset valley, a tributary of the Noguera Ribagorzana river) (Vilaplana, 1983) and the fluvio-glacial complex of Cerdaña-Capir in the western Pyrenees (Calvet, 1994, 1998). Those in the NW of the peninsula at Pias (Bibei valley, Galaico-Leonés Massif) (Pérez Alberti & Covelo Abeleira, 1996).

The first, according Vilaplana (1983) is a palaeolake damned by glaciers and containing *"lateral moraines with subglacial till sometimes covered by supraglacial till, both in relation to lacustrine, fluvial or slope deposits* [Fig 3]. *In relation to the stratigraphy, we atribute these deposits to the stabilization phase after the maximun of the last Quaternary Glaciation* [MGA]. *All the radiocarbon dates samples give an age of 34,000 years."*

Another deposit of stratigraphic interest is the fluvio-glacial complex of Cerdaña-Capir (Calvet, 1994, 1997), in which three generations of moraines, and associated aluvial terraces are described. The study, based on topography, stratigraphy and intensity of weathering, confirms that the three units were developed during different glacial phases dating from the Middle and Late Pleistocene. The more recent ones correspond to the Rissian and Würmian Alpine sequence, whilst the older is Middle Pleistocene, close to the limit with the Early Pleistocene, as suggested by the degree of weathering (Marine Isotope Stages 16 or 22).

Fig 4. Folded lacustrine-glacial sediments overlained by glacio-fluvial sediments in Pias (Bibei valley, NW Spain).

Table 1: Location of the key sections (UTM Zone 30)

	X UTM	Y UTM
Pias	173844	4669766
Llestui	804979	4719400
Cerdaña-Capir	905308	4710006

In Pias (Bibei valley) (Fig 4) the deposits consist of till, glacio-fuvial and glacio-lacustrine sediments (Pérez Alberti & Covelo Abeleira, 1996) (Figs. 4). The three major facies associations are ice-contact (Dmm, Dcm/Dcs, Gm, Gt(c) [minor Gs/Gt, Sm, Sl]), glacio-fluvial (Gs/$Gt_{(f)}$, Gp, St, Sl) and glacio-lacustrine (Fl, Sr, [minor St, Gp]). The lacustrine association underlays the last two. Deformations such as microfaults microfolds, and laminae bent by ice-rafted cobbles are common. Multiple glacier advance and retreat are documented by several till layers. The sedimentation started in a lake formed by the ice damming of the Bibei valley at Pias. The fluvio-glacial facies formed in a valley train settelment over the lacustrine sediments. The glacier advanced over this material, and partially deformed it. A possible new glacio-lacustrine phase developed as indicated by a gravelly deltaic facies. The sediments were deposited during the Late Pleistocene, just after the maximum glacial advance.

5. Glacial phases and their dating

So far no agreement exists among researchers on the number and ages of glacial cycles that affected the Spanish mountain ranges. Most researchers recognize a maximum advance of the glaciers (MAG) in all mountainous massifs affected by Pleistocene glaciation, which is generally linked to the final glacial cycle, i.e.: Würmian of the Alpine terminology. In the absence of absolute dates, this age is interpreted from the state of preservation of the landforms and deposits, and from the application of global palaeoclimatic models (Ruddiman and McIntyre, 1981). Accordingly, the maximum glacial advance is interpreted to correspond to the Weichselian glaciation (LGM), linking it to the cold minimum of the last glacial cycle (18,000-20,000 yrs BP). However, in the few cases where dates are available, they indicate an older age for the glacial maximum. Thus, in the Pyrenees, based on dates obtained in peat bogs, Bordonau *et al.* (1992) calculated it to be about 60,000 yrs BP and Montserrat (1992) about 45,000 yrs BP. This is in agreement with data from the northern Spanish mountain ranges (Mardones, 1982; Mardones and Jalut, 1983), which place the maximum advance of the glaciers (MAG) between 70,000 and 38,400 yrs BP. For the northwestern mountains of the Iberian Peninsula, the MAG apparently lies between 36,000 and 31,000 yrs BP, as indicated by radiocarbon dates on peat bogs and the correlation with data supplied by studies of coastal deposits and inland karstic caves (Valcárcel Díaz, 1998; Pérez Alberti and Valcárcel Díaz, 1988).

Fig. 5. End morainic ridge in the Piornedo valley, Ancares Mountains (NW Spain).

As for cycles of glaciation, good information is generally available only for the last one, any evidence of previous glaciation have been mostly lost by erosion. However , local evidence of previous glacial cycles has been found in the Pyrenees (Vilaplana, 1983; Clotet *et al.*, 1984; Serrano, 1992), in the Cordillera Cantábrica (Menéndez Duarte and Marquínez, 1996) and in the Galaico massif (Valcárcel Díaz, 1998) They were distinguished from the more recent forms by the comparison of the landforms and by their degree of weathering. In the eastern sector of the Pyrenees, up to three different glacial cycles can be distinguished, the oldest of which may date from the Middle Pleistocene (Marine Isotope Stages 16 or 22).

6. Conclusions

After a century and a half of research on Pleistocene glaciation in Spain, it is still difficult to make generalizations regarding the maximum glacial limits and the age of the individual ice advances, because many glaciated areas have still to be analyzed in detail.

The glaciation was controlled not only by the pre-existing landforms and the altitude, but also by the available humidity, related to their distance from the sea. This explains the progressive rise of the ELA towards the east and south, as well as the extensive glaciation in the northwestern mountains.

From the avaliable data it can be concluded that, although the Würmian glaciation seems to have been the dominant one, there is some evidence of previous glacial cycles.

Aknowledgements

The authors want to aknowledge the revision of the text made by Prof. P.I. Martini.

Bibliography Consulted

Acaso, E. (1983). Estudio del cuaternario en el Macizo Central de Gredos. Tesis Doctoral, Facultad de Ciencias. Universidad de Alcalá de Henares.

Alonso, V. (1992). Geomorfología de las cabeceras de los ríos Narcea, Navia y Sil y del Parque Nacional de la Montaña de Covadonga (NO de la Península Ibérica. Tesis doctoral (unpublished) Universidad de Oviedo, 366 pp.

Baysselance, A. (1883). Quelques traces glaciaires en Espagne. *Annuaire du Club Alpin Français*, **X**, 415.

Bordonau, J. (1992). *Els complexos glacio-lacustres relacionats amb el darrer cicle glacial als Pirineus*, 251 pp. Geoforma Ediciones. Logroño.

Bru, M. J. (1985): Estudi geomorfologic: El modelatge glacial d'un sector del Pirineu Central (Vals Ferrera i de Cardós) Tesis Doctoral (unpublished). 487 pp.

Calvet, M. (1994) Morphogenèse d'une montagne méditerranéenne, les Pyrénées orientales. Thése Doct. Etat. Univ. de Paris I, 3t. 1.177 p.

Calvet, M. (1998). Los complejos fluvioglaciares de Cerdaya-Capcir (Pirineos Orientales) y sus enseñanzas. *In:* Gómez Ortiz, A. & Pérez Alberti, A. (eds), *Las Huellas Glaciares de las Montañas Españolas*, 263-290. Santiago de Compostela, Universidade de Santiago de Compostela.

Castañón, J. C. (1989). Las formas de relieve de origen glaciar en los sectores central y oriental del Macizo Asturiano. Tesis Doctoral microfilm. Universidad de Oviedo. 775 pp.

Chueca Cía, J. (1993). Geomorfología de la Alta Ribagorza: Análisis de la dinámica de procesos en el ámbito supraforestal. *Serie Estudios Geográficos*, **N1 1**. Asociación Guayente. 250 pp. Zaragoza.

Chueca Cía, J.; Peña Monné, J.L.; Lampre Vitaller, F.; Gracía Ruiz, J.M. & Martí Bono, C. (1998). *Los glaciares del Pirineo Aragonés: Estudio de su evolución y extensión actual.* Departamento de geografía y Ordenación del Territorio. Zaragoza. 104 pp.

Clotet, N.; Gallart, F. & Tricart, J. (1984). manifestations glaciaires quaternarias dans le hault bassin du Llobregat (Prépyrénées Catalanes). *Acta Geológica Hispánica*, **19**, **4**, 249-265. Barcelona.

Daveau, S. (1971). La glaciation de la Serra da Estrela. *Finisterra*, **11**, 5-40. *Alcobaça.*

de Prado, C. (1852). Notes sur les blocs erratiques de la chaîne cantabrique. *Bulletin de la Societé Géologique de France*, **9**, 171-175.

Frochoso, M. (1980). Geomorfología del valle del Nansa. Tesis Doctoral microfilm. Universidad de Cantabria. 286 pp.

Frochoso Sánchez, M. & Castañón Álvarez, J.C. (1992). La glaciación Würm en las montañas cantábricas. *In:* Cearreta, A. & Ugarte, F.M. (eds), *The Quaternary in the Western Pyrenean Region.* Servicio editorial Universidad del Pais Vasco-Argitarapen Zerbitzzua Euskal Herriko Unibertsitatea. 319-332. Bilbao.

García Ruiz, J.M. (1979) El glaciarismo cuaternario en la Sierra de la Demanda. *Cuadernos de Investigación Geográfica,* **5 (2)**, 3-25.

Gómez Ortiz, A. (1980). Estudio geomorfológico del Pirineo catalán: morfogénesis glacial y periglacial de los altos niveles y vertientes meridionales de los macizos de Calmquerdós, Tossa Plana de llés y Port Negre (Cerdanya-Alt Urgell). Tesis doctoral. Departamento de Geografía. Universidad de Barcelona.

Gómez Ortiz, A. (1987). Morfología glaciar en le vertiente meridional de Sierra Nevada (Area Veleta-Mulhacen). *Estudios Geográficos.* **193**, 527-558.

Gómez Ortiz, A.. & Slavador Franch, F. (1991/92). Aportaciones significativas al conocimiento del glaciarismo y periglaciarismo de Sierra Nevada. *Notes de Geografía Fisica*, **20-21**, 89-101. Barcelona.

Gomez Ortiz, A. & Pérez Alberti, A. (eds) (1998). *Las Huellas Glaciares de las Montañas Españolas*. Servicio de publicacions da Universidade de Santiago de Compostela. Santiago de Compostela. 430 pp.

Hernadez Pacheco, F. (1957). El glaciarismo cuaternario de la Sierra de Queija (Orense). *Boletin da la Real Sociedad Española de historia Natural.* **55**, 27-74.

Jiménez Sanchez, M. (1994). Geomorfología de la cuenca alta del Río Nalón (Cordillera Cantábrica, Asturias), Tesis doctoral (unpublished), Universidad de Oviedo. 298 pp.

Kossel, U. (1996). Problemas geomorfológicos acerca de la determinación del máximo avance glaciar en la Sierra de Ancares (León-Lugo-Asturias). *In:* Pérez Alberti, A.; Martini, P.; Chestworth, W. & Martinez Cortizas, A. (eds), *Dinamica y evolución de medios cuaternarios.* 131-142. Santiago.

Marcos, J. & Palacios, D. (1995). *Evolución del relieve glaciar en la Gargantea Blanca. In* Aleixandre. T. & Pérez-Gonzalez, A. (eds), *Reconstrucción de paleoambientes y cambios climáticos durante el Cuaternario.* Monografías, **3**, 215-225, Centro de Ciencias Medioambientales. C.S.I.C. Madrid.

Mardones, M. (1982). Le pléistocène suérieur et l'Holocène du piémont de Lourdes: Le gisement de Biscaye (Hautes Pyrénées, France). Etude palynologique. These de 3éme cycle. Université de Toulouse.

Mardones, M. & Jalut, G. (1983). La tourbière de Biscaye (alt. 409 m, Hautes Pyrénées), approche paléoécologique des 45,000 dernières années. *Pollen et Spores*, **25 (2)**, 163-212.

Martí Bono, C.E. (1996). El glaciarismo caternario en el Alto Aragón Occidental. Tesis Doctoral, Facultad de Geología. Universitat de Barcelona, 254 pp. Barcelona.

Martí Bono, C & J. M. García Ruiz (eds) (1994). *El glaciarismo surpirenaico: Nuevas aportaciones*. Geoforma Ediciones. 142 pp.

Martínez de Pisón, E. & Alonso Otero, F. (1992). Algunas reflexiones sobre el glaciarismo en las montañas españolas. (Homenage al investigador Félix M. Ugarte Elorza). *Cuadernos de sección. Historia,* **20**, 109-121. Donostia.

Martínez De Pisón, E. & Arenillas Parra, M. (1979). Algunos problemas de morfología glaciar en la España Atlántica. (Homenage a LLuís Solé i Sabarís), *Acta Geológica Hispánica*, **14**, 445-450.

Martínez de Pisón, E. & Muñoz Jimenez, J. (1972). Observaciones sobre la morfología del Alto Gredos. *Estudios geográficos,* **129**, 3-103

Menéndez Duarte, R. & Marquínez, J. (1996). Glaciarismo y evolución tardiglaciar de kas vertientes en el valle de Somiedo. Cordillera Cantábrica. *Cuaternario y Geomorfología*, **N1 10**, 21-31.

Monserrat Martí, J.M. (1992). Evolución glaciar y postglaciar del clima y la vegetación en la vertiente sur del pirineo: estudio palinológico. *Consejo Superior de Investigaciones Científicas. Monografías del Instituto Pirenaico de Ecología.* **N1 6**, 146 pp.

Nussbaum, F. (1956). Observations morphologiques dans la région de la Noguera Pallaresa. *Pirineos*, **39-42**, 57-97. Jaca.

Ortigosa, L. (1986). *Geomorfología glaciar de la Sierra Cebollera (Sistema Ibérico).* Institutio de Estudios Riojanos, 77 pp. Logroño.

Pedraza, J. (1983). *Geomorfología del sistema central. In* Gutierrez Elorza, M. & Peña Monné, J.L. (eds), *Geomorfología de España,* 63-100. Madrid, Edit Rueda.

Pellicer, F. (1980). El periglaciarismo del Moncayo. *Geographicalia*, **7**, 3-26.

Penck, A. (1884). Das Klima Spaniens während der jüngeren Tertiärperiode und der Diluvialperiode. *Zeitschrift der Gesellschaft für Erdkunde zu Berlin*, **3**, 130-134.

Peña Monné, J.L.; Chueca Cia, J, & Lampre Vitaller, F. (in print). Los complejos glaciares y fluvioglaciares de la vertiente sur. *Síntesis Geológica del Pirineo*. Instituto Tecnógico Geominero de España. Madrid.

Pérez Alberti, A. (1991). La Geomorfología de la Galicia Sudoriental. Tesis doctoral. Universidade de Santiago. 185 pp.

Pérez Alberti, A. & Covelo Abeleira, P. (1996) Reconstrucción paleoambiental de la dinámica glaciar del alto Bibei durante el Pleistoceno reciente a partir del estudio de los sedeimentos acumulados en Pias (NW de la Península Ibérica). *In:* Pérez Alberti, A., Martini, P., Chestworth, W. & Martinez Cortizas, A. (eds), *Dinamica y evolución de medios cuaternarios*, 115-130. Santiago.

Pérez Alberti, A., Rodríguez Guitián, M.A. & Valcárcel Díaz, M. (1993). Las formas y los depósitos glaciares en las sierras orientales y septentrionales de Galicia (NW Península Ibérica). In: Pérez Alberti, A., Guitián Rivera, L. & Ramil Rego, P. (eds), *La evolución del paisaje en las montañas del entorno de los Caminos Jacobeos*, 61-90. Santiago, Xunta de Galicia.

Pérez Alberti, A. & Valcárcel Díaz, M. (1998). Caracterización y distribución espacial del glaciarismo en el noroeste de la Península Ibérica. *In* Gomez Ortiz, A. & Pérez Alberti, A. (eds), *Las Huellas Glaciares de las Montañas Españolas.* Servicio de publicacions da Universidade de Santiago de Compostela, 17-54.

Serrano Cañadas, E. (1991). Geomorfología glaciar de las montañas y valles de Panticosa y de la Ribera de Biescas (Pirineo Aragonés). Tesis doctoral en microficha. Servicio de publicaciones de la Universidad de Cantabria. Santander. 957 pp.

Serrat, D. (1977). Estudio geomorfológico del Pirineo Oriental (Puigmal-Costabona). Tesis Doctoral, Dpt. De Geomorfologia i Tectònica. Universitat de Barcelona, 222 pp.

Schmitz, H. (1969). Glazialmorphologische Untersuchungen im Bergland Nordwestspaniens (Galicien/León). Geographisches Institut der Universität zu Köln im Selbstverlag. 157 pp.

Turu i Michels, V. (1994). Datos para la determinación de la máxima extensión glaciar en los valles de Andorra (Pirineo Central). *In* Arnaez Vadillo; J., García Ruiz, J.M. & Gómez Villar, A. (eds), *Geomorfología en España,* 265-273. Logroño, Sociedad Española de Geomorfología

Valcárcel Díaz, M. (1998). Evolución geomorfológica y dinámica de las vertientes en el NE de Galicia. Importancia de los procesos de origen frío en un sector de las montañas lucenses. Tesis doctoral. Universidade de Santiago de Compostela. Inédita. 496 pp.

Valcárcel Díaz, M. & Pérez Alberti, A. (1998). Límite máximo de la glaciación y línea de equilibrio glaciar en el noroeste de la Península Ibérica durante el último periodo frío. In Gómez Ortiz, A., & Salvador Franch, F. (eds), Investigaciones recientes de la geomorfología española. Barcelona. 455-462.

Vidal Romaní, J.R.; Santos Fidalgo, L. & Jalut, G. (1995). Cronología relativa del máximo glaciar finipleistoceno en el sector Nor-Oriental de la Serra de Queixa (Ourense, Galicia, España). *Actas de la 30 Reunião do Quaternario Ibérico,* 210-215. Coimbra, GTPEQ-AEQUA.

Vilaplana, J.M. (1979). Estudi del glaciarisme de les valls de la Valira d'Ordino i d'Arisal (Andorra). Tesi de Llicenciatura, Universitat de Barcelona (Published in 1984 In *Arxius de la Secció de Ciències,* **LXXII**, Institut d'Estudis Catalans, Barcelona, 84 pp.

Vilaplana, J.M. (1983). Estudi del glaciarisme quaternari de les altes valls de la Ribagorça. Tesis Doctoral. Dpt. de Geomorfología i Tectònica, Universitat de Barcelona, 322 pp.

Quaternary Glaciations – Extent and Chronology
Editors J. Ehlers and P.L. Gibbard

Glacial events in the western Iberian Mountains

Bertrand Lemartinel

Université de Perpignan, 52, av. de Villeneuve,
F-66860 Perpignan Cédex, France

E-mail: lemart@univ-perp.fr

1. Glacial cirques

In the eastern part of the Iberian peninsula, apart from the Pyrenees, only the Moncayo reaches a sufficient height (2313 m) for the development of Pleistocene glaciers (Fig. 1). Their cirques are located on the northern slopes. These features are restricted to a narrow zone and of a small size, never larger than 1 km^2. Their flanks are very smooth and rise gently above the planar cirque floor. The cirques end downstream with a moderate but well marked constriction. In contrast, the sides of the cirque are very worn and smooth. The distribution and shape of these narrow, deep forms (300 - 400 m deep) seems to be controlled by bedrock fracture patterns. The locally very active glacial erosion system seems to have removed entire blocks of quartzite without problems. The lower limit of the previously glaciated zone is at around 1800 m. The latter may seem low for the latitude, but there is a strong thermal gradient caused by the dryness of the air at low levels, trapped in the Aragon basin. Today it is 0.78°C/100 m between 300-600 m. At higher altitudes, where the air gets more humid, the gradient is closer to 0.6°/100 m. This climatic isolation resulted in a lowered 0° isotherm during the Quaternary and the development of glaciers. These cannot exist today. Calculations show that at present the mean 0°C isotherm would be found at 2800 m a.s.l.

In the west the imprint of glaciers appears to be more important but does not have the same climatic significance. This is because the western parts of the chain are over 1800 m high, which encourages the development of glaciers. The features which can be seen here have already been described from the Cebollera by Ortigosa Izquierdo (1986, pp. 23 ff.). They are hardly different from those at the Moncayo. They are always orientated towards the north. The cirques do not particularly increase in size. Their mean size is about 34 ha, the largest reaching less than 1 km^2. The armchair form remains the most common (Fig. 2). The lowest altitude they reach is still 1800 m a.s.l. Nivation hollows are also still visible on the more gentle, south-facing slopes, generally in mylonitised zones. They are circular, shallow and do not exceed 40 m in diameter. On the high southern slopes of the Cebollera and in the Demanda, to the south of the San Lorenzo peak (2271 m), the poorly preserved remains of what may have been a small rock glacier are found. The glacial landforms, particularly on the Moncayo, provide evidence of the factors that controlled the development of glaciers in these mountains. Without doubt, their occurrence on the north-facing slopes of the largest summits results from the humid winds that were able to supply the mountains with sufficient snow. However, structural influences are also evident. To the east of the Picos de Urbión, traces of a glacier tongue over 2 km long are found, the existence of which was dependent on a particularly favourable rock structure in this area (Fig. 3). The shape of the glacier adjusted to grooves between beds of very resistant conglomerates between the Urbion peak and the Laguna Negra. This long depression lies constantly above 2000 m, which explains the development of this exceptionally long glacier (Figs 5 and 6).

Overall the glacial erosion, as spectacular as it is, remains a marginal feature in Aragon and Rioja. The basal height of the cirques, above 1800 m in the east and 1700 m in the west, is too close to the summits to allow a full morphogenetic development during the Quaternary cold periods. Consequently, the landforms are of a reduced size. Additionally, the basins occur on steep slopes, which do not favour glaciation. Consequently, only minor traces of glacial erosion of the high northern slopes are found in some armchair-shaped areas. Any major glacial development requires suitable lithostructural conditions, such as at the Urbion. In order to develop a widely applicable deglaciation model, it is neccessary to study the accumulated moraines. More knowledge of the matter must be obtained.

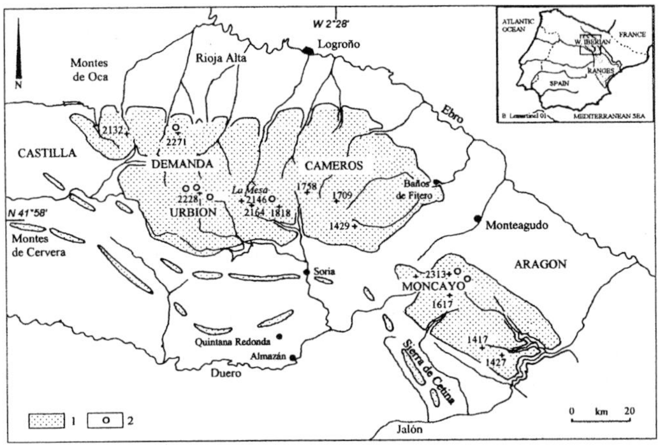

Fig. 1. Location map; 1 = Western Iberian Mountains; 2 = glacial features.

2. The poor development of moraines

The moraines are usually narrow features some 10-40 m wide, most of them single but some double ridges, forming the downhill limit of the former cirque glaciers. Thus, the Cucharón du Moncayo is blocked by two ridges of blocks (Fig. 2). The first crescent-shaped ridge represents the maximum glacial extent; it is composed of enormous boulders. The second, less spectacular ridge, somewhat doubled, occurs within the zone delimited by the external arc. In the Cebollera, the bedrock lithology influences the form of the moraines. Upstream of the Iregua, the morainic arc is composed of chaotic piles of grey sandstone. Then, below San Lorenzo the ridge visible above the ski-lift carpark is a mixture of sands and boulders; clays are naturally absent. The boulders and the matrix of the moraines are rarely altered. They are generally grey in

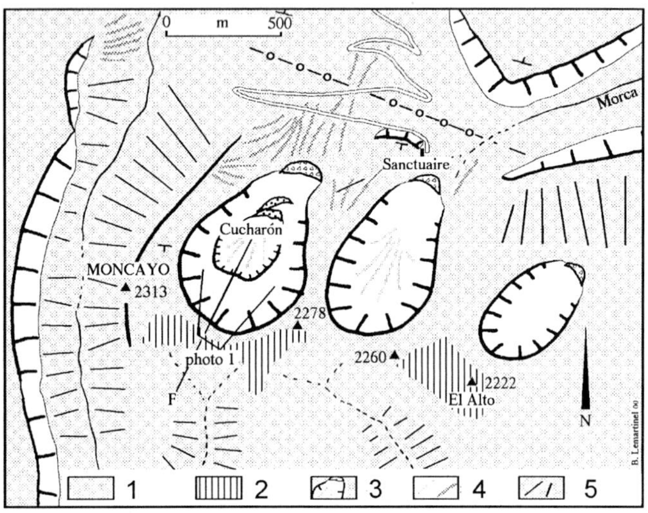

Fig. 2. Armchair-shaped cirques on the northern slope of the Moncayo. 1 = Quartzites; 2 = Oligocene erosional surface; 3 = Cirque; 4 = Scree; 5 = Slope direction

colour, even when the ridges are doubled. The few clay samples that have been collected showed that on both the Moncayo and the Urbión an association of kaolinite (peak as 2/3 in the diagram where 3/3 is the maximum value), illite (total) vermiculite (1/3) and quartz. Kaolinite can even comprise the whole clay mineral association in unweathered tills on the Urbión. The kaolinite originates from the Wealden (Lower Cretaceous) and not from recent soil formation, i.e. the minerals are inherited. It is tempting, therefore, to conclude that the Quaternary morphogenetic process was very recent, allowing little time for soil formation. But the very weak alteration of the siliceous material is difficult to interpret. One should not rely on this single criterion only to provide a relative date for the morainic deposits. Therefore, the fluvioglacial sequences downvalley have been used to try to confirm the preliminary chronology.

The latter are not very impressive. The moraines are perched on the constriction of the cirques and their connec-

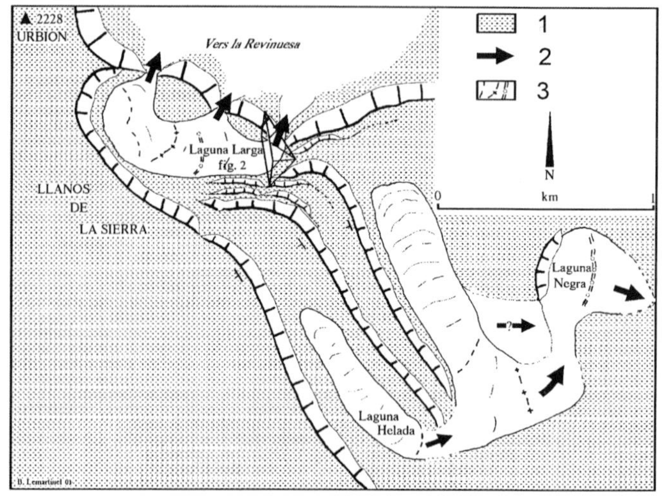

Fig. 3. Maximum extent of the glaciers on the Urbion. 1 = Wealden conglomerates; 2 = Ice flow directions; 3 = Retreat stages of the glaciers

tion to the drainage system is poorly established because there are no alluvial terraces present in the torrential upper courses of the rivers. In Demanda (Fig. 1) the glaciation was probably equivalent to the local T1 terrace. Unfortunately, it is not possible to establish more precisely the relations between the glacial and fluvial deposits. Nothing is known that can establish a link through the Gallego on the one hand from the retreat moraine at the higher terrace T2 at Senegüe and the lowest terrace, and on the other hand between that at Aurin and T2 (termed T1 in Barrère, 1975). Nevertheless, there are two opposing opinions. Astier & Latorre (1980) see a 'fluvio-glacial cone' 7 km long upstream of Razoncillo at the southern piedmont of the Cebollera. The present author disagrees with this interpretation. The landform in question comprises lower

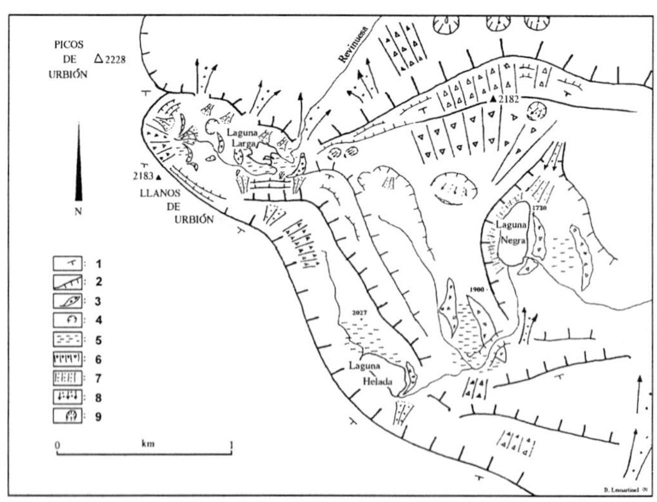

Fig. 4. Model of the Lagunas area after deglaciation. 1 = Dip direction; 2 = Structural scarp; 3 = Moraine; 4 = Nivation hollow; 5 = Peat bog; 6 = Direction of slope; 7 = Scree (relict);8 = Scree (active); 9 = Incisions in the slope.

Fig. 5. Multiple moraines and peat bogs at Lagunas Negra and Helada (Urbión), view towards the south

Neogene blocks of enormous size, which have been mistaken for cold climate landforms. The same error appears on the 1: 50,000 geological map sheet Villoslada de Cameros on which very gravelly deposits are recorded to the east of the village. The present author considers these to be of Miocene age. Certes in 1883 had already interpreted these Neogene formations as morainic accumulations. Unfortunately these old ideas have been revived by some modern authors.

Even if the myth of vast Sierra de Urbión fluvioglacial deposits is revived, the number of important morainic ridges in the region is still limited to four that reflect the decreasing levels of glaciation. In fact, they reflect successive ice advances of decreasing intensity. The morainic complex probably belongs to the same generation since the landforms are all perfectly preserved. The lack of differential alteration indicates that the features represent a single cold period. Relatively small temperature changes may have resulted in large consequences for the geometry of the glaciers in an area situated at the same altitudinal level. It is therefore not possible to reconstruct individual stages of ice retreat. Conversely, the final ridges are multiple, giving the impression that the glacier became fragmented into two or three segments (Figs 3 and 4). Once more this implies that the Urbión glacial complex is of relatively recent age, because the marks of the last cold event are perfectly conserved in the landscape and the no peat bogs have formed in the *lagunas* yet.

The moraines described here are only minor landforms, supporting the view that the effects of local glaciation were weak. Predominantly block accumulations are found that occur as chaotic ridges on the sides of the 'armchair' cirques particularly where the bedrock structure favoured their accumulation. The relationship of these ridges with downvalley fluvioglacial deposis was probably restricted by the limited supply of debris from the ice. Despite some authors' views nothing can be gained by comparison withthe slope sequences of the southern Pyrenees, even at the foot of the Urbión where the rock structure allowed relatively extensive glaciation. The higher number of morainic arcs found west of the Laguna Negra, although poorly formed, do not indicate formation over a long period of time. In general, the cirques and moraines were formed in areas protected from the sun during a relatively recent, possibly lengthy, cold episode during the Late Pleistocene. It is not totally clear, particularly in forested areas, if there is any evidence for older glacial events.

3. Evidence of a complex development

In the forest that covers the slopes beneath the Laguna Negra between 1550-1600 m a.s.l., enormous detached Wealden blocks are found. These rounded quartz conglomerates are rather atypical even if they were often locally derived. In contrast to the fine sandstones, that seem to be derived from upstream, they are striated. These rocks represent a moraine; they do not result from local mass movement. This interpretation is not new, it was already proposed in an article written in 1918 by Carandell & Gomez de Llarena (Ortigosa Izquierdo, 1985). Do these features represent a glacial episode older than that found at higher altitudes? That may also apply to deposits found in a stream cutting through the frontal moraine on the ledge in the area west of the Laguna Negra. Besides perfectly fresh grey material this exposure also includes some rubified sand bodies, orangy yellow in colour and containing oxidised pebbles. Clay mineral analysis shows no chemical differences between the altered and unaltered zones. In

Fig. 6. Laguna Urbión and morainic barrier. In the background: Sierra de Neila (view towards the southwest).

some other cases this analysis indicates abundant kaolinite (given as two thirds in the diagram) and illite (total) vermiculite (one third), and the predominance of quartz associated with feldspar. It is impossible to explain these isolated observations. Several hypotheses can be offered however.

Regarding the upper moraine, the remains of altered sediment are of doubtful significance when they are mixed with fresh deposits. But where are they derived from? A first possibility is that they are derived from hard-grounds that are found in the Wealden strata. Without entirely rejecting this view, this interpretation conflicts with the finds of residual oxides in all the accumulation sector of the moraines. The second possibility concerns the remobilisation of the inherited remains of a *Vallésien* – i.e. the terrestrial equivalent of the marine Seravallian and Tortonian stages - altered coating of the Urbión. The latter hypothesis is the reworking during the last cold period

Fig. 7. Cucharòn cirque (Moncayo), view towards the north.

of glacial material as well as old and altered material. It is likely that this scenario explains the lower moraine below that at the Laguna Negra. However, it is surprising in this case that no other evidence has been found.

There are still many questions concerning the lower morainic complex. One thing is certain, it was formed before that surrounding the Laguna Negra, but it is not known when. The surface alteration of the displaced blocks allows the development of a lithochromic colour scale. The removal of rock masses operates by fracturing along the network of cracks in conglomerates and sandstones. If fractures are locally stained by iron oxides, as suggested above, the block faces will be rubified, without any need to invoke subaerial evolution. Nevertheless, the striations are also coloured. Where the alteration of the sandstone surface is only a thin film, it indicates that the lower moraine has actually been altered *in situ*. Its relatively low altitude in the region implies that the glaciers which deposited it represent a stronger glacial advance than that represented in the cirques. However, it is very difficult to accept that an older and stronger glaciation existed based on the evidence of a single set of landforms only.

Fig. 8. Laguna Larga. In the background, to the right, Urbión Peak (2228 m).

Throughout the area, only one period of glacial morphogenesis can be demonstrated. The models described here, with one exception are relevant to all. The doubling of the moraines of limited size bordering the cirques like that north of Moncayo does not contradict the view that the features is constant. The greater number of ridges found in the Urbión is not additional evidence for a greater number of cold events. It simply reflects minor readvances in the retreat phase of the glaciers and, more importantly, the influence of local topography. It is the latter that, for the most part, conditions the nature of the glaciation in the western Iberian mountains. The exposure to the humid winds and shelter from the sun plays a decisive role. The glaciers are essentially confined to the northern slopes of the high crests, because the climatic conditions only marginally allow their development. It is however, not justified to think that earlier cold stages were unable to generate glaciers on the mountains, even though the traces found so far are insufficient to especially demonstrate that they existed.

Acknowledgement

This paper was translated by P. Gibbard. Many thanks from the author!

References

Astier, J. & Latorre, F. (1980). El glaciarismo cuaternario de la Sierra Cebollera, Obra cultural de la Caja de Ahorros y Préstamos de la provincia de Soria, Soria, 47 pp.

Carandell, J. & Gomez de Llarena, J. (1918). El glaciarismo cuaternario en los Montes Ibéricos, *Trabajos del Museo Nacional de Ciencias. Naturales, sección. Geología*, **22**, Junta para la Ampliación de Estudios e Investigaciones Científicas, Madrid, 62 pp.

Gonzalo Moreno, A. (1981). El relieve de la Rioja, analisis de geomorfología estructural, Instituto de Estudios Riojanos, Logroño, 508 pp.

Lemartinel, B. (1983). *La Demanda et ses piémonts. Etude géomorphologique*, Thèse de 3ème Cycle, Université de Paris I, 223 pp.

Lemartinel, B. (1985). Premières données sur le Quaternaire moyen de la région de la Demanda (Espagne), Physio-Géo, **14-15**, 53-60.

Lemartinel, B. (1993). Histoire d'une montagne: étude géomorphologique des Monts Celtibériques occidentaux, Thèse de Doctorat d'Etat, Université de Paris I, 680 pp.

Ortigosa Izquierdo, L. (1985). *Geomorfología glaciar cuaternaria de la Sierra Cebollera: (Sistema Ibérico)*; Memoria presentada para la obtención del grado de Licenciado en Filosofía y Letras por la Universidad de Zaragoza, Zaragoza , [s.n.], 1985, 222 p., 2h. pleg.

Ortigosa Izquierdo, L. (1986). *Geomorfología glaciar de la Sierra Cebollera*, Instituto de Estudios Riojanos, 1986, 77 pp.

Thornes, J.-B. (1968). Glacial and periglacial features in the Urbion Mountains (Spain), *Estudios Geologicos*, **24**: 245-258.

Quaternary Glaciations – Extent and Chronology
Editors J. Ehlers and P.L. Gibbard

Glacial history of Sweden

Jan Lundqvist
Stockholm University, Department of Physical Geography and Quaternary Geology, SE-106 91 Stockholm, Sweden

E-mail: jan.lundqvist@geo.su.se

Pre-Weichselian glaciations

Very little is known about the pre-Weichselian glacial history of Sweden. The landforms, with deep valleys and premontane lakes indicate that the mountain area of Sweden has experienced repeated glaciations throughout the Quaternary, probably even since late Tertiary time. Land surfaces fairly unaffected by glaciation only exist in the mountain range (Kleman & Stroeven, 1997). However, no accumulations are known from the earliest glacial stages. Subsequent glaciations have depleted all such deposits.

Tills representing early glaciations occur beneath organic-bearing sediments that are dated to the Holsteinian and Eemian interglacials (Fig. 1). None of them has been sufficiently dated. In the Alnarp depression traversing southernmost Sweden sediments have been tentatively assigned to the Holsteinian, or at least pre-Eemian. Below them, there is till with a northerly provenance. This till may be Elsterian or older but an Early Saalian age is not out of question (Miller, 1977).

At Öje in Dalarna, south-central Sweden, a sequence originally described by G. Lundqvist (1955) has been ascribed a Holsteinian age (Garcia Ambrosiani & Robertsson, 1998). Consequently, the lowermost till there is Elsterian or older.

The Pleistocene stratigraphy in northern Finland provides clear evidence that northern Sweden must have been covered by ice in pre-Holstenian time. The Naakenavaara Interglacial is referred to the Holsteinian and stratigraphically beneath it there are two till beds (V and VI) which accordingly must be Elsterian or older (Hirvas *et al.*, 1977; Hirvas, 1991; Aalto *et al.*, 1992).

From what is known about the extension of the Saalian and Elsterian ice sheets on the European continent there can be no doubt that the whole of Sweden was ice-covered during at least parts of these glacials. There is no evidence so far indicating any non-glacial phases in the Saalian and Elsterian. However, as evidence is scanty and age determinations are uncertain it is still possible that one or more complexes of stadials and interstadials might be present within the pre-Eemian glacial stratigraphy.

Early to Mid Weichselian glaciation

In northernmost Sweden intermorainic organic sediments above the Eemian deposits provide a reliable post-Eemian stratigraphy. Since the till beds show different characteristics and directions of transport the same stratigraphy can be identified at several other sites where Eemian deposits are absent. Therefore we have a good idea about the Early Weichselian glacial history of the region. Unfortunately, the stratigraphy differs considerably from the stratigraphy in northern Finland.

In Swedish Lapland two Weichselian interstadials have been identified (Lagerbäck & Robertsson, 1988). They were called the Peräpohjola and Tärendö Interstadials, then using the Finnish name for the first-mentioned stage. Correlation with the North-European Brørup and Odderade interstadials appears likely, although no finite ages have been obtained from the Swedish sites. This means that they correspond to Marine Isotope Substages (MIS) 5c and 5a, respectively (*cf.* Shackleton, 1969; Mangerud, 1991). Consequently, the stadials between them and separating them from the Eemian correspond to the cold phases 5d and 5b. In Finland, however, only one Early Weichselian interstadial has been identified (Hirvas *et al.*, 1977; Hirvas, 1991). Possibly only the area west of the River Torneälven was affected by the 5b ice sheet and the interstadials 5a and 5c occur as one single unit east of it.

First Weichselian glaciation. The effects of the first Weichselian ice sheet (MIS 5d, 117 – 105 ka BP) are well known in northernmost Sweden. This ice sheet was warm-based and had the greatest influence on the land surface. Its erosive effect was considerable and all the large-scale morphology of the Quaternary deposits is an effect of that glaciation. Within Sweden it moved in general from the

Fig. 1. Block of Eemian peat from the Leveäniemi deposit, Swedish Lapland, with numerous fragments of pine wood, leaves etc. Photo J. Lundqvist 1968.

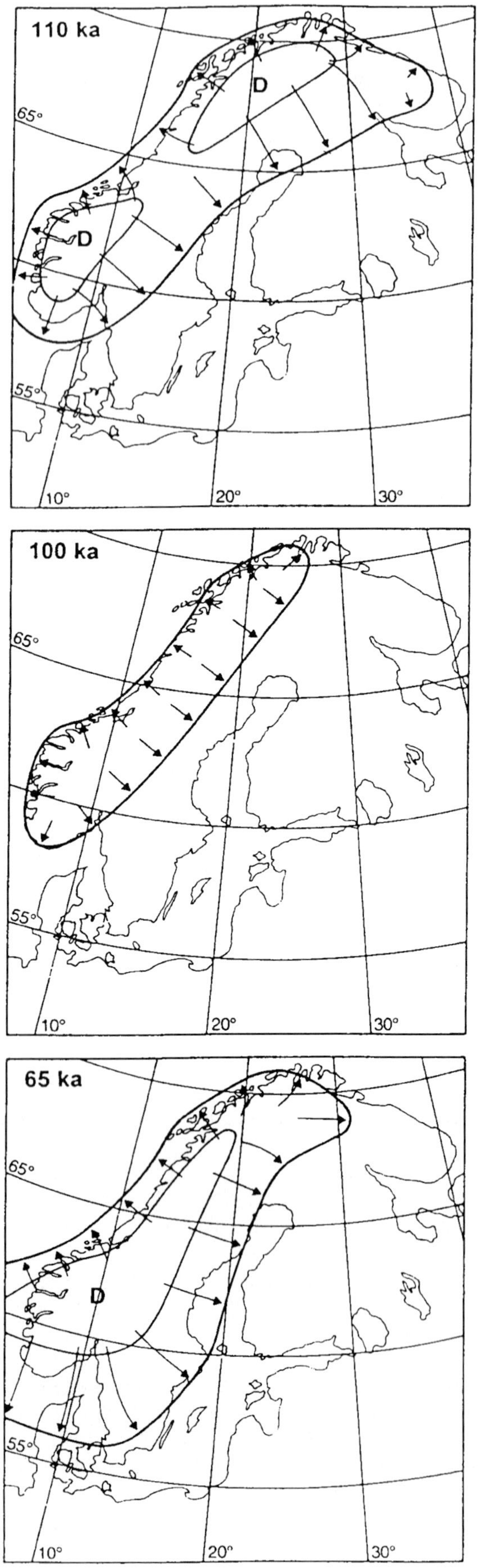

Fig. 2. Interpretation of ice margins and dispersal centres for Early and Mid Weichselian ice sheets according to Kleman et al. (1997).

mountains towards the Baltic Sea basin. In northern and central Lapland this means from the northwest, and this is the trend of the large drumlins (size of the order of several kilometres). Eskers and erosional features show the same trend. In this area these landforms are easily distinguished from the forms of the last glaciation, which are smaller and trending southwest – northeast. Farther to the south the difference between different generations of forms is not so easily seen since these forms trend approximately in the same direction.

The MIS 5d ice sheet appears to have been centred in the mountain region, and possibly even at its western margin (Fig. 2; Kleman *et al.*, 1997). It was controlled by a maritime climate, probably an effect of a relatively warm ocean at the end of the Eemian. This is in agreement with the near-coast centred Early Weichselian glaciation farther east, in northern Siberia (Astakhov, 1992; Svendsen *et al.*, 1999).

The extension of the MIS 5d ice sheet is more or less unknown. Maps showing a possible extension are rather theoretical. Kleman *et al.* (1997) visualised a northeast – southwest-trending ice margin through central Sweden on the basis of pre-Late Weichselian morphological features. In the west the ice should have reached down into the Skagerrak. The evidence is weak and no deposits really support this picture. However, the very westerly extension, out on the Norwegian shelf, is probably correct.

J. Lundqvist (1992a) showed a similar picture. According to this, the ice covered the Scandinavian peninsula only down to middle Sweden. This, also speculative, picture was based on the extension of a till cover which is considerably thicker north of that boundary and becomes much thinner south of a fairly well-defined zone.

The general outline of the deglaciation at the end of MIS 5d has been reconstructed in the north. Kleman *et al.* (1997) have identified five zones of recessional moraines, the last three of which are located in Sweden, the older ones in Finland. The oldest of the three Swedish zones was defined by Lagerbäck (1988) and is marked by the hummocky landscapes of Veiki moraine (Fig. 3; Hoppe, 1952). The Veiki moraine line is stratigraphically dated to MIS 5d by Lagerbäck (1988). The eastern boundary of this terrain shows several lobes. The most distinct one is the so-called Lainio Arc in Swedish Lapland (G. Lundqvist, 1961). Along the boundary there is often a high end-moraine-like ridge. A short distance east of the boundary there is in some places another less distinct zone, mostly not separated from the main line. Further to the west, lateral moraines in the eastern part of the mountain range have also been interpreted as corresponding to a MIS 5d recessional stage (Borgström, 1989; Kleman, 1992). All these moraines appear to reflect the retreat of the eastern MIS 5d ice margin towards the centre of the mountain range.

Second Weichselian glaciation. The glaciation that occurred between the Peräpohjola and Tärendö Inter-

Fig. 3. Veiki moraine south of Gällivare, Swedish Lapland, formed at the down-wasting of the first Weichselian ice sheet. The site is situated in the centre of the square marked 'Fig. 3' in Fig. 4. The square refers to a figure in Kleman et al. (1999). Photo J. Lundqvist 1975.

stadials, that is, MIS 5b (93 – 85 ka BP) is incompletely known. Lagerbäck (1988; Lagerbäck & Robertsson, 1988) identified a thin till unit between organic-bearing sediments from these interstadials in northern Swedish Lapland. Strangely, the corresponding ice-free stages are not identified in northern Finland, that is, on the eastern side of the River Torne älv (Hirvas, 1991). Consequently, a MIS 5b till unit cannot be defined in Finland. On the other hand, Lundqvist & Miller (1992) found stratigraphic evidence for this glaciation in Ångermanland, central Sweden. Even though the age control is poor, it appears likely that part of the morphological sequences found by Kleman *et al.* (1992) at the eastern margin of the mountains in Dalarna, south-central Sweden, could also have the same age.

In northern Lapland the 5b till is thin and of westerly provenance. The erosive effect of the glacier was insignificant. Even delicate landforms like Veiki moraine from the 5d deglaciation are preserved (Lagerbäck, 1988). The 5b ice must have been cold-based in wide areas. It may have been formed partly by instantaneous glacierization under the influence of a cool phase in the North Atlantic (Gard, 1988). Also in the more southern areas the corresponding till bed is thin and fragmentary.

The extension of the MIS 5b ice sheet is totally unknown. Possibly, it covered only the mountain range and near-mountain areas. Kleman *et al.* (1997) expressed some doubt about its existence. If it did exist, according to them, MIS 5b "was characterized by a restricted glaciation, with build-up of a thin, entirely cold-based ice sheet of limited size" (Fig. 2).

A different picture was presented by J. Lundqvist (1992a). This picture shows a west-easterly ice margin over south-central Sweden, and an ice sheet extending even to the coast of Finland. It was based on occurrences of a dark grey or bluish grey clayey till. This till is easily recognized and was interpreted by J. Lundqvist (1973a) and Björnbom (1979) as a stratigraphic unit. Mostly it is considered just a facies, but in Sweden it appears to occur at the same stratigraphic level wherever there is a clear stratigraphy. This till occurs as far south as Stockholm (Möller & Stålhös, 1965). It is of a westerly provenance as supported also by striae from the west, locally even from the WSW. If the correlation is correct, it shows that the ice was strongly mountain-dependent, in the Stockholm area originating from the mountains in southern Norway. The picture would be similar to the build-up phase of the third Weichselian ice sheet according to Kleman *et al.* (1997, fig. 11). Thus it seems possible that the dark till belongs to that stage instead. Obviously, the wide extension of the MIS 5b ice sheet is based on very weak evidence and must be considered highly speculative. Considering the conditions in Finland, a more restricted extension as shown on the main map appears more likely.

Late Weichselian glaciation

Different aspects of the last glaciation and its down-wasting have been published by J. Lundqvist (1997, 1998a, b). In this context Late Weichselian is defined according to Mangerud & Berglund (1978). This means that the boundary Mid/Late Weichselian is placed around 25 000 BP, or better defined, at the boundary between MIS 3 and 2. The dates in the following text are given in uncalibrated radiocarbon years BP but on the main map calibrated dates are also shown.

Early Late Weichselian stage. From the end of the Tärendö (Odderade) Interstadial through the Last Glaciation Maximum (LGM) virtually all of Sweden was glaciated. This corresponds to MIS 4 – 2. Probably glaciation started in the mountains, influenced by a warmer phase in the North Atlantic (Gard, 1988). East of the mountains instantaneous glacierization must have played an important

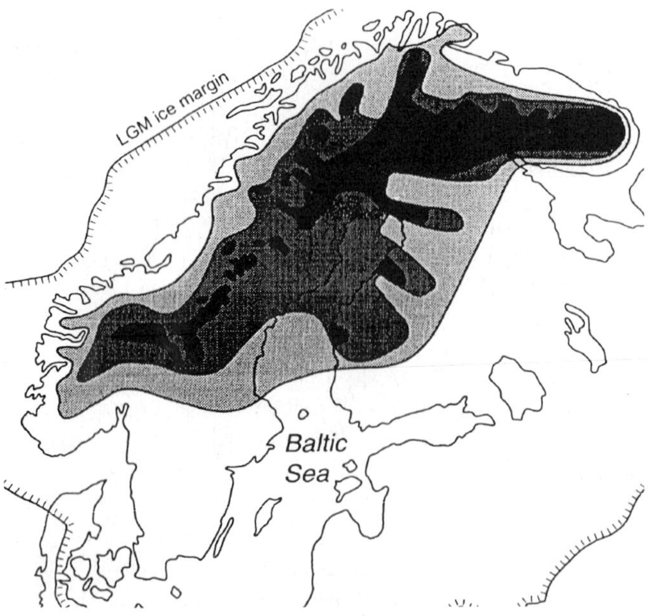

Fig. 4. Black areas show relict landscapes reflecting frozen-bed conditions during Mid and Late Weichselian. Light shading indicates minimum extent of frozen-bed areas during the LGM, and intermediate shading inferred extent of frozen bed in the Younger Dryas cold event. From Kleman et al. (1999).

role (J. Lundqvist, 1978). Conditions during the very cold Tärendö Interstadial favoured such a process in the continental areas. Whether this happened only in a marginal zone or more regionally is not known. Well preserved landforms, interstadial strata, frost-shattered bedrock and other ground features (Lagerbäck, 1988; Lagerbäck & Robertsson, 1988) are clear evidence that the advancing ice cannot possibly have been wet-based. Hättestrand (1997) has shown that cold-based conditions prevailed at least as far south as Lat. 60°30′ N. This does not only apply to a marginal zone, but to the entire central part of the ice sheet. During deglaciation the base gradually but only partly changed to temperate conditions (Fig. 4).

The advance of the ice front is not dated, but may have reached southern Sweden already in MIS 4. In the Baltic depression, where the sea-floor was covered by soft sediments, the advance was probably more rapid (*cf.* Boulton & Jones, 1979) than inland. Lagerlund (1987a) proposed that surges upon a substratum with low shear-stress provided the base for growth of thicker ice – according to him even ice domes. Modelling does not support the dome theory (Boulton *et al.*, 1985; Holmlund & Fastook, 1993) and it has not gained general acceptance (*cf.* Ringberg, 1989; Boulton & Payne, 1994; Ruszczińska-Szenajch, 1999). However, this theory is not a prerequisite for a rapid Baltic advance. There is evidence that the Baltic ice stream reached the European continent in MIS 4 (Drozdowski & Fedorowicz, 1987) or southeastern Denmark in late MIS 3 (Houmark-Nielsen, 1987). On the contrary, the western part of southernmost Sweden (Skåne) was never covered by Weichselian ice until in MIS 2 (Berglund & Lagerlund, 1981).

The hypothetical extension of the MIS 4 ice was illustrated by J. Lundqvist (1992a). According to him, entire Sweden except parts of the southwestern coast and western Skåne were ice-covered. This model is based on negative evidence only. Interstadial deposits along this coast are restricted to the Mid-Weichselian, whereas Early Weichselian interstadials are lacking (Hillefors, 1969). Coastal deposits in Sweden and Norway are radiocarbon-dated to ages between 24 000 and about 40 000 BP (Hillefors, 1969; Mangerud, 1981). Although such dates cannot be considered very accurate they indicate, together with their content of microfossils, a correlation with some of the cool interstadials on the continent (*cf.* Behre, 1989). They also indicate proximity of the ice margin. J. Lundqvist (1992a) thinks that the ice margin stayed along the coast throughout MIS 4 and 3, making limited retreats and readvances on several occasions.

Kleman *et al.* (1997) have presented a model in which the Early Weichelian glaciation after the MIS 5a interstadial started from a very westerly centre. At 65 ka the ice should have moved from southern Norway southeastwards across southern Sweden (Fig. 2). This model is supported by morphological features. It is in agreement with observations of an oldest ice movement from the northwest at some places on the west coast (Påsse *et al.*, 1988; Påsse, 1992).

Whether one model or the other is correct, it is clear that there was no major deglaciation after MIS 4. Directly following this stage the ice margin advanced to its maximum extension at the Brandenburg and other moraines on the continent and to the Main Stationary Line in Denmark in MIS 2. The whole of Sweden was ice-covered then. During the growth of the ice its divide moved eastwards to a culmination over the Gulf of Bothnia (Ljungner, 1949).

The classical model shows ice movement from north to south over southernmost Sweden. When deglaciation proceeded south of the Baltic basin and in Denmark the movement turned to a more east-westerly direction in the southernmost part of the basin. Ice streams on successively lower levels developed south of Sweden together with a northeast-southwesterly movement over the mainland (Holmström, 1904; and summary by Lagerlund, 1977).

This model has so far failed to account for some features in the glacial stratigraphy in Skåne. To overcome these difficulties Lagerlund (1987a) developed his ice-dome model. According to that model marginal domes should have formed due to higher precipitation along the southern front of the ice sheet. As mentioned above, this model has been criticized (Ringberg, 1989; Ruszczińska-Szenajch, 1999) and there is not yet a generally accepted solution. Possibly a compromise between the two models can be elaborated.

Late Weichselian deglaciation. The first part of Sweden to become ice-free was the Kullen peninsula in northwestern Skåne. This happened first before 15 500 BP but at the East Jylland readvance the area became again ice-covered (Lagerlund & Houmark-Nielsen, 1993). The final deglaciation there took place about 14 600 BP (17 200 cal. yr; Sandgren *et al.*, 1999; Sandgren & Snowball, 2001).

When the northwestern part of Skåne was free of ice, the sea inundated the coast. Its highest level is found on the Kullen peninsula at 88 - 89 m a.s.l. (Sandgren & Snowball, 2001). This is some 50 m higher than at other places. Possibly Mt. Kullen has been raised along fault lines after deglaciation (Lagerlund, 1977).

The picture of the following deglaciation in southernmost Sweden is depending on which of the two models we apply. The classical ice-stream model implies retreat of a lobate ice front from the Danish straits southwards and from Denmark eastwards. The Baltic ice lobe became gradually thinner, a High Baltic Ice Stream changing to a Middle Baltic and finally to a Low Baltic stream (Holmström, 1904). Its front receded eastwards simultaneously with the retreat northeastwards by the NE Ice. Between the two fronts there was a wedge-shaped ice-free area (Sandgren, 1983). Glacial lakes were ponded up there (Nilsson, 1958). At a readvance of the Low Baltic Ice about 13 500 BP to the Öresund region sequences including till-covered proglacial sediments were formed.

The dome model implies a north-southerly front receding eastwards across Skåne. Baltic lobes were less significant. However, according to Lagerlund (1987b) a Simrishamn readvance of a lobe took place about 13 300 BP.

Irrespective of which model we prefer, isolation of stagnant ice bodies was probably part of the process. Hummocky moraine, contrasting with flat areas of Low Baltic till, is an evidence of this. Due to isostatic rebound in combination with a eustatically low sea-level the Danish straits lay dry. The sea-level there was situated about 20 m below its present level. In the southern part of the Baltic basin a glacial lake, the Baltic Ice Lake, was ponded up. In this lake, ice retreat proceeded rapidly towards the south, and then further eastwards (*cf.* Stay, 1979; Duphorn *et al.*, 1979). When the ice front lay in eastern Skåne there was still ice south of Sweden. When the ice margin had receded less than 10 km inland, the ice sheet in the southern Baltic Ice Lake disappeared. Probably it broke up quickly when it became thin enough to float. The result of this upbreaking was a turn of the front from north-southerly trend in eastern Skåne to a west-easterly ice front across southern Sweden, from the southern Kattegat to south of the Isle of Öland. A possible correlation between ice fronts in southern Sweden and the the region south and east of the southern Baltic basin has been shown lately by Uścinowicz (1999).

The following deglaciation until the Younger Dryas time was characterized by retreat in three directions (*cf.* B.E. Berglund, 1979; J. Lundqvist, 1998a, b). The west-east trending ice margin in the south receded northwards. Along the west coast the margin moved inland, that is, towards the northeast, and in the Baltic basin towards the northwest. Along the western front moraines were formed. They were deposited in a marine environment and can be dated by means of shells or through correlation with radiocarbon-dated moraines in the Oslofjord area in Norway (Sørensen, 1979, 1992).

Series of small moraines, the Halland Coastal Moraines, developed already when the ice margin in the south passed across Skåne but their dating offers great problems. Ages ranging from 12 200 to more than 14 000 BP have been obtained (Fernlund, 1988, 1993; Påsse, 1992; M. Berglund, 1995) but around 13 000 appears most likely. Partly they are true marginal formations, but to some extent they may have been formed by extramarginal glaciotectonic folding (Fernlund, 1988).

Further inland, there are series of large moraines (Fig. 5; see Hillefors, 1975; Lundqvist & Wohlfarth, 2001). They indicate variations of the climate but may to some extent be topographically controlled. The Göteborg moraine forms very large ridges in valleys as well as on the higher ground. It has been dated to 12 400 – 12 800 BP and indicates a readvance at about 12 600, probably climatically induced. The age of the following Berghem moraine is between 12 000 BP (M. Berglund, 1995) and 12 300 (Fredén, 1988). The next moraine, the Trollhättan moraine, offers some problems. The line has been dated by means of lake sediments that were supposed to be situated inside it, (Björck & Digerfeldt, 1981). However, according to Lundqvist & Wohlfarth (2001) the line actually passes north of the studied lake. According to these authors an age around 11 800 is most likely.

In the west, where the more maritime climate appears to have had a greater influence on variations in the downwasting of the ice, some of the moraines indicate minor readvances of the ice margin. This applies to the Göteborg, Berghem and Trollhättan moraines. The next moraine in this sequence, the Levene moraine, shows considerable glaciotectonic disturbances with infolded clay beds. It represents a larger readvance in Allerød time, probably the beginning of the Younger Dryas climatic deterioration (Lundqvist & Wohlfarth, 2001).

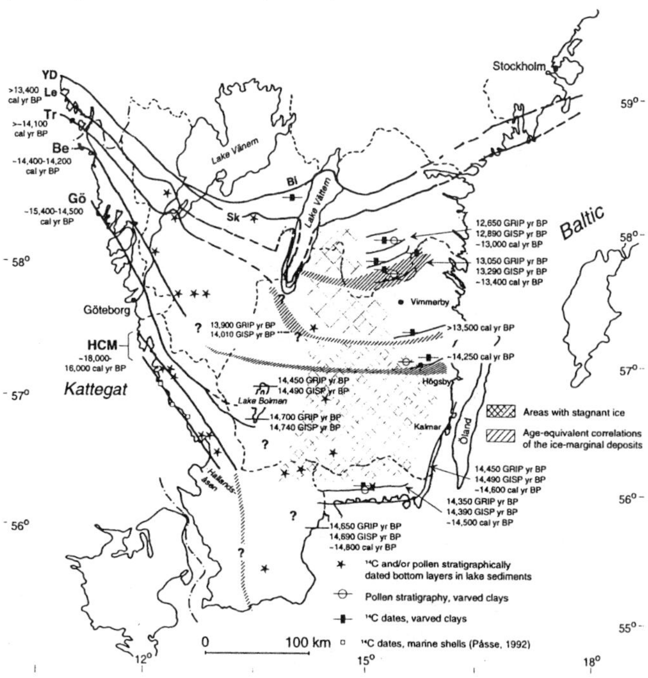

Fig. 5. Summary of datings of ice-marginal lines in southern Sweden according to Lundqvist & Wohlfarth (2001).

Fig. 6. Varved clay at Uppsala, Sweden. Dark layers correspond to winter sedimentation, and light layers to summer sedimentation. The wavy appearance of the layering is caused by the excavator. Photo J. Lundqvist 1958.

Towards the southeast all these moraines become less well developed and disappear gradually into hummocky moraine landscapes (Johnsson, 1956; Hillefors, 1975). Further to the east there are no end moraines. An alternation between landscapes of hummocky moraine and smoother moraine surfaces is the only evidence for climate variations. Marginal parts of the ice sheet wasted down as dead-ice, separated from the active ice front (Lagerlund *et al.*, 1983; Möller, 1987). In the eastern part of southern Sweden the retreat of the ice margin is dated biostratigraphically and with the clay-varve method (Fig. 6; Ringberg, 1991; Wohlfarth *et al.*, 1998; Björck, 1999). The differences in moraine morphology and the datings with methods that are not quite compatible have caused great difficulties in the correlation between the western and eastern parts of southern Sweden (Fig. 5; Lundqvist & Wohlfarth, 2001).

After the Levene readvance the ice margin retreated beyond the position of the Younger Dryas moraines before it readvanced up to these moraines. This is also known from Norway (Sørensen, 1992) and is indicated as well by a drop of the corresponding coastline in western Sweden (De Geer, 1909; Johansson, 1982). While the ice margin retreated an unknown distance and readvanced to the Younger Dryas position the isostatic rebound raised the land surface some 10 – 20 m. The magnitude of the drop implies a time of at least some 100 years, which would allow a retreat of a few tens of kilometres.

When the ice margin passed Mount Billingen, south-central Sweden, the Baltic Ice Lake was lowered to the level of the ocean and a direct connection to the sea was opened. Glacial clay and other sediments folded into the Younger Dryas moraines after an early drainage, traced also in the marine record (Bodén *et al.*, 1997), provide further evi- dence of the Younger Dryas readvance (Björck & Diger-feldt, 1984).

Around the southern end of Lake Vättern there is a thick sequence of till beds interlayered by glaciolacustrine and glaciofluvial sediments which represent readvances corresponding to both the Levene and Younger Dryas

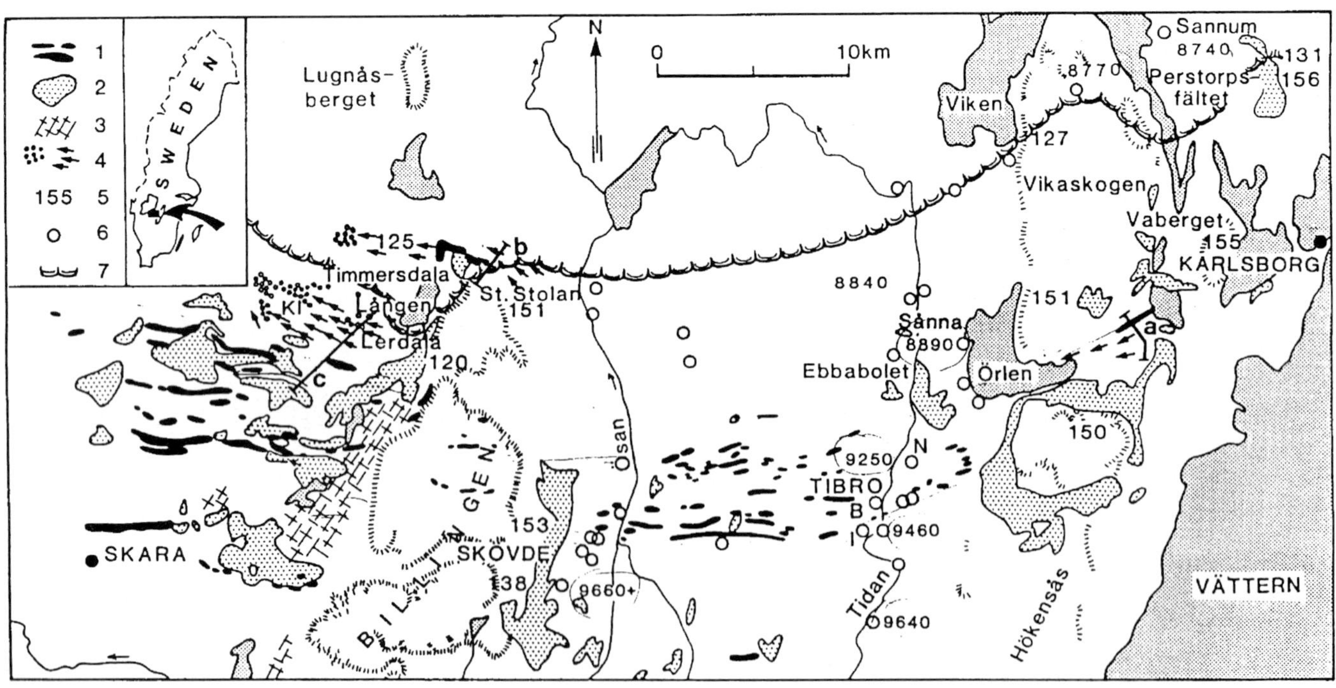

Fig. 7. The Billingen – Lake Vättern area according to Strömberg (1985). 1 = moraines; 2 = glaciofluvial deposits; 3 = kames; 4 = erosion and deposition from the catastrophic lowering of the Baltic Ice Lake; 5 = shore levels of the ocean and the Baltic Ice Lake; 6 = varve measurement with date in varve years B.C.; 7 = ice margin at the lowering event.

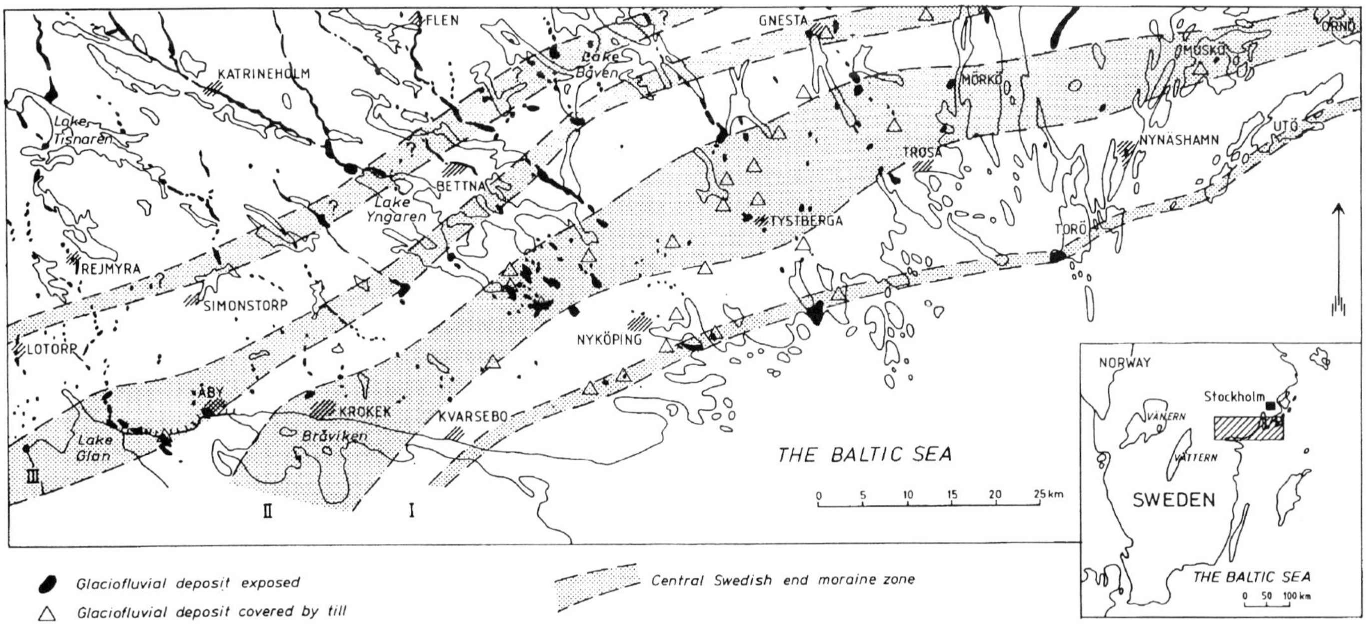

Fig. 8. Inferred ice-marginal lines belonging to the Central Swedish Ice Marginal Zone according to Persson (1983).

moraines (Waldemarson, 1986; J. Lundqvist, 1997). East of this, there is no evidence of significant readvances within Sweden (Kristiansson, 1986; Brunnberg, 1995).

The difference in activity of the ice in the west and east is reflected by the glacial morphology along the Younger Dryas line through Sweden. In the west there are big moraines and large marginal deltas. At Mt. Billingen a large delta of extremely coarse boulder-rich gravel is an effect of the second, final, catastrophic lowering of the Baltic Ice Lake to sea-level (Strömberg, 1977, 1992). Many small end moraines occur in that area (Fig. 7). The formations in western Sweden correspond to two distinct lines, the Skövde and Billingen moraines. The Skövde moraine represents the Younger Dryas advance while the Billingen moraine corresponds to the drainage event.

East of Lake Vättern there are no moraines except groups of rather small ones in the Linköping – Norrköping area (Bergström, 1992). The Younger Dryas line is marked by series of large deltas and kames and a zone of till, slightly thicker than immediately north and south of it (J. Lundqvist, 1989). Parallel to this, Persson (1983) distinguished two zones with till-covered sediments which he ascribed to oscillations corresponding to the Skövde and Billingen moraines. This pattern continues eastwards to the Baltic Sea (Fig. 8).

The Younger Dryas formations in Sweden, like in Norway and Finland, are evidence for strong activity of the ice sheet. North of them there are few such indications. Only in the west, some 80 – 100 km north of the Younger Dryas moraines, there are short moraines in valleys (J. Lundqvist, 1988, 1992b). They are not well dated but can be correlated with the Ås, Ski and Aker moraines farther west, in Norway (Sørensen, 1979, 1992), which shows that they are of late Younger Dryas to Preboreal age. Their continuation eastwards is marked by large glaciofluvial deposits north of Lake Vänern (Fredén, 1988).

Further to the east, as well as northwards, there are no true ice-marginal deposits in Sweden. Small De Geer moraines (Fig. 9) form a zone eastwards from the area at Lake Vänern to the Baltic coast. They were earlier considered end moraines (De Geer, 1940) but are nowadays interpreted as formed some distance behind the ice margin (Hoppe, 1948; Strömberg, 1981). Also large glaciofluvial delta-like deposits in the Stockholm area, earlier interpreted as marginal deposits comparable with the ones north of Lake Vänern, are now shown to represent increased meltwater discharge caused by more rapid retreat of the ice (Brunnberg, 1995). Thus there is no correspondence in eastern Sweden to the very large Salpausselkä moraines in Finland, nor to the Central Finland Ice-Marginal Formation, corresponding to the Jyväskylä readvance (Rainio *et al.*, 1986; Nenonen, 1992). Neither a readvance nor a slow-down of the recession of the ice margin is seen in the clay-varve sequences (Strömberg, 1989).

The deglaciation of Sweden after the Younger Dryas time has proceeded evenly without any breaks or readvances. The low parts of the area, up to a level rising from about 160 m above present-day sea-level in the north and south to 285 m near the central coast of the Gulf of Bothnia (Hörnsten, 1964), were covered by early stages of the Baltic Sea. Below the sea-level of the corresponding time the ice margin receded with a well-defined front. Calving played an important role in that environment. Calving bays formed at the lowest parts of the terrain and around the mouths of meltwater streams. Above the highest coast-line hummocky dead-ice moraine shows that down-wasting of stagnant ice was important. According to an earlier opinion dead-ice deglaciation was the dominating

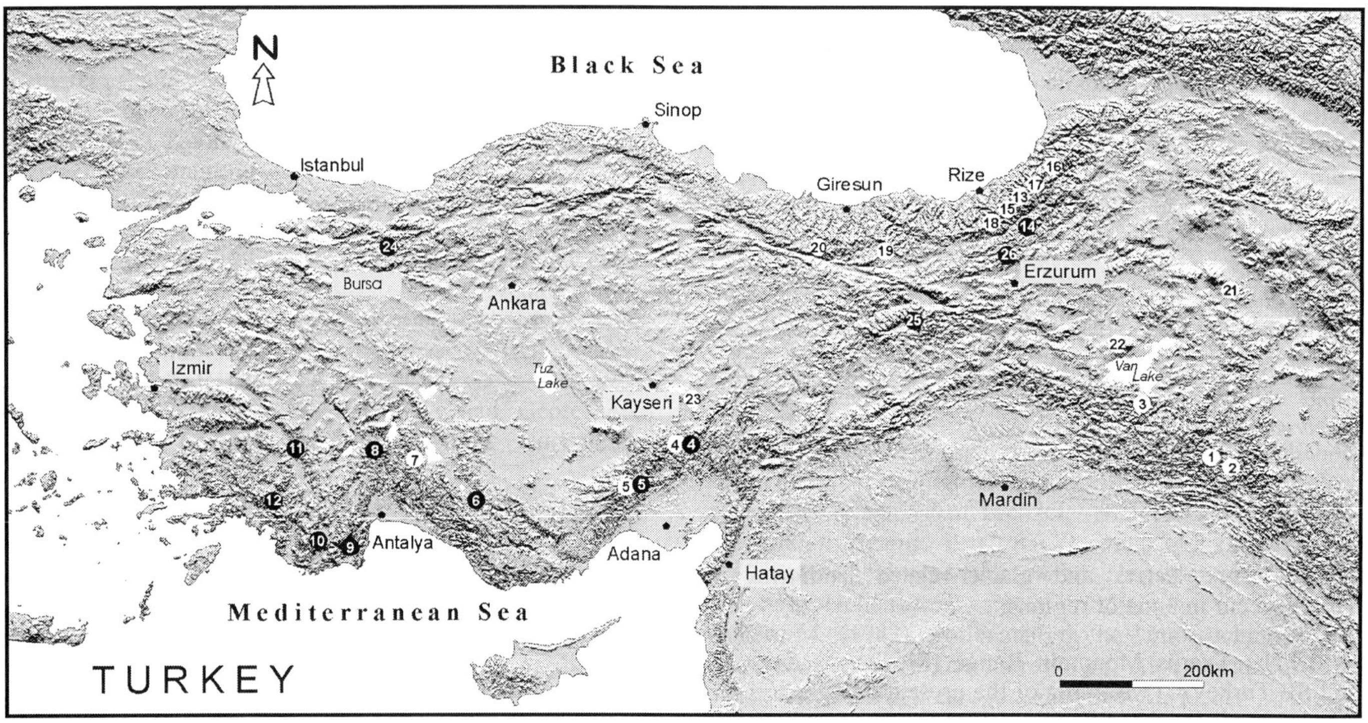

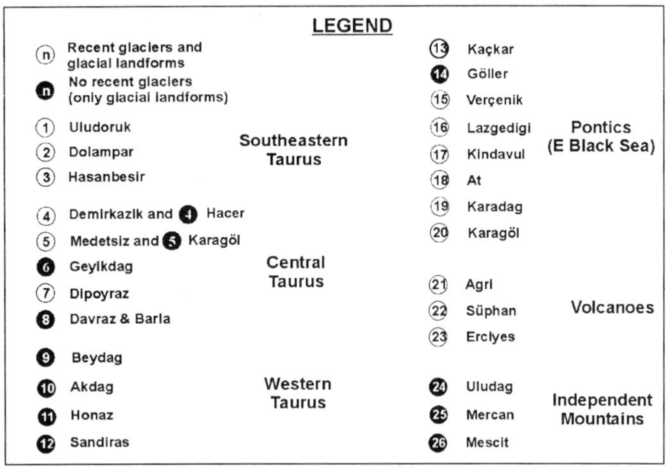

Fig. 1. Location map indicating glaciers and glacier related landforms (drawn by Dr. Onur Köse).

Glaciers and glacial landforms

Present-day glaciers and glacier-related landforms occur in 3 major regions in Turkey (Fig. 1 and Table 1): 1. The Taurus Mountains (Mediterranean coast and SE Turkey). 2. The Pontic Mountains (Eastern Black Sea coast). 3. Volcanoes and independent mountain chains scattered throughout the Anatolian plateau.

1. Taurus Mountains

The Taurus Mountains (Southeastern, Central and Southwestern Taurus) extend from the SW to the SE of Turkey, up to the Iranian border, running parallel to the Mediterranean coast. The southeastern part of the Taurus Mountains contains two thirds of the country's present day glaciers. Further to the west, no active glaciers exist probably because of the higher snowline altitude, high degree of erosional dissection and the very limited number of small peaks that reach above the snowline (Kurter, 1991).

a) Southeastern Taurus: This is the most important recently glaciated region in Turkey from which more than 20 glaciers have been reported (Erinç, 1952b). The glaciers are especially very well developed on Mount Cilo (4135 m). Here, the Uludoruk (Reşko) valley glacier, which is almost 4 km long, covers an area of 8 km^2 (Table 1). Maunsell first photographed this glacier in 1901. This picture, although not very clear, shows a thicker and broader glacial tongue, that reached down to lower levels than today. Bobek, who visited the area in 1937 found the glacier terminus at 2600 m (Bobek, 1940). The same glacier tongue was observed in 1948 to have retreated to the 2900 m level by Erinç (1952a). Finally, the satellite image observations made by Kurter (1991) indicate that the glacier has continued to retreat up to an elevation of nearly 3000 m. These observations clearly indicate thinning and shrinking of the glacier, at least since the beginning of the 20th century. Bobek (1940) describes traces of strongly developed Pleistocene glaciations from the northern flank.

Another glacier complex on Mount Cilo, the so-called Mia Hvara glacier that comprises three isolated glaciers in the Mia Hvara valley, probably originally formed one single glacier. In fact, the terminus of the middle Mia Hvara glacier, that reached down to 2550 m in 1937 (Bobek, 1940), was observed at the 2800 m level in 1948

Table 1. Location and types of modern glaciers and glacial deposits in Turkey. Modern and Last-Glacial snowlines are also indicated where available. Modified from Kurter & Sungur (1980) and Kurter (1991)

Mountain Range Name	Mountain or Peak Name	Peak Elevation (m)	Location (latitude and longitude)	Glacier Names	Type of Glacier	Area (km²)	Length (km)	Modern Snowline Elevation (m) *(Würmian Snowline Elevation (m))*	Type of Glacial Deposits
Cilo (Buzuldağ) (SE Taurus)	Uludoruk (Reşko)	4135	37°26' - 37°32' N 43°56' - 44°04' E	Uludoruk	Valley	8.0	4.0	3600	Lateral/terminal moraines
				Mia Hvara	Valley	2.5	1.5		
				5 small glaciers	Valley to mountain	0.3 to 1.0	0.2 to 0.5 each		
Sat (SE Taurus)	Dolampar	3794	37°18' - 37°24' N 44°10' - 44°20' E	Geverok	Valley	0.8	1.0	3500	Terminal moraines
				Unnamed	Valley	0.1	0.4		
Kavussahap (SE Taurus)	Hasanbeşir	3503	38°12' - 38°16' N 42°48' - 42°°54' E	Northwest	Mountain	0.06	0.3	3400	Terminal moraines
Aladağ (Central Taurus)	Demirkazık	3756	37°49' - 37°53' N 35°06' - 35°11' E	Lolut	Valley	0.5	1.0	3450 *(2200-1900)*	Terminal moraines
Aladağ (Central Taurus)	Mamerdiğin (Hacer Valley)	3407	37°47' - 37°49' N 35°13' - 35°21' E	No recent glacier				3450 *(2200-1900)*	Lateral/terminal and hummocky moraines
Bolkardağ (Central Taurus)	Gökboyun (Karagöl)	3524	37°26' - 37°33' N 34°36' - 34°50' E	No recent glacier				3450-3700 *(2200-2000 on south face, 1900-2075 North face)*	Lateral/terminal moraines and glacial lakes
Bolkardağ (Central Taurus)	Medetsiz	3524	37°26' - 37°33' N 34°36' - 34°50' E	North	Mountain	0.06	0.3	3450-3700 *(2200-2000 on south face, 1900-2075 North face)*	Terminal moraines
Geyikdağ (Central Taurus)	Geyikdağ (Namaras valley)	2850	36°45' - 36°50' N 32°09' - 32°14' E	No recent glacier				3200 *(2000)*	Lateral/terminal and hummocky moraines
Dedegöldağ (Central Taurus)	Dipoyraz	2997	37°40' - 37°45' N 31°19' - 31°24' E	Several glacierets towards the South	Cirque	0.3	up to 0.2	3300-3500 *(2350-2400)*	Terminal moraines
Isparta (Central Taurus)	Davras and Barla	2700	37°33' - 37°38' N 30°43' - 30°48' E	No recent glacier				*(2400)*	Lateral/terminal moraines
Beydağları (W Taurus)	Beydağ	3086	36°33' - 36°38' N 30°12' - 30°17' E	No recent glacier				3600 *(2400-2600)*	Terminal moraines
Akdağ (W Taurus)	Akdağ	3016	36°30' - 36°35' N 29°33' - 29°38' E	No recent glacier				3500 *(2200-2400)*	Terminal moraines
Honaz (W Taurus)	Honaz	2571	37°40' - 37°45' N 29°18' - 29°23' E	No recent glacier				3600 *(2600)*	Terminal moraines
Gölgelidağ (W Taurus)	Sandıras	2295	37°10' - 37°15' N 28°45' - 28°50' E	No recent glacier				*(2050-2000)*	Lateral/terminal moraines

Mountain Range Name	Mountain or Peak Name	Peak Elevation (m)	Location (latitude and longitude)	Glacier Names	Type of Glacier	Area (km^2)	Length (km)	Modern Snowline Elevation (m) *(Würmian Snowline Elevation (m))*	Type of Glacial Deposits
Rize (Pontics)	Kaçkar	3932	40°50' - 41°00' N 41°08' - 41°20' E	Kaçkar I Kaçkar II Kaçkar III Krenek I, II Dübe	Valley Valley Valley Cirque Cirque	0.8 0.5 0.3 0.3 0.01	1.3 0.7 0.5 0.5 0.1	3100-3200 *(2300-2500)* on N face, 3550 *(2600-2700)* on S face	Ablation, terminal, lateral, ground and hummocky moraines. Roches moutonnées, Glacial lakes
Rize (Pontics)	Göller (Hunut)	3560	40°40' - 40°55' N 41°03' - 41°13' E	No recent glacier				*(2650)*	Terminal moraines
Rize (Pontics)	Verçenik (Uçdoruk)	3710	40°40' - 40°46' N 40°52' - 41°05' E	Sinançor Dilektepe	Mountain Valley	0.05 0.14	0.3 0.7	3500 *(2700)*	Middle and terminal moraines, moraine-dammed glacial lakes
Altiparmak (Pontics)	Lazgedigi	3353	40°57' - 41°10' N 41°25' - 41°32' E	Kırmızıgedik	Cirque	0.3	0.5	*(2650)*	Terminal and ground moraines
Bulut (Pontics)	Kindevul	3562	40°53' - 41°00' N 41°15' - 41°23' E	Avucur	Cirque	0.015	0.15	*(2650)*	Lateral moraines
Soğanlı (W of Pontics)	At	3395	40°25' - 40°45' N 40°45' - 40°52' E	Few glacierets				*(2650)*	Terminal moraines
Gavur (SW of Pontics)	Karadağ (Aptalmusa)	3331	40°22' - 40°26' N 39°02' - 39°07' E	Avliyana	Mountain	0.045	0.15	3500 *(2600-2850)*	Terminal and hummocky moraines
Giresun (W of Pontics)	Karagöl	3107	40°30' - 40°32' N 38°08' - 38°13' E	Northwest Few small glaciers	Mountain	0.08	0.4	2900 *(2600-2700)*	Terminal moraines
Strato-volcano S of Iğdır (E Anatolia)	Ağrı (Ararat)	5165	39°41' - 39°44' N 44°15' - 44°19' E	11 glaciers	Ice cap	10.0	1.5 to 3.0	4300 *(3000)*	Terminal moraines
Strato-volcano N of Lake Van (E Anatolia)	Süphan	4058	38°53' - 38°55' N 42°47' - 42°52' E	South Few glaciers on N	Valley	3.0	1.5	3700-4000	Terminal moraines
Strato-volcano in Kayseri (Cappadocia)	Erciyes (Argeous)	3916	38°31' - 38°34' N 35°24' - 35°28' E	Northwest	Valley	0.11	0.38	3800 *(2700)* on North face 3400 *(3000)* on South face	Terminal, ablation and lateral moraines. Moraine covered dead ice masses. Sandur plains
Bursa (NW Anatolia)	Uludağ	2543	40°10' – 40°15' N 29°11' - 29°16' E	No recent glacier				*(2200-2330)*	Terminal moraines
Erzincan (E Anatolia)	Mercan	3368	39°25' - 39°30' N 39°15' - 39°10' E	No recent glacier				3600-3700 *(2750)*	Terminal and ground moraines
Erzurum (E Anatolia)	Mescid	3239	40°20' - 40°25' N 41°13' - 41°18' E	No recent glacier				3600-3700 *(2750)*	Glacial lakes

Fig. 2. Very thick lateral moraine developed on the northern flank of the Hacer valley (Aladağ, Central Taurus Mountains).

by Erinç (1952a). Satellite images confirm this height (Kurter, 1991). Besides, fresh terminal moraines indicate that the ice margin continued to retreat (Erinç, 1952a). The maximum extent of the Pleistocene (Würmian?) glaciers is well marked by prominent end moraines. The Mia Hvara glacier reached at least a length of 9 km. Neighbouring glaciers were 7, 5 and 6 km long, but in contrast to, for instance, the Alps, they did not coalesce to form an ice-stream network (Bobek, 1940).

In the southeastern Taurus Mountains, less important glaciers also exist on Mount Sat (3794 m) (Table 1). The most important, the Geverok glacier has a length of nearly 1 km. In the northern part of SE Taurus Mountains, a small glacier, which is only 300 m long and 200 m wide, was first described by Klaer (1965) and later by Schweizer (1972, 1975). This glacier occurs on Mount Hasanbeşir (3503 m), to the south of Lake Van, at an elevation of 3300 m, as deduced from satellite images (Kurter, 1991). Although Mount Sat is much lower than Mount Cilo, it bore impressive glaciers. In the Pleistocene, the Geverok glacier on Mount Sat reached a maximum length of 10 km (Bobek, 1940).

The actual snowline in the SE Taurus Mountains varies between 3400-3600 m and the Last Glacial snowline is estimated to be around 2800 m (Messerli, 1967). Therefore the presence of glaciers is easily explained by favourable local climatic and physiographic conditions.

b) Central Taurus: The smaller and less extensive, glaciers and glacier-related landforms of the Central Taurus Mountains also attracted scientific interest, starting from the German Alpine expedition of 1927 (Künne, 1928), probably because of their easier access. Aladağ and Bolkardağ constitute the two most important mountains of the Central Taurus where some small valley and mountain glaciers can be seen (Table 1). On Mount Aladağ, the southern flank of Demirkazık peak (3756 m) bears a 1 km long glacier (Lolut glacier) (Kurter, 1991). However, glacial landforms at much lower altitudes suggest that Pleistocene glaciations were far more extensive. To the S and SE of the Demirkazık and Kaldı peaks several valleys contain terminal and lateral moraines reaching down to altitudes of about 2100-2200 m (Blumenthal, 1952; Spreitzer, 1939, 1956, 1957, 1958, 1959, 1960, 1969, 1971a, 1971b; Birman, 1968). A field survey carried out by a team from Hacettepe University in 1997 confirmed the presence of morainic deposits in valleys to the E and S of the Demirkazık and Kaldı peaks. For instance, a lateral moraine that occupies the northern flank of Hacer valley (2000 m) is several hundred metres high and a few kilometres long. It is composed of large limestone blocks (up to 20 m in diameter), now partly covered by trees (Fig. 2). Glaciofluvial deposits to the E of the Hacer valley are also preserved down to an altitude of ca. 1100 m.

On the Medetsiz peak (3524 m) of Mount Bolkardağ, Kurter (1991) observed a mountain glacier of about 300 m long to descend from an altitude of 3350 down to 3000 m from satellite images. However, Catherine Kuzucuoğlu (personal communication), who visited the area in 1998, noticed only permanent snows and few stationary remains of glaciers in this locality. Contrary to the present-day tendency towards glacial retreat, glacial landforms in this area indicate more severe glaciations in Pleistocene times. Several very distinct moraines are present along the Maden and Ganimet streams, situated around 1750 m (Blumenthal, 1956a; Birman, 1968). Klaer (1969) and Messerli (1967) even suggest the presence of possible morainic deposits as low as 1700-1650 m in the Maden valley, descending from the Medetsiz peak.

Apart from the Bolkardağ and Aladağ Mountains, only the northern face of the Dedegöldağ (Dipoyraz peak, 2997 m) bears a few very small glaciers today (Delannoy & Mairie, 1983). Well-developed moraines around 2000 m are also present on the eastern face of the Dipoyraz peak (Olivier Monod, personal communication). All other mountains in the Central Taurus Range show no sign of modern glaciers. However, small morainic ridges on the Davras and Barla Mountains, near Isparta, are also reported

Fig 3. Typical hummocky moraines (m) developed between Cretaceous limestones (Kelçe Mountains; 2850 m) (a) and Cretaceous bedrock ridge (b). Note the glacial cirques and arêtes on the north-facing slope (c). Vegetation-covered surface is micritic limestone of Eocene age (d). (Namaras valley near Mount Geyikdağ, Central Taurus Mountains).

Fig. 4. Typical hummocky moraines (m) and flat-lying outwash fan (f) (approx. 500 m long) developed on the Namaras valley floor (2100m) (Central Taurus Mountains). Note part of a lateral morainic ridge (b) nearly 100 m high on the lower left corner.

(Atalay, 1987; Monod, 1977 and Olivier Monod, personal communication).

Another area situated 100 km NE of Antalya in the Central Taurus Mountains, is characterised by the presence of a peculiar hummocky topography that covers an area of approximately 30 km^2 in the Namaras and Susam valleys (2000 m), near Geyikdağ (Fig. 3 and 4) (Arpat & Özgül, 1972; Çiner *et al.*, 1999). In these valleys, coarse, loose materials form a chaotic 'knob-and-kettle' topography with hillocks up to 10 m high and 30 m wide, separated by irregular depressions (Fig. 5). These landforms are interpreted as hummocky disintegration moraines from former active glaciers (Çiner *et al.*, 1999). Several glaciers related landforms, such as kettle holes and lakes formed by lateral moraines are also very commonly observed in the area (Fig. 6 and 7).

The modern snowline in the central Taurus Mountains varies between 3200-3700 m and the Last-Glacial snowline is estimated to have been around 2200 m on average (Messerli, 1967).

Fig. 5. Section in a hummocky morainic mound showing a typical matrix-supported diamicton (Namaras valley near Mount Geyikdağ, Central Taurus Mountains).

c) Western Taurus: The mountains of the Western Taurus do not support recent glaciers. However, Beydağ (3086 m) and Akdağ (3016 m) show several cirques and well-developed morainic landforms, especially on their NE facing slopes (de Planhol, 1953; Onde, 1954; Messerli, 1967; Doğu *et al.*, 1999). A few moraines also occur on the NE flanks of Honaz (2571 m) (Yalçınlar, 1954, 1955; Darkot & Erinç, 1954; Erinç, 1955a, 1955b) and Sandıras Mountains (2295 m) (de Planhol, 1953; Doğu, 1993). The modern snowline is estimated to be around 3500 m (Doğu, 1993). On the other hand, the Last Glacial snowline elevation was approximately 2400 m in the Western Taurus Mountains, except for on Mount Sandıras where a snowline at 2000 m elevation was determined. This lower snowline reflects probably its vicinity to the sea and favourable sufficiently humid climatic conditions (de Planhol, 1953; Messerli, 1967; Doğu, 1993).

Presumably, all the recently glaciated mountains mentioned, must have supported Pleistocene glaciers. However, apart from calculations of the Last-Glacial snowline altitude, very little is known about the actual size and dynamics of glaciers in the central and southeastern Taurus Mountains. The distribution of glaciers, as shown in Fig. 1 and in digital map is thus largely based on the snowline altitude and height of the individual mountains.

2. Pontic (Eastern Black Sea) Mountains

The Pontic Mountains trend W-E along the Black Sea coast of Turkey. The mountain range increases in height towards the east, reaching altitudes of more than 3900 m (Stratil-Sauer 1961, 1964, 1965; Gall, 1966). The actual snowline elevation is much lower on the north facing slopes (3100-3200 m) as compared to the south facing ones (3550 m) because of the effect of humid air masses (Erinç, 1952a). The Last Glacial snowline elevation is estimated at 2600 m on average (Messerli, 1967).

The highest peak of the Pontic Range is Mount Kaçkar (3932 m). Five glaciers are developed on its northern flank (Table 1). Erinç (1949a), who first described the largest and named it Kaçkar I glacier, indicates that the glacier tongue descends down to 2850 m. On Landsat MSS images taken in 1975, the same glacier is seen to originate from 3650 m and terminate at an elevation of 2900 m with a total length of 1500 m (Kurter, 1991). Doğu *et al.* (1993), who carried out a detailed field survey in the area, give the length as 1250-1300 m (between 3600-3000 m altitude). Other glaciers present in the region are the Kaçkar II and III glaciers. Erinç (1949a) indicates that the tongues of these two glaciers descended down to 3000 and 2940 m respectively. According to Kurter (1991), Landsat MSS satellite images (taken in 1975) show two glaciers originating from 3650 m that extend down to 2990 (1 km long) and 3130 m respectively. The lower limits given by Doğu *et al.* (1993) are again somewhat higher, being 3080 and 3100 m respectively. From their map, glacier lengths of 700 and 500 m can be calculated. These three studies confirm the general tendency of glacial retreat in Turkey that has already been observed in the SE Taurus (Cilo) Mountains. Three other small cirque glaciers termed Krenek I and II (Krenek, 1932) and Dübe, are also present near Mount Kaçkar. According to Doğu *et al.* (1993), this most mountainous part of the Pontic Range contains four large U-shaped valleys where different glacier-related landforms such as ablation, terminal, lateral and ground moraines, roches moutonnées and glacial lakes (up to 750 m^2) are abundantly preserved.

The second highest peak in the Pontic Range is Mount Verçenik (3709 m). According to Erinç's (1949a) map three glaciers are present on its northern slope. However on the Landsat MSS images taken in 1975 only two glaciers, the Dilektepe (700 m) and Sinançor glaciers (300 m) remain (Kurter, 1991). Doğu *et al.* (1996) who carried out a field survey of the valleys near Mount Verçenik do not show those two glaciers on their map. They indicate, however, the presence of several cirque and moraine-dammed lakes and medial, terminal and lateral moraines. They also identified two different sets of morainic ridges that are several kilometres long.

Although less high, the easternmost parts of the Pontic Mountains (the Bulut-Altıparmak Mountains) bear some recent glaciers. The largest, the Kırmızıgedik glacier (500 m), occupies a cirque to the east of Lazgediği peak (3353 m) (Table 1). Another small glacier, Avucur glacier (only 150 m long) is found to the east of Kindevul peak (3562). To the west, Mount Karagöl (3107 m) bears several small glaciers (de Planhol & Bilgin, 1964). According to these authors, only one, the NW glacier (400 m long) descended down to 2850 m and deposited a set of terminal moraines probably during the Little Ice Age.

Most of the landforms in the Pontic Range, as everywhere else in Turkey, have yet to be studied in detail, and their exact age (Last Glacial or older) has not been established.

3. Volcanoes and independent mountain chains on the Anatolian plateau

The volcanoes in the interior of the country show signs of glacial activity and active glaciers (Table 1). Among them, Mount Ağrı (5165 m), located close to the Iranian and Armenian borders, is not only the highest mountain in Turkey, but is also the only mountain on which a recent ice cap is developed (Imhof, 1956). According to Blumenthal (1956b; 1958), eleven glaciers emerged from the summit, descending down to 3900 m on the north-facing slope and 4200 m on the south-facing slope, and covering an area of approximately 10 km^2. Birman (1968) who saw the glaciers from a distance during his visit in 1963 stated that the lowest glacier descends down to 4500 m on the western slope and to 3700 m on the northern slope. Landsat MSS images taken in 1976 clearly indicate a prominent northwesterly tilt of the ice cap, reaching down to an altitude of 4100 m (Kurter & Sungur, 1980). On Mount Ağrı, the actual snowline elevation is estimated as 4300 m (Klaer, 1965; Arkel, 1973; Kurter & Sungur, 1980). So far, no traces of any older, more extensive stages of ice cover could be identified. Blumenthal (1958) calculated a Pleistocene snowline elevation of 3000 m for an ice cap of presumably 100 km^2. He explains the absence of moraines by the lack of confining ridges to control valley glaciers, by insufficient debris load in the ice to form moraines and by volcanic eruptions that later covered the pre-existing moraines.

Fig. 6. Kettle hole (20 m in diameter) developed on a morainic ridge (Susam valley, Central Taurus Mountains). Person (encircled) for scale.

Another volcano, Mount Süphan (4058 m) is situated to the north of Lake Van in SE Turkey. Several small glaciers are developed on the northern slope of the crater (Kurter & Sungur, 1980). The largest is 2 km wide and 1.5 km long and descends down to 3400 m (Kurter, 1991).

The last volcano is Mount Erciyes (3916 m), where a glacier exists on its northwestern slope, located in Central Anatolia near the town of Kayseri. It was first visited by Penther (1905) who observed that the glacier descended to the 3100 m level, with a total length of 700 m (Blumenthal, 1938). Later, Erinç (1952a) described the glacier as having a length of 550 m reaching down only to 3380 m. He also noticed that the glacier was strongly covered by debris,

Fig. 7. Eğrigöl Lake formed by the damming of a lateral morainic ridge (r) (Eastern part of Namaras valley, Central Taurus Mountains). The village on the lakefront for scale.

interpreted as ablation moraine. Together with the observations of Bartsch (1935), Erinç (1952a) roughly calculated that the glacier had been shrinking at an average rate of 3 m per year over a period of 22 years. The most recent survey carried out by Güner & Emre (1983) on Mount Erciyes, indicates a glacier length of only 380 m and hence confirms the overall tendency of glacier retreat already observed in other Turkish glaciers. Five large and several smaller cirques indicate much stronger glaciations during the Quaternary. The largest Würmian glacier had a length of *c.* 5 km. Messerli (1967) estimates that the snowline occurred at 2700 m on the northern and 3000 m on the southern side. Erratic blocks have been interpreted as traces of older, more extensive glaciations. They have been found about 2 km beyond the maximum limits of the Würmian glaciers, suggesting a snowline some 200 m lower than in the Würmian.

Apart from the volcanoes, few mountain chains and massifs in Central Anatolia bear signs of past glacial activity. Among them Mount Uludağ (2543 m) is situated to the SE of the Marmara Sea near Bursa (Table 1). Birman (1968) who conducted a very rapid survey in the area described a small glacier, about 100 m in length and width, on the northwestern slope. Although more extensive past glaciations could be interpreted from the form of valleys and by the existence of lateral moraines several kilometres long, more recent studies (Atalay, 1987) do not indicate the presence of actual recent glaciers.

In eastern Anatolia, the Munzur Mountain chain (3368 m), near the town of Erzincan, experienced glacial activity during the Pleistocene when the Mercan valley glacier descended down to an altitude of 1650 m (Atalay, 1987; Türkünal, 1990). Another area glaciated during the Pleistocene is Mount Mescid (3239 m), to the north of Erzurum (Yalçınlar, 1951). Here, as on Mount Mercan, the Last Glacial snowline elevation is estimated to have been at around 2750 m, as compared to the modern snowline elevation of 3600-3700 m.

Conclusion

The presence of U-shaped valleys and morainic deposits and recent glaciers in the Turkish Mountains indicate the existence of past and present glacial activity. Unfortunately, the absence of dating of the morainic landforms makes it difficult to assign a precise age to the past glacial periods. However a general consensus seems to exist between scientists, that there must have been several ice advances and retreat phases (Louis, 1944; Erinç, 1952a; Blumenthal, 1958; Klaer, 1969, 1977, 1978; Messerli, 1967; Schweizer, 1975; Atalay, 1987; Kurter, 1991). For instance, Schweizer (1975) in his work on Mount Hasanbeşir tentatively attributed the oldest identified and most strongly weathered moraines to the second-but-last (Rissian) cold stage.

Most of the works carried out also concluded that, at the time of the maximum extension of glaciers during the Last Glacial, the snowline was 1000 to 1500 m lower than present (Kuzucuoğlu & Roberts, 1998), and numerous cirques and small valley glaciers developed down to an elevation of 2000 m. They all disappeared during the Last Glaciation.

The fact that moraines deposited during the postglacial readvance phases overlie a major part of the Pleistocene glacial deposits was interpreted by Erinç (1952a) as evidence for a more recent humid stage (Younger Dryas; Little Ice Age?) during which the valley glaciers descended lower than during the Last Glacial. Therefore, according to this author, the present-day glaciation cannot be considered as a continuation of that during the Pleistocene. Some very fresh morainic landforms are also observed in cirques that are not occupied by recent glaciers. In the Karagöl (de Planhol & Bilgin, 1964) and in the Mount Uludağ (Erinç, 1949b, 1952a, 1957; Pfannenstiel, 1956) these moraines have been attributed to the Little Ice Age.

The Pleistocene glaciations were ignored for a long time in Turkey. The pioneering works of Sırrı Erinç together with the indisputable presence of morainic deposits and glacier-related landforms observed on the mountain ranges over 2000 m indicate the need for a detailed study which will take into account the altitudes and orientations of those deposits and their relative chronologies. The study of glaciofluvial and lacustrine sediments deposited in the valleys and lakes will certainly yield interesting results regarding the advances and retreats of the glaciers and the fluctuations of climate at high altitudes.

The data available on Turkish glaciers indicate that the most recent glacier retreat probably started at the beginning of the 20th century, becoming faster since the 1930's (Erinç, 1952a; Erol, 1981; Güner & Emre, 1983; Kurter, 1991). This general shrinkage trend is yet to be quantified by additional field observations in order to understand the glacier evolution of Turkey. For this purpose a National Science Foundation (NSF) - Scientific and Technical Research Council of Turkey (TÜBITAK) joint project has been organised by Dr. Marek Zreda (University of Arizona) and by Dr. Attila Çiner (University of Hacettepe, Ankara) (Zreda *et al.*, 2001). During this 3 year project classical dating methods, such as historical documents, lichens, organic material dating in sediments, and more modern dating techniques such as cosmogenic ^{36}Cl method based on *in-situ* accumulation of ^{36}Cl in boulders exposed to cosmic radiation will be carried in order to determine the age of the landforms. It is hoped that this study will help to better understand the magnitude and timing of Quaternary glaciations in Turkey.

Acknowledgements

I thank Dr. Jürgen Ehlers for inviting me to write this overview paper about the Turkish Quaternary glaciations. His permanent help, critics and encouragement are greatly appreciated. I am also grateful to Dr. Catherine Kuzucuoğlu (CNRS, Meudon-France) who critically reviewed this paper. Thanks also go to Dr. Olivier Monod (CNRS, Orléans-France) and to Dr. Ali Fuat Doğu (University of Ankara) for indicating some of the moraines they observed in the Taurus and Pontic Mountains. Dr. Max Deynoux (CNRS, Strasbourg-France), Dr. Erdal Koşun (University of Antalya), Dr. Serdar Bayarı (University of Hacettepe) and Erdem Çörekçioğlu (MTA) are thanked for their field assistance.

References

Ainsworth, W. F. (1842). *Travels and researches in Asia Minor, Mesopotamia, Chaldea and Armenia*. London, J. W. Parker, **1**, 346 pp.; **2**, 399 pp.

Arpat, E. & Özgül, N. (1972). Rock glaciers around Geyikdağ, Central Taurus, Turkey (in Turkish) *Bulletin of Mineral Research and Exploration,* **80**, 30-35.

Atalay, İ. (1984). Glacial morphology of the Mescid Mountain (in Turkish) *Aegean Geographical Journal,* **2**, 129-138.

Atalay, İ. (1987). *Introduction to the Geomorphology of Turkey (in Turkish),* 2nd ed. İzmir, Eagean University Press, 456 pp.

Bartsch, G. (1935). Das Gebiet des Erciyes Dagi und die Stadt Kayseri in Mittel-Anatolien. *Jahrbuch der Geographischen Gesellschaft zu Hannover für 1934 und 1935*, 87-202.

Birman, J.H. (1968). Glacial reconnaissance in Turkey. *Geological Society of America Bulletin,* **79**, 1009-1026.

Blumenthal, M.M. (1938). Der Erdchias-Dagh, 3916m. *Die Alpen,* **14, 3**, 82-87.

Blumenthal, M.M. (1941). Un aperçu de la géologie des Taurus dans les Vilayets de Nigde et d'Adana. *Bulletin of Mineral Research and Exploration,* **6**, 45-95.

Blumenthal, M.M. (1952). Das taurische Hochgebirge des Aladağ, neuere Forschungen zu seiner Geographie, Stratigraphie und Tektonik. *Bulletin of Mineral Research and Exploration,* **6**, 136 pp.

Blumenthal, M.M. (1954). Im südostanatolischen Hochland zwischen dem Van-See und den Cilo-Ketten. *Die Alpen,* **30**, 223-240.

Blumenthal, M.M. (1956a). Geology of northern and western Bolkardağ region (in Turkish). *Bulletin of Mineral Research and Exploration,* **7**, 153p.

Blumenthal, M.M. (1956b). Die Vergletscherung des Ararat (Nordöstliche Türkei). *Geographica Helvetica,* **11, 4**, 263-264.

Blumenthal, M.M. (1958). Vom Ağrı Dag (Ararat) zum Kaçkar Dag. Bergfahrten in nordostanatolischen Grenzlanden. *Die Alpen,* **34**, 125-137.

Bobek, H. (1940). Die gegenwärtige und eiszeitliche Vergletscherung im Zentralkurdischen Hochgebirge. *Zeitschrift für Gletscherkunde,* **27, 1-2**, 50-87.

Çiner, A., Deynoux, M. & Çörekçioğlu, E. (1999). Hummocky moraines in the Namaras and Susam valleys, Central Taurids, SW Turkey. *Quaternary Science Reviews,* **18**, 659-669.

Darkot, B. & Erinç, S. (1954). Observations géographiques dans le sud - ouest de l'Anatolie. *Review of the Geographical Institute of the University of İstanbul*, **1**, 149-167.

Delannoy, J.J. & Maire, R. (1983). Le Massif de Dedegöl dağ (Taurus Occidental, Turquie). Recherches de géomorphologie glaciaire et karstique. *Bulletin de l'Association de Géographie. Française,* **491**, 43-53.

de Planhol, X. (1953). Les formes glaciaires du Sandıras Dag et la limite des neiges éternelles quaternaires dans le SO de l'Anatolie. *Compte Rendu Sommaire de la Société Géologique de France*, 263-265.

de Planhol, X. & Bilgin, T. (1964). Glaciaire et périglaciaire quaternaires et actuels dans le massif du Karagöl (chaines pontiques, Turquie). *Revue de Géographie Alpine*, 497-512.

Doğu, A.F. (1993). Glacier shapes on the Mount Sandıras (in Turkish). *Turkish Geography Bulletin, Ankara University*, 263-274.

Doğu, A.F., Çiçek, I, Gürgen, G. & Tuncel, H. (1996). Glacier shapes, yaylas and tourism on the Mount Üçdoruk (Verçenik) (in Turkish). *Turkish Geography Bulletin, Ankara University,* **5**, 29-51.

Doğu, A.F., Çiçek, I, Gürgen, G. & Tuncel, H. (1999). Geomorphology of Akdağ and its effect on human activities (in Turkish). *Turkish Geography Bulletin, Ankara University,* **7**, 95-120.

Doğu, A.F., Çiçek, I, Gürgen, G., Tuncel, H. & Somuncu, M. (1994). Glacier shapes, yaylas and tourism on the Mount Göller (in Turkish). *Turkish Geography Bulletin, Ankara University,* **3**, 193-218.

Doğu, A.F., Gürgen, G., Tuncel, H. & Çiçek, I. (1997). Glacier shapes, yaylas and tourism on the Bulut-Altiparmak Mountains (in Turkish). *Turkish Geography Bulletin, Ankara University,* **6**, 63-91.

Doğu, A.F., Somuncu, M., Çiçek, I, Tuncel, H. & Gürgen, G. (1993). Glacier shapes, yaylas and tourism on the Kaçkar Mountains (in Turkish). *Turkish Geography Bulletin, Ankara University*, 157-183.

Erinç, S. (1944). Glazialmorphologhie Untersuchungen im Nordostanatolischen Randgebirge. *İstanbul University Geography Inst. Pub., Ph.D dissertation Series.* **1**, 56 pp.

Erinç, S. (1949a). Eiszeitliche Formen und gegenwärtige Vergletscherung im nordostanatolischen Randgebirge. *Geologische Rundschau,* **37**, 75-83.

Erinç, S. (1949b). Research on glacial morphology of Mount Uludağ (in Turkish). *Review of the Geographical Institute of the University of İstanbul.* **11-12**, 79-94.

Erinç, S. (1951). The glacier of Erciyes in Pleistocene and Post-glacial epochs (in Turkish). *Review of the Geographical Institute of the University of İstanbul.* **1, 2**, 82-90.

Erinç, S. (1952a). Glacial evidences of the climatic variations in Turkey. *Geografiska Annaler,* **34, 1-2**, 89-98.

Erinç, S. (1952b). The present day glaciation in Turkey. General assembly and 17th International Congress of the Int. Geographical Union, 8th proceedings, Washington D.C., 326-330.

Erinç, S. (1953). From Lake Van to Mount Cilo (in Turkish). *Review of the Geographical Institute of the University of İstanbul.* **2, 3-4**, 84-106.

Erinç, S. (1954). The Pleistocene history of the Black Sea and adjacent countries with special reference to climatic changes. *Review of the Geographical Institute of the University of İstanbul.* **1**, 84-133.

Erinç, S. (1955a). Periglacial features on the Mount Honaz (SW Anatolia). *Review of the Geographical Institute of the University of İstanbul.* **2**, 185-187.

Erinç, S. (1955b). A propos de la note de İ. Yalçınlar intitulée "Sur la présence de formes glaciaires quaternaires au Honaz-Dag et au Boz-Dag (Ouest de la Turquie)". *Compte Rendu Sommaire de la Société Géologique de France*, **13-14**, 259-261.

Erinç, S. (1957). About the Uludağ periglacial (in Turkish). *Review of the Geographical Institute of the University of İstanbul.* **8**, 91-94.

Erinç, S. (1978). Changes in the physical environment in Turkey since the end of the Last Glacial. *In:* Brice, W. C. (ed.), *The Environmental History of the Near and Middle East since the Last Ice Age.* Academic Press, London, 87-110.

Erol, O. (1981). Quaternary pluvial and interpluvial conditions in Anatolia and environmental changes especially in south-central Anatolia since the last glaciation. *In:* Frey, W. & Uerpmann, H.P. (eds.), Beiträge zur Umweltgeschichte des vorderen Orients (Contributions to the environmental history of Southwest Asia). *Beihefte zum Tübinger Atlas des Vorderen Orients. Reihe A, Naturwissenschaften*, **8**, 101-109.

Gall, H. (1966). Gletscherkundliche Beobachtungen im Hochgebirge von Lasistan (nordostanatolisches Randgebirge) (Glaciologic observations in the high mountains of Lasistan, border ranges of northeastern Anatolia). *Mitteilungen der Oesterreichischen Geographischen Gesellschaft*, **108, 2-3**, 261-286.

Güner, Y. & Emre, Ö. (1983). Pleistocene glaciation on Mount Erciyes and its relation to volcanism (in Turkish). *Bulletin of Geomorphology,* **11**, 23-34.

Horvarth, E. (1975). Glaciers of Turkey, Armenian S.S.R. and Iran. In: W.O. Field (ed.), *Mountain Glaciers of the Northern Hemisphere, pt. III, Glaciers of Southern Asia*, US Army Cold Regions Research and Engineering Laboratory, 343-359.

Imhof, E. (1956). Der Ararat. *Die Alpen,* **32, 1**, 1-14.

İzbırak, R. (1951). Geographical research in Lake Van and in the Hakkari and Cilo Mountains (in Turkish). *Turkish Geography Bulletin, Ankara University,* **67, 4**, 149 pp.

Klaer, W. (1965). Geomorphologische Untersuchungen in den Randgebirgen des Van-See (Ostanatolien). *Zeitschrift für Geomorphologie,* **N.F., 9, 3**, 346-355.

Klaer, W. (1969). Glacialmorphologische Probleme in den Hochgebirgen Vorderasiens. *Erdkunde,* **23, 3**, 192-200.

Klaer, W. (1977). Zur Morphogenese und Formungsdynamik innerhalb des periglazialen Stockwerkes der Hochgebirge Vorderasiens (The morphogenesis and dynamics within the periglacial tiers in the Near East). *In:* Poser, H. (ed.), Formen, Formengesellschaften und Untergrenzen in den heutigen periglazialen Höhenstufen der Hochgebirge Europas und Afrikas zwischen Arktis und Aequator. *Abhandlungen der Akademie der Wissenschaften in Göttingen, Mathematisch - Physikalische Klasse,* **31**, 261-274.

Klaer, W. (1978). Periglaziale Hangformung in den Hochgebirgen vorderasiens (Periglacial slopes in the mountains of the Near East). In: Colloque sur le périglaciaire d'altitude du domaine méditérranéen et abords. *Association Géographique d'Alsace, France,* 253-261.

Krenek, L. (1932). Gletscher im Pontischen Gebirge. *Zeitschrift für Gletscherkunde,* **20, 1-3**, 129-131.

Künne, G. (1928). Die deutsche alpine Taurus-Expedition 1927 (Ala Dagh in Zilizien). *Petermanns Geographische Mitteilungen,* **74**, 273-276.

Kurter, A. & Sungur, K. (1980). Present glaciation in Turkey. In: World Glacier Inventory. Proceedings of the workshop at Riederalp, Switzerland, 17-22 September 1978. *International Association of Hydrological Sciences,* **126**, 155-160.

Kurter, A. (1991). Glaciers of Middle East and Africa - Glaciers of Turkey. *In:* Williams, R.S. & Ferrigno, J.G. (eds), Satellite Image Atlas of the World, *USGS Professional Paper*, **1386-G-1**, 1-30.

Kuzucuoğlu, C. & Roberts, N. (1998). Evolution de l'environnement en Anatolie de 20000 - 6000 BP. *Paléorient,* **23/2**, 7-24.

Lembke, H. (1939). Klima und Höhenstufen im Nordostanatolischen Randgebirge. *Zeitschrift der Gesellschaft für Erdkunde zu Berlin*, **74**, 171-184.

Leutelt, R. (1935). Glazialgeologische Bebachtungen im Lazistanischen Hochgebirge. *Zeitschrift für Gletscherkunde,* **23**, 67-80.

Louis, H. (1938). Eiszeitliche Seen in Anatolien. *Zeitschrift der Gesellschaft für Erdkunde zu Berlin,* **73, 7-8**, 267-285.

Louis, H. (1944). Die Spuren eiszeitlicher Vergletscherung in Anatolien. *Geologische Rundschau,* **34**, 447-481.

Maunsell, F. R. (1901). Central Kurdistan. *Geographical Journal,* **18, 2**, 121-144.

Messerli, B. (1964). Der Gletscher am Erciyas Dagh und das Problem der rezenten Schneegrenze im anatolischen und mediterranen Raum. *Geographica Helvetica,* **19, 1**, 19-34.

Messerli, B. (1967). Die eiszeitliche und die gegenwärtige Vergletscherung in Mittelmeerraum. *Geographica Helvetica,* **22**, 105-228.

Messerli, B. (1980). Mountain glaciers in the Mediterranean area and in Africa. In: World glacier inventory; workshop, *IAHS-AISH Publications*, **126**, 197-211.

Monod, O. (1977). Recherches géologiques dans les Taurides occidentales au sud de Beyşehir (Turquie). Unpublished Thesis, Univ. Paris Sud (Orsay), 442 pp.

Onde, H. (1954). Formes glaciaires dans le Massif Lycien de l'Akdağ (Turquie du Sud-Ouest). *Congrès Géologique International*, **XV**, 327-335.

Palgrave, W. G. (1872). Vestiges of the glacial period in northeastern Anatolia. *Nature,* **5**, 444-445.

Penther, A. (1905). Eine Reise in das Gebiet des Erdschias-Dagh (Kleinasien), 1902. *Abhandlungen der k. k. Geographischen Gesellschaft in Wien,* **6, 1**.

Pfannenstiel, A. (1956). Rezente Froststrukturböden und Karstformen des Ulu Dagh. *Akademie der Wissenschaften und der Literatur, Math.-Nat. Kl., Abh.*, **5**, 329-340.

Philippson, A. (1906). Ein Gletscher am Erdschias-dagh (Argaeus) in Kleinasien. *Zeitschrift für Gletscherkunde,* **1, 1**, 66-68.

Schweiser, G. (1972). Klimatisch bedingte geomorphologische und glaziologische Züge der Hochregion vorderasiatischer Gebirge (Iran und Ostanatolien). In: Geoecology of the high mountain regions of Eurasia. *Erdwissenschaftliche Forschung,* **4**, 221-236.

Schweiser, G. (1975). Untersuchungen zur Physiogeographie von Ostanatolien und Nordwestiran. *Tübinger Geographische Studien,* **60**, *Special Issue,* **9**, 145 pp.

Spreitzer, H. (1939). Bergfahrten und Forschungen im Ala Dag (Südostanatolien), Deutsche Taurus-Bergfahrt 1938 des Zweiges Klagenfurt, Die Kundfahrt und ihre Ergebnisse. *Zeitschrift des Deutschen Alpenvereins,* **70**, 50-52.

Spreitzer, H. (1956). Untersuchungen im Kilikischen Ala Dag im Taurus. *Mitteilungen der Österreichischen Geographischen Gesellschaft,* **98, 1**, 57-64.

Spreitzer, H. (1957). Zur Geographie des Kilikischen Aladağ im Taurus. *Festschrift zur Hundertjahrfeier der Geographischen Gesellschaft in Wien*, 414-459.

Spreitzer, H. (1958). Frührezente und rezente Hochstände der Gletscher des Kilikischen Ala Dag im Taurus. Innsbruck, *Geographische Forschungen,* **190**, 265-281.

Spreitzer, H. (1959). Fußflächen am Kilikischen Ala Dag im Taurus. *Mitteilungen der Österreichischen Geographischen Gesellschaft,* **101, II**, 183-201.

Spreitzer, H. (1960). Hangformung und Asymmetrie der Bergrücken in den Alpen und im Taurus. *Zeitschrift für Geomorphologie*, 211-236.

Spreitzer, H. (1969). Die eiszeitliche und gegenwärtige Vergletscherung des kikilschen Ala Dag im Taurus (Pleistocene geology and glaciation of Ala Dag, Taurus Mountains, Turkey). Actas - *Congreso Internacional del I. N. Q. U. A. (International Union for Quaternary Research)* **5, 1**, 339-347.

Spreitzer, H. (1971a). Rezente und eiszeitlische Grenzen der glazialen und periglazialen Höhenstufen im Zentralen Taurus (vornehmlich am Beispiel des Kilikischen Ala Dag). *Mitteilungen des Naturwissenschaftlichen Vereines für Steiermark,* **101**, 139-162.

Spreitzer, H. (1971b). Die eiszeitliche Depression der glazialen und periglazialen Region und die Pedimentbildung im Zentralen Taurus (bes. im Kilikischen Ala Dag) (The ice-age depression of the glacial and periglacial region, and pedimentation in the central Taurus Mountains, Cilician Ala Dag). *In:* Etudes sur le Quaternaire dans le Monde, Vol. 1. *Bulletin de l'Information, Association Française pour l'Etude du Quaternaire,* **4**, 105-111.

Stratil-Sauer, G. (1927). Der östliche Pontus. *Geographische Zeitschrift,* **33, 9**, 497-519.

Stratil-Sauer, G. (1961). Beobachtungen im Ostpontischen Gebirge unter besonderer Berücksichtigung der Kaltzeitformen. *Mitteilungen der Österreichischen Geographischen Gesellschaft.* **103, 1**, 1-24.

Stratil-Sauer, G. (1964). Beobachtungen im Ostpontischen Gebirge unter besonderer Berücksichtigung der Kaltzeitformen. Teil II. *Mitteilungen der Österreichischen Geographischen Gesellschaft.* **106, 1**, 16-44.

Stratil-Sauer, G. (1965). Beobachtungen im Ostpontischen Gebirge unter besonderer Berücksichtigung der Kaltzeitformen. Teil III. *Mitteilungen der Österreichischen Geographischen Gesellschaft.* **107, 1-2**, 20-44.

Türkünal, S. (1990). Mountain chains and mountains of Turkey, (in Turkish). *Bulletin of the Chamber of Geological Engineers of Turkey,* 30, 42 pp.

van Arkel, N.A. (1973). Die gegenwärtige Vergletscherung des Ararat. *Zeitschrift für Gletscherkunde und Glazialgeologie*, **9**, 89-103.

Wright, H.E. (1962). Pleistocene glaciation in Kurdistan. *Eiszeitalter und Gegenwart,* **12**, 131-164.

Yalçınlar, İ. (1951). La glaciation des chaines de Soğanlı - Kaçkar et de Mescid Dağ (NE de la Turquie). *Review of the Geographical Institute of the University of Istanbul,* **1, 2**, 50-55.

Yalçınlar, İ. (1954). Sur la présence de formes glaciaires quaternaires au Honaz-Dag et au Boz-Dag (Ouest de la Turquie). *Compte Rendu Sommaire de la Société Géologique de France,* **13**, 296-298.

Yalçınlar, İ. (1955). Etudes morphologiques sur la glaciation du Honaz-Dag et de la chaine de Boz-Dag (Turquie occidentale). *Review of the Geographical Institute of the University of Istanbul,* **2**, 45-55.

Zreda, M., Çiner, A. & Bayarı, S. (2001). Magnitude of Quaternary glaciers and glaciations from low to high latitudes: global or local dominant controlling factors? An ongoing project between the Scientific and Technical Research Council of Turkey (TÜBİTAK) and the National Science Foundation (NSF)

Quaternary Glaciations – Extent and Chronology
Editors J. Ehlers and P.L. Gibbard

Pleistocene glaciations in the Ukraine

Andrei V. Matoshko

Institute of Geography of the Ukrainian National Academy of Sciences, 44, Volodymyrska Street, 01034 Kyiv, Ukraine;

E-mail: matoshko@iptelecom.net.ua

Introduction

Investigations of Quaternary glaciations in the Ukraine, including the problem of their extent, have continued for over 190 years. Ukrainian, Polish, Belorussian, Russian and Lithuanian scientists have participated in this work. Evidence of erratics on the interfluve of Teterev and Zdvizch Rivers (Fig. 1) was first reported by Staszic (1806). Later, Kalinichenko (1840) described erratic boulders of Scandinavian origin from the Sula Valley and noted that their outermost southern extent is within the Poltava region. Since then, over 150 works have been published on the position of the glacial boundaries (GB). The most comprehensive reviews of these were given by Dmitriev (1937) and Bondarchuk (1959) for the north-eastern Ukraine, and by Tutkovsky (1901), Volossovich (1924), Ruhle (1937), Przepiorski (1938), Krygowski (1947) and Dorofeev (1972) for the northwestern Ukraine. A general review of the Ukrainian glacial history is given by Gozhik (1995).

The database contains the most representative sources reflecting the main opinions on the GB problem including the author's map based on the latest evidence. This version in part is published for the first time. The map includes: the GB of each ice sheet proper; the areas of mountain glaciations; sources and type of evidence on which the boundary lines are based; key sites on which the age of glaciation is established, the directions of proglacial meltwater streams, distribution of proglacial lakes and the occurrence of formations, the glacial origin of which is as yet uncertain.

After the Second World War, Raskatov (1954) and Zamory (1961) proposed that the Dnieper Glaciation was the only Quaternary glacial event in the Ukraine and that it had occupied both the north-eastern and the north-Western parts of the country, including the area around L'viv (digital map, boundary 8). However, this point of view was not generally accepted. Today, the traces of two glaciations, the Oka and the Dnieper, can be clearly distinguished. Also the presence of traces of the Narev (Lower Pleistocene) and Moscow (Middle Pleistocene) glaciations in the Ukraine is suggested by some authors. The problems associated with the glacial limits are considered separately for each glaciation and each region: the Middle Dnieper area, West (Volyn') Polissia, the L'viv region and the Carpathian Mountains.

Lower Pleistocene

Narev Glaciation

Isolated traces of the Narev Glaciation are only associated with deep incisions infilled by Quaternary deposits in West Polissia. Supposedly Narev age glacial deposits occur at the base of the incisions. They are represented by marl-chalk-bearing sediments with an admixture of crystalline material in some places and overlain by unstratified glaciofluvial sands of the Narev-Oka Horizons. On the basis of this evidence, Gruzman (1976) suggests that the Narev ice sheet occupied the northern part of Volyn' Polissia. However, these marl-chalk-bearing sediments can also be considered as the basal facies of Oka glaciofluvial deposits or normal alluvium. According to Gozhik (1995, p.213) "the lack of detailed evidence prevents the determination of an accurate age and correlation of these glaciofluvial deposits. It is quite possible that they may represent the early stage of the south Polish (i.e. Oka) glaciation". Thus, the presence of an ancient 'Narev Glaciation' in the Ukraine cannot be taken for granted.

Middle Pleistocene

Oka glaciation

L'viv lobe

The occurrence of glacial formations in the L'viv region was first noticed during the compilation of the Atlas of Galiciya at the end of 19th and beginning of the 20th century (Dorofeev, 1972). The boundary of 'maximum glaciation' was drawn on the Carpathian foothills on the basis of the distribution of so-called the 'mixed pebble gravel' which includes components of crystalline rocks. Later Przepiorski (1938) reproduced its position on his map (boundary 1) and it has since been reproduced on other maps with minor modifications (Krasnov, 1971; Buraczynski, 1997) (boundaries 2, 19).

Another group of researchers held the view that the 'mixed pebble gravel' does not reflect the position of the ice sheet front. In their version the GB took up a more northward position. One of them, Lomnicki (1897, 1898) drew the maximum glacial limit between the towns Rudki,

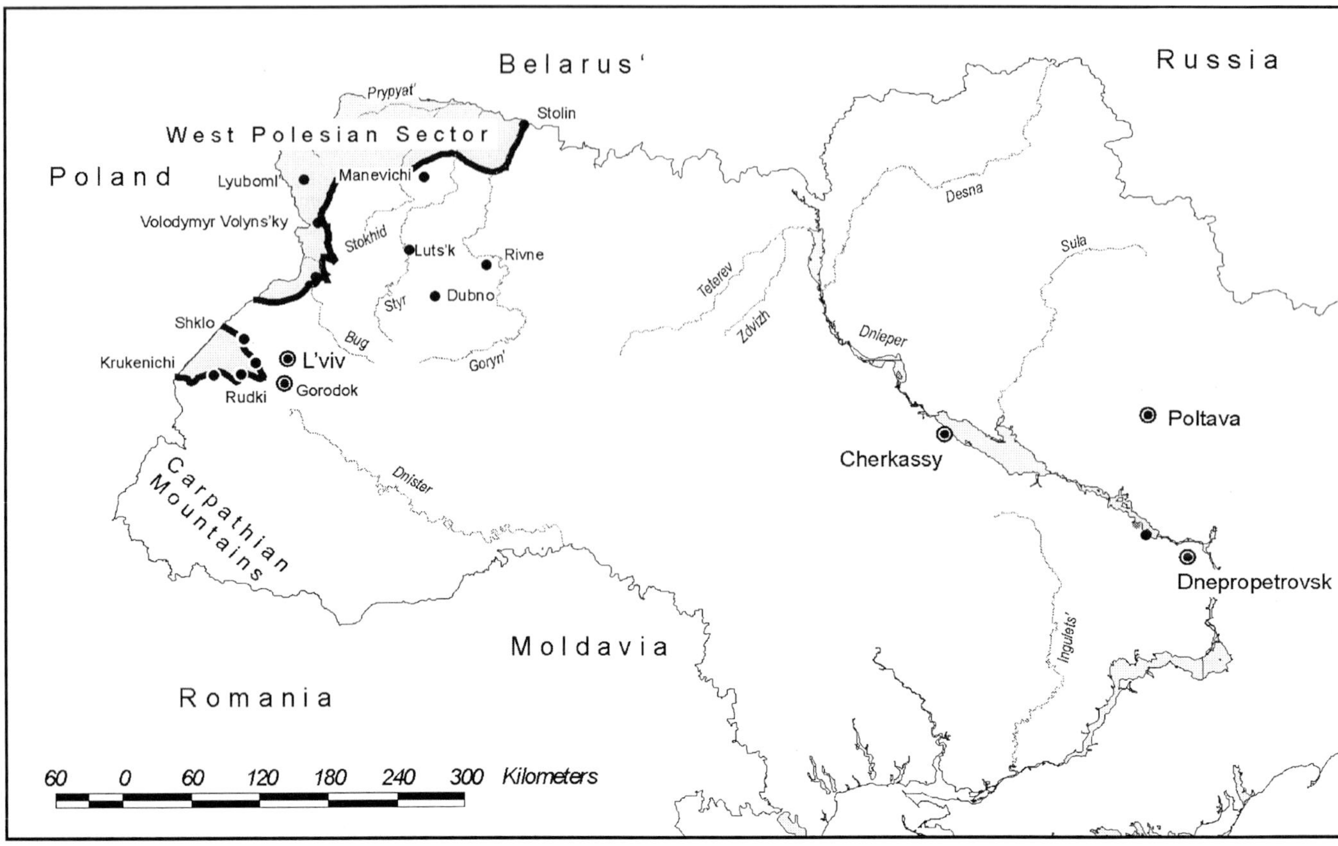

Fig. 1. The extent of the Oka Glaciation in the Ukraine.

Gorodock and Shklo (Fig. 1). The results of his investigation have been confirmed and supplemented by Gerenchuk *et al.* (1966), Dorofeev (1972) and Demedyuk & Demedyuk (1988, 1991).

The author's version of the GB (boundary 4) is similar to the latter. All well-known finds of erratic boulders, till remnants and glacial dislocations are within the area enclosed by this GB. To the west of L'viv City, the GB takes the form of a tongue that follows the scalloped contour of the L'viv lobe (Fig. 1). This lobe was restricted by the Roztochia upland in the northeast and by the Carpathian Mountains in the southwest. Likewise, the front of the tongue is controlled by the position of local heights on the San and Dniester river interfluve. Meltwater runoff valleys, recognised by Demedyuk & Demedyuk (1988), cross this interfluve. The occurence of gravel and pebbles of crystalline rocks in Quaternary alluvial deposits of the Upper Dniester depression are well known (Dorofeev, 1972). Both types of evidence indicate proglacial runoff towards the proto-Dniester valley.

The special problem of the GB in the L'viv region is the Roztochia rampart-like upland (300-400 m a.s.l.). On maps of Przepiorski (1938), Krasnov (1971) and Buraczynski (1997) (boundaries 1, 2, 19 respectively), the GB crosses the upland implying that its greater part was glaciated. However, Dorofeev (1972) pointed out that there are no erratics within the Roztochia upland (Fig. 1). On the contrary Gruzman (1976) indicates the presence of erratics at some localities in the Roztochia, but neither gives a detailed description nor their precise position. Buraczynski (1997) also showed till of the San Glaciation on one of his cross sections at two places 300 m a.s.l. (see map). It should be noted however, that no borehole posi-tions are indicated in his map in both these cases, and no source of information is given in the text. Moreover, nume-rous cross sections of Roztochia from Buraczynski's mono-graph show that San Stage till only occurs on the flanks of the Roztochia ridge in the valleys up to a level 250-270 m a.s.l. Only in the southwestern part of Roztochia (approxi-mately 150 km from L'viv City) does the till occur higher than 300 m a.s.l.

Today the majority of researchers assume that the L'viv region glacial formations should be referred to the Oka (Berezina, San) Horizon of Middle Pleistocene. Apart from long-distance geological correlation with adjacent regions (Poland, Volyn' Polissia), this assumption is based on the comprehensive study of the Kruckenichi key section (see map). At this site peat and glaciolacustrine silts are found overlying till. The results of spore and pollen analysis (Gerenchuk *et al.*, 1966; Artyshenko *et al.*, 1967; Kondratene, 1976; Grichuk & Gurtovaya, 1976, Velichkevich, 1982), indicate that the organic sediments date from the Likhvin (Holsteinian) Stage.

According to Gaigalas (1974), the assemblage of far-transported indicator boulders and pebbles from the till at this locality is similar to that from the equivalent Dainava and Upper Berezina horizons (Middle Pleistocene) of Lithuania and Belarus. The approximate age of till samples dated using the TL method at the Lyublin University laboratory (Butrym *et al.*, 1988) was found to be 510-520,000 years B.P.

It must be admitted that the problem of the GB in the L'viv region is not completely solved. The complete lack of morphological expression of the Oka glacial formation in the modern topography and the small quantity of available exposures pose major difficulties. Moreover, drilling materials from the area have also not been analyzed to the appropriate degree. A key problem awaiting further investigations is the question of the glaciation of the Roztochia upland.

The West Polesian sector

The West Polesian sector (Fig. 2) was traditionally considered by Soviet researchers as the area of a single, Dnieper glaciation (Zamory, 1956; Bondarchuck, 1961a, b; Krasnov, 1971). However, Polish scientists Ruhle (1937) and Krygowski (1947) pointed out the existence of two till horizons within this area. In 1966, Bogutskyi reported the occurrence of glacial formations (till remnants, erratic boulders and glacial dislocations) on the right bank of Bug River and close to the town of Sokal' (to the north of Roztochia within Male Polissia (digital map and Fig. 2). His glacial limit for this area (boundary 5) is still valid today. However, Bogutskyi's (1966) publication remained un-noticed for a long time.

Additional information was obtained in the course of a large-scale geological survey carried out in the 1970s. A network of deep incisions infilled by several glacial and interglacial horizons has been discovered beneath the Goryn', Styr, Stohkid and Upper Prypyat' river interfluves (Fig. 2). In most sections of the incisions, Lihkvin (Alexandriya, Mazovian) deposits were clearly distinguished by lithological and paleontological evidence (Dorofeev & Zalessky, 1978; Khursevich & Loginova, 1980). Zalessky & Mel'nichuk (1975) have also described the most ancient, the Nalibocksky (Beylovyezhian, Cromerian) interglacial horizon and reported its spore and pollen characteristics. Consequently, glacial deposits found underlying the Likhvin and overlying the Beylovyezhian deposits were referred to the Oka (Berezina) Horizon. Dorofeev & Zalessky (1978) recognised two pre-Likhvin till units: the Lower and Upper Berezina. The same distinct stratigraphical position is occupied by the analogues of Oka glacial deposits within the adjacent regions of Belorus and Poland, where they occur right up to the Ukrainian border. So the proposed Oka Stage glacial limit in this region follows the southern and south-eastern ends of the incisions in question (boundary 11). Such a plot must be regarded as a minimum version, the real Oka GB probably lying further to the south. Therefore only the approximate proglacial drainage pattern along the margin of the Oka ice sheet in this region can be given.

According to Gruzman (1978) the Dnieper ice stayed in the north of Volyn' Polissia, and glacial deposits of the Oka Stage not only form the marginal Lyuboml' - Stolin ridge but also occur right up to the line: Volodymyr Volyns'ki - Luts'k - Rivne. Besides that, the ice sheet invaded the territory of Male Polissia in a narrow lobe that stretched up to 175 km from the Polish border in a west-easterly direction. Gruzman also considers that the glacial deposits in the vast areas to the south of the Lyuboml'-Stolin ridge and within Male Polissia are represented predominantly by marly sands. Two isolated rafts of chalk blocks to the south of Dubno town are also shown in Gruzman's scheme (1978). These ideas are reflected in the unusual Oka glacial limit (boundary 3) on the Geomorphological map of the Ukrainian SSR and the Moldavian SSR (Sokolovsky, 1979).

Thus the existence of Lower Pleistocene glacial deposits in the West Polesian sector is apparent, but the identification of the corresponding glacial limit remains highly complicated. The problems are associated with deficient knowledge of the Quaternary of the middle part of the Goryn' and Styr river basins, the upper part of the Bug River basin and of the Male Polissia area as a whole.

Middle Dnieper area

Various researchers have indicated the possibility of the existence of 'Mindel' glacial deposits in the Dnieper valley and in adjacent areas (see review by Gozhik, *et al.*, 1985). The discussion in principle developed after detailed drilling of the Pereyaslav-Cherkass'ka depression (the so-called Shevchenko 'exaration' valley) in the 1960s. As a result of this drilling programme a thick glacial-alluvial (meltwater sediment) formation was distinguished, the Shevchenko Formation (Goretsky, 1970). The deposits of the Shevchenko Formation are underlain by basal till in many places. Several boreholes also penetrated an additional horizon of till in the middle part of the sequence (map for location). On the basis of pollen and plant macrofossil analyses, a group of researchers (Romodanova & Makhnach, 1968; Artyushenko *et al.*, 1969) suggested a Likhvin age for the deposits between the till horizons. Correspondingly, they believed that the upper till unit was of Dnieper age, while the lower one was older. Later Romodanova *et al.* (1969) concluded that both till units and the Shevchenko Formation as a whole were of Early Pleistocene age, and that during this interval an ice sheet penetrated the Middle Dnieper area in the form of a narrow tongue that advanced southeastwards to the latitude of Cherkassy. These ideas, as well as the interpretation of palynology and other paleontological evidence have been criticized by Goretsky (1970), Gubonina (1980) and Vozgrin (1985). In particular Goretsky convincingly showed in his detailed cross-sections how the till lying at the base of the Shevchenko Formation rises gradually on the depression flanks and joins with the Dnieper horizon till of the watershed areas. Lithological features and the local

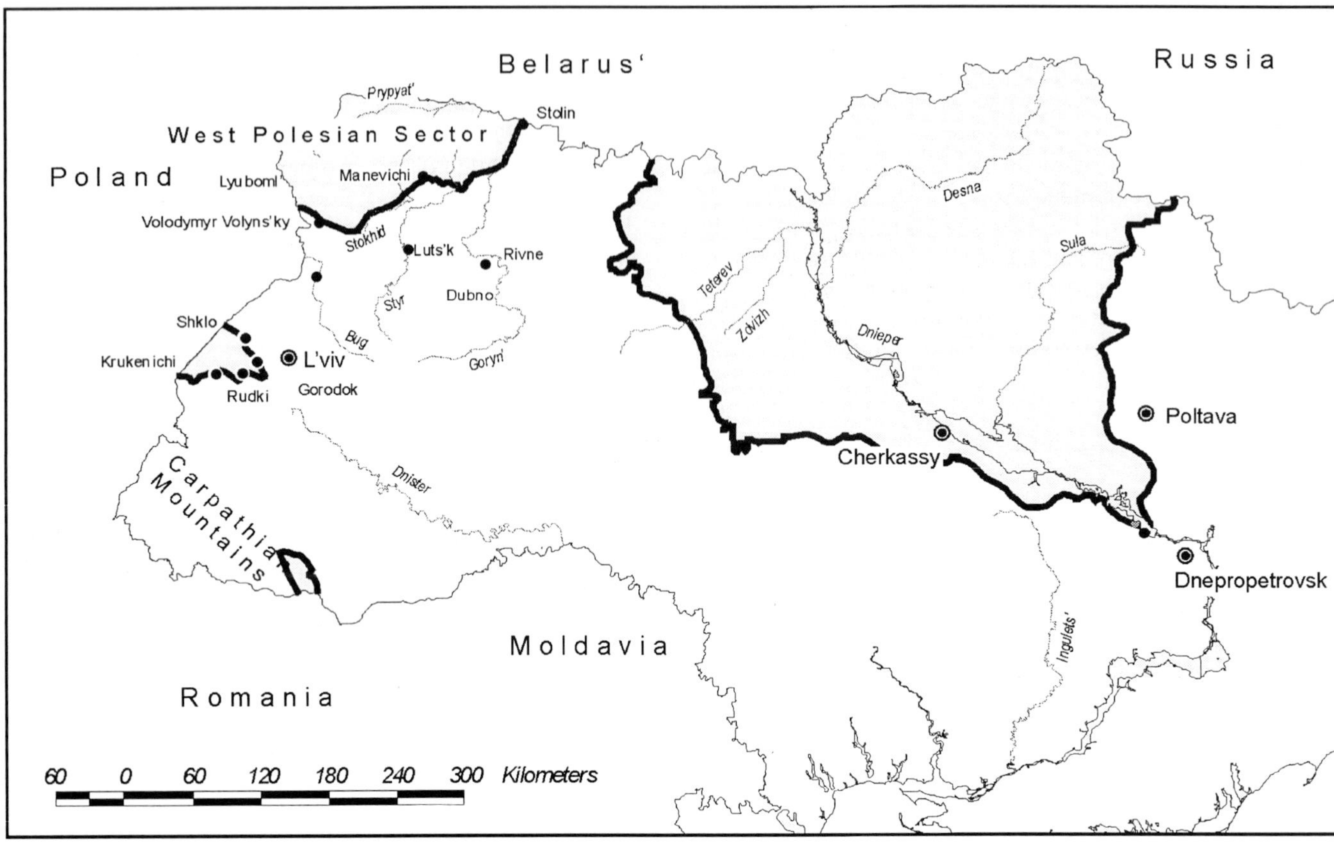

Fig. 2. The Dnieper Glaciation in the Ukraine and the glaciation in the Carpathian Mountains.

spread of the upper till unit indicate its ablation or waterlain origin. Investigations by Matoshko & Chugunny (1993) have confirmed the existence of only one Dnieper glacial horizon in the Middle Dnieper area.

Dnieper Glaciation

The Dnieper Stage glacial boundaries have been interpreted controversary over a long period of time. P.A.Tutkovsky (from Dmitriev, 1937) drew one of the first and best maps of their position. He already distinguished a Polessian driftless area between a 'Dnieper ice tongue' (Dnieper ice stream in modern terminology; Matoshko & Chugunny, 1995) that occurs in the Middle Dnieper area and a 'West Polessian lobe' (Fig. 2). Not all workers agreed with him, some denied the existence of the driftless area. In particular, this is reflected in the schematic map (boundary 15) of Bondarchuk (1961b). However, in a large-scale geological survey the presence of any glacial formations on the right bank of Prypyat' River (between the Ovruch Heights and Goryn' River) could not be established.

Dnieper Ice Stream

Apart from the plot by Bondarchuk (1961b), all other versions of the GB (boundaries: 7, 8, 12, 13, 16, 17) are similar. There are several small areas that have been changed where new data has been obtained. The present version (boundary 17) primarily follows the basal till distribution (Matoshko & Chugunny, 1993; Matoshko, 1995). It is based on investigations of both boreholes and exposures. In most cases, no traces of glacial activity can be found beyond the boundary of basal till distribution. Only in some localities ice-marginal landforms, such as ridges and hills of marginal ablation moraines, esker fans and valley sandar, mark the GB position.

Proglacial drainage was different for the eastern and western margins of the Dnieper ice stream. It is assumed that runoff along the eastern margin towards the south was relatively free and partially coincided with the proto-Psiol valley (Fig. 3). Only in its northern part two local spillway valleys are found, infilled by meltwater deposits. These gaps connect the drainage basins of the present Seim and Psiol Rivers (see map). Runoff from the Western margin was more complicated because the GB was situated in front of the highest part of the Prydniprovs'ka Upland. At the north-Western margin runoff was probably oriented to the north to the proglacial lake, which occupied the Polessian Driftless Area. Further to the south several spillway valleys have been determined by geological evidence whilst the existence of others has been postulated. They connect the present basins of Ros' and Pivdenyi Bug Rivers, Ros' and Gorny Tickitch Rivers, Ol'shanka and Shpolka Rivers, Tyasmin and Ingulets Rivers. Some form a part of the

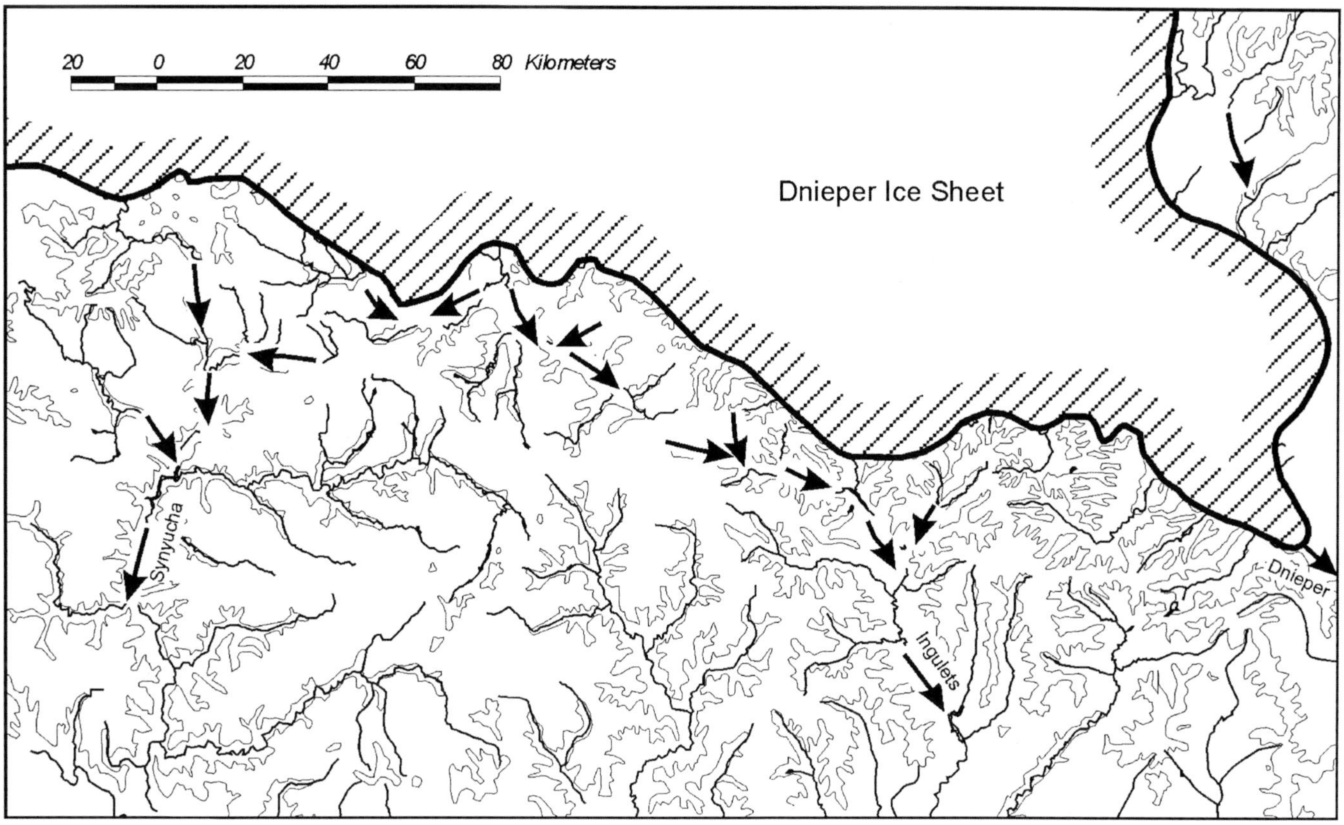

Fig. 3. Drainage of the Dnieper Ice Sheet.

longest valley sandar system of the Ukraine. This system extends through the present valleys of Gnylyi Tickitch - Synyucha, Ol'shanka Shpolka, Tyasmin - Gnylyi Tashlyk - Shpolka - Gnylyi Tickitch, Tyasmin - Ingulets (Matoshko & Chugunny, 1993). The main drainage lines on the western flank were the proto-Pivdenyi Bug, proto-Synyucha and proto-Ingulets valleys. At the same time the proto-Dnieper drained the western flank of the ice stream and probably carried the subglacial runoff of the ice stream's axial zone. All these water-ways took the meltwater to the Black Sea basin.

The outer boundaries of oscillations and recessional phases of the retreating Dnieper Glaciation according to Chugunny & Matoshko (1995) are also shown based on the position of marginal formations (push moraines, ablation moraines, ice-marginal landforms, esker fans, proximal slopes of sandar and valley sandar in some cases). An interesting problem concerning the Dnieper GB is its outermost southern margin. According to the present version, the GB is located near the town of Verkhnedneprovsk, where glacial dislocations occur on the right bank of the Dnieper River. However, some suggest that the Dnieper Ice Sheet penetrated southwards to Dnepropetrovsk City and possibly even further (see map). Single finds of supposedly erratic blocks, large erratic boulders and fragments of flow till (Karlov & Boyko, 1938; Karlov, 1951, 1955) provide evidence in support of this assertion. Unfortunately, nearly all these localities are now flooded and there is no possibility of carrying out additional research. It should be noted that the border regions with Russia on the right bank of the Desna River and also the Dnieper - Ingulets' river interfluve area are not well studied. Thus future changes to the Dnieper GB there can be expected.

The Dnieper till dates from the second part of the Middle Pleistocene. The Dnieper Glaciation was the only one that extended into the Middle Dnieper area and the Dnieper till unit is the main indicator horizon within the Quaternary sequence in this area. Its exact age is not known. The numerous thermoluminescence dates on the Dnieper till, as well as supposedly Oka and Moscow tills, carried out in the laboratory of the Institute of Geological Sciences (Kyiv) by V.N. Shelkoplyas are not accepted as accurate by the majority of researchers.

West Polessian Lobe

As well as within the Middle Dnieper area, the Dnieper glacial limit here (boundaries 7, 8, 12, 13) gradually approached its modern position (boundaries 11, 14). The latter is based on the results of Tutkowsky (1901), Dorofeev (1972), Palienko (1982), Matoshko & Chugunny (1993, 1995). However, in contrast to the previous region, the GB of the West Polessian lobe (Fig. 2) is drawn along morphologically defined end moraines (marginal thrust and ablation moraines), glaciofluvial fans, proximal margins of ice-dammed lakes, glacial dislocations and rare fragments of basal till. The latter is characterized by a high sand content, and very often in boreholes the till cannot be distinguished from meltwater deposits. This is one of the main problems concerning the establishment of the GB in this area.

To the west of the town of Manevichi, the Dnieper GB is differently defined (boundary 6, 7, 10). In the latest version (boundary 6) the GB passes over esker fans on the left bank of the Stokhid River and then over possible end moraines to the north of Volodymyr Volynskiy town (Matoshko & Chugunny, 1993, 1995). There are also two versions of the position of the Dnieper Glaciation retreat phases and oscillations: by Palienko (1982) and by Matoshko & Chugunny (1993, 1995). In the latter case they are reflected predominantly by marginal thrust features and push moraines.

Information on the proglacial drainage of this area is very contradictory because of the lack of reliable geological data. Only in the watershed area, between the basins of the Turiya and Styr Rivers, the Western Bug and the Styr Rivers were former spillways identified, now infilled by meltwater deposits (Matoshko & Chugunny, 1993). The analysis of the sub-Quaternary surface has provided evidence that the main proglacial runoff was orientated along the ice sheet margin to the north-east. Here, in the Polessian driftless area was a vast depression, which was blocked to the south by the slopes of the Pridniprovs'ka upland and to the north by the ice sheet margin. This closed area was probably occupied by a large proglacial lake (see digital map). However, it should be noted that in only some points in the possibly glacial lake do lacustrine deposits occur, and that no lake terraces have been found.

The age of the Dnieper glacial formations is defined by their stratigraphical position overlying the deposits of Likhvin age (see above).

Moscow Glaciation

At first Bondarchuk (1961a) and Marinich (1963) assumed the existence of two Middle Pleistocene till units in the Polissia area. Bondarchuk suggested an advance of the Moscow or Prypyat' Glaciaton into the Middle Dnieper area as far as the town Zolotonosha. Later, Khristoforova & Shelkoplyas (1976) and Vozgrin (1975, 1985) developed this idea further and found evidence for an expansion of the Moscow (Tyasmin, Sozh, Wartanian) Ice Sheet into the northern part of the Middle Dnieper area. In particular Vozgrin indicated the position of the Tyasmin GB on the right bank part of the Middle Dnieper area (boundary 18). Their main argument is the presence of two till horizons separated by deposits with organic inclusions at individual localities. However, closer inspection has shown that, in many cases, thrust moraines and phasial tills have been identified erroneously as stratigraphically different till units, and methodological mistakes were made in the identification of till units (Gozhik & Chugunny, 1981; Gozhik *et al.*, 1985). According to Gozhik (1995, p. 215), "the available evidence does not support unequivocally that the Ukraine was overridden during the subsequent Moscow Glaciation. ... Consequently, it seems that the Ukraine was glaciated twice during the Quaternary. In contrast to the Oka Glaciation, which was limited to the northwestern part of the Ukraine, the Dnieper Glaciation covered a considerable part of the middle course of the Dnieper River."

Mountain glaciation

Geomorphological observations from the beginning of the 20th century (Romer, 1906; Sviderski, 1938; Ivanov, 1950; Tsys', 1961; Tykhanych, 1967; Voropai & Kunytsya, 1969) indicated the existence of ancient cirque and cirque-valley glaciers within the Charnogora, Svydovets', Chyvchyny and Gorgany ridges and the Rakhivs'kyi massif (see map) in the Ukrainian Carpathian Mountains. Apart from this evidence, the author suggests the existence of slope, nivation glaciers and rock glaciers. Traces of glacier activity such as cirques, fragments of glacial troughs, valley basins and small end moraine ridges occur at altitudes in the range of 1350-1600 m. a.s.l.

The age of this glaciation is being discussed at present. Sviderski (1938) has suggested two Carpathian glaciations equivalent to the Mindelian and Rissian. In contrast, Tsys' (1961) thought a more recent glaciation had occurred. New evi-dence concerning the age of the glaciation has been obtained by Tretyak & Kuleshko (1982) from a prospection pit near the Breskul summit (Charnogora ridge). This pit, in the base of a short glacial valley, bordered with a clearly defined series of end moraine ridges, comprises downwards: two horizons of peat divided by a layer of rock glacier deposits and morainic material. A Late Holocene age of the peat has been established by ^{14}C dating. The age of the moraine is thus considered to be Late Pleistocene - Holocene. Undoubtedly, the problem of Carpathian Mountains glaciation during the Quaternary requires serious further study including modern dating methods.

References

Artyushenko, O.T., Romodanova, A.P. & Khristoforova, T.F. (1968). Istoriya rozvytku roslynnosti i stratigrafiya alluvialnykh vidkladiv dolyny Dnipra v Kanivskomu rayoni (History of vegatation evolution and stratigraphy of alluvial deposits in Kaniv region). *Ukrainskyii botanichnyi zhurnal*, **4**, 1115-1119.

Artyushenko, O.T., Khristoforova, T.F. & Kareva, O.V. (1967). Mindelske mizhlyodovikove torfovysche v okolicyah s.Krukenichi L'vivskoi oblasti (Mindel interglacial peat from environs of Krukenichi village in L'viv region). *Ukrainskyi botanchnyi zhurnal*, **4, t.24,** 76-84.

Bogutskyi, A.B. (1967). Morennye otlozheniya na teritorii Volynskoy vozvyshennosti (Till deposits within the Volyn' upland territory). *Doklady i soobscheniya L'vovskogo otdela Geographicheskogo obschestva UkrSSR za 1965* g. L'vov, Izdatelstvo L'vovskogo Universiteta.

Bondarchuk V.G. (1959). *Geologiya Ukrainy* (Geology of Ukraine): 832 pp. Kyiv: Vydavnytstvo AN UkrSSR.

Bondarchuk, V.G. (1961a). Structure of the Quaternary (Anthropogene) deposits. *Quaternary Period* **13, 14, 15**, 7-41.

Bondarchuk, V.G. (1961b). Glacial dislocations of the Middle Dnieper area. *Quaternary Period* **13, 14, 15**, 67-86.

Buraczynski, J. (1997). *Roztocze (budova - rzezba - krajobraz)*. 189 pp. Lublin: Zaklad Geografii Regionalnej, Universytet Marii Curie-Sklodowskiej.

Butrym, J., H. Maruszczak & Wojtanowicz, J. (1988). Chronologia termoluminescencyjna osadow ladolodu Sanian (= Elsterian II) w dorzeczu sanu i gornego Dniestru (Thermoluminescence chronology of the Sanian (=Elsterian II) inland-ice deposits in the San and Upper Dniester River basins). *Annales Societatis Geologorum Poloniae*, **58**: 191-205.

Chugunny, Yu.G. & Matoshko, A.V. (1995). Dnieper Glaciation - till deposits of ice marginal zones. *In:* Ehlers, J., Kozarski, S. & Gibbard, P. (eds.), *Glacial deposits in North-East Europe*, 249-256. Rotterdam, Brookfield, A.A.Balkema.

Demedyuk, N.S. & Demedyuk, Yu.N. (1988). *Dnestrovsky lednikovy complex Predkarpatiya* (Dniester glacial complex of Fore-part area of Carpathians): 55 pp. Kiev: Institut Geologicheskih Nauk AN UkrSSR, preprint 88-27.

Demedyuk, N.S. & Demedyuk, Yu.N. (1991). Paleogeographiya antropogena Dnestr-Sanskogo mezhdurechiya (paleogeography of the Dniester-San interfluve area in Anthropogene). *In:* Shnyukov, E.F. (ed.), *Litologiya osadochnogo chehla UkrSSR (paleogeographicheskie aspecty)*, 133-136. Kiev, "Naukova dumka".

Dmitriev, N.I. (1937). Granitsa morennyh otlozheny Dneprovskogo oledeneniya (On the boundary line of the Dnieper glaciation morainic debris). *Uchenye zapiski Kharkovskogo Universiteta*, **8-9**, 167-191.

Dorofeev, L.M. (1972). *Rasprostranenie i stratigraphicheskoe polozhenie lednikovyh otlozheny v zapadnoy chasti Ukrainy* (Spread and stratigraphic position of glacial deposits in the western part of Ukraine). Kraevye obrazovaniya materikovyh oledeneny, 246-253. Moskva, "Nauka".

Dorofeev, L.M. & Zalessky, I.I.. (1978). Stratigraphiya nizhne- i srednepleistocenovyh otlozheny mezhdurechiya Pripyat'-Turiya (Stratigraphy of Lower- and Middle Pleistocene deposits of the Prypyat'- Turiya interfluve). *Tektonika i stratigraflya*, **14**, 111-121.

Gaigalas, A.I. (1974). Petrographicheski sostav i orientirovka galek v lednikovyh otlozheniyah okrestnostey s.Kruckenichi (Petrographic content and pebbles orientation in glacial deposits from environs of Kruckenichi village). *In:* V.A.Kuznetsov (ed.), *Geologiya i geohimiya antropogena Belorussii*, 147-152. Minsk: "Nauka i tekhnika".

Gerenchuk, K.I., Demedyuk, N.S. & Zdenyuk, M.V. (1966). *Do chetvertynnoi paleogeographii Sansko-Dnistrovskogo mezhyrichia* (On Quaternary paleogeography of San-Dniester interfluve area). Paleogeographichni umovy teritorii Ukrainy v plioceni ta antropogeni. Kyiv: "Naukova dumka".

Goretsky, G.I. (1970). Alluvialnaya letopis' Velikogo Pra-Dnepra (Alluvial record of Great Pra-Dnieper): 492 pp. Moscow: "Nauka".

Gozhik, P. (1995). Glacial history of the Ukraine. *In:* Ehlers, J., Kozarski, S. & Gibbard, P. (eds.), *Glacial deposits in North-East Europe*, 213-215. Rotterdam, Brookfield, A.A.Balkema.

Gozhik, P.F. & Chugunny, Yu.G. (1981). O srednepleistocenovom oledenenii na teritorii Ukrainy (On the Middle Pleistocene Glaciation in the Ukraine territory). *In:* Velichko, A.A. (ed.), *Pleistocene glaciations at the East European Plain*, 86-91. Moscow, "Nauka".

Gozhik, P.F., Matoshko, A.V. & Chugunny, Yu.G. (1985). Vozrast lednikovyh obrazovany Srednego Pridneproviya (The age of Middle Dnieper area glacial formations): 35 pp. Kiev. Institut Geologicheskih Nauk, preprint 85-21.

Grichuk, V.P. & Gurtovaya, E.E. (1976). Mezhlednikovaya flora ozerno-bolotnyh otlozheny u s.Krukenichi i ee stretigraphicheskoe polozhenie (Interglacial flora of lacustrine-bog deposits near Krukenichi village and its stratigrafic position). *In:* Bondarchuk, V.G. (ed.), *5-e Vsesoyuznoe soveschanie po izucheniyu kraevyh obrazoveny maerikovyh oledeneny Tezisy dokladov*, 96-97. Kiev, Institut Geologicheskih Nauk.

Gruzman, G.G. (1976). Lednikovaya formatsiya Volyni (Glacial Formation of Volyn' region). *In:* Bondarchuk, V.G. (ed.), *Kraevye obrazoveniya materikovyh oledeneny*, 73-8 1. Kiev, Naukova dumka.

Gubonina, Z.P. (1980). Palinologicheskaya harakteristike podmorennyh otlozheny v besseine Dnepra (po dannym razreza "Gun'ki") (Pollen and spores characteristic of submoraine deposits in Dnieper basin (by data of "Gun'ki" section). *In:* Velichko, A.A. (ed.), *Vozrast i rasprostranenie maksimal'nogo oledeneniya Vostochnoy Evropy*. 153-168. Moskva, "Nauka".

Ivanov, B.N. (1950). Sledy oledeneniya Ukrainskih Carpat (The traces of Ukrainian Carpathians glaciation). *Uchenye zapiski Chernovitskogo Universiteta*, **2**, **8**, series - geol.- geomorphol. 14-21.

Kalinichenko, N. (1840). O kostyah drevnih zhivotnyh vnov' otkrytyh v Malorossiii (About bones of ancient animals - new discovery in Malorossiye). *Gorny zhurnal*.

Karlov, N.N. (1951). Valun finlyandskogo granite v okrestnostyah g. Pavlograda (The boulder of Finnish granite in environs of Pavlograd town). *Byulleten' komissii po izucheniyu chetvertichnogo perioda*, **16**, 73-74.

Karlov, N.N. (1955). Ob ostatkah osnovnoy moreny v g.Dnepropetrobske (About the remnants of basal till in

Dnepropetrovsk City). *Bulleten' komisii po izucheniyu chetvertichnogo perioda*, **20**, 82-86.

Karlov, M.M. & Boiko, K.A. (1938). Slidy liodoviokovoyi moreny v m.Dneprodzerzhinc'ku Dniprovs'koi oblasti (The traces of glacial moraine in Dneprodzerzhins'k City of Dnieper region). *Gelogichnyi zhurnal AN UkrRSR*, **3**, **5**, 201-208.

Khristoforova, T.F. & Shelkoplyas, V.N. (1976). O stratigraphicheskom raschlenenii srednepleistotsenovyh lednikovyh otlozheniy severnoy chasti pravoberezhiya UkrSSR (On stratigraphical subdivision of the glacial deposits of the northern part of the UkrSSR right bank area). *Chetvertychny period*, **16**, 107-113.

Khursevich, G.K. & Loginova, L.P. (1980). Vozrast i paleogeographicheskie usloviya nakopleniya drevne-ozernyh diatomovyh porod Severo-Zapada Volyni (The age and paleogeographic conditions of accumulation of ancient lacustrine diatomaceous rocks in North-West Volyn'). *In:* Bondarchuk, V.G. (ed.), *Paleolandshafty, fauna i flora lednikovyh i perygliacial'nykh zon pleistocena*, 32-34. Kiev, Institut Geologicheskih Nauk, preprint 80-15.

Kondratiėne, O.P. (1976). Osobennosti obrazovaniya mezh-lednikovyh otlozheny v s.Krukenichi i osnovnye zakono-mernosti razvitiya rastitel'nosti zapadnoy chasti Russkoy Ravniny v likhvinskom mezhlednikovie (Features of interglacial deposits formation in Krukenichi village and main regularities of vegetation evolution in Western part of Russian Plain in Likhvin Interglacial). *In:* Bondarchuk, V.G. (ed.), *5-e Vsesoyuznoe soveschanie po izucheniyu kraevyh obrazovany materikovyh oledeneny Tezisy dokladov*, 97-98. Kiev, Institut Geologicheskih Nauk.

Krasnov, I.I. (ed.) (1971). *Karta chetvertichnyh otlozheny Evropeiskoy chasti SSSR, masshtab 1: 1 500 000* (Map of Quaternary deposits of European part of USSR, scale 1: 1,500,000). GUGK USSR

Krygowski, B. (1947). Zarys geologiczno-morfologiczny poludniowego Polesie. Poznanskie tow. przyjac. nauk. Prace komisji matern.-pryrodn. zesz.1, t.5, seria A.

Lomnicki, A.M. (1897). *Atlas geologiczny Galicji. Tekst do zeszytu 10, SZ. 1. PAU*, 208 pp. Krakow.

Lomnicki, A.M. (1898). *Atlas geologiczny Galicji. Tekst do zeszytu 10, SZ. 2. PAU*, 167 pp. Krakow.

Marinich, A.M. (1963). *Geomorphologiya yuzhnogo Polessiya* (Geomorphology of south Polissia). 252 pp. Kiev, Izdatel'stvo Kievskogo Universiteta.

Matoshko, A.V. (1995). Dnieper Glaciation basal till deposits. *In:* Ehlers, J., Kozarski, S. & Gibbard, P. (eds.), *Glacial deposits in North-East Europe*, 231-239. Rotterdam, Brookfield, A.A. Balkema.

Matoshko, A.V. & Chugunny, YuG. (1993). Dneprovskoe oledenenie teritorii Ukrainy (geologichesky aspect) (Dnieper Glaciation of Ukraine (geological aspect): 192 pp. Kiev: "Naukova dumka".

Matoshko, A.V. & Chugunny, Yu.G. (1995). Geological activity and dynamic evolution of the Dnieper glaciation. *In:* Ehlers, J., Kozarski, S. & Gibbard, P.L. (eds), *Glacial deposits in North-East Europe*, 225-230. Rotterdam, Brookfield, A.A. Balkema,.

Palienko, V.P. (1982). Osobennosti glyatsiorel'efa kraevoy zony dneprovskogo lednika v predelah Volynskogo Polessya (Features of glacial relief of the Dnieper glacier marginal zone within Volyn' Polessie area). *In:* Bon-darchuk, V.G. (ed.), *Materialy po izucheniyu chetvertich-nogo perioda na teritorii Ukrainy*, 203-211. Kiev, Naukova dumka.

Przepiorski, W. (1938). Dyluwium na plaskowyzu Chyrowsko-Lwowskim. Rocznik 63, sesz. 11: 183-245. "Kosmos".

Raskatov, G.I. (1954). K voprosu o drevnechetvertichnyh oledeneniyah zapadnyh oblastey Ukrainy (On the problem about ancient Quaternary glaciations of the Ukrainian west regions). *Trudy Voronezhskogo Universiteta t*, **31**, zbornik rabot geolog. fakul'teta, 154-173.

Romer, E. (1906). Epoka lodowa na Swidowcu. *Rozpravy Ak. Um., Krakow.* **A 46** (Seria **III**, **6**), 11-80.

Romodanova, A.P. & Makhnach, N.A. (1968). Do ob-gruntuvannya viku antropogenovykh vidkladiv na perepoglyblenny chastyni Serednyogo Dnipra (On the basis of Anthropogene deposits age in overdeepening part of Middle Dnieper). *Doklady Akademii Nauk UkrSSR*, **5**, 135-142.

Romodanova, A.P., Vozgrin, B.D., Golubitskaya, A.A., Lipkina-Kuchinskaya, N.E. & Khristoforova, T.F. (1969). Old Pleistocene deposits of the Middle Dnieper Valley. *In:* Bondarchuk, V.G. (ed.), *Materials on the Quaternary period of the Ukraine*, 70-87 Kiev, "Naukova dumka" .

Ruhle, E. (1937). Utwory ludowcowe zachodniej csesci Polesia Wolynskiego. *Rocznik,* **A 62**, sesz. **1-2**, 81-109 "Kosmos".

Sokolovsky, I.L. (ed.) (1979). *Geomorphologicheskaya karta Ukrainskoy SSR i Moldavskoy SSR, masshtab 1: 1 000 000* (Geomorphologic map of Ukrainian SSR and Moldavian SSR, scale 1: l,000,000). GUGK USSR.

Staszic, S. (1806). *Carta geologica totius Poloniae, Moldaviae, Transilvaniae parttis Hungariae et Vala-chiae.* Warszawa.

Sviderski, B. (1938). *Geomorfologia* Czarnogory : Warsza-wa.

Tretyak, P.R. & Kuleshko, M.P. (1982). Degradatsiya poslednego oledeneniya v Ukrainskih Carpatakh (De-gradation of the last glaciation in Ukrainian Carpathians). *Doklady Akademii Nauk Ukr SSR*, **B (8)**, 26-31.

Tsys', P.N. (1961). Some Problems of the Quaternary morphogenesis of the Soviet Carpathians *Quaternary Period*, **13, 14, 15**, 231-239.

Tutkovsky, P. (1901). *Konechnye moreny, valunnye polosy i ozy b Yuzhnom Polessie. S kartoy* (End moraines, boulder belts and eskers in south Polissia. With map): 108 pp. Separate reprint. Kiev, Tipografiya Impera-torskogo Universiteta SV. Vladimira.

Tykhanich, V.V. (1967). K voprosu geomorfologii Svi-dovtsa (On the Problem of Svidovets geomorphology). *Doklady i soobscheniya L'vovskogo otdela Geographi-*

cheskogo obschestva UkrSSR za 1965g, 97-102. Izdatel'stvo L'vovskogo Universiteta.

Vevichkevich, F.Yu. (1982). *Pleistotsenovye flory lednikovyh oblastei Vostochno-Evropeiskoi ravniny* (The floras of the glaciated areas of the East-European plain). 240 pp. Minsk.

Volossovich, S.T. (1924). K voprosu o vozraste konechnyh moren Yuzhnogo Polessiya (On the problem about the age of south Polissia end moraines), *Kosmos*, **20**. Lvov.

Voropay, L.I. & Kunytsya, M.O. (1969). Drevneliodoviokovi formy rel'iefu masyvu Svidovets' v Ukrayins'kykh Carpatakh (Ancient glacial landforms of Svidovets' massif in Ukrainian Carpathians). *Geographichni doslidzhennya na Ukrayini*, **1**, 111-125.

Vozgrin, B.D. (1975). Kraiovi lyodovykovi utvorennya Serednyogo Prydniproviya ta pov'yazani z nymy korysni kopalyny (Marginal glacial formations of the Middle Dnieper area and associated mineral deposits). *Fizychna geografiya ta geomorfologiya,* **14**. 119-125.

Vozgrin, B.D. (1985). Lednikovye obrazovaniya tiligul'skogo i tyas'minskogo etapov v Srednem Pridneprovie (Glacial formations of the Tiligul and Tyasmin stages in the Middle Dnieper area). *In:* Velichko, A.A. (ed.), *Kraevye obrazovaniya materikovyh oledeneny*, 115-116. Moskva, "Nauka".

Zalessky, I.I. & Mel'nichuk, I.V. (1975). O paleogeographicheskih usloviyah formirovaniya antropogenovyh otlozheny Prip'yat' -Turiyskogo mezhdurechiya (On paleogeographic conditions of Quaternary deposits formation in Pripyat' -Turiya interfluve area). *Fizychna geographiya ta geomorphologiya*, **14**, 80-87.

Zamory, P.K. (1961). *Chetvertynni vidklady Ukrains'koi SSR* (Quaternary deposits of Ukrainian SSR), Part 1. 549 pp. Kyiv, Vydavnitstvo KDU.

Quaternary Glaciations – Extent and Chronology
Editors J. Ehlers and P.L. Gibbard

Evidence of European ice sheet fluctuation during the last glacial cycle

G.S. Boulton, P. Dongelmans, M. Punkari and M. Broadgate

Department of Geology and Geophysics, University of Edinburgh, Grant Institute, Kings Buildings, Edinburgh EH9 3JW, UK

Abstract

Satellite images provide unique means of identifying large scale flow-generated lineations produced by former ice sheets. They can be interpreted to reconstruct the major elements which make up the integrated, large scale structure of ice sheets: ice divides; ice streams; interstream ridges; ice shelves; calving bays. The evolving form of the European ice sheet during its decay after the Last Glacial Maximum (LGM) is reconstructed by reference to these components and in the context of a new map showing isochrons of retreat. During the retreat phase in particular the time dependent dynamic evolution of the ice sheet and the pattern of ice stream development are reconstructed.

Crossing lineations are widespread. The older ones are suggested to have formed during molten bed phases of ice sheet growth and preserved by frozen bed conditions during the glacial maximum, particularly in areas which lay, during deglaciation, beneath ice divides and inter-ice stream ridges, both areas of slow flow and frozen bed conditions.

Four phases of growth (A1 to A4) and five phases of decay (R1 to R5) are used to describe the major climatically and dynamically determined stages in the evolution of the ice sheet through the last glacial cycle. The growth and decay patterns are quite different, and associated with major shifts in the ice divide, reflecting growth from the Fennoscandian mountains and decay away from marine influenced margins.

1.0 Introduction

Reconstructions of the form and flow of former ice sheets have largely been based on geological evidence of features such as drumlins, striations and till fabrics which reflect the direction of ice sheet movement, rather than from the more sparse moraines which indicate the locations of former ice margins. The purpose of this study is to improve the glacial geological reconstruction of the European ice sheet presented by Boulton *et al.* (1985) by deducing palaeo-flows from a compilation of the distribution, primarily of drumlins, but also of moraines from the area of the ice sheet. The results have been presented in full and interpreted in terms of palaeoglaciological changes by Boulton *et al.* (2001).

2.0 Data

Two principal data sources have been used to plot the distribution of large scale glacigenic landforms within the area of northwest Europe believed to have been occupied by an ice sheet during the last glacial cycle. Satellite images have been used for the inner region of the ice sheet (Fennoscandia, northwestern Russia and much of the area of Estonia, Latvia and Lithuania), and existing fieldwork-based compilations for the peripheral areas (Krasnov, 1971; Chebotareva, 1977; Liedtke, 1981; Kozarski, 1986, 1988; Ehlers, 1990). In the former, MSS images (resolution 79 m) were used, with TM images (resolution 30 m) and SPOT images (resolution 10-20 m) for some more detailed interpretations. Interpretations utilise the capacity of spectral and spatial information to discriminate between drift deposits and bedrock and between glacial landforms and bedrock features (Punkari, 1982, 1985, 1993; Boulton *et al.,* 1985; Boulton & Clark, 1990a, b). The validity of the interpretations of glacial landforms has been checked, in selected areas, against field surveys where they exist but particularly through studies of conventional aerial photography, where there is rarely any doubt whether particular features are drumlins, ice smoothed rock surfaces or moraines. Given the scale of this study however, this can only be done for a very limited number of examples.

Three principle landform types have been mapped:

Eskers. Narrow winding ridges have been identified, approximately parallel to directions of ice flow, which coincide with the locations of eskers marked on the excellent detailed maps of the Norwegian, Swedish and Finnish Geological Surveys (Nordkalott Project, 1986). However, the narrowness and the sinuous or relatively disordered form of eskers makes them difficult to identify on the images used, unless they are long and continuous, in which case their continuity aids their identification. The current mapping therefore underestimates their frequency compared with the results of ground-based survey. Kleman *et al.* (1997) used the distribution of eskers on glaciated surfaces as an index of conditions at the ice/bed interface.

Moraines. Moraines form at and parallel to glacier margins. They are less sinuous than eskers, and can vary markedly in width along their length. This irregularity makes them difficult to identify on our the images. Relatively minor and discontinuous moraines which have been mapped on ground surveys, such as the Middle

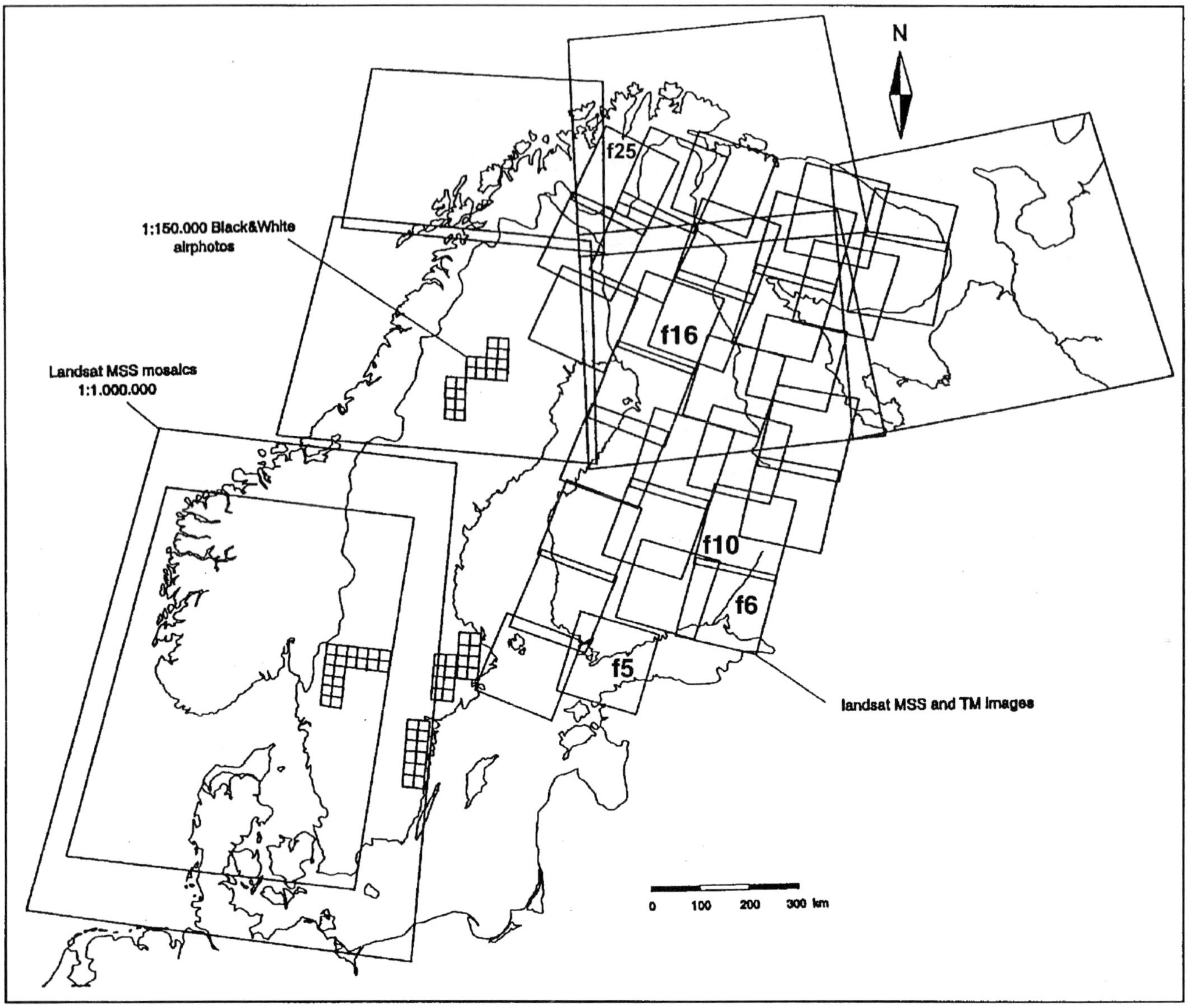

Fig. 1.*Location of the satellite images used to map glacial lineations, together with locations of conventional aerial photographs used to check satellite interpretations. Satellite imagery has been used for the whole of the shield areas, where patterns of arable agriculture do not obscure glacial features. Published ground observations have been used to cover areas beyond the shield, and limited areas the Baltic states have been studied using satellite-derived data. Numbers identify images referred to in figures 2 and 3 and in the text.*

Swedish Moraines, are difficult to identify, but major or continuous moraines, such as the Salpausselkä Moraines of Finland, are easily recognised.

Drumlins. Satellite survey comes into its own in identifying straight, flow-parallel, streamlined landforms which occur in fields of mutually parallel ridges. They are quite different from the more complex forms of moraines and from eskers. It is thought that almost all such flow-parallel features that we have mapped are sediment ridges, 10s-1000s of metres in length, metres to 10s of metres in height, and 10s to 100s of metres in width. They show great diversity of size and of detailed form and can all be reasonably grouped together as 'drumlins'. Satellite image-based maps tend to show a greater density of drumlins than maps of drumlin fields compiled from field surveys. Although such comparisons show that small drumlins are missed by satellite surveys, we are often able to identify drumlins of great extent and low relief which are too large to be discerned by field surveys, suggesting that in many drumlin fields, satellite surveys produce a better estimate of the total drumlin population. Satellite survey is also an effective means of identifying where changing directions of ice flow have created two populations of drumlins, in which older drumlins have been partially reworked to create a second, cross-cutting drumlin set (Boulton & Clark, 1990b).

The locations of the images, on which the present interpretations are based, are shown in Fig. 1. Examples of images and interpretations made from them are shown in Figs 2-4. Satellite image survey has the advantages that a

Fig. 2a-b. a) Conventional aerial photograph showing crossing lineations at Utsjoki in northern Finland. Drumlins directed towards the NNE (trend shown by large open arrows) have smaller flutes and drumlins directed towards the NE superimposed on their surfaces (trend shown by smaller filled arrows), suggesting a clockwise rotation of ice movement of about 25°.

b) A satellite image immediately to the south of a), from f25 on Fig. 1. The upward direction is towards 310° and the width of the image is 130 km. A strong S-N lineation defined by an extensive drumlin field is shown in the central part of the image, where it overprints an older NNE-trending lineation.

single operator or a small group can apply a well-defined set of criteria on a continental scale; that large features that are often unresolveable by ground or air photo survey can be identified and that groups of individually indistinct linear elements can be recomposed visually into coherent patterns to infer integrated patterns of ice flow. The level of coherent detail about patterns of lineation over wide areas that can be mapped from satellite imagery is generally greater than has hitherto been obtained from ground surveys such as those used by Kleman *et al.* (1997; see their Fig. 3). Satellite-derived data contrasts with stratigraphi-cal and sedimentological evidence of palaeoglaciology, which tends only to yield information from very small areas about continental-scale glacier behaviour. In this study continent-wide spatial patterns of flow parallel lineations have been plotted as a basis for reconstructing spatial patterns of ice sheet behaviour. It has the disadvantage that it only permits, at best, relative dating and has the danger that temporally separated flow events are grouped together. It should ultimately be integrated with stratigraphic and sedimentological data (Dongelmans, 1997).

To guard against subjective interpretations, the authors have individually and independently produced interpretations of the same areas and cross-checked the results. In general, substantively different interpretations have not been found. In some areas, bedrock structures show linear patterns, but many such areas can be eliminated by using existing geological and geophysical maps.

Figs 3a, c and e show details of flow-parallel lineations observed on satellite images at a scale of 1: 300,000. We term these first order lineations. In some cases it is clear that different parts of an the image show flow lineations which could not reflect synchronous patterns of basal ice flow. In some cases, lineations cross, and in some of these, it is possible to establish the relative ages of lineations from satellite images (Boulton, 1987). In some, relative ages have been checked by reference to aerial photographs, but in others no clear interpretations of relative age could be made. It is not possible to show first order interpretations on an ice sheet wide basis in figures of the scale used in this article. In order to show ice sheet wide patterns, three successive approximations have been made:

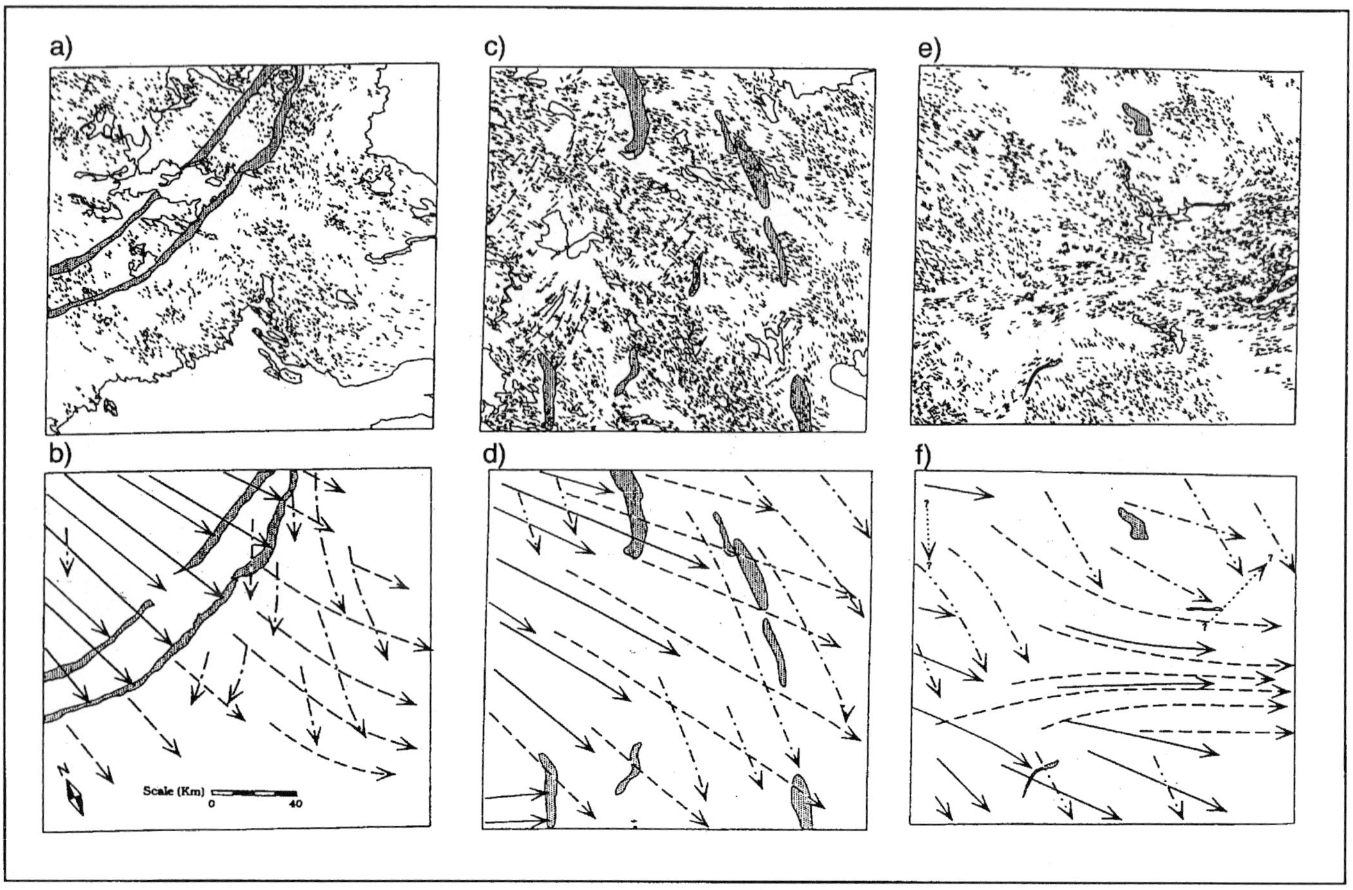

Fig. 3 a-f: Examples of interpretations of satellite images. a), c), and e) show first-order interpretations directly from satellite images. b), d) and f) show interpretations of the pattern in terms of integrated, second-order lineation sets. Different second order sets are shown by different arrow types. ***a-b)*** *Image f5 (Fig. 1) in the area of the First and Second Salpausselkä Moraines in eastern Finland. The interpretation suggests an oldest fanning lineation pattern (possibly an ice stream) reflecting a NNE-SSW flow, overlain by successively younger lineations reflecting NNW-SSE flow. The NNE-SSW sets south of the First Salpausselkä Moraine are separated because of the long ice sheet still-stand at the moraine.* ***c-d)*** *Image fl0b (Fig. 1) in eastern Finland. An older NS lineation with a superimposed NW-SE lineation. Earlier (dashed line) and younger integrated sets are distinguished.* ***e-f)*** *Image fl6 (Fig. 1) from northern Finland showing an early N-S lineation superimposed by an older and a younger WNW-ESE lineation, convergent towards the SE, reflecting a change from non-streaming flow to streaming flow towards the head of the White Sea during ice-sheet retreat.*

a) Maps of first order interpretations have been generalised to produce second-order interpretations of lineation sets which clearly form parts of spatially integrated groups and can be summarised by a series of lines which are longer than the original lines representing individual geomorphological features (Fig. 3b, d, f). Many show several crossing lineation sets.

b) The second order interpretations have been further compiled to produce third-order lineation maps of large sectors of the ice sheet (Fig. 4).

c) These have been further simplified to produce fourth-order maps of lineation trends over the whole area of the ice sheet (Fig. 5) by selecting representative third order lineations trends.

3.0 The Last Deglaciation

3.1 Geological reconstruction - the pattern of retreat

Two types of data are used in reconstructing the pattern of retreat. From the area of the Shield, flow lineations and a relatively small number of ice-marginal moraines observed on satellite imagery are used. In the flanking area of younger sediments, a compilation of field observations from a variety of authors and a relatively small cover of satellite images from the Baltic states and north-west Russia and Belorussia are utilised.

Figs 3-5 show that from the area in which satellite images have been used, flow lineations from different phases of ice sheet history are preserved and superimposed. In this section, those lineations are identified which are thought to have been produced during the last deglaciation. Section 5 shows how they can be stripped away in order to infer ice sheet flow patterns from earlier stages. Ice-marginal landforms (end moraines, outwash plains, etc.), produced during deglaciation, are direct evidence of the pattern of final retreat (Fig. 6). However, except for the Younger Dryas moraines, only a few significant end moraines occur in the shield areas to guide reconstruction of deglaciation. Following rules set out in section 3, it is expected

- that the strongest lineation to be generated in the sub-marginal zone of the ice sheet where velocities were highest,

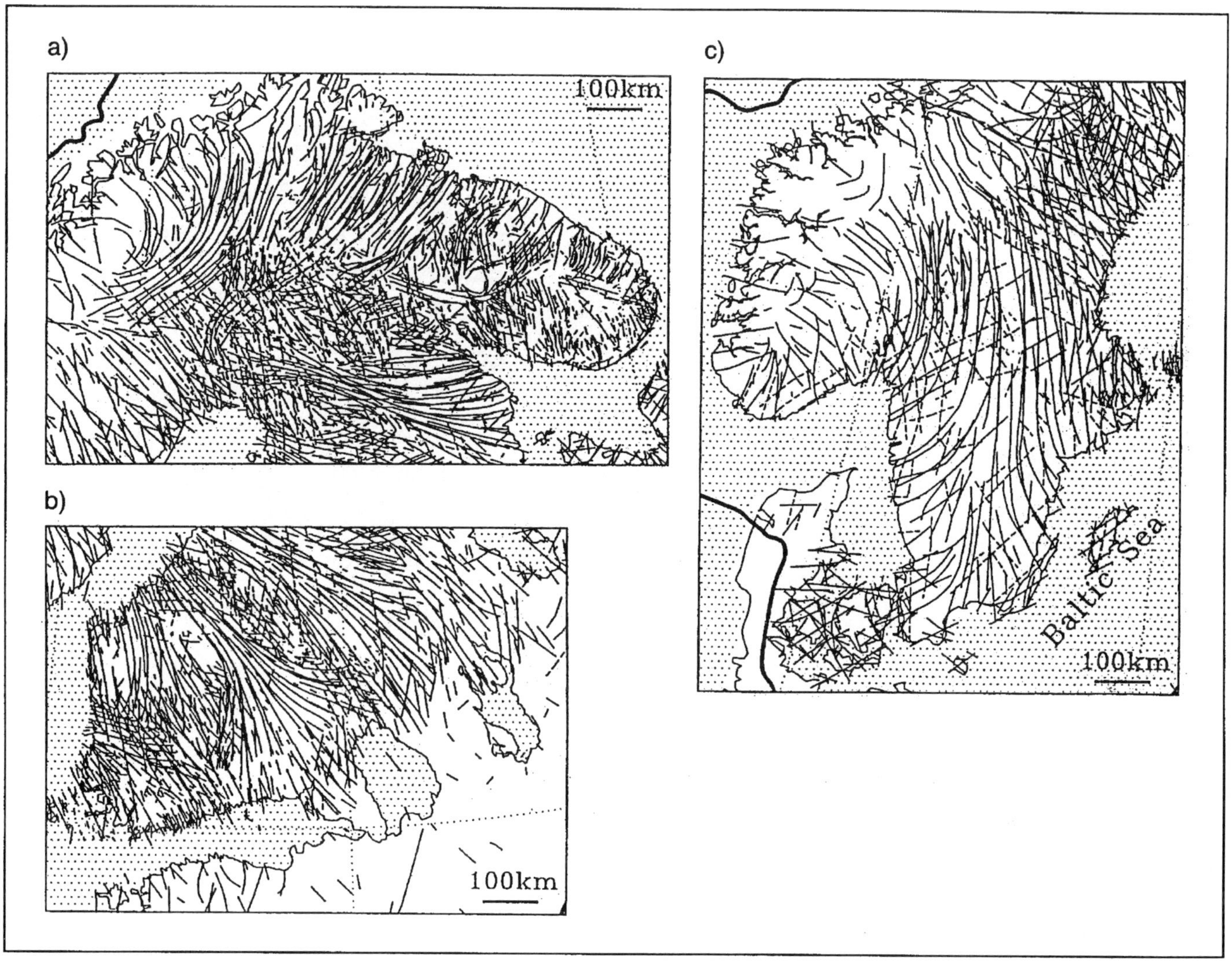

Fig. 4a-c. Third-order interpretations, generalised from second order lineations such as those shown in Fig. 3, showing lineation patterns over the shield area of the ice sheet. a) Northern Scandinavia. The lineation convergences into the head of a very strong lineation sets running ESE towards the White Sea, which shows divergence along the White Sea shore are particularly interesting. b) Southern Finland and the area around the Gulf of Finland. Note the crossing lineations in the inter-ice stream areas and reflecting interactions between ice streams during their lifetimes. c) Southern Sweden, southern Norway and part of Denmark.

- that in most cases, the greatest lineation density will reflect flow in this zone during the retreat phase,
- and that, on a horizontal bed, the trend of these lineations will lie normal to the trend of the ice margin.

The prime exceptions to this will be:

- where fast ice streams have produced strong lineations far from the ice sheet margin and which are preserved because streaming flow ceases prior to deglaciation of the area.
- where lineations are created in a zone of melting and which are preserved because of onset of ice/bed freezing prior to deglaciation.

If these rules are applied to lineations in areas where ice-marginal moraines have survived, it is found that the ice-marginal retreat pattern inferred from the densest lineations coincides with that inferred from moraines (eg. Fig. 3a-b). This approach permits:

- a definition of the general pattern of retreat with much greater precision than by using ice margin-parallel forms alone;
- an identification of those lineations which are not normal to the retreating ice margin, and which thus reflect either flow patterns far from the retreating ice margin or patterns produced during earlier phases of glaciation.

In Fig. 7, following Boulton *et al.* (1985), ice margin trends during retreat have been determined by assuming that they lie normal to the dominant flow lineations. This gives a clear picture of the trajectory of retreat of the ice margin and of the sub-marginal flow patterns associated

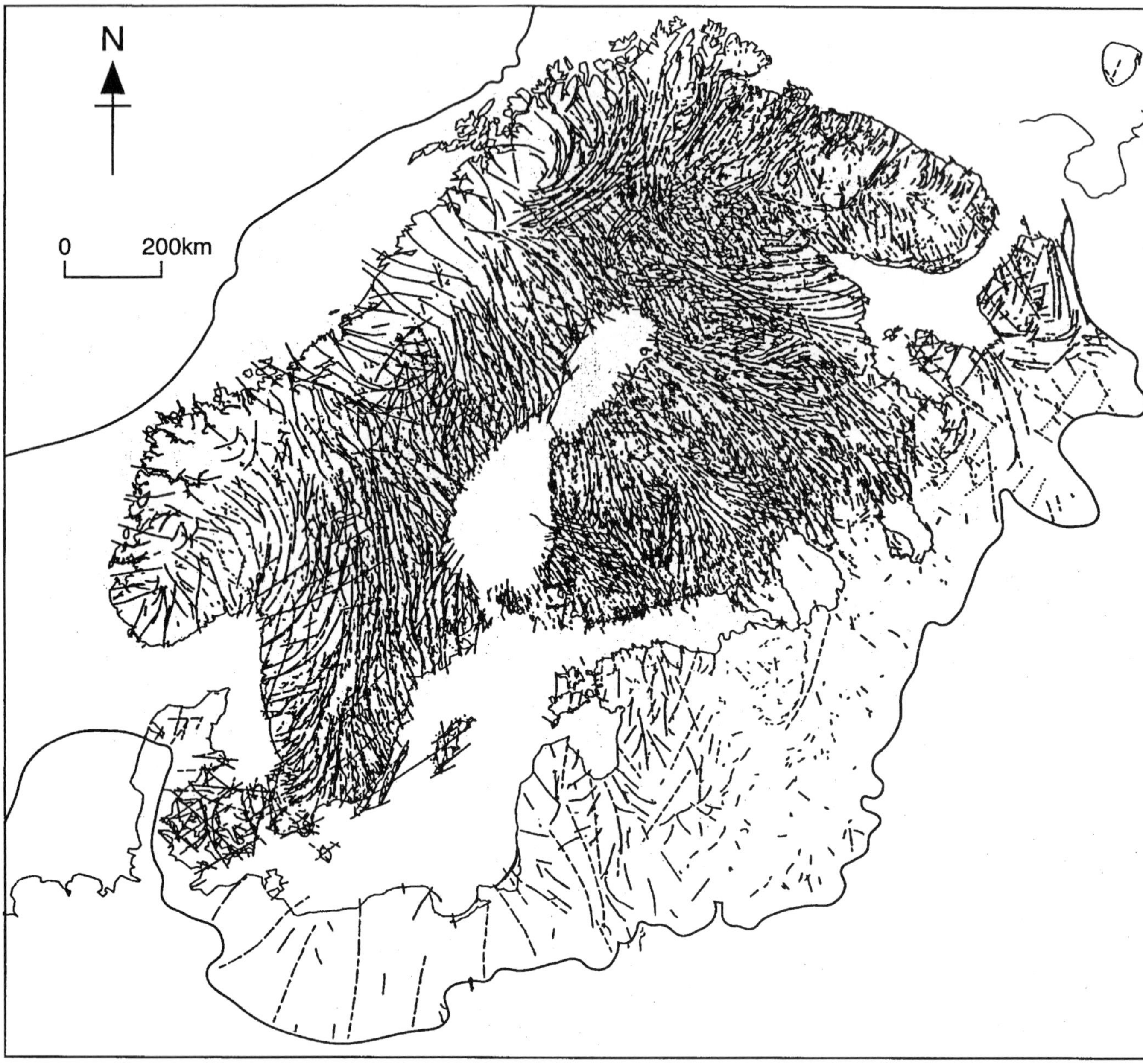

Fig. 5. Fourth-order longitudinal lineation sets plotted on a European scale. Note the areas of abundant crossing lineations between the strongly lineated zones which are interpreted as ice streams, particularly clearly seen in southern Finland (also Fig. 4b). They reflect the location of sluggish inter-stream ridges. They may also be characterised by basal freezing. Crossing lineations also occur in former ice divide areas on the eastern side of the Scandinavian mountain chain.

with it. It permits interpretation of aspects of the dynamic behaviour of the ice sheet during retreat.

Beyond the Shield, in the area south of the Baltic Sea, much of the evidence of ice sheet behaviour during retreat comes from the distribution of major morainic ridges formed at the maximum and during retreat. The maximum extent of the ice sheet and its retreat in the northeastern North Sea and on the Norwegian continental shelf have been summarised by Holtedahl (1993) and Sejrup *et al.* (1994).

3.2 Geological reconstruction – the tempo of retreat

In this section an attempt is made to translate the pattern of retreat shown in Fig. 7 into one showing specific iso-chrons (Fig. 8). This is relatively simple in the shield area where a unique record of the tempo of ice-sheet retreat is contained in the annually laminated varve chronology of Sweden, Finland and the area around the Gulf of Finland.

Beyond this zone however, lineation data is sparse and chronology is much more difficult to establish. This latter

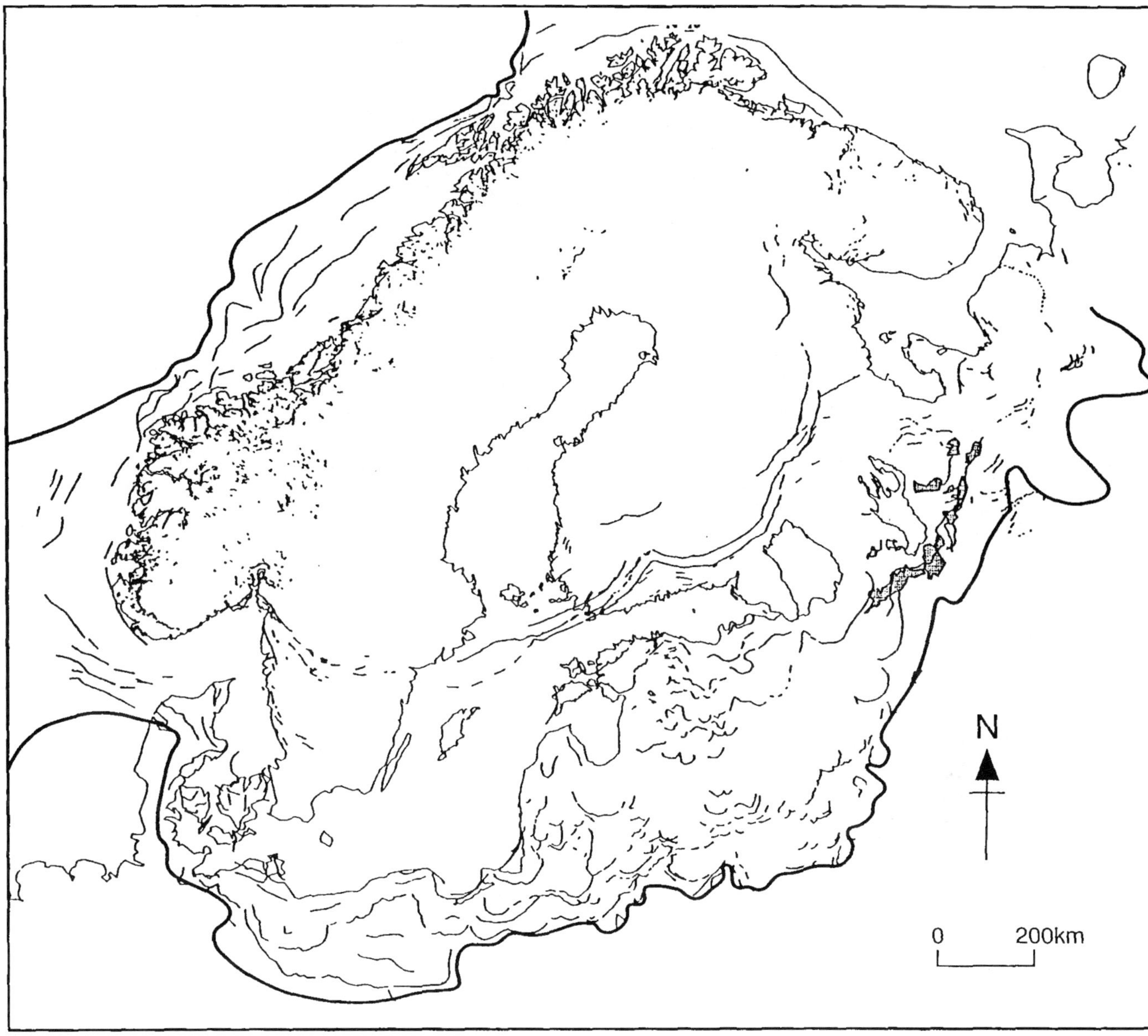

Fig. 6. The principal zones of ice-marginal moraines and other ice-marginal deposits within the area of the Weichselian ice sheet. Apart from the Younger Dryas Salpausselkä moraines and their correlatives, most moraines occur within the zone of relatively slow retreat lying beyond the shield (cf. Fig. 7).

area is discussed before retreat chronology within the varve zone is analysed.

3.2.1 'Pre-varve' retreat

There is strong evidence that the maximum extent of the Late Weichselian Scandinavian ice sheet was time-transgressive. The earliest advance to the maximum for which evidence is available is in the southwest, where the ice sheet, flowing from a centre in southwest Norway, crossed the North Sea to become confluent with an ice sheet over the British Isles at about 28 ka (Sejrup *et al.*, 1994, 2000) (Fig. 8). There is evidence that during this phase, the Norwegian Channel, which extends from the Skagerrak and skirts the southern and southwestern coast of Norway as far as the continental shelf edge off southwest Norway, was the location of a major ice stream (King *et al.*, 1996; Sejrup *et al.*, 1996, 1997).

The connection between the European and British ice sheets was broken shortly after 23 ka, and the European ice sheet margin retreated to the vicinity of the coast of southwest Norway by about 20ka (Valen *et al.*, 1996). Between 20 ka and 18 ka, the ice sheet readvanced from the vicinity of the Norwegian coast to reach a maximum just beyond the southwest margin of the Norwegian Channel (the Tampen advance; Sejrup *et al.*, 1994). It seems most likely that the sharp re-entrant in the maximum extent of

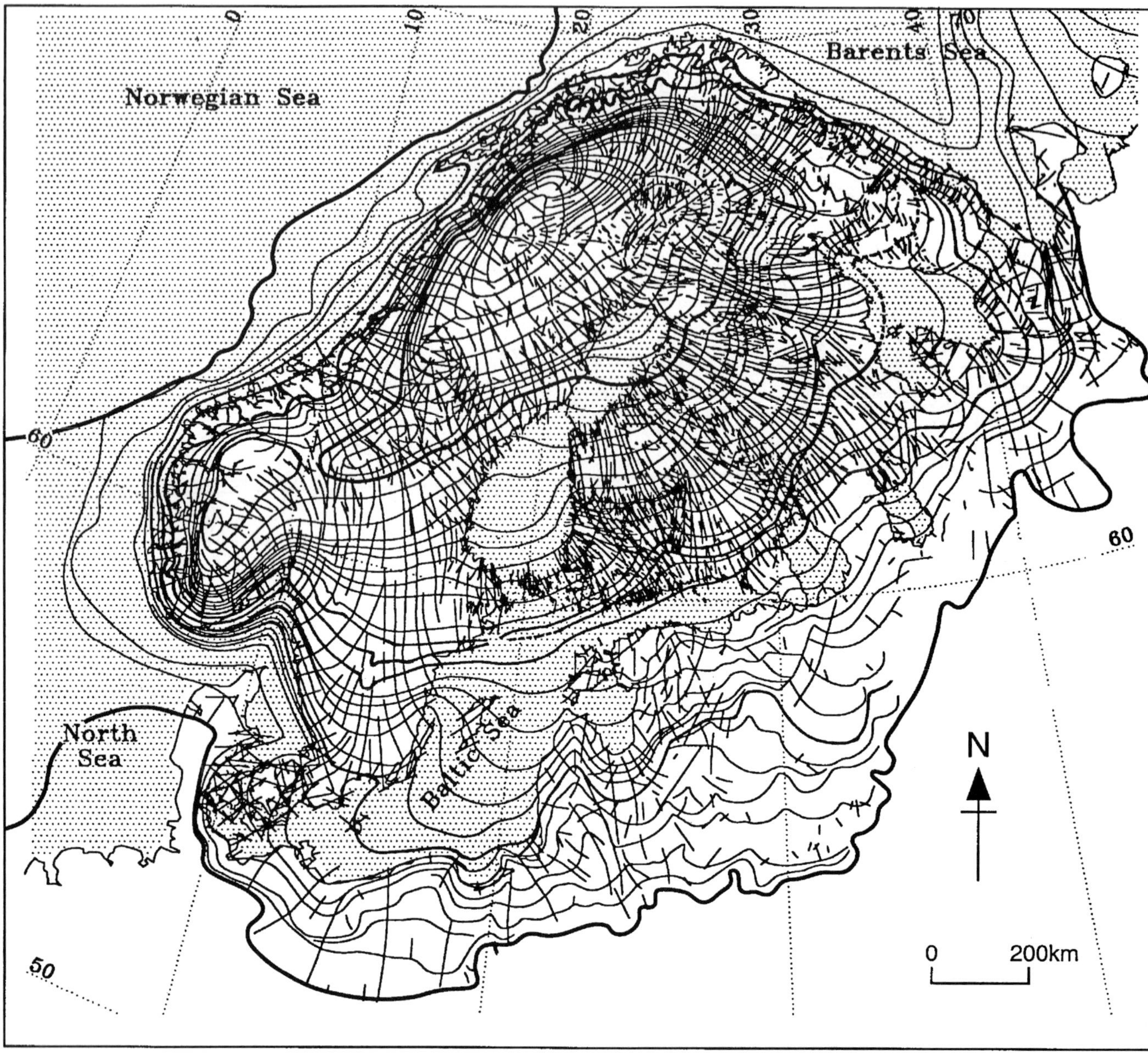

Fig. 7. Inferred pattern of ice-front retreat of the last Northern European ice sheet, superimposed on the dominant patterns of longitudinal lineation. Within the area of the shield, retreat lines are constructed from normals to fourth-order, flow-generated lineations (Fig. 5). Beyond the shield they are largely derived from the distribution of moraines (Fig. 6). Within this latter area, some longitudinal lineations are also shown, derived from published field mapping of drumlins. Note that there is a strong correspondence between inferred ice margin positions on the shield and the few moraines, such as the Salpausselkä, which occur there.

the Weichselian ice sheet margin in northern Jutland (Fig. 8) represents an overstepping by this advance of the margin of the older North Sea ice sheet, such that the maximum extent in much of Jutland and northwest Germany, is roughly contemporary with the Tampen advance. Furthermore, the distribution of moraines in northern Germany suggests that the eastward continuation of this advance, represented by the north-south Weichselian ice margin through Denmark, also oversteps the line of the older, Brandenburg moraines and continues into the Poznan Moraines (Fig. 8).

The configuration of moraines in northern Germany and Poland (Fig. 8) suggests that an early advance occurred in this sector to the glacial maximum along the Brandenburg-Lezno moraines, and that this was followed by a retreat and then a readvance which created the Poznan stage moraines (Kozarski, 1986, 1988).

Further to the east and west, the readvance formed the glacial maximum. It is possible that the early, Brandenburg-Lezno advance was a consequence of stronger ice sheet flow down the axis of the southern Baltic Sea. It may reflect the same phase of ice sheet growth which led to

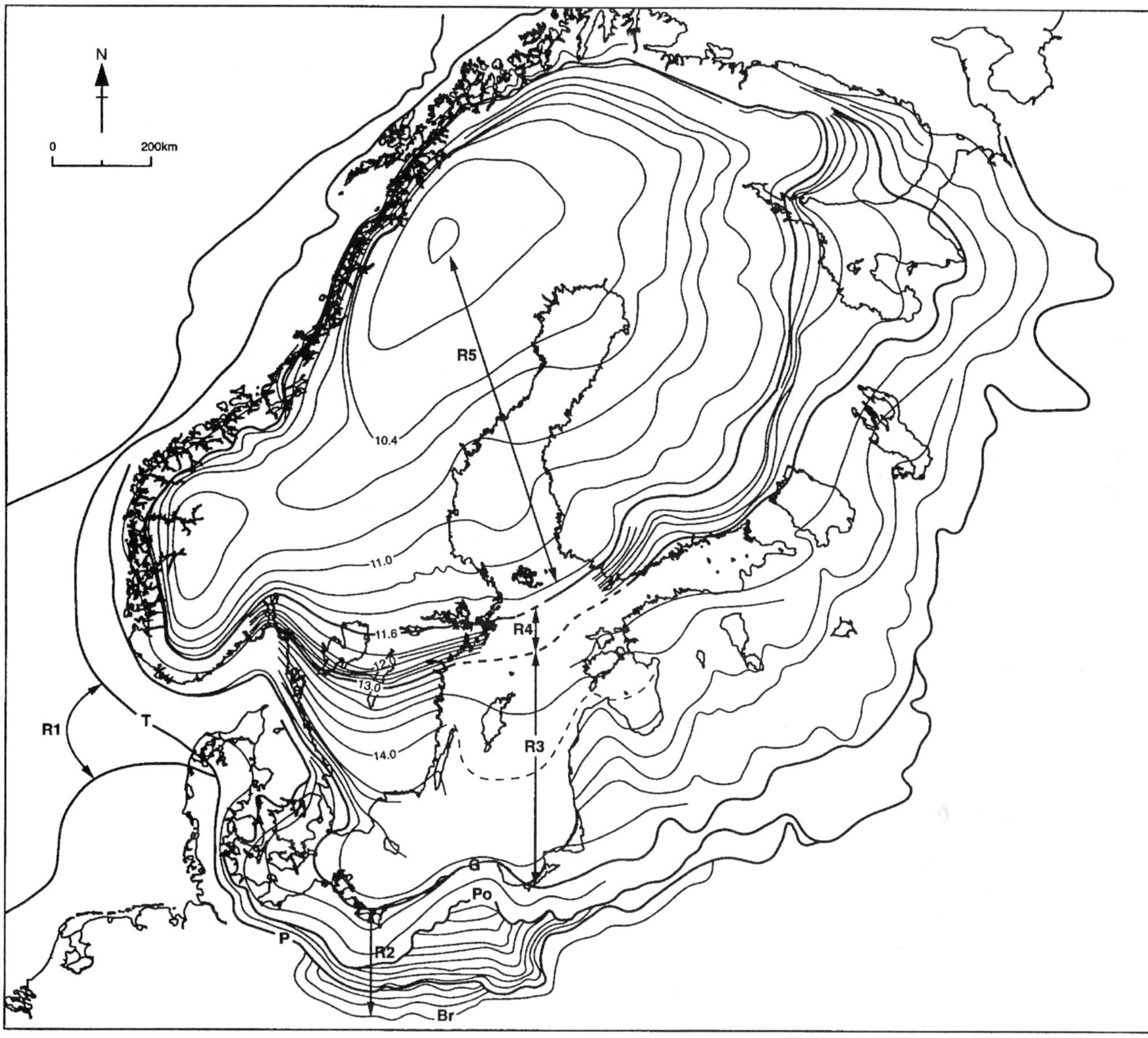

Fig. 8. Isochrons of retreat of the last Northern European ice sheet. The retreat pattern shown in Fig. 7 is translated into isochrons of retreat by calibration with the Swedish and Finnish varve timescales and available corrected ^{14}C determinations. Retreat is subdivided into a series of principal phases R1-R5, with dates of R1 = 28 ka-20 ka; R2 = 20 ka to 15.2 ka; R3 = 15.2 to 13 ka; R4 = 13 ka to 11.5/6 ka; R5 = 11.5/6 to final decay. Dating of the early stages of retreat is speculative. The major LGM and early retreat moraine phases are shown as Br = Brandenburg-Lezno (22-20 ka); P = Poznan (18.0 ka); Pm = Pomeranian (16.5 ka); G = Gardno (15.4 ka). The pattern of retreat in Poland follows Kozarski (1986). It is clear that the maximum glacial extent was reached earliest in the southeast (at about 28 ka – Sejrup et al., 1994) and last in the east (between 17 and 15 ka). The ice sheet was still advancing in the east whilst retreating in the west.

extension of ice from the southern Norwegian centre of ice sheet growth into the North Sea.)

The later, Pomeranian Moraines (Fig. 8), have been shown by Kozarski (1986) to cut across the earlier, post-Poznan Moraines in Poland, representing yet another major advance of the ice sheet margin in this sector. The Gardno Moraine in northernmost Poland (Kozarski, 1986) may represent a relatively small readvance, as there is no clear evidence of major cross-cutting relationships with the major moraines to the south.

There is very little independent evidence of the age of the glacial maximum or of ice-marginal positions in the southern (German-Polish) sector of the ice sheet. Prior to its maximum extent, the expanding ice sheet moved across the southern Norwegian and southwest Swedish coasts between 30 and 24 ka (Hillefors, 1974; Andersen, 1987) and crossed the Polish coast slightly before 22ka BP (Kozarski, 1988). The ice sheet is suggested to have advanced to the maximum extent in Poland shortly after 20 ka BP (TL yr - Ralska-Jasiewiczowa & Rzetkowska, 1987; Mojski, 1992). The time window for the glacial maximum in Poland is closed by an uncalibrated ^{14}C date of 13,500 BP on the northern coast of Poland on the earliest organic deposits after deglaciation (Ralska-Jasiewiczowa & Rzetkowska,

1987). Adjusting this date using the correction of Bard *et al.* (1993), gives an age of 15,950 BP (from here on ^{14}C ages are given as corrected ages unless otherwise stated). Although there is considerable uncertainty about the timing of the glacial maximum and of deglaciation of the area to the south of the Baltic Sea, the above estimates are not inconsistent with other records. The isotopic minimum in deep-ocean stratigraphy, thought to record the global ice volume maximum during the last glacial period, occurred at 22 ka (Imbrie et al., 1984), as did the period of maximum cooling in the Greenland ice-core record (Johnsen et al., 1992; Bond et al., 1993).

Support for the general distribution of retreat isochrons in the southeast sector of the ice sheet during the early stages of retreat is given by Sandgren *et al.* (1997), who have used palaeomagnetic correlations with varve-dated sites in southern Sweden and Karelia to suggest deglaciation of Lake Tamula in Estonia at 14,400 calendar years BP. It indicates that the proposed isochrons are consistent with the time of deglaciation of northern Poland and of southern Estonia (Fig. 8).

If the ice sheet readvances in the southern sector margin (Kozarski, 1988) are climatically driven rather than dynamic, it is tempting to suggest the following correlations with post-LGM cooling events in the GISP/GRIP records: Poznan Moraine - 18,500 BP; Pomeranian Moraine - 16,500 BP; Gardno Moraine- 15,400 BP; compared with Kozarski's (1986) estimates of 20,400 for the Lezno, 18,400 for the Poznan, 15,200 for the Pommeranian and 13,300 BP for the Gardno moraines.

The trend of morainic features in the southeastern sector of the ice sheet suggests that the features which are co-linear with the Gardno Moraine ultimately converge, about 300 km south of Lake Ladoga, with the maximum extent of the Weichselian glaciation (Fig. 8), just as the Poznan and Pommeranian moraines appear to further west. If this is so, it suggests that whilst the southern and south western margins of the ice sheet were retreating by up to 300 km, the eastern and southeastern margins of the ice sheet were stationary or still advancing. This conclusion is supported by evidence that in the area to the southeast of the White Sea, the glacial maximum was not reached until after 17 ka and that retreat from the maximum position did not begin until about 15 ka (Larsen *et al.*, 1999).

In the northern sector, the ice sheet appears to have reached a maximum in northern Norway between 19-18.5 ka BP (Vorren et al., 1988), whilst an ice sheet over the Barents Sea, believed to have been confluent with the European ice sheet, is thought to have reached its maximum extent and begun to collapse rapidly by 15 ka and had largely disappeared by 12 ka (Landvik *et al.*, 1998).

3.2.2 The varve zone

The most important record of the tempo of deglaciation is in the varve-based 'Swedish Timescale' (De Geer, 1940), which has been extended back continously from final deglaciation at about 8,700 varve years B.P. to about 11,500 varve years BP (Strömberg, 1985). However, Younger Dryas-age advances of the ice sheet in the Middle Swedish End Moraine zone make it very difficult to connect this chronology with that represented by about 2000 varve years in southern Sweden. The latter is a floating chronology because of dislocation in the Middle Swedish End Moraine zone, but if this dislocation is ignored, appears to extend back to about 13,300 varve years BP (e.g. Wohlfarth *et al.*, 1995; Strömberg, 1994; Björck *et al.*, 1995). The work of Strömberg (1994) appears to indicate that the maximum of the Younger Dryas readvance in the Middle Swedish zone was reached at 11,410 varve years BP, and that ice front positions representing a period of 180 years were later overridden by the Younger Dryas readvance.

The interpretation summarised by Björck *et al.* (1995) and Wohlfarth and others (1995; and Wohlfarth personal communication, 1995) can in principle be used to determine the rate of ice-sheet retreat from 13,258 varve years BP, in eastern Scania, to 9200 varve years BP in Västernorrland. Although correlation between varve and ^{14}C years has been difficult to achieve, a direct comparison of varve and AMS ^{14}C chronology (Björck *et al.*, 1995) shows the latter to be systematically older by 500-1,500 years during the period 9,500-12,500 varve years BP, although there is good agreement between the varve chronology and the ^{14}C chronology calibrated (Stuiver & Reimer, 1993) against the German oak-pine chronology (Kromer & Becker, 1993) and U/Th ages on corals (Bard et al., 1993) up to about 12,000 varve years BP. The U/Th-calibrated ^{14}C record beyond 12,000 varve years BP gives much greater ages than the varve record.

A means of confirming the precise magnitude of the mis-match between varve years and calendar years has been developed by Andren *et al* (1999). They have suggested that varve thickness will largely reflect ice sheet melting rate as a direct result of changing summer temperatures. They have compared varve thickness data for the period 10,250-11,550 varve years with the GRIP ice core ^{18}O record (Johnsen *et al.*, 1992), which is probably the best proxy, high resolution guide to the trend of climatic variations in western Europe during the period, reflected for example in the excellent match between the GRIP record (Fig. 9a) and the sea surface temperatures in the North Atlantic (Fig. 9b; Kroon *et al.*, 1997). They show that a very good correlation exists between the two records for the highly distinctive pattern of change during the Younger Dryas/Preboreal climatic shift, provided that 875 years are added to the varve ages.

We have therefore added 875 years to the whole varve chronology, and added 180 years to that part lying south of the Middle Swedish End Moraines, to produce the pattern of ice-sheet retreat rate shown in Fig. 9c. The pattern is very similar to that shown by the GRIP and NE Atlantic records, but lags behind it. The best fit is obtained by assuming that the retreat rate lags 250 to 500 years behind phases of climatic warming.

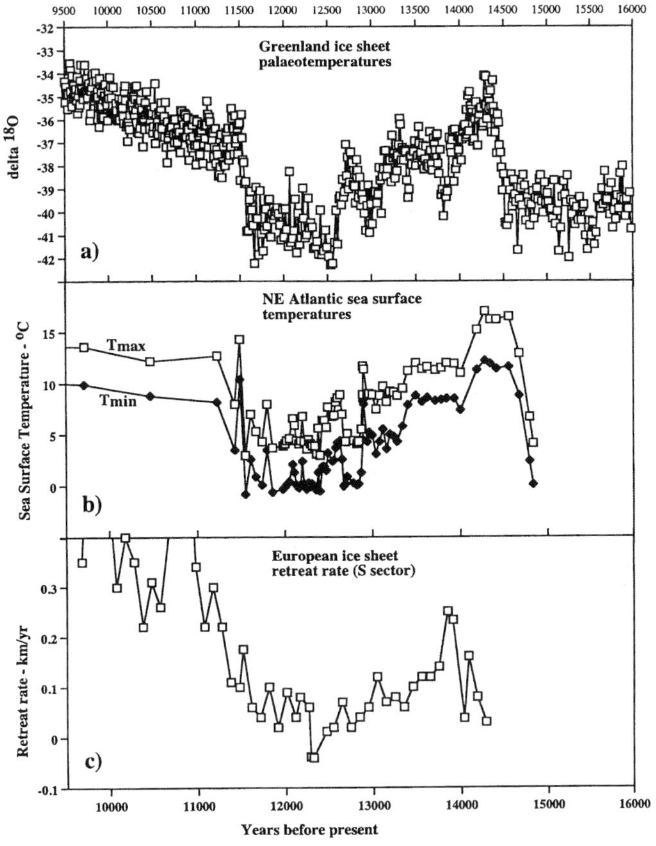

Fig. 9 a-c. Correlation of the corrected varve zone retreat of the ice sheet with other contemporary records of climate change. a) The GRIP $\delta^{18}O$ record (Johnsen et al., 1992) between 16,000 and 9,500 BP. b) Sea surface temperture record in the NE Atlantic, off western Scotland (Kroon et al., 1997) showing the match between a European proxy climate indicator and the GRIP record. The upper curve shows maximum and lower curve minimum temperatures. c) Retreat rate of the European ice sheet in the Swedish varve zone. The timescale is derived from a correlation with the GRIP core based on variations of varve thickness (Andren et al., 1999). It suggests that the retreat rate lags climate forcing by 250-500 years. This is suggested to be an ice-sheet dynamic effect.

This corrected varve-based chronology is now applied to the ice-sheet retreat pattern shown in Fig. 8 by ascribing ages to retreat isochrons at the points at which they cross the varve transect series in eastern Sweden. Fig. 10a shows the inferred pattern of time dependent retreat of the southern sector of the ice sheet, not only for the area to which the corrected varve timescale can be applied, but also for the earlier phase of retreat by using the tentative correlations offered between the Poznan, Pommeranian and Gardno readvances and Kozarski's (1986, 1988) estimate of retreat rate between them.

A varve chronology for Finland was first published by Sauramo (1918), but has been a 'floating chronology' until recently. Strömberg (1990) has suggested that the zero varve of the Finnish chronology (Sauramo, 1923) was equivalent to 10,643 varve years in the Swedish chronology. This has been used as the basis of the chronology in Finland. It has now been extended to the area of the eastern Gulf of Finland and as far as Lake Ladoga (Hang, 1997).

Although it is still premature to create an unequivocal reconstruction of European ice sheet deglaciation, in view of much poor or heterogeneous dating evidence, the absence of evidence of ice margin trends in marine areas, and the fragmentary evidence of trends in areas such as Belorussia, it is helpful to create a deglaciation hypothesis

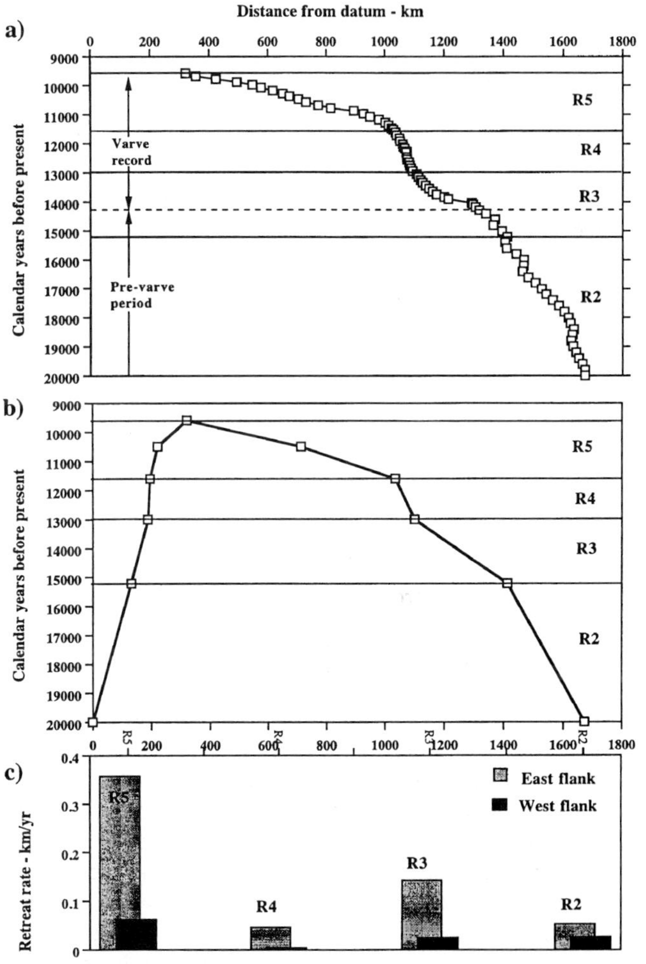

Fig. 10 a-c. Tempo of retreat of the ice sheet along a transect from northern Germany, along the southeast coast of Sweden and west to the Norwegian continental shelf (Boulton et al., 2001). a) Retreat along the transect on the southern flank of the ice sheet. The early part of the retreat is based on estimates of age by Kozarski (1986, 1988) and speculative correlations with the GRIP record. After 15.2 ka, the retreat chronology is based on the varve chronology as far as the north of Stockholm. Beyond this point, varve dates are projected onto the line of section using the isochron map (Fig. 8). b) Net retreat during each of phases R2-R5 along the transect on the eastern (right) and western (left) sides of the final ice divide. The retreat on the western flank of the ice sheet prior to 13 ka (Younger Dryas Stadial) is poorly known. c) Retreat rates during phases R2-R5 on the eastern (light shading) and western (dark shading) flanks of the ice sheet. Retreat rates are invariably less on the western side of the divide, reflecting less negative mass balances on the more maritime and mountainous flank.

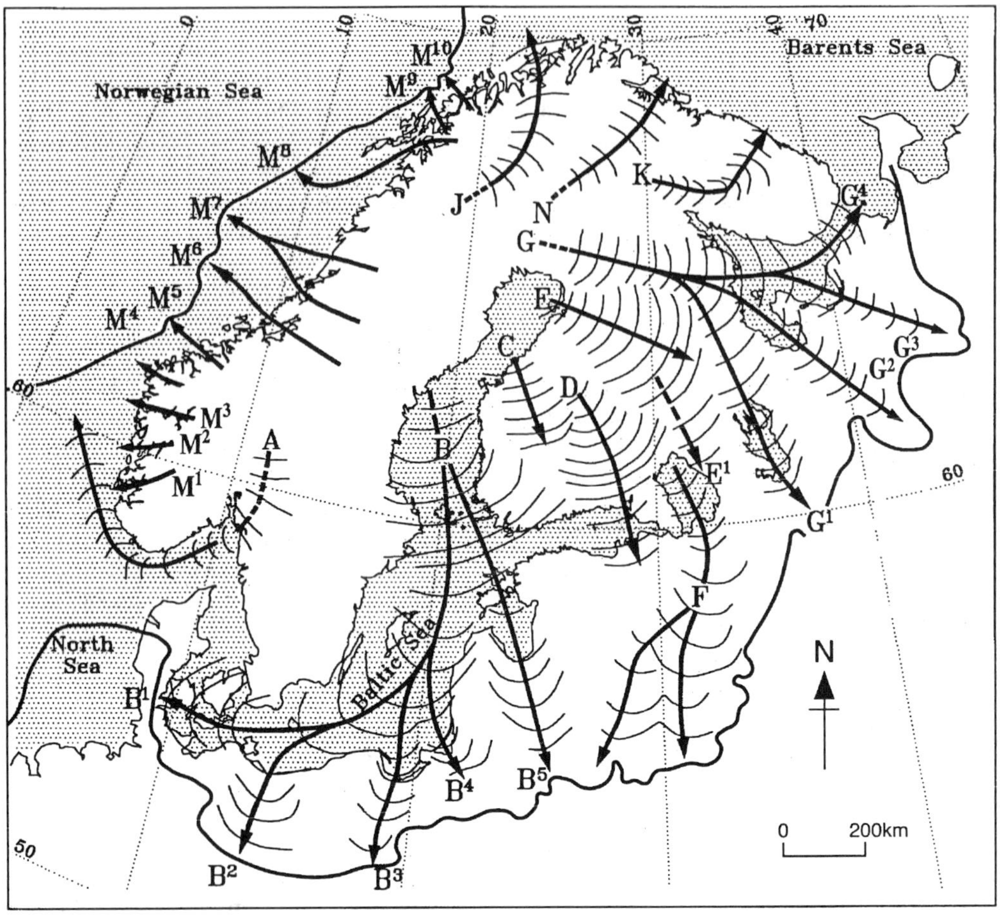

Fig. 11. Time-dependent trajectories of land-based major ice streams which existed during decay of the ice sheet, marked A-J (named in the text). The stippled zones show the swathes along which ice streams were active during deglaciation. The swathes are presumed to reflect time-transgressive activity, not the length of an ice stream at any one time. Marine ice streams are shown on the western margin of the ice sheet. Evidence for these elements include shelf troughs, arcuate moraines at the mouths of shelf troughs and major accumulations of glacial sediment at the shelf edge or on the upper continental slope.

as a spur to testing and to provide a best estimate of deglaciation for quantitive simulation models. The evidence of trends of ice margins and the varve chronology in much of Sweden, and to a lesser extent in Finland, make it possible to produce a persuasive reconstruction of deglaciation in those areas. South of the Baltic and the Gulf of Finland and in northern Finland and the Kola Peninsula, the dating evidence is sparse, and ice-marginal trends are uncertain east and southeast of the Gulf of Finland. However, combining the concentric trends normal to flow lines shown in Fig. 7 with the evidence of ice-marginal landforms and the chronological information discussed above, allows the reconstruction of the tempo and pattern of ice sheet decay during the last deglaciation in Fig. 8. The interpreted tempo of deglaciation along a transect from northern Poland to western Norway, through the centres of final ice sheet decay in Norrbotten, is show in Fig. 10a-c using the adjusted varve timescale based on the correlation with GRIP by Andren *et al.* (1999).

3.3 Glaciological implications – streaming within the retreating ice sheet

It has been assumed that, on a relatively flat surface, very strong flow-parallel lineation sets and/or evidence of lobation of the ice sheet margin indicate the fomer existence of an ice stream. However, lobate ice margins can be interpreted in two ways: as the termini of ice streams which protrude beyond the normal ice margin because of high ice velocities along their axes, or as expressions of radial flow from the end of an ice divide or ice dome. In the first case they represent a zone of relatively fast flow, in the second, a zone of relatively slow flow. Lagerlund (1980) has interpreted the southern Baltic Sea lobes as evidence of an ice dome in the southern Baltic rather than of ice streams. Lobes and associated lineations which occur entirely on land, for instance in Finland, show lineations at their margins which are not normal to the margins, as would occur if they were ice domes, but are aligned at angles acute to the margin, which is compatible with an ice stream origin. The fact that evidence for the southern Baltic lobes lies partly beneath the sea does not permit the two hypotheses to be tested as they can be in Finland. It is suggested however that the similarities of form should be taken as evidence of similarity of origin in the absence of strong evidence to the contrary.

The relationship between lobation and strong lineation can be used to infer the time transgressive history of an ice stream and to interpret its length (Boulton *et al.*, 2001). Using this approach, the pattern of time transgressive streaming in the ice sheet during its retreat is shown in Fig. 11. This diagram does not show the length of ice streams at any one stage of their history, but the periods during which ice streams reached the margin of the ice sheet. It shows a lettering scheme which identifies the ice streams on the terrestrial margin of the ice sheet. Ice streams B, F and G are deduced largely from lobation of the contemporary ice margin. Ice stream A is an outcome of the convergence of ice flowing from flanking land areas into the Skagerrak embayment, but there is seismic evidence that it extended along the axis of the Norwegian Channel (Sejrup *et al.*, 2000) and lineations on Jæren, in southwest Norway, where the ice stream transgressed onto land (Sejrup *et al.*, 1997). Others are deduced from locally strong lineation sets.

Lineation evidence for ice streams is restricted to the eastern, northern and southern margins of the retreating ice sheet. By analogy with modern ice sheets, ice streams would also be expected at the marine margins of the ice sheet. Ice stream J for example, is well marked by strong lineations on land and trends distally towards a well-defined, over-deepened shelf trough and a lobate glacial unit, presumed to be a moraine, at the shelf edge (Rokoengen *et al.*, 1979). In a tentative estimate of the locations of former ice streams at the western, marine margin, over-deepened troughs on the shelf and lobate moraines or thick glacial sediment units at the shelf edge (Rokoengen *et al.*, 1979; King *et al.*, 1987; Rokoengen *et al.*, 1988; Rise et al., 1988; Holtedahl, 1993) have been used as possible indicators of the palaeo-ice streams, marked on Fig. 11.

If lobation of the contemporary ice sheet margin, based on moraine distributions (e.g. Chebotareva, 1977) as an index of streaming in the south eastern flank of the ice sheet south of the Baltic and the Gulf of Finland is used, many seem to correspond with streams on the Shield which are clearly indicated by lineations patterns. They have therefore been identified, using suffixes, as ice stream complexes which appear to recur along the same axis and which are often succeeded by single major ice streams at a later stage of retreat. They are:

B1 - B5- Southern Baltic ice stream complex.
E - E1- Karelian ice stream complex.
G1 - G4- White Sea ice stream complex.

3.4 Glaciological implications – the phases of retreat

The structural evolution of the ice sheet during its decay is characterised by sub-dividing it into a series of phases as shown in Figs 8 and 10.

Phase R1 – 29 ka to 20 ka

The growth of the Northern European ice sheet culminated in its southwest, west and northwesterly sectors during this period, when it extended onto continental in the North Sea, the northwestern and northern. It then retreated from these shelves before the GRIP core indicates that a significant climatic warming had occurred.

Phase R2 – 20 ka to 15.2 ka

This is characterised by:

a) a large net retreat in the southern sector whilst the eastern sector was stationary at, or advancing to, the maximum extent;

b) major readvances separated by intervening retreats of the ice-sheet margin in the southwestern and southern sectors.

Phase R3 - 15.2 – 13 ka

This was characterised by general retreat around the whole of the ice-sheet margin, reflecting a strong climatic amelioration. The corrected varve record (Fig. 10a) shows a rapid early retreat followed by a decceleration towards the Younger Dryas moraines. Retreat at the western margin was much slower (Fig. 10b-c). Particularly dramatic rates of retreat occurred in the southern Baltic and White Sea regions (Fig. 8), which were probably enhanced by iceberg calving.

Phase R4 - 13 ka to 11.5/11.6 ka

The Younger Dryas Stadial, which occurred between about 12,860-11,640 yr BP, was associated with a sudden temperature drop of 7°C at the beginning of the period and a similar increase at the end (Johnsen et al., 1992; Bond et al., 1993). Large terminal moraines, the Fennoscandian moraines, were generated in Norway (Ra Moraines), Sweden (Middle-Swedish Moraines) and Finland (Salpausselkä I-III) with smaller moraines in Russian Karelia (Rugozero and Nyukozero Moraines - Punkari, 1985), although only discontinuous moraines were produced on the Kola Peninsula (Punkari, 1993). The ice margin readvanced in western Norway (Mangerud et al., 1979) and western Sweden between 12.8 and 12 ka, although it was either stationary or retreated slightly further east during the same time period.

Phase R5 - 11.5/6 ka to final retreat

Rapid retreat of the southeastern flank ice sheet was resumed and the western, southwestern and northern flanks withdrew from the marine margin more slowly as a consequence of substantial climatic amelioration. After 10.6 ka, the ice sheet broke up into separate domes, the main ice mass centred to the east of the modern ice cap of Svartisen, along the Norwegian-Swedish border, a dome centred over the Jotunheimen area in southwest Norway, and a small dome separating later in central Norway/ Sweden (Jämtland). The last ice caps were located further to the east than the initial growth centres. Some strong lineations entering into the northern extremity of the Gulf of Bothnia from the northwest (Fig. 4a) suggest that some streaming continued here at least into the early part of this phase, although it is not clear whether this belongs to a successor of streams E, D or C, or a newly-created stream.

4.0 OLDER FLOW SETS - THE PATTERN OF ICE SHEET GROWTH

4.1 The pattern of pre-retreat lineations

Crossing lineations have been observed in many areas, where flow-parallel lineations on satellite images and drumlins and striations observed on the ground clearly predate the flow features produced near to the retreating ice margin during final deglaciation (eg. Figs 3-5). Superimposed drumlins reflecting different flow directions were first identified by Rose & Letzer (1977). Boulton (1987) showed how they could be produced by changing ice-sheet flow directions, and Boulton & Clark (1990a, b) mapped crossing lineations which reflect such features over wide areas of Canada from satellite imagery. In Fig. 12, all lineations which lie parallel to the flow during final retreat have been removed from the total lineation population to reveal lineations which unequivocally pre-date final retreat. This is not to say that many lineations which happen to be parallel to marginal flow during final retreat do not pre-date that retreat, merely that our conservative approach does not recognise them. The residuals come predominantly from the areas of crossing lineations shown in Figs 4-5. They tend to form well-defined clusters, which are particularly strong in northern Fennoscandia, although residuals occur throughout central and southern Sweden, in particular in southwestern Sweden. In northern Finland/Sweden and in central Sweden (Fig. 12), these early lineations tend to form two sets, with the earlier set in the more northerly area oriented NW-SE and the later set oriented NNW-SSE. It is suggested that many residual lineations were protected from overprinting by the re-organisation of ice-sheet dynamics during the latest phase of glaciation, by temporary phases of basal freezing (Kleman, 1990), migration of the ice divide over the site (Boulton & Clark, 1990) or because they lay beneath zones of sluggish flow between ice streams when the area was in the near-terminal zone during deglaciation (Fig. 4b).

The pattern of residual lineations, shown in Fig. 12 could have formed far from the ice-sheet margin during the final retreat, or during an earlier stage of ice-sheet retreat or advance, either in the Weichselian or during an earlier glacial stage (Tanner, 1915; Punkari, 1985; Nordkalott Project, 1986).

In southwestern Sweden, the residuals show a north-south trend compared with the NE-SW trend during deglaciation. In central Sweden they show a NW-SE trend compared with a NNW-SSE trend during deglaciation. In northeastern Sweden, northern Finland and northwestern Russia they show a NW-SE to NNW-SSE trend compared to a dominant W-E trend during deglaciation; and in southern Finland they show a N-S trend compared with a NW-SE trend during deglaciation. In the area north of the Gulf of Bothnia, the pattern of residuals is more complex, progressively swinging round towards a north-south trend further to the south and east (Fig. 11). Studies of their interrelations suggest that the NW-SE lineation is the earlier, over-printed by a later lineation which fans out from northernmost Sweden in an arc from SW and W.

The pattern of final retreat on the SE flank of the ice sheet east of the Baltic Sea, with a margin oriented in a NE-SW direction, indicates that the dominant flow during retreat in this sector was NW-SE, from a principle ice divide oriented SW-NE. It is concluded therefore that neither the N-S orientation in southern Finland, nor the earlier NW-SE lineations in northeastern Sweden belong to the final decay phase. However, in the Gulf of Finland region, some of the older N and NE directions may represent flow far inside the receding margin during the last deglaciation, produced by convergent drawdown at the head of the Baltic Sea ice stream.

4.2 Reconstructing the pattern of ice sheet growth

The change in lineation trend between early and late lineations must reflect a shift in the pattern of ice flow. There is relative continuity and integration in the residual lineation pattern (Fig. 12) (apart from in southern Sweden, where a NW-SE lineation is replaced by a superimposed N-S lineation further south, and northern Finland / Sweden, where a NW-SE lineation is succeeded to the south by a N-S lineation superimposed upon it). There is also a systematic ice-sheet wide directional shift between pre-retreat and retreat patterns; clockwise in the southeast and central sector and anti-clockwise in the eastern and northeastern sectors. These observations support that the early lineation pattern resulted from a single glacial phase in which the overall patterns of flow and ice divide distribution were different from those during retreat.

Whilst it is possible that the different patterns were the product of different glacial periods in which the ice sheets had different geometries, Boulton *et al.* (2001) suggest it to be most likely that the early lineations reflect growth phases of the Weichselian ice sheet in which the pattern of flow was different from the pattern of flow during retreat. They also suggest that, apart from any possible streaming (see below), strong lineations able to survive a succeeding glacial phase with a different flow direction are most likely to be produced in the near-margin high velocity zone. If the residual lineations shown in Fig. 12 belong to the phase of ice sheet growth, the general time-transgressive trajectory of the expanding ice sheet margin can be determined (Fig. 12). This pattern is not surprising, for the following reasons:

- The advance along the land sector from northern Finland to the Gulf of Finland appears to have been faster than from central Sweden over the central Baltic Sea. This could reflect slowing of advance beyond the east coast of Sweden because of calving into contemporary deep water in the Gulf of Bothnia.

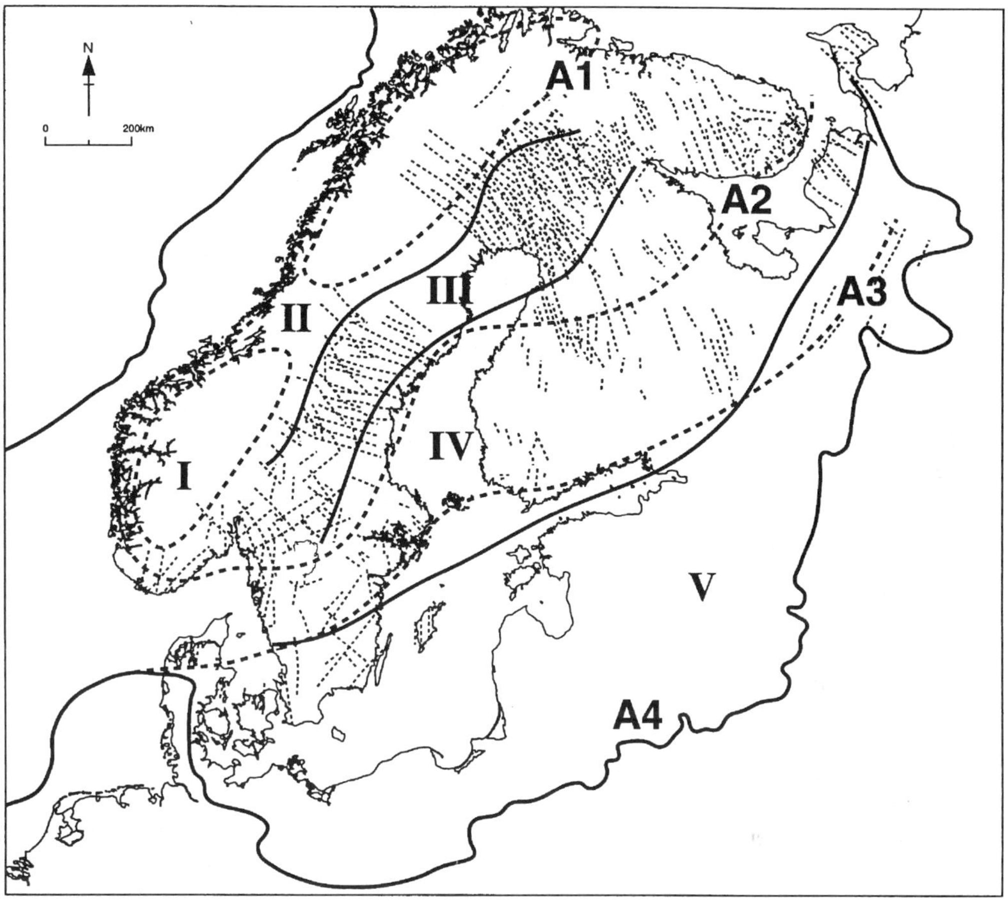

Fig. 12. The major pre-maximum lineation sets which differ in orientation from the retreat sets. They tend to be preserved in areas which lay beneath the principal ice divide at the maximum of glaciation, which protected them from erosion because of frozen bed conditions and low ice velocities (Boulton et al., 2001), and beneath inter-stream ridges (Figs 5, 7, 11) characterised by low ice velocities and possibly frozen bed conditions.

Assumed basal thermal regime. The solid lines identify different lineation zones, and are explained by reference to the time dependent thermal zonation (Boulton et al.,2001). Zone I: no major pre-maximum lineation sets; the area in which there was no basal melting through much of the glacial cycle. Zone II: a single major set of pre-maximum lineations; a zone in which there was basal melting during the Early Weichselian prior to 80 ka, but which was subsequently characterised by basal freezing. Zone III: two major sets of pre-maximum lineations; explained as a zone in which the same Early Weichselian lineation set as those in zone II is crossed by a lineation set formed in the zone of melting during the Middle Weichselian and in the phase of ice sheet growth prior to 25 ka. Its outer limit shows the limit of an Early Weichselian advance. Zone IV: only contains one major pre-maximum lineation set; explained as forming during ice sheet growth immediately prior to the last glacial maximum with a post-maximum lineation being superimposed on it after an intervening phase of basal freezing. Zone V: no major pre-maximum lineation set, explained as a zone which only experienced basal melting, so that lineations have been progressively modified to leave only one major lineation set.

Patterns of ice sheet growth. Pre-maximum lineation sets are used to determine patterns of growth of the last ice sheet based on the assumption that during ice advance the maximum rate of erosion/deposition will be in the terminal zone of the ice sheet (Boulton et al., 2001). It is also based on the view that on a horizontal surface, ice margin trends will lie normal to the flow direction. The general pattern of growth is illustrated by concentric dashed lines, which are arbitrarily used to subdivide growth into four main phases (A1-A4). Patterns of growth in western Norway are entirely speculative, as is the connection between phase 3 and the extension of the ice sheet into the central North Sea.

- The lineation pattern in northwestern Russia and northern Scandinavia suggests that an ice divide extended to the north east at least as far north as the northern coast, and that ice must have transgressed into the shallow waters of southern Barents Sea. This pattern of extension is more likely during an ice sheet growth stage when relative sea levels were low, compared to a retreat stage when they are high because of isostatic depression of the crust.
- A similar argument can be employed in southern Sweden, where it is reasonable to expect an advance into a

shallow early-glacial southern Baltic Sea from the mountainous area to the north. The retreat takes place however predominantly along the axis of the southern Baltic Sea, which would be compatible with strong calving into an isostatically-deepened basin.

- The pattern shows uniform flow directions without the lobate structures typical of the later retreat. During advance, remoulding of the substratum will commence as the ice sheet moves over a site and continue progressively beneath the ice sheet. During retreat, fast, marginal remoulding will be the last glacial event to which an area is subject. As a consequence, marginal features, such as lobes, will be well reflected in the glacial geology left by retreat, but will tend to be overprinted by deeper processes during advance (Boulton, 1996; Boulton *et al.*, 2001).

This interpretation depends upon the assumption that the rate at which geomorphological work will be done, and therefore the intensity of lineation, will depend upon the velocity of the ice sheet over its bed and the ice sheet residence time. The sub-marginal high velocity zone near to the equilibrium line would be the zone in which strongest lineations would be expected to develop. This zone of maximum geomorphic work would be expected to have been time-transgressive. With this in mind, it is thought that the broad arc described by the major residual lineation set is just the pattern which is likely to have been produced by the expansion of an ice sheet from its early nucleation and growth centres in the mountains of western Scandinavia during its expansion to the south and east if lineations were generated predominantly in a sub-marginal zone of strong geomorphic activity.

We speculatively suggest a simple glaciological rationale for the general pattern of ice sheet expansion suggested in Fig. 12 through a subdivision into four phases, though without any chronological implications except in the final, A4, phase.

Phase A1. *Ice-sheet initiation* begins as separate domes in the areas of Jotunheimen/Hardangervidda in southwest Norway and in the Scandinavian mountains north of Kebnekaise in northern Sweden/Lapland, from which they grew to coalesce. Ice-sheet initiation was a consequence of climatic deterioration but the location of nucleation centres was determined by topography. The two early growth centres, in south west Norway and northern Sweden/ Lapland, are currently the two highest and most heavily-glaciated areas along the Scandinavian mountain chain. These are the obvious areas for ice-sheet nucleation, although further growth is not merely depend upon the presence of high mountains, but also the existence of a large high altitude hinterland on which large ice domes can grow. The northern growth centre lay significantly further north than the northern centre during final decay, whereas the southern growth and decay centres seem to have been coincident.

Phase A2. *Radial expansion of the coalesced ice sheet and elongation of the ice divide towards the north and northeast and into the Kola Peninsula.* This expansion was driven by climate and occurred radially away from the centres of nucleation. The extension of the divide to the northeast must also have been climatically-determined, and must reflect intersection of the permanent snowline with the land surface along the Barents Sea coast of northern Norway. There is no high mountain topography in the northern Kola Peninsula. Coastal hills have a maximum elevation of 500m in the west and up to 350m further east. These low elevations must have been adequate to permit glacial initiation, and must reflect a strong moisture flux into the region, either from the northwest or southeast. Under such conditions, it is probable that ice-sheet initiation would have occurred in the Barents/Novaya Zemlya/Kara region to permit coalescence with the growing Northern European ice sheet in the region by late phase A2. The earlier NW-SE lineations in southern Sweden probably belong to this phase.

Phase A3. *Strong growth in the eastern sector of the ice sheet.* The trend reflected by the eastern extension of the divide in A2 is continued during phase A3 by extension to the south in the eastern sector. This was not matched by similar extension further west. This contrast may reflect the difficulty the ice sheet found in crossing the Baltic Sea and Gulf of Bothnia from the east coast of Sweden because of strong calving in deep water. It is only as the ice sheet advanced southwards on the eastern side of the Gulf of Bothnia that ice extended across the Gulf, possibly reflecting greater effectiveness in spanning the Gulf when ice flowed into it from both sides. This contrast between southeast and southwestern sectors may therefore be a dynamic rather than a climatic effect. The apparent failure of the ice sheet to cross from southern Sweden to Denmark in an area lacking large expanses of deep water might however indicate a strong climatic gradient between a cold but still moist eastern area, and a much warmer western area. It is presumed that the ice sheet advanced in the west at least as far as the coastal zone during phase A3 or even A2. The strong growth in the eastern sector and limited growth in the west during this phase implies a swinging of the ice divide about a pivot in southwest Norway, from a SW-NE orientation during phase 2 to a WSW-ENE orientation by the end of phase A3.

Phase A4. *Expansion in the central southern sector.* The ice sheet extended to, or close to the south west coast of Norway by the end of phase A3, and then expanded into the central North Sea at about 28 ka (Sejrup *et al.*, 1994) during A4, when it also expanded in the central southern sector. (Phase A4 and phase R1 overlap. The ice sheet dome over the central North Sea appears to have collapsed before the ice sheet reached its maximum extent in the southern and southeastern sectors.) This expansion would have been associated with a bowing towards the south of the primary ice divide, or the development of a secondary divide towards the south . This could account for some of the SW-directed lineations in the Stockholm area.

Further east, in the Baltic Sea and Finland, there was a very large advance, compared with further west in the North Sea area, where the ice sheet had already reached its maximum advance by early in A4, and further east in northwest Russia. This is consistent with recent AMS ^{14}C re-dating of mammoth remains (Ukkonen *et al.,* 1999) from six sites pre-dating the last, Late Weichselian ice sheet advance. These have produced ages ranging from 31,970 ± 950 BP from northwestern Finland, immediately west of the head of the Gulf of Bothnia, to 22,420 ± 315 BP in central Finland to 23,340 ± 350 BP at Helsinki in southern Finland. Notwithstanding the uncertainties associated with the dating of bone of this age (e.g.: Berglund *et al.,*1976), the evidence suggests that the ice sheet advanced by some 700-800 km to its maximum extent after about 22-23 ka. It is also consistent with the evidence from northwest Russia of a late advance to and retreat from the regional maximum, although the extent of the A4 advance there is less than further west. The N-S lineations in southern Sweden which post-date NW-SE lineations belong to this phase.

The moisture sources which fed the expansion in northwest Russia in phase A3 may have been cut off from the westerly moisture sources by ice-sheet growth. Further west, the central southern sector expansion may have reflected increasingly maritime conditions, which in the extreme west had become so warm as to inhibit further growth. Such changes may be explained by some of the relationships between storm tracks, maritime/continental contrasts and glacierisation pointed out by Kvasov (1978). The same highly maritime conditions in the south west would have created a relatively energetic ice sheet, which was able to cross the deep waters of the Norwegian Channel (300m depth at the present day) and extend into the central North Sea region to become confluent with a British ice sheet, when we assume that a primary ice divide would have extended from south west Norway into the central Norh Sea. It is thought that advance to the continental shelf edge in the west occured during phase A4.

Evidence from the Kattegat area of south Sweden/Denmark suggests two early stage ice advances (Sjørring, 1983; Houmark-Nielsen, 1987) a Norwegian advance and an Old Baltic advance. Sjørring (1983) regarded the former as the earlier of the two, whilst Houmark-Nielsen (1987) and Ringberg (1988) believed it to be the latter. The succeeding Main Danish Advance and later readvances are believed (Houmark-Nielsen, 1987) to have coincided with the LGM and early retreat stages from it. The authors suggest that advance from the north in the area of Jutland could belong to late phase A2, A3 or early phase A4 conditions, whilst in Sweden they are most likely to reflect A4 conditions.

The absence of evidence of flow from a Barents Sea ice mass in the NNE and from a Kara Sea mass in the NE, suggests that ice on the European mainland was dominantly Europe-centred and can largely be considered as a single dynamic entity.

4.3 The chronology of ice sheet fluctuation prior to the Last Glacial Maximum

There are currently two schools of thought about the tempo of glaciation in northern Fennoscandia. One group (Group A: e.g.: Korpela, 1969; Hirvas & Nenonen, 1985; Lagerbäck, 1988; Kleman, 1990; Sutinen, 1992) argues for the preservation of tills, eskers and major end moraines from the early Weichselian and Saalian glaciations, which were separated from the phase of advance to the Weichselain maximum by periods of deglaciation. This is based on discoveries of organic matter underlying the uppermost till unit and which are interpreted as evidence of deglaciation. They suggest that the organic units separate two glacial phases during which the patterns of ice flow (determined from till fabric) were quite different. Another group (Group B: Aario & Forsström, 1979; Punkari, 1984, 1993; Forsström & Eronen, 1991; Punkari & Forsström, 1995) argues that much of the organic material is redeposited, is not therefore evidence of deglaciation, and that the inter-till organic beds are seldom found between tills reflecting the two different flow patterns in question.

The views of group A imply strong ice-marginal oscillations through the Weichselian, with the ice-sheet margin withdrawing to northern Fennoscandia during interstadials (Mangerud, 1991; Lundqvist, 1986). Group B stress a longer, slower and more sustained ice sheet build up to its maximum Weichselian extent, with less dramatic pre-LGM oscillations on the eastern flank of the ice sheet.

Boulton et al. (2001) have used a glaciological flowline model to simulate the glaciological implications of the view of group A. It generates two well-integrated sets of time-transgressive glacial lineation. The earlier set would is best interpreted as a combination of sets formed in zone II, at about 90 ka, and in zone III, between about 50 ka and 20 ka, with the younger set representing deglaciation in zone III.

Acknowledgements

We are grateful for helpful discussions to Barbara Wohlfarth, Svante Björk, Nick Hulton, Anto Raukas, Jonathon Perry, Lars Forsström, Michael Houmark-Nielsen and Jürgen Ehlers.

References

Aario R. & Forsström L. (1979). Glacial stratigraphy of Koillismaa and North Kainuu, Finland. *Fennia,* **157:2**, 1-49.

Andersen, B.G. (1987). Quaternary research in Norway 1960-1986. *Boreas,* **16**, 327-338.

Andren, T., Björck, J. & Johnsen, S. (1999). Correlation of Swedish glacial varves with the Greenland (GRIP) oxygen isotope record. *Journal of Quaternary Science,* **14**, 361-371.

Bard, E., Arnold, M., Fairbanks, R.G. & Hamelin, B. (1993). $^{230}Th/^{234}U$ and ^{14}C ages obtained by mass spectrometry on corals. *Radiocarbon*, **35**, 191-199.

Berglund, B.E., Hakansson, S. & Lagerlund, E. (1976). Radiocarbon-dated mammoth finds in South Sweden. *Boreas,* **5**, 177-191.

Björck, S., Wohlfarth, B. & Possnert, G. (1995). ^{14}C AMS measurements from the Late Weichselian part of the Swedish Time Scale. *Quaternary International*, **27**, 11-18.

Bond, G., Broecker, W., Johnsen, S., McManus, J., Labeyrie, L., Jouzel, J. & Bonani, G. (1993). Correlations between climate records from North Atlantic sediments and Greenland ice. *Nature,* **365**, 143-147.

Boulton, G.S. (1987). A theory of drumlin formation by subglacial deformation. *In:* Rose, J. & Menzies, J. (eds), *Drumlins*, Rotterdam, Balkema, 25-80.

Boulton, G.S. (1996). Theory of glacial erosion, transport and deposition as a consequence of subglacial sediment deformation. *Journal of Glaciology*, **42**, 43-62.

Boulton, G.S. & Clark, C.D. (1990a). A highly mobile Laurentide ice sheet revealed by satellite images of glacial lineations. *Nature*, **346**, 813-7.

Boulton, G.S. & Clark, C.D. (1990b). The Laurentide ice sheet through the last glacial cycle: the topology of drift lineations as a key to the dynamic behaviour of former ice sheets. *Transactions of the Royal Society of Edinburgh: Earth Sciences*, **81**, 327-347.

Boulton, G.S., Smith, G.D., Jones, A.S. & Newsome, J. (1985). Glacial geology and glaciology of the last mid-latitude ice sheets. *Journal of the Geological Society, London,* **142**, 447-474.

Boulton, G.S., Dongelmans, M., Punkari, M. & Broadgate, M. (2001). Palaeoglaciology of an ice sheet through the Weichselian. *Quaternary Science Reviews,* **20**, 591-625.

Chebotareva, N.S. (ed.) (1977). *The structure and dynamics of the last ice sheet of Europe* (in Russian). Moscow, Nauka, 143 pp.

De Geer, G. (1940). Geochronologie Suecica Principles. *Kungliga Svenska Vetenskaps-Akademiens Handlingar* **3: 18:6**, 367 pp.

Dongelmans, P. (1997). Glacier dynamics of the Fenno-scandian ice sheet: a remote sensing study. Unpublished PhD thesis, University of Edinburgh.

Ehlers, J. (1990). Reconstructing the dynamics of the north west European Pleistocene ice sheets. *Quaternary Science Reviews,* **9**, 71-83.

Forsström, L. & Eronen, M. (1991). New information about the Eemian and Early Weichselian in Finland. *Striae*, **34**, 31-38.

Hang, T. (1997). Clay varve chronology in the Eastern Baltic area. *Geologiska Föreningens i Stockholm Förhandlingar,* **119**, 295-300.

Hillefors, A. (1974). The stratigraphy and genesis of the Dösebacka and Ellesbo drumlins. *Geologiska Föreningens i Stockholm Förhandlingar*, **96**, 355-374.

Hirvas, H. & Nenonen, K. (1985). The till stratigrphy of Finland. *INQUA Till Symposium, Finland 1985. Geological Survey of Finland, Special Paper*, **3**, 49-63.

Holtedahl, H. (1993). *Marine geology of the Norwegian continental margin.* Geological Survey of Norway, Trondheim, Norway. 150 pp.

Houmark-Nielsen, M. (1987). Pleistocene stratigraphy and glacial history of the central part of Denmark. *Bulletin of the Geological Society of Denmark,* **36, 1-2**, 189 pp.

Imbrie, J., Hays, J.D., Martinson, D.G., McIntyre, A., Mix, A.C., Morley, J.J., Pisias, N.G., Prell, W.L. & Shackleton, N.J. (1984). The orbital theory of Pleistocene climate: support from a revised chronology of the marine ^{18}O record. *In:* Berger, A., Imbrie, J., Hays, J., Kukla, G. & Saltzman B. (eds), *Milankovitch and climate*, Dordrecht, D. Reidel, 269-306.

Johnsen, S.J., Clausen, H.B., Dansgaard, W., Fuhrer, K., Gundestrup, N., Hammer, C.U., Iversen, P., Jouzel, J., Stauffer, B. & Steffensen, J.P. (1992). Irregular glacial interstadials recorded in a new Greenland ice core. *Nature,* **359**, 311-313.

King, E.L., Rokoengen, K. & Gunleiksrud, T. (1987). Quaternary seismostratigraphy of the mid Norwegian Shelf, 65°N-67°, 30'N. A till tongue stratigraphy. *IKU, Publication*, **114**, 58 pp.

King, E.L., Sejrup, H.P., Haflidason, H., Elverhøi, A. & Aarseth, I. (1996). Quaternary seismic stratigraphy of the North Sea Fan; glacially fed gravity flow aprons, hemipelagic sediments and large submarine slides. *Marine Geology,* **130**, 293-315.

Kleman, J. (1990). On the use of glacial striae for reconstruction of palaeo-ice sheet flow patterns. *Geografiska Annaler,* **72A**, 217-36.

Kleman, J., Hätterstrand, C., Borgström, I. & Stroeven, A. (1997). Fennoscandian palaeoglaciology reconstructed using a glacial geological inversion model. *Journal of Glaciology,* **43**, 283-299.

Korpela, K. (1969). Die Weichsel-Eiszeit und ihr Interstadial in Perapohjola im Licht von submoränen Sedimenten. *Annales Academiae Scientiarum Fennicae,* **A III, 99**, 108 pp.

Kozarski, S. (1986). Timescales and the rhythm of Vistulian geomorphic events in the Polish lowlands. *Czasopismo Geograficzne,* **57**, 247-270.

Kozarski, S. (1988). Time and dynamics of the last Scandinavian ice-sheet retreat from northwestern Poland. *Geographica Polonica,* **55**, 91 -101.

Krasnov, I.I. (1971). *The map of Quaternary deposits of Europe and territories surrounding it, scale 1:1,500,000.* Ministry of Geology SSSR.

Kromer, B. & Becker, B. (1993). German oak and pine ^{14}C calibration, 7200-9439 BC. *Radiocarbon*, **35**, 125-135.

Kroon, D., Austin, W.E.N., Chapman, M.R. & Ganssen, G.M. (1997). Deglacial surface circulation changes in the northeast Atlantic: temperature and salinity records off NW Scotland on a century scale. *Palaeoceanography*, **12**, 755-763.

Kvasov, D.D. (1978). The Barents ice sheet as a relay regulator of glacial-interglacial alternation. *Quaternary Research,* **9**, 288-299.

Lagerbäck, R. (1988). The Veiki moraines in northern Sweden - widespread evidence of an early Weichselian deglaciation. *Boreas,* **17**, 469-86.

Lagerlund, E. (1980). Litostratigrafisk indelning av Västskånes Pleistocen och en ny glaciationsmodell för Weichsel. *University of Lund, Department of Quaternary Geology, Thesis,* **5**, 106 pp.

Landvik, J.Y., Bondevik, S., Elvehøi, A., Fjeldskaar, W., Mangerud, J., Salvigsen, O., Siegert, M., Svendsen, J-I. & Vorren, T. (1998. The last glacial maximum of Svalbard and the Barents Sea area: ice sheet extent and configuration. *Quaternary Science Reviews,* **17**, 43-75.

Larsen, E., Lyså, A., Demidov, I., Funder, S., Houmark-Nielsen, M., Kjaer, K.H. & Murray, A.S. (1999). Age and extent of the Scandinavian ice sheet in northwest Russia. *Boreas* 28, 115-132.

Liedtke, H. (1981). Die nordischen Vereisungen in Mitteleuropa (2nd edition). *Forschungen zur deutschen Landeskunde*, **204**,. 307 pp.

Lundqvist, J. (1986). Stratigraphy of the central area of the Scandinavian glaciation. Quaternary Sciences Reviews, **5**, 251-268.

Mangerud, J. (1991). The Scandinavian ice sheet through the last interglacial/glacial cycle. *In:* Frenzel, B. (ed.), *Klimageschichtliche Probleme der letzten 130,000 Jahre.* Stuttgart / New York, G. Fischer, 307-330.

Mangerud, J., Larsen, E., Longva, O. & Sønstegaard, E. (1979). Glacial history of western Norway 15,000-10,000 B.P. *Boreas,* **8**, 179-187.

Mojski, J.E. (1992). Vistulian stratigraphy and TL dates in Poland. *Sveriges Geologiska Undersökning,* **Ca 81**, 195-200.

Nordkalott Project (1986). *Map of Quaternary geology, sheet 5: Ice flow directions, Northern Fennoscandia.* Geological Surveys of Finland, Norway and Sweden. Uppsala.

Punkari, M. (1982). Glacial geomorphology and dynamics in the eastern parts of the Baltic Shield interpreted using Landsat imagery. *Photogrammetric Journal of Finland,* **9**, 77-93.

Punkari, M. (1984), The relations between glacial dynamics and tills in the eastern part of the Baltic Shield. *Striae,* **20**, 49-54.

Punkari, M. (1985). Glacial geomorphology and dynamics in Soviet Karelia interpreted by means of satellite imagery. *Fennia,* **163**, 113-153.

Punkari, M. (1993). Modelling of the dynamics of the Scandinavian ice sheet using remote sensing and GIS methods. *In:* Aber, J. (ed.), Glaciotectonics and mapping glacial deposits. *Proceedings of the INQUA comission on formation and properties of glacial deposits. Canadian Plains Research Center, University of Regina*, 232-250.

Punkari, M. & Forsström, L. (1995). Organic remains in Finnish subglacial Sediments. *Quaternary Research,* **43**, 414-425.

Ralska-Jasiewiczowa, M. & Rzetkowska, A. (1987). Pollen macrofossil stratigraphy of fossil lake sediments at Niechorze I. *Acta Palaeobotanica,* **22**, 153-158.

Ringberg, B. (1988). Late Weichselian geology in southernmost Sweden. *Boreas,* **17**, 243-263.

Rise, L., Roekoengen, K., Sættem, J. & Bugge, T. (1988). *Thickness of Quaternary deposits on the mid-Norwegian Continental Shelf.* Map in scale 1:1.000.000. IKU Publication, **119**.

Roekoengen, K., Bugge, T., Lofaldi, M. & Vigran, J.O. (1979). Quaternary geology and deglaciation of the continental shelf off Troms, North Norway. *Boreas,* **8**, 217-227.

Roekoengen, K., Rise, L., Bugge, T. & Sættem, J. (1988). *Bedrock geology of the mid Norwegian continental shelf.* Map in scale 1:1.000.000. IKU Publication, **118**.

Rose, J. & Letzer, J.M. (1977). Superimposed drumlins. *Journal of Glaciology,* **18**, 471-480.

Sandgren, P., Hang, T. & Snowball, I.F. (1997). A late Weichselian geomagnetic record from Lake Tamula, SE Estonia. *Geologiska Föreningens i Stockholm Förhandlingar,* **119**, 279-284.

Sauramo, M. (1918). Geochronologische Studien über die spätglaziale Zeit in Südfinnland. *Fennia,* **41**, 1-44.

Sauramo, M. (1923). Studies on the Quaternary varve sediments in southern Finland. *Fennia,* **44:1**, 1-164.

Sejrup, H.P., Haflidason, H., Aarseth, I., King, E., Forsberg, C.F., Long, D. & Roekoengen, K. (1994). Late Weichselian glaciation history of the northern North Sea. *Boreas,* **23**, 1-13.

Sejrup, H.P., King, E.L., Aarseth, I., Haflidason, H. & Elverhøi, A. (1996). Quaternary erosional and depositional processes: western Norwegian fjords, Norwegian Channel and North Sea Fan. *Journal of the Geological Society of London,* **117**, 187-202.

Sejrup, H.P., Landvik, J.Y., Larsen, E., Janocko, J., Eiriksson, J. & King, E.L. (1997). The Jaeren area, a border zone of the Norwegian Channel ice stream. *Quaternary Science Reviews,* **17**, 801-812.

Sejrup, H.P., Larsen, E., Landvik, J., King, E.L., Haflidson, H. & Nesje, A. (2000). Quaternary glaciations in southern Fennoscandia: evidence form southwestern Norway and the northern North Sea region. *Quaternary Science Reviews,* **19**, 667-685.

Sjørring, S. (1983). The glacial history of Denmark. *In:* Ehlers, J. (ed.), *Glacial deposits in north-west Europe*, 163-179. Rotterdam, Balkema.

Strömberg, B. (1985). Revision of the lateglacial Swedish varve chronology. *Boreas,* **14**, 101-105.

Strömberg, B. (1990). A connection between the clay varve chronologies in Sweden and Finland. *Annales Academiae Scientiarum Fennicae,* **154 A:III**. 32 pp.

Strömberg, B. (1994). Younger Dryas deglaciation at Mt. Billingen, and clay varve dating of the Younger Dryas /Preboreal transition. *Boreas,* **23**, 177-193.

Stuiver, M. & Reimer, P.J. (1993). Extended ^{14}C data base and revised CALIB 3.0 ^{14}C age calibration program 1993. *Radiocarbon,* **35**, 215-230.

Sutinen, R. (1992). Glacial deposits, their electrical properties and surveying by image interpretation and ground penetrating radar. *Geological Survey of Finland, Bulletin,* **359**. 123 pp.

Tanner, V. (1915). Studier öfver kvartärsystemet i Fennoskandias nordliga delar. *Fennia,* **36**. 815 pp.

Ukkonen, P., Lunkka, J.P., Jungner, H. & Donner, J. (1999). New radiocarbon dates from Finnish mammoths indicating large ice-free areas in Fennoscandia during the middle Weichselian. *Journal of Quaternary Science,* **14**, 711-714.

Valen, V., Mangerud, J., Larsen, E. & Hufthammer, A.K. (1996). Sedimentology and stratigraphy in the cave Hamnsundhelleren, western Norway. *Journal of Quaternary Science,* **11**, 185-201.

Vorren, T.O., Vorren, K.-D., Alm, T., Gulliksen, S. & Løvlie, R. (1988). The last deglaciation (20,000 to 11,000 B.P.) on Andøya, northern Norway. *Boreas,* **17**, 41-77.

Wohlfarth, B., Björck, S. & Possner, G. (1995). The Swedish time scale - a potential calibration tool for the radiocarbon timescale during the Late Weichselian. *Radiocarbon,* **37**, 347-359.

Index